D0555382

ARLIS

Pettijohn · Potter · Siever

Sand and Sandstone

With 258 Figures

Springer-Verlag
New York · Heidelberg · Berlin 1973

F. J. PETTIJOHN, Professor of Geology, The Johns Hopkins University, Baltimore, MD 21218/U.S.A.

PAUL EDWIN POTTER, Professor of Geology, University of Cincinnati, Cincinnati, OH 45221/U.S.A.

RAYMOND SIEVER, Professor of Geology, Harvard University, Cambridge, MA 02138/U.S.A.

Library of Congress Catalog Card Number 79-168605.

Printed in the United States of America.

9 8 7 6 5 4

ISBN 0-387-90071-3 Springer-Verlag New York Heidelberg Berlin (soft cover)
ISBN 3-540-90071-3 Springer-Verlag Berlin Heidelberg New York (soft cover)
ISBN 0-387-05528-2 Springer-Verlag New York Heidelberg Berlin (hard cover)
ISBN 3-540-05528-2 Springer-Verlag Berlin Heidelberg New York (hard cover)

SAND as

grains
crystals
laminae
boxes
piles
shadows
beds
waves

dunes
bodies
banks
bars
barriers
beaches
belts
cheniers

lenses
pods
ribbons
sheets
shoestrings
basins

so surely there is plenty of it!

Plenty for you, for me, for everyone
to share its many sparkling facets.

Preface

This book is the outgrowth of a week-long conference on sandstone organized by the authors, first held at Banff, Alberta, in 1964 under the auspices of the Alberta Association of Petroleum Geologists and the University of Alberta, and again, in 1965, at Bloomington, Indiana, under the sponsorship of the Indiana Geological Survey and the Department of Geology, Indiana University. A 200-page syllabus was prepared for the second conference and published by the Indiana Geological Survey. Continuing interest in and demand for the syllabus prompted us to update and expand its contents. The result is this book.

We hope this work will be useful as a text or supplementary text for advanced undergraduate and graduate courses in sedimentation, sedimentary petrology, or general petrology and perhaps will be helpful to the teachers of such courses. Though we have focussed on sandstones we have necessarily included much of interest to students of all sediments. We hope also that it will be a useful reference work for the professional geologist, especially those concerned with petroleum, ground-water, and economic geology either in industry or government. Because the subject is so closely tied to surface processes it may also be of interest to geomorphologists and engineers who deal with beaches and rivers where sand is in transit.

This work presupposes a general knowledge of the elements of mineralogy, chemistry and statistics on the part of the reader. As no investigation of sediments — especially sandstones — can be considered adequate or complete without careful microscopical analysis, we also presume, therefore, that the user of this book has the knowledge and skills needed to study sands and sandstones under the microscope.

On the other hand, some cognate fields of knowledge are less familiar to geologists and while we did not include a section on statistics or thermodynamics, we did include a section on the principles underlying fluid flow and the propulsion of granular materials. We feel that some knowledge of this subject will become increasingly important in understanding physical sedimentation and the resulting textures and structures of sands.

The book is organized in such a manner as to lead the reader from consideration of the component grains in a sandstone to the analysis of sandstones in the sedimentary basin as a whole. The first half is largely descriptive, a summary of what is known about sandstones beginning with the components, their composition (Chapter 2) and geometrical properties (Chapter 3), progressing to the larger organization and structure (Chapter 4) to the whole rock itself (Chapters 5, 6 and 7). The second half of the book is more largely interpretative and process-oriented. It includes the processes of sand formation (Chapter 8), transportation and deposition (Chapter 9), and post-depositional alteration (Chapter 10). The book concludes with a résume of the relation of sands to their environment of deposition and to other sediments (Chapter 11) and a summary of their distribution in space and time (Chapter 12). We have included a synoptic review of several better-known sedimentary basins in which an integrated approach — involving stratigraphy, sedimentary petrology, and paleocurrents — was used to unravel geologic history.

For the most part, analytical techniques are omitted. They are adequately covered in several modern texts and manuals (see references, p. 19). Exceptionally, however, we have included a short appendix on the art of petrographical description and analysis which, like field work, is best learned perhaps from experience under the guidance of a skilled master of the subject. We felt it worthwhile, however, to set down some guiding principles as these are seldom made explicit in most published works.

We did not include many "case histories" because, unlike in law or psychiatry, we feel that the student can turn to no better source of instruction than the rocks themselves. No course on this subject can be considered adequate or complete without a well-integrated program of field and laboratory studies. The student, under the supervision of his teacher, should work out his own problems. The clinic is a better guide to practice than the case book.

References to the literature are of two kinds — actual citations in the text to specific papers and a collection of annotated references. The latter for the most part supplement rather than repeat the text citations. Both are placed at the end of the appropriate chapters. In general, our references are selective, that is, although they include some older classic papers emphasis is on the more recent ones. In many cases, such as the chapter on sedimentary structures, we did not feel the need of an in-depth review of the literature inasmuch as several specialized modern works which contain an extensive bibliography are readily available.

As is inevitable in a work of this kind, much of what is contained therein is a compilation from many sources which transcend and go beyond the immediate and direct experience of the authors. We have tried to acknowledge our debt to these sources at the appropriate places. We also wish to acknowledge the helpful criticism of those who read sections of this work when it was in manuscript form. In particular, we are indebted to Earle McBride, University of Texas, for checking our glossary of rock names applied to sandstones, to Robert L. Smith, U.S. Geological Survey, William F. Jenks, University of Cincinnati, and Richard V. Fisher, University of California at Santa Barbara, for reading the chapter on volcaniclastic sands, to Lee Suttner for criticism of Chapters 4 and 6, to Gerald V. Middleton of McMaster University, Yaron M. Sternberg of the University of Maryland and John B. Southard of the Massachusetts Institute of Technology for their help with the chapter on transport and deposition, to S. V. Hrabar of the Humble Oil Company for reading all of Chapter 11 and Donald A. Holm of Williams, Arizona, and Richard Mast of the Illinois Geological Survey for reading parts of it, to D. A. Pretorius of the University of Witwatersrand and R. W. Ojakangas of the University of Minnesota at Duluth for their comments on portions of Chapter 12, and to Miriam Kastner for help in the X-ray and electron probe analysis of the Trivoli Sandstone. Alan S. Horowitz of Indiana University read and helpfully edited many of the chapters. We wish to thank Mrs. Susan Berson, Miss Kathleen Feinour, Miss Jean Dell'Uomo, Mrs. Debby Powell, and Miss Cynthia Worswick for the final typing of the manuscript and our publishers for their help in the preparation of the illustrations and seeing the work through the press.

To emphasize our spirit of teamwork we have listed our names in alphabetical order.

January 1, 1972

F. J. Pettijohn
Paul Edwin Potter
Raymond Siever

Table of Contents

Chapter 1. Introduction and Source Materials

Sand and Sandstone Defined

Sands and clastic sediments, in general, differ from the igneous and other crystalline rocks in possessing a framework of grains – a framework stable in the earth's gravitational field. Unlike the grains of the igneous and related rocks which are in continuous contact with their neighbors, the grains in a sand are generally in tangential contact only and thus form an open, three-dimensional network. As a consequence, sands have a high porosity – have a fluid-filled pore system. The unequal distribution of stress along grain boundaries may lead to solution at points of pressure and deposition elsewhere increasing the surfaces of contact and decreasing the pore space. Such action, coupled in some cases with the introduction of cementing materials, leads to the ultimate end-product – a rock with grains in continuous contact and without porosity. In this manner, a sand with tangential contacts and a porosity of 35 to 40 percent is converted to an interlocking crystalline mosaic with zero porosity.

Sand is loose, non-cohesive granular material, the grains or framework elements of which must by definition be sand-sized. Various attempts have been made to define sand more precisely. These attempts are largely directed toward expressing grain size in terms of grain "diameter" of some specified magnitude. Inasmuch as sand grains are non-regular solids, it is first necessary to define the term "diameter" as applied to such solids (see Chap. 3). Attempts to codify the meaning of "sand" as a size-term are many. The effort to do so is usually part of a larger effort to codify all size terms and to construct a "grade scale" (see Chap. 3). The various choices made for the size class "sand" in some of these grade scales are shown in Fig. 1-1. We shall here adopt the diameter limits 0.0625 (1/16) and 2.0 mm for "sand" – limits which have become generally accepted among sedimentologists.

Sand, although restricted by definition to the 0.0625 to 2.00 mm range of diameters, actually encompasses a vast range in grain size. A grain 2 mm in diameter, as a sphere, has a volume of about 4.2 mm^3. A grain 1/16 (0.0625) mm in diameter, has a volume of about 0.00012 mm^3. The larger volume is 34,688 times greater than the smaller. In short, while sand has a thirty-two fold range in diameter, it has nearly a 35,000-fold range in volume – the truest measure of size.

Definitions of "sand" as a deposit – as distinct from a size term – are diverse. No generally accepted usage is apparent from a review of the literature. The questions are: should the average, median or modal size of the material designated "sand" fall in the sand range? Or must 50 percent or some other specified proportion of the material be within these limits? Or to put the

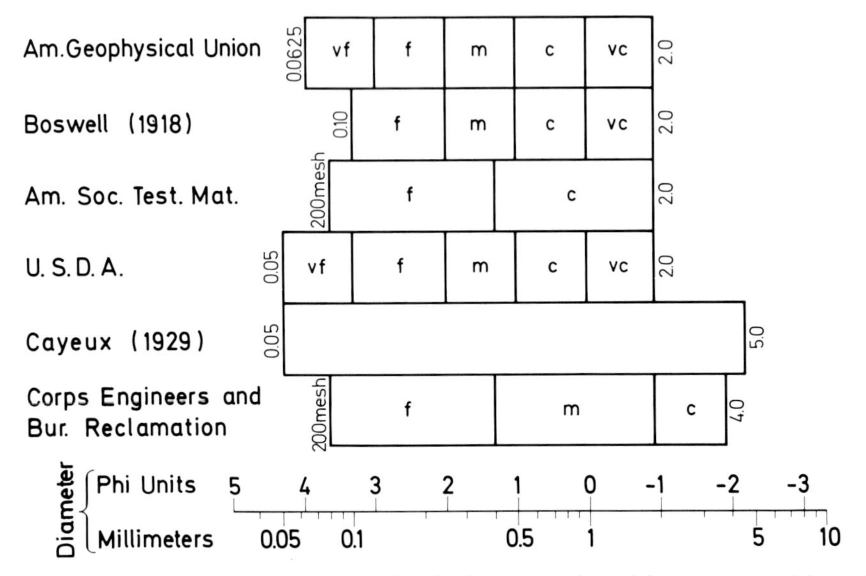

Fig. 1-1. Grade scales and size limits of sand. Note the diverse meanings of the size terms and the variations in the limits of sand. *vf*—very fine; *f*—fine; *m*—medium; *c*—coarse; *vc*—very coarse

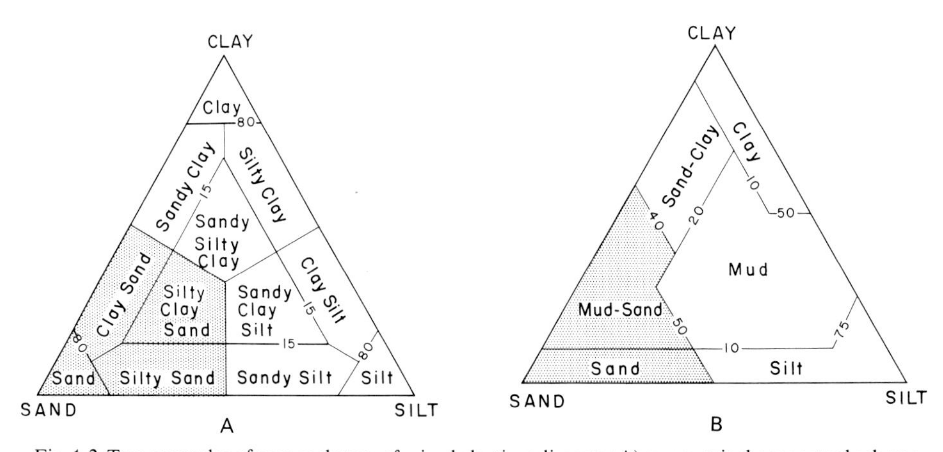

Fig. 1-2. Two examples of nomenclature of mixed clastic sediments: A) symmetrical conceptual scheme and B) asymmetrical scheme, the latter based on actual usage of marine geologists (Modified from Shepard, 1954, Fig. 4 and 5)

question another way, what are the permissible proportions of oversized or undersized material in "sand"? Two of the various alternatives are shown in Fig. 1-2, together with suggested nomenclatural solutions to the problem of sands with various admixtures of other grades.

It is clear from the above discussion that "sand," both as a size term and as a deposit, is defined without reference to composition or to genesis. It could be a quartz sand or a carbonate sand. It could arise (Fig. 1-3) as the indestructible residue from decomposition of a granite (quartz sand) or the product of chemical precipitation (oolitic sand). In practice, we tend to call all these materials of diverse compositions and origins "sand." However, many tend to restrict the term "sandstone" to those indurated sands of siliceous character. The lithified carbonate sands would be termed limestones – not sandstones. Even the term "sand" without adjectival modifiers tends to imply a siliceous composition. The terms "arenite" and "psammite" have been proposed as size terms devoid of any compositional connotations to avoid this ambiguity. In this book we largely exclude the carbonate and non-siliceous sands from our consideration.

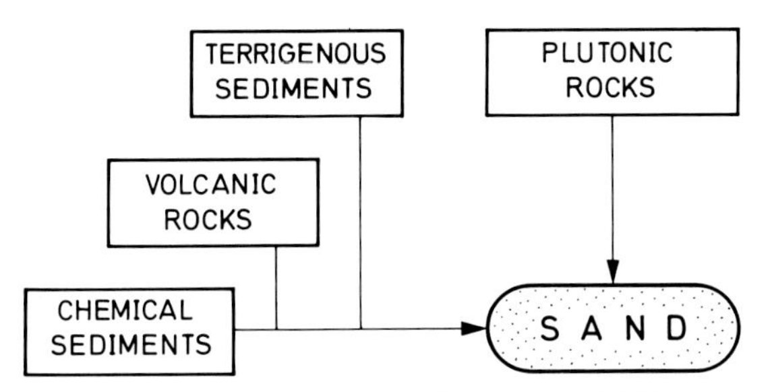

Fig. 1-3. Genesis of sand. Plutonic rocks supply mostly quartz and feldspar, terrigenous sediments mostly quartz and rock fragments, volcanic rocks mostly rock fragments and glass, and chemical sediments mostly carbonate debris

Sand may be defined in terms of certain arbitrarily agreed upon size limits as indicated above. Some investigators, however, have supposed that there are some "natural" limits which set sand apart from other materials. Wentworth (1933), for example, presumed that the size limits of the several principal classes of clastic sediments, sand, silt, clay, were genetically circumscribed because of the mode of derivation from the parent rock and because of certain fundamental modes of transport by running water. Sand is a product of breakdown of coarse-grained source rocks and the range of sizes (primarily of the quartz) is limited by the original texture of the source rock. There is a presumed dearth of material in those size grades transitional to gravel and to silt (for a review of this problem see Russell, 1968). Sand, unlike the finer materials, is largely transported by rolling and sliding along the bottom or by saltation and only to a smaller extent by turbulent suspension.

Bagnold (1941, p. 6) places the lower limit of sand as that at which the terminal fall velocity is less than the upward eddy currents and the upper limit as that size such that a grain resting on the surface ceases to be moveable either by direct pressure of the fluid or by the impact of other moving grains. This kind of behavioral definition of sand depends on the nature of the flowing medium (air or water) and must be valid for "average" conditions of flow. The size limits thus defined approximate those set by tradition. Bagnold notes further that materials designated "sand" have one peculiar characteristic which is not shared by coarser or finer materials – namely, the power of self-accumulation – of utilizing the energy of the transporting medium to collect their scattered components together in definite heaps, leaving the intervening surface free of grains. The common mode of transport of sand is by the migration of such heaps or "dunes" be they subaerial or subaqueous.

Relative and Absolute Abundance

Several methods have been employed to estimate the abundance of sandstone relative to the other common sedimentary rocks. These consist either of actual measurement of many stratigraphic sections (Table 1-1) or of calculation of the relative proportion of sandstone based on some geochemical considerations (Table 1-2). The results obtained by these two approaches are somewhat different. In general, the proportions of sandstone and limestone determined by actual measurement are greater than those derived by calculations. This may in part be due to loss of the finer clayey fraction to the deep sea so that shales are

Table 1-1. *Percentage of common sedimentary rocks based on stratigraphic measurements*

	Leith and Mead (1915)[1]	Schuchert (1931)[2]	Kuenen (1941)[3]	Krynine (1948)[4]	Horn and Adams (1966)[5]		Ronov (1968)[6]	
					Continent-Shield	Mobile belt-Shelf	Platforms	Geo-synclines
Shale	46	44	57	42	53	59	49	39
Sandstone	32	37	14	40	28	36	24	19
Limestone	22	19	29	18	19[a]	5[a]	21	16

[a] Includes 3 percent evaporite.

1. Leith and Mead (1915, p. 60) based these figures on the average of North American sections aggregating 520,000 feet; an average of sections totalling 188,000 in Eurasia gives the proportions 49, 32 and 19 respectively.
2. Schuchert (1931, p. 12) based his figures on measurement of North American Paleozoic maximum of 259,000 feet.
3. Kuenen (1941, p. 168) derives his figures from measurements in the East Indies.
4. Krynine (1948, p. 156) did not indicate how these estimates were obtained.
5. Horn and Adams (1966, p. 282) utilize the data from several sources, both published and unpublished. They do not explain how their estimates were made.
6. Summarized in Ronov, 1968, p. 30. Volcanic rocks are estimated to form an additional 25 percent of the geosynclinal fill. Of the total volume of sediments, 75 percent are geosynclinal and 25 percent platform. The weighted proportion of sandstone is 20 or, if volcanics are excluded, 25 percent.

Table 1-2. *Percentage of common sedimentary rocks based on geochemical and other calculations*

	Mead[1] (1907)	Clarke[2] (1924)	Holmes[3] (1913)	Wickman[a 4] (1954)
Shale	82	80	70	83
Sandstone	12	15	16	8
Limestone	6	5	14	9[b]

[a] Percentage values calculated from Wickman's data.
[b] "Carbonate rock."

1. Mead's figures were derived from bulk chemical analyses by calculating the proportions of average shale, sandstone and limestone which combined would be as nearly like the average igneous rock as possible.
2. Clarke (1924, p. 34) obtained his figures by assigning all of the free quartz of the average crystalline rock to the production of sandstone and half of the calcium to the formation of limestone. The figure for quartz was obtained from a statistical examination of 700 igneous rocks.
3. Holmes' (1913, p. 60) estimates are the proportions of sediment now deposited annually.
4. Wickman's calculations are "a modernized version of Mead's ideas."

under-represented in the stratigraphic column with a compensatory increase in sandstone and limestone. In summary, sandstones constitute 14 percent of all sediments according to Kuenen (1941), about 20 percent according to Ronov and others (1963, p. 212), and about 32–37 percent according to Leith and Mead (1915) and to Schuchert (1931). Kuenen's estimate is based solely on Indonesian data; Ronov's most recent estimate, 26 percent, apparently applies only to the Russian platform (Ronov and others, 1969, p. 192).

Poldervaart (1955) considered the estimates of Leith and Mead, Schuchert, and Krynine to be applicable to the continental-shield areas, whereas Kuenen's Indonesian estimate was thought to be more characteristic of the younger folded belts of the world. Because the sediments of the former have a volume of about 52.5×10^6 km^3 and the latter a volume of about 126×10^6 km^3, the weighted proportion of sandstone is about 26 percent (based on 43 percent sand in continental-shield regions and 18 percent in the younger folded belt.) Horn and Adams (1966) likewise noted that the proportion of sand in the continental-shield area was different from that in the mobile belts. They presumed that the sand content in the latter was higher than in the former – a conclusion at variance with other estimates (Table 1-1). The volume of the continental-shield sediments was estimated to be at 127×10^6 km^3 and that of the mobile belts 395×10^6 km^3 – estimates considerably higher than those of Poldervaart. The weighted proportion of sandstone (based on 28 percent in the continental-shield areas and 36 percent in the sediments of the mobile belts and shelf) is, therefore, 34 percent. Ronov (1968, p. 30), on the other hand, estimated 25 percent of the total volume of sediment to be sand. Because Horn and Adams did not indicate how their estimates of the percentage of sandstone were obtained nor how the respective volumes were determined, it is difficult to evaluate their results. Hence we will adopt a rounded figure of 25 percent – a result more in accord with both the estimates of Poldervaart and Ronov. One fourth the total volume of sediment (the deep oceans and suboceanic materials excluded), is, therefore, sandstone.

What is the total volume of all sedimentary material in the earth's crust? Various estimates have been made (Table 1-3). These estimates have been arrived at in different ways.

Clarke (1924, p. 32) estimates the total volume of average igneous rock which must be weathered to provide the sodium in the sea plus that retained

Table 1-3. *Total volume of sedimentary deposits*

Authority	Kilometers^3
1. Clarke, F. W. (1924, p. 32)	3.7×10^8
2. Kuenen (1941, p. 188)	13.0×10^8
3. Goldschmidt (1933)	3.0×10^8
4. Wickman (1954)	$4.1 \pm 0.6 \times 10^8$
5. Poldervaart (1955, p. 126–130)	6.3×10^8
6. Horn and Adams (1966, p. 282)	10.8×10^8
7. Ronov (1968, p. 29)	9.0×10^8
8. Blatt (1970, p. 259)	4.8×10^8

in the sediments. He calculates this to be 84×10^6 mi^3 or 350×10^6 km^3. To this is added 10 percent for porosity and additions from the atmosphere to give the 3.7×10^8 km^3. Goldschmidt (1933, p. 866) likewise calculated the amount of weathered igneous rocks and that of the sediments formed therefrom based on the sodium content of the oceans. Kuenen (1941) applied various corrections to Clarke's data and obtained a figure of 8×10^8 km^3 to which he added an estimated 5×10^8 km^3 of disintegrated but undecomposed material (tuffs, graywackes, etc.) for a total of 13.0×10^8 km^3.

Wickman's (1954) calculations are similar to those of Mead, but based on better data.

Poldervaart's (1955, p. 124, etc.) estimates are based on the thickness of the sediments in the continental-shield areas, the younger folded belts, the ocean basins and thc suboceanic areas (shelves, etc.). Poldervaart used Kay's estimates (Kay, 1951, p. 92) for the first two and estimates based on geophysics and rates of sedimentation for the last two. Horn and Adams (1966) approached the problem in somewhat the same manner as Poldervaart, but using somewhat different data, arrived at a figure of 10.8×10^8 km^3 – an estimate surpassed only by those of Kuenen and of Blatt (1970). The estimates of Poldervaart and of Horn and Adams include deep oceanic and suboceanic sediments of which sand forms a negligible part. Sand is concentrated on the continental blocks rather than in the oceanic basins. Sand forms a fourth to a third of the sediments on the continental platform (including the mobile belts and continguous shelves). A recent estimate of total sediment volume is that of Blatt (1970, p. 259) who estimated 10×10^8 mi^3 or about 4.8×10^8 km^3. This estimate is by far the largest on record being about four times that of Poldervaart.

Using Poldervaart's estimate of volume for the continental sediments (176×10^6 km^3) and assuming one fourth of it to be sand, the total volume of sand in the world is 44.0×10^6 km^3. If we use the larger estimate of Horn and Adams (522×10^6 km^3) and assume one third of it is sand, the total volume of sand is about 174×10^6 km^3. The total *mass* of sediments on the continental block is estimated by Poldervaart to be 480×10^{15} tons (4800 Gg) of which, therefore 120×10^{15} metric tons (1200 Gg) is sand. If an individual sand grain (1 mm in diameter and density of 2.7) has a mass of about 0.0014 grams, there would be some 85.7×10^{24} grains of sand in the crust of the earth. As most sand grains are of smaller size, the total number would be much larger (eight times as many if the grains were 0.5 mm in diameter and sixteen times as many for grains 0.25 mm diameter).

Is this quantity a fixed sum or is it constantly being added to? Weathering breaks rocks down. One product of this breakdown is sand which, therefore, increases the total quantity of sand on earth. But sand grains are subject to abrasion and other size-reduction processes. Conceivably such action would so reduce the size of the material acted on so as to eliminate it from the sand grade. Moreover, sandstones are, in turn, owing to deep burial, being transferred to those zones within the earth where they undergo metasomatic transformation into granites, gneisses, and other rocks no longer recognizable as sediments. Do the processes of sand formation and sand destruction balance? Is there a steady state or is there a net increase in the total quantity of sand? Certainly these are fundamental questions that deserve our attention.

Kuenen (1959) has estimated that the yearly production of quartz sand is of the order of 0.05 cubic kilometers per year. He has shown also that the loss of sand by abrasion is incredibly slow. Such rounding of sand as has taken place is thought to be largely due to eolian action in desert areas. Balancing new sand production against losses, led Kuenen to the conclusion that the total quantity of sand is increasing. He estimated that during each and every second in the incredibly long past, the number of quartz grains on earth has increased by 1,000 million grains!

Distribution, Past and Present

Just where is sand in the world today and where did the sand accumulations of the past take place?

Where is sand found in the world today? The most obvious places are the rivers and beaches and to a lesser extent the dunes and shallow shelf seas. The fluvial sands include those found on alluvial fans, in river channels and on floodplains, and those of the deltas of both lakes and the oceans. A little sand also escapes the river channel and finds its way into the backwater swamps and bayous. Shoreline sand includes not only that of the beaches but also that found on offshore bars, in lagoons and on tidal flats. Many windblown dunes are closely associated with beaches and also with major rivers but the most impressive eolian sands are those of the dune fields of some desert basins. Marine sands are largely shelf sands though some sand occurs on the continental slopes, on most of the continental rises as well as on the oceanic abyssal plains and isolated sediment ponds that are found in hilly or mountainous subsea topography.

In short, there seems to be no large geomorphic region of the earth where sand is not found. The deep oceanic basins, the most extensive geomorphic elements, have the least, being almost devoid of sand – and containing only the scattered grains of eolian origin, the turbidite sands on the continental rise and the abyssal plains, and, in some cases, volcaniclastic sands generated by subaqueous eruptions. Clearly the principal environments of sand accumulation are on the continent. The absence of sand in any particular environment is probably due more to an absence of supply rather than to conditions unfavorable for its accumulation.

Not all of the environments of sand accumulation are of equal importance. Not only are some of lesser importance than others but many are places of temporary lodgement of sand which ultimately will be eroded and the sand retrieved to be redeposited elsewhere. Much of the present-day detritus from the western cordillera of North America, for example, is trapped in intermontaine basins and will, in time, be recycled and redeposited. One should not presume, however, that the ultimate destination of all sand is the sea. The ultimate destination of most sand is a geosyncline where it more commonly is deposited in an alluvial rather than a marine environment. It has been estimated that three quarters of all the sediments of the geologic past are in geosynclines and that only one fourth is found on the cratonic platform. Sands form a significant part of the accumulation in both these tectonic elements.

It is noteworthy that most common modern sites of sand accumulation – the beaches and rivers – are linear features and the sand associated with them is confined to a narrow zone. Yet the sands of the past commonly occur in areally extensive stratiform sheets. This discrepancy between the essentially linear loci of sand deposition in the present day and the extensive sand sheets of the past, suggests that the latter are the product of lateral shifting of the loci of sand deposition through time by the lateral migration of streams or by transgressive or regressive shift of the shorcline.

Exceptions to this rule are the more extensive dune fields of some deserts and the broad expanse of sand on some shelf areas. The sand on modern continental shelves, however, turns out to be a relict sand not in equilibrium with the present regime and is probably a fluvial deposit inherited from the low stand of the sea in glacial times (Emery, 1966, p. 12).

In summary, sand is the most continental of all sediments. It is produced in grains that are too large to be blown or washed far off the continents. Therefore, it remains as an ever increasing cover on the continental blocks. Sand is produced on the continent, is shifted from the higher places to the lower sites of accumulation. The only "leakage" from the continents is due to a trifling amount carried to the deep sea in dust storms or by turbidity currents which transport sand down the continental slope to the abyssal plains.

Where did sand accumulation take place in the geologic past? Presumably it could and did accumulate in the same environments then as now. But the relative importance of each of these environments was vastly different. In the central Appalachians, for example, sandstone constitutes about twenty three percent of the whole section (Colton, 1970, p. 11). Of this about fifty-five or sixty percent is believed to be alluvial, about twenty-five percent marine turbidite; the balance, no more than twenty percent, is probably littoral or shallow marine. None is identifiable as eolian. These figures emphasize the importance of the alluvial sediments in miogeosynclines, of which the Appalachians is perhaps a fairly typical example.

Although no ancient eolian sand accumulations are undisputably recognizable as such, some sands display an extraordinary roundness. If Kuenen is right in believing that such roundness is most likely acquired by eolian action, many of the sand grains in the world's accumulation have had a desert eolian stage at some time in their history. Kuenen (1959, p. 23) has estimated that 2×10^6 km^2 of desert is needed to keep the world average roundness constant (to offset the new, sharp-cornered sand added each year).

Nothing has been said here about the distribution of the several kinds of sand (arkosic, lithic, etc.) in the present day world or about their distribution in the geologic past. It may be that the bulk composition of sand has changed with time. These problems are treated in the last chapter in this book.

History of Investigation

The scientific investigation of sands and sandstones goes back nearly two centuries. The earliest work on sands – mainly river and beach sands – was directed toward determining their mineral composition – work inspired in part by the fact that sands may contain useful materials such as gold. This activity was greatly stimulated and advanced by the use of the polarizing microscope. These early researches were largely descriptive and attempted only to record what was there. The indurated sands – the sandstones – were looked on mainly as stratigraphic entities – formations in a geological column. They were described by the field geologist, named, and then placed in the proper position in the geologic section. Names such as the Millstone Grit, the Old Red Sandstone and the Buntsandstein appeared in the early literature and bear witness to the earliest field studies involving sandstones. Interpretations of their origin were based largely on field observations of sedimentary structures, such as ripple marks, and on the contained fossils, if any.

Not until the thin-section technique was available was there serious study of the fabric and composition of these sandstones and utilization of their microscopical characteristics to elucidate the natural history of the rock. Henry Clifton Sorby, whose Presidential address in 1879 before the Geological Society of London, "On the Structure and Origin of Non-calcareous Stratified Rocks," was a milestone, initiated the modern approach to the study of sandstones. One of the earliest papers on the petrography of sandstones was his paper on the Millstone Grit (Sorby, 1859b). In addition to his microscopical investigations, Sorby made many significant field observations on the structure of sandstones, particularly on crossbedding – researches which anticipated the paleocurrent analyses a century later.

Despite Sorby's brilliant demonstration of the usefulness of the polarizing microscope in the study of the sedimentary rocks, it was to the igneous rocks that the thin section technique was largely applied – especially by the German school of petrographers (Rosenbusch, Zirkel and others). A major exception was the work of Lucien Cayeux, whose monographs on the sedimentary rocks of France remain unsurpassed even to this day. Among these was his work on the Tertiary Sandstones of the Paris Basin (1906) and his volume on the siliceous rocks of France (Cayeux, 1929).

The work on the mineralogy of sands, much of it on the minor accessory minerals – the "heavy minerals" – was greatly expedited by the microscopic techniques using polarized light. The utility of heavy minerals in stratigraphic correlation, especially subsurface correlation in oil field exploration and development, led to widespread interest in these constituents of sand. This interest culminated in the appearance of H. B. Milner's "Introduction to Sedimentary Petrography", which first appeared in 1922 – a revised fourth edition appeared forty years later – and which was oriented largely toward the use of heavy minerals in subsurface correlation. For the most part, heavy minerals of sands now play only a minor role in correlation – having been superseded by microfossils and to an even greater extent by geophysical logging techniques.

The mineralogy of sands – ancient as well as Recent – continues to be of interest, however, in provenance studies. Studies of the mineralogy led further to paleogeographic analyses represented, for example, by the modern work of Füchtbauer (1964) on the Molasse of southern Germany. Such studies owe much to the pioneer work of Mackie on the sands and sandstones of Scotland. Mackie attempted to work out the principles of interpretation of sand mineralogy – such as the use of quartz varieties in provenance (1896), the climatic significance of feldspar (1899) and the use of mineral analyses of ancient sandstones to unravel their natural history.

Excepting, perhaps, Hadding's work on the sandstones of Sweden (1929), there have been few monographic studies of sandstones in the Cayeux tradition. There have been, however, noteworthy classical studies of particular sandstone formations. Well-known examples are Krynine's study of the Devonian Third Bradford Sand of Pennsylvania (1940), Dake's earlier study of the Ordovician St. Peter Sandstone of the Upper Mississippi Valley (1921), and Gilligan's study of the Millstone Grit (1920) in Great Britain. Mention should also be made of Marcus Goldman's study of the Miocene Catahoula Sandstone of Texas (1915), which is an unsurpassed example of how much information and understanding can be extracted from a small sample of sandstone.

Interest in the mineral composition of sands, particularly modern beach, dune and river sands, was superseded, or perhaps more properly supplemented, by an interest in their "mechanical" composition. Grain-size analyses, and later the measurement of grain shape and roundness, led to an era of "quantitative sedimentation" and to the study of sands as gross particulate systems. A major contribution to this approach was J.A. Udden's paper on "Mechanical Constitution of Clastic Sediments" published in 1914. Although including silts and gravels, this work was largely a compilation of grain-size analyses of sands. The concept of sand as a population of grains led to the application of statistical methods of population analysis. The pioneer work of Wentworth (1929) and Trask (1932) was followed by a flood of papers. Most influential in this approach, perhaps, are the many papers of W.C. Krumbein (beginning in 1936). Efforts continue to find ways of utilizing the grain size distribution to discriminate between differing environments and/or agents of deposition.

Most recent studies of sand deposits have emphasized the anisotropic fabric of such accumulations. Sand accumulations show a response to the earth's magnetic and gravitational fields and to fluid flow systems. Such response extends from the orientation of individual non-spherical grains through primary sedimentary structures of the deposit to the shape and orientation of the sand body itself. Such "paleocurrent analyses" early made by Sorby (1859a) have been much in vogue during the past decade (see Potter and Pettijohn, 1963).

With the growth and interest in geochemistry since the Second World War, there has been renewed interest in sediments as the products of a complex process of chemical differentiation and fractionation. The sands constitute such a differentiate – the "resistates." As a result of the pioneer efforts of geochemists such as F.W. Clarke, data on the chemical composition

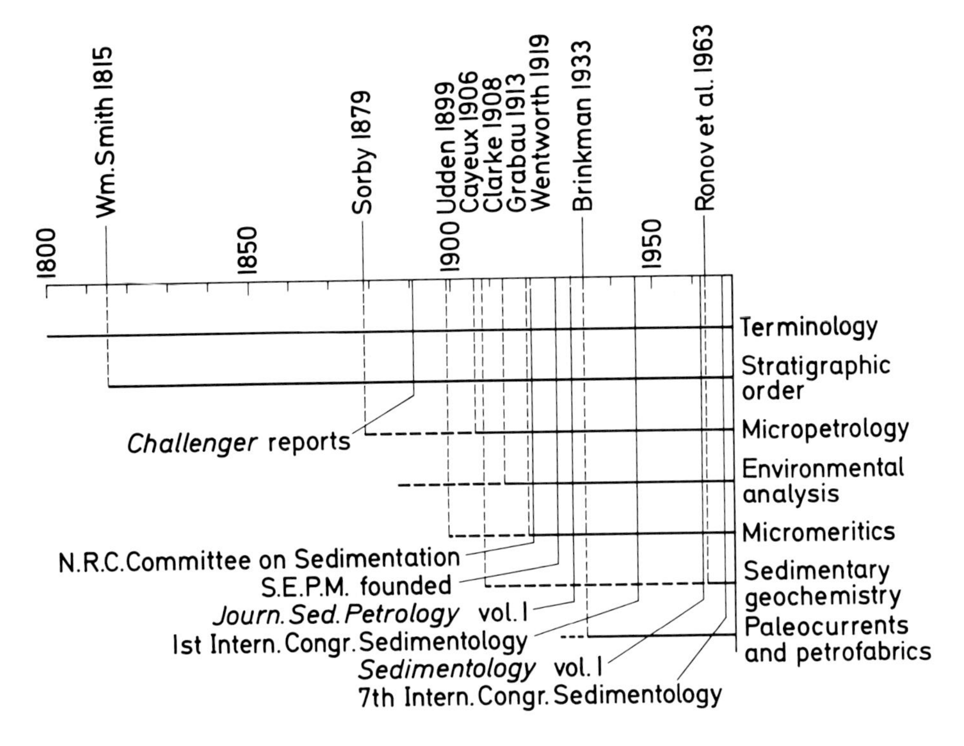

Fig. 1-4. Evolution of "arenology"

of sands and other sediments has accumulated (Clarke, 1924). Sandstones have a bulk composition reflecting their degree of maturity, that is, the degree to which they have evolved toward the stable end-product of the sand-forming processes.

It should be noted, in conclusion, that there is yet another way of looking at sandstones. Not only is a sandstone a petrographic entity, expressed by such designations as arkose or graywacke, and also a stratigraphic entity, as are the Old Red Sandstone and Millstone Grit, but a sandstone may be looked upon as a component of a sedimentary assemblage or association. Thus we may have a "flysch sandstone" or a "molasse sandstone." Such designations refer to a facies presumed to mark a particular stage in the filling of a geosyncline. The flysch assemblage is a thick marine sequence of interbedded turbidite sands, commonly graywackes in the older systems, and pelagic shales laid down during the final stages of filling of a geosyncline. The molasse, on the other hand, is a thick sequence, mainly alluvial but including some marine beds, of coarse conglomeratic sands and shales deposited in a foredeep marginal to the newly uplifted and deformed orogenic belt. Such terms as "flysch sandstone" or "molasse sandstone," therefore are neither stratigraphic nor petrographic terms in the strict sense, but are perhaps something of both coupled with a tectonic concept – that of the geosyncline. In a similar fashion we might speak of a "cratonic sandstone" – again with reference to a sandstone with certain petrographic characters generated and deposited in the cratonic environment.

In summary, the history of the investigations of sands and sandstones reveals the shifting emphasis in ways of looking at sandstones. It is our aim in this book to look at sands and sandstones from all of these points of view and to utilize the several approaches to an overall understanding and interpretation of the natural history of these deposits. The several points of view and the evolution of the science of "arenology" are summarized in Fig. 1-4.

Economic Value of Sand

Sand is an important economic resource. The uses of sand are many. For some purposes such as an abrasive to clean a skillet or mess kit most any sand will do; other uses require a particular kind and quality of sand (Boswell, 1919). Sands are a source of silica for making sodium silicate, for manufacture of carborundum (silicon carbide), for silica brick, for manufacture of both common and optical glass. Sand is an ingredient in plaster, in concrete, is added to clays to reduce shrinkage and cracking in brick manufacture, and is mixed with asphalt to make "road dressing." It is used in foundries as moulding and parting sand, is used as an abrasive (sand paper, sand blast), and employed in the grinding of marble, plate glass and metal. Some sands are used as soil conditioners (lime sand) or fertilizers (glauconitic greensand). Sand is used in filtration and as friction sand (on locomotives). And perhaps most important of all, it is what every child loves to play in.

Sands are exploited for rare minerals and rare elements which they contain. Some are gold bearing; others contains gems, platinum, uranium, tin (cassiterite) and tungsten (wolframite). Others are exploited for monazite (containing thorium and rare earths), zircon (for zirconium) rutile (for titanium) and others. The search for and exploitation of these sands is a special art – that of alluvial prospecting (Raeburn and Milner, 1927; Smirnov, 1965).

In addition to the uses to which sands are put and in addition to the rarer constituents which are recovered from sands, these materials constitute most important reservoirs for the storage of valuable fluids. Large volumes of fluid are contained in the pore systems of sands and sandstones. They are important reservoirs of fresh waters, of brines and of petroleum and natural gas. Sand strata are also the conduits for artesian flow. Knowledge of the shape and attitude of these sand reservoirs and their porosity and permeability is necessary for extraction of the fluids therein. Fluids may be injected into sands also. Sands are thus utilized for natural gas storage and for recharge with fresh waters for future use or injected with water to drive out the contained oil.

Sandstones are used for building stone construction, as flagstone, and, if crushed, as road fill, road metal, and railroad ballast.

The economic aspects of sands are not confined to its value as raw material and its various uses. Sand production, movement and deposition is of concern to the engineering geologist and to the geomorphologist, especially those concerned with shore erosion and harbour development. Some understanding of the sand budget – sand supply and accretion and sand deficit or removal – is necessary if the problems of shore engineering are to be solved. Similarly dune stabilization and prevention of sand encroachment on cultivated lands and forests or on roads and other structures requires an understanding of the geology of sand. Many of the problems of river management likewise involve sand.

To the three of us the study of sands is perhaps the most rewarding of all sediments. They are more amenable to study than the shales and have greater value in determination of provenance, paleocurrents, and paleogeographic reconstruction than either shales or limestones. Surely if one had to pick a *single* lithology from which he had to reconstruct the earth's geologic history, it would be sandstone: it goes back further in time than limestone, is less susceptible to diagenesis than shales and carbonate rocks, and is widespread all over the continents.

References

Bagnold, R. A.: The physics of blown sand and desert dunes, 265 p. London: Methuen 1941.

Blatt, Harvey: Determination of mean sediment thickness in the crust: a sedimentologic model. Geol. Soc. America. Bull. **81**, 255–262 (1970).

Boswell, P. G. H.: Sands; considered geologically and industrially, under war conditions. Inaugural Lecture, 38 p. Univ. Liverpool Press 1919.

Cayeux, Lucien: Structure et origine des grès du Tertiare parisien. Ètudes des gites minéraux de la France, 160 p. Paris: Impr. Nationale 1906.

— Les roches sédimentaires de France, Roches siliceuses, 774 p. Paris: Impr. Nationale 1929.

Clarke, F. W.: The data of geochemistry (5th ed.), 841 p. U.S. Geol. Survey Bull. **770**, (1924).

Colton, G. W.: The Appalachian basin – Its depositional sequences and their geologic relationships. In: Fisher, G. W., Pettijohn, F. J., Reed, J. C., Weaver, K. N., Eds.: Studies of Appalachian geology, central and southern. p. 5–47. New York: Interscience Publ. 1970.

Dake, C. L.: The problem of the St. Peter sandstone. Univ. Missouri School Mines and Metall. Bull. (tech. ser.) **6**, 158 p. (1921).

Emery, K. O.: Geologic background. In: The Atlantic Continental Shelf and Slope of the United States. U.S. Geol. Survey Prof. Paper **529**-A, A 1–A 23 (1966).

Füchtbauer, H.: Sedimentpetrographische Untersuchungen in der älteren Molasse nördlich der Alpen. Eclogae géol. Helvetiae, **57**, 157–298 (1964).

Gilligan, A.: The petrography of the Millstone Grit of Yorkshire. Geol. Soc. London Quart. Jour. **75**, 251–294 (1920).

Goldman, M. I.: Petrographic evidence on the origin of the Catahoula Sandstone of Texas. Am. Jour. Sci., ser. 4, **39**, 261–287 (1915).

Goldschmidt, V. M.: Grundlagen der quantitativen Geochemie. Fortschr. Min. Krist, Petrogr. **17**, 112–156 (1933).

Hadding, A.: The pre-Quaternary sedimentary rocks of Sweden: Part III, The Paleozoic and Mesozoic sandstones of Sweden. Lunds Univ. Årsskr., N. F., Avd. 2, **25**, nr. 3, 287 p. (1929).

Holmes, A.: The age of the earth, 196 p. London: Nelson 1913.

Horn, M. K., Adams, J. A. S.: Computer-derived geochemical balances and element abundances. Geochim. et Cosmochim. Acta **30**, 279–290 (1966).

Kay, Marshall: North American geosynclines. Geol. Soc. America Mem. **48**, 143 p. (1951).
Krynine, P. D.: Petrology and genesis of the Third Bradford Sand. Pennsylvania State College Bull. **29**, 134 p. (1940).
— The megascopic study and field classification of sedimentary rocks. Jour. Geology **56**, 130–165 (1948).
Krumbein, W. C.: Application of logarithmic moments to size frequency distributions of sediments. Jour. Sed. Petrology **6**, 35–47 (1936).
Kuenen, Ph. H.: Geochemical calculations concerning the total mass of sediments in the earth. Am. Jour. Sci., **239**, 161–190 (1941).
— Sand – its origin, transportation, abrasion and accumulation. Geol. Soc. S. Africa, Annexure **62**, 33 p. (1959).
Leith, C. K., Mead, W. J.: Metamorphic geology, 337 p. New York: Henry Holt 1915.
Mackie, W.: The sands and sandstones of eastern Moray. Trans. Edinburgh Geol. Soc. **7**, 148–172 (1896).
— The felspars present in sedimentary rocks as indicators of contemporaneous climates. Trans. Edinburgh Geol. Soc. **7**, 443–468 (1899).
Mead, W. J.: Redistribution of elements in the formation of sedimentary rocks. Jour. Geology **15**, 250 (1907).
Milner, H. B.: An introduction to sedimentary petrography, 125 p. London: Murby 1922.
— Sedimentary petrography, Vol. 1, 643 p. Methods in sedimentary petrography, Vol. 2, 715 p, Principles and applications. New York: Macmillan 1962.
Poldervaart, A.: Chemistry of the earth's crust. In: Poldervaart, A., Ed.: Crust of the earth – a symposium. Geol. Soc. America Spec. Paper **62**, 119–144 (1955).
Potter, P. E., Pettijohn, F. J.: Paleocurrents and basin analysis, 296 p. Berlin-Göttingen-Heidelberg: Springer 1963.
Raeburn, C., Milner, H. B.: Alluvial prospecting, 478 p. London: Murby 1927.
Ronov, A. B.: Probable changes in the composition of sea water during the course of geologic time. Sedimentology **10**, 25–43 (1968).
— Migdisov, A. A., Barskaya, N. V.: Tectonic cycles and regularities in the development of sedimentary rocks and paleogeographic environments of sedimentation of the Russian platform (an approach to a quantitative study). Sedimentology **13**, 179–212 (1969).
— Mikhailovskaya, M. S., Solodkova, I. I.: Evolution of the chemical and mineralogical composition of arenaceous rocks. In: Vinogradov, A. P., Ed.: Chemistry of the earth's crust, v. 1, Acad., Sci., U.S.S.R. Israel Progr. Sci., Translations **1966**, 212–262 (1963).
Russell, R. J.: Where most grains of very coarse sand and gravel are deposited. Sedimentology **11**, 31–38 (1968).
Schuchert, Chas.: Geochronology or the age of the earth on the basis of sediments and life. In: The age of the earth. Bull. Nat. Research Council **80**, 10–64 (1931).
Shepard, F. P.: Nomenclature based on sand-silt-clay ratios. Jour. Sed. Petrology **24**, 151–158 (1954).
Smirnov, V. I.: Geologiia rossypei (Geology of placer deposits). Div. Earth Sci., Sci. Council Ore-formation, Acad. Sci. U.S.S.R., Moscow, 400 p. (1965).
Sorby, H. C.: On the structure and origin of the Millstone-Grit of South Yorkshire. Proc. West Yorkshire Geol. Soc. **3**, 669–675 (1859a).
— On the structures produced by the current present during the deposition of stratified rocks. The Geologist **2**, 137–147 (1859b).
— On the structure and origin of non-calcareous stratified rocks. Pres. address, Geol. Soc. London Proc. **35**, 1879, 56–77 (1880).
Trask, P. D.: Origin and environment of source sediments of petroleum, 323 p. Houston: Gulf Publ. Co. 1932.
Trefethen, J. M.: Classification of sediments. Am. Jour. Sci. **248**, 55–62 (1950).
Udden, J. A.: Mechanical composition of clastic sediments. Geol. Soc. America Bull. **25**, 655–744 (1914).
Wentworth, C. K.: Method of computing mechanical composition types of sediments. Geol. Soc. America Bull. **40**, 771–790 (1929).
— Fundamental limits to the sizes of clastic grains. Science **77**, 633–634 (1933).
Wickman, F. F.: The "total" amount of sediment and the composition of the "average igneous rock". Geochim. et Cosmochim. Acta **5**, 97–110 (1954).

General Source Materials for the Study of Sand and Sandstone

References and Textbooks

Barth, T. F. W., Correns, C. W., Eskola, P.: Die Entstehung der Gesteine, 422 p. Berlin, Göttingen and Heidelberg, Springer 1960.

Part II, the sedimentary rocks, includes a brief discussion of weathering transport, texture and diagenesis. Strong emphasis on chemical composition and geochemistry. Very quantitative for its time. Reprinted without change from the original 1939 edition. A good source for older European references.

Baturin, V. P.: Petrograficheskii analiz geologicheskogo proshlogo po terrigennym komponentam (Petrographic analysis of terrigenous components of the geologic past), 335 p. Akad. Nauk SSSR, Institut Goriuchikh Iskopaemykh 1947.

A well-illustrated text in two parts: terrigenous components (3 chapters on mineralogy, grain size and shape, 124 pages) and geography and paleogeography (6 chapters, 195 pages). In part II there is strong emphasis on present physiography and sands as keys to ancient provenance studies plus some stress on stream hydraulics. This book is a further development of the author's 1937 work entitled, "Paleogeography Based on Terrigenous Components". Approximetely 300 references, about 100 of which are to the non-Russian literature.

Bilibin, Ju. A.: Osnovy geologii rossypei (Principles of the geology of placer deposits), 471 p. Moskow: Izdatel'stvo Akad. Nauk SSSR 1955.

Twenty-six chapters organized into four parts: formation of placer deposits, types, structure and lithology, and conclusions. Alluvial deposits are the center of interest but there is one chapter each on lake, lagoon, delta, and glacial placers plus one on beaches. Strong emphasis on geomorphology. Contains only a few illustrations and almost no references.

Bolewski Andrzej, and Turnau-Morawska, Maria: Petrografia, 811 p. Warsaw: Wydawnictwa Geologiczne 1963.

Igneous, sedimentary and metamorphic rocks in 33 chapters, 16 of which are devoted to sediments (p. 339–632). Historical introduction and brief section on methods. Well illustrated with many photomicrographs, many chemical analyses and some modal analyses. Mostly Polish references and examples.

Boswell, P. G. H.: On the mineralogy of the sedimentary rocks, 393 p. London: Murby 1932.

A collection of short essays on selected topics and a series of abstracts of more than 1,000 papers dealing with sedimentary mineralogy – mainly of sands. Classified indices.

Bouma, A. H., Brouwer, A.: Eds. Turbidites, 264 p. Amsterdam: Elsevier 1964.

Fifteen papers, mostly in English, on turbidite deposition.

Cailleux, A., Tricart, J.: Initiation à l'étude des sables et des galets, various paging. Paris: Centre de Documentation Univ. 1959.

Three volumes, the last two tables. Summarizes the properties of sands and pebbles in different environments throughout the world.

Carozzi, A.: Microscopic sedimentary petrography, 485 p. New York: John Wiley, Inc. 1960.

Qualitative description and 88 figures, mostly photomicrographs or drawings thereof. Good for western European references. Three parts (clastic, biochemical and chemical rocks) in nine chapters.

Cayeux, L.: Contribution à l'étude micrographique des terraines sédimentaires I. Étude de quelques dépôts siliceux secondaires et tertiares du Bassin de Paris, 206 p. Lille: Le Bigot Féres 1897.

Petrography of quartz, chert, and tuffaceous chalk. Detailed treatment of framework grains, organisms, cement and some chemical data. Chapter 4 devoted to glauconite; chapter 5 to radiolarites. The first "superdetailed" petrographic description and interpretation?

Cayeux, L.: Les roches sédimentaires de France, v. 1, Roches siliceuses, 696 p. Paris: Masson 1929.	A monographic, superbly illustrated, study of sandstones and cherts as seen in thin section. A classic landmark in the development of sedimentary petrology. Ten chapters, 30 plates, and chemical analyses.
Füchtbauer, Hans, Müller, German: Sedimente und Sedimentgesteine, Teil II, 726 p. Stuttgart: E. Schweizerbart'sche Verlagsbuchhandlung 1970.	Seven chapters emphasizing petrography. Chapter 3 (92 p.) most concerns sandstones. Many illustrations and references.
Garrels, Robert M., Mackenzie Fred T.: Evolution of sedimentary rocks, 397 p. New York: Norton 1971.	Sedimentary rocks as participants in the major geochemical cycle of the earth, using newest ideas of origin of sea water and sediments as chemical entities.
Grabau, Amadeus, W.: Principles of stratigraphy, Vols. 1 and 2, p. 1–581 and p. 582–1185. New York: Dover 1960.	Really sedimentology in its broadest sense. Thirty-two chapters that cover the field with much on sand and sandstone. First written in 1913 and revised in 1924. A classic far ahead of its time.
Hadding, Assar: The pre-Quaternary sedimentary rocks of Sweden, Pts. 1–6: Lund Univ. Årssk., N.F. Avd 2; A survey of the pre-Quaternary sedimentary rocks of Sweden and the Paleozoic and Mesozoic conglomerates of Sweden, v. 23, Pts. 1 and 2. 171 p.; The Paleozoic and Mesozoic sandstones of Sweden, v. 25, Pt. 3, 287 p., 1927, and later.	Classic studies, well ahead of their time, that are still worth consulting. Qualitative petrography, some chemical analyses and many very good photomicrographs, Part III has a classification of sandstones.
Hatch, F.H., Rastall, R.H.: Petrology of the sedimentary rocks, 4th ed., 408 p. London: Murby 1965.	A general textbook on sedimentary rocks – classification, description, and origin, revised by J.T. Greensmith. Eight chapters beginning with environments of deposition and ending with lithologic associations. Six deal with sandstones. Few illustrations and references. First published in 1913.
Jung, Jean: Précis de pétrographie, 314 p. Paris: Masson 1958.	Part II has a qualitative petrography of sedimentary rocks (p. 51–142). Mostly line drawings. No references. Has an index of sedimentary rock names and terms.
Krumbein, W.C., Sloss, L.L.: Stratigraphy and sedimentation, 2nd ed., 660 p. San Francisco: W.H. Freeman and Co. 1963.	A very good beginning text on a wide range of sedimentary topics. Chapters 11, 12, and 13, sedimentary tectonics, stratigraphic maps, and stratigraphic analysis are noteworthy, especially, Chapter 13, which includes an excellent discussion of lithologic associations.
Kukal, Zdeněk: Geologie recentnich sedimentů, 441 p. Praha: Nakladatelství Ceskoslovenske Akad. Věd 1964.	A well illustrated (121 figs.) and well-referenced compilation of modern sediments from the world over. Comprehensive in topical coverage with 22 chapters, including an initial one on classification of environments. Clays and carbonates as well as sands. Some geochemistry. Over 200 tables.
Logvinenko, N.W.: Petrografiia osadochnykh porod (Petrography of sedimentary rocks), 416 p. Moscow: Vysshaiashkola 1967.	Seventeen chapters arranged in four parts: Part 1, weathering, mineralogy, and textures and structures; Part 2, description of types of sedimentary rocks; Part 3, Larger concepts.
Lombard, A.P.: Géologie sédimentaire, les series marines, 722 p. Paris: Masson 1956.	A general treatise on sedimentary rocks of marine origin primarily based on field studies. In 4 parts: Part I, modern sediments; Part 2, marine sedimentary rocks; Part III, sedimentary associations; and Part IV, origin. Substantial bibliography.

Nalivkin, D. V.: Uchenie o fatsiiakh (Facies studies). Akad. Nauk. SSSR **1** and **2**, 534 p. and 393 p. (1956).	Volume one has five chapters; volume two has but two. A combination of stratigraphy and sedimentation, the latter organized by environment. Chapter 2 of volume 2 gives criteria for environmental determination. Few illustrations but many references.
Papin, Victor C.: Petrographia rocilor sedimentaire, 506 p. Bucuresti: Editura Stiintifică 1960.	A general text with 31 chapters and four parts: an introduction (53 p.), constituents (140 p.), lithogenesis (118 p.) and general petrography (178 p.). Contains a chapter on the pyroclastic rocks and 144 photomicrographs many of which are sandstones. Over 500 references.
Petránek, Jan: Usazené horniny – jejich složeni, vžnik a ložiska (Sedimentary rocks – their composition, origin, and deposition), 718 p. Prague: Naklad, Ces. Akad. Ved. 1963.	A general work on the nature and origin of sedimentary rocks with more than 1500 references, 350 figures and 150 photographs.
Pettijohn, F. J.: Sedimentary rocks, 2nd ed., 718 p. New York: Harper and Row 1957.	A general text and reference work on sedimentary rocks, their nature and origin; 173 line drawings and 119 tables. Excellent photomicrography and many references. Fifteen chapters beginning with individual particles and ending with the geological history of sediments.
Pustovalov, Leonid: Petrografiia osadochnykh porod (Petrography of sedimentary rocks). Vols. 1 and 2, 420 p. Moscow and Leningrad: Gostoptekhizdat 1940.	Structure, texture, color, and description of principal sedimentary rocks. Six chapters.
Radulesca, Dan.: Petrografia rocilor sedimentare, 2nd ed., 350 p. Bucuresti: Editura Didactică si Pedagogiă 1965.	Three major parts: petrogenesis (5 chapters) with emphasis on modern sediments and processes; petrography (13 chapters) includes composition, structures, and discussion of major groups; problems of petrology (6 chapters) with emphasis on petrographic provinces, lithic associations, and changes with time. Many references and illustrations.
Raeburn, C., Milner, H. B.: Alluvial prospecting. 478 p. London: Murby 1927.	Ten chapters including theory and practical methods plus diagnostic properties of important alluvial heavy minerals.
Ruchin, L. B.: Grundzüge der Lithologie, 806 p. (German translation of original Russian work). Berlin: Akademie-Verlag 1958.	A general treatise on the nature and origin of sedimentary deposits. Twenty eight chapters in five parts: sedimentary rocks, origin and processes of formation, facies and facies analysis, major sedimentary deposits, and Recent sedimentation. A good source for pre-1958 Russian references.
— Osnovy obshcheĭ paleogeografii (Principles of general paleogeography), 3rd. ed., 628 p. Leningrad: Gostoptekhizdat 1962.	Unified treatment of an old topic from a Russian viewpoint. Fourteen chapters in four parts: principles, methods, chief types of ancient landscapes, and practical significance of paleogeographical maps. Broad in scope and well referenced.
Selley, R. C.: Ancient sedimentary environments. A brief survey, 240 p. Ithaca: Cornell University Press 1970.	Ten major environments, five of which principally concern sand. Economic applications.

Shvetsov, M. S.: Petrografiia osadochnykh porod (Petrography of sedimentary rocks), 3rd ed., 416 p. Moscow: Gostoptekhizdat 1958.

Organized into 3 parts: origin of sedimentary rocks (2 chapters), their component parts and structures (3 chapters), and chief rock types (8 chapters). Weathering and common sediments is followed by common components and sedimentary structures and finally the types and origins of major sedimentary clans. Two interesting chapters conclude the book: sedimentary rocks as gatherers of precious metals (Ch. 12) and the development and contemporary state of the study of sedimentary rocks in the Soviet Union plus an elaboration of questions about the regularities of their formation (Ch. 13). Two hundred and forty four references and 40 plates.

Smirnov, V. I., Ed.: Geologiia rossypei (Geology of placer deposits), 400 p. Moscow: Acad. Sci. USSR Izd. Nauka 1965.

A collection of 55 papers arranged into five parts: general questions, placer deposits of gold, placer deposits of titanium and rare earths, placer deposits of diamonds, and methods of study. Contains a 40-page bibliography of Russian literature on commercial placer deposits. Paleogeography, paleocurrents, and some papers on experimental hydraulics of heavy minerals.

Strakov, N. M.: Osnovy teorii litogeneza, zakonomernosti sostava i razmeshcheniia gumidnykh otlozheniĭ (Theory of lithogenesis, Regularities in the composition and deposition of humid climate deposits), Vol. 2, 574 p. Moscow: Akad. Nauk SSSR, Geol. Inst. 1960.

Ten chapters: grain size; clay minerals; distribution of chemical elements; aluminum, iron, and manganese ores; accumulations of P, $CaCO_3$, $MgCO_3$ and SiO_2; organic matter; basin sediments as a physical-chemical system (diagenesis); properties of diagenetic mineral formation; carbonaceous concretions. Chapter 1 is most important to sandstones and has four parts: sorting in stable regions, sorting and climate, grain size in tectonically active areas, and sorting and its relation to environment of deposition.

— Osnovy teorii litogeneza (Theory of lithogenesis), Vol. 1, 245 p. English edition by Consultants Bureau. New York: 1962, 1967.

Volume 1 stresses the formation of sedimentary rocks in response to climate. Chemical analyses. Many world maps. Five chapters.

— Osnovy teorii litogeneza, sakonomernosti sostava i razmeshcheniia aridnyk otlotheniĭ (Theory of lithogenesis, regularities of the composition and deposition of arid deposits), Vol. 3, 550 p. Moscow: Akad. Nauk SSSR, Geol. Inst. 1962.

Three parts. The four chapters of part I are: general characteristics of arid zone terrigenous sedimentation; Cu, Pb and Zn in the arid zone; P, $CaCO_3$, $MgCO_3$ and SiO_2 in weakly mineralized arid zone basins; and accumulations of organic matter. Part II is a description of contemporary saline deposits (three chapters). In Part III ancient saline deposits of different types are described and explained (six chapters).

—, and others: Obrazovanie osadkov v sovremennykh vodoemakh (Formation of sediments in Recent basins), 791 p. Moscow: Akad. Nauk. SSSR 1954.

Three parts: methods of study, lake sediments, and processes (with chemical overtones). Fourteen chapters, the last compares modern and ancient.

Teodorovich, G. I.: Uchenie ob osadochnykh porodakh (Study of sedimentary rocks), 572 p. Leningrad: Gostoptekhizdat 1958.

A general text (not seen by us).

Trask, Parker, D., Ed.: Recent marine sediments, a symposium, 726 p. Tulsa: Soc. Econ. Paleon. Mineral. 1955.

A reprinting of the classic 1939 edition. Includes a bibliography of about 500 post-1939 articles.

Twenhofel, W. H.: Principles of sedimentation, 2nd ed., 673 p. New York: McGraw-Hill 1949.

A text, abridged and amended, based primarily on the material in the "Treatise on Sedimentation." No equations. Few illustrations.

Twenhofel, W. H., and others: Treatise on sedimentation, 926 p. Baltimore: Williams and Wilkens 1932.	The standard work covering the literature prior to 1932. Primarily a treatise on processes and environments. Qualitative. Good source for older English language literature.
Van Straaten, L. M. J. U., Ed.: Deltaic and shallow marine deposits, 464 p. Amsterdam: Elsevier 1964.	Contains many short papers given at the 6th International Sedimentological Congress.
Vatan, A.: Pétrographie sédimentaire, 279 p. Paris: Editions Technip 1954.	A general introductory text mostly for terrigenous sediments. Has a brief but interesting section on the history of sedimentary petrography.
— Manuel de sédimentologie, 397 p. Paris: Editions Technip 1967.	A general beginning text: fourteen chapters plus four appendices on methods. Covers all major rock types plus petroleum. Well illustrated. Good source of European references. Easy French.
Weller, J. M.: Stratigraphic principles and practice, 725 p. New York: Harper 1960.	The broader aspects of sedimentation as seen by a widely experienced stratigrapher. Four parts (introduction, materials of stratigraphy, stratigraphic bodies and relations, and an appendix) in 18 chapters, the last entitled "Field Work and Geologic Mapping." Well illustrated. Sparingly referenced.
Williams, Howel, Turner, F. J., Gilbert, C. M.: Petrography, 406 p. San Francisco: Freeman 1954.	More descriptive than interpretative, qualitative rather than quantitative. Informative line drawings. A beginning text.

Petrographic Manuals and Lexicons

Bonorino, F. G., Teruggi, M. E.: Lexico sedimentologico, 164 p. Buenos Aires: Inst. Nac. Inv. Cien. Nat. 1952.	A lexicon of terms: English, German, and Spanish.
Bouma, A. H.: Methods for the study of sedimentary structures, 458 p. New York: Wiley – Interscience 1969.	A comprehensive manual of methods applicable mainly to unconsolidated sediments. Sedimentary peels, impregnation, radiography, and other methods. Sampling.
Brajnikov, B., and others: Techniques d'étude des sédiments, Pt. II, 110 p. Paris: Hermann 1943.	Three parts (chemical, physical and mineralogical techniques) with a section on how to analyze associated water. Sparingly referenced.
Cayeux, L.: Introduction à l'étude petrographique des roches sedimentaires, 524 p. Paris: Imprimerie Nationale 1931.	The first monographic treatment of sedimentary petrography. One hundred and eighty seven pages devoted to physical and microchemical methods plus staining technique; the minerals of sedimentary rocks cover 194 pages. The text is illustrated by a supplementary atlas containing 56 full-page plates. Emphasis on microchemical methods for silicates is unique.
Commité des Technicians: Essai de nomenclature des roches sédimentaires, 78 p. Paris: Editions Technip 1961.	Primarily a glossary of terms but also contains some useful charts and tables.
Duplaiz, S.: Determination microscopique de mineraux des sable, 80 p. Paris-Liege: Librarie Polytech. Ch. Beranger 1948.	Contains a description of properties of some 70 species or varieties of detrital minerals, both light and heavy. Most important species illustrated. Contains identification tables.
Folk, R. L.: Petrology of sedimentary rocks, 170 p. Austin: Texas Hemphill's Book Store 1968.	A very useful syllabus, mostly for arenites. Good for teaching, although virtually no references. Reprinted frequently.

Griffiths, J. C.: Scientific method in analysis of sediments, 508 p. New York: McGraw-Hill 1967.	A personal philosophy on how to study sediments, especially sandstones, with emphasis on statistical treatment. Twenty-two chapters.
Jones, M. P., Fleming, M. G.: Identification of mineral grains, 102 p. Amsterdam: Elsevier 1965.	Selection, separation, and identification (optical, magnetic and chemical) procedures. Determinative tables.
Köster, Erhard: Granulometrische und morphometrische Meßmethoden an Mineralkörnern, Steinen und sonstigen Stoffen, 336 p. Stuttgart: F. Enke 1964.	An up-to-date treatment of many of the same topics covered by Krumbein and Pettijohn's manual.
Krumbein, W. C., Pettijohn, F. J.: Manual of sedimentary petrography, 549 p. New York: Appleton-Century-Crofts 1938.	An older standard work on laboratory methods with special emphasis on statistical summarization of quantitative data. Contains a long section on identification of detrital minerals, both light and heavy. Most important minerals illustrated. Identification tables.
Kummel, B., Raup, D. M., Eds.: Handbook of paleontological techniques, 825 p. San Francisco: Freeman 1965.	Part 4 is particularly useful for the sandstones.
Milner, H. B.: Sedimentary petrography; Vol. 1, 643 p. Methods in sedimentary petrography; Vol. 2, 715 p. Principles and applications. New York: Macmillan 1962.	Volume 1 contains a synoptic treatment of all laboratory methods of sedimentary analysis: sampling, sample preparation, size analysis, mineralogical analysis with microscope, chemical and microchemical analysis, X-ray, spectrographic, electrochemical, fluorescence, electron microscopy, and nuclear methods. Volume 2 has a major section on properties of detrital minerals (more than 200 p.) and shorter sections on clay minerals and consolidated sediments.
Müller, G.: Methoden der Sediment-Untersuchung, 303 p. Stuttgart: E. Schweizerbart'sche Verlagsbuchhandlung 1964.	A manual covering geophysical methods useful in the field, field methods of sedimentological analysis, laboratory methods (sample preparation, grain size analysis, fabric analysis, mineral separation, mineral determination by microscopic, X-ray, DTA, and infrared methods, determination of organic content, porosity and permeability). Now translated into English.
Plas, Leedert van der: The identification of detrital feldspars, 305 p. Amsterdam-New York: Elsevier 1966.	Exhaustive determinative tables and methods.
Prebrazhenskii, I. A., Sarkisyan, S. G.: Mineraly osadochnykh porod (Minerals of sedimentary rocks), 462 p. Moscow: Gostoptekhizdat 1954.	A treatise on the determinative mineralogy of sedimentary rocks.
Russell, R. D.: Tables for the determination of detrital minerals. In: Report of the Committee on Sedimentation, 1940–1941, Division of Geology and Geography, National Research Council, p. 6–8 (Available from Department of Earth and Planetary Sciences, The Johns Hopkins University, Baltimore, Maryland 21218) 1942.	Compact but very useful tables that cost $ 0.50 per set.

Strakhov, N. M.: Methode d'étude des roches sédimentaires, vols. 1 and 2. Service Inf. Géol., Annales, Nov. 1958, no. 35, 1007 p. (1957).	A very comprehensive work dealing with field observations, petrographic analysis in the laboratory, chemical methods (for organic material, trace elements, pH, bulk composition, etc.), and methods appropriate for special classes of sediments. Translated from the Russian.
Thoulet, J.: Précis d'analyse des fonds sous-marins actuels et anciens, 220 p. Paris: Libraire Militaire, R. Chapelot et Cie 1907.	A manual with 9 chapters including mechanical, physical, chemical and mineralogical methods as well as recording forms for the analysis of clays and sands. Tables for mineral identification, optical and chemical.
Tickell, R. G.: The examination of fragmental rocks, 127 p. Palo Alto: Stanford Univ. Press 1947.	A brief student manual dealing with measurement of grain size, porosity and permeability, preparation of samples, mineral identification and description of minerals. Includes tabulated physical and optical properties of each mineral species and identification tables.
— The techniques of sedimentary mineralogy, 220 p. Amsterdam: Elsevier 1965.	Size analysis, bulk properties, preparation of sediments, mineral identification, etc.
Twenhofel, W. H., Tyler, S. A.: Methods of study of sediments, 183 p. New York: McGraw-Hill 1941.	A brief, student manual on the laboratory study of sediments. Deals with field study, sampling, preparation of samples, mechanical analysis, mineral separation, quantitative determination of mineral content, graphic methods, chemical methods of mineral separation, and various physical properties. Lacks data on mineral species and identification tables.

Periodicals and Serials

American Association of Petroleum Geologists Bulletin. Tulsa, Oklahoma: American Association of Petroleum Geologists.	Stresses petroleum but contain articles relevant to sands each month.
Deep-Sea Research. Oxford: Pergamon.	Modern marine sediments from many points of view. Bimonthly.
Developments in Sedimentology. Amsterdam: Elsevier.	A series of books, some nine to date, on topics in sedimentation, so far primarily about sands.
Journal of Sedimentary Petrology. Tulsa: Society of Economic Paleontologists and Mineralogists.	All aspects of sediments and sedimentation. Contains more on sands than any other journal.
Limnology and Oceanography: Lawrence, Kansas: (Allen Press, Inc.) American Soc. of Limnology and Oceanography, Inc.	Occasional articles on the physical and chemical aspects of modern sands. Quarterly.
Marine Geology. Amsterdam: Elsevier.	Much of interest on geology, geochemistry and geophysics of sea bottom and shore and nearly always one or two articles on sands.
Maritime Sediments. Halifax and Fredericton.	A quarterly journal devoted to studies of modern and ancient sediments in the Maritime provinces of Canada and adjacent areas.
Palaeogeography, Palaeoclimatology, Palaeoecology. Amsterdam: Elsevier.	A quarterly journal very broad in scope with mostly English-language articles.

Sedimentary Geology. Amsterdam: Elsevier.	A new journal devoted to applied and regional sedimentology.
Sedimentology. Amsterdam: Elsevier.	Much on sands and sandstones. Four issues per volume, currently two volumes a year.
Senckenbergiana Lethaea: Frankfurt am Main.	Emphasis on biologic aspects of sedimentation – especially Lebenspuren – and modern tidal sediments of North Sea.
Senckenbergiana Maritima. Frankfurt am Main: Senckenbergische Naturforschende Gesellschaft.	Devoted to comparison of modern and ancient depositional environments and biotopes: marine geology, marine biology and "actual paleontology".

Bibliographies

General

Abstracts of North American Geology. Washington D.C.: U.S. Geological Survey.	Monthly. Generally short abstracts. Began 1966. Terminated at end of 1971.
Annotated Bibliography of Economic Geology. Urbana: The Economic Geology Publishing Co.	A general international bibliography with economic emphasis from the late 20's to the present. Annotated and indexed.
Bibliography and Index of Geology. Boulder, Colorado: Geological Society of America.	A monthly bibliography and index. Covers the earth science literature of the world including North America. No abstracts. Began in 1969.
Bibliography and index of geology exclusive of North America. Boulder, Colorado: Geological Society of America.	Comprehensive coverage with informative brief abstracts. Issued annually from 1934–1968. Now superseded by Bibliography and Index of Geology.
Bibliography of North American Geology. Washington: U.S. Geological Survey.	Comprehensive coverage of North American literature from 1785 to present. No abstracts. Issued annually as Bulletins of the U.S. Geol. Survey. Discontinued 1967.
Bulletin Signaletique. Paris: Centre National de la Recherche Scientifique.	Part 10, Sciences of the Earth, appears monthly and contains brief abstracts in French of current geologic happenings. Well indexed.
Canadian Index to Geoscience Data, Ed. 70-1, Information Science Industries, Ottawa 5, Ontario, Canada.	Eleven volumes of computer typed bibliographic geoscience data (over 6000 pages) for the ten provinces and territories of Canada plus a thesaurus. A good source to find out about sand and sandstone in Canada.
Geoscience Abstracts. Washington, D.C.: American Geological Institute.	Mostly North American literature plus Soviet literature in English translation. Largely author's abstracts. Discontinued in 1967.
Mineralogical Abstracts. London: The Mineralogical Society of Great Britain and the Mineralogical Society of America.	Quarterly with an index every two years.
Referativnyi Zhurnal. Geologiia: Moscow, Acad. Nauk SSSR.	Monthly issues plus yearly subject and author indexes. Some abstracts have illustrations.
Ward, D. C.: Geologic reference sources, 114 p. Univ. Colorado Studies, Ser. Earth Sci. No.5, 1967.	A good "where to find it" containing much of value – from guides to translation services to films and slides to geochronology and much more.

Zentralblatt für Geologie und Paläontologie, Teil 1. Stuttgart: E. Schweizerbart'sche Verlagsbuchhandlung.	Part I contains annotations for general, applied regional and historical geology. Special subheadings for marine geology and sedimentology. In addition, each issue contains a literature review on a special topic. Emphasizes the European literature. Since 1901 under this title.

Sedimentation

Kummel, B., Raup, D. M.: Handbook of paleontological techniques, 825 p. San Francisco: Freeman 1965.	Much of value for references to source materials, p. 767–832.
Institute Francais du Pétrole: Bibliographie europeene des progres recents de la sédimentologie. Rev. Inst. Francais Pétrole **10**, Suppl. 5, 78 p. (1955).	The literature of nine European countries and Japan.
International Association of Sedimentology: Bibliographie des travaux recents de sedimentologie, 164 p. Paris: Editions Technip 1960.	The bibliographies of 13 countries plus review articles by Shepard, F. P., Siever, R.
Inter-Agency Committee on Water Resources: 1952–64, Annotated bibliography on hydrology and sedimentation 1941–50, 1950–54, 1955–58, 1959–62 (United States and Canada), various paging. Washington: U.S. Government Printing Office.	Author and subject indices. Brief annotations.
Israel Program for Scientific Translations, 1965 to 1967, Sedimentation: annotated bibliography of foreign literature: Washington, U.S. Dept. of Agriculture and National Science Foundation, Survey No. 1 for 1959–1964 (1965), Survey No. 2 for 1959–1965 (1966) and Survey No. 3 for 1965–1967 (1967), various paging.	The 3 volumes contain a total of 2925 references to the technical literature (U.S. excluded) from articles written in more than 20 languages. Strong emphasis on modern sedimentation. Excellent subject index.
Kuenen, Ph. H., Humbert, F. L.: Bibliography of turbidity currents and turbidites. In: Bouma, A. H., Brouwer, A., Eds.: Turbidites, p. 222–246. Amsterdam: Elsevier 1964.	Over 600 references.
Oceanic Index: Oceanic Research Institute, La Jolla, California 1964 to present.	Twenty four major headings with biology, coastal engineering, chemical oceanography and geology of most interest to students of sandstones. Titles and some annotations. Annual index. Four to six issues a year.
Pugh, W. E., Preston, B. G.: Bibliography of stratigraphic traps, 195 p. Tulsa: Seismograph Service Corp. 1951.	Traps identified by location, age and type and very briefly annotated.
U.S. Geological Survey. Abstracts of North American Geology: Washington, U.S. Government Printing Office, various paging 1966 to date.	Monthly abstracts commonly of moderate length. See also Bibliography of North American Geology 1785–1964.

Part I

The Fundamental Properties of Sandstones

Mineralogy, texture and sedimentary structures are essential to the study of all sedimentary rocks, be they limestones, coal, evaporites or sandstones. In the broadest sense, this trilology represents what one needs to know to get started in the study of sediments. Hence we have grouped these three together and placed them first recognizing that the experienced reader may briefly pass them by whereas the less experienced may profit by a careful reading. Perhaps mineralogy, texture and sedimentary structures are not the most exciting aspects of sands and sandstones, but certainly they are the most fundamental.

Chapter 2. Mineral and Chemical Composition

Introduction

Sandstones are mixtures of mineral grains and rock fragments coming from naturally disaggregated products of erosion of rocks of all kinds. The total variety of rock types in any given eroding watershed may be represented in the sediment product. Theoretically, therefore, the number of mineral species to be found in all sandstones is as large as the total number of mineral species known. Even a given specific sandstone might be expected to have a large variety of minerals, since a glance at any geologic map will show the average watershed to have rocks with a large variety of minerals present. In fact the expectation proves to be untrue, for the abundant minerals of sandstones belong to a few major groups; many varieties of heavy minerals (most present in trace amounts) may be found, but the list is by no means very large. Obviously, the processes determining mineral composition of sandstones are more complex than simple mixing ones from source areas of different kinds. The discrepancy is great between observed and theoretically possible combinations of minerals.

Minerals may be lost or modified by weathering in the source area, by transportation to the site of sedimentation, and by diagenesis. Because the mineralogy of a sandstone is the inheritance from the source area as modified by sedimentary processes, it is one of the most practical studies that can be made to reconstruct the provenance, including tectonics and climate, for the effects of transportation, including distance and direction, and for chemical additions during sedimentation and diagenesis.

Minerals may be lost by weathering at the source in a variety of ways. The most prominent of these ways is by chemical decomposition or alteration. Thus in the course of weathering feldspars may alter to kaolinite or an intermediate product; pyroxenes and amphiboles are more likely to simply dissolve and be transported as dissolved ions. In contrast, some minerals, such as quartz, are very slightly soluble and will in general be transported unchanged in amount or character from the source rocks. Thus, very largely, all sandstones represent a residuum from surficial chemical weathering processes. The longer or more intense those processes, the less the residuum will resemble the original mixture of source rocks from which it came. We can, therefore, by noting the extent of loss of unstable minerals deduce the amount of weathering in the source area.

The transportation of minerals from the source area to the sedimentary basin where they are finally deposited can also result in attrition of minerals. Most important is *differential abrasion*, whereby the softer minerals may be decreased in size relative to the harder minerals, or simply become more rounded than the harder ones. Many minerals may be so soft as to not survive rigorous transportation as sand sized grains; these will then find their place in silts or muds rather than sands. The effect of transportation is also to sort on the basis of shape. To the extent that this factor operates we may find sands derived from the same source materials that will be of varying mineralogical as well as textural composition in different parts of the sedimentary basin.

Finally, the mineral composition of sandstones may be drastically altered by dissolution, precipitation, or alteration during diagenesis. In this process unstable minerals may be lost completely or partially. New minerals, carbonates in particular, may be added by precipitation from solution. If we can recognize these diagenetic effects and subtract them from the composition of the entire rock, we can arrive at an evaluation of what the original sand may have been.

Aside from those processes which result in the disappearance or appearance of minerals in a sandstone, the simple effects of mixing may give rise to compositions which differ drastically from the source rocks. Many terrains from which sand grains are eroded are composed of a variety of igneous, metamorphic, and sedimentary rocks; the minerals from all these rocks become mixed. Certain minerals, such as olivines and pyroxenes, though they may be abundant in specific igneous or metamorphic rock types, are not abundant in the overall composition of an average terrain. Thus where olivine may be a significant rock-forming mineral and constitute up to 30 percent of some basic igneous rocks, it is usually so diluted that it rarely makes up more than 1 to 2 percent of completely unweathered sandy detritus. In contrast to this, feldspar is such an abundant mineral in so many rock types, as is quartz, that it is unlikely that it will disappear by simple mixing and dilution.

When we consider that most sediments may contain materials that have been recycled from older sediments, we recognize that the processes of *sedimentary differentiation*, the segregation of different sizes and compositions by weathering and transportation, can extend through one or many sedimentary cycles. Some processes by which minerals are lost or gained may operate intensively during a single cycle. Others, such as the abrasion rounding of some minerals, may need a number of cycles before their effect can be noticed. The problem of interpretation of mineral composition of sediments lies in separating out the effects of the last sedimentary cycle of weathering, transport, deposition, and diagenesis from the long term effects of many cycles.

The size of mineral crystals in source rocks determines the presence of monomineralic or rock fragments in the sandstone produced therefrom. Finely crystalline source rocks, igneous, metamorphic, or sedimentary, whose grain or crystal size is less than fine sand size, 1/16 mm diameter, contribute only rock fragments to sandstones. The coarser grained source rocks will contribute mostly mineral grains, but, depending upon the ratio of grain size of source rock to grain size of sandstone, may also provide rock fragments. Thus the larger the grain of sandstone, the more likely the occurrence of rock fragments and the better the assessment of source terrain lithologies. Rock fragments made up of two or three crystals are rare. We either find rock fragments made up of a cluster of many mineral grains or the single mineral grains as sand grains. Just as minerals may be lost by decay or abraded to very fine grains and thus deposited with the muds during a sedimentary cycle, so may rock fragments. For example, if quartzite is the only rock fragment type in a coarse quartzose sandstone we cannot simply interpret rock fragment composition to mean that the source lands consisted only of quartzite. Other lithologic types may have been eroded from the source terrain only to be subsequently eliminated as sand-sized material by mechanical or chemical processes. Thus we face the same choice of interpretations as for mineral grains: the absence of a mineral or a rock type may be due either to (1) its absence at the source, or (2) its disappearance during weathering and transport. The better the general knowledge of sedimentary processes and the specific knowledge of the particular geologic problem, the better one can make a choice that is all important to the reconstruction of geologic history.

The interpretation of the mineralogy of a sandstone is dependent on a proper assignment of mineral species present to a meaningful genetic category. For example, a single list of minerals present is virtually meaningless; we need instead several lists by categories, such as primary detrital, precipitated cement, and post-depositional alteration products.

The distinction between "detrital" and "chemical" minerals has genetic significance by extension of the concept of detrital and chemical rocks. Objectively, the petrographer distinguishes rounded grains that indicate a detrital origin from interlocking crystal growth fabrics that indicate a chemical origin. We distinguish between minerals that are only slightly soluble in water, such as the detrital silicates, and minerals soluble in water, such as the chemical carbonates. The distinction cannot be made for any mineral species as such, for many minerals are both detrital and chemical in origin in the same rock. For example, rounded detrital quartz grains may be cemented by chemically precipitated secondary quartz cement. It is also possible to find carbonate and gypsum sands with all of the characters of detrital silicate sands. Clay minerals partake of some of the characteristics of both detrital and chemical groups. The distinctions between the detrital and chemical categories then are mainly based on the geometry of the particular grain. Fig. 2-1 shows diagrammatically the relationships between the detrital and the chemical mineral groups. We may sum up the utility of the two groups by noting that detrital minerals reflect the source contribution of preexisting rocks, removal by weathering and transport, and diagenesis. The chemical group allows us to interpret the environment of deposition and of diagenesis.

A different way of looking at minerals in sandstones is that used by geochronologists, who determine by radioactivity the average age of zircons or other datable minerals (Tatsumoto and Patterson, 1964). Because the youngest detrital mineral in the sediment can be no younger than the age of the sediment itself, and also because many ages of preexisting rocks may be represented in the source area, an apparent radioactive age determination of the detrital minerals will give us some sort of average age. The ages of source rocks that we refer to are in terms of "radioactive events," which may simply refer to metamorphic episodes rather than the first date of crystallization of the source rock. The dating of detrital minerals commonly involves many assumptions and difficult mineral separation problems. The major assumption that we know may be violated, is that there is no diagenetic effect on the detrital mineral assemblage. Hurley and others (1963) have shown that ages of micas and illites of Atlantic Ocean sedi-

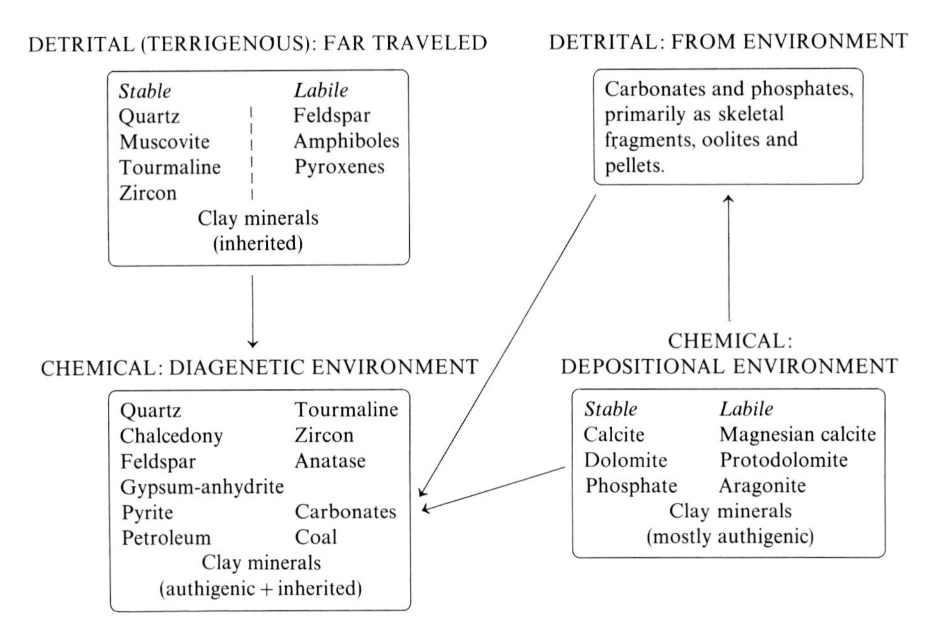

Fig. 2-1. Derivation of minerals in sandstones

ments are probably too young because of the presence of diagenetically formed illite. Such diagenetically formed materials will have some younger date than the date of the sediment, and will influence the total average of detrital and diagenetic in the younger age direction. There is some promise that with more refined methods we may begin to date diagenetic events in relation to the age of the formation. But there is also promise that age dating of sandstones can be used for provenance studies, as has been shown by Brookins and Voss (1970), who found a Middle to Lower Devonian age for both sheet and channel sandstones of the Pennsylvanian of Kansas.

Although up to now mineralogy of sandstones has been determined almost entirely by the polarizing microscope, the widespread availability of X-ray diffraction equipment makes possible the use of that equipment as a supplementary tool. Precise analyses in short times for varieties of feldspars, carbonates, and heavy minerals identifiable under the microscope only with painstaking effort are possible in this way (Pryor and Hester, 1969). In addition, mineralogy based on X-ray data can include structural and compositional information not given by optical methods. For much work in sedimentary geochemistry, we need more precise mineralogical determinations and precise chemical compositions of the mineral phases present. For that work one of the newest instruments, the electron microprobe, is available. The electron microprobe is an X-ray spectrography instrument in which the target of an electron beam is a thin section or polished section rather than the conventional copper or other metal target of X-ray diffraction instruments. The chemical elements of the rock are excited by the electron beam and give off X-rays. The characteristic X-ray radiation for the elements present in a particular tiny area as small as one micron in diameter of the thin section impinged by the electron beam is adequate to identify the elements quantitatively. Though the data we get from such analyses are in terms of the quantitative analysis of the elements, the mineralogy may be calculated from the chemical composition. The general use of the electron probe is covered by Birks (1963). For a full discussion of the use of the electron probe for analysis of rock forming minerals see Sweatman and Long (1969).

The surface appearance of sandstones is strongly influenced by mineral composition. The pure quartz arenites that contain virtually all quartz tend to be white or very light gray in color. The same types of sandstones, which may contain authigenic quartz overgrowths, many of them euhedral, are frequently called "sparkling" sandstones because of their glint in the light. On the other hand, sandstones rich in rock fragments and clay tend to be various shades of gray, greenish gray, and dark gray. The graywackes, again partly because of their fine grain, appear to be very dark. For a long time geologists have been familiar with reddish and brownish colors given to sandstones by the contained iron minerals, most frequently coatings of hematite or limonite on the silicate grains. Attractive as color may be for the identification of certain sandstones, in particular for stratigraphic marker beds, the analysis of color in terms of mineral composition, texture, and other fundamental properties of the rock is most difficult.

In the following description, specific mineral groups are classed as detrital or nondetrital on the basis of their dominant mode of occurrence; thus all of the silica minerals are discussed under detrital although it is recognized that some of them are of secondary chemical origin.

The Detrital Minerals

The Silica Minerals

Only one crystalline polymorph of SiO_2, *low quartz*, is thermodynamically stable under sedimentary conditions, and it is one of the most common minerals in sandstones (Siever, 1957, p. 822; Frondel, 1962, p. 3). Other polymorphs, such as tridymite and cristobalite, are rarely found. Materials in sediments have been identified as cristobalite on the basis of some similarities in X-ray diffraction pattern to that of crystallized cristobalite (Swineford and Franks, 1959); other similar materials, particularly those in very young or Recent sediments, usually prove to be a variety of amorphous silica or opal. Many opals and silica glasses, though they do have some similarities to the cristobalite structure over very short distances, do not display periodicity for more than a few unit cells and thus cannot properly be called cristobalite. Amorphous silica varieties, including opal, siliceous sinter, silica glass (lechatelierite) and others, are abundant in modern and ancient sands with volcanic affinities, but for most non-volcanic sandstones, we need consider only chert (chalcedony). In practically all ancient rocks older than Tertiary, chert is made up of microcrystalline quartz (Midgley, 1951; Folk and Weaver, 1952), as determined by X-ray diffraction. The anomalously low indices of refraction and birefringence of chalcedony noted under the microscope, stem from the fibrous nature of the quartz and from water-filled holes in the structure that are readily seen by electron microscopy. Some Tertiary rocks contain opaline (amorphous) silica (Frye and Swineford, 1946), which will presumably invert to the more stable microcrystalline quartz with time. Opaline silica may originate as a devitrification product of volcanic ash, as small irregular secretions of plant origin, opal phytoliths (Baker, 1959), or as the hard parts of diatoms, radiolarians and the siliceous sponges. There is only one report of the precipitation of crystalline quartz at the surface of the earth at low temperatures in any modern environments, that in a manganese nodule described by Harder and Menschel (1967).

Quartz grains are the common detrital constituents of most sandstones, and, because quartz occurs in so many igneous and metamorphic rocks, attempts were made long ago to use them for source rock determination. The terms *mono-* and *polycrystalline* are often used to describe quartz varities: monocrystalline quartz refers to grains consisting of single crystals and polycrystalline to aggregates of crystals. The terms *unit* and *composite* are also used.

The characters first used were the nature of undulatory extinction (Mackie, 1896) and types of inclusions (Gilligan, 1920, p. 259–260). There followed a history of refinement, culminating in Krynine's (1940, p. 13–20) enumeration of 10 different varieties, based on inclusions, shapes of grains, the nature of the extinction, and the form of the boundaries of the grains, as well as polycrystalline nature of some grains. The varieties were assigned to igneous, metamorphic, or sedimentary source rocks.

Blatt and Christie (1963) restudied the problem of undulatory extinction and concluded that this criterion is not valid for source determinations in the way that it had been used in the past. They stressed the importance of using the universal stage to determine true undulatory extinction and, by a study of various source rocks and sediments, concluded that the only rocks that contained a high proportion of non-undulose quartz were volcanic extrusives and the quartz arenites of Paleozoic and Precambrian age. They concluded that undulose extinction was simply the optical expression of a strained crystal and that most rocks had been subjected to some sort of deformation, either in the course of crystallization or after formation. Because the volcanic extrusives are a relatively small proportion of presumed source rocks of many arenites containing quartz, undulose extinction is not particularly useful, except possibly for the abundant volcaniclastic sands of eugeosynclinal belts. The explanation for the quartz of older quartz arenites being unstrained was sought by Blatt and Christie (p. 571) in the greater thermodynamic stability of the unstrained as opposed to the strained grains. This was correlated with the greater probability of strained grains wearing out during transportation in the course of multiple sedimentary cycles.

Polycrystalline quartz grains include those from igneous and metamorphic rocks, quartzite, sandstones, and cherts (Fig. 2-2), although chert, because of its fineness of grain, is commonly considered separately. Excluding many microcrystalline cherts, the genetic distinction between these possible origins is not fully established by optical criteria such as internal suturing, inclusions, grain shape and internal crystal size. An example of this type of study is Blatt's (1967) in which he suggests that internal crystal size and morphology may help

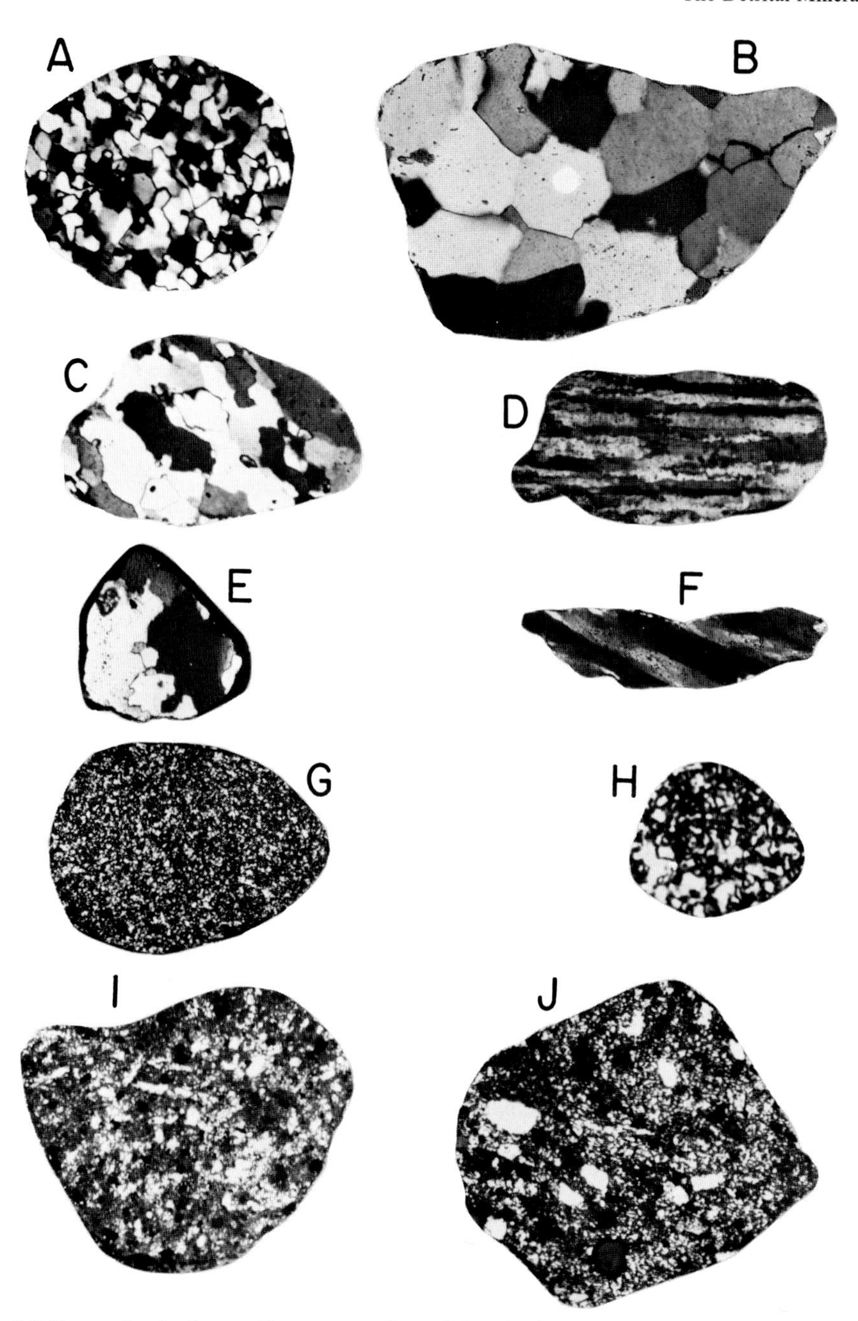

Fig. 2-2. Composite of polycrystalline quartz grains and chert (various sources): A) and B) polycrystalline quartz with uniformly sized grains having mostly straight contacts (polygonized quartz), × 80 and 120; C) Polycrystalline quartz grain consisting of elongate, slightly sutured grains, × 40; D) polycrystalline quartz with almost perfect orientation of elongate crystals some of which have sutured contacts, × 40; E) Polycrystalline grain with coarse interlocking mosaic of crystals, × 40; F) "pseudo"-polycrystalline grain that in reality is moncrystalline but has distinct zones of undulatory extinction, × 40; G) fine grained chert × 40; H) coarse-grained chert, × 80; I) spicular chert, × 120; and J) silty chert grain

distinguish gneissic from granitic polycrystalline quartz. Overlapping characteristics are the principal problem. Polycrystalline quartz is commonly less abundant in well rounded, well sorted quartz-rich sandstones then in poorly rounded sandstones rich in feldspar and rock fragments. Apparently it is more easily eliminated by abrasion then monocrystalline quartz. Nonetheless polycrystalline quartz is a most useful index for mapping petrographic provinces. Certainly it is a good example of how a varietal type may be useful even if its ultimate origin is not fully clear (see Chap. 8).

There has also been some study of quartz types by chemical methods, although rarely have they been used to systematically map varietal types across a basin. In general, the chemical methods are slower, more difficult and more expensive than optical ones.

The study of inclusions may be useful, but is time-consuming and difficult. The trace elements of quartz varieties have been investigated and seem to show some differences as determined by spectrography (Dennen, 1964). The application of cathodo-luminescence to petrography (Smith and Stenstrom, 1965; Sippel, 1968) promises to be most useful in distinguishing the quality and quantity of trace elements in various quartz grains in sandstones. The luminescence is largely a function of the amount and kind of trace elements incorporated in the quartz at the time of crystallization. Because this trace element composition does seem to vary with different source rocks, though not necessarily always constant within a particular source rock type, it can be mapped quickly and simply with relatively inexpensive equipment. Differences in luminescence between igneous and varieties of metamorphic and sedimentary quartz can be pronounced. The fundamental interpretation of trace element incorporation in quartz proceeds from the knowledge that the incorporation of trace elements in the crystal structure is primarily a function of temperature. The higher temperature varieties are those that contain greater numbers and concentrations of trace elements.

The varieties of quartz most useful for source evaluation are polycrystalline quartz grains, chert grains, and those grains made up of quartz plus rounded secondary overgrowths that can be recognized as second-cycle grains. Though it may be theoretically possible to see multicycle grains, those with two or more generations of rounded overgrowths, the probabilities of proving such an origin are very low.

Quartz is ubiquitous as a chemical cement, almost everywhere deposited in optical continuity with the original detrital grain. Such secondary quartz is relatively free of inclusions and trace elements and thus is obviously related to a low temperature origin. Chert and opaline silica are also found as pore filling cements or as rims around detrital quartz grains. Crystal growth experiments have shown that secondary overgrowths on strained crystals may become strained simply as a function of inheriting a crystal structure of the seed crystal. In view of this it may not always be simply assumed that the fact that an overgrowth is strained necessarily means that the rock was subjected to tectonic stress after the authigenic quartz was precipitated. Diagenetic processes resulting in silica precipitation are discussed in detail in Chap. 10. The ways in which provenance and diagenesis combine to form the many species of the silica minerals in sandstones is shown in Fig. 2-3.

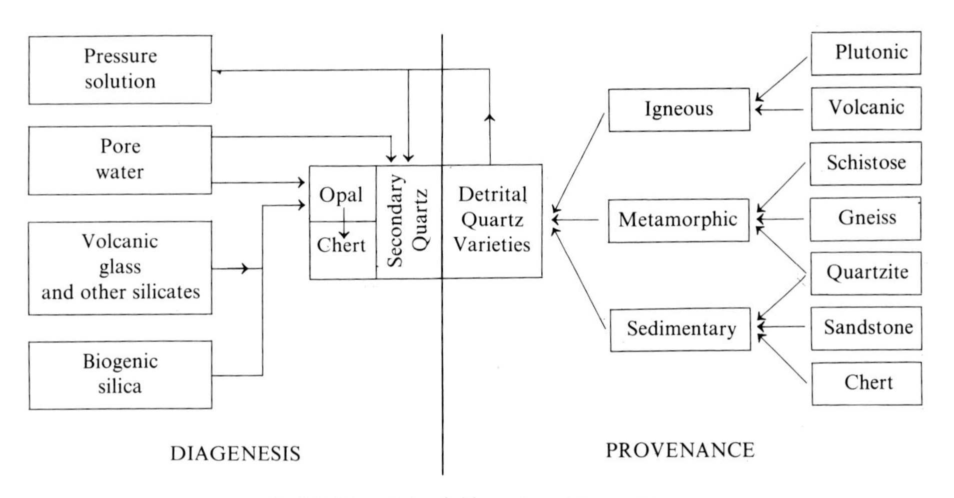

Fig. 2-3. The origin of silica minerals in sandstones

Feldspars

All varieties of feldspar have been noted as detrital minerals. The feldspar minerals have been studied intensively in the past two decades from mineralogical, crystallographic, and petrologic points of view; an extended treatment is given by Deer and others (1963, p. 1–178). Although relatively few petrographic analyses give a breakdown of feldspar varieties, a survey of those that do suggest that K-feldspar (orthoclase, microcline) is most abundant and that sodic plagioclases far outweigh calcic ones. Potassium feldspar, especially microcline, is a characteristic feldspar of many arkoses but the few published analyses of all sandstones in which varieties of K-feldspar are quantitatively distinguished show less microcline than orthoclase. This is not an identification problem, for microcline normally shows its characteristic twinning and so is easily recognized. Many authors who do not give quantitative measurements nevertheless state that microcline is the most abundant species. There is little doubt from the published literature that sodic varieties predominate; they are the only feldspars in many graywackes. No quantitative estimates have been made of the proportion of the most calcic plagioclases, bytownite and anorthite. In general the amount of untwinned K-feldspar is probably underestimated relative to plagioclase, because it is that feldspar which is most easily not identified as such in thin sections, if staining techniques are not used. It is also well known that many different feldspar compositions may occur in the same thin section (a situation far different from metamorphic or igneous rocks, which at equilibrium will normally contain two, or if perthite is included, three varieties, not counting zoned crystals). The fact that there are many different compositions present together is proof of detrital origin, for any rock that is presumably approaching equilibrium, such as most metamorphic and igneous rocks, would not have such a mixed assemblage. Electron microprobe analysis of feldspar compositions promises to give us much needed information on the composition of feldspars in sandstones. It is possible with this instrument to make accurate analyses of relative proportions of potassium, sodium, and calcium of individual feldspar grains in thin sections.

The relative proportions of K-feldspar, albitic, and anorthitic plagioclases may be due either to the relative abundance of those feldspars in the igneous and metamorphic source rocks, or to a differential stability in earth surface environments. It certainly is true that the anorthitic component of plagioclases is far less frequently encountered in source rocks than is the albitic component, though this may be solely a function of the commonness of acidic rocks that we see at the surface of the earth. No such arguments can be made with respect to potassium and sodium rich feldspars, for the relative alkali abundance can be specified only with respect to the particular terrain that is being eroded. If there is a difference of abundance of K-feldspar and Na-feldspar, we must have recourse to stability arguments. Recent studies of weathering of igneous rocks by Feth and others (1964) and Garrels and Mackenzie (1967) appear to confirm the impression gained from petrographic analysis that the order of stability in the weathering environment is K-feldspar – most stable, albite – less stable, and anorthite – least stable. These studies are based on non-equilibrium studies of the kinetics of weathering in the Sierra Nevada mountains, but do seem to confirm laboratory equilibrium data by Orville (1963) that suggest the same order of stability is to be expected in equilibrium at low temperatures in surface environments. All of these studies confirm the ideas of Goldich (1938, p. 36 and 54) based on soil and rock weathering studies. Confirming this trend is the relatively frequently noted composition of authigenic plagioclases as sodic rather than calcium-rich. The authigenic sodic plagioclases are almost always extremely pure albite.

Feldspar in Recent and Modern Sands. Feldspar is a very abundant constituent in modern sands of many diverse origins. As can be seen from Table 2-1, the feldspar content varies from 1 percent or less to 77 percent. And although some of these sands contain 25 or more percent feldspar, few, if any, would be termed "arkose" or even "subarkose." In nearly all, the percentage of rock particles exceeds feldspar and the sands, if lithified, would more resemble lithic arenites. The reason few have the composition arkose is that these sands are far-travelled and have a mixed provenance. Arkoses, usually being derived almost wholly from a highly feldspathic igneous or metamorphic terrain, almost always form in close proximity to their sources and, in many cases, have a very limited distribution. These accumulations, though local, may be very thick.

The seven "gruss" samples taken from the drainage area of the South Platte River, Colorado, are representative of the ultimate source material of most sand. They contain an average of 69.0 percent feldspar, about equally divided between K-feldspar and plagioclase (Hayes, 1962).

River sands seem to be more feldspathic than either dune or beach sands. They average twice as much as either of the latter. This may be related to some attrition of feldspar on beaches and dunes or to the fact that many of the latter are reworked from or transported from less feldspathic relict Pleistocene continental shelf sands or have a large admixture of that origin. It is certainly worth noting that even beaches and dunes have appreciable quantities of feldspar, a fact that militates against the origin of the pure quartz arenites as the products of disappearance of feldspar in such environments. Field and Pilkey (1969) have determined the abundance of feldspars in continental shelf sediments off the southeastern United States and concluded that the low amounts, averaging

Table 2-1. *Feldspar content of Recent and Pleistocene North American sands*

Type	Locality	No. of samples	Range	Average	Reference
Gruss	Colorado	7	—	69	Hayes (1962)
Glacial, Outwash	Illinois	24	—	14	Willman (1942)
Glacial, Till (sand portion)	Illinois	5	15 to 20	17	Willman (1942)
River	Missouri River	26	26 to 45	32	Hayes (1962)
Rivers	South Platte and Platte, Colorado	11	—	50	Hayes (1962)
Rivers, Small	Virginia	3	—	14	Giles and Pilkey (1965)
Rivers, Small	North Carolina	12	—	7	Giles and Pilkey (1965)
Rivers, Small	South Carolina	17	—	5	Giles and Pilkey (1965)
Rivers, Small	Georgia	8	—	7	Giles and Pilkey (1965)
River	Illinois River	3	9 to 12	10	Willman (1942)
River	Ohio River	4	6 to 21	15	Willman (1942)
River	Wabash River	2	17 to 19	18	Willman (1942)
River	Mississippi River (Illinois)	13	16 to 34	25	Willman (1942)
River	Mississippi (Illinois to Gulf)	62	15 to 26	21	Russell (1937)
River	Appalachicola River (Florida)	2	—	6	Hsu (1960)
River	Mobile River, Alabama	1	—	1	Hsu (1960)
River	Brazos River, Texas	2	—	11	Hsu (1960)
River	Colorado River, Texas	1	—	18	Hsu (1960)
River	Columbia River	4	—	23	Whetten (1966)
Rivers, Small	Mexican streams	7	13 to 28	18	Webb and Potter (1969)
River, Small	Jacalitos Creek, Calif.	1	—	25	Gilbert (1954) (p. 285)
Beach, Lake	Lake Erie	7	22 to 28	25	Pettijohn and Lundahl (1943)
Beach, Lake	Lake Michigan	4	—	14	Willman (1942)
Beach, Sea	Quebec, Labrador (Greenland)	9	27 to 77	49	Martens (1929)
Beach, Sea	Massachusetts	17	1 to 26	8	Trowbridge and Shepard (1932)
Beach, Sea	Gulf Coast, Louisiana	15	12 to 19	17	Hsu (1960)
Beach, Sea	East Gulf	13	—	<1	Hsu (1960)
Beach, Sea	Texas Gulf	14	8 to 9	8	Hsu (1960)
Beach, Sea	North Carolina	17	—	5	Giles and Pilkey (1965)
Beach, Sea	South Carolina	17	—	5	Giles and Pilkey (1965)
Beach, Sea	Georgia	11	—	4	Giles and Pilkey (1965)
Beach, Sea	Florida	18	—	2	Giles and Pilkey (1965)
Dunes	Illinois	47	8 to 29	18	Willman (1942)
Dunes	North Carolina	17	—	5	Giles and Pilkey (1965)
Dunes	South Carolina	7	—	3	Giles and Pilkey (1965)
Dunes	Georgia	11	—	2	Giles and Pilkey (1965)
Dunes	Florida	15	—	2	Giles and Pilkey (1965)
Continental Shelf	Southeastern U.S.	80	—	3.3	Field and Pilkey (1969)[a]
Continental Shelf	Southeastern U.S.	67	—	5.2	Field and Pilkey (1969)[b]
Summary					
	Average River	168	—	22.0	
	Average Beach	142	—	10.1	
	Average Dune	95	—	10.6	
	Grand Average	404	1 to 77	15.3	

[a] 0.250–0.364 mm size fraction.
[b] 0.125–0.177 mm size fraction.

3 to 5 percent, reflect provenance from Piedmont rivers draining deeply weathered source areas. This is in contrast to the much higher proportions, 11 to 30 percent, in New England shelf sediments reported by McMaster and Garrison (1966). The latter are derived from a much less weathered source overlain by feldspathic Pleistocene till.

Inspection of Table 2-1 shows that, independent of their depositional environment, the sands of the northern United States are more feldspathic than those of the southern or southeastern states. This difference is probably due to the fact that the sands of the northern areas were formed by reworking of the glacial drift, the sand fraction of which is quite feldspathic.

The average of 404 river, beach, and dune samples is 15.3 percent feldspar, which is exactly the same as the average of 435 sandstones from the Russian platform (see Table 2-3). No claims can be made for the adequacy of either sample; that of Recent and Pleistocene sands is probable overweighted with sands of glacial derivation and hence more feldspathic than would otherwise be the case.

Feldspar in Ancient Sandstones. The average sandstone is calculated (from the average bulk chemical composition) to contain 11.5 percent feldspar (Clarke, 1924, p. 33). Leith and Mead (1915, p. 76) estimate by calculation 8.4 percent. Krynine (1948, Fig. 11) gives a figure of 11 to 12 percent. Actual modal analyses show that some sandstones are essentially feldspar-free whereas in others feldspar constitutes over 90 percent of the framework fraction (Tables 2-2 and 2-3). The

Table 2-2. *Feldspar content of North American sands and sandstones*

Age	Number of formations	Percent feldspar
Pre-Devonian	35	5.1
Devonian-Permian	29	5.8
Mesozoic	12	25.0
Tertiary	22	21.0
Pleistocent-Recent		15.3
Unweighted mean		14.4

Table 2-3. *Feldspar content of sandstones of Russian platform (Ronov and others, 1963)*

Age	No. of samples	Feldspar
Precambrian	65	30.5
Cambrian	18	16.6
Silurian	14	9.6
Devonian	177	8.9
Carboniferous	95	4.8
Triassic	5	61.6
Jurassic	23	42.8
Cretaceous	20	15.0
Tertiary	10	31.1
Quaternary	8	22.6
	Average	15.3

average feldspar content of North American sands and sandstones determined from a random selection of published analyses is about 14 percent; the average sandstone of the Russian platform is estimated to be about 15 percent.

The cause of the variations shown in the two tables is not readily apparent. In the North American data the highest values are for sandstones in the orogenic belts, the lowest in the sandstones of the continental interior. Most of the Mesozoic and Tertiary sandstones are from the Coast Ranges and the Rocky Mountain areas whereas the Paleozoic sandstones are from the continental interior and are generally texturally and mineralogically more mature and hence feldspar-poor. However even the Paleozoic sandstones of the Appalachians and Ouachita orogenic belts are poor in feldspar though rich in rock fragments. The high feldspar content of some of the Russian sandstones is attributed to tectonism, i.e. rapid uplift and erosion of incompletely weathered debris; the low feldspar content of others is correlated with the end-stages of erosion and low relief and crustal stability (Ronov and others, 1963, p. 225).

The Origin of Feldspar in Sandstones. Because feldspars, as quartz, are so ubiquitous in metamorphic and igneous rocks, the presence of undifferentiated feldspar as such has little interpretative value for precise source rock assessment unless the precise composition is specified. The composition can be determined most precisely by optical or X-ray diffraction methods, but staining techniques are rapid and convenient for distinguishing plagioclase from K-feldspar (van der Plas, 1966, p. 49–52). The value of feldspars of different types for provenance is partly based on the limited distribution of certain types, as, for example, sanidine, which is associated with high temperature contact metamorphic or volcanic rocks. Microcline, in contrast, is widely distributed in metamorphic and plutonic igneous rocks but not in volcanics. Plagioclase compositions correlate with the chemical compositions of metamorphic and igneous rocks but are not diagnostic of specific rock types. Optical heterogeneity of feldspars was studied by Rimsaite (1967), who related such characters as oscillatory or normal zoning and various kinds of intergrowths to composition and origin of source rock types. But because feldspars are so much more characteristic of and abundant in metamorphic and igneous than in sedimentary terrains, their presence in a sandstone is most useful in a general way. Their abundance is related to (1) source rock composition (see Chap. 8), (2) chemical weathering in the source area, (3) abrasion and solution during transportation, and (4) solution during diagenesis (Fig. 2-4).

We may relate chemical weathering in the source area to the concept of sedimentary tectonics in the following way. The survival of unstable minerals in the weathering environment, and thus their transportation to the sedimentary basin, is a function of the ratio of chemical to mechanical weathering in the source area. The ratio of chemical to mechanical weathering is related to the topography of the source area in the main. The topography, of course, is largely a function of tectonics. Thus we may go from the quartz-feldspar ratio to an evaluation of the tectonic state of the source area, having corrected for other things such as source rock composition and changes during transportation and diagenesis. The argument of sedimentary tectonics tends to underplay the role of climate or at least climate that is not directly controlled by topography. There has been discussion of the relative importance of climate and tectonics (that is, topo-

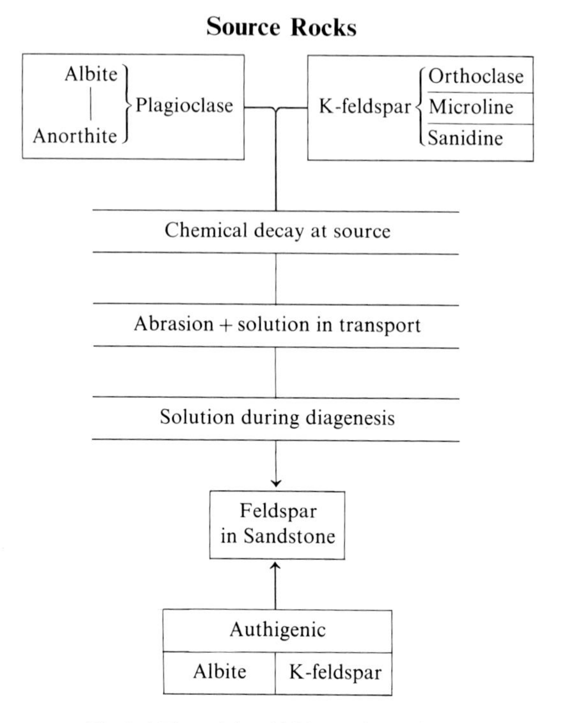

Fig. 2-4. The origin of feldspars in sandstones

graphy) in the chemical weathering and formation of soils (Strakov, 1967, p. 3–23). It is true that both may be important in particular cases. It appears yet to be demonstrated, however, which one or the other is of overriding importance. To the extent that this is so, there is a certain indeterminacy in the interpretation of the ratio of feldspar to quartz or generally of stable to unstable minerals. A somewhat similar discussion has gone on with respect to the disappearance of feldspar and other unstable minerals as a result of abrasion and solution in the course of transportation as discussed on p. 56 of this chapter.

Authigenic feldspars in sandstones have been reported so frequently that there is little doubt of their common occurrence (Pettijohn, 1957, p. 664–667; Baskin, 1956). The chemical conditions for the precipitation of feldspar are determined by the relative amounts of their components in solution, that is, K^+, Na^+, Ca^{++}, and SiO_2, as well as pH. These variables as related to feldspar precipitation have been discussed by Garrels and Christ (1965, p. 359–363). A measure of the stability of feldspar with respect to a solution and thus of the probability of its precipitation may be given by the appropriate alkali metal-hydrogen ion ratio, given a minimum amount of silica present in solution. A consideration of the temperature coefficients of the reaction as given by Hemley and Jones (1964) as well as the geological facts of the occurrence of authigenic feldspars suggests that slightly elevated temperatures, presumably associated with moderate to deep burial, are an important factor in their precipitation.

Micas, Chlorites, and Clay Minerals

These minerals can be considered together because they are all closely related in chemical composition (hydrous aluminosilicates) and crystal structure (sheet structures). They also are related to each other by virtue of their occurrence in sandstones. It is frequently impossible to note any discontinuity in size between large flakes, easily identifiable as detrital micas, and the very fine grained interstitial clay characteristic of many sandstones. Nevertheless there does seem to be some virtue in considering separately the two end members, the large detrital grains, and the fine grained clays. The two can occur together without discontinuity in the size distribution; as two size varieties separated by a discontinuity in the size distribution; and each size separately without the other size present.

Large Detrital Grains. Large detrital grains of muscovite, biotite, and chlorite are common in sandstones, either fresh or altered, and species can be readily identified under the microscope. In most sandstones they are a minor constituent, except along fine grained partings and some shaly sandstones, in which they may be abundant. Because of their thin sheet shape and consequent lower settling velocity, they are associated with quartz and feldspar grains of smaller sand or silt size. Submarine fans on the continental rise below submarine canyons show a greater abundance of micas in the finer grain fractions farther out on the fan, while the more proximal parts of the fan, coarser grained, have less mica. This is in general true of silts in comparison with sands. The Eocene Tyee sandstone of Oregon mapped by Lovell (1969, p. 950) shows much the same pattern for a turbidite sandstone, where the micas are associated with the more distal portions of the formation farther down the depositional slope. Here too the coarser nearer shore facies have less mica in them. The well crystallized detrital micas can be interpreted in much the same fashion as feldspars, but may serve the purpose most usefully in the study of very fine grained sandstones and siltstones. Muscovite is much more resistant to chemical weathering than biotite or chlorite, but there are not enough data to permit an evaluation of the stability of biotite relative to the several varieties of chlorite during weathering. There appears to be a greater tendency of chlorite to degrade to finer particles than biotite. So it is reasonable that chlorite is frequently lumped with the fine clay fraction in sandstones. The alteration of biotite to chlorite in place is also common. Frequently one can see the pattern of alteration in a single grain. The alteration of grains of biotite to glauconite was observed by Galliher (1939).

Fine Grained Clays. The much finer grained clay found in sandstones as the essential constituent of matrix and argillaceous rock fragments includes all major groups of clays: the kaolin group (kaolinite, dickite, halloysite), the micas (muscovite, illite, glauconite), the montmorillonite or smectite group (nontronite, saponite, and many others), the chlorite group, the mixed-layer group (corrensite and others) (Grim, 1968, p. 558, 564–65; Brown, 1961). The structures of these groups are made up of simple building blocks of alumina octahedra and silica tetrahedra, each of which forms sheets in the a–b crystallographic plane (Fig. 2-5). The division of the clay minerals into the various groups is primarily on a structural, rather than a compositional basis. Thus the minerals of the kaolin group are those in which one octahedral layer is linked to one tetrahedral layer

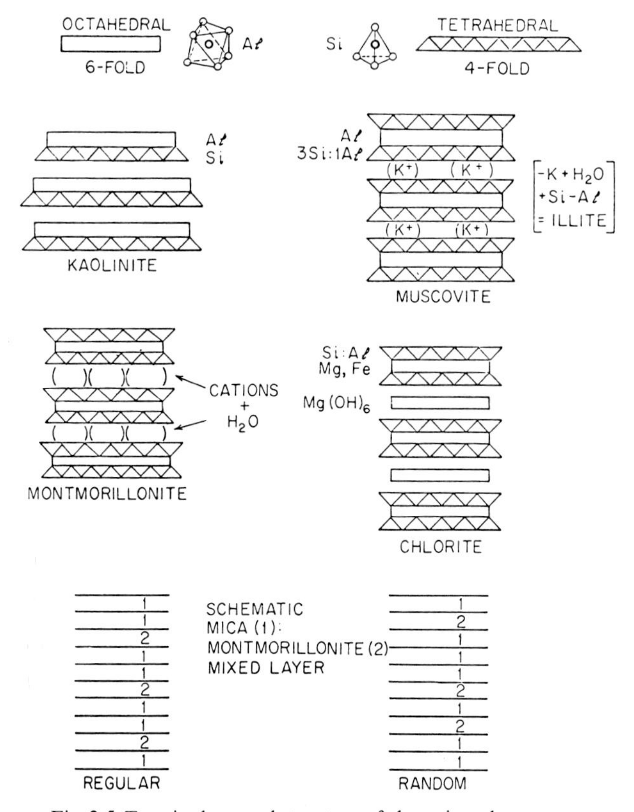

Fig. 2-5. Terminology and structure of clay minerals

and the packet is a 1 : 1 group. Varieties of stacking polymorphs are possible in the kaolinite group giving rise to the minerals kaolinite, dickite, and nacrite. The micas are the packets made up of an octahedral layer between two tetrahedral layers, the packets being joined together with interlayer cations, such as potassium in biotite and muscovite. The chlorite group is characterized by much the same variety of packet but the interlayer position between packets is occupied by a brucite, $Mg(OH)_2$, layer. The montmorillonites are those which, basically similar to the mica packet, have water and cations between the layers. Many compositional varieties are possible within each structural group. Though many of these have been given specific names as varieties based on composition, one can generally consider each of the groups to contain a wide and indefinite spectrum of compositions. The X-ray diffraction identification of clay minerals proceeds primarily on the basis of these different structural groups (Brown, 1961).

The structure of the clay minerals is basic to the concept of ion exchange (base exchange), a process whereby cations in the interlayer positions, and to a lesser extent in other sites, may equilibrate with the external environment. Thus a sodium montmorillonite can be put in a calcium-rich solution and become a calcium montmorillonite. The extent of exchangeability (exchange capacity) is greatest for the montmorillonites, intermediate for the mixed-layer, illite, and chlorite groups, and lowest for the kaolinite group. Cations in octahedral positions are not generally exchangeable in short term laboratory experiments, and so for relatively short periods of time, clays will act as stable detrital minerals insofar as octahedral and tetrahedral sheets are concerned, but as labile chemical components insofar as interlayer positions are concerned. Thus we may expect that over the short period of time it takes a clay mineral to be transported from river water to sea water and buried it will exchange only interlayer positions as well as all those ions available on edges and corners on the outside of crystallites. Over the time scale of millions of years, characteristic of diagenesis, all parts of the clay mineral structure may react with the environment.

Recent laboratory work confirms thin section studies that have shown the probability of direct precipitation of clay minerals from aqueous solutions at low temperatures and pressures (see, for example, Siffert, 1962). In addition, laboratory experiments have shown that clay minerals will partly dissolve and alter in hours or days (Mackenzie and Garrels, 1965; Siever, 1968). Over periods up to six months clay minerals come to some quasi-equilibrium with respect to the solutions that bathe them. Work on the clay minerals in river waters and a variety of sea waters has likewise shown that the data are compatible with relatively quick equilibration of some sort, probably the clay surface only, with the surrounding waters. There is some question as to whether this means that the clays that react with their surrounding waters are those well crystallized clays that are identified by X-ray diffraction procedures, or rather are that proportion of the clay mineral fraction which is in the very fine grained and relatively amorphous state characteristic of many soil clays. If the latter, they are likely not to be identified by X-ray diffraction means and at the same time are likely to be more chemically reactive.

Precise identification of clay minerals is primarily by X-ray diffraction methods though rough approximate identification by optical methods is necessary for thin-section study. But precise identification by optical methods in thin-section is often difficult (Grim, 1968, p. 432) though there are extensive treatments of optical identification procedures in immersion media available (Correns and Piller, 1955; Grim, 1968, p. 412–433). Procedures for quantitative estimation of the relative abundance of the major clay mineral groups (mica, kaolinite, chlorite, montmorillonite) include separation and size fractionation of clay from sandstone, making diffraction patterns with and without addition of ethylene glycol (for determination of montmorillonite or other expandable clays), measuring X-ray peak areas, and applying correction factors for X-ray scattering efficiency (Johns, Grim, and Bradley, 1954). The difficulties are such that analyses do not have a precision better than 10 percent unless extraordinary care is taken to insure reproducibility. Electron microscopy (Bates and Comer, 1955; Beutelspacher and van der Marel, 1968), electron diffraction (Zvyagin, 1967), and infrared spectroscopy (Farmer and Russell, 1964) are finding increasing use for detailed studies of clay minerals. The electron probe is also useful for investigations of clay mineral compositions.

Interpretation of clay mineralogy proceeds from a knowledge of (1) how weathering processes form clays from primary silicate minerals and alter them from pre-existing clays, (2) the nature of ion-exchange and other short term chemical changes that take place in the environment of deposition, and (3) the kinds of overall structural and chemical changes that take place over a long time. Thus (1) is related to the source rock, climate, and topography; (2) is related to the environment of deposition and later (groundwater) environments; and (3) is a function of long term diagenesis. This might make clay mineralogy a powerful tool were it not for the fact that each type of change partially obliterates the record of earlier changes. Nonetheless, in comparative petrologic studies, clay minerals may find some utility for mapping petrographic provinces, for evaluating environmental differences, and for assessing the effects of diagenesis.

Recent studies by Biscaye (1965) and others have shown that the clay mineral distributions of Atlantic and Pacific sediments can be understood most easily in terms of detrital petrographic provinces, where the transport is by a combination of wind and ocean currents. Though nearshore differential flocculation has often been invoked, the oceanic province boundaries do not fit such an origin, which would be expected to result in a different pattern in nearshore and estuarine areas than farther out to sea. There is a clear correspondence between the climate, weathering, and source rock compositions of the adjacent continents and the distribution of clay minerals on the sea floor. This does not rule out certain diagenetic changes that can be found in oceanic sediments, such as the alteration of volcanic ash to montmorillonitic clay. Nevertheless the major factor determining the observable clay mineral composition is the source contribution. Studies of ancient rocks can also be interpreted in the same way. In ancient rocks also a number of studies have pointed out the presence of authigenic kaolinite or dickite in sandstones. One study (Glass and others, 1957) has definitely shown that the kaolinite is authigenic and could not be a different source contribution because the adjacent shales do not contain kaolinite. Much of the clay matrix of graywackes is apparently secondary, a diagenetic product, rather than originally deposited as a fine-grained clay. The difficulty in understanding how ordinary currents, including most turbid currents, can introduce the extraordinarily high proportions of fine-grained clay with medium- and coarse-grained sand fragments, tends to force the conclusion that much clay matrix may be a diagenetic or metamorphic product.

The conditions for the formation and chemical stability of the various clay minerals, at least in idealized form, are considered by Garrels and Christ (1965, p. 352–362) and Krauskopf (1967, p. 176–203). As in the case of the feldspars, the pertinent parameters are the concentrations of alkali metals, alkaline earths, hydrogen ion, silica, and alumina in the environment (see Chap. 10). Thus it appears that the clue to understanding the origin of clay minerals in a particular rock will come from areal mapping of the clay mineral composition rather than an attempt only to infer origin from a particular composition in a particular place.

Heavy Minerals

In this group we include various silicates and oxides that are found in small quantities in sandstones, the total quantity of such constituents rarely making up more than 1 percent of the rock. They range from tourmaline and zircon, which do not occur in large amounts in any source rock but are resistant to mechanical and chemical attack, to the amphiboles and pyroxenes, which may be abundant constituents of some source rocks but show little resistance to decay. To the extent that the heavy minerals survive the hazards of weathering, transport, and diagenesis and to the degree that they occur in a restricted range of provenance types, they are most useful for source rock interpretation (see Chap. 8). Because it is most unusual for a source area-sedimentary basin couple to change appreciably and synchronously over short periods of geologic time, thus giving rise to a different and laterally constant mineral assemblage for each stratigraphic unit, heavy minerals are not of great use for ordinary detailed time-stratigraphic correlation, just as lithologic facies in general are not useful as time-stratigraphic markers. To the extent, however, that heavy minerals can be used as a facies correlation indicating sedimentary dispersal from particular source areas which are undergoing tectonic evolution, they can be extraordinarily useful. Thus one can map the progress of an orogenic episode in a source area which led to gradual unroofing from a sedimentary to a metamorphic to an igneous terrane by noting the change in heavy mineral suites going upwards in the sandstones derived from that source. The stratigraphy of the heavy mineral zones of the sandstones will be the reverse of the sequence in the source area.

When used in conjunction with textural analysis and light mineralogy, heavy mineralogy becomes most valuable for establishing petrographic provinces and source rock types. Again it is the mapping that proves the utility of the heavy minerals. Because the size distributions of each mineral variety may be different, one must be careful to either include the entire size range to cover the entire heavy mineral assemblage, or to count the heavy minerals within a particular group. Hydraulic ratios (Rittenhouse, 1943; Briggs and others, 1962) can be used with a proper understanding of the density-size relationships of heavy minerals to evaluate more carefully the total heavy mineral fraction. Hubert (1960, p. 208–221) has proposed a zircon-tourmaline-rutile maturity index of heavy mineral assemblages that he relates to a source contribution modified by sedimentary differentiation in the environment.

Special attention has been given to two of the commonest stable heavy minerals, tourmaline and zircon, in much the same way that quartz varieties have been studied in the hope of gaining greater effectiveness in provenance determinations. Tomita (1954) surveyed the literature on zircons and concluded that purple or rose-pink zircon comes only from Archean gneisses or granites; the color was considered radiogenic, its depth or intensity increasing with age. Poldervaart (1955) studied zircons from sedimentary rocks in comparison with their distribution in igneous rocks and found that shape in relation to size was more important than other variables, the larger the zircon the more rounded, if sedimentary; the smaller the more rounded in the rare occurrences of resorbed zircons in granites. Callender and Folk (1958) measured a variety of properties of zircons in Tertiary sand of central Texas, including color, elongation, and inclusions, but found only the degree of idiomorphism to be correlated with another petrologic variable, the relative amount of volcanogenic detritus. Krynine (1946) attempted to distinguish color varieties of tourmaline as a basis for provenance, but there is no simple relationship between color and composition or between composition and the type of igneous or metamorphic rock in which tourmaline is found (Deer and others, 1962, p. 313).

As covered in Chap. 3, the roundness of heavy minerals, as of any detrital grains, may indicate petrographic provinces by showing different abrasion histories. The stability and persistence of heavy minerals with age is covered in Chap. 10.

Rock Fragments

The major types of rock fragments in sandstones are (1) the argillaceous group including shale, slate, phyllite, and schist, (2) volcanic rocks including glass, and (3) silica group of quartz and cherts (Fig. 2-6). Of lesser importance but locally important are the carbonate rock fragments.

The relation between rock fragment content and age has only recently been fully appreciated. Rock fragments are clearly recognized and identifiable in thin sections of modern sands. The argillaceous rock fragments are probably not identified as such in many older sandstones because they may be squashed, deformed, and molded about the more competent grains so that they blend into or appear as clay matrix (Allen, 1962, p. 678). All types of such squashing can be seen in suitable thin sections. It is likely that in a great many of the oldest sandstones the squashing is almost complete and the former rock fragments are confused with a fine-grained clay matrix. It is almost impossible to give a nonarbitrary and completely reproducible way of recognizing squashed rock fragments when all outlines are gone. Clay aggregates formed during diagenesis may also be confused with rock fragments. The tougher varieties of argillaceous material (more highly metamorphosed) are more likely to be identified as rock fragments; soft shales rarely survive as detrital fragments of sand size and, even if they do, are more likely to lose their identity after deposition. There is some question as to the viability of argillaceous rock fragments during stream transport; most seem to be quickly degraded to silt or clay size in modern streams. Some tumbler barrel experiments also indicate a lack of durability of shale fragments with transport. In consequence it may be possible that abundant argillaceous rock fragments in a sandstone indicate a nearby source of that material. Thus one might speak of a "cannibalism" index in terms of the proportion of soft argillaceous rock fragments preserved in a sandstone.

Volcanic rock fragments, including pyroclastic debris, are abundant in some sandstones. In fact, they may be the major or only constituent of some. The fragments may be of older volcanic terrains or of contemporaneous origin from volcanic activity within or near the depositional environment. When altered diagenetically, such rocks are likely to contain zeolite minerals. Devitrification of glasses within the sandstones can lead to the production of opal cement. Some formations can equally well be considered as volcanic rocks or as sandstones made up of volcanic rock fragments. There is probably only a semantic difference between the consideration of a welded tuff as a volcanic rock or as a sediment in which the conditions for cementation are especially related to the temperature of formation.

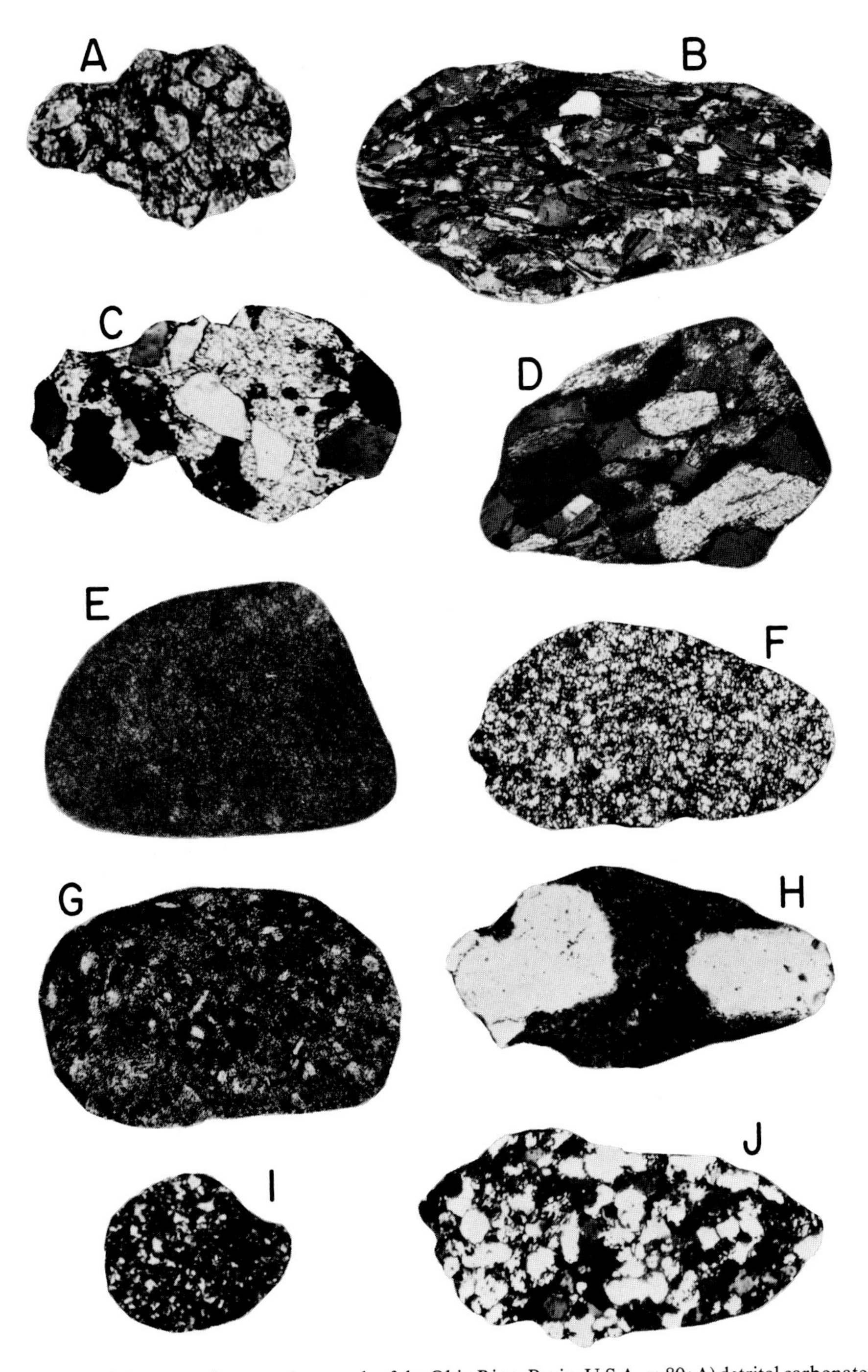

Fig. 2-6. Rock fragments from modern sands of the Ohio River Basin, U.S.A. × 80; A) detrital carbonate, B) muscovite schist; C) sandy carbonate; D) hornblende schist, E) micritic limestone; F) microcrystalline carbonate, G) silty shale; H) sandstone cemented by chert; I) chert, and J) coarse siltstone. Photographs courtsey of James F. Friberg

Table 2-4. *Rock fragment content of North American Recent and Pleistocene sands*

Type	Locality	No. of samples	Range	Average	Reference
River	Mississippi River	62	9 to 19	13	Russell (1937)
River	Ohio River	1	—	11	Hunter (1967)[a]
River	Mississippi River (Below junction of Missouri)	2	22 to 27	24	Hunter (1967)[a]
River	Columbia River	11	—	31	Whetten (1966)[b]
Rivers, small	Mexican streams	7	45 to 69	57	Webb and Potter (1970)
River, small	Jacalitos Creek, California	1	—	40	Gilbert *in* Williams and others (1954, p. 285)
River	Rio Grande, Texas	1	—	40	Nanz (1954)[c]
Average river		85		20	
Beach	Lake Michigan	2	—	17	Hunter (1967)[a]
Beach	Texas	2	—	15	Nanz (1954)[c]

[a] Data from 1.25–2.75ϕ range only.
[b] Includes some coarse silt grades also.
[c] Estimated from published plot.

Rock Particles in Recent and Ancient Sands. Rock fragments are widely distributed in modern sands but the kinds and even the total quantity present are seldom specified. Such data on their occurrence as could be obtained by a search of the literature is given in Table 2-4. These data are, in some cases, incomplete and for certain size-grades only. To the extent that only the finer grades were examined, the rock fragment content is too low. The original analyses in such cases show that, unlike feldspar, the rock fragment content tends to be markedly size dependent, the proportion of such fragments increasing with increasing size (Hunter, 1967; Okada, 1966, Table I). There is also some ambiguity about what is reported as a "rock fragment." Polycrystalline quartz, chert, and ordinary rock fragments may be placed together or reported separately. In Table 2-4 chert is included with the rock fragments.

As was the case with feldspar, some small streams draining areas of restricted lithology are apt to be highest in rock fragments. The sands of large rivers are perhaps more "normal." Data are too few to draw any far-reaching conclusions about the rock fragment content of the several environments of sand accumulation. The data of Nanz (1954, Fig. 19) suggest that the rock fragment content is less on beaches and dunes than in rivers of the same geographic province.

If sand carried by large modern rivers is a guide, most sandstones would be subarkoses or sublithic arenites (subgraywackes), the average feldspar and rock fragment contents being 15 and 19 percent respectively. The relative importance of feldspar and rock fragments seems to be a matter of provenance.

It is obvious from this review of the meager literature on the rock fragment content of modern sands that more and better data are needed. Petrographers have been prone to concentrate on the minor accessory minerals of sand, the heavy minerals, and neglect the light minerals. Moreover, study of the constituents as grains is not adequate. Most rock fragments are opaque to transmitted light and unless sectioned are unidentifiable. They have been generally ignored.

What is needed are thin section studies of modern sands – sands artificially indurated in place and then examined in thin section exactly as ancient sandstones are examined. We may then have adequate data to make some generalizations regarding the production and distribution of rock particles. It is particularly disappointing not to have this information because, unlike the study of modern marine sediments, the cost of data collection and analysis is very small.

The abundance and kind of rock fragments in ancient sands is perhaps better known than in modern sands due primarily to thin section analysis of these indurated sands. Table 12-3 will give the reader some indication of the range and average rock fragment content of Appalachian sandstones.

Several factors determine the rock fragment content, namely (1) grain size, (2) provenance, (3) maturity, and (4) age. In general, the rock fragment content is a function of grain size, the coarser the sand the higher the rock fragment content, other things being equal (Allen, 1962, p. 673; Shiki, 1959).

As noted by Potter and Pryor (1961, p. 1224–1226) the sands derived from the Canadian Shield contain more feldspar than rock particles; those of Appalachian derivation show the reverse. The Shield area is more largely "granitic" whereas the Appalachians, including the crystalline Piedmont, is a sedimentary-metamorphic terrane. The texturally mature sands – those better rounded and better sorted, tend to have a lower rock fragment content (as well as a lower feldspar content).

The Chemical Minerals

Carbonates

The carbonate minerals in sandstones are both detrital and chemically precipitated, the latter by far predominant.

Detrital. Detrital carbonate is abundant in calcareous sands as skeletal fragments, oolites and faecal pellets, none of which are traveled from a distant source. Such debris may also be admixed in varying, but usually minor amounts, in terrigenous sandstones. Majewske (1969) and Horowitz and Potter (1971) provide guides to the recognition of skeletal debris in thin section.

Detrital carbonate grains derived from source regions outside the basin are also present in some sandstones (Fig. 2-6). Such grains have rounded, abraded boundaries and are significant in some molasse sandstones (Füchtbauer, 1964, Fig. 16). High relief and arid climate both favor terrigenous carbonate. Of the two, high relief with a resulting high ratio of mechanical to chemical erosion appears to be the more important. But the dryer the climate, the lower relief need be. In any case, the solubility of carbonate plus its softness and cleavage all decrease the probability of its survival by stream transport. Acid etching and/or staining are useful when studying thin sections rich in carbonate rock fragments.

Calcite and Dolomite. Calcite and dolomite are abundant in sandstones as pore filling and replacement cements of postdepositional origin. Grains of detrital and primary dolomite have been identified by Sabins (1962). Some pore filling cement may be recrystallized from originally detrital carbonate grains or may be a precipitate from an aqueous solution in an originally empty pore. In contrast to the carbonate of many modern sediments, the calcite and dolomite of sandstone cements are apparently relatively pure, with no excess Mg in the calcite and no excess Ca in the dolomite, although the number of accurate analyses are few (Graf, 1960, p. 24–42). Aragonite has not been reported as a precipitated cement in ancient sandstones, though it is known in beach-rock in modern carbonate environments and aragonite cemented modern sands (Allen and others, 1969). Calcite may be a primary cement or recrystallized from earlier detrital origin but dolomite appears almost always as a replacement of a calcite precursor. The clue to this difference lies in replacement textures (see Chap. 10) and crystal habit in particular. Calcite is commonly anhedral, made up of an interlocking crystal mosaic whose individual crystals may vary from a few microns to centimeters (as in sand crystals). In contrast dolomite and ferroandolomite most commonly assume a rhombohedral form (Fig. 6-5). Though this is not exclusively so, the presence of rhombohedra is one of the quickest clues to the identification of dolomite, as the distinction between calcite and dolomite by ordinary optical criteria is difficult. Differentiation between calcite and dolomite is most quickly achieved by one of a variety of staining tests, the stains usually taking advantage of the presence of small amounts of iron in all dolomites (Friedman, 1959; Warne, 1962).

Fe-Mn-Carbonates. The iron-rich carbonates are less common than calcite and dolomite in most sandstones, but are abundant constituents of some. Investigations by Goldsmith and Graf (1960) and Goldsmith and others (1962) have elucidated the phase relations in the system Ca–Mg–Fe–Mn–CO_2 (Fig. 2-7).

Of the various minerals in this chemical system, the ones that are important in sandstones in addition to calcite and dolomite are siderite and the iron-rich dolomites or ankerites. Rhodochrosite is known in some concretionary forms in sandstones, but has rarely been reported as cement.

The precipitation of calcite and dolomite is primarily influenced by the variables of Ca and Mg abundance and the pH of the environment, whereas the iron and manganese carbonates have an additional dependence on the oxidation-reduction balance of the environment. Because iron and manganese can be in carbonate minerals only in their reduced state, a redox potential is immediately established. The redox potential and carbon dioxide pressure necessary for the precipitation of siderite has been discussed by Garrels and Christ (1965, p. 201–211). For any given amount of dissolved carbonate these relations can be shown on an Eh-pH diagram (Fig. 2-8). The most probable reducing agent responsible for the reduction of iron is organic matter. Thus we may look for clues to the formation of siderite and other reduced metal carbonates in terms of a linkage to environments in which organic matter, either in soluble or solid form are abundant; such places are stagnant basins, or tidal or estuarine environments where organic productivity is high. Recent work by Hallam (1966, 1967) suggests that sideritic concretions in the Jurassic of England are tied to shoreline conditions and indirectly to the presence of organic matter which supplied the reducing capacity. As is shown in the phase diagram of the carbonates (Fig. 2-7), it is possible at low temperatures to have a coexisting

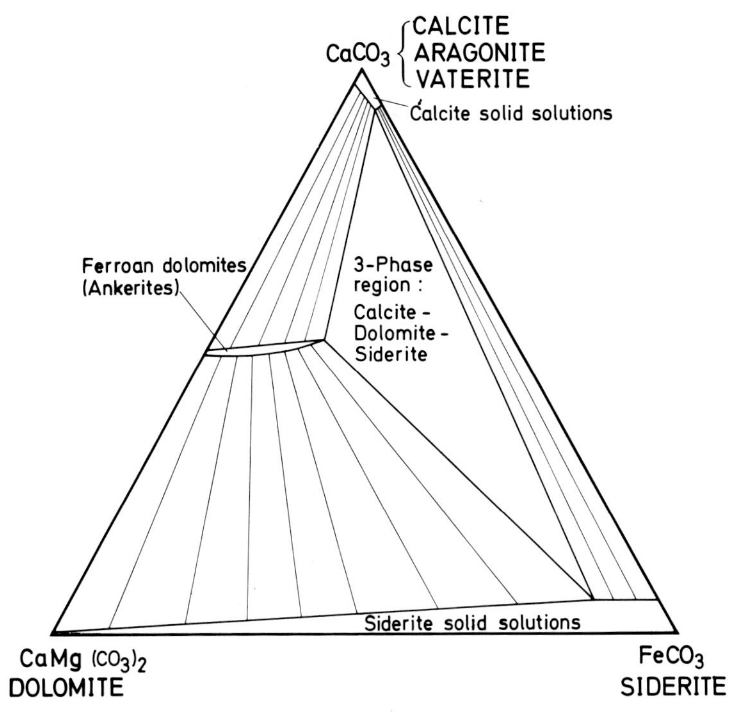

Fig. 2-7. Schematic phase diagram of the Ca-Mg-Fe-CO_2 system, estimated, at 25° C. Estimate based on data in Goldsmith and others (1962)

equilibrium assemblage of calcite, an iron-rich dolomite, and siderite. These three cements are found in what appears to be this kind of equilibrium in certain Carboniferous sandstones of the mid-continent of the United States. If other such associations are noted we may have a better clue to the composition of the subsurface fluids from which the cements were precipitated.

The iron-rich magnesium carbonates are frequently found as concretionary accumulations in sandstones; calcite and dolomite concretions are also well known. These concretions have not been studied a great deal, but do seem to be related to bedding planes, which may be either the result of special permeability along certain bedding planes, or a stratigraphic expression of particular compositions of the water in the environment. The concretions appear to be pure segregations of carbonate cement enclosing sand grains, which speaks for their post-depositional origin. The alignment along bedding planes may also be related to diagenetic redistribution of originally detrital carbonate.

Sulfates

Gypsum, anhydrite, and barite are the three common sulfates found as cementing agents in sandstones. Though modern gypsum sand dunes are known, there have been no ancient gypsum rich rocks reported that have been interpreted as lithified gypsum sand dunes. The sulfate of these cements is ultimately derived from that

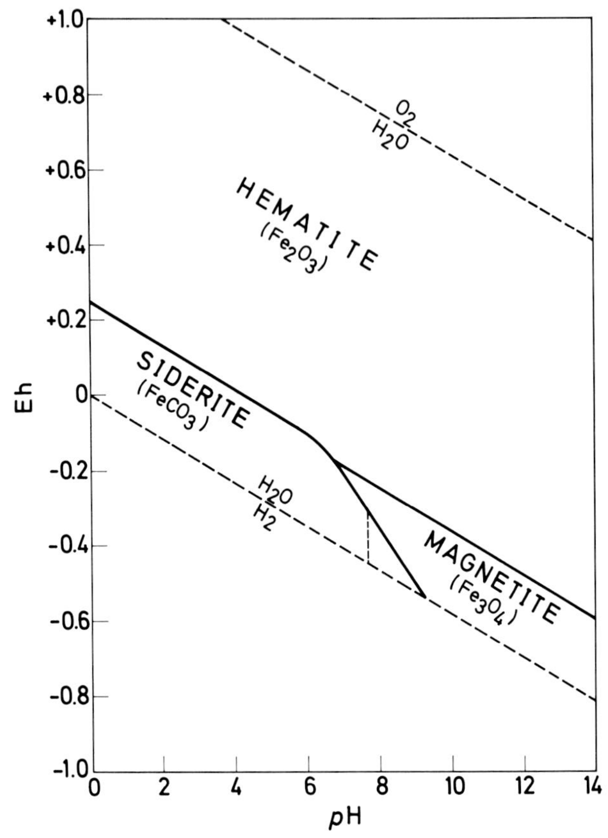

Fig. 2-8. Eh-pH diagram of stability field of siderite in relation to hematite and magnetite in water at 25° C., 1 atmosphere total pressure, and total dissolved carbonate of 10^{-2} moles per liter. Dotted line is for activity of Fe^{++} in solution $= 10^{-6}$ (Modified from Garrels and Christ, 1965, Fig. 7.13)

dissolved in sea water or in subsurface brines, but the cements, in contrast to bedded deposits of gypsum and anhydrite, are not to be taken as certain indicators of evaporite formation. A suggestion, however, that much of the cement may indeed be related to evaporite formation has come from recent work on the Persian Gulf sediments of supratidal flats (Kinsman, 1964). His work seems to indicate that downward migration of evaporated sea water from the supratidal flats may be responsible for much of the gypsum-anhydrite that has been deposited in pore spaces of pre-existing sediment below the flats. The migration of the water is in response to a density gradient induced by the evaporation of the brines at the surface.

The conditions for the presence of gypsum as opposed to anhydrite as functions of pressure, temperature, and salinity have been studied by a number of workers, the last two most recently by Hardie (1967). Gypsum, containing two molecules of water, will convert to anhydrite as the temperature is increased, the pressure is increased, or the salinity is increased (Fig. 2-9). All of these effects are predictable from the volume and density relationships of the two minerals. It

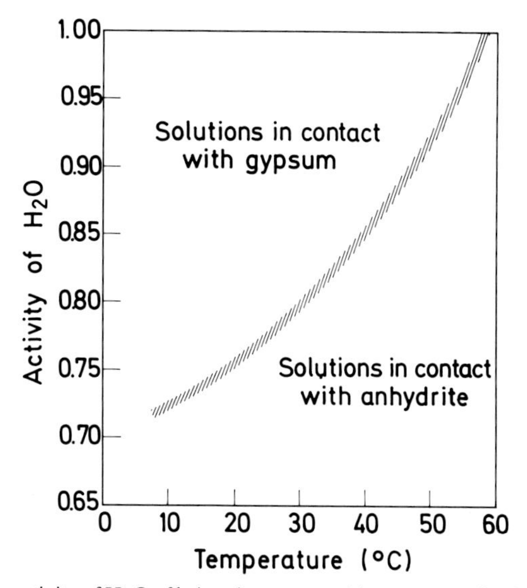

Fig. 2-9. Temperature activity of H_2O of brines in contact with gypsum and anhydrite from experimental data and natural occurrences (Modified from Hardie, 1967, Fig. 6). Activity of H_2O is a thermodynamic measure related to salinity; the activity of pure water = 1

now appears that the depth of burial is minimal, 2000 to 3000 feet (645 to 970 m), for the pressure and temperature to be reached at which gypsum converts to anhydrite, depending on the salinity of the brine.

Barite occurs as concretions as well as cement. The origin of barite in modern sediments seems to be related to vulcanism on the sea floor according to Arrhenius and Bonatti (1965). This has been deduced from studies of barite formation in modern sediments in the deep ocean; barite appears to be localized with respect to volcanic provinces in the sea. There is also evidence that much of the barite formed in economically workable deposits as well as sandstone cements and concretions is the result of barite precipitation from ascending brines from deeper seated barium containing basement rocks, according to Hanor (1967). Because of the very low solubility product of barium sulfate, the presence of barite is more to be ascribed to the presence of barium than that of sulfate, since sulfate is almost always present either from sea water or the typical subsurface brine.

Sulfides

Pyrite is the principal sulfide found in sandstones. Lesser amounts of marcasite have been reported, and in the occasionally economically valuable deposit, other sulfides as well. The sulfide appears to be almost entirely diagenetic in origin, though certain Precambrian deposits such as those of the Mississagi Quartzite of the Blind River area in Ontario and the Witwatersrand uranium deposits in South Africa have large amounts of apparently detrital rounded pyrite grains. These detrital rounded grains in the Precambrian deposits have been used to indicate a non-oxygenic atmosphere roughly two billion years ago (Rutten, 1962, p. 100).

Pyrite is genetically related to a more or less amorphous common mineral, FeS, "hydrotroilite," now properly called mackinawite, that is found in modern sediments under reducing conditions. FeS forms first by the bacterial reduction of sulfate in sea water and during early diagenesis is converted to FeS_2, pyrite (Berner, 1964). Thus pyrite in a sandstone, if it is not detrital, is presumptive evidence of at least locally reducing conditions. The reducing conditions have often been tied to a silled anoxygenic basin, but in the case of sandstones it is more likely that it is the presence of large amounts of organic matter on tidal flats that is responsible for the reduction. There is some evidence from modern sediment studies and ancient sediments that pyrite directly precipitated by bacteria can also be found, though no examples have been quoted from sandstones.

Because the pyrite that is formed diagenetically may commonly be found as rounded aggregates it is not always easy to distinguish such diagenetic pyrite from possible rounded detrital grains of pyrite. It is presumed that in an oxygen-rich environment, such as the average weathering environment of the earth today, the oxidation of the sulfide will proceed so rapidly that pyrite, though a hard, abrasion resistant mineral, will be destroyed in large quantities. Thus we find pyrite as a common but very minor constituent of the heavy mineral suite of most modern sands. But pyrite may persist for many miles downstream of areas where mine wastes containing pyrite have been dumped into a river. The persistence of pyrite as a detrital mineral is a subject that needs much more investigation.

Other Minerals

A variety of other minerals are of chemical origin, the most important being phosphates ("collophane" or carbonate-apatite), iron silicates (such as chamosite, greenalite, and glauconite), the zeolites, and the iron and titanium oxides.

The precipitation of phosphate in most sandstones appears to be related to the biological world, as reworked products of bone or shell formation. Phosphate also occurs as surficial or soil segregations and concretions. The mineralogy of these phosphates is complex, most of the minerals being some mixture of hydroxy-fluor-carbonate-apatite. Both detrital bone and shell fragments and cements appear to be poorly crystallized and have only vaguely defined optical properties. In both modern and ancient sediments, phosphate abundance has been related to organic productivity and, in particular, to upwelling nutrients on a relatively shallow marine shelf.

The iron silicates can be considered to be varieties of clay minerals that are iron rich. Glauconite appears to be the most common such mineral throughout the geologic column. Glauconite is a mica related to illite in structure, but with a great deal of both ferrous and ferric iron. Glauconite formation has been studied by Takahashi (1939), Cloud (1955), and Burst (1958). The glauconites appear to be related to mildly reducing conditions, perhaps locally induced by the presence of abundant organic matter, related to relatively near shore and shallow water environments. It is probable that there are a variety of conditions suitable for the formation of glauconite, as for the other iron silicates that are found in sediments, all of them relating to conditions which allow the entrance of ferrous iron into the clay structure. Almost always glauconite appears to be of marine origin, though some few occurrences of nonmarine origin have been reported.

The chamosite minerals are a variety of chlorite, and in general appear to need some of the same conditions for their formation as glauconite. An example of chamosite beds in a sandy formation is the Jurassic Northampton formation of England (Taylor, 1949). It is not clear whether the minerals greenalite and stilpnomelane are primary or diagenetic sedimentary minerals; they may primarily be metamorphic equivalents of iron-rich clay minerals. The most likely origin of the iron-rich clays and iron silicates in sedimentary environments is the diagenetic transformation of ferric hydroxide rims or accumulations that come down from rivers into estuarine or near shore environments. The ferric hydroxide, when intimately mixed with detrital clay minerals such as chlorites or illites, may spontaneously reorganize, and if sufficient organic matter is in the environment to provide a slightly reducing environment, to enter the clay mineral structure and form the iron silicates. The sandstones that contain these minerals are most frequently associated with the ironstone shales or carbonates which have been used as iron ores.

The same general scheme of origin has been suggested for the hematitic and limonitic iron oxide coatings of red sandstones. The iron in these formations is contained either in hematite or in the mixture of goethite and lepidocrocite that we call by the field name limonite. It is doubtful that much of this iron was transported to the sedimentary environment as detritus in the form of hematite or limonite. Most of it was probably transported as a finely divided ferric hydroxide gel, either as a suspension or adsorbed on the surfaces of clay minerals or sand grains. After deposition the ferric hydroxides spontaneously dewater to form limonite, or if complete dewatering occurs, to form hematite. It has recently been suggested by Walker (1967) that many of the red sandstones of desert environments owe their hematite to precipitation by ground waters that get their iron from the weathering of iron rich primary silicates, such as the hornblendes. Much of the recent work on the remanent magnetism of sandstones seems to support the general view of the hematite-limonite being of immediately post-depositional origin.

The titanium and iron-titanium oxides are quantitatively unimportant in sandstones, found mainly as minor components of the heavy mineral suites. These oxides are resistant to chemical and mechanical decomposition and so tend to persist for long times throughout the sedimentary cycle.

The zeolite minerals have assumed increasing importance in work on sandstones in recent years. These minerals, formerly considered in the nature of curiosities, and rarely identified as such, have now been shown to be wide-spread and abundant in many of the sandstones of eugeosynclinal tracts. Analcime is found in and associated with sandstones of other kinds, such as the Triassic Lockatong formation of the Newark Series of the Appalachians (Van Houten, 1962). The Lockatong rocks are thought to have been deposited in a lacustrine environment as groups of detrital and chemical cycles related to climate. Zeolite minerals most commonly found in sandstones are analcime, laumontite, heulandite, clinoptilolite, and mordenite. A complete review of the mineralogy and diagenetic origin of the zeolite minerals has been given by Hay (1966).

The zeolite minerals are closely linked to the presence of volcanic glass. It now appears that the zeolites form as early decomposition products of volcanic glasses, either in weathering or sedimentary environments. Other zeolites form in response to very high pH, high silica environments such as the alkaline lakes of the western United States. Sequences of zeolite and other silicate minerals that have been found with increasing depth in eugeosynclinal accumulations have been noted by Coombs and others (1959) and Packham and Crook (1960) and have been designated the zeolite facies, a true metamorphic facies, by Turner and Verhoogen (1960, p. 532). It seems likely that the zeolites will receive much more attention in the interpretation of the volcanic rich sands in the future.

Organic Matter

Organic matter in sandstones is a biochemical precipitate characteristic of the environment. The organic matter always has been profoundly altered by diagenesis. The varieties of organic matter in sandstones range from oil and gas to solid particles of high carbon content material. None are as easily characterized as minerals; they are identified and classified according to their chemical composition. They can also be partly characterized by their physical properties just as are minerals, but analytical chemical descriptions are much too difficult and time consuming to permit quick and easy identification. Much recent work in organic geochemistry has been devoted to specifying the kinds and amounts of organic compounds in recent and ancient sediments (Breger, 1963; Columbo and Hobson, 1964; Degens, 1965, p. 202–312; Manskaya and Drozdova, 1968; Eglinton, 1969).

We can easily distinguish the black opaque from the brownish translucent matter and the yellowish and brownish transparent materials that are noted in sandstones. The darkness of color is related to increasing carbon content. Most petrographic descriptions have simply lumped all of these materials into organic matter – if they are mentioned at all. We can identify woody and leafy materials as precursors of coaly fragments, but such fossils cannot be used to specify the origin of kerogen, oil and gas.

The organic compounds of sandstones are derived by the breakdown of plant and animal tissues in the sedimentary environment; later the decomposition products may be solubilized, transported, and reprecipitated during stages of diagenesis. The ultimate chemical product of all these processes is apparently more linked to a complex diagenetic history than to the particular pathways linked to specific environments. The existence of organic matter is obviously the result of either high organic productivity in comparison to the oxidative breakdown of such organic material, or to the presence of particular biologic entities that are highly resistant to oxidation, such as some of the plant cutins, spores, and other epidermal materials. Mineral charcoal, fusain, is also a prominent constituent of sandstones and is probably of the same origin as in coal, most likely of forest fire origin.

Sandstones in general are characterized by extremely small amounts of organic matter, less than 0.1 percent as opposed to shales and limestones. Some graywackes are an exception and may contain up to several percent organic matter. Most of the fluid organic matter, oil and gas, has been introduced into the sandstone after deposition by migration from the beds in which the organic matter was indigenous. The organic matter that is native to sandstones is dominantly detrital. Even tidal flat sandstones, where organic productivity is high, have a relatively small residue of organic matter left after the early biological diagenetic processes by which organisms, micro- and macro-, oxidize the original organic component. Modern sands, whether alluvial, deltaic, nearshore marine, or deep sea turbidite, tend to be associated with aerated environments; the currents that bring in the sand also tend to mix the water and keep it oxygenated. The presence of oxygen and the high permeability of the sand promotes oxidation of much of the organic matter.

Relation of Mineralogy to Texture

Mineralogy and Size

Because mineral and rock fragments break into smaller pieces and become reduced in size by abrasion as a function of cleavage, fracture, brittleness, and hardness they may be expected to have size distributions in sandstones that are related to the relevant mineralogical properties. Most work has been done on the effects of abrasion on mineral composition (Pettijohn, 1957, p. 588), which could be used to predict the differential rate of size decrease of minerals, but few analyses of size distributions of different minerals have been made that reflect abrasion resistance. One such is that of Füchtbauer (1964, Fig. 25), who compared quartz, rock fragments, and feldspar and showed that the quartz was appreciably larger in modal and median size than the others (see Chap. 5, Figs. 5-1 and 5-2). Long known is the tendency for finer grained and silty sands to contain more clay than coarser ones. But it is not possible to generalize yet on the data of Füchtbauer, who pointed out that a coarse grained sand might be classified as a different sandstone type from a fine grained sand from the same stratigraphic unit, just because of the differential size distribution of minerals.

The size of heavy minerals has been investigated extensively by van Andel (1955, p. 530–535) who showed how the different size distributions were related to hydraulic ratios and thus to the densities of the minerals. It appears that the size effect on heavy minerals is a complex of both the hydraulic ratio and the differential resistance to abrasion, for the analyses are not easily predictable on the basis of either alone. A further complicating factor in the interpretation of size distributions of heavy minerals is the original size distribution in source rocks, which might be vastly different for different minerals, as, for example, between small accessory zircons in a granite and large tourmaline crystals from pegmatites. Faced with the possibility of significant differences in mineralogy with size, the petrologist is well advised to sample both coarser and finer beds to establish the range of compositions.

Mineralogy and Resistance to Abrasion

The possibility that selective abrasion would drastically affect mineral composition by loss of finer and softer minerals has been argued for strongly by many who believe that quartz arenites may be derived by abrasion from lithic arenites rich in feldspar and rock fragments. In spite of the attractiveness of this possibility, there is no great amount of evidence that feldspar on beaches, dunes, or in nearshore marine environments decreases radically from selective abrasion. Few careful studies have been made of the disappearance of rock fragments by abrasion, but qualitative observations suggest that the softer shale fragments do degrade quickly in most sand transporting environments. But the harder metamorphic and igneous rock fragments appear to persist for longer distances of transport.

The disappearance of soft minerals that are minor components of source contributions accounts in part for the general absence of minerals like talc, gypsum, and most carbonates, though there is no doubt that solution plays an important role for the carbonates and sulfates. What complicates this problem is the association of chemical alteration with mechanical degradation, so that when minerals such as talc become extremely fine grained, they become more prone to sorb cations and transform to clay minerals. If mechanical degradation were the only important effect, then we should see the same overall mineral composition but softer minerals should be more abundant in fine silt and clay sizes. Thus the feldspar that supposedly is worn away from a quartz sand should appear as an important constituent of the associated shales and this apparently does not happen. Though this subject is an old one, there is little basis for revising the conclusions of Pettijohn (1957, p.561) that the abrasion effect exists, but it may be obscured by more important effects and is not necessarily the major cause of changes in composition.

Chemical Composition

It is obvious that the chemical composition of sandstones is a different way of designating the mineralogical composition and vice versa. If all of the minerals of a given sample could be enumerated, an accurate modal analysis made, and the chemical compositions of all of the minerals were accurately known, we could simply calculate the bulk chemical composition. This is rarely, if ever, possible because of the wide range of compositions in individual mineral species, particularly the clay minerals. Such compositional variations cannot easily be analyzed by optical, X-ray diffraction, or other methods. Such an attempt would be extraordinarily difficult if it were to be made with respect to minor or trace elements. To avoid these difficulties and yet at the same time to study the distribution of the elements that otherwise might not be revealed, we use bulk chemical analysis. Bulk chemical composition of metamorphic and igneous rocks are generally more simply interpreted than those of sandstones, both because the crystallines are usually treated as equilibrium mineral assemblages and because they do not involve the complex mechanical mixing and segregation processes inherent in the making of a sandstone. Bulk chemical composition, for example, does not distinguish between a detrital component and cement, a most important distinction. This lack of discrimination is a serious drawback to chemical analyses. Nonetheless, chemical composition is useful in establishing a baseline for the study of metamorphic derivatives of sandstones.

Ordinarily the chemical analysis is given as the oxides, commonly implying a precision of three or four significant figures, that is, to the nearest 0.01 percent. It is doubtful that this level of precision is warranted as a basis for much geological work on sandstones, for though the analysis of the particular specimen given to the analyst may be that precise, there is little reason to believe that such specimens are chosen randomly from a homogeneous population of rocks in the field. In addition to the probable lack of meaning of the last significant figure in the analysis that results from haphazard sampling, there is some doubt that differences in hundreths of a percent will prove to be geologically significant, for sandstones have such variable compositions, reflecting their coming from such diverse origins (Pettijohn, 1963).

Some analysts have attempted to chemically fractionate fine-grained sediments in order to better interpret bulk chemical analyses (Hirst and Nicholls, 1958; Goldberg and Griffin, 1964). The approach has been to treat the sediment with successively stronger acids so as to leach soluble minerals. This divides the sediments into water-soluble, weak acid-soluble and strong acid-soluble fractions. Oversimply put, the interpretation assumes a detrital origin for the least soluble fractions and authigenic origin for the more soluble ones. The fact that very insoluble minerals such as zircon or tourmaline are sometimes identifiable as authigenic and that soluble minerals such as dolomite may occur as detrital grains casts doubt on the complete validity of this idea. The chemical fractionation approach has not been explored with sandstones; a study of some carbonate deep sea sands in which the carbonate fraction was digested with weak acid to concentrate the detrital silicate fraction is one exception (Siever and Kastner, 1967).

Chemical Composition as a Function of Mineral Constituents

Alkalies (Group I). The abundant elements of this group in sandstones are as they are in most other crustal rocks, Na and K, though in sandstones neither exceeds a few percent of the total. In the non-clayey sandstones the bulk of the Na and K is in alkali feldspars and muscovite, the only igneous or metamorphic alkali metal minerals that survive as detritals to any significant extent. The Na and K in the argillaceous sandstones may be mainly in the clay minerals illite and montmorillonite. K commonly exceeds Na in both argillaceous and non-argillaceous sandstones, a reflection of dominance of mica over feldspar in most of the latter and illite over montmorillonite in the former. The rocks in which Na is the more abundant are primarily the graywackes, in which Na-rich feldspars are the main source of the high Na-content. It is not certain if this is related to a dominance of detrital albitic feldspar over K-feldspar or calcic plagioclases; it is more likely that it is related to the albitization of feldspars that is common in the diagenesis or low grade metamorphism of these rocks. It may in some graywackes come from Na-rich volcanics. Little is known of the distribution of Cs and Rb in sandstones; it is likely that they are held primarily in the clays and that any fractionation is governed by ionic radius and potential (Goldschmidt, 1954, p. 163–173). Li, Na, and K are all found in zeolite minerals, either as major components or as impurities.

Alkaline Earths (Group II). Carbonates and clay minerals account for most of the elements of this group. The only other mineral containing a Group II element in significant amounts is anorthitic (calcic) plagioclase. Ca and Mg are the only two elements of quantitative importance. Little is known of the distribution of Sr and Ba and practically nothing about Be is known in sandstones.

Ca is characteristically present as carbonate, either in calcite or dolomite, rarely in aragonite, or gypsum or anhydrite. Minor contributions to total Ca-content may come from montmorillonitic clays. Sandstones that contain anorthite plagioclase may owe their Ca to this mineral. Mg is partitioned between dolomite, chloritic and montmorillonitic clays. Ca is more abundant than Mg in most sandstone types, a reflection of the generally greater abundance of calcite over dolomite. In graywackes the two are more nearly equal as a result of the large amount of chloritic clay that is typical of the matrix of many of these rocks.

Ba and Sr, when present in more than trace amounts, are found as sulfates in the barite-celestite solid-solution system. In mixed carbonate-silicate sands, particularly the very young ones, Ba and Sr are in the carbonate, proxying for Ca or Mg.

The elements of Group II originate as chemical precipitates in sandstones, primarily as the carbonates, and much may be diagenetic. The exception is the Mg in chloritic clays and this too may be of the same origin as discussed in the section of origin of graywacke matrix in Chap. 10. Because of the relatively high solubility of the carbonates and sulfates and the relative ease of leaching of Mg from chlorite clays, one can see some justification for referring to acid-soluble constituents as correlative with diagenetic origin, but only at the cost of great oversimplification.

Aluminium (Group III). Al is the only abundant element of this group present in sandstones. It is present almost exclusively in aluminosilicates, though gibbsite and boehmite may be significant in some and minor amounts can be attributed to heavy minerals such as spinels, pyroxenes, epidote, and micas. Abundance of Al is in general, then, related to the abundance of feldspars, micas and clays, regardless of whether the clay is in argillaceous rock fragments or in interstitial clay, either detrital matrix or diagenetic. The variability of Al content in the different clay minerals is probably less important in affecting the bulk composition than is the quantity of clay. This is illustrated by the fact that the more aluminous clays are associated with quartz arenites but the quartz arenites as a group are low in Al, because they contain so little clay. At the other end of the sandstone compositional spectrum, the graywackes, Al is high not because the clays are aluminous, for they are not, but because both clays and feldspars are abundant.

Boron is present in trace quantities in the clay fraction of sandstones, although little is known about it. For the shales and muds, where it has been chiefly studied, there is a large literature concerning its possible significance. Some have tried to relate boron concentration directly to salinity (Degens and others, 1957 and 1957), to clay mineral composition (Lerman, 1966) and still others have correlated it with the percentage of less than 2 μ clay (Shimp and others, 1969). Complicating the interpretation of boron in the clay fraction of sandstones is good permeability, which favors postdepositional diagenesis of the clay minerals (Glass and others, 1957).

Silica. Silica abundance is rather simply related to the ratio of silicate to non-silicate minerals. Clearly, those sandstones with large amounts of carbonate, sulfate, or oxide, all of which are most commonly present as cements, will be low in silica. Next in importance are the relative amounts of quartz and chert in relation to other silicates or aluminosilicates. In this way silica content matches roughly with maturity factors in mineralogical composition. But to the extent that relative amounts of feldspar and clay minerals have some effect, feldspar being more siliceous than most clays, silica content would not correlate well with maturity.

Others. Both ferrous and ferric iron are present in sandstones as components of many minerals. Fe^{++} is in clays, mainly chlorites with lesser amounts in montmorillonites and illites; in carbonates as siderite and in ferroan dolomites (ankerites); in sulfides, primarily pyrite; and as a minor constituent of feldspars as other silicates. Fe^{3+} is normally present as the hydrous and anhydrous oxides, hematite, goethite, and lepidocrocite; and in the glauconites. Volcanic rock fragments may be rich in Fe^{++}; to the extent that they or Fe-rich glasses dominate the volcaniclastic sands, those sands will have a high Fe^{++} content. The Fe-oxides, though they may give the sandstone a pronounced red appearance, are normally dispersed as fine grained pigment and are not quantitatively abundant enough to strongly affect total iron composition.

Titanium is present in small amounts, mostly in clays, but some is in the heavy minerals rutile, ilmenite, brookite, and anatase. Though these minerals may be authigenic it is doubtful that the whole rock will be enriched in Ti by this means. Sulfur is present as sulfate, mainly as gypsum and anhydrite, and sulfide, pyrite and marcasite. Thus, with the exception of some few sandstones in which gypsum or pyrite can be demonstrated to be detrital, S is restricted to the chemical precipitates, and, within that group, largely to those of diagenetic origin. Phosphorus is present as detrital apatite and as biochemically precipitated carbonate-apatite (collophane) as well as minor impurities in a host of other minerals.

Isotopic Composition

Though much work has been done on the isotopic composition of sedimentary rocks and minerals, practically no effort has been devoted to sandstones. Studies of C^{12}/C^{13} and O^{18}/O^{16} ratios of carbonate minerals have generally not included the cements of sandstones. The same has been true of S^{32}/S^{34} and other isotopic studies. Any guesses that we might essay would only be inferences by extension from the studies of other rock types. To that extent we suggest that the study of the stable isotopes of oxygen, carbon, and sulfur in the cements or other chemical components of sandstones can be petrologically useful for the determination of temperature-salinity relationships that may be related to depth of burial, replacement versus void filling origin of carbonate cements, biogenic versus abiogenic formation, primary versus diagenetic formation, and probably many others. Savin and Epstein (1970) have reported oxygen isotope ratios of 7 modern sands and 5 Precambrian sandstones and concluded that O^{18}/O^{16} ratios may indicate high temperature rock provenance of detrital quartz and feldspar as opposed to low temperature origin of authigenic quartz and feldspar. They also suggested how the temperature of authigenic feldspar formations might be calculated.

Chemical Composition as a Function of Sandstone Type

The range of chemical composition of the common sandstone types and the average sandstone has been summarized most recently by Pettijohn (1963, Tables 12 and 13), from which Table 2-5 has been taken. The quartz arenites are nearly pure SiO_2, the Al_2O_3 coming from clay and the CaO from calcite cement. The lithic arenites show much more Al_2O_3, and the other major elements, most deriving from the argillaceous rock fragments. Graywackes have lower SiO_2

Table 2-5. *Mean composition of principal sandstone classes and average sandstone (Pettijohn, 1963, p. 15)*

	Quartz Arenite	Lithic Arenite	Gray-Wacke	Arkose	Average sandstone		
					A	B	C
SiO_2	95.4	66.1	66.7	77.1	78.66	84.86	77.6
Al_2O_3	1.1	8.1	13.5	8.7	4.78	5.96	7.1
Fe_2O_3	0.4	3.8	1.6	1.5	1.08	1.39	1.7
FeO	0.2	1.4	3.5	0.7	0.30	0.84	1.5
MgO	0.1	2.4	2.1	0.5	1.17	0.52	1.2
CaO	1.6	6.2	2.5	2.7	5.52	1.05	3.1
Na_2O	0.1	0.9	2.9	1.5	0.45	0.76	1.2
K_2O	0.2	1.3	2.0	2.8	1.32	1.16	1.3
H_2O+	0.3	3.6	2.4	0.9	1.33	1.47	1.7
H_2O-		0.7	0.6		0.31	0.27	0.4
TiO_2	0.2	0.3	0.6	0.3	0.25	0.41	0.4
P_2O_5		0.1	0.2	0.1	0.08	0.06	0.1
MnO		0.1	0.1	0.2	trace	trace	0.1
CO_2	1.1	5.0	1.2	3.0	5.04	1.01	2.5
SO_3			0.3		0.07	0.09	0.1
Cl					trace	trace	trace
F							trace
S			0.1				trace
BaO					0.05	0.01	trace
SrO					trace	none	trace
C			0.1				trace
Ignition Loss							
Total	100.7	100.0	100.4	100.0	100.41	99.86	100.0

A. Composite analysis of 253 sandstones.
B. Composite analysis of 371 sandstones used for building purposes.
C. Computed by taking 26 parts average graywacke, 25 parts average lithic arenite, 15 parts average arkose, and 34 parts average quartz arenite.

than most sandstones, more Al_2O_3, and dominance of Na_2O over K_2O and MgO over CaO. Arkose have high Al_2O_3, K_2O, and Na_2O contents predictable from the high proportion of feldspars.

Chemical Classification of Sandstones

Probably the simplest way to categorize sandstones by chemical composition rather than by mineralogical-textural composition is one that takes into account the general geochemical behavior of the elements by following the Periodic Table of the elements in association with their crystal chemistry, as shown in Fig. 2-10. This classification is not designed for and is not particularly useful for naming purposes but simply shows some of the relationships between elemental composition, mineralogy, and rock type.

The classification differentiates mature and immature as those terms are used in both mineralogical and textural classifications (see Chap. 3 and 5). It does so by using the ratio of SiO_2 to Al_2O_3 rather than by using either alone. The mature quartz rich sandstones have high SiO_2/Al_2O_3 ratios by virtue of the absence of aluminosilicates, either clays (a textural attribute) or primary silicates (a mineralogical attribute). This kind of division should not be confused with weathering maturity in the geochemical sense in which low SiO_2/Al_2O_3 ratios are characteristic of the most maturely weathered soils in which SiO_2 is leached before Al_2O_3.

The differences between alkali metal-rich and alkali metal-poor sandstones can almost equally well be taken as corresponding to a maturity index defined in chemical compositional terms (Fig. 2-11), where the contours are drawn for the ratio of $SiO_2 + Al_2O_3$ to $Na_2O + K_2O$ and the individual analyses are plotted by the ratio of Na_2O to K_2O and the ratio of SiO_2 to Al_2O_3. Either way of looking at the chemical analysis is bound to reveal the same maturation concept that is

Structural frame work: Anion Groups	Exchangeable cations		Associated Petrographic Type
High SiO_2/Al_2O_3 (mature) (little clay or detrital Al-silicate)	Alkaline-earth-rich (carbonate cement) Alkaline-earth-poor (silica cement)		Quartz arenites
Low SiO_2/Al_2O_3 (immature) (clay + detrital Al-silicate)	Alkali-metal-rich (feldspar and clays)	$Na_2O > K_2O$	Feldspathic graywackes
		$Na_2O < K_2O$	Arkoses; Lithic graywackes
	Alkali-metal-poor (aluminous clays)		Lithic arenites

Fig. 2-10. A chemical classification of sandstones

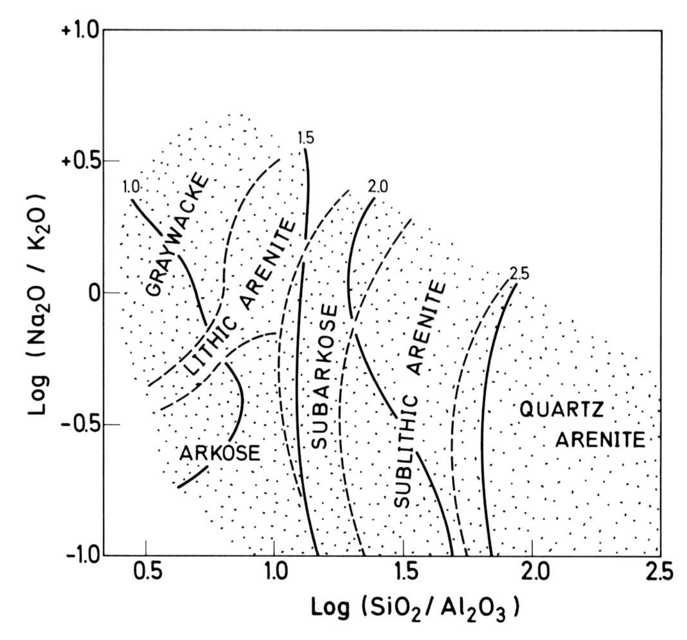

Fig. 2-11. SiO_2/Al_2O_3 vs. Na_2O/K_2O ratios of sandstones. Stipple outlines area of most analyses. Contours (heavy lines) show values of $\log\left(\frac{SiO_2 + Al_2O_3}{Na_2O + K_2O}\right)$. Analyses drawn largely from Pettijohn (1963)

expressed by modal analyses; the longer a sand has been in the sedimentary "mill" the more it tends to be enriched in quartz at the expense of everything else.

The differences between immature sandstones, all of which are low in SiO_2/Al_2O_3 ratios, can be expressed as further subdivisions of that ratio into "low" and "very low" or, as is done here, into alkali metal-rich and alkali metal-poor. Though not a neat division it does separate most lithic sandstones from other immature sandstones, primarily reflecting the fact that many of the lithic sandstones tend to have more aluminous, and therefore less alkali metal-rich, clays in their rock fragments and matrices. The exceptions are those lithic sandstones close in nature to lithic graywackes, in which rock fragments may be of volcanic or other alkali metal-rich rocks.

Alkali metal-rich immature sandstones are here divided into two classes on the basis of a long recognized difference in K_2O/Na_2O ratios. The high Na_2O/K_2O ratios of graywackes are attributable to a dominance of albitic plagioclase feldspars over K-feldspars and K-micas. To the extent that some, primarily the younger, graywackes contain abundant montmorillonite, the ratio is further increased. The frequency with which albitic rims, either diagenetic or low-grade metamorphic, have been observed in many graywackes is petrographic evidence of a secondary source of Na that is not related to provenance. Thus the ratio of feldspars is not purely nor may not even be largely due to distinctive source areas with more plagioclase, but comes from the typical secondary growth of Na-feldspars in strongly deformed and metamorphic belts of which graywackes are so typical.

The division of the mature sandstones takes recognition of the fact that the chemical analysis is strongly affected by the type of cementing agent. All gradations from pure silica to pure carbonate cement are known, sometimes in a single stratigraphic unit. One might erect another pigeonhole for analyses that reflect other cements, such as Ba-rich for barite cemented sandstones, and P-rich for phosphate cemented sandstones. Regardless of the scheme, it will reflect the diagenetic additions to the mature sandstones.

Thus, in contrast to the igneous and metamorphic rocks, where bulk chemical analyses are all important in any classification scheme and can be more or less neatly harmonized with modal mineralogical analyses, sandstones' chemical analyses do not fit so neatly with mineralogical analyses. The bulk chemistry of sandstones reflects the sometimes divergent and sometimes reinforcing effects of mineralogical versus textural sorting and differentiation, and the widely differing effects of source, environment, and diagenesis in shaping the rock's present composition.

References

Allen, J. R. L.: Petrology, origin and deposition of the highest Lower Old Red sandstone of Shropshire, England. Jour. Sed. Petrology **32**, 657–697 (1962).

Allen, R. C., Gavish, E., Friedman, G. M., Sanders, J. E.: Aragonite-cemented sandstone from outer continental shelf off Delaware Bay: submarine lithification mechanism yields product resembling beachrock. Jour. Sed. Petrology **39**, 136–149 (1969).

Andel, Tj. H. van: Sediments of the Rhone delta, pt. II, Sources and deposition of heavy minerals. Koninkl. Nederlandsch. Geol. Mijnb. Genoot. Geol. Ser. **15**, 357–556 (1955).

Arrhenius, G., Bonatti, E.: Neptunism and volcanism in the ocean. In: Sears, M., Ed.: Progress in Oceanography, Vol. 3, p. 7–22. London: Pergamon 1965.

Baker, G.: Opal phytoliths in some Victorian soils and "red rain" residues. Australian Jour. Botany **7**, 64–87 (1959).

Baskin, Yehuda: A study of authigenic feldspars. Jour. Geology **64**, 132–155 (1956).

Bates, T. F., Comer, J. J.: Electron microscopy of clay surfaces. Clays and clay minerals: 3rd Nat'l. Conf. on Clays and Clay Minerals. Proc. U.S. Nat'l. Acad. Sci. Pub. **395**, 1–25 (1955).

Berner, R. A.: Iron sulfides formed from aqueous solution at low temperatures and atmospheric pressure. Jour. Geology **72**, 293–306 (1964).

Beutelspacher, H., van der Marel, H. W.: Atlas of electron microscopy of clay minerals and their admixtures, 333 p. Amsterdam: Elsevier 1968.

Birks, L. S.: Electron probe microanalysis, 253 p. New York: Interscience 1963.

Biscaye, Pierre: Mineralogy and sedimentation of recent deep-sea clay in the Atlantic Ocean and adjacent seas and ocean. Geol. Soc. America Bull. **76**, 803–832 (1965).

Blatt, H.: Original characters of clastic quartz. Jour. Sed. Petrology **37**, 401–424 (1967).

— Christie, J. M.: Undulatory extinction in quartz of igneous and metamorphic rocks and its significance in provenance studies of sedimentary rocks. Jour. Sed. Petrology **33**, 559–579 (1963).

Breger, I. A., Ed.: Organic geochemistry, 658 p. New York: Macmillan 1963.

Briggs, L. I., McCulloch, D. S., Moser, Frank: The hydraulic shape of sand particles. Jour. Sed. Petrology **32**, 645–656 (1962).

Brookins, D. G., Voss, J. D.: Age dating of muscovites from Pennsylvanian sandstones near Wamego, Kansas. Am. Assoc. Petroleum Geologists Bull. **54**, 353–356 (1970).

Brown, George, Ed.: The X-ray identification and crystal structures of clay minerals, 544 p. London: Mineralog. Soc. Great Britain 1961.

Burst, J. F.: "Glauconite" pellets: their mineral nature and application to stratigraphic problems. Amer. Assoc. Petroleum Geologists Bull. **42**, 310–327 (1958).

Callender, D. L., Folk, R. L.: Idiomorphic zircon, key to volcanism in the lower Tertiary sands of central Texas. Am. Jour. Sci. **256**, 257–269 (1958).

Clarke, F. W.: The data of geochemistry, 5th Ed. U.S. Geol. Survey Bull. **770**, 841 p. (1924).
Cloud, P. E.: Physical limits of glauconite formation. Amer. Assoc. Petroleum Geologists Bull. **39**, 484–492 (1955).
Coombs, D. S., Ellis, A. J., Fyfe, W. S., Gaylor, A. M.: The zeolite facies, with comments on the interpretation of hydrothermal syntheses. Geochim. et Cosmochim. Acta **17**, 53–107 (1959).
Columbo, Umberto, Hobson, G. C., Eds.: Advances in organic geochemistry, 488 p. New York: Macmillan 1964.
Correns, C. W., Piller, H.: Mikroskopie der feinkornigen Silikatminerale. In: Mikroskopie der Silikate; Pt. 1, Handbuch der Mikroskopie in der Technik, Vol. IV, p. 699–780. Mikroskopie der Gesteine. 796 p. Frankfurt: Umschau Verlag 1955.
Deer, W. A., Howie, R. A., Zussman, J.: Rock-forming minerals, Vol. 1, 333 p. Ortho- and ring silicates. New York: Wiley 1962.
— — — Rock-forming minerals, Vol. 4, 435 p. Framework silicates. New York: Wiley 1963.
Degens, Egon: Geochemistry of sediments: a brief survey, 342 p. Englewood Cliffs, New Jersey: Prentice-Hall 1965.
Degens, E. T., Williams, E. G., Keith, M. L.: Environmental studies of Carboniferous sediments, pt. 1, Geochemical criteria for differentiating marine and freshwater shales. Am. Assoc. Petroleum. Geologists Bull. **41**, 2427–2455 (1957).
— — — Environmental studies of Carboniferous sediments, pt. 2, Application of geochemical criteria. Am. Assoc. Petroleum Geologists Bull. **42**, 981–997 (1958).
Dennen, W. H.: Impurities in quartz. Geol. Soc. America Bull. **75**, 241–246 (1964).
Eglinton, Geoffrey, Ed.: Organic geochemistry, 720 p. Berlin-Heidelberg-New York: 1969.
Farmer, V. C., Russell, J. D.: The infra-red spectra of layer silicates. Spectrochim. Acta **20**, 1149–1173 (1964).
Feth, J. H., Roberson, C. E., Polzer, W. L.: Sources of mineral constituents in water from granitic rocks, Sierra Nevada, California and Nevada, 70 p. U.S. Geol. Survey Water-Supply Paper **1535**-I 1964.
Field, M. E., Pilkey, O. H.: Feldspar in Atlantic continental margin sands off the southeastern United States. Geol. Soc. America Bull. **80**, 2097–2102 (1969).
Folk, R. L., Weaver, C. E.: A study of the texture and composition of chert. Am. Jour. Sci. **250**, 498–510 (1952).
Friedman, G. M.: Identification of carbonate minerals by staining methods. Jour. Sed. Petrology **29**, 87–97 (1959).
Frondel, C.: Dana's System of Mineralogy, 7th Ed., 334 p. New York: Wiley 1962.
Frye, J. C., Swineford, A.: Silicified rock in the Ogallala formation. Kansas Geol. Survey Bull. **64**, pt. 2, 33–76 (1946).
Füchtbauer, H.: Sedimentpetrographische Untersuchungen in der älteren Molasse nördlich der Alpen. Eclogae Geol. Helv. **57**, 157–298 (1964).
Galliher, E. W.: Biotite-glauconite transformation and associated minerals. In: Recent Marine Sediments. Tulsa: Am. Assoc. Petroleum Geologists 513–515 (1939).
Garrels, R. M., Christ, C. L.: Solutions, minerals, and equilibria, 435 p. New York: Harper and Row 1965.
— Mackenzie, F. T.: Origin of the chemical compositions of some springs and lakes. In: Equilibrium concepts in natural water systems. American Chemical Society Advances in Chemistry Ser. **67**, 222–242 (1967).
— — Siever, R.: Sedimentary cycling in relation to the history of the continents and oceans. In: Robertson, E. D., Ed.: The nature of the solid earth. 93–121. New York: McGraw-Hill 1971.
Giles, R. T., Pilkey, O. H.: Atlantic beach and dune sediments of the southern United States. Jour. Sed. Petrology **35**, 900–910 (1965).
Gilligan, A.: The petrography of the Millstone Grit of Yorkshire. Geol. Soc. London Quart. Jour. **75**, 260–262 (1920).
Glass, H. D., Potter, P. E., Siever, R.: Clay mineralogy of some basal Pennsylvanian sandstones, clays and shales. Amer. Assoc. Petroleum Geologists Bull. **40**, 750–754 (1957).
Goldberg, E. D., Griffin, J. J.: Sedimentation rates and mineralogy in the South Atlantic. Jour. Geophys. Research **69**, 4293–4309 (1964).
Goldich, S. S.: A study in rock weathering. Jour. Geology **46**, 17–58 (1938).
Goldschmidt, V. M.: In: Muir, A., Ed.: Geochemistry, 730 p. Oxford: Oxford Univ. Press 1954.

Goldsmith, J. R., Graf, D. L.: Subsolidus relations in the system $CaCO_3$–$MgCO_3$–$MnCO_3$. Jour. Geology **68**, 324–335 (1960).

Goldsmith, J. R., Graf, D. L., Witters, Juanita, Northrop, D. A.: Studies in the system $CaCO_3$–$MgCO_3$–$MnCO_3$–$FeCO_3$. Jour. Geology **70**, 659–688 (1962).

Graf, D. L.: Carbonate mineralogy, carbonate sediments, pt. 1 of Geochemistry of carbonate sediments and sedimentary carbonate rocks. Illinois Geol. Survey Circ. **297**, 39 p. (1960).

Grim, R. E.: Clay mineralogy, 2nd Ed., 596 p. New York: McGraw-Hill 1968.

Hanor, J. S.: Regional control and zoning of barite in eastern North America. Econ. Geology **62**, 870 (1967).

Harder, H., Menschel, G.: Quarzbildungen am Meeresboden. Die Naturwissenschaften **54**, 561 (1967).

Hardie, L. A.: The gypsum-anhydrite equilibrium at one atmosphere pressure. Am. Mineralogist **52**, 171–200 (1967).

Hallam, A.: Depositional environment of British Liassic ironstones considered in the context of their facies relationships. Nature **209**, 1306–1307 (1966).

— Siderite- and calcite-bearing concretionary nodules in the Lias of Yorkshire. Geol. Mag. [Great Britain] **104**, 222–227 (1967).

Hay, R. L.: Zeolites and zeolite reactions in sedimentary rocks. Geol. Soc. America Spec. Paper **85**, 130 p. (1966).

Hayes, J. R.: Quartz and feldspar content in South Platte, Platte and Missouri river sands. Jour. Sed. Petrology **32**, 793–800 (1962).

Hemley, J. J., Jones, W. R.: Chemical aspects of hydrothermal alteration with emphasis on hydrogen metasomatism. Econ. Geology **59**, 538–569 (1964).

Hirst, D. M., Nicholls, G. D.: Techniques in sedimentary geochemistry, pt. 1, separation of the detrital and non-detrital fractions of limestones. Jour. Sed. Petrology **28**, 468–481 (1958).

Horowitz, Alan, Potter, Paul Edwin: Introductory Petrography of Fossils, 325 p. Berlin-Heidelberg-New York: Springer 1971.

Hubert, J. F.: Petrology of the Fountain and Lyons Formations. Front Range, Colorado. Colorado School Mines Quart. **55**, no. 1, 1–242 (1960).

Hunter, R. E.: The petrography of some Illinois Pleistocene and Recent sands. Sed. Geology **1**, 57–75 (1967).

Hurley, P. M., Hunt, J. M., Pinson, W. H., Jr., Fairbairn, H. W.: K–Ar age values on the clay fractions in dated shales. Geochim. et Cosmochim. Acta **27**, 279–284 (1963).

Hsu, K. Jinghwa: Texture and mineralogy of the Recent sands of the Gulf Coast. Jour. Sed. Petrology **30**, 380–403 (1960).

Johns, W. D., Grim, R. E., Bradley, W. F.: Quantitative estimations of clay minerals. Jour. Sed. Petrology **24**, 242–251 (1954).

Kinsman, D. J. J.: The Recent carbonate sediments near Halat el Bahrani Trucial Coast, Persian Gulf. In: Deltaic and shallow marine deposits. Developments in Sedimentology **1**, 185–192 (1964).

Krauskopf, Konrad: Introduction to geochemistry, 721 p. New York: McGraw-Hill 1967.

Krynine, P. D.: Petrology and genesis of the Third Bradford Sand. Pennsylvania State Coll. Bull. **29**, 134 p. (1940).

— The tourmaline group in sediments. Jour. Geology **54**, 65–87 (1946).

— The megascopic study and field classification of sedimentary rocks. Jour. Geology **56**, 130–165 (1948).

Leith, C. K., Mead, W. J.: Metamorphic geology, 337 p. New York: Henry Holt 1915.

Lerman, A.: Boron in clays and estimation of paleosalinities. Sedimentology **6**, 267–286 (1966).

Lovell, J. P. B.: Tyee formation: a study of proximality in turbidites. Jour. Sed. Petrology **39**, 935–953 (1969).

Mackenzie, F. T., Garrels, R. M.: Silicates: reactivity with sea water. Science **150**, 57–58 (1965).

Mackie, W.: The sands and sandstones of Eastern Moray. Edinburgh Geol. Soc. Trans. **7**, 148–172 (1896).

Manskaya, S. M., Drozdova, T. V.; Shapiro, L., Breger, I. A., trans. and eds.: Geochemistry of organic substances, 345 p. London: Pergamon 1968.

Majewske, Otto P.: Recognition of invertebrate fossil fragments in rocks and thin sections, 101 p. Leiden: Brill 1971.

McMaster, R. L., Garrison, L.: Mineralogy and origin of southern New England shelf sediments. Jour. Sed. Petrology **36**, 1131–1142 (1966).

Midgley, H. G.: Chalcedony and flint. Geol. Mag. [Great Britain] **88**, 179–184 (1951).

Nanz, R. H., Jr.: Genesis of Oligocene sandstone reservoir, Seeligson field, Jim Wells and Kleberg Counties, Texas: Am. Assoc. Petroleum Geologists Bull. **38**, 96–117 (1954).

Okada, H.: Non-graywacke "turbidite" sandstones in the Welsh geosyncline. Sedimentology **7**, 211–232 (1966).

Orville, P.: Alkali ion exchange between vapor and feldspar phases. Am. Jour. Sci. **261**, 201–237 (1963).

Packham, G. H., Crook, K. A. W.: The principle of diagenetic facies and some of its implications. Jour. Geology **68**, 392–407 (1960).

Pettijohn, F. J.: Sedimentary rocks, 2nd Ed., 718 p. New York: Harper 1957.

— Chemical composition of sandstones – excluding carbonate and volcanic sands. U.S. Geol. Survey Prof. Paper **440**-S, 19 p., 1963.

Plas, L. van der: The identification of detrital feldspars, 305 p. Amsterdam: Elsevier 1966.

Poldervaart, A.: Zircons in rocks, pt. 1, sedimentary rocks. Am. Jour. Sci. **253**, 433–461 (1955).

Potter, P. E., Pryor, W. A.: Dispersal centers of Paleozoic and later clastics of the upper Mississippi valley and adjacent areas. Geol. Soc. America Bull. **72**, 1195–1250 (1961).

Pryor, W. A., Hester, N. C.: X-ray diffraction analysis of heavy minerals. Jour. Sed. Petrology **39**, 1384–1389 (1969).

Rimsaite, J.: Optical heterogeneity of feldspars observed in diverse Canadian rocks. Schweiz. Mineralog. Petrog. Mitt. **47**, 61–76 (1967).

Rittenhouse, G. A.: Transportation and deposition of heavy minerals. Geol. Soc. America Bull. **54**, 1725–1780 (1943).

Ronov, A. B., Mikhailovskaya, M. S., Solodkova, I. I.: Evolution of the chemical and mineralogical composition of arenaceous rocks. In: Chemistry of the earth's crust, Vol. 1. U.S.S.R. Acad. Sci., Israel Progr. Sci., Translations, 1966, p. 212–262 (1963).

Russell, R. D.: Mineral composition of Mississippi River sands. Geol. Soc. America Bull. **48**, 1307 – 1348 (1937).

Rutten, M. G.: The geological aspects of the origin of life on earth, 146 p. Amsterdam: Elsevier (1962).

Sabins, F. F., Jr.: Grains of detrital, secondary, and primary dolomite from Cretaceous strata of the western interior. Geol. Soc. America Bull. **73**, 1183–1196 (1962).

Savin, S. M., Epstein, Samuel: The oxygen isotopic compositions of coarse grained sedimentary rocks and minerals. Geochim. et Cosmochim. Acta **34**, 323 – 329 (1970).

Shiki, T.: Studies on sandstone in the Maizuru Zone, Southwest Japan, pt. 1, Importance of some relations between mineral composition and grain size. Mem. Coll. Sci., Univ. Kyoto, ser. B **29**, 291 – 324 (1959).

Shimp, N. F., Witter, J., Potter, P. E., Schleicher, J. A.: Distinguishing marine and freshwater muds. Jour. Geology **77**, 566 – 580 (1969).

Siever, R.: The silica budget in the sedimentary cycle. Am. Mineralogist **42**, 821 – 841 (1957).

– Establishment of equilibrium between clays and sea water. Earth Planet. Sci. Letters, **5**, 106 – 110 (1968).

Siever, Raymond, Kastner, Miriam: Mineralogy and petrology of some Mid-Atlantic Ridge sediments. Jour. Marine Research **25**. 263 – 278 (1967).

Siffert, Bernard: Quelques reactions de la silice en solution: la formation des argiles. Service Carte Geol. Alsace Lorraine Memoires No. **21**, 86 p. (1962).

Sippel, R. F.: Sandstone petrology, evidence from luminescence petrography. J. Sed. Petrology **38**, 530 – 554 (1968).

Smith, J. V., Stenstrom, R. C.: Electron excited luminescence as a petrologic tool. Jour. Geol. **73**, 627 – 635 (1965).

Strakov, N. M.: Principles of lithogenesis, Vol. 1, 245 p. New York: Consultants Bureau 1967.

Sweatman, T. R., Long, J. V. P.: Quantitative electron-probe microanalysis of rock-forming minerals. Jour. Petrology **10**, 332 – 379 (1969).

Swineford, A., Franks, P. O.: Opal in the Ogallala formation, Kansas. In: Silica in Sediments. Soc. Econ. Paleontologists and Mineralogists Spec. Pub. **7**, 111–120 (1959).

Takahashi, J. I.: Synopsis of glauconitization. In: Recent Marine Sediments. Tulsa, Oklahoma: Amer. Assoc. Petroleum Geologists 503–512 (1939).

Tatsumoto, M., Patterson, C.: Age studies of zircon and feldspar concentrates from the Franconia sandstone: Jour. Geol. **72**, 232 – 242 (1964).

Taylor, J. H.: Petrology of the Northampton sand ironstone formation. Geol. Survey Great Britain Mem., 111 p. 1949.

Tomita, Toru: Geologic significance of the color of granite zircon, and the discovery of the Precambrian in Japan. Kyushu Univ. Fac. Sci. Memoir **4**, 135 – 161 (1954).

Trowbridge, A. C., Shepard, F. P.: Sedimentation in Massachusetts Bay: Jour. Sed. Petrology, **2**, 3 – 37 (1932).

Turner, F. J., Verhoogen, John: Igneous and metamorphic petrology, 694 p. New York: McGraw Hill 1960.

Van Houten, F. B.: Cyclic sedimentation and the origin of analcime-rich Upper Triassic Lockatong formation, west-central New Jersey and adjacent Pennsylvania. Am. Jour. Sci. **260**, 561 – 576 (1962).

Walker, T. R.: Formation of red beds in modern and ancient deserts. Geol. Soc. America Bull. **78**, 353 – 368 (1967).

Warne, S. St. J.: A quick field or laboratory staining scheme for the differentation of the major carbonate minerals. Jour. Sed. Petrology, **32**, 29 – 38 (1962).

Webb, W. M., Potter, P. E.: Petrology and geochemistry of modern sands derived from a volcanic terrain, western Chihuahua. Bol: Soc. Mexicana **32**, 45–61 (1969).

Whetten, J. T.: Sediments from the lower Columbia River and origin of graywacke. Science **152**, 1057 – 1058 (1966).

Willman, H. B.: Feldspar in Illinois sands; a study in resources: Illinois Geol. Sur., Rept. Inv. **79**, 87 p. (1942).

Williams, Howell, Turner, F. J., Gilbert, C. M.: Petrography, 406 p. San Francisco: Freeman 1954.

Zvyagin, B. B., Lyse, S., trans.: Electrondiffraction analysis of clay mineral structure (Translated from the Russian by S. Lyse), 364 p. New York: Plenum Press 1967.

Chapter 3. Texture

Introduction

Texture includes the shape, roundness, surface features, grain size, and fabric of the components – principally the detrital ones – of a sandstone. Except for fabric and surface features, all have been studied exhaustively so that there is a great literature.

What has been the object of this research? Identification of the environment of deposition has been the principal objective and represents perhaps 80 percent of the effort, the underlying assumption being that the physical processes at the site of deposition impart a distinctive textural "fingerprint" to the sand. In other words, what effects do currents, wave height, and water depth, to name but a few, have on grain size, surface texture, roundness, and fabric? One of the fundamental problems of identification of environment of sands and sandstones by textural measurements has been to determine the extent to which grain size, rounding and surface texture are inherited from a previous history – or are really only the response to processes that acted immediately before burial in the last environment.

Other objectives of textural study include 1) the determination of useful physical properties such as porosity, permeability and crushing strength, 2) the mapping of dispersal patterns by textural measurements and 3) making distinctions among stratigraphic units.

Three recent comprehensive summaries of texture include those by Rosenfelder (1961), Köster (1964), and Müller (1964). All three emphasize methods and techniques. Folk's syllabus (1968, p. 3–64) is also very useful.

Grain Size

Geologists have long been fascinated by the problem of extracting geologic information from a grain size (granulometric) analysis of a sand or sandstone. As a result, there is a larger literature on the techniques and interpretation of grain size than on any other aspect of texture. Although the ratio of the output of important geological results to the input of measurement and statistical analysis is at best only modest, there is no sign of slackening of rate of publication.

Meaning of Size

Although it is one of the most widely used terms in sedimentology, the "size" of a particle is not uniquely defined except perhaps for only the most simple of geometric objects such as a sphere (diameter) or a cube (length of an edge). But

Table 3-1. *Differing definitions of particle size*[a] *(Allen, 1968, Table 2.1)*

Symbol	Name	Definition
d_s	Surface diameter	The diameter of a sphere having the same surface area as the particle.
d_v	Volume diameter	The diameter of a sphere having the same volume as the particle.
d_d	Drag diameter	The diameter of a sphere having the same resistance to motion as the particle in a fluid of the same viscosity and at the same velocity.
d_a	Projected area diameter	The diameter of a sphere having the same projected area as the particle when viewed in a direction perpendicular to a plane of stability.
d_f	Free-falling diameter	The diameter of a sphere having the same density and the same free-falling speed as the particle in a fluid of the same density and viscosity.
d_{St}	Stokes' diameter $d_{St} = (d_v^3/d_d)^{\frac{1}{2}}$	The free-falling diameter in the laminar flow region ($Re < 0.2$).
d_A	Sieve diameter	The width of the minimum square aperture through which the particle will pass.
d_{vs}	Specific surface diameter $d_{vs} = d_v^3/d_s^2$	The diameter of a sphere having the same ratio of surface area to volume as the particle.

[a] Reproduced by permission of the author and Chapman and Hall, Ltd.

for irregular particles such as sand grains, size commonly depends on the method of measurement, which in turn depends on the object of study. Moreover, as the particle becomes less equant, discrepancy between the measures becomes progressively greater. Table 3-1 summarizes the different definitions of particle size. A sedimentologist studying behavior of grains in a fluid might find their drag or Stokes diameter, measures related to fluid flow (see Chap. 9), much more meaningful than volume diameter. On the other hand, a petrographer studying the downcurrent gradient of abundance of carbonate rock fragments in an alluvial sand might consider their postdepositional solubility to be important and, consequently, specific surface diameter might be most useful to him. Thus the measure chosen will reflect the object of study as well as the technique used. In general, one measure of size converts to another by expressing it in terms of a corresponding equivalent diameter.

Techniques

Modern summaries of particle size methodology, written from a nongeologic point of view, are given by A.S.T.M. (1959), Herdan (1960), Irani and Callis (1963), and Allen (1968). Brewer (1964, p. 18–40) and Folk (1966) provide excellent geologically oriented summaries of techniques, Folk's summary being especially recommended.

Measurement. The different available size measurement techniques apply to widely different size ranges (Fig. 3-1). With the exception of the settling tube and use of the thin section, measurement techniques in sedimentology have been stabilized

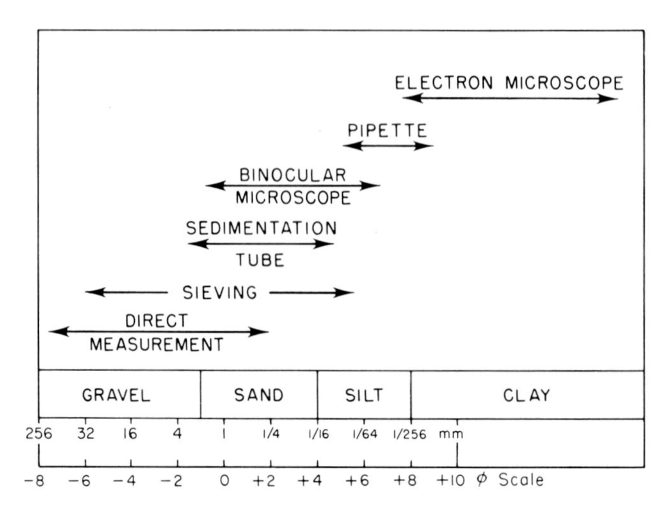

Fig. 3-1. Range of applicability of different techniques of size analysis

for more than two decades. Sieving remains the preferred method for unconsolidated sands and friable sandstones. It is relatively rapid, trouble free, cheap, and the oldest, most tested method. But particularly since Friedman (1958 and 1962) resolved the problem of correlating thin section and sieve size estimates, direct measurement in thin section is more widely used; it is the only one possible for completely cemented sandstones even though only the mean and sorting can be obtained by this method. The binocular microscope can also be used to estimate means and sorting quickly and easily using standard comparison sets (Swann and others, 1959; Emrich and Wobber, 1963). The settling tube, formerly used only for clay sized sediment, is beginning to be used more widely for unconsolidated sands and employs modern instrumentation to measure settling velocities rapidly and accurately. It yields an automatically recorded result that can be empirically related to a standard size distribution curve (Zeigler and others, 1964). Like measurement in thin section, only the mean and sorting can be obtained.

Choice of method depends both on the degree of consolidation and the objectives of the study. If the sand is unconsolidated and only the mean and sorting are of interest and their range of variability is large, the binocular microscope may suffice. But if skewness and kurtosis are desired or if the range of variation of the mean and sorting is small, very careful sizing with sieves of smaller class intervals is necessary.

Description. Because size is a continuous variable and many particles are needed for a good estimate, a grade scale is necessary. A *grade scale* consists of a series of class intervals having some constant relationship to one another. Udden (1898, p. 6–7) early recognized the need for a grade scale and based his on powers of 2. A natural scale to use is a logarithmic one, because a difference of a millimeter between sand grains is significant but between boulders is trivial. In addition, samples from single populations of grains tend to plot as straight lines on arithmetic probability graph paper. Kolmogorov (1941) has

Table 3-2. *Terminology and class intervals for grade scales*

	U.S. Standard sieve mesh	Millimeters		Phi (ϕ) units	Wentworth size class
GRAVEL	Use wire	4096		–12	
	squares	1024		–10	Boulder
		256	256	– 8	
		64	64	– 6	Cobble
		16		– 4	Pebble
	5	4	4	– 2	
	6	3.36		– 1.75	
	7	2.83		– 1.5	Granule
	8	2.38		– 1.25	
	10	2.00	2	– 1.0	
SAND	12	1.68		– 0.75	
	14	1.41		– 0.5	Very coarse sand
	16	1.19		– 0.25	
	18	1.00	1	0.0	
	20	0.84		0.25	
	25	0.71		0.5	Coarse sand
	30	0.59		0.75	
	35	0.50	1/2	1.0	
	40	0.42		1.25	
	45	0.35		1.5	Medium sand
	50	0.30		1.75	
	60	0.25	1/4	2.0	
	70	0.210		2.25	
	80	0.177		2.5	Fine sand
	100	0.149		2.75	
	120	0.125	1/8	3.0	
	140	0.105		3.25	
	170	0.088		3.5	Very fine sand
	200	0.074		3.75	
	230	0.0625	1/16	4.0	
SILT	270	0.053		4.25	
	325	0.044		4.5	Coarse silt
		0.037		4.75	
		0.031	1/32	5.0	
		0.0156	1/64	6.0	Medium silt
	Use	0.0078	1/128	7.0	Fine silt
	pipette	0.0039	1/256	8.0	Very fine silt
MUD	or	0.0020		9.0	
	hydro-	0.00098		10.0	Clay
	meter	0.00049		11.0	
		0.00024		12.0	
		0.00012		13.0	
		0.00006		14.0	

shown that the log-normal distribution is the theoretically expected one under certain conditions of grinding and crushing, if one assumes homogeneous material. Wentworth (1922) modified Udden's scale as Krumbein did later (1934), when he proposed his phi scale which is defined as

$$\phi = -\log_2 S$$

where S is size in millimeters. Tables for conversion of millimeters to phi units are available (Page, 1955) or the slide rule or log table can be used keeping in mind than the base is 2. Table 3-2 shows the relationships between size in millimeters, mesh, and phi units and also gives terminology. Quarter phi units should be used for the best estimates of size parameters and shape of the cumulative curve. Pettijohn (1957, p. 21–27) reviewed grain size terminology, which is far from standardized. The principal problem of size analysis is, however, not the lack of standardized nomenclature, but rather its geologic significance.

A size distribution is usually described either by employing some form of the cumulative frequency curve or by numerically specifying its central tendency by mean, mode, or median and its form or shape by sorting, skewness, and kurtosis. These can be obtained either from the cumulative curve (graphic measures) or calculated directly from the sieve analysis (moment measures). Because almost every size analysis is open ended – it is rarely practical to use sieves smaller than 30 microns – most graphic measures are superior to moment measures even though the latter can be calculated rapidly by computer (Pierce and Good, 1966). Moreover, visual inspection of the cumulative curve may show whether two or more populations are present and reveal possible errors in the analytical method. Hence it always pays to see the cumulative curve.

The starting point for the description of the cumulative curve is the *percentile*, which is obtained by arranging all the obervations in order, smallest to largest, accumulating them to 100 percent and counting off p percent of them (Fig. 3-2). For example, if x is grain size in phi units, then an arbitrary pth percentile is the value of x for which p percent of the distribution is smaller, ϕ_p. When using phi units, one should always plot the cumulative curve on arithmetic probability paper, because it permits more accurate extrapolation between sieve sizes. Alternatively, one can plot millimeter data directly on logarithmic probability paper. If one wishes to dissect a cumulative curve into its parts or subpopulations (Spencer, 1963), a probability plot is essential. Subpopulations are most easily identified by sharp angular discontinuities in the cumulative curve.

To specify both central tendency and shape of the curve, linear combinations of percentiles are used, a *linear combination* being nothing more than the addition or subtraction of a series of terms all raised to the first power, such as, for example, this estimate of the mean

$$(\phi_{10} + \phi_{30} + \phi_{50} + \phi_{70} + \phi_{90})/5$$

where the ϕ's are percentiles. The more percentiles that are read from the cumulative curve, the more accurately one can specify the desired parameter, such as the mean, mode, or sorting. Thus, initially, only two percentiles, ϕ_{25} and

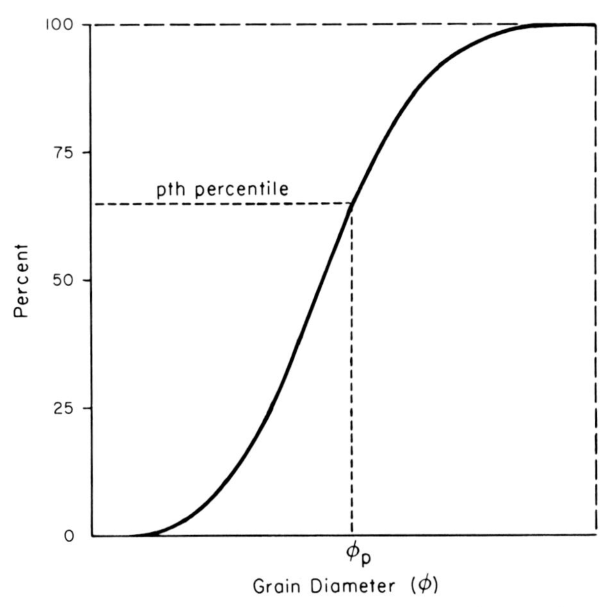

Fig. 3-2. Cumulative curve and definition of percentile

ϕ_{75}, called the first and third quartiles, were used to determine sorting but subsequently others seeking more information from the cumulative curve have used as many as five or more (Folk and Ward, 1957, p. 13–14; McCammon, 1962). Table 3-3 summarizes the most commonly used graphic and moment measures.

Other methods of description are essentially graphic and include plots of sand-silt-clay ratios, which are very popular with modern sediment workers, *CM* diagrams (Passega, 1957, 1964), the arithmetic-cumulative probability plots of the Dutch school (Doeglas, 1946; van Andel and Postma, 1954, p. 80–94), and logarithmic probability plots (Sindowski, 1957).

Ratios of sand, silt, and clay are usually plotted on triangles and used to define and standardize the naming of mixtures of sand, silt and clay, especially in unconsolidated sediments. The *CM* diagram plots maximum size (99th percentile) against median size (50th percentile), a simultaneous specification of central tendency and sorting. *CM* diagrams have been used for both modern sands and ancient sandstones. Graphic plots on arithmetic-cumulative probability paper are claimed to reveal the existence of a limited number, called *F*, *F-S*, *M*, *M-S*, *M-C*, and *C-S-C* types of size distributions that are identified on the basis of the shape rather than the position of the curve (van Andel and Postma, 1954, Table 6). The *F*, *M*, and *B* curves are for sands, and the *C* and *S* curves are for clays. The curves are believed to more clearly reveal size differentiation and mixing than other methods. Both *CM* diagrams and arithmetic-cumulative probability plots have enjoyed much greater popularity in Europe than in America. Sindowski (1957, Fig. 4) recognized 10 types of cumulative curves on the basis of their shapes. Gradations between types of curves are common and the difficulty of assignment of these intermediates is a shortcoming of these methods.

Table 3-3. *Graphic and moment measures*

Name	Graphic formula[a]	Moment formula[b]	Remarks
Mean	$Me_\phi = \frac{(\phi_{16} + \phi_{50} + \phi_{84})}{3}$	First moment $\bar{x}_\phi = \sum_{i=1}^{n} f_i m_{i\phi}$	All three measures of central tendency reflect average kinetic energy of depositing medium plus size distribution of available sediment.
Median	$Md_\phi = \phi_{50}$		
Mode	$M_\phi =$ Midpoint of most abundant class interval		
Bimodality Index	$Mi_\phi = 1 + \frac{(\phi_f - \phi_c)}{2\phi}$		Measure of bimodality, if present: ϕ_f is midpoint of finest mode; ϕ_c is midpoint of coarsest.
Sorting	Inclusive graphic standard deviation $s_I = \frac{\phi_{84} - \phi_{16}}{4} + \frac{\phi_{95} - \phi_5}{6.6}$	Second moment $s_\phi = \left[\sum_{i=1}^{n} f_i (M_{i\phi} - \bar{x}_\phi)^2\right]^{1/2}$	Measures dispersion, which is dependent upon velocity variations plus bimodality.
Skewness	Inclusive graphic skewness $SK_I = \frac{\phi_{84} + \phi_{16} - 2\phi_{50}}{2(\phi_{84} - \phi_{16})} + \frac{\phi_{95} + \phi_5 - 2\phi_{50}}{2(\phi_{95} - \phi_5)}$	Third moment $3_\phi = \sum_{i=1}^{n} \frac{f_i (M_{i\phi} - \bar{x}_\phi)^3}{s_\phi^3}$	Measures asymmetry, the direction of "tails," which are widely believed to have environmental significance. Skewness varies from +1.0 (positive) to 0.0 (symmetrical) to −1.0 (negative).
Kurtosis	$K_G = \frac{(\phi_{95} - \phi_5)}{2.44(\phi_{75} - \phi_{25})}$	Fourth moment $4_\phi = \sum_{i=1}^{n} \frac{f_i (M_{i\phi} - \bar{x}_\phi)^4}{s_\phi^4}$	Measures peakedness. Graphic measure is ratio of sorting of centered 90 percent to centered 50 percent. $K_G = 1.0$ for normal curve, $K_G > 1.0$ for peaked curve, and $K_G < 1.0$ for flattened curve.

[a] All after Folk and Ward (1957) except the bimodality index after Sahu (1964).

[b] f_i = fraction of total weight in each class interval; $m_{i\phi}$ = midpoint of each class interval in phi units.

Sampling. Proper sampling is vital for grain size studies. Determination of average grain size of a deposit at a given point requires either a channel sample or a composite obtained from spot samples taken across the deposit. Values of sorting, skewness, and kurtosis obtained from such samples are more valuable in determining regional variations than in environmental interpretation. If size gradients of the pebbles of a pebbly sandstone are to be established, one can record from outcrop only the largest pebble; a more statistically stable estimate can be obtained by recording the average of a small number, such as the largest ten. Rigorously, number of pebbles observed at each point should be the same, but in practice this has been hard to maintain. To minimize differential abrasion effects, pebble lithology should always be the same. Fifty to one hundred grams of sand are ample for either sieve or sedimentation tube analyses. Two to four hundred grains are usually measured in thin section.

A different procedure is needed for environmental studies. Here it is best to obtain samples from a single bed or sedimentation unit defined by Otto (1938, p. 575) as "that thickness of sediment which was deposited under essentially constant physical conditions." Attention to sedimentary structures now assumes importance because ripple mark, crossbedding, parting lamination, graded beds, and other structures constitute natural units of sampling. Because sedimentary structures are the product of fluid dynamics, there is a probable correlation between the sedimentary structure and the grain size distribution. When sampled in this way, the size parameters yield a maximum of environmental information. As before, either single, spot samples may be analyzed or composites from the same structure, the former probably being superior.

Statistical Measures

Central Tendency. The grain size of a sand is commonly clustered around an average value termed the *mean, mode* or *median* (Table 3-3), which is controlled by a combination of two factors: the average competence of the depositing medium and the initial size of the source materials. Contrasting methods of transport are responsible for the almost universal sharp segregation of clay from silt and sand: clay particles are transported entirely as turbulent suspensions while most silt, almost all sand, and all larger particles are transported along the bottom as bed load by a combination of sliding, saltation and rolling. The lateral sequence sand-silt-clay found over distances of tens to hundreds of kilometers in many basins is almost always related to declining current strength rather than abrasion. Supporting evidence for a current strength interpretation is the negligible rate of abrasion of sand (Kuenen, 1959). Because the mean of a distribution depends on the entire size curve it is a better estimate of central tendency than the median. In a statistically normal distribution mean, mode, and median all coincide.

Bimodality. An index (Table 3-3) was proposed by Sahu (1964, p. 77) to measure bimodality. Bimodality may result from a variety of causes: a combination of bed and suspension load transport, infiltration, post-depositional diagenesis, or lack of certain size grades in some source materials. The presence of two or more modes complicates interpretation of all statistical measures.

Sorting. Dispersion around central tendency determines sorting. Because the tails of a distribution have been thought to be environmentally sensitive, estimates of sorting have been designed to reflect them. By this view, an estimate of sorting based on the 84th and 16th percentiles should be superior to one based on the 75th and 25th percentiles, ones closer to the central tendency. Folk and Ward (1957, p. 13) combined each of these measures in their inclusive graphic standard deviation, σ_I, which is the average of sorting based on both the central and exterior parts of the size distribution (Table 3-3). One can also estimate sorting visually or measure it semiquantitatively in thin section (Folk, 1968, p. 26). Sharp and Fan (1963) proposed an entropy measure of sorting which has maximum value of 100 percent when all grains are of a single size class and zero percent when each class has the same number of grains. Whether a distribution is normal or not, this function measures the evenness of sorting.

Fluctuations in velocity, contributions from suspension as well as traction transport to the same bed, sampling more than one sedimentation unit, and source sands with more than one mode all contribute to sorting variation. The effect of inheritance on sorting has been emphasized by Folk and Robles (1964, p. 290–291). Their point of view may be stated as follows: commonly sands contain several grain populations, each with its own size distribution, so that best sorting results when the different modes are close together and worst when they are far apart. Thus in polymodal sediments, sorting, as well as other measures, simply reflects the relative magnitude and separation of the different modes. It is widely believed that sorting is best when sand is repeatedly available for reworking by currents of moderate intensity. Probably the worst sorting is found in sands subject to one brief episode of mass transport, such as a submarine slide or a turbidity current, and deposited below wave base in deep water. Kuenen (1964, p. 213–215) suggested the term "repository" for an environment where grains become unavailable for transport once deposited. The sand of a barchan dune moving over bedrock provides a good example of an environment lacking repositories and thus one with excellent sorting.

What are the limiting values of sorting that have been observed? Folk and Robles (1964, p. 290) suggest that sands deposited in the surf zone have values of σ_I of 0.3–0.6ϕ. Little other information is available on limiting values of sorting in other environments. Even though it is hard to estimate limiting values, such an inquiry strikes us as worthwhile.

Skewness. The asymmetry of a distribution is measured by skewness and is determined by the relative importance of the tails of the distribution (Fig. 3-3). As with sorting a number of skewness formulas have been proposed. Folk and Ward (1957, p. 14) provide the most comprehensive graphic formula, one that uses six percentiles (Table 3-3). Those who have attempted to draw conclusions about either transport processes or the environmental significance of the size distribution have focused attention on the tails of the distribution and paid particular attention to skewness.

Kurtosis. Roughly speaking, kurtosis (Table 3-3) is a measure of the peakedness of the distribution; if a distribution is flatter than a normal one it is called platykurtic but, if more peaked, it is called leptokurtic. Few geologic conclusions have been deduced from kurtosis values alone.

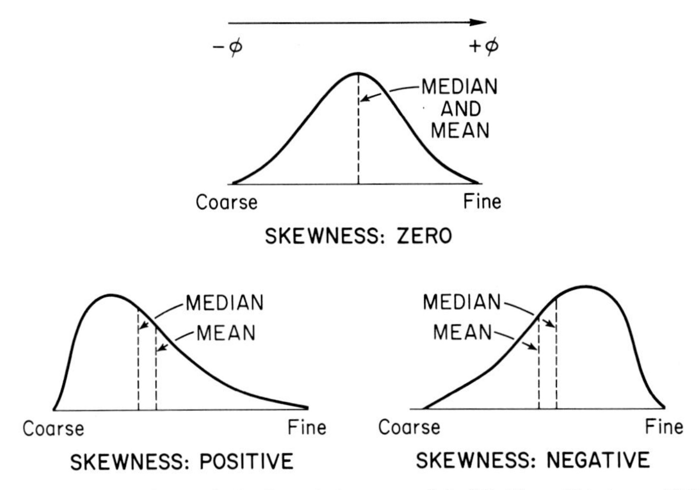

Fig. 3-3. Positively and negatively skewed size curves (Modified from Friedman, 1961, Fig. 7)

The moment measure formulas for skewness and kurtosis both illustrate why, if meaningful conclusions are to be drawn about them, much care must be used in obtaining the original size data. Because the deviations from the mean as well as the inclusive sample standard deviation are raised to the 3rd and 4th powers, errors in the original data can produce widely misleading estimates of both skewness and kurtosis.

Mapping Dispersal Patterns

One of the important parts of inferring the geologic history of a sandstone is the reconstruction of the *dispersal pattern* in which sand was distributed by water or wind currents from the source areas to and in the sedimentary basin. Because a dispersal patters is two-dimensional on the face of the earth, it is expressed as a map of the areal distribution of one or more variables whose distribution was controlled by the current system. In this way we infer a paleoslope or paleowind and build towards environmental recognition. Textural parameters can be used for this; a measure of central tendency can be mapped or, less commonly, the maximum observed size, the latter being more subject to sampling fluctuations than central tendency. Some have smoothed these results to emphasize major trends. Size gradients are usually easy to establish in boulder trains, loess, and ash falls (Potter and Pettijohn, 1963, Chap. 8). Loess (Smith, 1942) and ash falls (Eaton, 1964) show especially good gradients because, as air velocities decline and a particle is deposited, the land surface acts as an absorbing repository.

Valley train gravels usually display marked down-current size decline as do alluvial fan deposits (Blissenbach, 1954) for which Sternberg's law (1875, p. 486–487) serves as a first approximation. He related size decline to distance of transport by a simple, negative exponential decay function, $W = W_0 e^{-as}$, where W is weight of the largest observed pebble at some point downstream or downdip, W_0 is the largest initial weight in the source area, a is a constant for a particular

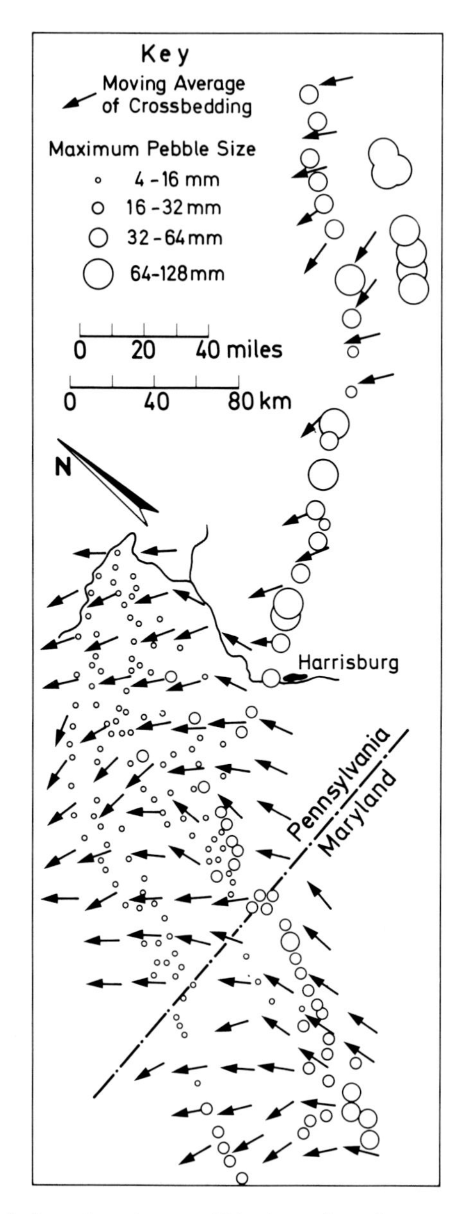

Fig. 3-4. Downcurrent decline of maximum pebble size and moving average of crossbedding in the fluvial Tuscarora Quartzite (Silurian) in the Appalachian Basin, U.S.A. (Modified from Yeakel, 1962, Figs. 8 and 14)

stream and s is transport distance from the source. Instead of weight, the diameter, d, can be used. In ancient deposits one must estimate W_0 or d_0 to determine s. A number of workers (McDowell, 1957, p. 27–28 and Pelletier, 1958, p. 1056) have applied Sternberg's law to the pebbles of conglomeratic sandstone with good results (Fig. 3-4). Selective sorting and abrasion during transport both

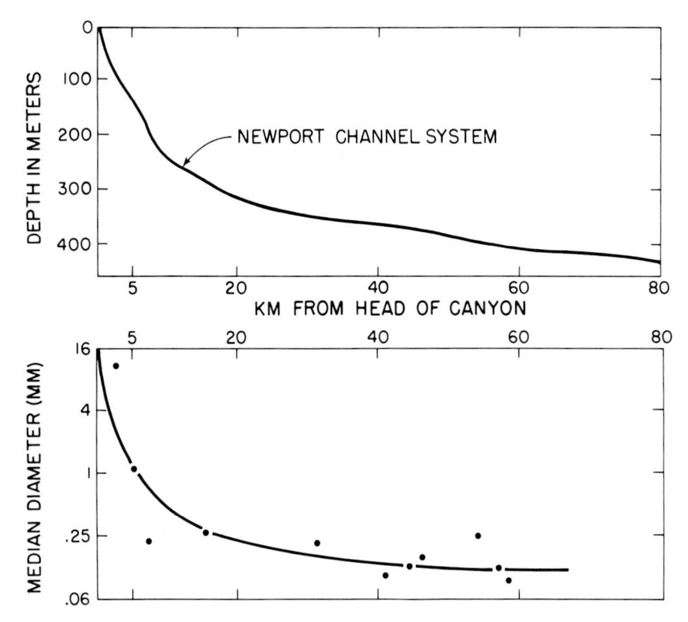

Fig. 3-5. Downfan decline of grain size in surface samples from modern turbidites. Note correlation between grain size and slope (Modified from Hand and Emery, 1964, Fig. 7)

contribute to down-current size decline in waterlaid deposits with the former probably much the more important for sand. In wind deposits selective sorting is probably dominant. Some modern turbidite sands also have good size decline curves that correlate well with slope on a submarine fan (Fig. 3-5) much as on subaerial fans. Ancient turbidites have similar relations, but difficulty of

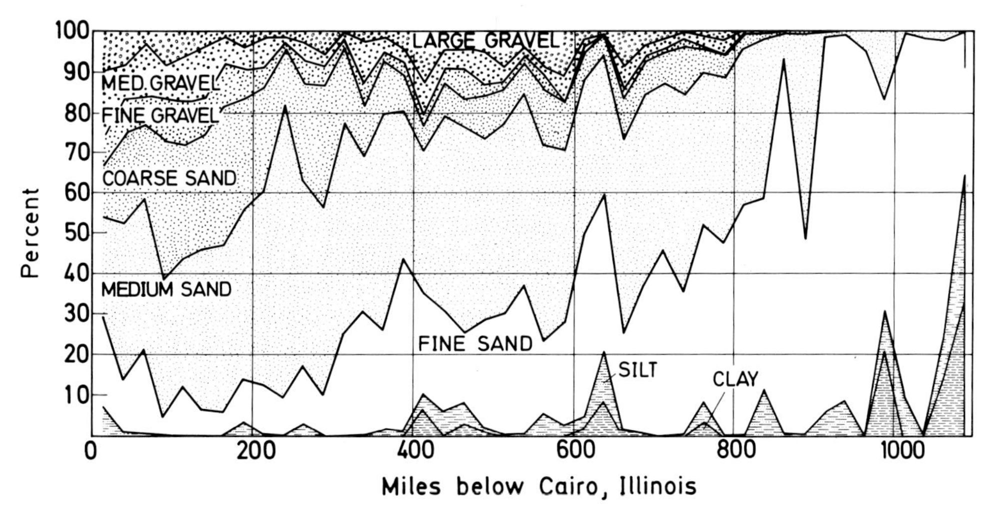

Fig. 3-6. Grain size of Mississippi River sands, Cairo, Illinois to Gulf of Mexico (Modified from U.S. Corps of Engineers, 1935, Pl. 58)

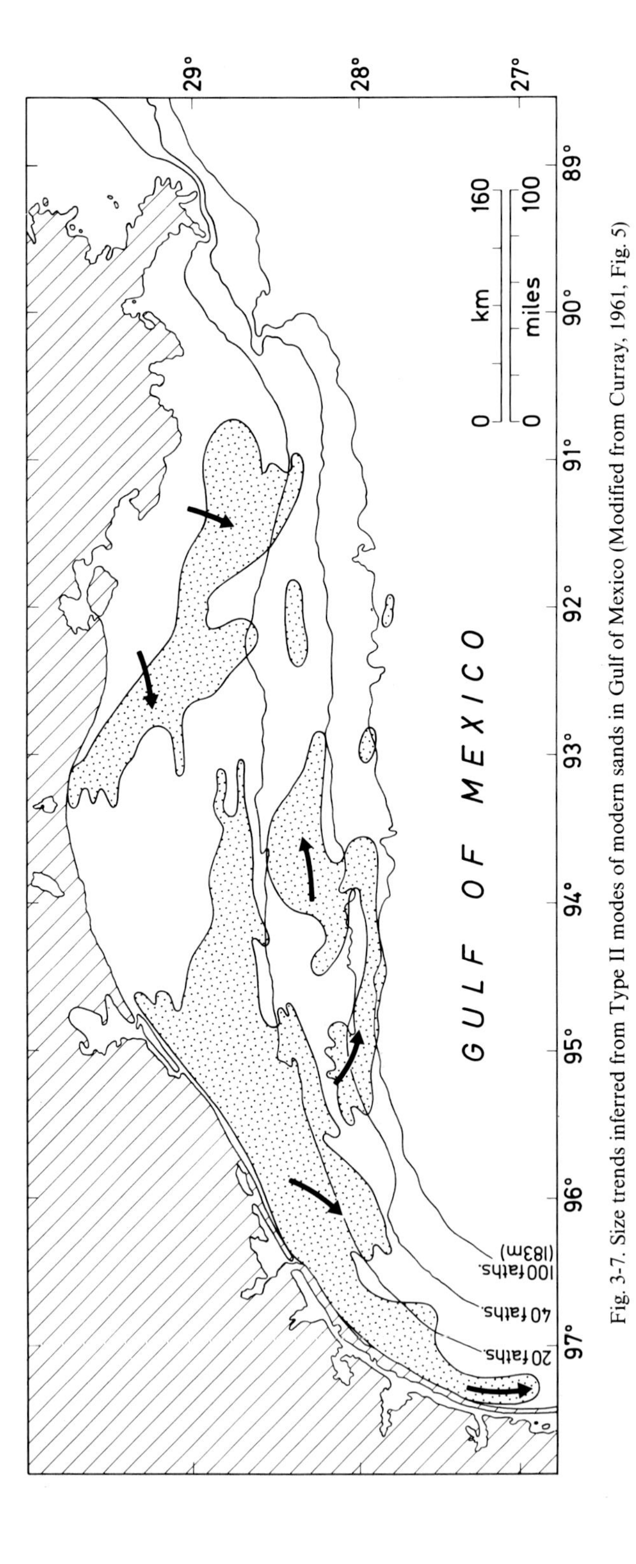

Fig. 3-7. Size trends inferred from Type II modes of modern sands in Gulf of Mexico (Modified from Curray, 1961, Fig. 5)

correlation in deformed sequences that lack good marker beds generally complicates lateral comparison. Usually it is easier to see vertical rather than lateral grain size variations in a turbidite sequence.

But are there size gradients in the typically fine to medium grained sandstones deposited on broad, low gradient coastal plains and on shallow marine shelves? If the lower portions of the Mississippi River from Cairo, Illinois, to the Gulf of Mexico are accepted as a reasonable model for a large alluvial river (Fig. 3-6), strong gradients may be present in alluvial sandstones deposited on coastal plains. Unfortunately, the search for size trends has not been very successful in most ancient medium- to fine-grained sandstones. Sandstones of many formations probably represent several interbedded environments and may be mixtures of transgressive and regressive sandstones. In addition, in a formation 50 or more feet thick, local vertical textural variation is high and hence many samples are commonly necessary to obtain a good estimate. Pelletier, however, did demonstrate systematic textural gradients in the sands of the Triassic Grey Beds of British Columbia (1965, Figs. 4 and 6). Curray (1961) achieved some success by dissecting into subpopulations, each with its own mode, the size curves of approximately 750 samples of modern shelf sands from the Gulf of Mexico (Fig. 3-7).

Eolian dune sheets, because they lack repositories, may have the weakest of all gradients, but almost no data is available.

Shape and Roundness

Shape and roundness are the properties of sand grains that have great significance for the study of the effect of the transport process on the debris furnished by the source area. These properties reveal the modification of angular grains of many shapes by abrasion, solution, and current sorting. Though they are so important, shape and roundness of sand grains have received but a fraction of the effort that has been devoted to size studies, possibly because most methods involve individual grain measurement that is long and tedious as well as imprecise.

Shape is defined by various ratios of a particle's long (L), intermediate (I) and short (S) axes. There are two aspects of shape: sphericity and form. *Sphericity* is a quantitative parameter measuring the departure of a body from equidimensionality. Sneed and Folk (1958, p. 117–125) review the various measures of sphericity and proposed a new one, *maximum projection sphericity* (also called effective settling sphericity), which is the ratio of the cross section area of a sphere of the same volume as the particle divided by its maximum projection area and is quantitatively defined as

$$\psi_p = \sqrt[3]{S^2/LI}\,.$$

They believe this quantity to be a better measure of a particle's behavior in a fluid medium than previous ones. One can calculate maximum projection sphericity graphically (Fig. 3-8). Briggs and others (1962, p. 654), however, found most of the measures of geometric shape to correlate about equally well with the effect of

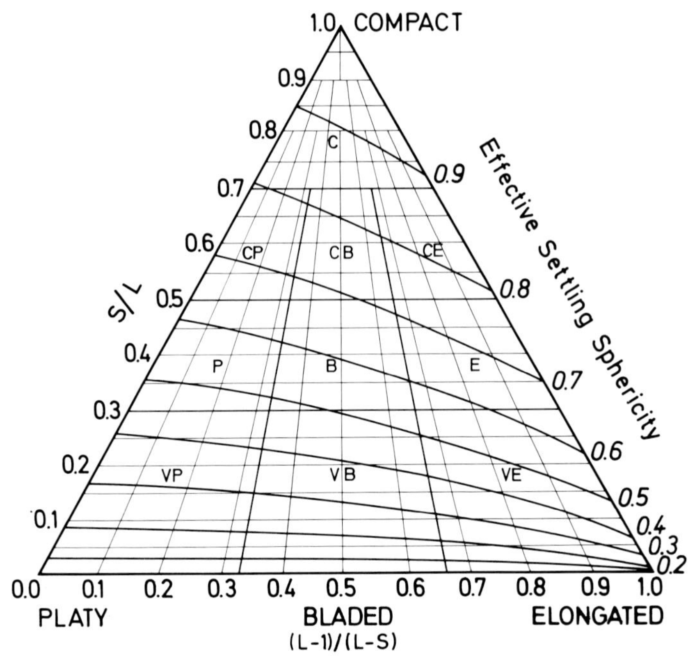

Fig. 3-8. Graphic calculation of maximum projection sphericity. To determine sphericity, first compute S/L and then $(L\text{-}I)/(L\text{-}S)$. The intersection of the ratios S/L and $(L\text{-}I)/(L\text{-}S)$ determines a point in the triangles, the sphericity of which is obtained by interpolating between the curved lines. L = long diameter; I = intermediate diameter; S = short diameter. The ten form classes are: *C*, compact, *CP*, compact platy; *CB*, compact-bladed; *CE*, compact-elongate; *P*, platy; *B*, bladed; *E*, elongate; *VP*, very platy; *VB*, very bladed, and *VE*, very elongate (Modified from Sneed and Folk, 1958, Fig. 2)

shape of sand grains on fluid flow characteristics. *Form* will distinguish a prolate (one long and two short axes) from an oblate spheroid (two long and one short axes) whereas sphericity will not (Fig. 3-8). Form is defined by two pairs of ratios of axis lengths, short (S), intermediate (I), and long (L), S/L and $(L\text{-}I)/(L\text{-}S)$, rather than a single ratio as is sphericity. Zingg (1935, Fig. 7) proposed four classes based on the ratios I/L and S/I and subsequently Folk and Ward (1957) proposed 10 form classes.

In most shape and roundness studies 25 to 100 particles per sample have been used to estimate average values. Although rarely done, one can also calculate the "sorting" of such distributions.

Few three dimensional studies of sand grain shape have been made, largely because of difficulties of measurement. In thin section, shape is usually approximated by the ratio of the apparent longest to shortest axis. In contrast, shape studies of gravels are fairly numerous.

Shape is generally regarded as having little environmental sensitivity even though hydrodynamic and aerodynamic experience indicate it may be of some importance, particularly as it affects settling behavior. Various shape factors have been proposed to more effectively take into account the effect of shape on settling velocity (Schultz and others, 1954). Perhaps the dependence of shape on original form and internal vectorial properties is chiefly responsible for its apparent lack of environmental sensitivity. Briggs and others (1962, p. 654) found, however, that sorting of heavy minerals by shape is just as important as sorting by density.

Roundness is geometrically distinct from shape and is concerned with the curvature of corners. It is quantitatively defined as $\frac{\sum_i^n r_i/R}{n}$, where r_i is the radius of a circle inscribed in the ith corner of a grain, n the number of corners and R the smallest radius that will circumscribe the grain. One uses the silhouette of the maximum projection plane of the particle to determine r_i and R. Standard sets of images are commonly used to determine roundness and differing degrees of roundness from angular to well rounded as quantitatively defined. Wadell (1932) and Krumbein (1934) are largely responsible for our present day concepts of roundness and the use of a chart to estimate it. Later Powers (1953) proposed a scale based on two sets of images for grains of different sphericity (see Appendix, Fig. A-2). The ϱ scale of Folk (1955, p. 294) is based on logarithmic intervals of the Powers Scale.

The degree of rounding of a detrital particle depends on its size, physical characteristics, and history of abrasion. Laboratory studies and measurement of sand roundness in modern sediments have shown grain rounding to be a very slow process, one that rapidly becomes much slower as size decreases. Hence after a given amount of transport, larger grains are better rounded than smaller ones. Kuenen's experiments (1959, p. 186) showed that 20,000 km of transport would cause no more than 1 percent loss of weight of an angular, medium-grained quartz sand thus confirming earlier experimental studies. Although one might wish to estimate quantitatively the distance of travel for sand from its average roundness or percentage of angular grains, this is as yet impossible, because the processes of rounding are still poorly understood. One of the difficulties is that the rate of rounding may vary greatly with depositional environment or rate of sedimentation. Chemical solution appears to play a secondary role for most silicates but is no doubt of some importance for more soluble detrital carbonate grains.

Well-rounded grains are the result of either many cycles of transport, each contributing its small share of rounding, or the result of intensive abrasion in a special environment where rounding was accomplished very rapidly. Petrographers appear to be about equally divided between these two, some favoring the "inheritance" viewpoint and others the "environment" viewpoint. The beach and, to some extent, the dune are the environments that have been suspected as producing well-rounded grains, especially in the geologic past, where greater tectonic stability on ancient cratons may have been responsible for greater reworking of sand than in present-day sediments. To date, however, most studies of modern beach deposits generally do not show strikingly more rounded grains than their presumed continental source sands.

Because of the slow rate of rounding of sand-size particles, roundness trends in most sands are difficult to establish. Some investigators have simply used it to help define mineral associations or dispersal patterns based on the

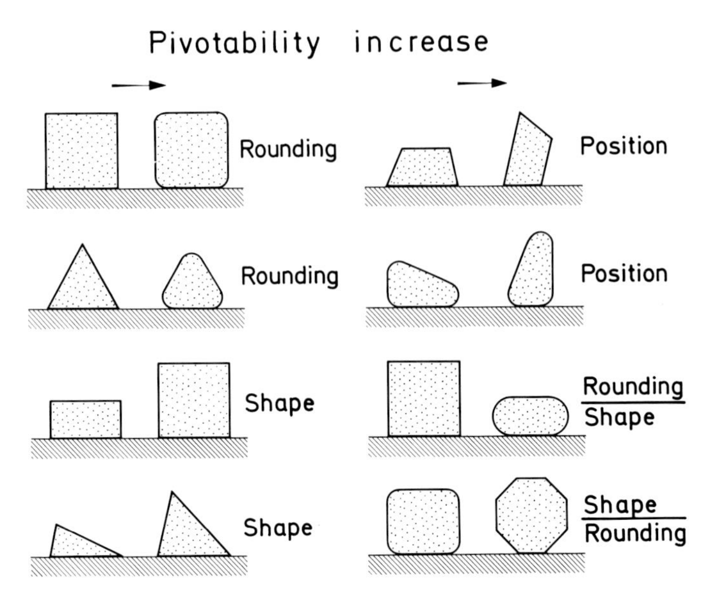

Fig. 3-9. Pivotability or rollability (Modified from Kuenen, 1964, Fig. 1)

presence or absence or an arbitrary abundance of angular grains. When determining the percentage of angular or round grains, comparisons should only be made with the same size fraction and mineral. Roundness provinces may also be defined by the angularity of heavy minerals, generally either zircon or tourmaline, as well as quartz. Although rounded overgrowths are decisive evidence of multicycle grains, they are usually so small in number as to be impractical for petrographic mapping. Because it makes them easier to recognize, cathodoluminescence petrography (Sipple, 1968) promises to increase the utility of multicycle grains.

Briggs and others (1962, p. 654) found that visual estimates of grain roundness had negligible correlation with the fluid dynamic behavior of sand grains. Boggs (1967) details the use of an electronic particle size analyzer to determine roundness and sphericity of sand grains. A minor use of grain roundness has been stratigraphic – to help distinguish between stratigraphic units.

A new "shape" concept – pivotability – was recently introduced by Shepard and Young (1961) and elaborated by Kuenen (1964). It is defined as the minimum angle of slope which will cause a particle to roll or fall. Winkelmolen (1969) has made an exhaustive study of this aspect of particle shape and introduced the term *rollability*. Rollability is a property derived from roundness and shape as shown in Fig. 3-9 and is in essence a *functional* shape description – in contrast to a purely geometric one – and is determined for loose sand in a machine. Winkelmolen used such a device to investigate various beach and dune sands and concluded that plots of rollability against grain size were helpful in discriminating between them. His work is one of the most complete studies on record of the sedimentological aspects of grain shape.

Surface Texture

The idea that surface textures of sand grains might be useful for discovering the effects of erosion, transport, and the sedimentary environment has been with us for a long time, for it is apparent that the physical and chemical effects involved may produce a variety of recognizable textures. Kuenen and Perdok (1962, p. 648) concisely describe the appearance of the varying surface textures of sand grains in reflected light:

> "... all transitions can be found, from grains that are beautifully polished to slightly pitted or scratched ones or those that have a milky surface but are still translucent, to ones with a dull or even quite opaque surface."

A large literature exists on the possible environmental significance of surface textures, usually under the general heading of frosting. No doubt this literature must have come from early observations that quartz grains of dunes, especially desert dunes, tend to have a dull opaque surface, whereas beach and fluvial sand grains have a shiny, polished appearance. Kuenen and Perdok (1962, p. 648) believe wind and water to be minor factors that cause frosting and think that chemical action is of principal importance – alternate solution and precipitation by dew in a desert environment. Electron microscope studies of sand grain surfaces could confirm this. To a lesser extent, frosting may result from corrosive solutions that promote replacement of quartz after deposition, commonly by calcite. This would give rise to frosted grains after decementation. River abrasion may actually "defrost" grains, reconverting a frosted surface into a polished one. Although occasionally noted, surface textures have not been of much help as environmental indicators in ancient sandstones. Perhaps Cailleux (1962) has been the most ardent advocate of frosting as an indicator of the eolian origin of ancient sandstones.

Recently the electron microscope has been used to study surface texture (Porter, 1962; Krinsley and Margolis, 1969) and there is promise of data of interpretative value to come from these studies, particularly for provenance studies of recent sands. Ancient cemented sandstones are likely to be less easily interpreted. The scanning electron microscope is particularly well suited for this type of study because of its great depth of field (Stieglitz, 1969).

Textural Maturity

Sedimentologists, recognizing the wide range of sandstone textures from clay rich and angular to well sorted and rounded, have extended the concept of maturity from the mineralogy of sands to the description of textural variations. Plumley (1948, p. 571–579) defined a mature gravel or sand as one consisting of a well sorted, mineralogically mature framework of grains. Folk (1951) defined *textural maturity* as the degree to which a sand is free of interstitial clay and is well sorted and well rounded. He gives a clear procedure for determination of three stages of textural maturity (see Appendix, Table A-1). These three stages are based on the idea that in transport clay is first removed, then framework grains are sorted, and that much later they are finally rounded. In short, proponents of the idea of textural maturity attempt to relate winnowing and abrasion by currents to sorting and rounding. The *final* depositional environment, however, controls textural maturity. Thus during a storm well rounded and well sorted marine shelf sand might be mixed with much clay in deeper water. Or bioturbation can produce a like mixture. Appreciable clay in either a well sorted or well rounded framework is called a *textural inversion.* A poorly sorted framework of very well rounded grains provides another example of a textural inversion. The relationships between textural and mineralogical maturity are not fully established largely because they have not been systematically and quantitatively studied.

The concept of textural maturity has been applied primarily to ancient sandstones where it is used for much the same purposes as a size analysis. Both rounding and sorting of framework grains are easily determined in thin section. But clay content is much more troublesome for it is commonly difficult to distinguish detrital from authigenic clay in many ancient sandstones, a distinction that is vital if textural maturity is to be meaningful. As a means of easy and rapid textural description in consolidated sandstones, however, the concept of textural maturity can be useful.

Environmental Recognition

Can the size distribution of a sandstone be used to identify its environment of deposition? The straightforward approach has been tried at least since the time of Udden (1914) and is a classic example of uniformitarianism: sample modern sands of known environments, hope to establish differences between them, and then use these differences to identify an ancient equivalent. This is the basic foundation upon which almost all environmental size analysis is founded.

The classification schemes include one or more variables and may use graphic plots of parameters obtained from the size curve as well as more formal statistical methods. For example, Keller (1945) used the ratio of the size of the larger of the two class intervals proximal to the modal class – his $F:C$ ratios – to distinguish between wind and water currents. This is a rough measure of skewness. Earlier von Engelhardt (1940) suggested the ratio of mean radius of quartz to mean radius of heavy mineral as a way of distinguishing between wind and water currents. Recently, Hand (1967) confirmed this suggestion for some New Jersey beach and dune sands. More complicated are methods such as those used by Friedman (1961 and 1967), who plotted different pairs of parameters, such as median versus sorting, sorting versus skewness, and skewness versus kurtosis. Friedman used parameters based on data obtained with quarter ϕ units and only reported on modern sands. Later Moiola and Weiser (1968) obtained essentially similar results for some of their plots using parameters estimated from half or even whole ϕ units.

Even more refined analysis may be obtained with a linear discriminant function, as first applied to size analysis of modern sands by Sahu (1964). The discriminant function is a multivariate classifying function that assigns an unknown sample specified by several variables, for example, median (x_1), sorting (x_2), and skewness (x_3), to one of two or more populations of different environments, such as modern fluvial, beach, and dune sands. For three variables the discriminant has the form

$$D_3 = a_1 x_1 + a_2 x_2 + a_3 x_3$$

where the a's are estimated from the data and D_3 is the discriminatory index. By determining x_1, x_2, and x_3 on the unknown, it can then be assigned to one of the parent populations. Another possibility, although it has been but little used, is to combine size parameters with textural ones, such as roundness or shape, in a discriminant function. Krumbein and Graybill (1965) give details and examples of discriminant functions.

Attempts to relate the size distribution of a sandstone to its environment of deposition have had but limited success even with modern sands where results have not always been consistent, some finding distinctions between environments where others failed. Consideration of some of the assumptions behind the "fingerprinting" approach suggests why this is so.

Fingerprinting assumes that the sizing processes in the final environment have effectively modified the size distribution of the sand supplied to it. In short, it *assumes* that the latest environmental processes eliminate the effects of inheritance. Consider a sand in a stream draining a source region underlain by a very well sorted quartz arenite. Could the textural parameters of the sand in the stream significantly differ from those of the source sand itself? We regard inheritance as particularly important when studying cratonic multicyclic sandstones, for such sands characteristically have had long histories of abrasion and hence their size distributions may reflect sedimentary processes or events quite different from those of the final depositional environment. Another aspect of inheritance that argues against universal solutions to the size analysis-environment problem is that different source sands have different modes and, as a consequence, criteria that may separate beach from fluvial sands in one basin will not separate them in another. Finally, nearly all the fingerprinting methods ignore sedimentary structures as a sampling unit, one that might very well influence sizing. Is it correct to compare the size distribution of parting lineation on a beach to that of crossbedding in a stream? Which is more important to sand sizing – the generalized fluid dynamics of an overall environment or the specific fluid flow regimes that produce the different sedimentary structures that are so ubiquitous in sandstones the world over?

Primarily for the above reasons we doubt the existence of a universal solution to environmental identification and discrimination by size analysis, especially by either single or groups of samples selected without regard to sedimentary structures, bedding facies, or position within the sand body. Much more promising to us is the analysis of vertical profiles of mean and maximum size in a sand body. Fluvial and tidal sand bodies commonly have upward declining size curves, barrier islands have upward increasing size curves, and marine sheet sands may increase or decrease depending upon whether they migrate shoreward or seaward or they may not show any vertical trends (see Chap. 11). When combined with study of sedimentary structures, we believe that the increasing fineness or coarseness upwards of sand bodies can be most informative.

Several other approaches to size analysis deserve mention.

Passega (1957 and 1964) believes that if one makes *CM* diagrams for 30 or more samples from a sand section one can distinguish between traction and suspension transport of sand. Water depth – at least comparatively within a basin – has also been inferred by him from the maximum size transported by bottom currents.

Q-mode factor analysis will also probably play an increasing role in our search for a better understanding of size analysis. With the help of *Q*-mode factor analysis, Klovan (1964) classified 69 Recent sediment samples from Barataria Bay into three groups and concluded that causal process factors for each group were surf and bottom currents and gravitational (quiet water) settling. Solohub and Klovan (1970) using *Q*-mode factor analysis tested most of the grain-size parameters or combinations that have been proposed for environmental discrimination in lacustrine environments in Lake Winnipeg, Manitoba, Canada. None of the measures could reliably identify the depositional environment when treated as unknowns, but factor analysis gave a map consistent with energy conditions in the various environments. Imbrie and van Andel (1964) and Krumbein and Graybill (1965, p. 368–375 and 400–406) give introductions to *Q*-mode factor analysis.

A rather different point of view has been developed by Spencer (1963), who like Folk and Robles (1964, p. 290), believes that variations in mean sorting, skewness, and kurtosis only reflect the relative importance of the three size populations of gravel sand and clay, each of which is log normally distributed. Spencer suggests that in most sands "sorting" is really only a measure of relative importance of framework grains to interstitial matrix. He considers 30 μ as the most logical upper limit of matrix. As we have noted earlier, however, diagenesis may convert argillaceous rock fragments into matrix, so that the grain-matrix ratio of an ancient sandstone may differ significantly from its original one. Visher (1969), in an approach somewhat similar to that of Spencer, dissects the cumulative curve into its components and, using modern sands as a guide, relates them to seven major processes of sand deposition. His Table 1 compactly summarizes his many cumulative curves.

Control of Physical Properties

A considerable portion of all size analysis studies has been made in an effort to relate physical properties, such as permeability, sonic transmissibility, tensile strength, and thermal conductivity, to the size distribution. This effort has usually proved to be moderately successful, but care is often necessary to have a carefully designed experiment, since, in general, any aggregate physical property, *P*, is a complex function of size, grain shape, fabric, and composition, with the result that these physical properties may have to be sorted out via regression analysis (Krumbein and Graybill, 1965, Chaps. 10 and 12). This is especially so if the property is a "sensitive" one such as permeability. Erfroymson (1960) describes a stepwise regression procedure that permits one to test the effect of adding successive independent variables, $x_1, x_2, x_3 \ldots$ to the dependent variable, *P*.

By adding one independent variable at a time, one obtains a series of intermediate equations

$$P' = a_0' + a_1' x_1$$
$$P'' = a_0'' + a_1'' x_1 + a_2'' x_2$$
$$P''' = a_0''' + a_1''' x_1 + a_2''' x_2 + a_3''' x_3$$
$$\dots\dots\dots\dots\dots\dots\dots\dots$$

where the a's are estimated from the data. One accepts or rejects a particular variable, such as x_3, if it improves or worsens the "goodness of fit" of the equation. Griffiths (1961, p. 494–497) gives a full discussion of the geologic rationale behind the regression equation approach.

Fabric

The fabric of sand and sandstones, the ways in which the grains are put together to make an aggregate, should be related in some way to the ways that currents deposit large numbers of grains of diverse size, shape, and roundness, and to the ways in which the aggregate is compacted by physical and chemical processes. To the extent that we can define and measure fabric by parameters independent of the other textural parameters, we may succeed in using such properties to infer current and compaction regimes. We may also use those properties for practical purposes in engineering geology as a guide to crushing and bearing strength.

Grain to Grain Relations

Grain to grain relations have been described in thin section by a combination of qualitative terms and quantitative indices, both of which attempt to infer three-dimensional relationships in the plane of the thin section. At present we have the terminology and a little of the methodology, but we are desperately short of systematic, mapping studies. Knowledge of the spatial distribution of fabric types in a sandstone body is needed for a better understanding of the origin and relative ages of their cementing agents and diagenetic processes. One impediment to progress is lack of reliable instrumentation that could specify grain to grain relations more quickly than grain-by-grain microscopic counting methods and perhaps do so in three dimensions as well. Cathodo-luminescence petrography (Sippel, 1968) is an example of how improved instrumentation may be helpful. Another difficulty in mapping fabric type is that it is highly variable – much like permeability – as can be seen in many single thin sections. Some of this variation is linked to primary deposition, subtle differences in original grain to grain relations being characteristic of each lamination and bed. These differences influence, of course, later diagenetic processes. What is needed is a better way of increasing the "signal-to-noise ratio" of fabric studies.

Table 3-4 defines the qualitative and quantitative terms used in studies of grain to grain relations in sandstones and Fig. 3-10 illustrates the application of

Table 3-4. *Terms used to specify sandstone fabric*

QUALITATIVE

Concavo-convex contact (Taylor, 1950, p. 707): one that appears as a curved line in the plane of section.

Fixed margin (Allen, 1962, p. 678): that part of a grain in contact with another in the plane of section.

Fixed grain (Allen, 1962, p. 678): fixed margin exceeds free margin.

Floating grain: no contacts with other grains in the plane of section.

Framework fraction: the stress-transmitting portion of a sand.

Free margin (Allen, 1962, p. 678): that part of grain not in contact with other grains in the plane of section.

Free grain (Allen, 1962, p. 678): free margin exceeds fixed margin.

Long contact (Taylor, 1950, p. 707). a contact that appears as a straight line in the plane of section.

Packing (Kahn, 1956, p. 390): mutual spatial relationships among grains.

Sutured contact: mutual stylolitic interpenetration of two or more grains.

Tangential contact (Taylor, 1950, p. 707): one that appears as a point in the plane of section.

QUANTITATIVE

Condensation index (Allen, 1962, p. 678): ratio of percentage of fixed rock fragments to percentage of free grains.

Contact index: number of contacts per grain.

Horizontal packing intercept (Mellon, 1964, Fig. 7): average horizontal distance between framework grains.

Packing density (Kahn, 1956, p. 390): length of grains intercepted divided by length of traverse × 100.

Packing index (Emery and Griffiths, 1954, p. 71): the product of the number of quartz to quartz contacts per traverse and the average quartz diameter, the product being divided by the total length of traverse.

Packing proximity (Kahn, 1956, p. 390): number of grain to grain contacts divided by total number of contacts of all kinds (grain to matrix and grain to cement) × 100.

Vertical packing intercept (Mellon, 1964, Fig. 7): average vertical distance between framework grains.

the quantitative fabric indices listed in the table. Table 3-5 indicates possible relationships between fabric elements, pressure solution and cementing agents.

Orientation

More progress has been made with framework grain shape-orientation studies, in which methodology, at least those methods of measurement of the long axes or apparent long axes of individual grains in thin section, is well standardized and applications are beginning to be made (Table 3-5B). Shape fabric of framework grains has two major modes: a principal one of long axes parallel to current flow and imbricated 15 to 18° upcurrent and a secondary one at right angles to current flow (Fig. 3-11). The secondary mode, not always present, is believed to result from transverse rolling. Hamilton and others (1968) confirmed orientation of the major mode parallel to the dip of sloping sand surfaces, both wet and dry. When making orientation studies, care should be taken to relate grain orientation to bedding, for grains always accumulate with respect to a depositional surface – and not necessarily a horizontal one.

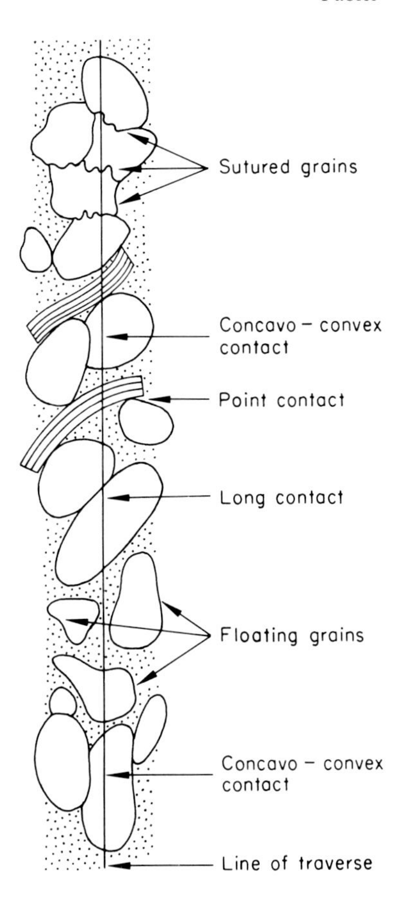

Fig. 3-10. Definition sketch of fabric terminology: quartz (white), mica (lined), and matrix (stippled)

A current field of interest is the relationship between sedimentary structures, such as crossbedding, parting lineation, and flute casts, and their constituent grains. These studies suggest that there are persistent correlations between the constructional forms, such as sand waves and parting lineation, and the orientation of their constituent sand grains. Correlation of flute mark and grain fabric has also been observed, although correlation between erosional structures such as flute marks and fabric, may be less perfect because erosion and deposition are separated in time.

Usually there is little need for regional studies of grain orientation, because sedimentary structures are generally available to determine paleocurrents. Sestini (1964), however, used fabric to infer current direction in calcarenites of turbidite origin, because directional structures were not readily measurable.

Progress in the development of aggregate or bulk methods of fabric specification by an appropriate "black box" has been made, but they have not as yet been widely applied. Magnetic susceptibility appears to be promising (Rees, 1965). Earlier Zimmerle and Bonham (1962) used an electronic flying spot scanner to establish fabric.

Table 3-5. *Sandstone fabric (modified from Adams, 1964, Table 1)*

A. Grain to grain relations specified chiefly by qualitative observation supplemented by some quantitative indices on nonoriented samples. Chiefly used to interpret and predict reservoir porosity.

1. Much pressure solution.

   Many sutured contacts, grain to grain contacts large, packing density high, and no porosity and cement.

2. Moderate pressure solution.

   Some sutured contacts, principally equidimensional grains, chiefly with concave-convex contacts. Cement is mostly quartz overgrowths with or without minor amounts of carbonate or clay. Little porosity.

3. Minor pressure solution.

   Mostly original grain outlines with long and tangential contacts. Low to moderate number of grain to grain contacts and moderate packing density. May be either well or poorly cemented by quartz overgrowths, clay, and carbonate. May have moderate porosity, if poorly cemented.

4. No pressure solution.

   Chiefly original grain outlines with either tangential contacts or floating grains. Number of grain to grain contacts is low as is packing density. Cements are mostly carbonate or clay. Porosity is high when cementation is limited.

B. Orientation of framework specified by quantitative measurement of oriented samples. Porosity and cementation are independent of orientation. Chiefly used to determine current direction in undeformed sediments.

1. Particulate methods.

   Visual, direct measurement of either long axes or apparent long axes of framework grains usually in thin section.

2. Aggregate methods.

   Measurement of a bulk geophysical property by an appropriate black box that can be correlated with the orientation of the framework grains.

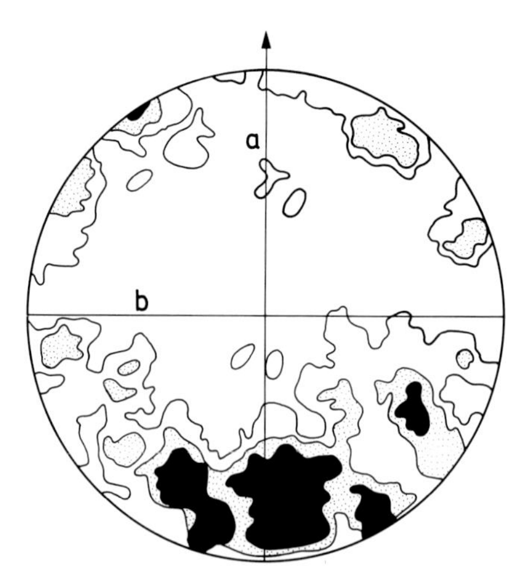

Fig. 3-11. Orientation of sand grains in flume. Long axes of quartz grains plotted on equal area net. Note imbrication. Arrow shows current (Modified from Rusnak, 1957, Fig. 9d)

Potter and Pettijohn (1963, p. 23–61) summarize the fabric literature of sands through early 1963. Johansson (1965) gives an excellent digest of orientation analysis.

Porosity and Permeability

A sandstone consists of a framework of grains, interstitial detrital silt and clay, chemical cements and an interconnecting network of void space or pores. The study of the pore system of rocks is included in the area of petrophysics. Contributions to this field have been made by petroleum engineers, groundwater hydrologists, soil scientists and some geologists. Very much still remains to be learned.

It is the pore system that permits a sandstone to store and transmit fluid – groundwater, oil, gas, and mineralizing solutions. The size, shape and pattern of this pore system in a sand or sandstone is very difficult to specify. Lack of geometric regularity and small size are the chief obstacles. The scanning electron microscope with its high depth of field beautifully illustrates the intricate surfaces of pores in sandstone (Fig. 3-12).

Porosity and permeability are two concepts central to the analysis of flow in pore systems. Like density, both are mass properties.

Void space is that portion of a sandstone not occupied by its solid components. Voids may either be connected or isolated. Porosity, a scalar quantity, is expressed

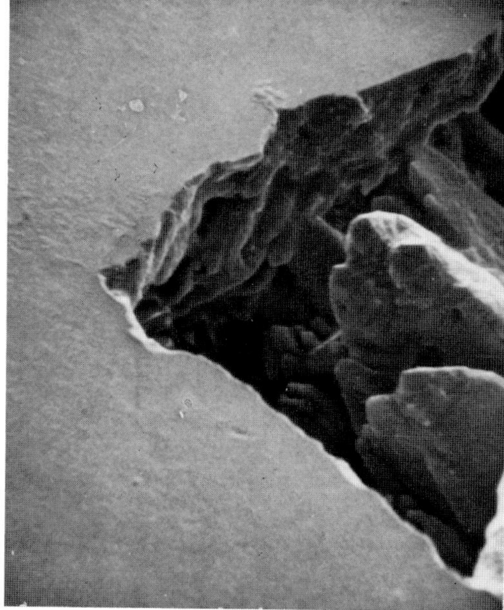

Fig. 3-12. Pores of Berea Sandstone (Mississippian) as seen by scanning electron microscope. Left × 170, and right × 510 (Weinbrandt and Fatt, 1969, Figs. 10 and 11). Reproduced by permission of the Journal of Petroleum Technology

as a percent. *Absolute* or *total porosity* is defined as

$$P_t = \left(\frac{\text{bulk volume} - \text{solid volume}}{\text{bulk volume}} \right) 100 \,,$$

and *effective porosity* as

$$P_e = \left(\frac{\text{interconnected pore volume}}{\text{bulk volume}} \right) 100 \,.$$

Absolute or total porosity is the percentage of all void space in a sand, whether the pores are connected or not whereas effective porosity is the percentage of interconnected pore space. Effective porosity, commonly less than total porosity, is what is ordinarily measured in most reservoir studies. Pumice and foam plastic are two substances with very large absolute but very small effective porosity. Pirson (1958, p. 31–40) summarizes the many different laboratory measurement techniques used to determine porosity. Different methods are used to determine total and effective porosities. Semi-quantitative estimates can also be obtained from electric logs, micro-logs, and sonic logs.

The packing of equal sized spheres has been studied in order to approximate porosities found in granular materials. Scheidegger (1957, p. 16–19) reviewed this literature and concluded that for a given mode of packing of uniformly sized spheres porosity is independent of size of the spheres. Tightest packing (rhombohedral) has a porosity of 25.9 percent and loosest packing has porosity of 87.5 percent. The abundance of grains of mixed sizes and non-spherical framework grains, the filling of pores by cementation and interstitial clay minimizes the significance of these studies for sandstones, however. In addition, it has been virtually impossible to specify the irregular packing of framework grains of sandstones by crystallographic terms such as rhombohedral, isometric, etc.

Compaction and cementation can reduce porosity from 50 percent or more in modern sands to virtually zero in quartzites and other holocrystalline sandstones. Most Phanerozoic sandstones that form petroleum reservoirs commonly have porosities of 5 to 20 percent. As cementation increases, particularly silica cementation, fracture porosity is likely to increase with respect to intergranular porosity, because the sandstone is more brittle and thus fractures more readily.

Permeability or *hydraulic conductivity*, K, measures the ability of a sandstone to transmit fluids and is defined as a constant of proportionality for the laminar flow of a particular fluid in a particular porous substance by the Darcy equation:

$$Q = KA(dp/dl)$$

where Q is volume of transmitted flow per unit time, A is cross sectional area and dp/dl is the dimensionless hydraulic gradient (change of pressure p in flow direction, l). Thus the rate of flow is directly proportional to K, cross sectional area and hydraulic gradient. The dimensions of K are L/T. This measure of permeability depends on the fluid as well as the pore system of the medium and the direction of permeability measurement in the medium. Thus it is a vectorial or tensor property in contrast to porosity. *Specific permeability*, k, is a measure of permeability depending only on the medium and is defined as

$$Q = \frac{kA\gamma}{\mu}\frac{dp}{dl}$$

where γ is the specific weight of the fluid and μ its viscosity. Specific permeability is measured in Darcys and has dimensions of L^2; it can be thought of as the characteristic pore area controlling the flow.

Pirson (1958, p. 62–87) describes laboratory techniques for measurement of permeability. Pottier and others (1964) summarize different petrophysical methods and give resulting data for the Hassi Messaoud Sandstone (Cambrian) in Algeria.

Among the rock properties affecting permeability are grain size, sorting, orientation, and packing of framework grains and cementation and bedding, all associated in a complicated and as yet not fully understood way, even though much study has been done. Experimentally, it has been found that the finer the grain size and the poorer the sorting of loose sand, the smaller its permeability, as shown by Krumbein and Monk (1942, p. 10) in the formula

$$K = Cd^2 e^{-1.35\sigma}$$

where C is a constant, d is geometric mean grain diameter, e is the base of natural logarithms, and σ is standard deviation of the sand. Thus in a graded turbidite bed that fines upward permeability will have a parallel decline. The tighter the packing density of the sand the lower its effective porosity and hence the lower its permeability, all other factors remaining equal. Orientation and packing of the framework grains of a sandstone appear to have a weak control on permeability in the plane of the bedding and a much stronger control in vertical sections parallel to sand transport direction (Mast and Potter, 1963, p. 558–559). At deposition, framework sand grains accumulate with their long axes parallel to current flow and imbricated upcurrent 15 to 18 degrees from the depositional interface. As a consequence, an anisotropy is imposed on the pore system. Bedding too has an important effect on permeability – one that probably appreciably exceeds that of fabric – for slight pauses in sand deposition are commonly marked by an accumulation of thin mud laminations of low permeability and consequently vertical flow is inhibited. Thus both grain fabric and laminations cause vertical permeabilities to be smaller than horizontal ones while grain orientation imparts weak anisotropy to permeability in the plane of the bedding. As a consequence, permeability is a directional property and is correctly described in three dimensions by an ellipsoid or mathematically as a tensor.

In modern sands permeabilities of 10 to more than 100 darcys have been reported. But in most consolidated sandstones values of more than one or two darcys are unusual and many of the sandstones in petroleum reservoirs have permeabilities of a few to a few hundred millidarcys. Statistical distributions of permeability in reservoirs, especially from the same bedding facies in a reservoir, tend to be lognormally distributed, whereas porosity typically has a normal distribution. Moreover, variance of permeability is much greater than that of porosity (Fig. 3-13) and can vary markedly vertically, centimeter by centimeter.

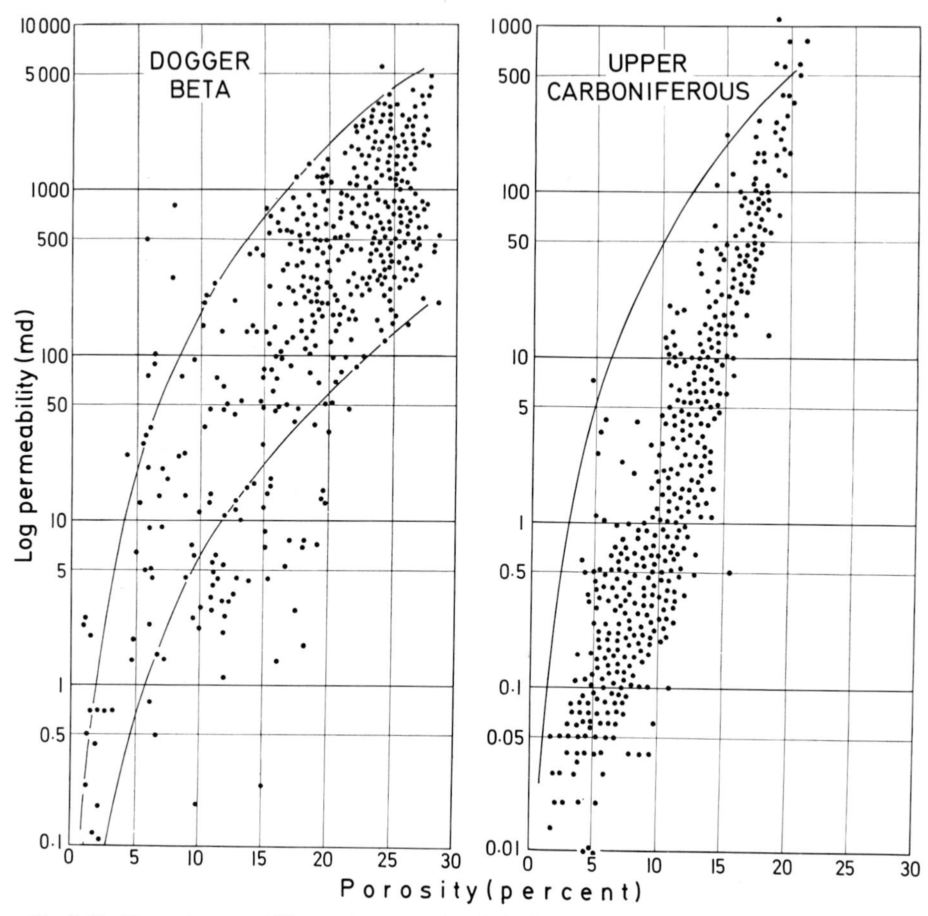

Fig. 3-13. Plots of permeability against porosity. Left, Dogger-β (Jurassic) and right, Upper Carboniferous sandstone. Note correlation (Füchtbauer, 1967, Figs. 7 and 10). Reproduced by permission of the Elsevier Publishing Company

It is also clear from Fig. 3-13 that there can be good correlation between effective porosity and permeability in a sandstone reservoir. The Kozeny-Carman equation (Scheidegger, 1957, p. 100–108) for permeability,

$$K = \frac{e^3}{5S_v^2(1-e)^2}$$

where S_v is the specific surface exposed to the fluid (surface exposed to fluid per unit volume of solid) and e is the base of natural logarithms. The Kozeny-Carman equation gives some insight into the dependence of permeability on effective porosity, especially in unconsolidated sands. As a first approximation, permeability is proportional to the first power of porosity and inversely proportional to the second power of specific surface. The Kozeny-Carman equation is useful in that it explains why a fine grained sand or silt with effective porosity identical to that of a coarser grained sand has smaller permeability: as grain size decreases, specific surface increases and consequently also resistance to flow (Fig. 3-14).

And why is permeability so much more variable than the porosity of a sandstone? There appear to be two principal reasons. A given porosity defines neither the drag imparted to a fluid by the small scale roughness of the walls of the pore system (Fig. 3-12) nor the path length that the fluid must flow to go between any two points in the sandstone. Change in either or both can alter permeability without change of porosity. As a consequence, permeability is much

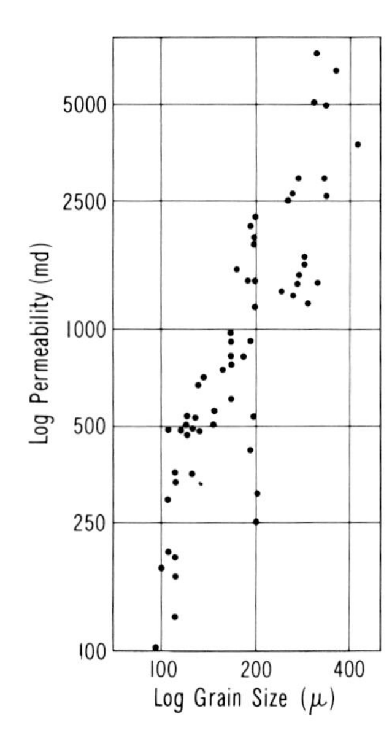

Fig. 3-14. Dependence of permeability on grain size in the Bentheimer Sandstone of the Scherrhorn oilfield near Lingen, Germany. Regression equation is $\log_{10} y = -2.1007 + 2.221 \log_{10} x$, where y is permeability in millidacrys and x is grain size in millimeters. Scatter diagram based on random selection of data from Fig. 49 of von Engelhardt (1960)

more variable than porosity. The Kozeny-Carman equation may also shed some light on the much greater spatial variability of permeability than porosity. For example, while progressive cementation decreases both specific surface and porosity, perhaps both decrease at variable rates so that permeability, being inversely proportional to the *square* of specific surface, is much more variable than porosity. Von Engelhardt and Pitter (1951) give much quantitative data on the relationships between porosity, permeability and grain size. Von Engelhardt (1960) gives a thorough treatment of the porosity, transmissive characteristics, and chemical composition of the pore system in sediments. Scheidegger (1957) has a more mathematical treatment.

References Cited

Adams, W. L.: Diagenetic aspects of Lower Morrowan, Pennsylvanian sandstones, northwestern Oklahoma. Am. Assoc. Petroleum Geologists Bull. **48**, 1568–1580 (1964).

Allen, J. R. L.: Petrology, origin and deposition of the highest Lower Old Red Sandstone of Shropshire, England. Jour. Sed. Petrology **32**, 657–697 (1962).

Allen, Terence: Particle size measurement, 248 p. London: Chapman and Hall 1968.

van Andel, Tj. H., Postma, H.: Recent sediments of the Gulf of Paria, Vol. 1. Koninkl. Nederlandsche Akad. Wetensch. Verh. **20**, 245 p. (1954).
A.S.T.M.: Symposium on particle size measurement. Am. Soc. Testing Materials, Sp. Tech. Pub. **234**, 303 p. (1959).
Blissenbach, Erich: Geology of alluvial fans in semiarid regions. Geol. Soc. America Bull. **65**, 175–189 (1954).
Boggs, Sam, Jr.: Measurement of roundness and sphericity parameters using an electronic particle size analyzer: Jour. Sed. Petrology **37**, 908–913 (1967).
Brewer, Roy: Fabric and mineral analysis of soils, 470 p. New York: Wiley 1964.
Briggs, L. I., McCulloch, D. S., Moser, Frank: The hydraulic shape of sand particles. Jour. Sed. Petrology **32**, 645–656 (1962).
Cailleux, Andre: Aspects mats des grains de quartz: Koninkl. Nederlandsche Akad. Wetensch. Proc., ser. B., **65**, 393–394 (1962).
Curray, J. R.: Tracing sediment masses by grain size modes: Internat. Geol. Cong., 21st, Copenhagen 1960, Proc., pt. 23, 119–130 (1961).
Doeglas, D. J.: Interpretation of the results of mechanical analysis. Jour. Sed. Petrology **16**, 19–40 (1946).
Eaton, G. P.: Windborne volcanic ash: A possible index to polar wandering. Jour. Geology **72**, 1–35 (1964).
Emery, J. R., Griffiths, J. C.: Reconaissance investigation into relationships between behavior and petrographic properties of some Mississippian sediments. Pennsylvania State Univ., Min. Ind. Expt. Sta. Bull. **62**, 67–80 (1954).
Emrich, Grover, Wobber, F. J.: A rapid method for estimating sedimentary parameters: Jour. Sed. Petrology **33**, 831–841 (1963).
Engelhardt, Wolf von: Die Unterscheidung wasser- und windsortierter Sande auf Grund der Korngrößenverteilung ihrer leichten und schweren Gemengteile. Chemie Erde **12**, 451–465 (1940).
— Der Porenraum der Sedimente, 207 p. Berlin-Göttingen-Heidelberg: Springer 1960.
—, Pitter, H.: Über die Zusammenhänge zwischen Porosität, Permeabilität und Korngröße bei Sand und Sandsteinen. Heidelberger Beitr. Min. Petrogr. **2**, 477–491 (1951).
Erfroymson, M. A.: Multiple regression analysis. In: A. Ralston and H. S. Wilf, Eds.: Mathematical methods for digital computers, Vol. 1, p. 191–203. New York: Wiley 1960.
Folk, R. L.: Stages of textural maturity in sedimentary rocks. Jour. Sed. Petrology **21**, 127–130 (1951).
— Student operator error in determination of roundness, sphericity, and grain size. Jour. Sed. Petrology **25**, 297–301 (1955).
— A review of grain-size parameters. Sedimentology **6**, 73–93 (1966).
— Petrology of sedimentary rocks, 170 p. Austin, Texas: Hemphill's Book Store 1968.
—, Robles, Rogelio: Carbonate sands of Isla Perez, Alacran Reef complex, Yucatan. Jour. Geology **72**, 255–292 (1964).
—, Ward, W. C.: Brazos River bar: a study in the significance of grain size parameters. Jour. Sed. Petrology **27**, 3–26 (1957).
Friedman, G. M.: Determination of sieve-size distribution from thin section data for sedimentary petrological studies. Jour. Geology **66**, 394–416 (1958).
— Distinction between dune, beach, and river sands from their textural characteristics. Jour. Sed. Petrology **31**, 514–529 (1961).
— Comparison of moment measures for sieving and thin section data in sedimentary petrologic studies. Jour. Sed. Petrology **32**, 15–25 (1962).
— Dynamic processes and statistical parameters compared for size frequency distribution of beach and river sands. Jour. Sed. Petrology **37**, 327–354 (1967).
Füchtbauer, Hans: Influence of different types of diagenesis on sandstone porosity: World Petroleum Cong., 7th, Mexico, Proc. **2**, 353–369 (1967).
Griffiths, J. C.: Measurement of the properties of sediments. Jour. Geology **69**, 487–498 (1961).
Hamilton, N., Owens, W. H., Rees, A. I.: Laboratory experiments on the production of grain orientation in shearing sand. Jour. Geology **76**, 465–472 (1968).
Hand, B. M.: Differentation of beach and dune sands, using settling velocities of light and heavy minerals. Jour. Sed. Petrology **37**, 514–520 (1967).
—, Emery, K. O.: Turbidites and topography of north end of San Diego Trough, California. Jour. Geology **72**, 526–542 (1964).

Herdan, G.: Small particle statistics, 418 p. New York: Academic Press 1960.
Imbrie, J. M., van Andel, Tj. H.: Vector analysis of heavy mineral data. Geol. Soc. America Bull. **75**, 1131–1156 (1964).
Irani, R. R., Callis, C. F.: Particle size: measurement, interpretation and application, 165 p. New York: Wiley 1963.
Johansson, C. E.: Structural studies of sedimentary deposits. Geol. Fören. Stockholm Förh. **87**, 3–61 (1965).
Kahn, J. S.: The analysis and distribution of the properties of packing in sand size sediments, 1. On the measurement of packing in sandstones. Jour. Geology, **64**, 385–395 (1956).
Keller, W. D.: Size distributions of sand in some dunes, beaches, and sandstones. Am. Assoc. Petroleum Geologists Bull. **29**, 215–221 (1945).
Klovan, J. E.: The use of factor analysis in determining depositional environments from grain-size distributions. Jour. Sed. Petrology **36**, 115–125 (1966).
Kolmogorov, A. N.: Über das logarithmische Verteilungsgesetz der Teilchen bei Zerstückelung: Dokl. Akad. Nauk S.S.S.R. **31**, 99–101 (1941).
Köster, Erhard: Granulometrische und morphometrische Messmethoden an Mineralkörnern, Steinen, und sonstigen Stoffen, 336 p. Stuttgart: Enke 1964.
Krinsley, David, Margolis, S.: A study of quartz sand grain surface textures with the scanning electron microscope. New York Acad. Sci. Trans. Ser. **31**, 457–477 (1969).
Krumbein, W. C.: Size frequency distribution of sediments: Jour. Sed. Petrology, **4**, p. 65–77 (1934).
—, Graybill, F. A.: An introduction to statistical models in geology, 475 p. New York: McGraw-Hill 1965.
—, Monk, G. D.: Permeability as a function of the size parameters of unconsolidated sands, 11 p. Am. Inst. Mining Metall. Engineers. Tech. Pub. 1492, 1942.
Kuenen, Ph. H.: Experimental abrasion: 3. Fluviatile action on sand. Am. Jour. Sci. **257**, 172–190 (1959).
— Pivotability studies of sand by a shapesorter. In: L. M. J. U. van Straaten, Ed.: Developments in sedimentology, Vol. 1, p. 208–215. Amsterdam: Elsevier 1964.
—, Perdok, W. G.: Experimental abrasion: 5. Frosting and defrosting of quartz grains. Jour. Geology **70**, 648–658 (1962).
Mast, R. F., Potter, P. E.: Sedimentary structures, sand shape fabrics, and permeability, pt. 2. Jour. Geology **71**, 548–565 (1963).
McCammon, R. B.: Efficiencies of percentile measures for describing the mean size and sorting of sedimentary particles. Jour. Geology **70**, 453–465 (1962).
McDowell, J. P.: The sedimentary petrology of the Mississagi quartzite in the Blind River area. Ontario Dept. Mines Geol. Circ. **6**, 31 p. (1957).
Mellon, G. B.: Discriminatory analysis of calcite- and silicate-cemented phases of the Mountain Park Sandstone. Jour. Geology **72**, 786–809 (1964).
Moiola, R. J., Weiser, D.: Textural parameters: An evaluation. Jour. Sed. Petrology **38**, 45–53 (1968).
Müller, G.: Methoden der Sediment-Untersuchung, 303 p. Stuttgart: Schweizerbart 1964.
Otto, George: The sedimentation unit and its use in field sampling. Jour. Geology **46**, 569–582 (1938).
Page, H. G.: Phi-millimeter conversion table. Jour. Sed. Petrology **25**, 285–292 (1955).
Passega, Renato: Texture as characteristic of clastic deposition. Am. Assoc. Petroleum Geologists Bull. **41**, 1952–1984 (1957).
— Grain size representation by CM patterns as a geological tool. Jour. Sed. Petrology **34**, 830–847 (1964).
Pelletier, B. R.: Pocono paleocurrents in Pennsylvania and Maryland. Geol. Soc. America Bull. **69**, 1033–1064 (1958).
— Paleocurrents in the Triassic of northeastern British Columbia. In: Middleton, G. V., Ed.: Primary sedimentary structures and their hydrodynamic interpretation. Soc. Econ. Paleontologists and Mineralogists Spec. Pub. **12**, 233–245 (1965).
Pettijohn, F. J.: Sedimentary rocks, 2nd Ed. 718 p. New York: Harper 1957.
Pierce, J. W., Good, D. I.: Fortran II program for standard-size analysis of unconsolidated sediments using an IBM 1620 computer. Kansas Geol. Survey, Spec. Distrib. Pub. **28**, 18 (1966).
Pirson, S. J.: Oil reservoir engineering, 2nd Ed., 735 p. New York: McGraw-Hill 1958.
Plumley, W. J.: Black Hills terrace gravels: A study in sediment transport. Jour. Geology **48**, 526–577 (1948).

Porter, J. J.: Electron microscopy of sand surface texture. Jour. Sed. Petrology **32**, 124–135 (1962).

Pottier, J., Jacguin, C., Marle, E., Montadert, L.: Méthodes et moyens pour l'étude des milieux poreux naturels. Rev. Inst. Francais du Pétrole **19**, 872–900 (1964).

Potter, P. E., Pettijohn, F. J.: Paleocurrents and basin analysis, 296 p. Berlin-Göttingen-Heidelberg: Springer 1963.

Powers, M. C.: A new roundness scale for sedimentary particles. Jour. Sed. Petrology **23**, 117–119 (1953).

Rees, A. I.: The use of anisotropy of magnetic susceptibility in the formation of sedimentary fabric. Sedimentology **4**, 257–271 (1965).

Rosenfelder, A.: Contribution à l'analyse textural des sediments. Ser. Carte Geol. Algerie, Bull. **29**, 310 (1961).

Rusnak, G. A.: The orientation of sand grains under conditions of "unidirectional" fluid flow, 1. Theory and experiment. Jour. Geology **65**, 384–409 (1957).

Sahu, Basanta K.: Depositional mechanisms from the size analysis of clastic sediments. Jour. Sed. Petrology **34**, 73–84 (1964).

Scheidegger, A. E.: The physics of flow through porous media, 236 p. New York: Macmillan 1957.

Schultz, E. F., Wilde, R. F., Albertson, M. L.: Influence of shape on the fall velocity of sedimentary particles. Colorado Agr. and Mech. Research Found. Rept. to Missouri River Div., Corps of Engineers, U.S. Army, Omaha, M. D. Sed. Ser. **5**, 161 p. (1954).

Sestini, G.: Paleocorrenti eoceniche nell'area tosco-umbra: Soc. Geol. Italiana Boll. **83**, 1–54 (1964).

Sharp, W. E., Fan, Pow-Foong: A sorting index: Jour. Geology **71**, 76–84 (1963).

Shepard, F. P., Young, R.: Distinguishing between beach and dune sands. Jour. Sed. Petrology **31**, 196–214 (1961).

Sindowski, Karl-Heinz: Die synoptische Methode des Kornkurven-Vergleiches zur Ausdeutung fossiler Sedimentationsräume. Geol. Jahrb. **73**, 235–275 (1957).

Sippel, R. F.: Sandstone petrology, evidence from luminescence petrography. Jour. Sed. Petrology **38**, 530–554 (1968).

Smith, G. C.: Illinois loess – variation in its properties and distribution: A pedologic interpretation. Illinois Agr. Exp. Sta. Bull. **490**, 139–183 (1942).

Sneed, E. D., Folk, R. L.: Pebbles in the lower Colorado River, Texas, a study in particle morphogenesis. Jour. Geology **66**, 114–150 (1958).

Solohub, J. T., Klovan, J. E.: Evaluation of grain-size parameters in lacustrine environments: Jour. Sed. Petrology **40**, 81–101 (1970).

Spencer, D. W.: The interpretation of grain size distribution curves of clastic sediments. Jour. Sed. Petrology **33**, 180–190 (1963).

Sternberg, H.: Untersuchungen über Lang- und Querprofil geschiebeführender Flüsse. Zeitschr. Bauwesen **25**, 483–506 (1875).

Stieglitz, R. D.: Surface textures of quartz and heavy mineral grains from fresh-water environments: an application of scanning electron microscopy. Geol. Soc. America Bull. **80**, 2091–2094 (1969).

Swann, D. H., Fisher, R. W., Walters, M. J.: Visual estimates of grain size distribution in some Chester sandstones. Illinois Geol. Survey Circ. **280**, 43 p. (1959).

Taylor, J. M.: Pore-space reduction in sandstone: Am. Assoc. Petroleum Geologists Bull. **34**, 701–716 (1950).

Udden, J. A.: Mechanical composition of wind deposits. Augustana Library Pub. **1**, 69 p. (1898).

— Mechanical composition of clastic sediments. Geol. Soc. America Bull. **25**, 655–744 (1914).

U.S. Corps Engineers: Studies of river bed materials and their movement, with special reference to the Lower Mississippi River. Vicksburg, U.S. Waterways Expt. Sta. Paper **17**, 161 p. (1935).

Visher, G. S.: Grain size distributions and depositional processes: Jour. Sed. Petrology **39**, 1074–1106 (1969).

Wadell, Hakon: Volume, shape and roundness of rock particles. Jour. Geology **40**, 443–451 (1935).

Wentworth, C. K.: A scale of grade and class terms for clastic sediments. Jour. Geology **30**, 377–392 (1922).

Weinbrandt, R. M., Fatt, I.: A scanning electron microscope study of the pore structure of sandstone. Jour. Petroleum Tech. **21**, 543–548 (1969).

Winkelmolen, A. M.: Experimental rollability and natural shape sorting of sand. 141 p. Groningen: Rijksuniversiteit Groningen 1969.

Yeakel, L. S., Jr.: Tuscarora, Juniata and Bald Eagle paleocurrents and paleogeography in the Central Appalachians. Geol. Soc. America Bull. **73**, 1515–1540 (1962).

Zeigler, J. M., Hayes, C. R., Webb, D. C.: Direct readout of sediment analyses by settling tube for computer processing: Science **145**, 51 (1964).

Zimmerle, W., Bonham, L. C.: Rapid methods for dimensional grain orientation measurements. Jour. Sed. Petrology **32**, 751–763 (1962).

Zingg, Th.: Beiträge zur Schotteranalyse. Schweiz. Mineralog. Petrog. Mitt. **15**, 39–140 (1935).

Chapter 4. Sedimentary Structures and Bedding

Introduction

Like texture and composition, sedimentary structures and bedding are inherent in sedimentation. Both are made visible by variations in grain size and to a lesser extent by mineralogy (Fig. 4-1). Because the great majority of structures can be seen with the naked eye, their study is as old as geology itself and therefore most of what we know has arisen from observation of ancient sediments. Only very recently have flume and modern sediment studies contributed significantly to the study of structures. Structures have been used (1) as guides to determine the agent or environment of deposition, (2) as guides to stratigraphic order, by determination of top and bottom, (3) to map paleocurrent systems, (4) as indices of flow conditions and (5) to assess chemical changes after deposition.

Obviously then, sedimentary structures are as important to the study of sand and sandstones as are texture and mineralogy. But unlike texture and miner-

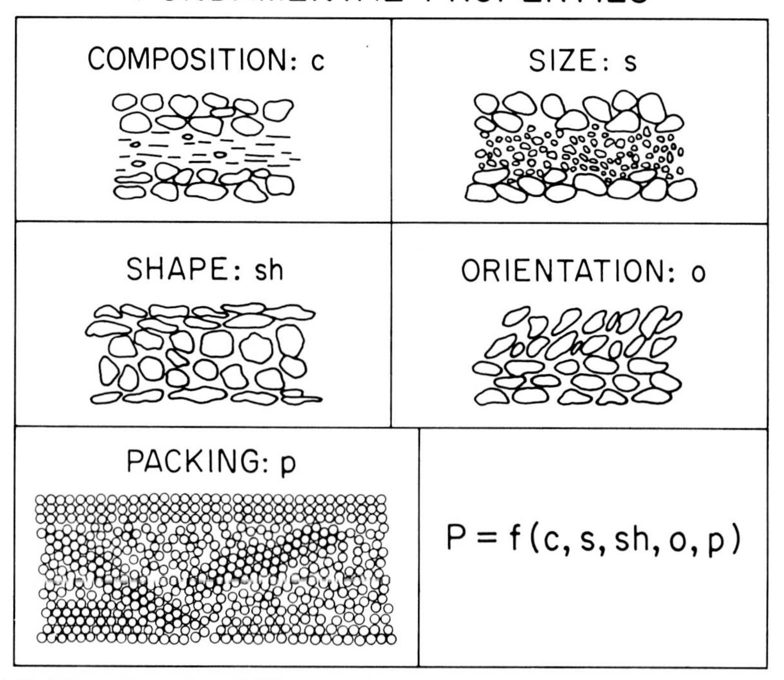

Fig. 4-1. Bedding as the product of different combinations of composition, size, shape, orientation and packing (Modified from Griffiths, 1961, Fig. 3)

alogy, most sedimentary structures can be studied only in outcrops and cores. Because of their scale, they constitute a "higher organization" involving 10^6 to 10^{10} grains so that microscopic study is largely inappropriate.

The study of sedimentary structures has generated a very large literature. Indicative of present interest, especially in the current structures and their usefulness in paleocurrent analysis and as indices of flow conditions, are the larger monographic works. Because these summarize most of the earlier literature, we have restricted our citations to the most recent papers and, exceptionally, to earlier classic studies. The reader is referred to such compilations as that by Shrock (1948) on the utility of structures in determining stratigraphic order, and to those of Khabakov (1962), Pettijohn and Potter (1964), Gubler and others (1966), and Conybeare and Crook (1968) which deal with all types of structures. Dzulynski and Sanders (1962) and Dzulynski and Walton (1965) give detailed coverage of sole marks of sandstones. Allen (1969) covers current ripples and the structures produced by their migration. The relation between primary current structures and the hydrodynamics is the topic of a symposium volume (Middleton, 1965). The reader is referred to the annotated references at the end of this chapter.

It is not our task to deal with all sedimentary structures. We do, however, summarize what is known about those which characterize sandstones in order to make our treatment complete.

Table 4-1. *Sedimentary structures: process and structure*

CURRENT

Depositional
- beach cusps
- graded bedding
- parallel lamination (parting lineation)
- sand waves (ripple mark and crossbedding)
- wave and swash marks

CURRENT

Erosional
- channels
- obstacle scours
- rill marks
- scour marks

Tool marks
- bounce, brush, prod and skip marks
- roll marks
- slide and groove marks
- striations and grooves

DEFORMATIONAL
- desiccation (mud crack casts)
- eruption (sand volcanoes and spring pits)
- founder and load structures
- impact (spray, hail, and rain pits)
- injection (neptunian dikes and sills)
- slump (folds, faults, and breccia)

BIOGENIC

Animal
- crawling trails
- feeding trails
- grazing trails
- residence structures
- resting trails

Plant
- impressions
- rootlets

CHEMICAL
- cementation (sand crystals)
- crystallization (salt and ice)
- diffusion (color banding)
- pressure solution (stylolites)
- replacement (nodules)

It is worth noting that some structures, such as ripple marks and cross-bedding are readily observable in both ancient and modern sands. Others such as the sole marks of which flute casts are an example, are seen only in those ancient sandstones which separate along bedding planes; the unconsolidated nature of modern sands precludes examination of the underside of sand layers.

A structure cannot be defined in the same precise manner as a geometric object, such as a cube or a cylinder. As with fossils and organic forms in general, a picture is essential for its description.

There are four broad types of sedimentary structures (Table 4-1): (1) *current structures* formed by currents of water, air, and even ice as sediment is transported and deposited, (2) *deformational structures* formed shortly after deposition and before consolidation mostly by slumping and foundering but also by escaping fluid and gas, (3) *biogenic structures* of organic origin such as tracks, trails and burrows plus a few formed by plants, and (4) *chemical structures* formed by chemical processes during and after lithification of the sand. Current, deformational, and most organic structures are all made very early in a sand's history before effective consolidation, but only the structures of current origin are strictly "primary." It is chiefly these structures that have been used to help interpret ancient environments and map paleocurrent systems. Practically all the current structures and many of the deformational and biogenic structures can form in a matter of hours, some even in a matter of minutes. In contrast, the chemical structures develop over much longer intervals – perhaps hundreds or thousands of years. Some structures, of course, are a mixture of more than one origin and hence they can be and have been classified in many different ways.

Terminology is abundant, perhaps too much so. There are probably over 400 English-language names for current and deformational structures (Pettijohn and Potter, 1964, p. 275), many others for those structures of biologic origin, commonly designated by generic and specific names from the Greek and Latin, and a modest number for chemical structures. Fortunately much of value can be learned from sedimentary structures without recourse to excessive terminology, though those who explore further in this field will need to become familiar with a good many terms and a glossary is helpful (Pettijohn and Potter, 1964).

The objectives of this chapter are to summarize very concisely what is known about the "geometry" and utility of sedimentary structures; their processes of formation – current, chemical, and rheologic – being presented mainly in Chap. 9.

Current and Deformational Structures

Bedding

Review of most sedimentary structures soon discloses that they can be classified and defined as some aspect of *bedding*. Our approach is here mainly descriptive, is based mainly on form and geometry, and is discussed under four headings (Table 4-2). The genetic aspects of bedding and related structures are covered in Chap. 9.

Table 4-2. *Classification of primary sedimentary structures (Pettijohn and Potter, 1964, p. 5)*

BEDDING, EXTERNAL FORM

1. Beds *equal* or *subequal* in thickness; beds laterally uniform in thickness; beds continuous
2. Beds *unequal* in thickness; beds laterally uniform in thickness; beds continuous
3. Beds *unequal* in thickness; beds laterally variable in thickness; beds continuous
4. Beds *unequal* in thickness; beds laterally variable in thickness; beds discontinuous

BEDDING, INTERNAL ORGANIZATION AND STRUCTURE

1. Massive (structureless)
2. Laminated (horizontally-laminated; crosslaminated)
3. Graded
4. Imbricated and other oriented internal fabrics
5. Growth structures (stromatolites, etc.)

BEDDING PLANE MARKINGS AND IRREGULARITIES

1. On base of bed
   (a) Load structures (load casts)
   (b) Current structures (scour marks and tool marks)
   (c) Organic markings (ichnofossils)
2. Within the bed
   (a) Parting lineation
   (b) Organic markings
3. On top of the bed
   (a) Ripple marks
   (b) Erosional marks (rill marks; current crescents)
   (c) Pits and small impressions (bubble and rain prints)
   (d) Mud cracks, mud-crack casts, ice-crystal casts, salt-crystal casts
   (e) Organic markings (ichnofossils)

BEDDING DEFORMED BY PENECONTEMPORANEOUS PROCESSES

1. Founder and load structures (ball-and-pillow structures, load casts)
2. Convolute bedding
3. Slump structures (folds, faults, and breccias)
4. Injection structures (sandstone dikes, etc.)
5. Organic structures (burrows, "churned" beds, etc.)

External Form. Bedding has been recognized and described for many years. Nonetheless, attempts to classify it have generally been unsuccessful primarily because of failure to describe it meaningfully. Perhaps the best summary has been made by Bokman (1956) and later by Campbell (1967), whose Table 1 is particularly complete. Bokman defined a lamination as the smallest recognizable unit layer of particles in a sediment. It may vary from microscopic size in silts to many centimeters in coarse gravels. Following Otto, (1938, p. 575), Bokman defined a *bed* or *sedimentation unit* as, "that thickness of sediment which appears to have been deposited under essentially constant physical conditions." Otto had recognized that random deviations around a mean will be present and hence he added the qualifier, "under essentially constant physical conditions."

A *set* is a lithologic unit composed of two or more consecutive beds of the same lithology. All three terms are relative ones and have no thickness connotations.

What are the basic properties of a bed? They are its thickness and lateral continuity. Several attempts have been made to describe thickness and redefine the terms which denote thickness (Grumbt, 1969, Fig. 7). Lateral continuity is a variant of the thickness problem. Some beds persist virtually without change in thickness even in large outcrops whereas others, such as ripples, pinch and

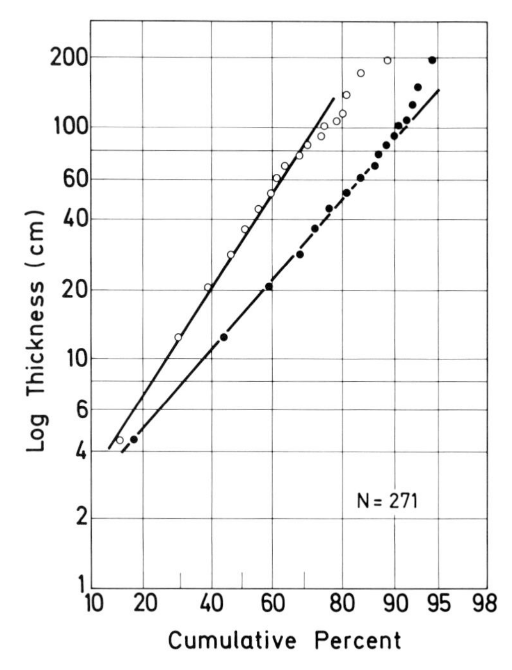

Fig. 4-2. Approach to lognormality of thickness of turbidite sandstone beds (Scott, 1966, Fig. 13). Whether lognormal or not, virtually all such distributions are strongly skewed toward the thinner beds

swell (wavy and flaser bedding) and some pinch out (Reineck and Wunderlich, 1968).

Bedding, like other rock properties, is susceptible to measurement, particularly in vertical sections. Measurement of bedding thickness is an elemental part of all stratigraphic studies. Bed thickness is related to current competence, stronger currents producing thicker and commonly coarser grained beds. Many bedding thicknesses seem to be log normal (Fig. 4-2) and practically all are strongly skewed toward the thinner beds. Kelley (1956, p. 299) proposed a *stratification index*, defined as the number of beds $\times$ 100 divided by the thickness of the section measured, which although not widely used, is essentially the reciprocal of the average thickness of the beds. Bokman (1957) proposed a geometric *theta scale* to facilitate statistical computation. A geometric scale such as Bokman's tends to normalize originally skewed thickness distributions much the same as the phi scale does for size distributions (see, for example, Enos,

1969, p. 706). As with any statistical distribution, mean, mode and some measure of dispersion are needed to specify any particular bedding sequence. Plots of vertical profiles of bed thickness (rhythmograms) have been used for precise correlation or to search for cycles (Dean and Anderson, 1967). Most striking success has been with glacial varves and evaporites. Rhythmograms may also be useful in attempting correlation in facies lacking good marker beds such as turbidites.

Fig. 4-3. X-ray and normal (inset) photograph of fine-grained, laminated sandstone (Hamblin, 1965, Fig. 14)

The four bedding types of Table 4-2 represent a progression from maximum order (uniform thickness within and between beds) to minimum order (variable thickness within and between beds plus discontinuous beds). This is a progression from uniform to very variable flow conditions – from deposition virtually without erosion to deposition with appreciable erosional scour and fill.

Internal Organization and Structure. Internally beds may be (1) massive or structureless, (2) may be either "horizontally"-laminated or show diagonal or cross-laminations, (3) may be graded, (4) may display internal imbrication, (5) or may exhibit "growth" bedding produced by rhythmic precipitation or by organisms such as stromatolites. All occur in sands and sandstones, although algal stromatolites are very rare (Davis, 1968).

Massive Bedding is bedding seemingly without internal structure. Such lack of internal structure can be deceptive, however, as Hamblin (1965) and others have shown (Fig. 4-3). Truly massive beds of sand appear to be very rare which is indeed fortunate, for if they were common, we would be hard pressed to explain them.

Laminated Bedding is moderately common and many such laminated beds yield excellent flagstone (Fig. 4-4). Laminated bedding forms a small part of many sands and occurs in virtually every major environment. Its origin – whether the product of weak or strong currents – is as yet not well established.

Crossbedding is one of the most characteristics structures of sands. This structure, also known as current bedding, cross-lamination, cross-stratification, diagonal bedding, and inclined bedding, may be seen in both modern and ancient sands. As here defined, it is a structure confined to a single sedimentation unit characterized by internal bedding or laminations, called foreset

Fig. 4-4. Laminated bedding of fine-grained Sappington Formation (Mississippian-Devonian). Bedding planes have weak parting lineation. NW$\frac{1}{4}$SW$\frac{1}{4}$ sec. 31, T. 2 N., R. 1 E., Broadwater County, Montana, U.S.A.

bedding, inclined to the principal surface of accumulation. This definition is restrictive and excludes the inclined bedding of talus, of lateral accretion deposits such as the slip slope of a point bar, and other stratification with a high initial dip such as that formed by the basinward growth of delta front as well as the lower angle stratification related to a prograding beach.

Several classifications of crossbedding, dependent upon the geometry of the structure, have been proposed of which the most widely followed is that of McKee and Weir (1953). Allen (1963) recognized some fifteen varieties. In actual practice, it is difficult to apply these classifications owing to the fact that exposures are seldom adequate or complete enough to determine to which class a given cross-bed set belongs. It is even difficult, in small outcrops, to distinguish between planar and trough crossbedding (see Fig. 4-5).

What kind of meaningful observations can one make on crossbedding? Normally one is restricted to two situations: a vertical cross-section or a bedding-plane exposure. The vertical section may have any orientation with respect to the crossbedding. The most useful, perhaps, is a longitudinal section, that is, one parallel to the direction of current flow. In such a section, one can see whether or not the traces of the planes which bound the set are *parallel* or *convergent*, whether the traces of the foreset surfaces are straight or curved and whether they are *tangent* to the base of the bed or not. One can also determine the *scale* of the cross-bedding, that is the thickness of the set (Figs.

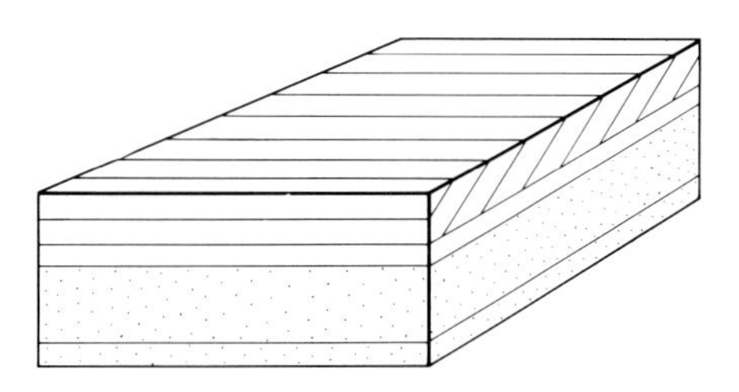

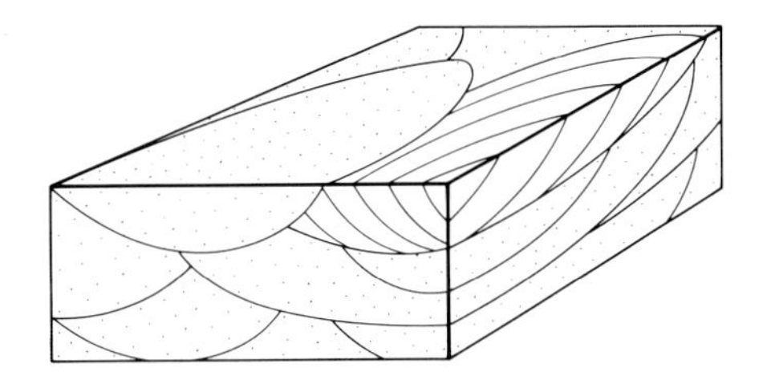

Fig. 4-5. Principal types of crossbedding: planar (above) and trough (below)

4-6 and 4-7). Crossbedding varies from sets but a centimeter or two thick (small scale) to those thirty or more meters thick (large scale). In most sandstones the average is 15 to 60 cm. The observer can also measure the dihedral angle between the upper bounding surface of the set and the foreset planes. This angle is often referred to as the *angle of repose* or *inclination*, a designation approximately but not strictly correct. Most important is to measure the *azimuth* of the dip of the foreset planes which is the presumed down-current direction. In tectonically tilted strata, the dip of the foreset planes and that of the true bedding are both recorded and the down-current azimuth determined after correction for tectonic tilt – a correction made by use of the stereographic projection or by computation.

Vertical sections normal to the current flow yield very little information except in the case of trough crossbedding. If a complete section of a trough is exposed one can measure the depth of the trough (maximum thickness of the set) and its width to compute a width-depth ratio. In general there seems to be a fixed width/depth ratio which is independent of scale.

Most useful are bedding-plane sections, especially of the trough cross-bedded units. Such sections enable one to determine the bisectrix of the trough –

Fig. 4-6. Small-scale crossbedding near top of a fine-grained turbidite bed. Approximately one half natural size. Paleocene flysch near east end of San Telmo beach, Zumaja, Vascongadas Province, Spain

Fig. 4-7. Crossbedding in Lower Devonian sandstone, transport from left to right. Oued Tairi, Tassili Externe, Department of Oasis, Algeria

the down-current direction. Bedding-plane sections are the most useful for distinguishing between trough crossbedding and the ordinary planar-tabular sets. The horizontal traces of the foreset surfaces are sharply curved in the trough crossbedding; they are straight or nearly so in the planar crossbedding.

It is important to observe the relation between cross-bedded sets. They may be superposed one on the other. Commonly one set is separated from the next by ordinary bedded strata. The latter may be, in fact, "top-set" or "back-set" beds which, though appearing horizontal in outcrop, in fact dip up-current at very low angles. Ripple cross-lamination presents special situations. Ripples that are superimposed slightly out of phase and regularly displaced form what has been termed "climbing ripples" or "ripple-drift lamination."

Cross-laminations may be deformed – either at the time of deposition by some kind of "soft-sediment deformation" or by shear during tectonic movements. Normal penecontemporaneous deformation results in over-steepening of the foresets, in some cases even overturning; tectonic deformation leads to either oversteepening or flattening and other distortions (Ramsey, 1961).

Crossbedding, as here defined, seems to be the product of down-current migration of a sand wave of some kind. The very smallest scale cross-lamination is the product of ripple migration. That of most fluvial sandstones is formed by migration of subaqueous "dunes" forming medium-scale structures. Large-scale crossbedding is the product of migration of large dunes, either subaqueous or eolian. Such factors as scale, angle of inclination, tangency or lack of it, planar or trough structure, or trough dimensions and depth-width ratios have been in

Fig. 4-8. Graded bedding, Archean turbidite, Minnitaki Lake, Ontario, Canada, photograph by R. G. Walker

Fig. 4-9. Varieties of graded bedding (Kuenen, 1953, Fig. 1)

part investigated but they are not as yet fully understood (Jopling, 1965; Harms and others, 1963).

At this writing there are no certain criteria, based either on experimental or on theoretical considerations or on empirical observations, to relate the geometry of the crossbedding to either the agent or the environment of deposition. Nonetheless crossbedding has great value in paleocurrent analysis and paleogeographic interpretation as outlined later in this chapter.

Graded bedding is defined by an upward decline of grain size *within* a bed and is of several varieties (Figs. 4-8 and 4-9). Graded beds form by deposition from a decaying current and may range from a centimeter or less to several meters in thickness. The graded materials may be silt, sand or even gravels. In a general way, the coarser the material, the thicker the graded unit. In general, most graded sandstones, commonly the graywackes, range from a few centimeters to a meter in thickness. Much thicker beds are likely to be composites. Graded sequences display a log normal thickness distribution. Graded beds commonly have a distinct internal, vertical sequence of structures, which like grain size, is the response to a decaying current (Fig. 4-10). Typically, graded bedding is found in thick sections of immature sandstones – the graywackes – of geosynclines. Here density or turbidity currents – turbid mixtures of mud and sand and water – are believed by most sedimentologists to flow periodically downslope and transport sand into deep water, where normally only muds would accumulate. Unlike crossbedding, the graded beds of turbidites can

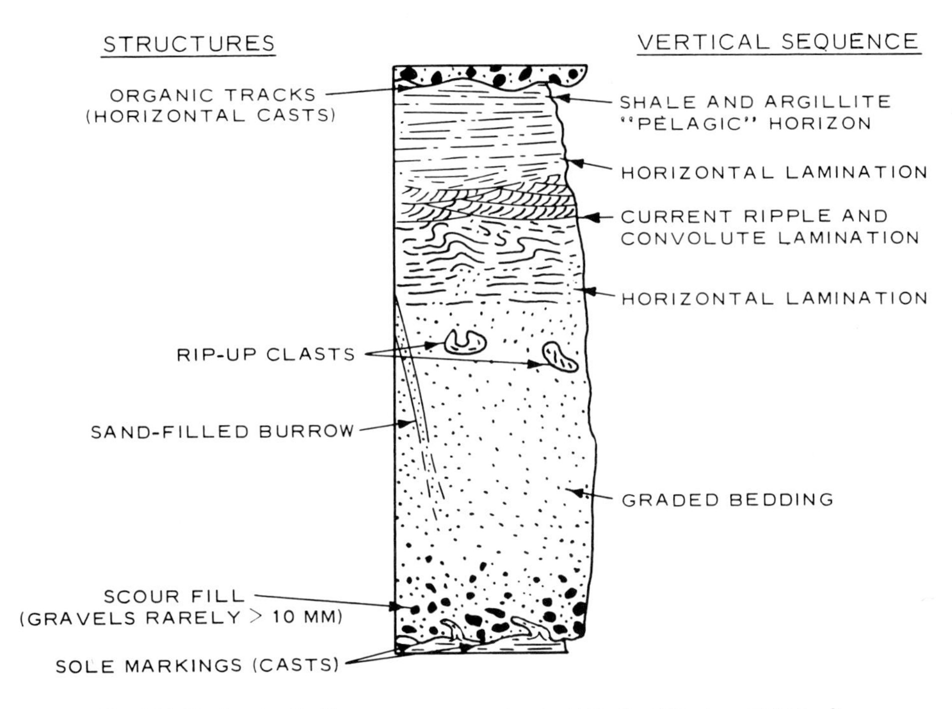

Fig. 4-10. Ideal cycle of sedimentary structures in a turbidite bed (Stanley, 1963, Fig. 2)

have wide lateral extent, some individual ones having been correlated for many miles or kilometers.

Bedding Plane Markings and Irregularities. Many bedding planes of sandstones, if examined closely, show a variety of structures. They may be divided into those on the *base* or sole of the bed, those on *top* of the bed, and those on planes *within* the bed.

Sole markings are features characteristic of the underside of sandstone and some limestone beds that rest on siltstone and shales (Table 4-3). These features have been known for many years but only recently studied intensively. Most of these structures are the "negatives" or "casts" of depressions or markings originally produced in the mud over which the sand was spread. The structures originate by (1) the action of currents on the mud surface, (2) unequal loading of the soft hydroplastic mud, or by (3) the activity of organisms on this surface. Although they occur in almost all sands, sole marks are parti-

Table 4-3. *Inorganic sole markings and deformation structures (modified from Potter and Pettijohn, 1963, Table 5-1)*

Agent	Process	Name of structure
Produced by current	Current scour	Flute (casts)
	Engraved by moving objects (tools)	Tool marks
	(a) Drag	Drag marks (groove and striation casts)
	(b) Saltation	Bounce, brush and prod marks (and casts)
	(c) Rolling	Roll marks
	Unknown	Channels
Produced by gravity	Unequal loading	Load pockets (load casts)
	Foundering (thixotropic transformation)	Ball-and-pillow structure
	Slump or slide	Slide marks (and casts); slump folds and faults and breccias
Produced by "liquidation"	Injection and other processes	Sandstone dikes and sills; sand volcanoes; mud volcanoes
Complex current and gravitational interaction		Convolute laminations

cularly abundant in turbidites where they provide the best means of determining current flow.

Of the various structures produced by the action of currents, the most common is the *flute* formed on a somewhat firm mud surface and filled with sand and hence preserved as a raised structure or *flute cast* on the underside or sole of the overlying sand bed (Figs. 4-11 and 4-12).

Flute casts are subconical structures with a rounded or bulbous upcurrent nose, the other end flaring out and merging with the bedding plane. The structure has also been designated as flute molds, flow marks, scour casts, scour finger, vortex cast and turboglyph. Flute casts vary from a few centimeters in length to giant structures a meter or even two meters in length. Solitary flutes are rare; most commonly they occur in swarms within which the individual flute casts may be widely spaced, close-spaced, or even overlapping. It is common for successive sandstone beds to show flute casts. In other words, when conditions were right for the production of a flute swarm during the deposition of one sand bed, these conditions persisted during the formation of subsequent beds.

Flute casts vary in shape; those of a given swarm being more or less alike. Some are elongate, relatively narrow structures, others have a broader deltoid form. Some have good bilateral symmetry, others show less regular form commonly with a twisted "beak." There seems to be a transition from well-defined flute casts to more irregular transverse scour casts and to relatively weak greatly elongate furrow casts.

The less regular forms resemble load casts but have a more regular shape and also show clear evidence of their erosional origin such as the transection of laminations in the subjacent mud or silt. The laminations may be differentially eroded so that the sand filling shows "terraces" forming a sculptured flute cast. In contrast, the laminations associated with load casts are not transected by the structure but are, instead, distorted or deformed by it. Some flute fillings were the cause of loading and deformation. Such "load-casted" flute marks are good examples of hybrid structure – the type that are difficult to classify.

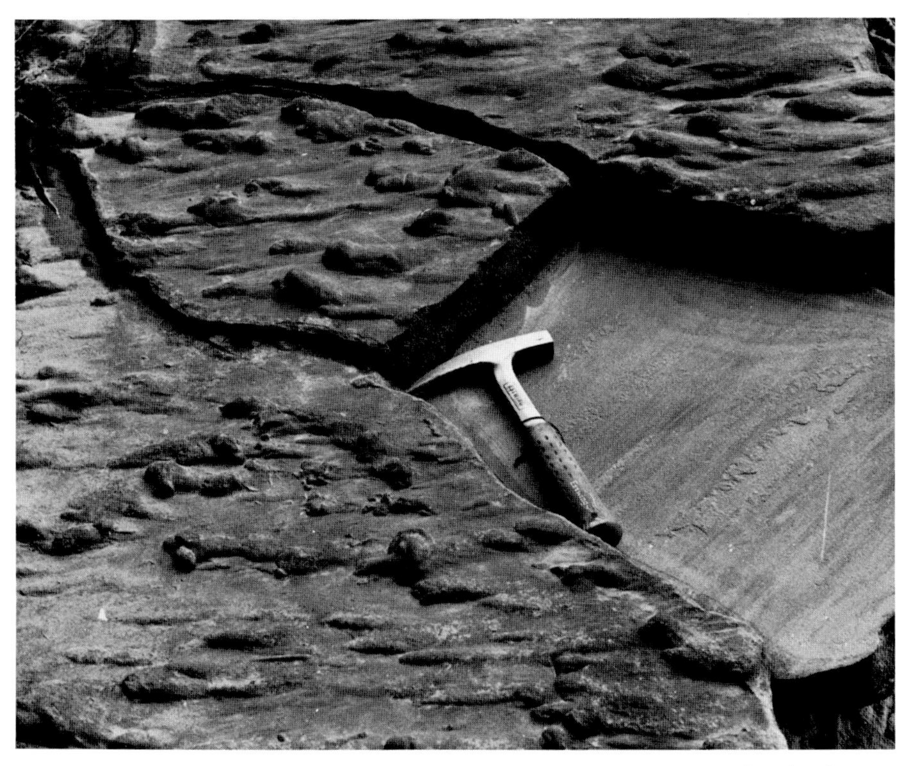

Fig. 4-11. Partial section through a turbidite sandstone: flute casts and parting lineation in overturned bed of Marnoso Arenacea (Miocene). Current from right to left. Note divergence in orientation between direction indicated by flutes and parting lineation. Near Marradi, Tuscany, Italy

Although flutes may be associated with grooves, such is not the rule. In general they are mutually exclusive.

Flutes appear to be the product of local eddies. Their size is dependent on the size of the eddy, the latter being perhaps a function of the current strength. When the flow conditions are right for the production of one eddy, they are right for the generation of a field of such vortices. There are many unknown factors which govern the size, shape, and spacing of flutes.

Flutes are the most common and most useful of the current-produced sole marks. Their form is a sure guide to the direction of current flow and, although not exclusively the product of turbidity flows, they are most characteristic of the flysch facies.

Fig. 4-12. Flute casts from Denbigh Grits, Salopian (Middle Silurian) at Penstrowed Quarry, Wales. Photograph by Norman McIver

A structure produced by current scour, and hence akin to flutes, is the *current crescent* or as the Germans call it Hufeisenwülste. It is indeed a horseshoe-shaped depression developed by current scour around an obstacle lying on the sand surface. These commonly form around shale intraclasts in fluvial sandstones. The clast and moat are buried with additional sand. Ultimately the shale weathers out leaving a hole. The moat appears as a raised ridge on the underside of the sandstone bed surrounding the hole.

Groove casts (Shrock, 1948, p. 162) are raised, rectilinear, rounded to sharp-crested features found on the underside of some sandstone beds. They are particularly characteristic of turbidite sands. They are presumed to originate by the filling of a corresponding furrow in the underlying shale (mud) and were in fact called "mud furrows" by Hall who observed them in the Devonian of New York State over 100 years ago (Hall, 1843, p. 424). They have also been called "drag marks" and "drag casts" from their presumed formation by objects being dragged over the mud bottom (Kuenen, 1957, p. 243).

Groove casts seldom appear singly – they generally occur in sets, commonly as two sets intersecting at an acute angle on the same surface (Fig. 4-13). The individual ridge displays a relief of only a millimeter or two, rarely more than a centimeter. They are remarkably straight and in most exposures show neither a beginning nor an end. Some are multiple, and are ornamented with a second order series of microgrooves or ridges. Within a given set there is little or no deviation in azimuths. Groove casts may be few or many; the latter sets partially effacing the earlier ones. Groove marks should be distinguished from *slide marks* or casts (Fig. 4-14) formed by the movement of a large object – such as a shale raft – over the bottom. Such sliding objects tend to pivot or rotate so that the marks they produce all curve alike. Normal grooves, on the other hand, being produced by many individual objects show no such coordinated behaviour. Groove casts may be associated with prod casts, skip and bounce casts and brush marks but rarely with flute casts.

Terminations are seldom seen; rarer still are terminations marked by a shell fragment or other recognizable tool presumed to be responsible for the original groove.

Groove casts are observable only on the base of the indurated sands resting on shale which has weathered away exposing the bottom surface of the sandstone bed. Like flute casts, they are most abundant on the base of turbidite sands and are the most common heiroglyphic structure of the flysch facies.

Their origin was long unsuspected. It now seems clear that they are the product of tools swept along by the current which engrave the surface of a relatively firm mud bottom. This view is supported by the finding of such tools – shells, large sand grains, mud lumps – at the downcurrent end of the groove and the parallelism of the grooves with the direction of current flow as shown by other criteria. The exact dynamics is not yet clear. Most objects moved by a current proceed by rolling and saltatory leaps and are constantly rotating or twisting about. To form a groove requires continuous contact, even pressure, and nonrotary movement. Eddy motion produces flutes, not grooves. What conditions control each and which of the two is the more proximal?

Groove casts, by reason of their abundance, are among the most useful paleocurrent indicators especially if used in conjunction with structures which yield the sense of the motion. If the currents responsible for them were indeed turbidity or density currents moving down slope, then some interpretative problems arise. The divergence in direction shown by intersecting sets or by aberrant directions on a particular bed within a more regular sequence, raises questions about the paleoslope. Clearly not all grooves are cut by currents moving down slope.

As may be seen in Table 4-3, there are other marks or structures in addition to the more common flute and groove casts. Included here are those produced by objects which touch bottom intermittently and those which are generated by rolling objects. The first group includes bounce, brush and prod casts. *Bounce casts*, also *skip casts*, are marks spaced at rather regular intervals and are impact structures made by an object pursuing a saltatory path. The

Fig. 4-13. Bottom markings in Denbigh Grits, Penstrowed Quarry, Wales. Note two sets of groove casts. N. L. McIver photograph

Fig. 4-14. Large, multiple subparallel slide marks on base of near-vertical turbidite. *Arensica numulitica* flysch (Eocene) along the Zumaja-Guetaria Highway (Km 34), Vascongadas Province, Spain

brush cast differs in that the bottom contacts are accidental and not regularly repeated and moreover are prolonged enough for the constructions of a slight mound of material pushed up by the forward moving object. The *prod cast* is, as the name implies, generated by an object such as a partially water-logged stick, impinging on the bottom, being forced downward into it and then being rotated in the forward direction and lifted free of it. The more prominent down-current terminal point of the mark lies at the end of a short groove.

Roll marks are diverse. Common in some flysch sequences are those formed by rolling, in wheel or hoop-fashion, of planar coiled shells, especially cephalopods. These leave a characteristic "signature" or track (Seilacher, 1963b).

Mudcracks develop in cohesive materials which undergo shrinkage upon loss of water. Common muds illustrate this very well. Noncohesive granular materials, such as sand cannot be expected to shows mudcracks. Nevertheless a polygonal pattern of raised ridges is present on the *underside* of some sandstones. This structure is in reality produced by an input of sand over a mudcracked surface, the sand filling of the cracks becoming a part of the overlying sand layer itself. The shale ultimately weathers away leaving the "casts" of the crack fillings welded to the sandstone layer itself.

Structures characteristic of the top of sandstone beds include ripple marks, rill marks, pits and prints and, in the finer sands and silts; ice-crystal casts and molds. Biogenic structures are also common on some beds. All of the above marks can also occur as casts on the bottom of the bed.

Historically, one of the earlier observed structures of sand and sandstone and one about which the most has been written are *ripples* and *ripple marks*. There is an enormous literature on the subject – not only by geologists but also by those interested in the physics of grain movement and ripple or wave phenomena. Early classic papers by geologists are those of Kindle (1917) and Bucher (1919). One of the most recent comprehensive reviews of the subject is that of Allen (1969).

Most of the earlier work dealt with the form or morphology of ripples as seen in modern sands or as displayed on bedding planes of the older sandstones. During the past decade interest has shifted to the structure of ripples and to features, such as ripple-drift, best observed in cross section.

Ripple marks are sand waves of the smallest scale and thus form from weaker currents than those that produce dunes which generate the common and large-scale crossbedding. Ripple marks are characteristic of noncohesive granular materials of sand size. They may develop in either siliceous or carbonate sands but do not form in either coarse materials such as gravel or the finer silts and muds.

Classification has been particularly troublesome because of the great variety of ripples and the gradation from one type to another. Figure 4-15 illustrates but one of the many different varieties of ripple mark. Very broadly, ripples are of two types: those with symmetrical cross section generally attributed to wave-generated oscillatory currents and those with asymmetrical cross sections produced by unidirectional currents of wind or water. Small-scale current ripples are of many varieties and forms and belong to a sequence such that they either grade or intermingle with one another (Fig. 4-16). Ripple marks display

Fig. 4-15. Fossil ripple mark. Horton Group (Mississippian), Minas Basin, Walton, Nova Scotia. Note two sets of ripples. Photograph by H. P. Eugster

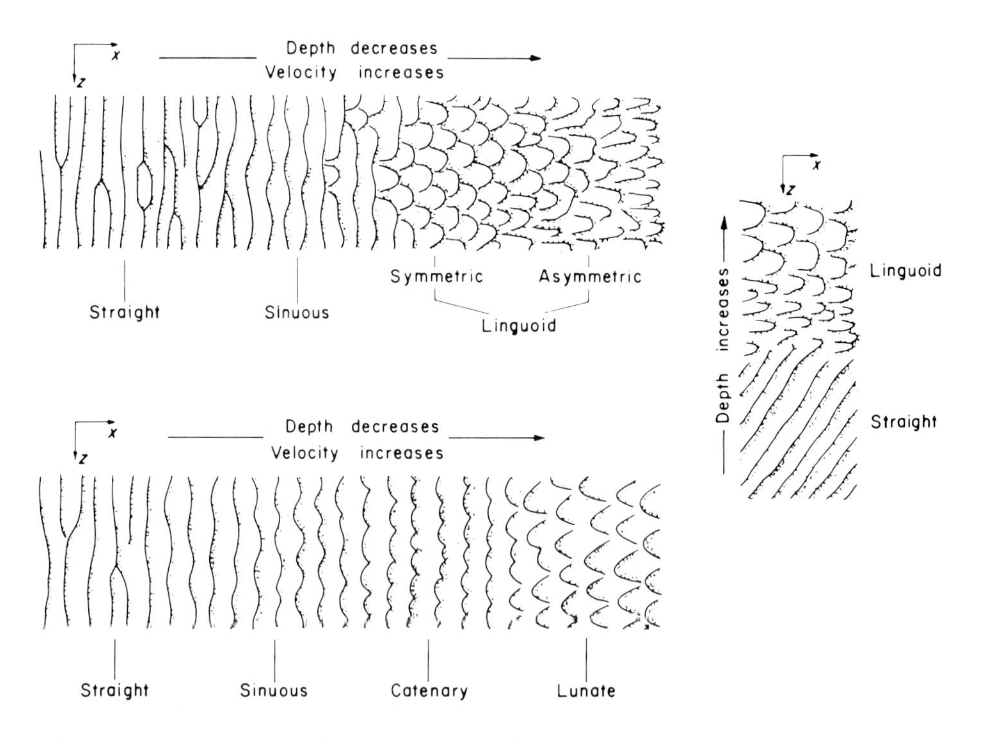

Fig. 4-16. Nomenclature and relationship between different patterns of current ripples (Allen, 1969, Fig. 4-61)

patterns termed *rectilinear*, if continuous and normal to the current, or *sinuous*, if continuous but with some sinuousity. Many current ripples do not form a continuous pattern but are instead broken up into laterally compressed crescentic structures of U-shape in plan. They are termed *crescentic* or barchanoid if the extremities point down-current and *lunate* if they point up-stream. Several indices based on such parameters as length and height have been thought to discriminate between eolian and aqueous ripples. Since undoubted eolian ripples are rarely, if ever, found in the geologic record, this distinction has little meaning.

Because ripples form wherever sand is in motion, they occur in a wide range of environments from the intertidal zone to the deep-sea. In shallow seas they tend to be parallel to the shoreline and this tendency has useful paleogeographic significance. Their orientation should be systematically recorded.

In many ripple systems sand is transported up the stoss side and avalanches down the lee side causing the ripple to migrate downstream without any change in the level of sedimentation. If, however, sand is *added* to the system, the rate of accumulation on the lee side increases and each ripple climbs the backslope of the one in front of it to form *climbing* ripples or ripple-drift cross-lamination (Sorby, 1859, 1908, p. 184; Walker, 1963, 1969; Jopling and Walker, 1968).

In detail the morphology is variable and several subtypes of ripple-drift structure are recognizable. In one, all the deposition is on the lee side; another is characterized by deposition on both stoss and lee sides with individual laminations continuous across the system. A third type is characterized by a sort of grading, mud being collected in the ripple trough and silt and sand being segregated on the stoss side. Although laminations are continuous in this case they change composition from the stoss to the lee sides. This type of ripple-drift seems to be characteristic of turbidite sedimentation (Walker, 1963).

The angle of climb is controlled by the rates of deposition from suspension relative to the rate of bed-load movement. At high-flow regimes (Chap. 9), the sediment is swept to the lee side and the angle of climb is small (Walker, 1969).

The "rib-and-furrow" structure of some sandstones is simply the bedding-plane expression of micro-cross-lamination generated by migration of crescentic ripples. It is trough cross-stratification in miniature.

The ripple structure, as seen in cross section, may be deformed. Local oversteepening and overturning of the foresets are common, especially in rippled silts and fine sand. There seems to be some relation between these deformed ripples and convolute bedding – the latter perhaps an extreme end-product of ripple deformation. Isolated or "starved" ripples which migrate over a mud bottom may show subsidence into the mud, that is, initiate load casts.

Some sands separate along bedding planes forming very regular flags. These parting surfaces may display a faint lineation, *parting lineation*, which is also called primary current lineation, related to grain alignment. Commonly the parting is slightly imperfect so that the surface has plasterlike remnants clinging to the bedding surface. These irregular patches are elongate in the direction of the

Fig. 4-17. Parting lineation in Annot Sandstone (Tertiary), a turbidite. Current parallels parting. Menu gives scale. Near Peira-Cava, Maritime Alps, France

lineation and are best seen in intense low-angle illumination. The term *parting step lineation* (Fig. 4-17) has been applied to this structure (McBride and Yeakel, 1963).

Contemporaneously Deformed Bedding

Following deposition, or concurrent with it and before consolidation, a sand deposit may undergo deformation. Various structures are produced. These may be attributed to one or another of three different processes. One is a convective-like pattern of motion which results in a vertical transfer of material. Such motion is initiated by an unstable density stratification such as occurs where a bed of sand is deposited on a less dense, water-saturated finer silt or clay. If the underlying material undergoes a thixotropic transformation (Chap. 9), with loss of strength, a series of convective cells may be set up with downward movement of sand and compensatory upward movement of silt or clay. The motion may be slow or it may be rapid and catastrophic.

A second process of deformation is the result of an instability due to over-steepened depositional slopes. The resulting movement has a large lateral component and hence a near-horizontal transfer of material. This may be a slow creep or rapid slump or slide.

A third process may involve a pseudoliquifaction of the sand – development of a "quicksand" capable of injection as sills, dikes, or diapirs.

Fig. 4-18. Bottom view of small load casts at base of fine-grained, thin-bedded Sample Sandstone (Mississippian). NW$\frac{1}{4}$ SE$\frac{1}{4}$ NW$\frac{1}{4}$ sec. 25, T. 8 N., R. 4 W., Owen County, Indiana, U.S.A.

Of the various structures due to vertical readjustments, *load casts* or load pockets are the most common (Fig. 4-18). These are bulbous, mammillary or papilliform downward protrusions of sand produced by unequal loading of the underlying hydroplastic mud. These structures may become sack-like, attached to the mother bed by a constricted neck; in rare cases they become detached and sink downward in the underlying materials. These are *load pouches*, and if detached, *load balls.*

As indicated by their process of formation, load casts are not "casts" at all. Unlike flute casts they did not fill a pre-existing cavity or "mold." Laminations in the underlying mud are deformed and the mud is forced upward in a thin tongue-like fashion between the downward protruding load casts.

Load casts generally lack symmetry or plan of arrangement and may be very irregular in size and form. They appear as swellings on the underside of a sand bed, varying from slight bulges to deep or shallow rounded sacks. Load structures may have been initiated as flutes, or even grooves, and by "starved" or isolated sand ripples. Usually with the load-casted ripples there is a pattern of arrangement and an internal structure inherited from the parent structure.

Load casts are indicative of no particular environment. The only requirement is deposition of sand on a water-saturated hydroplastic layer. They seem to be more common in turbidite sequences but even here some beds are devoid of load casts – others will display them. Their growth in one case and not in the other seems to indicate the condition of the underlying bed. If one turbidite flow follows on the heels of another, conditions are more favorable

Fig. 4-19. Ball and pillow structure. Annot Sandstone (Oligocene), Peira Cava, Maritime Alps, France

for load-casting whereas if there is a significant time-lapse and dewatering of the subjacent bed, no load-casting will take place.

Some sandstone beds, like some subaqueous lava flows, display an ellipsoidal or *pillow structure* (Fig. 4-19). The sand is broken up into numerous, and generally closely-packed, ball- or pillow-form masses. These have been called "hassocks" and "pseudo-nodules" which they resemble. The less satisfactory term "flow rolls" has also been used.

The closely-packed balls or pillows vary from objects a few centimeters to bodies several meters in diameter. They are rarely spherical, are more commonly ellipsoidal or kidney-shaped. If the sandstone were laminated, such laminations within the balls or pillows are deformed, generally conforming to the rounded lower half of the ball or pillow, perhaps contorted in the central part. In structure, the pillows are like basins with rims slightly recurved inward. Or they may be likened in form to an inverted mushroom without a stem. The cup-like or basin-structure generally is convex downward, concave upward, in some cases tilted but not overturned.

Despite earlier views to the contrary, the pillows are not concretions nor the products of spheroidal weathering, both of which are known in sandstones. Nor are they the products of slump as commonly stated. Their symmetry and orientation is a response to downward not lateral movement. That such saucer- or kidney-shaped structures can be produced by foundering of a sand bed was shown experimentally by Kuenen (1958, p. 18). The conviction has grown that some such mechanism is responsible for their formation (Sorauf, 1965; Howard and Lohrengel, 1969). Perhaps this action was sudden or catastrophic. There is clearly an affinity between these structures and load pockets, and especially with the detached load balls. All are related to vertical transfer of materials. Perhaps the differences are related only to rates – one being slow, the other rapid. Just like load casts, an underlying hydroplastic layer – a layer of mud that has not had time to dewater – is necessary.

Under some conditions sands are deformed by gravity-induced movements with a large lateral component while the sediment is still unconsolidated and still in the environment of deposition (we thus exclude tectonic and other later movements). The terms "slump" and "slide" have been applied to such movements and "slump structures" to the resulting features produced. In general "slump" conveys the idea of a local phenomenon; slides denote movement of greater horizontal displacement – miles in some cases. "Creep" also denotes down-slope movement of a slow imperceptible nature.

All sediments may be involved in these movements. They are not peculiar to sands, although sand deposits may be much deformed by them. One structure, involving only sand, is deformed crossbedding (Jones, 1962). This is made manifest by over-steepening of the foreset planes – which may display inclinations up to 90 degrees or even be overturned – the overturning being always in the downcurrent direction.

In rare cases the laminations are crumpled. Clearly there has been creep or slumping or some kind of shear imposed on the sand by drag of the depositing current or by lowering of water level, or earthquake shock. Tectonic folding will also distort crossbedding (Ramsey, 1961) producing either steepened foresets or very much flattened foresets. The deformation described here, however, is penecontemporaneous as the overlying or underlying strata may be unaffected.

Large scale slumps and slides involving many beds, including interbedded shales and other lithologies, are known (Potter and Pettijohn, 1963, p. 155). The effects vary from a chaotic mixture of the blocks of the more resistant beds caught up in a matrix formed from the least competent materials to a well-defined décollement above which the strata are distorted, folded or brecciated. Where the movement involved interbedded but unconsolidated sand and clay, the latter, being tenacious, yields fragments which become enveloped in a sand matrix – the unconsolidated sand yielding by grain flow. The shale fragments may be bent or otherwise distorted. If the muds are hydroplastic then both sand and mud flow and a streaked "migmatitic" mixture results. If there is a great deal of clay and relatively little sand or silt, the strata may retain integrity and be thrown into a series of disharmonic folds above a glide surface.

Fig. 4-20. Convolute bedding at top of a fine-grained turbidite bed: cross section at bottom shows convoluted bed following lamination (see Fig. 4-10) whereas top is a view of convoluted upper surface of the same bed. Paleocene flysch. San Telmo Beach, near Zumaja, Vascongadas Province, Spain

Various types of penecontemporaneous slump have been described by Kuenen (1949) and Gregory (1969).

The circumstances which initiate slumps are many and varied. Apparently slides, once initiated, can move over a very flat surface, even a level area, and can, if they have enough momentum, move up slope. Subaerial slides have been reputed to ride on a "cushion of air."

Convolute bedding, also known as convolute lamination, slip bedding, intrastratal contortions and crinkled bedding, is one of the most difficult structures to define or explain (Fig. 4-20). Rich's expression, "intrastratal contortions," is perhaps the best as it emphasizes the intrastratal nature, i.e. the observation that these folds or convolutions affect the laminations *within* a bed but *not* the bed itself.

Convolute bedding seems to characterize relatively thin-beds – 2 to 25 cm thick – of coarse silt or fine sand. Within such a bed – either siliceous or carbonate – one observes a complex set of folds. The individual laminations are continuous and traceable from fold to fold, although micro-unconformities occur within the structure. The synclines tend to be broad and open; the anticlines tight and peaked. These structures tend to die out toward the base of the bed, and commonly toward the top of the bed also. In some beds the anticlines appear truncated as if eroded.

Clearly the distorted laminations are not involved in folds of the ordinary kind, that is, they do not record lateral compression or telescoping as is true of slump folds nor are they elongated as are ordinary folds. The bedding plane section shows them to be a series of sharp domes and basins. These observations suggest that they are due to a complex system of vertical motions.

All evidence – confinement to very fine sand or coarse silt, restriction to a single bed, symmetry indicating vertical transfer of material – indicates some kind of internal readjustment of material in a quick or near-quick condition. Various theories have been put forward (Potter and Pettijohn, 1963, p. 152; Davies, 1965, p. 308). None are wholly satisfactory.

Under some conditions sands become "quick," that is, have such a loose structure and contain so much water that they are capable of injection into fissures to form *sandstone dikes*, or to be injected along bedding planes to form *sandstone sills*. Although some dikes are simple fillings, grain by grain, of widened joints from above, most bear evidence of instantaneous, forcible injection inasmuch as many follow the bedding for a short distance or even pass into sills.

Sandstone dikes and the problems of the origin have been reviewed elsewhere (Potter and Pettijohn, 1963, p. 162). They vary in size from a centimeter or two in width to several meters. They may be injected after consolidation or they may be injected early. – prior to compaction. The former are sharp-edged, straight-walled; the latter are contorted.

Biogenic Structures

As early as 1850, James Hall commented on tracks and trails in the Silurian sandstones of New York State but for many years afterward biogenic structures were regarded by nearly all geologists as little more than exotic curiosities. Although their study began in ancient sandstones, modern sediment study early played an important role in the better understanding of biogenic structures with the Germans being foremost in the investigation of these structures. Important early progress was made by Walther in the late 19th century at a marine station on the Bay of Naples and in the early 20th century by Rudolph Richter in the North Sea region at Senckenberg-am-Meer. Other no-

Fig. 4-21. Bedding of siltstone and shale disrupted by burrowing. Tradewater Formation (Pennsylvanian) near Dawson Springs, Hopkins County, Kentucky U.S.A.

ted pre-World War II German contributors were Abel, Krejci-Graf, and Häntzschel. But it was only in the early fifties that the interpretative value of biogenic structures began to be more fully exploited by Seilacher and others. Today, interest in biogenic structures is greater than ever. The study of biogenic structures is called *ichnology*. As might be expected, much of the literature on this subject is in German.

Terrigenous silts and sands are rarely rich in fossil remains but they can contain good records of animal activity called *trace fossils*, *ichnofossils*, or *Lebensspuren*. Trace fossils are in fact best preserved in silts and sands, because they require textural contrast to be evident – burrowers, crawlers, and creepers in silt and sand-free muds leave little obvious evidence of their existence other than the disruption of lamination and distinctive textures (Fig. 4-21). Trace fossils may be found on the top of the bed, within it, on its base, or outside of it. It should be kept in mind, however, that trails on the bottom of a bed can be made by planned mining as well as by free grazing with later infilling

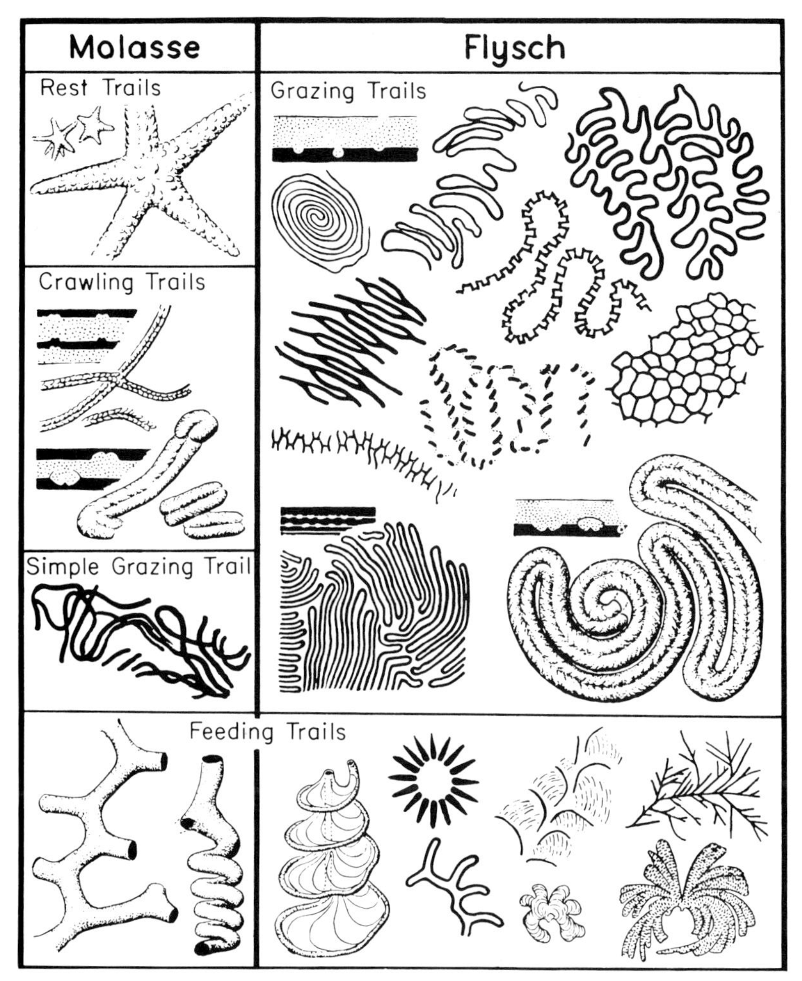

Fig. 4-22. Five functional classes of trace fossils and their suggested occurrence in flysch and molasse deposits according to Seilacher (1955, Fig. 2)

of sand. Seilacher (1964) recognized five functional classes of trace fossils (Fig. 4-22).

What general statements can we make about trace fossils? According to Seilacher and others, trace fossils have a long time span but a restricted facies distribution, are practically never reworked, and are, unlike body fossils, greatly enhanced by diagenesis so that like alcoholic beverages they improve with age.

Table 4-4 gives a functional classification. Seilacher (1964, Fig. 7) recognized three broad facies each defined by different proportions of the five trace fossil types shown in Table 4-4; one facies *Nereites* is associated with flysch deposition and the other two *Zoophycos* and *Cruziana* with shelf deposition. Fig. 4-23 is typical of a grazing trail in flysch, Fig. 4-21 represents an intensively burrowed 40 cm zone where molasse deposition was at a near standstill and Fig. 4-24 shows *Taonurus*, a feeding structure.

Table 4-4. *Functional classification of ichnofossils (modified from Seilacher, 1953, Fig. 6)*

CRAWLING TRAILS (Kriechspuren; *Repichnia*)
Trails or burrows produced by mobile benthonic organisms while moving.

FEEDING STRUCTURES (Fressbauten; *Fodincha*)
Burrows made by semi-sessile bottom feeders; structure radiates from a place of origin.

GRAZING TRAILS (Weidespuren; *Pasichnia*)
Sinuous trails or burrows of mobile mud-eating organisms, "grazing trails" at or below the interface.

RESIDENCE STRUCTURES (Wohnbauten; *Domichnia*)
Permanent shelters made by either mobile or semi-attached organisms.

RESTING MARKS (Ruhespuren; *Cubichnia*)
Shallow resting tracks made by mobile animals while resting.

Fig. 4-23. Systematic feeding trail on underside of a flysch bed. Pietraforte Formation (Cretaceous). Near Florence, Tuscany, Italy

Of what use are trace fossils? Their narrow facies distribution may enable one to zone a shelf as it passes into deeper water (Farrow, 1966, Fig. 11), help establish progressive changes in water depth in vertical profiles and thus sharpen our perception of tectonic events (Seilacher, 1963a), or perhaps help pinpoint a strand line (Weimer and Hoyt, 1964). They also give good information on relative rates of sedimentation, slow deposition exposing the interface to intensive attack by burrowing organisms forming a bioturbate in which all laminations are destroyed. Faecal pellets may be especially abundant in such zones.

In contrast, rapid sedimentation forms well-laminated sands. In the absence of other structures, Lebensspuren can be used to determine top and bottom of a sequence. A very minor use is their contribution to paleocurrent analysis.

To obtain maximum information from trace fossils, one should systematically make comparison between different known environments and *map* trace fossil distribution in the field.

Fig. 4-24. Systematic feeding burrow (*Taonurus*) in siltstone of Borden Formation (Mississippian)

Plant impressions and rootlets have value for identification of fresh and brackish water as opposed to marine environments. They are especially helpful if their occurrence is correlated with other characters, such as color, the rootlets being associated with red sediments, the marine fossils with green sands, etc.

Chemical Structures

Sedimentary structures of secondary origin are the result of both precipitation and solution. By and large, such structures have received but little attention since World War II. In sandstones, one finds structures due to solution and structures due to precipitation. Stylolites are an example of the first class; crystal aggregates such as sand crystals and barite rosettes are peculiar to sandstones and belong to the second class. Concretions, especially carbonate concretions, characterize both sands and silts.

Carbonate is the most common material that is precipitated to form chemical structures. Calcium carbonate may produce euhedral *sand crystals*. The luster mottling of some sandstones reflects such poikiloblastic calcite. Barite *rosettes* are of similar origin and character.

The *concretions* of sandstones are commonly larger than those of silts and clays, the greater permeability of the sand permitting greater transport of material in solution (Fig. 4-25). Laminations may pass undisturbed through

Fig. 4-25. Large concretions weathered out of crossbedded sandstone, "Rock City", southwest of Minneapolis, Ottawa County, Kansas, U.S.A. Photograph by Alvin Hornbaker, Kansas Geological Survey

the concretions. Such concretions may form quite early so that they may be the principal component of some intraformational conglomerates. Iron carbonate concretions may replace fine sands and silts; they are normally spherical and ellipsoidal, being flattened parallel to bedding, and may or may not have a nucleus. They vary in size from a few centimeters to bodies as much as 3 meters in diameter. Preferred shapes and orientation have been noted and may well reflect the anisotropic permeability of the host. Voidal concretions are hollow, variously shaped bodies with a hard rim of iron oxide or limonite. Extended tube-like shapes are common. These form above the water table principally in friable sands.

Although it has but rarely been done, systematic mapping of concretionary facies should be most rewarding primarily because it would show relationship to associated facies thus facilitating their integration with the entire section.

Replacement features such as chert and phosphate nodules – irregular tuberous bodies of matter unlike the host – are rare in sandstone.

Fig. 4-26. Stylolites in sandstone, Alleghany Formation (Pennsylvanian), Garret County, Maryland. Note two stylolite seams; core has parted or separated along the upper stylolitic surface, photograph by H. P. Eugster

Stylolites resulting from later pressure solution (Fig. 4-26) are rather common in clean sandstones but become less abundant as clay content increases (Heald, 1955). They are best observed in cores.

Color-banding is a rhythmic precipitation of iron-oxide in thin, closely spaced, generally curved layers. It closely mimics bedding laminations for which it may be mistaken.

Obtaining Maximum Value from Sedimentary Structures

Maximum information is usually obtained by relating the kind and abundance of sedimentary structures to position within the sand accumulation – be it a sand bank in a tidal estuary, a beach or a single bed. Is a particular structure more common at the base, at the top, or near the margins of a sand? What are the associations of structures? How do such associations correlate with bedding thickness, grain size, and body fossils? What is the vertical sequence of structures? How many bedding facies does a sand body have? *Field mapping* will nearly always show a systematic distribution of kinds, abundance and magnitude of sedimentary structures in all sands so that one can speak of the cross-bedded, ripple-bedded or horizontally-bedded facies of a sand body. Commonly, the relationship of such facies to each other is the key to the origin of the sand. Why? Because a sand deposit is the integrated response to a unified, over-all process such that the whole is the ordered sum of its parts. Only by systematic mapping, can the geologist effectively see this unified response – defined by the different bedding facies – and so best identify the ancient environment.

Because current structures are not as such restricted to single environments – practically every such structure can be found wherever sand is transported in volume – so it is very rare indeed that one can deduce a specific environment from the *single* occurrence of a particular current structure. Hence attempts to relate specific structures to a particular environment such as fore-beach, fluvial point-bar, etc. have to our knowledge rarely if ever been successful.

To facilitate recording field observations, several schedules have been proposed (Bouma, 1962; Gubler and others, 1966, p. 240). If the investigator has some preliminary knowledge of the sandstones with which he will be working, he may want to design a specialized notebook form, or even better, design a special punch card for field use so as to directly translate field data to computer tape for subsequent statistical analyses. Punch card recording of data seems especially suitable for studies of core materials where one might wish to zone a sandstone body and relate the resultant zones to reservoir characteristics.

An important additional feature that should never be overlooked is the use of current structures for determining paleocurrent systems.

Directional Structures: Use and Interpretation

Sedimentary structures which provide information on current direction are called *directional structures.* Henry Sorby in England first utilized crossbedding to determine current direction in 1853 and Ruedemann made the first paleocurrent map utilizing linear fossil debris and "mud furrows" in the Utica shale of New York in 1897. By *paleocurrent* is meant the current system at time of deposition. In spite of these and a few other early efforts by Hyde, Rubey and Bass, and Reiche in America and Brinkman, Forche, and Bausch van Bertsbergh in Europe, paleocurrent mapping did not receive general acceptance until the early nineteen fifties, when it was suddenly rediscovered and became popular so that today its use is commonplace.

For a directional structure to be useful, it must be easy to measure and widespread. It should also correlate with the direction of the principal current. Commonly-used directional structures are summarized in Table 4-5. The vast majority of maps have been made with crossbedding, in either fluvial-deltaic,

Table 4-5. *Common directional structures: measurement and occurrence*

STRUCTURE	MEASUREMENT	OCCURRENCE
Crossbedding	Trough axis and maximum dip direction of foreset are direction of current flow. One reading per bed. The most widely measured directional structure.	Present in almost all traction-transported sands. Beds thicker than 30 cm present in all but turbidite sands. Rarely solitary.
Flutes and grooves	Long dimension is parallel to current; blunt end of flute points up-current. Measure trend on each bed. Second most commonly measured structure.	Only abundant in turbidite sands but present in all except eolianites.
Ripple mark	Direction of steep slope is direction of current in asymmetrical ripples; in symmetrical ripples strike of crest is strike of shoreline.	Found everywhere but perhaps most abundant and most varied types occur near the strand line.
Parting lineation	Trend of parting parallels flow; measure trend of each set.	In all environments but rarely systematically measured because of infrequent occurrence.

eolian, and shelf sands or with sole marks in turbidites. Ripple mark has played a very minor role.

What benefits can result from paleocurrent measurement? Knowledge of current direction can (1) predict direction of elongation of a sandstone body, (2) outline the paleocurrent system of a basin and thus contribute to better understanding of the arrangement of its sedimentary fill, (3) help to determine if a structural feature in a basin was active or not during deposition, (4) help locate source regions that lie beyond a basin margin, (5) aid paleoecology by establishing the direction of supply of nutrient carrying currents, (6) contribute to some types of stratigraphic correlation problems, and (7) can exercise the dominant control, through grain fabric, on bulk geophysical properties such as electrical and thermal conductivity. As we have already seen, current systems determine the distribution of mineral associations in a basin and some size patterns as well. For effort expended, measurement of directional structures is probably more rewarding to the field geologist than anything else. Crimes (1970) gives a good example of the role of sedimentary structures, particularly directional structures, in his facies analysis of the Cambrian sediments in the Caledonian trough in Wales.

Directional structures are usually measured "as they come" in an outcrop as one moves either vertically up section or horizontally along it as in a road cut or along a creek bed. Usually half a dozen or so measurements of crossbedding per outcrop will suffice; but if orientation is very variable, as in some tidally influenced sands, twenty or more readings may be necessary to establish a pattern of bimodality. Vertical profiles of current direction can be informative in assessing current stability; pronounced back and forth alternation of ripple mark and crossbedding suggests a strong tidal influence. In order to specify the convergence, divergence, or uniformity of paleocurrents most effectively, it is generally best to have more outcrops with fewer readings than a smaller number of outcrops each with numerous readings. It is best to compute and plot an average for each outcrop. The vector mean is a satisfactory average (Steinmetz, 1962, and Jones, 1967) for all except markedly bimodal distributions. For the latter, either midpoints of the two modes should be plotted or special computations are needed (Jones and James, 1969).

How well do directional structures correlate with the direction of the local current? In a general way, one can say that the magnitude (volume or thick-

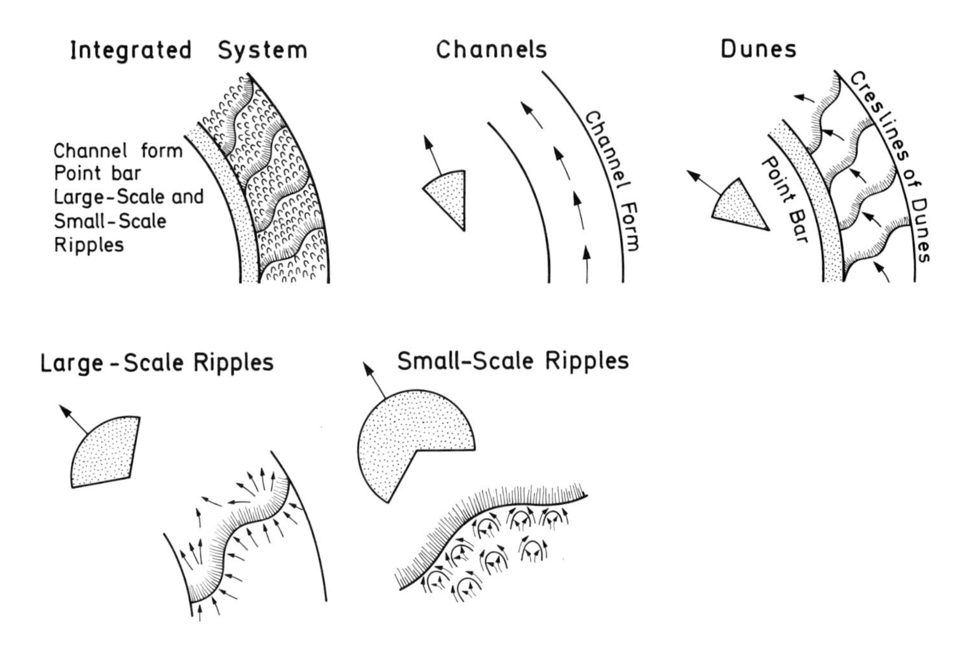

Fig. 4-27. Hierarchical order of flow systems: channel, dunes, and large and small scale ripples (Modified from Allen, 1966, Fig. 22)

ness) of a current structure is a reflection of the magnitude and importance of the current. For example, most ripple mark in alluvial or tidal sands develops in response to weak currents, ones usually associated with secondary flow induced by microrelief along the bottom whereas crossbeds develop from dunes migrating subparallel to the principal current of the channel. Under such conditions, crossbedding readings will be obviously superior to ripple mark. Allen (1966) first explicitly illustrated the hierarchical nature of flow systems and associated bed forms for alluvial sands (Fig. 4-27). In essence, the hierarchy is based on the variability or variance of the directional structures, channels having the least and small scale ripples the most. Mean direction is essentially unaffected, however. Thus, as a rule, careful measurement of many different directional structures will in general yield an integrated movement plan (Fig. 4-28). Grain fabric commonly has good correlation with directional structures, although a few exceptions exist.

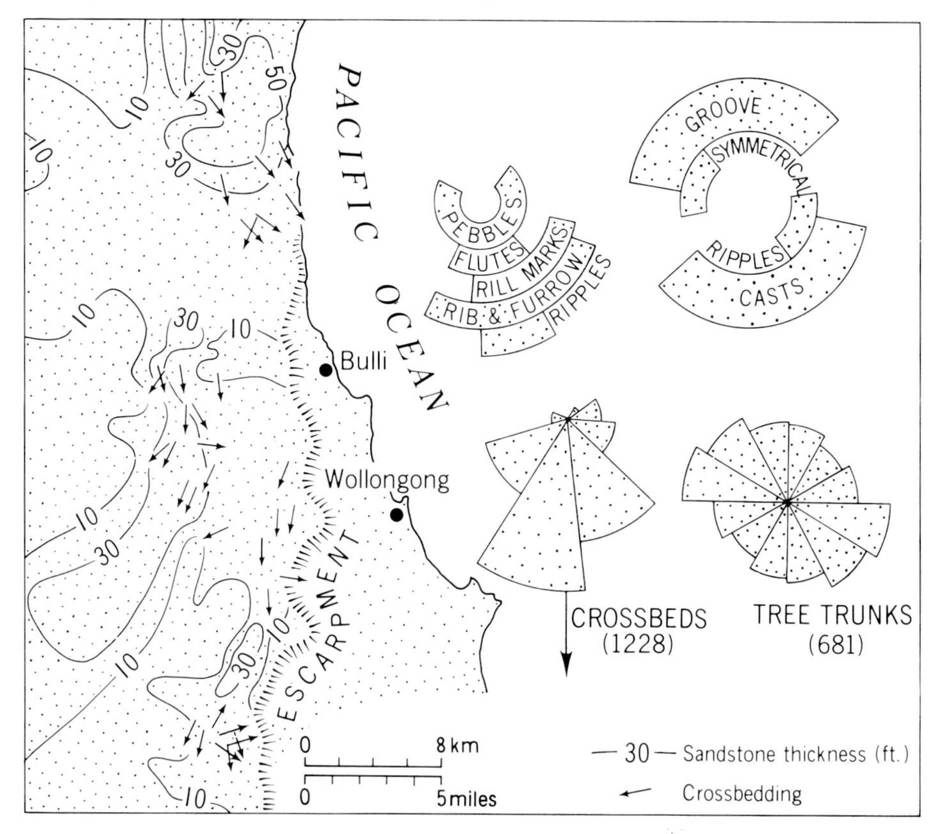

Fig. 4-28. Fluvial sandstone in Triassic coal measures and directional structures. Note similarity of general flow direction and correlation with sand body trend (Modified from Diessal and others, 1967, Figs. 3 and 5)

Figs. 4-29 and 4-30 illustrate but two of the many paleocurrent maps that have been made. One can also smooth such maps either by hand or with computer-derived trend surfaces (Fox, 1967). Interpretation of paleocurrent maps is most reliable when integrated with other aspects of the geology: petrographic provinces, facies distribution, thickness, directions of on- and off-lap, paleochannel systems, etc. The internal consistency that usually results from such integration generally resolves the question whether the regional current system reflects paleoslope or not. By *paleoslope* is meant the down-dip slope – into either deeper water (marine sands) or lower elevation above sea level (fluvial sands) – of a former depositional basin.

Paleocurrent patterns are of three types: unimodal, bimodal, and polymodal or random (Fig. 4-31). By far the most commonly reported – perhaps ninety percent of all studies – is the unimodal pattern which is characteristic of fluvial and deltaic and most turbidite sands. Many ancient sandstones claimed to be eolian also have unimodal patterns. Bipolar current roses are characteristic of tidally influenced sands, be they estuarine or wholly marine. Such sands may also have either on- or off-shore (or shoal) unimodal current

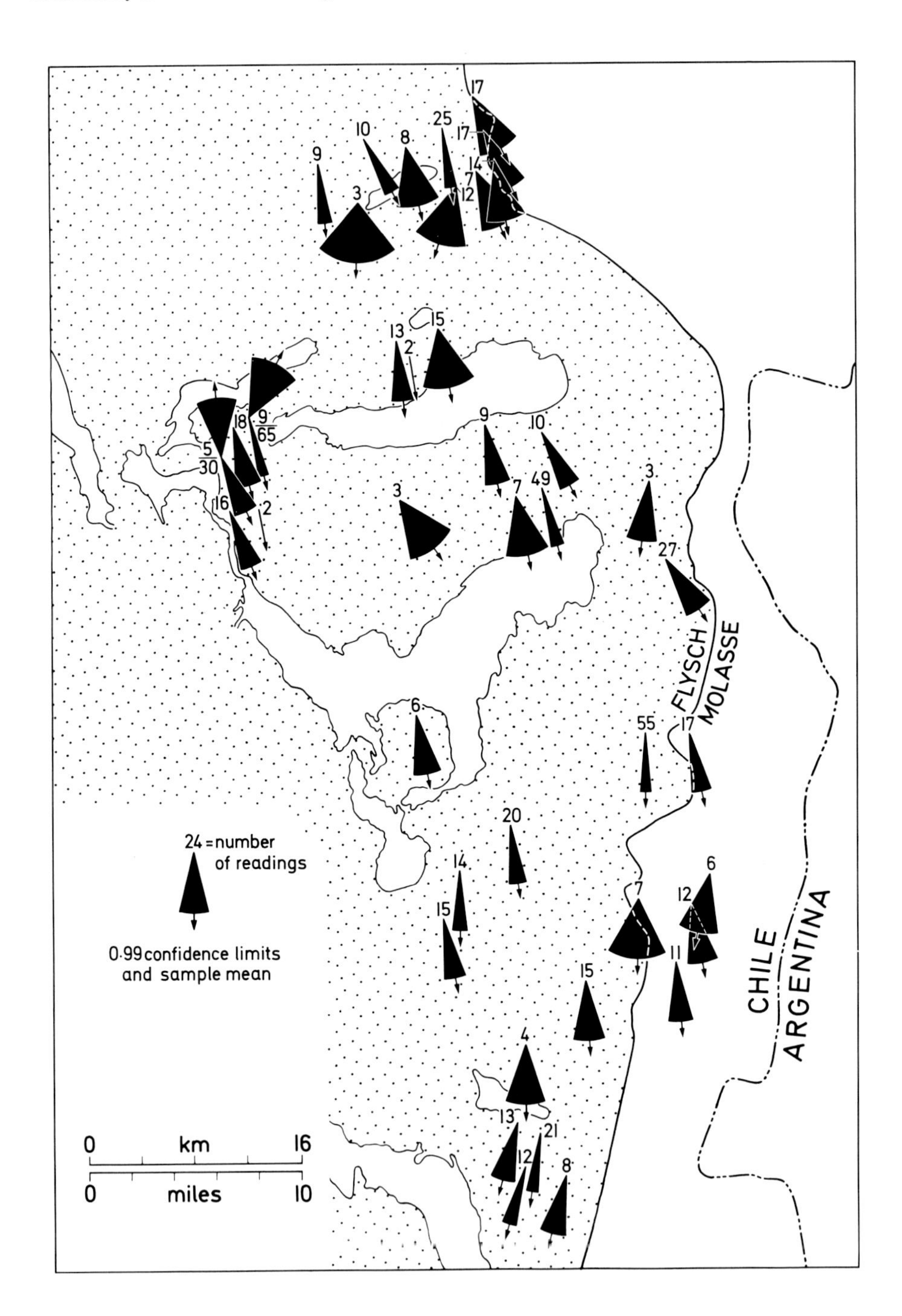

Fig. 4-29. Paleocurrent pattern of 641 flutes in flysch sandstone. Cerro Toro and Tres Pasas Formations (Cretaceous) in southern Chile (Modified from Scott, 1966, Fig. 6)

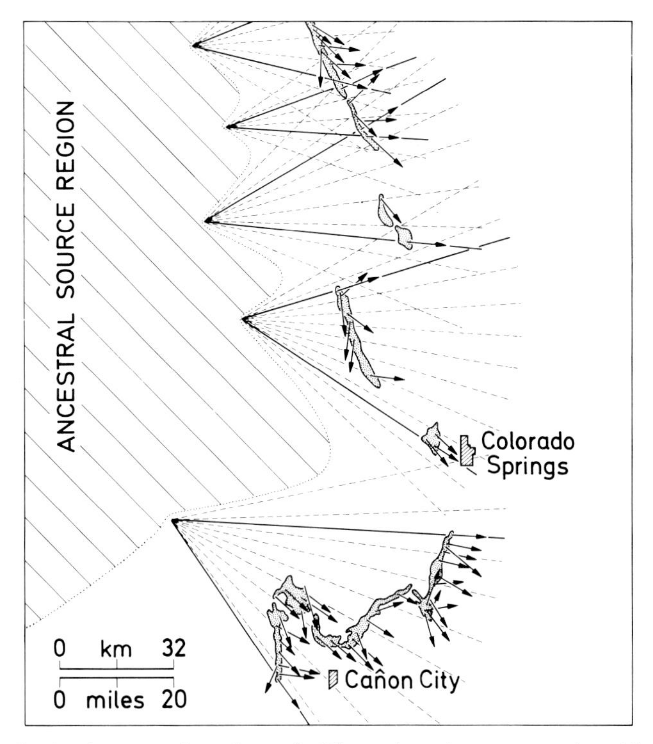

Fig. 4-30. Imaginative interpretation of crossbedding orientation suggests deposition from point sources of alluvial fans at base of mountain front. Fountain Sandstone (Pennsylvanian) of central Colorado, U.S.A. (Modified from Howard, 1966, Fig. 3)

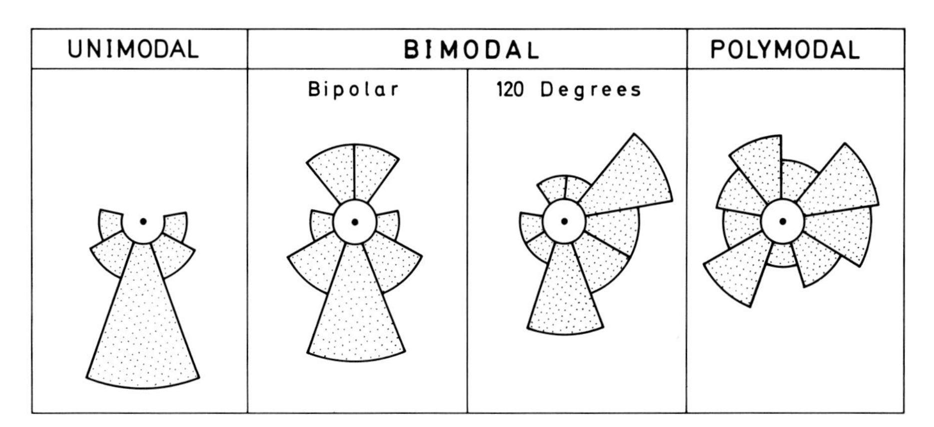

Fig. 4-31. Paleocurrent windroses and environment (Modified from Selley, 1968, Fig. 2)

roses depending upon the relative strengths of ebb and flood tides. Bimodal-perpendicular patterns have been reported but are exceedingly rare. Polymodal distributions may represent mixtures of different current systems or simply the random pattern of a very variable current system. A number of marine shelf sands have essentially random current patterns.

Table 4-6. *Depositional environments, characteristic directional structure and dispersal pattern*

ENVIRONMENT	DIRECTIONAL STRUCTURES	DISPERSAL PATTERN
Eolian	Crossbedding	Independent of paleoslope. Relationship of crossbedding orientation in continental eolianites to prevailing winds not well established.
Alluvial	Crossbedding	Unimodal and down slope; upcurrent direction is direction of source, if not too distant.
Beach	Crossbedding	Bimodal and some unimodal. Variable orientation with respect to strand line.
Delta	Crossbedding	Mostly unimodal and down slope. Very minor bimodality resulting from tidal action.
Estuary	Crossbedding	Unimodal and bimodal with modes parallel to estuary axis and hence at right angles to strand line.
Shallow Shelf	Crossbedding	Most variable: some unimodal but more commonly bimodal and random patterns. Care needed to infer paleoscope.
Turbidites	Sole and ripple marks and some slide structures	Unimodal. Commonly considered down paleoslope and formed by gravity controlled turbidity currents.

When considering different current-rose patterns, one should keep in mind that many observations are needed to distinguish, for example, a trimodal current rose from a random one. A simple test to determine the reality of any particular pattern is to randomly discard half of the data. If the two new distributions are similar to the original one, the current pattern can, as a first approximation, be considered an adequate sample of the parent population. Potter and Pettijohn (1963, p. 86–89) review the relation between variance of crossbedding and depositional environment as does Selley (1968), the latter suggesting a number of different paleocurrent models. Table 4-6 summarizes presently known relations between directional structures, environment and dispersal pattern.

References

Allen, J. R. L.: The classification of cross-stratified units with notes on their origin. Sedimentology **2**, 93–114 (1963).

— On bed forms and paleocurrents. Sedimentology, **6**, 153–190 (1966).

— Current ripples, 433 p. Amsterdam: North Holland Publ. Co. 1969.

Bokman, J. W.: Terminology for stratification. Geol. Soc. America Bull. **67**, 125–126 (1956).

— Suggested use of bed-thickness measurements in stratigraphic descriptions. Jour. Sed. Petrology **27**, 333–335 (1957).

Bouma, A. H.: Sedimentology of some flysch deposits, 169 p. Amsterdam: Elsevier Publishing Co. 1962.

Bucher, W. H.: On ripples and related sedimentary surface forms and their paleogeographic interpretations. Am. Jour. Sci. (ser. 4), **47**, 149–210, 241–269 (1919).

Campbell, C. V.: Lamina, laminaset, bed, and bedset. Sedimentology, **8**, 7–26 (1967).

Cannon, J. L.: Outcrop examination and interpretation of paleocurrent patterns of the Blackleaf formation near Great Falls, Montana. In: 17th Annual Field Conference, Billings Geol. Soc. **1966**, 71–111.

Conybeare, C. E. B., Crook, K. A. W.: Manual of sedimentary structures. Australian Dept. Natl. Development, Bur. Min. Res., Geol., and Geophysics, Bull. **102**, 327 p. (1968).

Crimes, T. P.: A facies analysis of the Cambrian of Wales. Paleogeog., Palaeoclimatol., and Palaeoecol. **7**, 113–170 (1970).

Davies, H. G.: Convolute lamination and other structures from the Lower Coal Measures of Yorkshire. Sedimentology **5**, 305–326 (1965).

Davis, R. A., Jr.: Algal stromatolites composed of quartz sandstone: Jour. Sed. Petrology **38**, 953–955 (1968).

Dean, W. E., Jr., Anderson, R. Y.: Correlation of turbidite strata in the Pennsylvanian Haymond Formation, Marathon Region, Texas. Jour. Geology **75**, 59–75 (1967).

Diessel, C. F. K., Driver, R. C., Moelle, K. H. R.: Some geological investigations into a Triassic river system in the roof strata of the Bulli Seam, southern Coalfield N.S.W.: Australian Inst. Min. Metal. Proc., No. **221**, 19–37 (1967).

Dzulynski, S., Sanders, J. E.: Current marks on firm mud bottoms. Connecticut Acad. Arts. and Sci., Trans. **42**, 57–96 (1962).

— Walton, E. K.: Sedimentary features of flysch and greywackes. Developments in Sedimentology No. 7, 300 p. Amsterdam: Elsevier (1965).

Enos, Paul: Anatomy of a flysch. Jour. Sed. Petrology **39**, 680–723 (1969).

Farrow, G. F.: Bathymetric zonation of Jurassic trace fossils from the coast of Yorkshire, England: Palaeogeog., Palaeoclimatol., and Paleoecol. **2**, 103–151 (1966).

Fox, W. T.: Fortran IV program for vector trend analysis of directional data. Kansas Geol. Survey, Computer Contr. **11**, 36 p. (1967).

Gregory, M. R.: Sedimentary features and penecontemporaneous slumping in the Waitemata Group Whangaparaoa Peninsula, North Auckland, New Zealand. New Zealand Jour. Geol. Geophysics **12**, 248–282 (1969).

Griffiths, J. C.: Measurement of the properties of sediments. Jour. Geology **69**, 487–498 (1961).

Grumbt, Eberhard: Beziehungen zwischen Korngröbe, Schichtung, Materialbestand und anderen sedimentologischen Merkmalen in feinklastischen Sedimenten. Berlin, Geologie **18**, 151–167 (1969).

Gubler, Y., Bugnicourt, D., Faber, S., Kubler, B., Nyssen, R.: Essai de nomenclature et caractérisation des principales structures sédimentaires, 291 p. Paris: Editions Technip, 1966.

Hall, James: Remarks upon casts of mud furrows, wave lines, and other markings upon rocks of the New York System. Assoc. Amer. Geol. Rept. **1843**, 422–432.

— On the trails and tracks in the sandstones of the Clinton Group of New York. Am. Assoc. Adv. Sci. Proc. **2**, 256–260 (1850).

Hamblin, W. K.: Internal structures of "homogeneous" sandstones: Kansas Geol. Survey, Bull. **175**, 569–582 (1965).

Harms, J. C., MacKenzie, D. B., McCubbin, D. G.: Stratification in modern sands of the Red River, Louisiana. Jour. Geology **71**, 566–580 (1963).

Heald, M. T.: Stylolites in sandstone. Jour. Geology **63**, 101–114 (1955).

Howard, J. D.: Patterns of sediment dispersal in the Fountain Formation of Colorado. Mountain Geologist **3**, 147–153 (1966).

— Lohrengel, C. F., II.: Large nontectonic deformational structures from Upper Cretaceous rocks of Utah. Jour. Sed. Petrology **39**, 1032–1039 (1969).

Jones, G. P.: Deformed cross-stratification in Cretaceous Bima sandstone, Nigeria. Jour. Sed. Petrology **32**, 231–239 (1962).

Jones, T. A.: Estimation and testing procedures for circular normally distributed data: Office Naval Research, ONR Task No. 388-078, Contract Nonr-1228 (36), 61 p., (1967).

— James, W. R.: Analysis of bimodal orientation data. Mathematical Geology **1**, 129–135 (1969).

Jopling, A. V.: Hydraulic factors controlling the shape of laminae in laboratory deltas. Jour. Sed. Petrology **35**, 777–791 (1965).

Jopling, A. V., Walker, R. G.: Morphology and origin of ripple-drift lamination, with examples from the Pleistocene of Massachusetts: Jour. Sed. Petrology **38**, 971–984 (1968).

Kelley, V. C.: Thickness of strata. Jour. Sed. Petrology **26**, 289–300 (1956).

Khabakov, A. V., Ed.: Atlas tekstur: struktur osadochyhk gornykh porod (An atlas of textures and structures of sedimentary rocks, pt. 1, clastic and argillaceous rocks), 578 p. Moscow: VSEGEI, 1962.

Kindle, E. M.: Recent and fossil ripple mark. Canada Geol. Survey Mus. Bull. **25**, 1–56 (1917).

Kuenen, Ph. H.: Slumping in the Carboniferous rocks of Pembrokeshire: Geol. Soc. London, Quart. Jour. **104**, 365–385 (1949).

— Graded bedding, with observations on Lower Paleozoic rocks of Britain. Koninkl. Nederlandse Akad. Wetensch. Afd. Nat. Verh., 1st Ser. **20**, 1–47 (1953).

— Sole markings of graded graywacke beds. Jour. Geology **65**, 231–258 (1957).

— Experiments in geology. Glasgow Geol. Soc. Trans. **23**, 1–28 (1958).

McBride, E. G., Yeakel, L. S.: Relationship between parting lineation and rock fabric. Jour. Sed. Petrology **33**, 779–782 (1963).

McKee, E. D., Wier, G. W.: Terminology of stratification and cross-stratification: Geol. Soc. America Bull. **64**, 381–390 (1953).

Middleton, G. V. (Ed.): Primary sedimentary structures and their hydrodynamic interpretation. Soc. Econ. Paleon. Mineral. Spec. Public. No. **12**, 265 p. (1965).

Otto, G. H.: The sedimentation unit and its use in field sampling. Jour. Geology **46**, 569–581 (1938).

Pettijohn, F. J., Potter, P. E.: Atlas and glossary of primary sedimentary structures, 117 pl., 370 p. New York: Springer 1964.

Potter, P. E., Pettijohn, F. J.: Paleocurrents and basin analysis, 296 p. Berlin-Göttingen-Heidelberg: Springer 1963.

Ramsey, J. G.: The effects of folding upon the orientation of sedimentation structures. Jour. Geology **69**, 84–100 (1961).

Reineck, H.-E., Wunderlich, F.: Classification and origin of flaser and lenticular bedding. Sedimentology **11**, 99–104 (1968).

Ruedemann, R.: Evidence of current action in the Ordovician of New York. Amer. Geologist **19**, 367–391 (1897).

Scott, K. M.: Sedimentology and dispersal pattern of a Cretaceous flysch sequence, Patagonian Andes, Southern Chile. Am. Assoc. Petroleum Geologists Bull. **50**, 72–107 (1966).

Seilacher, Adolf: Die geologische Bedeutung fossiler Lebensspuren: Deutsche geol. Gesell. Zeitschr. **105**, 214–227 (1953).

— Studien zur Palichnologie I. Über die Methoden der Palichnologie. Neues Jahrb. Geologie u. Paläontologie Abh. **96**, 421–452 (1955).

— Kaledonischer Unterbau der Irakiden. Neues Jahrb. Geologie u. Paläontologie Mh. **1963**a, 527–542.

— Umlagerung und Rolltransport von Cephalopoden-Gehäusen: Neues Jahrb. Geologie u. Paläontologie **11**, 593–615 (1963b).

— Biogenic sedimentary structures. In: Imbrie, John and Newell, Norman, Eds. Approaches to paleoecology, 296–316. New York: John Wiley and Sons, Inc. 1964.

Selley, R. C.: A classification of paleocurrent models. Jour Geology **76**, 99–110 (1968).

Shrock, R. R.: Sequences in layered rocks, 507 p. New York: McGraw-Hill 1948.

Sorauf, J. E.: Flow rolls of Upper Devonian rocks of south-central New York State. Jour. Sed. Petrology **35**, 553–563 (1965).

Sorby, H. C.: On the oscillation of the current drifting sandstone beds of the southeast of Northumberland, and on their general direction in the coal field in the neighborhood of Edinburgh. Repts. Proc. Geol. Polytechnic Soc. of the West Riding of Yorkshire **1853**, 225–231.

— On the structures produced by the currents present during the deposition of stratified rocks: Geologist **2**, 137–147 (1859).

— On the application of quantitative methods to the study of the structure and history of rocks. Geol. Soc. London Quart. Jour. **64**, 171–232 (1908).

Stanley, D. J.: Vertical petrographic variability in Annot sandstone turbidites: some preliminary observations and generalizations. Jour. Sed. Petrology **33**, 783–788 (1963).

Steinmetz, Richard: Analysis of vectorial data. Jour. Sed. Petrology **32**, 801–812 (1962).

Walker, R. G.: Distinctive types of ripple-drift cross-lamination. Sedimentology **2**, 173–188 (1963).
— Geometrical analysis of ripple-drift cross-lamination: Canadian Jour. Earth Sci. **6**, 383–392 (1969).
Weimer, R. J., Hoyt, J. H.: Burrows of *Callianassa major* Say, geologic indicators of littoral and shallow neritic environments. Jour. Paleontology **38**, 761–767 (1964).

Annotated References

General

Khabakov, A. V., Ed.: Atlas tekstur i struktur osadochyhk gornykh porod (An atlas of textures and structures of sedimentary rocks, pt. 1, Clastic and argillaceous rocks), 578 p. Moscow: VSEGEI 1962 (Russian, with French translation of plate captions).

A well-illustrated book with 268 plates; emphasis on textures and structures as seen in the outcrop or under the microscope, primarily in the older rocks. Some illustrations from modern sediments.

Pettijohn, F. J., Potter, P. E.: Atlas and glossary of primary sedimentary structures, 370 p. Berlin-Göttingen-Heidelberg-New York: Springer 1964.

Primarily a collection of 198 outstanding photographs of sedimentary structures, mainly from older rocks, with a short introductory text and an extended glossary. All subject matter including glossary and plate captions in English, German, French, and Spanish.

Shrock, R. R.: Sequence in layered rocks, 507 p. New York: McGraw-Hill 1948.

Special emphasis on the utility of primary structures to ascertain stratigraphic order. Includes a long section on structures in layered volcanic rocks. Mainly English references.

Conybeare, C. E. B., Crook, K. A. W.: Manual of sedimentary structures. Australian Dept. Nat. Development, Bur. Min. Res., Geol. and Geophysics, Bull. **102**, 327 (1968).

Contains ten pages on classification and 52 pages on description and interpretation of sedimentary structures. The main feature are the 108 plates. Bibliography of about 250 references.

Gubler, Y., Bugnicourt, D., Faber, J., Kubler, B., Nyssen, R.: Essai de nomenclature et caractérisation des principales structures sédimentaires, 291 p. Paris: Editions Technip. 1966.

A well-illustrated, very systematic treatment and well-referenced work dealing with gross geometry of sediments, their stratification and structures. Structures are grouped into those which are directional, of organic origin, due to escape of gas, of climatic origin, and diagenetic. There is a chapter on applications. One hundred and ninety five figures.

Bouma, A. H.: Methods for the study of sedimentary structures, 458 p. New York: John Wiley 1969.

A large monographic work primarily on methods applicable in the field including use of box coring for study of structures in unconsolidated subaqueous sediments and use of peel and replica techniques.

Classification

Andersen, S. A.: Om Aase og Terrasser inden for Susaa's Vandomraade og deres Vidnesbyrd om Isafsmeltningens Forlöb (The eskers and terraces in the basin of the River Susaa and their evidences of the process of the ice-waning): Danmarks geol. Undersögelse Raekke II, no. 45, 201 p. (Danish; English summ.) 1931.

Primarily a study of fluvioglacial deposits but contains one of the first organized discussions of the primary sedimentary structures in relation to direction of current flow and flow regime (See in particular, p. 173–177 and Fig. 38).

Andrée, K.: Ursachen und Arten der Schichtung. Geol. Rundschau **6**, 351–397 (1915).

One of the first papers to attempt a classification of bedding including a discussion of cyclical bedding and a classification of various types of crossbedding or diagonal bedding.

Birkenmajer, K.: Systematyka warstwawan w utworach fliszowych i podobynch (Classification of bedding in flysch and similar graded deposits). Studia Geol. Polonica **3**, 133 p. (1959) (Polish; English summary).

A major effort to classify bedding types and sedimentary structures found in flysch and similar graded deposits. Contains an elaborate scheme for symbolic representation of each class.

Botvinkina, L. N.: Morfologicheskaia klassificatsiiâ sloistosti osadochaykh porod: Akad. Nauk. S.S.S.R., Ser. Geol., no. **6**, 16–33 (1959). (Morphological classification of bedding in sedimentary rocks. Amer. Geol. Inst. translation issued 1961, p. 13–30.)

A detailed analysis of the problem of classification of bedding. Three main types, crossbedding, rippled bedding, and horizontal bedding, are recognized. Each type is subdivided in detail, and the classification schemes are presented in tabular form. The divisions are based on magnitude or scale, on relationships of contact surfaces, on shape, and on directions (that is, in same or opposed directions).

— Slosistot osadochaykh porod (Bedding of sedimentary rocks). Akad. Nauk. U.S.S.R., Trudy Geol. Inst., no. **59**, 542 p. (1962).

A monographic treatment in three parts: morphology (14 tables), environmental and genetic significance, and applications (to facies analysis, prospecting, paleoclimatology, engineering, etc.) The last chapter is devoted to methods of study. Fifteen pages of references, eleven of which are to the Russian works. The culmination of a geologist's life long interest in sedimentary structures and bedding.

— Metodicheskoe rukovodstvo po izucheniiû stoistostiv (Manual on the methods of studying bedding). Akad. Nauk. U.S.S.R. Geol. Inst., Trans. **119**, 253 p. (1965).

Methods, classification, and environmental interpretation. Some overlap with 1962 effort. English table of contents.

Botvinkina, L. N., and others: Atlas litogeneticheskikh tipov uglenosnykh otlozhenii srednego Karbona Donetskogo basseina (Atlas of lithogenic types of coal-bearing deposits of the Middle Carboniferous of the Donetz Basin). Akad. Nauk. U.S.S.R., Inst. Geol. Nauk., Moscow, **1956**, 368 p.

A collection of photographs primarily showing structures characteristic of coal measures strata. A must for the coal measure geologist.

Krejci-Graf, K.: Definition der Begriffe Marken, Spuren, Fahrten, Bauten, Hieroglyfen und Fucoiden. Senckenbergiana **4**, 19–39 (1932).

An early effort to make a comprehensive detailed classification of sedimentary structures.

Crossbedding, Ripplemark, and Sand Waves

Allen, J. R. L.: The classification of cross-stratified units with notes on their origin. Sedimentology **2**, 93–114 (1963).

A descriptive classification based on six attributes (grouping, magnitude, erosional or nonerosional bounding surfaces, shape of lower bounding surface, relation of foresets to bounding surface, and homogeneity or heterogeneity of materials). Fifteen types of crossbedding (with Greek-letter designations) are described. A three-fold genetic classification is also proposed.

Harms, J. C., MacKenzie, D. B., McCubbin, D. G.: Stratification in modern sands of the Red River, Louisiana. Jour. Geology **71**, 566–580 (1963).
A short but well-documented study of modern fluvial crossbedding, primarily large-scale trough crossbedding and small-scale "rib-and-furrow."

Jordan, G. F.: Large submarine sand waves. Science **136**, 839–847 (1962).
A short review of what is known about rhythmic submarine sand ridges or waves which are presumed to be transport forms. (The author does not discuss internal structure, which, as crossbedding, is all that would be available in geologic record).

McKee, E. D.: Dune structures: Sedimentology, Special Edition **7**, 1–69 (1966).
The details of crossbedding, principally in the White Sands of New Mexico.

— Wier, G. W.: Terminology of stratification and cross-stratification. Geol. Soc. America Bull. **64**, 381–390 (1953).
A much-cited geometrical classification of crossbedding and related terminology.

Niehoff, W.: Die Primär gerichteten Sedimentstrukturen insbesondere die Schrägschichtung im Koblenzquarzit am Mittelrhein. Geol. Rundschau, **47**, 252–321 (1958).
An extended description and analysis of large-scale crossbedding and small-scale ripple cross-lamination in the Coblenz quartzite.

Reineck, Hans-Erich: Sedimentgefüge im Bereich der südlichen Nordsee. Senckenberg. naturf. Gesell. Abh. **505**, 138 p. (1963).
A very comprehensive study of structures of a modern shallow marine shelf and tidal flat, particularly sand waves and their internal structure, crossbedding. Certainly the best study of crossbedding in the shallow marine-tidal realm yet made. Twenty illustrations, 12 plates, and 8 large maps.

Walker, R. G.: Distinctive types of ripple-drift cross-lamination. Sedimentology **2**, 173–188 (1963).
Recognizes three types of ripple-drift lamination based on internal geometry.

Wurster, Paul: Geometrie und Geologie von Kreuzschichtungs-Körpern. Geol. Rundschau **58**, 322–359 (1958).
Detailed analysis of crossbedding with special emphasis on the three-dimensional form of cross-bedded units.

Hieroglyphs and Sole Marks

Crowell, J. C.: Directional current structures from the Pre-Alpine Flysch, Switzerland. Geol. Soc. America Bull. **66**, 1351–1384 (1955).
This paper contains one of the first efforts to classify hieroglyphs and sole marks.

Dzulynski, S.: Directional structures in flysch. Studia Geol. Polonica, **12**, 136 p. (1963).
An extended treatment of the sedimentary structures in flysch well illustrated with 61 plates. Text both in Polish and English. Contains a substantial bibliography.

— Sanders, J. E.: Current marks on firm mud bottoms. Connecticut Acad. Arts and Sci. Trans. **42**, 57–96 (1962).
A short concise summary, description, classification, and genesis of the sole markings of the flysch facies. Twenty-two excellent plates.

— Walton, E. K.: Sedimentary features of flysch and greywackes, 300 p. Amsterdam: Elsevier Publishing Co. 1965.
The most comprehensive work on the sedimentary structures of the flysch facies. Excellent.

Kuenen, Ph. H.: Sole markings of graded graywacke beds. Jour. Geology **65**, 231–259 (1957).

One of the first "modern" papers dealing with sole markings and their origin.

Lanteaume, Marcel, Beaudoin, Bernard, Campredon, Robert: Figures sédimentaires du Flysch "Gres D'Annot" du synclinal de Peira-Cava, 99 p. Paris: Editions du Centre Nat. Recherche Sci. 1967.

Sixty one, high-quality plates of sole markings on sandstones in a flysch sequence from the French Maritime Alps. Contains a classification of sole marks (Fig. 5).

Vassoevich, N. B.: O nekotorykh flishevykh tekturakh (Znakakh) (On some flysch textures). Trudy Lvovs. Geol. Obsh. Univ. Ivan Franko, Geol. Ser., no. 3, 17–85 (1953) (Russian).

Contains one of the most comprehensive classifications schemes. Vassoevich coined a large number of new terms, derived from the Greek, only a few of which are generally used. The English definitions of these terms are given in Pettijohn and Potter (1964).

Biogenic Sedimentary Structures

Abel, Othenio: Vorzeitliche Lebensspuren, 644 p. Jena: Gustav Fischer 1935.

The most massive treatise on the subject written by the author of some 19 texts and monographs. Contains 530 illustrations. Emphasizes biological aspects.

Bajard, Jacques: Figures et structures sédimentaires dans la zone intertidal de la partie orientale de la Baie du Mont-Saint-Michel. Rev. Géog. Phy. et Géol. Dyn. **8**, 39–111 (1966).

One hundred and sixty informative illustrations of biogenic and hydraulic structures found on a famous intertidal flat.

Crimes, T. P., Harper, J. C., Eds.: Trace fossils. Geological Jour., Special Issue **3**, 547 p. (1970).

Thirty five articles by many of the world's experts. Many high-quality illustrations. Helpful trace fossil index.

Häntzschel, W.: Trace fossils and problematica. In: R. C. Moore, Ed. Treatise on invertebrate paleontology, Part W, Miscellanea, p. W 177–W 245. Boulder, Colo.: Geol. Soc. America and Univ. of Kansas Press 1962.

A brief but very good introduction that includes definitions of terms, classification, and nomenclature. Contains all the generic names that have been used. Well illustrated. Recognizes trace fossils, body fossils (impressions), and borings and reviews names of "fossils" of probable inorganic origin.

— Fossilum catalogus, 1: Animalia Pars 108: s'Gravenhage, Dr. W. Junk, 142 p.

The synonomy and legality of over 700 names plus large bibliography.

Lessertisseur, J.: Traces fossiles d'activité animale et leur signification paleobiologique. Soc. geol. France, Mem. **74**, 142 p. (1955).

Tracks and trails in French.

Martinsson, A.: Aspects of a Middle Cambrian thanatotope on Öland. Geol. fören Stockholm Förh. **87**, 181–230 (1965).

Contains an extended discussion of trace fossils (p. 201–226) illustrated by photographs. A good example of the interrelations between sedimentological (structures and paleocurrents), diagenetic, and biological aspects of a deposit.

Schäfer, Wilhelm: Aktuo-Paläontologie, 666 p. Frankfurt am Main: Kramer 1962.

Devoted to the modern, but diagrams and illustrations are very applicable to the ancient. A landmark study that summarizes the work of many years of the Forschungsanstalt für Meeresgeologie und Meeresbiologie in Wilhelmshaven. Many references including even a listing of films available from the institute.

Seilacher, A.: Biogenic sedimentary structures. In: Imbrie, John, Newell, Norman, Eds.: Approaches to paleoecology, p. 296–316. New York: John Wiley 1964.

The best summary in English of "trace fossils" including description, classification, origin, and environmental significance. Well illustrated by sketches; includes tables and a bibliography.

— Fossil behavior. Scientific American **217**, 72–80 (1967).

Semipopular account by Herr Lebenspuren. A good place to begin.

Directional Structures

Brinkman, R.: Über Kreuzschichtung in deutschen Buntsandsteinbecken. Göttingen Nachr., Math.-physik. Kl. IV, Fachgr. IV, No. 32, 1–12 (1933).

Probably the first modern study. Over 4,000 measurements; relates variability to type stream pattern. A classic.

Klein, George de Vries: Paleocurrent analysis in relation to modern marine sediment dispersal patterns. Am. Assoc. Petroleum Geologists Bull. **51**, 366–382 (1967).

Many references to directional studies in the modern marine realm plus a critical view of "downslope" in turbidites.

Nagahama, H., Hirokawa, O., Enda, T.: History of researches on paleocurrents in reference to sedimentary structures – with paleocurrent maps and photographs of sedimentary structures. Japan Geol. Survey Bull **19**, 1–17 (1968).

Bibliography of Japanese work plus maps and 22 superb plates.

Pelletier, B. R.: Paleocurrents in the Triassic of northeastern British Columbia. In: Middleton, G. W., Ed.: Primary sedimentary structures and their hydrodynamic interpretation. Tulsa, Soc. Econ. Paleont. Mineral., Spec. Pub. No. 12, **1965**, 233–245.

A short presentation of ripple marks and cross-bedding and how they relate to grain size in a marine shelf sequence. Paleocurrents go down paleoslope.

Potter, P. E., Pettijohn, F. J.: Paleocurrents and basin analysis, 296 p. Berlin-Göttingen-Heidelberg: Springer 1963.

Directional structures from all aspects. Over 800 references.

Scott, Kevin M.: Sedimentology and dispersal pattern of a Cretaceous flysch sequence, Patagonian Andes, southern Chile. Am. Assoc. Petroleum Geologists Bull. **50**, 72–107 (1966).

A "second stage" article on a turbidity sequence with a good discussion of the downslope (flute marks) versus across the slope (gravity slides) problem.

Part II

The Petrography of Sandstones

Utilizing mineralogy and texture, petrography is the first major step toward putting the parts of a sandstone together to form an integrated, meaningful description which is the basis for almost all subsequent inferences about a sandstone's origin and economic potential. Classification is a natural consequence of the systematic petrographic description of sandstones for, like all natural materials, sandstones differ from one another and one needs a convenient practical shorthand to quickly and conveniently label and classify them. The two petrographic chapters that follow describe major petrographic types, their occurrence, and origin. Together these three chapters form the basis for the serious study of sandstones, a work that was petrographically begun by Henry Clifton Sorby in 1877 and 1880 in his papers "The Application of the Microscope to Geology" and "On the Structure and Origin of Non-calcareous Stratified Rocks." The appendix offers guides to and gives an example of a detailed petrographic description and its interpretation.

Chapter 5. Petrographic Classification and Glossary

Introduction

There are three major types of sand: terrigenous, carbonate, and pyroclastic. *Terrigenous* sands are most abundant and are all ultimately derived from outside the basin of deposition by erosion of pre-existing crystalline, volcanic, and sedimentary rocks and, except for eolianites, are all deposited by water. Silicates predominate. *Carbonate* sands are virtually all deposited in marine waters and consist primarily of skeletal grains, oolites plus other coated grains, some locally derived detrital carbonate called *intraclasts*, and *terrigenous* carbonate. The latter is really a terrigenous sand, and not abundant except where there is very rapid erosion of thick carbonate sections in orogenic belts or in some glacial situations. *Pyroclastic* sands are those derived directly from volcanic explosion as, for example, ash, lapilli, and bombs. They may be deposited on either land or in water. Pyroclastic sands are less abundant than either terrigenous or carbonate sands. The term *volcaniclastic* is also applied to some sediments and refers to clastic materials rich in volcanic debris which may be either of pyroclastic origin or a normal terrigenous (epiclastic) sand derived from an older volcanic terrane.

It is noteworthy that the above classification, although simple and fitting readily into our ordinary geologic thinking, is not internally consistent as it is based on a variety of concepts: terrigenous sands are those classified by their source, carbonate sands are defined by composition, and pyroclastic sands by their agent of formation, vulcanism. Nonetheless the three-fold division is a valid one because of the unlike modes of sand production represented. This classification is based only on the detrital fraction and hence minimizes the complications of diagenesis.

In nature there are gradations between all three types. Pyroclastic and carbonate sand may occur together, for example, in an oceanic setting where skeletal debris is accumulating on a shallow shelf marginal to islands of volcanic origin. Or terrigenous sand may be carried into the area by longshore currents to become a minor constituent of the indigenous oolitic and skeletal sands.

We are primarily concerned with the terrigenous sands. But before considering their classification, let us first take a look at the more general problems of petrographic nomenclature and classification.

Nomenclature and Classification

Nomenclature and classification are parts of the same problem. To give something a name is to set it apart from all other things – to put it in a class by itself. The need for classification and nomenclature is both to facilitate communication and to organize our thinking. Our problem is to consider the petrographic classification of sandstone – itself a name designating a class of sedimentary rocks.

Classification is a mental discipline of value because it forces us to think clearly and to express our ideas about things in a logical and orderly way. But as ideas are subject to evolution and change, so, too, will classifications evolve and change. There can be, therefore, no final "correct" classification – only passing agreement. Classification is, thus, an attempt to express in a concise form our present concepts and is a means to facilitate communication. It is not an end in itself.

To classify is to define and to define is to draw limits based on some tangible property. Of all the many properties of sandstones, which shall be the defining parameters? Without doubt, it is manifestly impossible to utilize all properties. How then shall the choice be made? We think the answer depends on the point of view and the use to be made of the classification. An engineer might select properties meaningful to him in view of his interest in the behaviour of sandstones relative to engineering problems. A geologist might choose a different set of parameters related to pertinent geological questions. The ultimate geological question is origin and, therefore, the defining parameters should be as genetically significant as possible. This implies that a simple inventory of the minerals in a sand contributes but little to an understanding of its origin – what is really required is that they be arranged in meaningful genetic groups. One needs to discriminate, for example, between the original detrital components, those introduced as cements, and those produced by post-depositional metamorphism or weathering.

It is often said that there are two kinds of classifications: descriptive and genetic. And usually it is added that, although we ultimately want a genetic classification, we first need one which is purely descriptive. No wholly descriptive one is in fact possible. Such a classification, were it possible, would be largely meaningless and hence useless. Virtually all geologic "facts" are in truth interpretations based on origin. "Sandstone" is a good illustration. It is difficult to define independently of its origin. Even a simple term such as "pebble" has strong genetic connotations (which distinguish it from other round objects such as concretions).

Geological terms, such as rock names, present a motley array. Some are derived from common speech and have been redefined. These include such terms as "sand" and "clay." Others have been coined to fill a need. Unfortunately, usage tends to vary and terms come to be used quite differently than intended. This has led some to abandon some older terms and to the invention of new ones. A few are tied to a particular place – a place which is a first usage – a "type" locality. The term "graywacke" is an example. Abandonment of the term, as advocated by some, would not solve our problem – the problem of defining and naming the rock.

To summarize, classification to be geologically meaningful should be based on easily observed, genetically significant variables that are susceptible to quantitative analysis; the names applied to the classes defined are more acceptable if they correspond to current usage and do not depart greatly from original meaning.

Defining Parameters

All of the three types of sands – terrigenous, carbonate, and pyroclastic – can be considered in terms of their relative proportions of framework grains, detrital matrix (called *micrite* if carbonate rather than argillaceous mud) and chemical cement. Framework grains are those that support the sand, whereas matrix and chemical cement either wholly or partially fill the pore space.

The framework grains of noncarbonate sands are mainly quartz, feldspar, and fine-grained rock fragments (chert, limestone, siltstone, argillite, slate, glass, rhyolite and other volcanic fragments, etc.). Micas commonly play but a subordinate role as framework elements. Others such as heavy minerals and glauconite are negligible except locally (as in placers and greensands). The three basic framework components of terrigenous and pyroclastic sands are thus quartz, feldspar, and rock fragments.

Should a classification be based on the framework grains alone, or on framework grains plus detrital matrix, or should the chemically precipitated cements also be included? The independence of mineralogy and texture bears upon this question. In igneous petrology both are used to define rock clans and are treated as independent variables. Can we do the same for sands and sandstones? To state the question differently, is mineralogical composition dependent or independent of grain size?

Scatter diagrams show, with few exceptions, a relation between grain size and composition (Fig. 5-1). Grain size is, for example, correlated with the amount of detrital matrix and also with proportion of polycrystalline quartz and rock fragments. The coarser the sand the greater the proportion of rock particles and polycrystalline quartz. This latter relationship may lead to classifying the lower and coarser portion of a graded sandstone differently than the finer top (Fig. 5-2). Because of this observation and because grain size variation is the rule rather than the exception in sands, we believe that a classification with too many subdivisions can indeed be misleading. To minimize the effects of granular variation, nomenclature and subdivision should not become too complicated.

Regardless of the defining parameters finally selected, there is always the problem of subdivision and the naming of mixtures. How many subdivisions, what limits, and what names? The possibilities are really endless and consequently there is plenty of opportunity for individual expression.

A major sandstone type is distinguished by a dominant constituent. For example, if more than 50 percent of a sand consists of detritus derived from pre-existing rocks, it is a terrigenous sand; or if 50 or more percent is derived directly from volcanic explosion, it is a pyroclastic sand.

But, as subdivisions within these major types are made, disagreement develops as where to place exact limits even if the basic parameters are similarly chosen. It has been suggested that such problems can be avoided by not using a name at all but relying instead on only a few relevant numbers (Rodgers, 1950, p. 308; Füchtbauer, 1964, p. 165–166; Boggs, 1967) or perhaps simply reporting the entire composition. This suggestion presumes that one has made a quantitative petrographic modal analysis. The latter is, of course, necessary

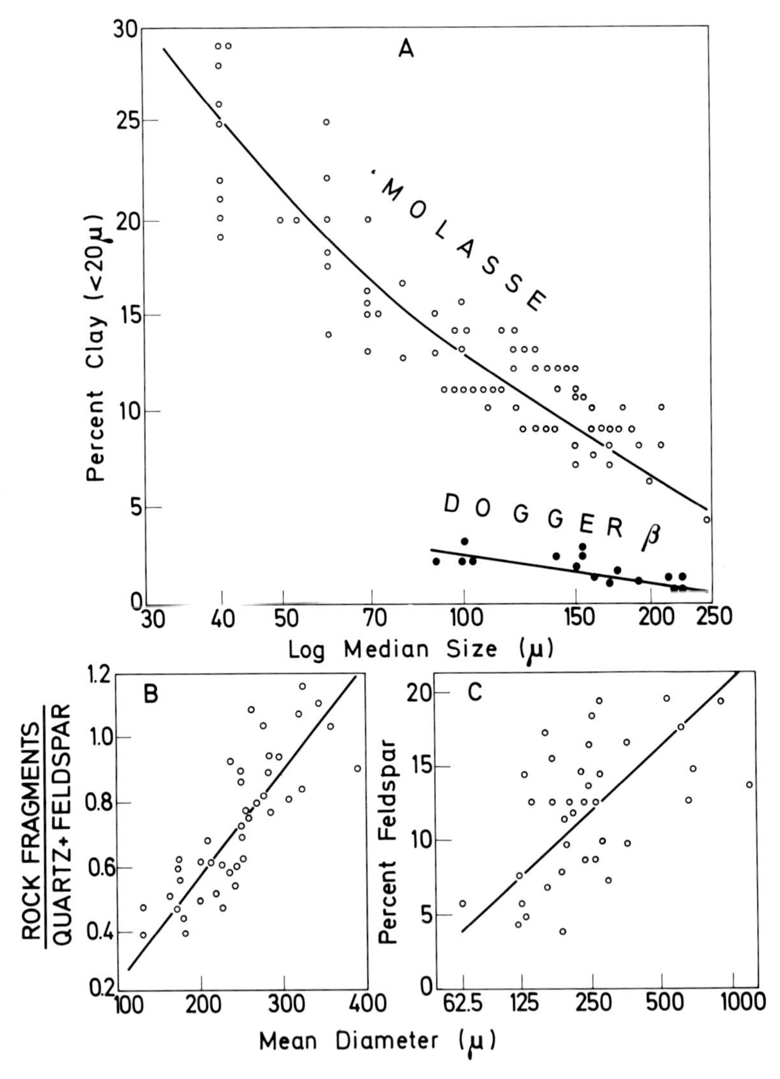

Fig. 5-1. Scatter diagrams of composition versus grain size: A) Strong (Molasse) and weak (Dogger β) dependence of clay content on grain size. Most sandstones lie between these extremes (Modified from Füchtbauer, 1964, Fig. 23); B) Dependence in Clee Sandstone (Devonian) (after Allen, 1962, Fig. 9); C) Feldspar becomes less abundant with finer grain in Silurian and Ordovician turbidites (Modified from Okada, 1966, Fig. 8)

for many statistical purposes, such as factor analysis to identify petrographic end members. Moreover, no quantitative classification of sandstones makes sense without the modal analyses to support it. Petrologic interpretation of a given sandstone is often based on its quantitative analysis – an analysis that also allows it to be classified. Though we might envision a literature that quoted only modal analyses and used no classification or nomenclature, such a procedure is all right up to the point of either discussion or final report writing when a name does become necessary for effective communication.

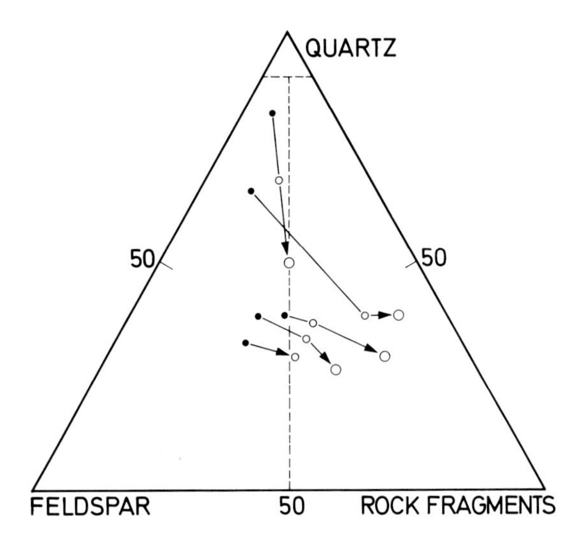

Fig. 5-2. Compositional plots change with grain size in graywackes. Each arrow connects samples from the same sequence (Modified from Füchtbauer, 1967, Fig. 1). Arrows go from fine to coarse

Major Trends in Sandstone Classification

The last twenty-five years have seen an almost continuous flow of papers on sandstone classification. Much of this literature has been summarized by Klein (1963). Without reviewing the history of the classification controversy in complete detail, it is worthwhile to note the major trends in the discussion. Some of the classification schemes have received wide usage; others have been proposed and hardly used except by the proposer. In a sense a good many petrographers have voted on their preference by how such classifications work when actually using the microscope rather than on abstract or theoretical grounds. We have been mindful of that kind of choice, for a classification is truly of value only if it can be and is used.

The use of the microscope is central, even though Krynine titled his 1948 paper, "The Megascopic Study and Field Classification of Sedimentary Rocks." Though careful field observation, including the use of a hand lens, makes a large number of distinctions, it is doubtful that any of the proposed classification schemes can be applied in any precise fashion without a careful and quantitative microscopical analysis. Thus a field classification is bound to be limited to general types, without quantitative limits, and is likely to be affected by variables of lesser interest, such as color or friability. There should be no difficulty in recognizing the pure quartz sandstones, the highly feldspathic arkoses, or the dark graywackes that are the end members of the classification; there is less point in trying to discriminate precisely among rocks of mixed composition.

The major impetus to sandstone classification came from the proposals of P. D. Krynine in the years 1940–1948 and F. J. Pettijohn from 1943–1957. Both recognized the importance of mineralogy as a clue to source-rock composition and source-area tectonism but put different emphasis on the role of texture. It is clear that all of the other schemes proposed derive from one or the other or both of them. A major concept used – borrowed from igneous petrology – was the composition triangle for representation of modal analyses and the blocking out of fields within that triangle. Equally important was the recognition that the relative amounts of feldspar (or other igneous minerals) and quartz (or other siliceous sedimentary minerals) was a clue to the composition of source rocks, the tectonics of the source terrain, and the weathering processes that produced detritus.

Krynine (1948) produced a composition triangle that considered only the detrital fraction and divided the rocks into classes based on the three-component system: *quartz* (+chert) – *feldspar* (+kaolin) – *phyllosilicates* (micas+chlorite). Though he noted that much of the mica and chlorite was very fine grained matrix (as was the kaolin), it was the mineral composition rather than the texture that was the important criterion. He explicitly recognized the importance of micaceous low rank metamorphic rocks and other types of rock fragments as part of the mica-chlorite component. Central to Krynine's ideas was the quantitative and interpretive significance of the graywackes, those sandstones with high proportions of the mica+chlorite component and varying amounts of the feldspar+kaolin component. He divided "low" and "high" rank graywackes on the relative amounts of the feldspar + kaolin and mica + chlorite components.

A somewhat different emphasis was given by Pettijohn to the classification of the rock fragment – mica+chlorite+clay matrix complex that Krynine had treated as only a mineralogical problem. In the fullest statement of his ideas (1954) Pettijohn erected a four-component system in which the matrix, a fine grained mixture of various kinds of clay (commonly altered to mica-chlorite by diagenesis) was a component by itself, differentiated from the framework rock-fragment components. The rock fragment component included the same micaceous low rank metamorphic rocks that Krynine assigned to his mica+chlorite component; the difference was that Pettijohn essentially took that component of Krynine's and divided it into two on the basis of texture. Where Krynine distinguished between high and low rank graywacke, Pettijohn used graywacke and subgraywacke. There proved to be little significant difference in the names and the way they were applied between the two systems of classification when high rank, "true," or "typical" graywackes, arkoses, and orthoquartzites were being considered. The major difficulties came with sandstones that were only slightly feldspathic and had only about 10 to 15 percent detrital matrix of an uncertain mineralogy. The importance of the detrital matrix was recognized by both Krynine and Pettijohn but it was elevated to a classification variable by Pettijohn, because he reasoned that it was a fluidity index of the transporting medium and closely related to the mechanics of turbidity flows of dense clay suspensions.

Many other attempts were made to resolve the ambiguities in the treatment of both texture and mineralogy in the same classification. Folk (1954, 1956) redefined Krynine's classification by introducing grain size end members with all gradations between gravel, sand, silt, and mud, as well as using the abundance of clay as an index of what he called "textural maturity," an extension of the idea of mineralogical maturity. Depending on the amount of clay and the sorting of the framework fraction, sandstones would be classed as immature, submature, mature, or supermature. Gilbert (Williams, Turner, and Gilbert, 1954, p. 290) used a textural criterion, the degree of sorting and detrital matrix, to divide all sandstones into suites, the wackes and the arenites. Each of these was classified according to the same general three component system used by Pettijohn. The division between the two suites was made at the 10 percent matrix point, whereas Pettijohn had used primary sedimentary structures to divide marine sandstones into two general groups, the graywacke, characterized by graded bedding and sole markings, and the arkose-quartzose sandstone suite, characterized by crossbedding and other related current structures. Many other proposals have been made that amplify, amend, extend, or modify the earlier classifications (Table 5-1).

Though there has been no overwhelming acceptance of any one classification, most petrographers have tended to clump about the classification schemes of Krynine as modified by Folk, Pettijohn, or Gilbert. The impression from the literature of the past decade is that most practising microscopists prefer to recognize texture explicitly, and favor dividing sandstones into two great groups, the poorly sorted, matrix-rich, and the well-sorted matrix-poor. Within that conceptual framework, there are minor variations in such matters as the choice between classifying a metamorphic quartzite grain as a particle of quartz or a metamorphic rock fragment. Such choices are probably always going to vary, for different sandstones may need "tailormade" decisions on how to classify most meaningfully.

Making a Choice

The classification we use for terrigenous sands and sandstones (Fig. 5-3) is a very simple one and one that, as indicated above, is generally consistent with current usage. It is appropriate for both ancient and modern sands. Basically, it uses only framework grains of quartz, feldspar, and rock fragments of sand size. As a secondary criterion, the classification distinguishes between the "clean" sands or *arenites* – sands with less than 15 percent matrix – and the "dirty" sands or *wackes* – those with more than 15 percent matrix. Among the matrix-poor sands, those with no more than five percent of either feldspar or

Table 5-1. *Summary of classifications of terrigenous sandstone (modified from McBride, 1963, Table 1)*

Reference	Basis of classification	End-members of classification			Comments
Fischer (1933)	Mineralogy	Quartz	Feldspar	Rock fragments	First use of triangular diagram for sandstone composition?
Krynine (1948)		Quartz	Feldspar and kaolin	Micas and chlorite	Ignores rock fragments, Graywacke based solely on mica and chlorite.
Folk (1954)		Quartz and chert	Feldspar and volcanic rock fragments	Metamorphic rock fragments, micas, metamorphic quartz	Graywacke based solely on metamorphic constituents. Sedimentary rock fragments ignored.
van Andel (1958)		Quartz	Feldspar	Rock fragments and chert	Graywacke based solely on rock fragments and chert.
Füchtbauer (1959)		Quartz	Feldspar	Rock fragments and chert	Recognizes clay-rich and clay-poor sandstone types.
Hubert (1960)		Quartz, chert and metaquartzite	Feldspar and feldspathic crystalline rock fragments	Micas and micaceous rock fragments	Non-micaceous rock fragments are not treated as a major constituent. Classification designed originally for feldspathic rocks.
Fujii (1962)		Quartz plus chert	Feldspar	Rock fragments	Divides triangle into five fields.
McBride (1963)		Quartz, chert and quartzite	Feldspar	Rock fragments	Ignores large micas. Has 8 classes.
Shutov (1967)		Quartz	Feldspar	Rock fragments	Divides triangle into 12 fields forming three major groups; graywacke based solely on rock fragment content.
Teodorovich (1967)		Quartz	Feldspar and mica/chlorite	Rock fragments	Eighteen subdivisions of triangle; addition of pyroclastic material requires tetrahedral representation and 11 additional subclasses.
Tallman (1949)	Texture and Mineralogy	Quartz	Feldspar		Graywacke based solely on matrix content. Rock fragments ignored.
Dapples, Krumbein, and Sloss (1953)		Quartz and chert	K and Na feldspar	Rock fragments and matrix	Graywacke based on sum of rock fragments and matrix. Ca feldspar ignored.

Table 5-1 (continued)

Reference	Basis of classification	End-members of classification			Comments
Williams and others (1954), Dott (1964)		Quartz, chert, quartzite	Feldspar	Unstable fine-grained rock fragments	Recognizes two suites on basis of > or < 10% matrix. Graywacke used as special rock type and not part of classification.
Bokman (1955)		Quartz	Feldspar and rock fragments	Clay	Graywacke based solely on clay content. Feldspar and rock fragments not differentiated.
Pettijohn (1957)		Quartz and chert	Feldspar	Rock fragments	Clay matrix is most important property of graywacke.
Sahu (1965)		Stable grains (Quartz, chert, quartzite, plus tourmaline, etc.)	Unstable grains (Feldspar, rock fragments, and micas)	Matrix	Eight clans of sandstone.
Krumbein and Sloss (1966)		Quartz	Feldspar	Clay, sericite, and chlorite	Ignores rock fragments. Graywacke based on clay, sericite, chlorite, and feldspar content.
Boggs (1967)		Siliceous resistates	Feldspar	Labile grains	Ten principal and 10 subclasses.
Packham (1954)	Structure, Texture, and Mineralogy	Quartz and chert	Unstable minerals and rock fragments	Matrix	Recognizes two suites. Graywacke based on deposition by turbidity current.
Crook (1960)		Quartz and chert	Unstable minerals and rock fragments	Matrix	Recognizes three suites. Graywacke based on deposition by turbidity current.

Others papers dealing with sandstone classifications, not represented in this table, include those of Michot (1958), Kossovskaya (1962), Shutov (1965), Chab (1967), Chang (1967), Wang (1967), Chen (1968), Konta (1968), Peykh (1969), and Travis (1970).

rock particles are called *quartz arenite*; commonly in the past we have called these sandstones *orthoquartzites*. Those with 25 percent or more feldspar, which exceeds rock fragments, are the *arkosic arenites*. *Arkoses* belong to this clan. Those with 25 or more percent of rock fragments, but a lesser amount of feldspar, are the *lithic arenites* (some may choose to contract this to *litharenites*). Transitional classes, subarkose and sublithwacke (protoquartzite) may be recognized (Fig. 5-3).

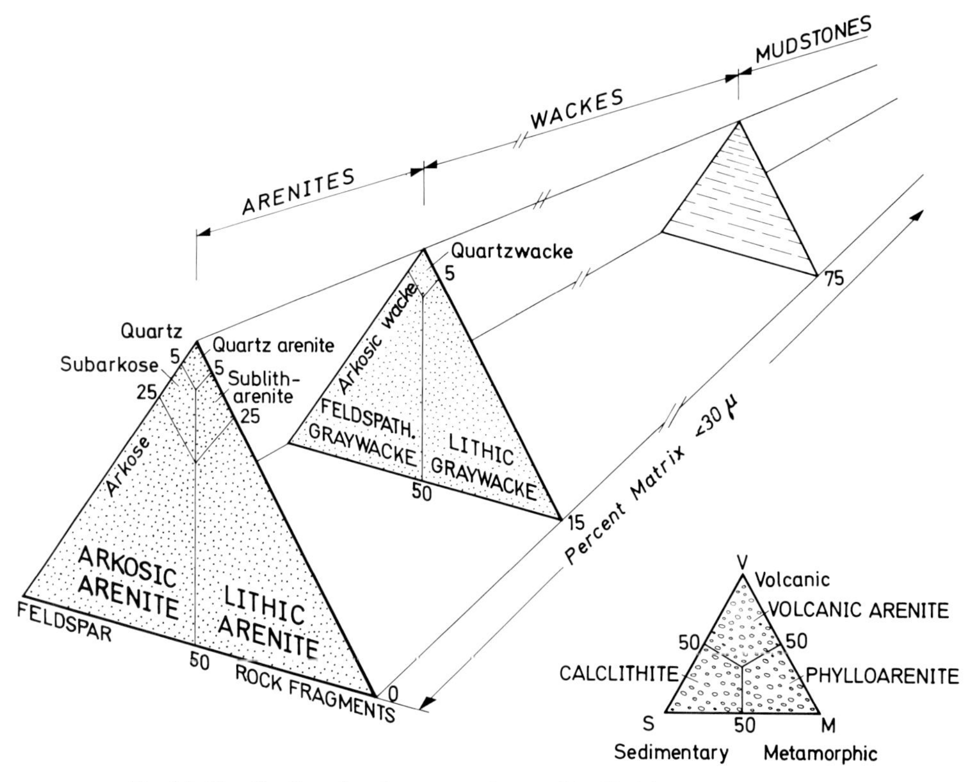

Fig. 5-3. Classification of terrigenous sandstones (Modified from Dott, 1964, Fig. 3)

The term *feldspathic arenite* is a loosely-used, all-encompassing term. Many arenites contain some feldspar. Some contain a great deal. As here used, any sandstone with five or more percent feldspar is "feldspathic." As thus defined, feldspathic sandstone includes subarkose, arkose, some litharenites, and many graywackes. The reader will note that "arkose" is here defined in a narrower sense. Commonly it is said that any sandstone with 25 or more percent feldspar is an arkose. This is usually true but as here defined an arkose must also have more feldspar than rock fragments. If it does not, it is a *feldspathic lithic arenite*. Also we have followed Gilbert in dividing the arkoses into *arkosic arenite* (a "clean" arkose) and *arkosic wacke* (a "dirty" arkose) – the latter having a significant matrix content.

The *litharenites* (a contraction of lithic arenite) are those arenites with 25 or more percent of rock particles and a minimal matrix content. Most commonly these rock particles are pelitic in character (shale, siltstone, slate, phyllite, and mica schist). Consequently the litharenites with such phylloid fragments have been termed *phyllarenites* (Folk, 1968, p. 131). Most lithic arenites are phyllarenites. Other important types include those in which the dominant rock particles are limestone or dolomite. The term *calclithite* (Folk, 1968, p. 141) has been proposed to distinguish this carbonate sand from *calcarenite* – a carbonate sand produced by biochemical or chemical precipitation. Other lithic arenites include *chert arenite*, if chert is the dominant detrital rock particle. Some lithic arenites may be *volcanic arenites* if the rock detritus is derived from disintegration of extrusive or flow rocks.

The transitional class, *sublitharenite* has a lesser rock particle content. The term *protoquartzite* has been used by Krynine (Payne, 1951) for these sandstones. Likewise the term *subarkose* is used for a transitional class of arenites with less feldspar than an arkose and with few or no rock particles.

For those sandstones with a significant matrix content (15 percent) the general term *wacke* (Fischer, 1933, p. 366; Williams, Turner, and Gilbert, 1954, p. 290) has been used. These are dominantly the *graywackes*, a term which has engendered considerable controversy. We retain the term and recognize two main classes: *feldspathic graywacke* and *lithic graywacke* depending on whether detrital feldspar or detrital rock particles dominate the rock particle fraction. *Quartz wackes* constitute a minor and relatively small group within the wacke clan.

In summary, we have two main groups of sandstone on the basis of their matrix content. The common sands or arenites, devoid of much matrix, are divided into three main families (quartz arenite, arkosic arenite, lithic arenite) and two subfamilies (subarkose and sublitharenite). The investigator has the option of recognizing some lesser sand types within the lithicarenite family base on the dominant rock particle present. It is noteworthy that the mineralogy or composition of the cement does not enter into the classification. The petrographer can, and should, indicate the character of the cement by a proper adjectival designation, for example, calcareous subarkose.

The above classification is a classification based on mineral composition and generally has minimal dependence on the environment of deposition. For example, a quartz arenite may have been deposited as a subaerial dune, in a stream or on a beach. Or an arkosic arenite may have been deposited on an alluvial piedmont fan or as a marine shelf sand. Because the character of the source rocks very largely determines mineral composition, this classification is largely relatable to source area composition and ultimately to tectonics.

We have used detrital matrix as a classifying criterion despite the fact that, as in the case of "micrite" in the carbonate rocks, there is little agreement among petrographers as to the upper size limit of matrix. Should it be 4 microns (1/256 mm) as in the Wentworth scale? Most have set it higher, some using 20 microns, some 30, and some even 62.5. We have selected 30 microns as the upper limit of matrix. This seems to us most satisfactory, although we recognize that others may alter this limit.

It is becoming increasingly clear that matrix does not all result from primary deposition. In addition to primary entrapment, matrix may be introduced by infiltration shortly after deposition, and it may be of diagenetic origin. Proof of the diagenetic origin of much of the matrix of ancient turbidites lies in their Recent and Tertiary counterparts – thick sections of rhythmically bedded sands that have abundant graded bedding and sole marks but which are singularly free of detrital matrix (Cummins, 1962; Hubert, 1964, p. 776). Such observations suggest that much matrix may be produced by transformation of argillaceous and volcanic rock fragments after deposition. For example, squashing can destroy the original structure of argillaceous rock fragments so that they become indistinguishable from fine grained detrital matrix. And finally matrix is also a product of precipitation in the void space between grains. Unfortunately, it is not easy to consistently distinguish these different origins of clay-sized materials. Matrix can also be replaced by chemically precipitated carbonate cement. Following Fischer (1933), Gilbert (Williams, Turner, and Gilbert, 1954, p. 290) and Dott (1964), we think the term *wacke* is a useful one to substitue for arenite when the sandstone has more than 15 percent matrix, as long as one remembers that the term itself is descriptive and implies nothing about the mechanism of transport or origin of the matrix. This usage of wacke differs from that of some who equate it with a sandstone presumed to have been deposited by a density current, one that flowed, impelled by gravity, downhill along the bottom. As used by us, wacke is a synonym for muddy or clayey sand or sandstone. If working with modern sands, the term "muddy" is probably preferable to wacke, should more than 15 percent matrix be present.

If, as pointed out above, some or much of this matrix is diagenetic or post-depositional in origin, this utilization of a diagenetic feature is at variance with our disregard of the precipitated mineral cements. These do not enter into our classification except as adjectival modifiers. We are aware of this inconsistency but we justify our decision by the prevailing treatment of the graywackes – that class of sandstones with a significant matrix content. The graywacke problem is discussed at length in the following chapter.

Our approach to classification recognizes the sensitivity of mineralogical composition on grain size and the role that diagenesis can play in origin of matrix. Because of these factors, we believe that the conservative scheme of Fig. 5-3 represents about as many petrographic types as can be meaningfully used. With appropriate adjectival modifiers this classification can also be used for sands rich in unusual constituents. Should one encounter, for example, an arenite in which olivine is the dominant mineral, the rock can be called an olivine arenite. Or should the framework be glass shards the sandstone can be described as a vitric arenite.

Complete description should, of course, include textural terms (Table 3-1) so that the final rock name might be, "medium grained, well sorted lithic arenite" the order being size, sorting and rock name. To this one might add degree of cementation, porosity, and possibly color, if deemed necessary.

And what of carbonate or calcarenite sands? Like their terrigenous counterparts, the basic components are framework grains, micrite, and chemical cement. Framework grains that are derived from within the basin of deposition, and commonly have experienced some transportation, are called *allochems*.

Because these allochems, which include skeletal debris, are capable of extensive subdivision, there are many more components to use in defining and mapping the calcarenites than are commonly available for terrigenous sands. Carbonate classifications are many and varied (Ham, 1962).

In carbonate rocks, the problems of microcrystalline calcite are comparable to those of matrix in the terrigenous sandstone. Distinguishing between detrital micrite and later secondary recrystallization that produces microspar can be just as difficult as unravelling the transformations experienced by matrix in a lithic arenite. Folk (1965) emphasizes this complexity for the carbonate rocks.

Glossary of Rock Names Applied to Sands and Sandstones

The great interest in the related problems of classification and nomenclature of sandstones, which began with the 1948 issue of the *Journal of Geology* on classification of sediments, especially Krynine's paper which contained a rational classification of the sandstones, has led to a proliferation of new terms and many varied redefinitions of older ones. This proliferation is a burden to the average reader and a source of confusion and misunderstanding. We felt the need, therefore, for a glossary which contains both the new and old terms and which provides a reference to the original definition in the case of the new terms and references to the several variants in usage and redefinitions of the older terms.

The only other glossary restricted to petrographic terminology of sandstones that we know of is that prepared by V. T. Allen and published in the *Report of the Committee on Sedimentation* of the National Research Council in 1936. This has proved of great help with the older terms.

We have included every term known to us, including many, which in our judgment might best be forgotten, as well as some old and now obsolete terms. We did not elect to pass judgment on the terms but felt rather that there was need for a glossary as complete as we could make it. The reader will observe, in the chapter on the petrography of sandstones, that we find need for relatively few rock names – no more than a dozen or fifteen of the one hundred or more contained in the glossary. The reader, however, will encounter many terms, good or bad, useful or not, and will need a definition.

We have included only rock names. Terms denoting textures or structures of sandstones have been omitted. These are covered to some extent in general dictionaries and, in the case of structures, in special glossaries (Pettijohn and Potter, 1964).

Where possible, we have given the reference to the first use of the term; such reference being placed immediately after the term. But inasmuch as many of the terms, such as grit, sand, brownstone and the like were common language or trade terms appropriated by the petrologist, it is not always possible to cite an original definition.

It is probably not practical to give all the variants of many well-known terms. It would be difficult, for example, to run down and give a reference to all the size-definitions of "sand." The effort at quantification of many rock names has led to a wide variation in the definitions of these terms. Not only is there a great variation in the choice of the mineralogical parameters used in classification (whether, for example, chert should be included with quartz or put with rock particles), but there is a wide variation in the percentage limits of these components. The combinations and permutations that are possible – and that, in fact, exist – are very large. A complete tabulation would be tedious to make and probably rather cumbersome to use. Many of these variations have been shown in graphic form in Klein's 1963 review of the classification question.

The reader will note that the glossary is alphabetical. Moreover, the adjectival modifier, if there is one, such as feldspathic graywacke, is listed first, not afterwards. This arrangement is simplest to construct; any other raises troublesome problems. For example, we believe it is easier to list and find "coral sand" or "lithic graywacke" than "sand, coral" or "graywacke, lithic."

We have restricted our glossary to terms found in the English-language literature.

ANEMOSILICARENITE (-YTE) (Grabau, 1904; 1913, p. 293)	An eolian sand of siliceous composition.
ARENITE (ARENYTE) (Grabau, 1904; 1913, p. 285)	1. A term of Latin derivation (*arena*: sand) used to denote a consolidated or lithified sand without respect to composition. Composition may be indicated by suitable prefix, thus: calcarenite, silicarenite. Adj.: *arenaceous*. 2. A sandstone with less than ten percent matrix material, as opposed to "wacke" (Williams, Turner and Gilbert, 1954, p. 290). Synonyms: Sandstone, psammite.
ARGILLACEOUS SANDSTONE	The term "argillaceous sandstone" may be loosely applied to impure sandstones containing an indefinite amount of fine silt and clay (Williams, Turner, and Gilbert, 1954, p. 290); considered by some as a synonym for "wacke." Used synonymously with the field term "dirty" sandstone.
ARKOSE (Brongiart, 1826)	1. A sandstone, formed principally by mechanical disintegration of a granitic rock, consisting of coarse grains of quartz and feldspar (Brongiart). 2. Any feldspathic sandstone. 3. A highly feldspathic sandstone with 30 percent or more feldspar (Krynine, 1940, p. 50). 4. An arenite or sandstone consisting of 80 percent or more quartz and feldspar, feldspar (and rock particles) over 25 percent, and feldspar exceeding rock particles (Pettijohn, 1957, p. 332). 5. A general term including (a) "arkosic arenites," clean sands with 25 percent or more feldspar, and (b) "arkosic wackes," sands with over 10 percent matrix and 25 percent or more feldspar (Williams, Turner, and Gilbert, 1954, p. 294).
ARKOSIC ARENITE (Williams, Turner, and Gilbert, 1954, p. 294, Fig. 97)	A matrix-free (under 10 percent) arenite with feldspar exceeding rock fragments; an arkose.
ARKOSIC SANDSTONE	An arkose; a feldspathic sandstone.

ARKOSIC GRAYWACKE (Williams, Turner, and Gilbert, 1954, p. 294, 314)	A graywacke with abundance of feldspar; a hard, dark-colored arkosic wacke.
ARKOSIC WACKE (Williams, Turner, and Gilbert, 1954, p. 291)	A wacke containing more feldspar grains than rock fragments; a feldspathic graywacke.
ARKOSITE (Tieje, 1921, p. 655)	1. Well-cemented arkose but without interlocked grains (see arkositite). 2. An arkose which is the lithified equivalent of a quartzite (Grout, 1932, p. 367). Synonyms: Quartzitic arkose; arkosic quartzite.
ARKOSITITE (Tieje, 1921, p. 655)	An arkose cemented *with* grains interlocked; bears same relation to arkosite as quartz sandstone does to quartzite.
ARTICULITE	According to Holmes (1920, p. 36) the term articulite was used by Wetherell in 1867 for flexible sandstone. It does not seem to be in general use.
ASPHALTIC SAND	A natural mixture of asphalt with varying proportions of sand, a tar sand; a bituminous sand.
ATMOSILICARENITE (-YTE) (Grabau, 1904; 1913, p. 296)	A siliceous sand resulting from atmospheric action (weathering) leading to disintegration of parent rock; example: gruss. Results from passive action of atmosphere in contrast to atmosphere in motion; see anemosilicarenite.
AUTOARENITE (-YTE) (Grabau, 1904; 1913, p. 296)	A sand produced by crushing due to earth movements or tectonic pressures; the sand-size equivalent of an autoclastic breccia.
BASAL ARKOSE	An arkosic sandstone basal to a sedimentary sequence resting unconformably on a granitic terrane; the arkosic equivalent of a granitic basal conglomerate. May grade downward into sedentary or residual arkose.
BIOARENITE (-YTE) (Grabau, 1904; 1913, p. 296)	A sand produced by the activity of organisms, including man, for example, roofing granules prepared by crushing of quartzite, greenstones, and other rocks.
BITUMINOUS SAND	A sand-asphalt mixture. Synonyms: Asphaltic sand; tar sand.
BLACK SAND	Usually refers to a magnetite or ilmenite sand; a common placerlike accumulation especially on beaches very much enriched in magnetite or ilmenite.
BROWNSTONE	A ferruginous quartz sandstone in which the grains are generally coated with iron oxide (Grabau, 1920, p. 579). The term was once widely used for the reddish-brown sandstone from the Triassic of the eastern United States, especially the Portland stone of Connecticut, once extensively quarried as a building stone. It has been little used for other sandstones.
BURRSTONE (also BUHRSTONE)	Given as "Burrh-stone" in Humble (1840, p. 35); also "burystone." A term used for a porous siliceous sandstone of angular grain suitable for millstones. Many geologists consider that the term has no special geological significance and should be discarded (Allen, 1936, p. 24).
CALCARENACEOUS ORTHOQUARTZITE (Pettijohn, 1957, p. 405)	A term suggested for a sandstone consisting of subequal proportions of *detrital* carbonate and quartz.

CALCARENITE (-YTE) (Grabau, 1904; 1913, p. 290)	A term for a sandstone composed of grains of calcium carbonate with or without carbonate cement. Commonly considered to be a limestone and commonly held to be intrabasinal in origin (composed of oolites, skeletal debris and intraclasts) rather than terrestrial. A terrestrial sand derived from pre-existing carbonate rocks is *calclithite*, which see.
CALCLITHITE (Folk, 1959, p. 36)	A terrigenous sandstone containing over 50 percent carbonate particles derived by weathering and erosion of pre-existing limestones and dolomites. The 50 percent figure is probably not generally adhered to. Twenty five percent of rock particles defines a lithic arenite; of these 50 percent or more must be terrigenous carbonate detritus.
CARBONACEOUS SAND	Any sandstone that contains an appreciable amount of *detrital* carbonaceous or woody particles.
CHERT ARENITE (Folk, 1968, p. 125)	A lithic arenite in which chert particles are the dominant rock-particle constituent.
CLASMOSCHIST	A term attributed by Roberts (1839, p. 71) to Coneybeare to be used in place of graywacke. Obsolete.
CORAL SAND	A calcareous sand formed by break-down of reef rock. Coral detritus forms a significant but not necessarily dominant part of this debris.
CRYSTAL SANDSTONE	1. A sandstone in which calcite has been deposited in the pores in large patches or units having a single crystallographic orientation resulting in a "poikiloblastic" or "luster-mottling" effect. In some rare sandstones with incomplete cementation the carbonate occurs as sand-filled scalenohedra of calcite – "sand crystals." 2. Term more commonly applied to sandstones in which the quartz grains have been enlarged by deposition of silica so that the grains show regenerated crystal facets and in some cases nearly perfect quartz euhedra. Crystal sandstones of this nature sparkle in bright sunlight. Sometimes called "sparkling sandstone."
FELDSPATHIC ARENITE (Williams, Turner, and Gilbert, 1954, p. 316)	A feldspathic sandstone, with 10 to 25 percent detrital feldspar, usually also containing a smaller proportion of rock particles.
FELDSPATHIC GRAYWACKE	1. A graywacke containing appreciable feldspar. 2. A graywacke in which feldspar exceeds rock particles (Pettijohn, 1954, p. 363). Synonyms: "High-rank graywacke" (Krynine, 1945); "feldspathic wacke" and "arkosic wacke" are arkoses with a significant proportion of matrix (Williams, Turner, and Gilbert, 1954, p. 292), not graywackes.
FELDSPATHIC LITHARENITE (McBride, 1963, p. 667)	A lithic arenite containing appreciable (over 10 percent) feldspar.
FELDSPATHIC LITHWACKE (Casshyap, 1967)	Essentially a lithic graywacke (over 15 percent matrix) in which rock fragments exceed feldspar but the latter forms 10 percent or more of the sand fraction.
FELDSPATHIC POLY-LITHARENITE (Folk, 1968, p. 135)	A polylitharenite containing appreciable (over 10 percent) feldspar.

FELDSPATHIC SANDSTONE

1. Any sandstone containing appreciable feldspar; would include arkose, some graywackes, etc.
2. A sandstone with 10 to 25 percent feldspar but without appreciable matrix (Pettijohn, 1949, Table 55).
Synonym: Subarkose.

FELDSPATHIC SUBLITH-ARENITE (McBride, 1963, p. 667)

A lithic subarkose.

FELDSPATHIC WACKE (Williams, Turner, and Gilbert, 1954, p. 292)

A wacke with 10 to 25 percent feldspar (A wacke with over 25 percent feldspar is an "arkosic wacke").

FLAGSTONE

A name applied to thin-bedded sandstones capable of being split or parted to form flags suitable for paving.
A calcareous sandstone which splits readily along micaceous layers (Tyrrell, 1926, p. 210). Thin-bedded argillaceous sandstone used chiefly for paving (Ries, 1910, p. 117).

FLEXIBLE SANDSTONE

See *Itacolumite.*

FLYSCH SANDSTONE

A sandstone characteristic of the flysch facies; generally a turbidite, commonly graywacke in the older systems.

FRANGITE (from Latin: *frango* – to break up; Bastin, 1909, p. 450)

All sedimentary rocks formed by the disintegration but without decomposition of igneous rocks, and without extensive mechanical sorting; includes arkoses, graywackes, grits and their metamorphic derivatives.

FREESTONE

1. Any kind of stone, the texture of which is so free or loose that it may be easily worked (Humble, 1840, p. 100).
2. Any stone, such as sandstone, which can be freely worked or quarried, especially one that cuts well in all directions.
3. A uniform, thick-bedded sandstone with few divisional planes; rarely a similar limestone (Tyrrell, 1926, p. 210).

GANISTER (Also gannister)

A hard, compact, highly siliceous sedimentary rock, with fine, uniform granular texture sand composed essentially of angular quartz grains cemented with secondary silica. A fine-grained quartzose sandstone consisting of angular grains, cemented with silica (Williams, Turner, and Gilbert, 1954, p. 283; see also Thomas, 1918, p. 3). Originally a local name for a variety of sandstone found in the Yorkshire and Derbyshire coal fields (Roberts, 1839, p. 66).

GLAUCONARENITE (-YTE) (Grabau, 1904, p. 245)

A glauconitic sand.

GLAUCONITIC SANDSTONE

See *Greensand.*

GRANITE WASH

1. A term loosely applied to tongues of poorly-sorted, little-rounded arkose intercalated with ordinary sediments derived from near-by buried hills of granite.
2. May also be applied to modern alluvium derived locally from exposed granite hills.

GRANULITE (-YTE) (Grabau, 1904; 1913, p. 283)

A sand-sized deposit of constructional but nonclastic origin (ex. oolitic sand); corresponds to the term *arenite*, a sand of clastic origin (Commonly now used for a large number of coarse-grained metamorphic rocks such as "pyroxene granulite").

GRAYWACKE (also GREY-WACKE) (Lasius, 1789)

Term was originally applied to the dark, tough Paleozoic Kulm sandstones in the Harz Mountains of Germany by Lasius. From the German "grauewacke."
1. Defined by Geike (1885, p. 162) as "a compact aggregate of rounded or subangular grains of quartz, feldspar, slate, or other minerals or rocks cemented by a paste ... gray, as its name denotes, is the prevailing color .... The rock is distinguished from ordinary sandstone by its darker hue, its hardness, the variety of its component grains, and above all, by the compact cement in which the grains are embedded." Various attempts have been made to redefine the term more precisely. These have involved (1) the proportion of matrix ("paste") and (2) the proportions of the labile components (feldspars and rock particles). Pettijohn (1954), for example, required 15 or more percent matrix and 25 or more percent labile constituents.
2. Other workers have departed from the classical definition and expanded it to encompass sandstones differing materially from the type graywacke. Some ignore the matrix; others exclude feldspar. For example: Krynine (1940, p. 51) defined graywacke simply as "a clastic rock containing a substantial amount (20 percent and over) of dark rock fragments or dark-colored ferromagnesian minerals." Later he (1945) stressed only the rock particles – primarily "chert, slate, schists, phyllites, etc.", that is, largely low-grade metapelites. This usage was followed by Folk (1954). Graywacke as thus defined is essentially a lithic arenite (var. phyllarenite). This is also the usage of Shutov (1967, Fig. 6). Recently (1968, p. 125) Folk has dropped the term "graywacke" and replaced it by "lithic arenite."
3. Graywacke has also been defined as "a variety of sandstone composed of material derived from the disintegration of basic igneous rocks of granular texture .... Thus defined it is the ferromagnesian equivalent of arkose" (Twenhofel, 1932, p. 231). This usage has gained little or no support.
For a review of the problem of defining graywacke see Dott (1964).

GREENSAND

Sands containing appreciable glauconite are referred to as green sands, or greensand. They are usually mixtures of quartz and glauconite in all possible proportions.
Synonyms: Glauconarenite, glauconitic sandstone.

GRIT

First used as a provincial term for a coarse-grained sand or sandstone (Humble, 1840, p. 35), for example: Millstone Grit. Later sharpness of grain became a part of the definition (Holmes, 1920, p. 50).

GRUSS (also GRUS)

The fragmental products of *in situ* disintegration (with little or no decomposition) of granite.
Synonyms: Residual arkose, sedentary arkose.
Also used for *any* rock that is finely granulated but not decomposed by weathering.

HIGH-RANK GRAYWACKE (Krynine, 1945)

Essentially a lithic arenite or sandstone with appreciable detrital feldspar; may also be a classical graywacke rich in feldspar. Synonym: "Feldspathic graywacke" (Folk, 1954); "Feldspathic litharenite" (Folk, 1968, p. 124).

HYDRARENITE (-YTE) (Grabau, 1904; 1913, p. 294)	A water-laid sandstone of variable composition.
HYDROSILICARENITE (-YTE) (Grabau, 1904; 1913, p. 294)	A water-laid siliceous sandstone, that is, one consisting of siliceous detritus (quartz, feldspar, etc.).
ITACOLUMITE	A peculiar quartz schist named for Itacolumi, a mountain in Brazil, composed of interlocking quartz grains with some mica, which exhibits some flexibility (Page, 1865, p. 204). Also known in North Carolina and elsewhere. Is considered by some to be a metamorphic rock rather than a sandstone. A term attributed to Humboldt by Holmes (1928, p. 126). Synonym: Flexible sandstone.
LITHARENITE (McBride, 1963)	A contraction of the term "lithic arenite." Defined as a sandstone with over 25 percent rock particles and less than 10 percent feldspar.
LITHIC ARENITE (Williams, Turner, and Gilbert, 1954, p. 304)	Sandstones, with less than 10 percent matrix in which rock particles (unstable, fine-grained) are an important constituent. Synonym: "Low-rank graywacke" of Krynine and "lithic sandstone" of Pettijohn.
LITHIC ARKOSE (Folk, 1968, p. 124; McBride, 1963)	An arkose with a feldspar/rock fragment ratio between 1 : 1 and 3 : 1 (Folk, p. 124); an arkose with over 10 percent rock fragments (McBride, Fig. 1).
LITHIC ARKOSIC WACKE (Casshyap, 1967)	A graywacke in which feldspar exceeds rock particles. Synonym: "Feldspathic graywacke" (Pettijohn, 1954).
LITHIC GRAYWACKE (Pettijohn, 1954)	A graywacke in which rock particles exceed feldspar.
LITHIC SANDSTONE (Pettijohn, 1954)	1. A sandstone, with less than 15 percent matrix, and 5 percent or more feldspar and rock particles but with rock particles exceeding feldspar. Subdivided into subgraywackes and protoquartzites. 2. Now (1970) used interchangeably with lithic arenite and defined as having 25 or more percent labile constituents but with rock particles exceeding feldspar. Synonym: Lithic arenite of Gilbert (Williams, Turner, and Gilbert, 1954, p. 304).
LITHIC SUBARKOSE (McBride, 1963)	A sandstone or arenite composed of abundant subequal amounts of detrital feldspar and rock fragments (more than 10 percent but less than 25 percent). The term "lithic subarkose" is preferred to the more cumbersome but equally appropriate "feldspathic sublitharenite."
LITHIC SUBARKOSIC WACKE (Casshyap, 1967)	A wacke with subequal proportion of feldspar and rock fragments but no more than 25 percent of either.
LITHIC WACKE (Williams, Turner, and Gilbert, 1954, p. 301)	A sandstone with 10 percent or more of matrix and a significant proportion of detrital rock particles – rock particles exceeding feldspar; "lithic graywacke" of Gilbert is a lithic wacke characterized by great hardness and gray color. Synonym: Essentially same, in most cases, as lithic graywacke of Pettijohn (1954).

LOW-RANK GRAYWACKE (Krynine, 1945)	A lithic arenite in most cases; also a graywacke as defined in the classical sense with little or no feldspar.
MENGWACKE (Fischer, 1933, p. 336)	A wacke with 33–90 percent unstable mineral constituents.
METAQUARTZITE (Krynine, 1945)	A quartzite of metamorphic origin in contrast to orthoquartzite which is a primary sedimentary quartzite. Synonym: Paraquartzite (Tieje, 1921).
MICROBRECCIA	Ill-sorted sandy rocks in which the grains are sharply angular (Williams, Turner, and Gilbert, 1954, p. 283).
MOLASSE SANDSTONE (Cayeux, 1929, p. 164)	A sandstone of the molasse facies; characterized by Cayeux as poorly rounded, poorly sorted, coarse sand rich in rock fragments and generally calcareous. Probably generally a lithic arenite, in places arkosic. The product formed by the demolition of a newly elevated orogenic belt.
OIL SAND	Oil-saturated sand.
ORTHOARENITE (Marchese and Garrasino, 1969, p. 283)	An arenite with detrital matrix under 15 percent.
ORTHOQUARTZITE (Tieje, 1921, p. 655)	1. A sandstone converted to a quartzite with interlocking grains "cemented only through infiltration and pressure" as opposed to "paraquartzite," a quartzite originating mainly through contact metamorphism (Tieje, 1921, p. 655). 2. A term revived by Krynine (1941, 1945, 1948) for sandstones consisting almost exclusively of detrital quartz. Although commonly cemented by quartz and truly "quartzitic" in character, most contain some carbonate cement and as the proportion of this nonsiliceous material increases, the orthoquartzites become less cohesive. The term has, by usage, been extended to all sandstones, indurated or friable, with 95 or more percent detrital quartz (Pettijohn, 1957, p. 295). Synonyms: *Quartz arenite*, *quartzitic sandstone* (Krynine, 1940, p. 51) a sedimentary quartzite, *quartzose sandstone* (Krynine, 1940, p. 51) 95 percent or more quartz but *not* cemented by silica.
PARARENITE (Marchese and Garrasino, 1969, p. 283)	An arenite with detrital matrix between 15 and 70 percent; prefix "para-" may be used with other terms, such as *paralithite*. We would suggest that the prefix be used for any matrix-rich sandstones whether the matrix is detrital or diagenetic.
PARAQUARTZITE (Tieje, 1921, p. 655)	A truly metamorphic quartzite derived from a sandstone by action of heat and pressure. Synonym: Metaquartzite.
PHYLLARENITE (Folk, 1968, p. 131)	Essentially a lithic arenite characterized by an abundance of detrital rock particles of low-grade metamorphic pelitic rocks: slate, phyllite, mica schist. Synonym: Most so-called "low-rank graywackes" of Krynine and Folk.
PLAGIOCLASE ARKOSE (Folk, 1968, p. 130)	An arkose, the chief feldspar of which is plagioclase.
POLYLITHARENITE (Folk, 1968, p. 135)	A lithic arenite with a diversity of sand-sized rock particles – volcanic, sedimentary, and metamorphic.

PROTOQUARTZITE (Krynine, 1952)

A term first used by Krynine (Payne and others, 1952) but never really defined. A member of the series "low-rank graywacke-quartzose graywacke-protoquartzite-orthoquartzite" and hence by implication a sandstone with an appreciable but not excessive quantity of detrital rock particles and little or no feldspar.
Defined by Pettijohn (1954) as a sandstone with 5 to 25 percent rock particles, a variety of lithic sandstone. Detrital feldspar less than detrital rock particles.
Synonym: Sublitharenite (McBride, 1963).

PSAMMITE (also PSAMMYTE)

From the Greek *psammos*: sand; adj. *psammitic*.
1. Formerly used in Europe for fine-grained clayey sandstone in which the component grains are scarcely distinguishable with naked eye (Oldham, 1879, p. 44).
2. Generally used to denote sandstones without reference to composition.
Synonym: Arenite, sandstone.
Note: Tyrrell (1921) would apply "psammitic" only to metamorphic rocks; the Latin term "arenaceous" would be applied to sedimentary rocks; the Greek terms would be used when the rocks were hardened and altered beyond the limits implied by the Latin terms.

QUARTZ ARENITE (Williams, Turner, and Gilbert, 1954, p. 292)

Sandstones consisting essentially of quartz without appreciable matrix (under 10 percent), containing no more than 10 percent of either feldspar and rock particles.
Most authors would restrict allowable "contaminants" to 5 percent or less, or to 5 percent or less of either feldspar or rock particles.
Synonym: Quartzose sandstone, orthoquartzite, quartz-arenite.

QUARTZ GRAYWACKE (Williams, Turner, and Gilbert, 1954, p. 294)

An exceptionally quartz-rich graywacke; quartz wacke.

QUARTZARENITE (McBride, 1963)

A contraction of quartz arenite.

QUARTZ-FREE WACKE (Quarz-frei wacke of Fischer, 1933)

A wacke with over 90 percent unstable mineral constituents.

QUARTZITE

The term "quartzite" is commonly used for a metamorphic rock produced by recrystallization of a sandstone. It has been defined by Holmes (1920, p. 194) as "a granulose metamorphic rock, representing a recrystallized sandstone, consisting predominantly of quartz." The term "quartzite" has come to include also sedimentary quartzites (orthoquartzites) or sandstones cemented with silica which has grown in optical continuity around each detrital quartz grain. The term "metaquartzite" or "paraquartzite" is reserved for the truly metamorphic quartzite.
Quartzites have been defined as rocks which fracture through rather than around the constituent grains.

QUARTZITIC GRIT

A grit with properties of a quartzite.

QUARTZITIC SANDSTONE

A sedimentary quartzite (Krynine, 1940, p. 51); a sandstone approaching the character of a quartzite; an orthoquartzite.

QUARTZOSE SANDSTONE	1. A sandstone of 95 or more percent detrital quartz but not cemented with silica as opposed to quartzitic sandstone so cemented (Krynine, 1940, p. 51). 2. A sandstone of 95 or more percent detrital quartz, essentially synonymous with *quartz arenite* or *orthoquartzite*.
QUARZMENGWACKE (Fischer, 1933, p. 336)	A wacke with 10 to 33 percent unstable mineral constituents.
QUARZWACKE (Fischer, 1933, p. 336)	1. A wacke with less than 10 percent unstable mineral constituents. 2. Redefined as a sandstone with 10 percent or more matrix and no more than 10 percent each detrital feldspar and rock particles (Williams, Turner, and Gilbert, 1954, p. 292). 3. Used by Krumbein and Sloss (1963, p. 172) as synonym for subgraywacke (Pettijohn, 1949, p. 255).
QUICKSAND	Medium to fine-grained sand containing much water which yields readily to pressure or weight and hence apt to engulf persons and animals coming upon it. Capable of injection into fissures.
RESIDUAL ARKOSE	An arkose formed *in situ* by disintegration of a granite; an untransported arkose. Also called *sedentary arkose*, commonly grading into the underlying granite (Barton, 1916, p. 447); related to gruss.
REDSTONE (Krynine, 1950, p. 103)	A brick-red, clayey sandstone consisting of angular quartz, mica and feldspar grains embedded in a matrix of red hematitic clay. Named for Redstone Hill, southwest of Southington, Connecticut, U.S.A. A ferruginous arkosic wacke.
SAND	1. Noncohesive granular material of specified size (commonly 1/16 to 2 mm in diameter). May be of organic, chemical, volcanic or clastic origin and of widely varying composition such as $CaCO_3$, SiO_2, $CaSO_4 \cdot 2H_2O$, etc. 2, *Detrital* material of specified size range. 3. A term applied to a siliceous detrital deposit composed mainly of quartz particles. 4. Material having a terminal fall velocity less than the upward eddy currents and an upper limit such that a grain resting on the surface ceases to be moveable either by direct pressure of the fluid or by impact of other moving grains (Bagnold, 1941, p. 6). 5. A drilling term for an oil-bearing horizon. Many definitions emphasize "water-worn," "fragments," "clastic," "siliceous," and "detrital." But many sands are not water-worn, for example, gruss, nor fragmental, for example, oolitic sand, nor siliceous, for example, gypsum, nor detrital, for example, coral sand.
SAND ROCK (Tieje, 1921, p. 655)	A weakly or poorly cemented sand may be called *sand rock*. Also rural terminology in Appalachia.
SANDSTONE	Defined by Lyell (1833, p. 79) as "any stone which is composed of an agglutionation of grains of sand, whether calcareous, siliceous, or any other mineral nature." In practice, only consolidated sands which are dominantly siliceous are designated sandstone. A consolidated calcareous sand would be called limestone.

SCHIST ARENITE — A term attributed to Adolph Knopf (Krynine, 1937, p. 427) for sandstones with a significant proportion of metamorphic rock particles (schists and phyllites); a particular kind of lithic arenite.

SCHIST WACKE (Williams, Turner, and Gilbert, 1954, p. 292) — A lithic wacke characterized by an abundance of metapelitic rock particles (slate, phyllite, schist).

SEDARENITE (Folk, 1968, fig. on p. 124) — A lithic arenite, the rock particles of which are of sedimentary origin in contrast to volcanic arenite and phyllarenite (of metamorphic rock particles).

SILICARENITE (-YTE) (Grabau, 1904; 1913, p. 290) — A purely siliceous sand or arenite such as the St. Peter Sandstone.

SILICEOUS SANDSTONE — 1. A sandstone cemented with silica; a hard quartizitic sandstone.
2. A sandstone, the detrital components of which are siliceous (quartz, feldspar, etc.).

SILICINATE QUARTZOSE SANDSTONE (Allen, 1936, p. 40) — The adjective "silicinate" (also "calcarinate" and "ferruginate") are used to denote the composition of the *cement* of the sandstone.

SPARAGMITE (Blaas, 1898, p. 18) — A term applied to coarse Eocambrian arkoses in Norway and Sweden.

SUBARKOSE (Pettijohn, 1954) — A sandstone with 5 to 25 percent labile components of which feldspar exceeds rock particles; a feldspathic sandstone with less feldspar than that of a normal arkose.

SUBARKOSE WACKE (Casshyap, 1967) — Essentially a wacke (over 15 percent matrix) with 5 to 25 percent feldspar; a species of feldspathic graywacke.
Synonym: Feldspathic wacke (Williams, Turner, and Gilbert, 1954, p. 292).

SUBFELDSPATHIC LITHIC ARENITE (Williams, Turner, and Gilbert, 1954, Fig. 97) — An arenite with 10 or less percent feldspar and a larger quantity of rock fragments.

SUBFELDSPATHIC LITHIC WACKE (Williams, Turner, and Gilbert, 1954, Fig. 96) — A lithic wacke containing less than 10 percent feldspar. A species of feldspathic graywacke.
Synonym: Subarkose wacke (Casshyap, 1967).

SUBGRAYWACKE (Pettijohn, 1949, p. 255) — 1. A sandstone (wacke) with over 20 percent matrix and less than 10 percent feldspar (Pettijohn, 1949, p. 255); essentially a quartzwacke.
2. A sandstone, with less than 15 percent matrix and over 25 percent labile grains in which rock particles exceed feldspar; essentially a lithic arenite (Pettijohn, 1954).
Subgraywacke has a superficial resemblance to a graywacke especially in color and rock particle content, but as defined under No. 2 above, it is without matrix.

SUBLITHARENITE (McBride, 1963, p. 667) — A rock analogous to subarkose but containing rock fragments instead of feldspar; with 5 to 25 percent rock fragments and 0 to 10 percent feldspar, and 65 to 95 percent quartz.
Synonym: Protoquartzite.

SUBLITHWACKE (Casshyap, 1967) — A wacke with 5 to 25 percent detrital rock particles; a sublitharenite with over 15 percent matrix.

SUBPHYLLARENITE (Folk, 1968, p. 132, 134) — A sublitharenite with metapelitic rock particles.

TUFFACEOUS SANDSTONE — A sandstone composed of or containing an appreciable proportion of material of pyroclastic origin.

VOLCANIC ARENITE — 1. A lithic arenite consisting of terrigenous volcanic detritus of epiclastic origin.
2. An arenite of pyroclastic debris, including rock particles, crystal debris and glass fragments; a tuff.
Synonym: Volcanic sandstone.

VOLCANIC SANDSTONE — See *volcanic arenite* above.

VOLCANIC WACKE (Williams, Turner, and Gilbert, 1954, p. 303) — A volcanic sand with a matrix of fine volcanic debris; a species of graywacke of strictly volcanic origin.

VOLCANICLASTIC SANDSTONE (Fisher, 1961) — A sandstone consisting of either (a) pyroclastic debris or (b) terrigenous volcanic detritus of epiclastic origin.

WACKE — 1. An unsorted mixture of sand, silt, and clay; a "loam."
2. Rocks in which the clastic grains are approximately evenly divided between the several size grades (Fischer, 1933).
3. A sandstone with 10 or more percent argillaceous matrix materials (Williams, Turner, and Gilbert, 1954, p. 290).
4. A dirty green to brownish-black clay arising as end-product of *in situ* decomposition of basalt (Geike, 1882, p. 161).

References

Allen, J. R. L.: Petrology, origin, and deposition of the highest Lower Old Red Sandstone of Shropshire, England. Jour. Sed. Petrology, **32**, 657–697 (1962).

Allen, V. T.: Terminology of medium-grained sediments. In: Rept. Comm. Sedimentation. Natl. Research Council **1935-36**, 18–47 (1936).

van Andel, T. H.: Origin and classification of Cretaceous, Paleocene, and Eocene sandstones of western Venezuela. Am. Assoc. Petroleum Geologists Bull., 42, p. 734–763 (1958).

Bagnold, R. A.: The physics of blown sand and desert dunes, 265 p. London: Methuen 1941.

Barton, D. C.: The geological significance and classification of arkose deposits. Jour. Geology **24**, 417–449 (1916).

Bastin, E. S.: Chemical composition as a criterion in identifying metamorphosed sediments. Jour. Geology **17**, 445–472 (1909).

* Blaas, J.: Katechismus der Petrographie, 2nd ed, 242 p.: Leipzig: J. J. Weber 1898.

Boggs, Sam, Jr.: A numerical method for sandstone classification. Jour. Sed. Petrology **37**, 548–555 (1967).

Bokman, John: Sandstone classification in relation to composition and texture. Jour. Sed. Petrology **25**, 201–206 (1955).

* Brongiart, A.: L'arkose, caractères minéralogiques et histoire géognostique de cette roche. Paris: Ann. Sci. naturelles **8**, 113–163 (1826).

Cassyhap, S. M.: On the classification of argillaceous sandstone. Ann. Geol. Dept., Aligarh Muslim Univ. **3**, 48–50 (1967).

Cayeux, L.: Les roches sédimentaries de France – roches siliceuses. 774 p. Paris, Impr. Nationale 1929.

Cháb, Jan: Poznámka ki klasifikaci psamitů (Note on the classification of psammites). Včst. Ustrcd. ùst. geol. Ceshost **42**, 225–227 (1967).

Chang, Shih-Chicao: A new sandstone classification scheme. Geol. Soc. China, Proc. No. 10, p. 107–114 (1967).

Chen, P. Y.: A modification of sandstone classification. Jour. Sed. Petrology **38**, 54–60 (1968).

Crook, A. W.: Classification of arenites. Am. Jour. Sci. **258**, 419–428 (1960).

* Original reference not seen by authors.

Cummins, W. A.: The greywacke problem. Liverpool and Manchester Geol. Jour. **3**, 51–72 (1962).
Dapples, E. C., Krumbein, W. C., Sloss, L. L.: Petrographic and lithologic attributes of sandstones. Jour. Geology **61**, p. 291–317 (1953).
Dott, R. L., Jr.: Wacke, graywacke and matrix – What approach to immature sandstone classification? Jour. Sed. Petrology **34**, 625–632 (1964).
Eisbacher, Gerhard: Über Merkmalsabhängigkeit bei der Klassifikation von Sandsteinen (gezeigt am Beispiel des alpinen Buntsandsteins). Neues Jahrb. Mineral. Monat., no. 6, 161–165 (1964).
Fischer, Georg: Die Petrographie der Grauwacken. Jahr. preuss. geol. Landesanstalt **54**, 320–343 (1933).
Fisher, R. V.: Proposed classification of volcaniclastic sediments and rocks. Geol. Soc. America, Bull. **72**, 1409–1414 (1961).
Folk, R. L.: The distinction between grain size and mineral composition in sedimentary-rock nomenclature. Jour. Geology **62**, 344–359 (1954).
— The role of texture and composition in sandstone classification. Jour. Sed. Petrology **26**, 166–171 (1956).
— Practical petrographic classification of limestones: Am. Assoc. Petroleum Geologists Bull. **43**, 1–38 (1959).
— Some aspects of recrystallization in ancient limestones. In: Pray, L. C. and Murray, R. C., eds., Dolomitization and limestone diagenesis: Soc. Econ. Paleon. Mineral. Spec. Pub. **13**, 14–48 (1965).
— Petrology of sedimentary rocks. 170 p. Austin, Texas: Hemphill's, 1968.
Füchtbauer, Hans: Zur Nomenklatur der Sedimentgesteine. Erdöl und Kohle **12**, 605–613 (1959).
— Sedimentpetrographische Untersuchungen in der älteren Molasse nördlich der Alpen. Eclogae geol. Helvetiae **57**, 157–298 (1964).
— Die Sandsteine in der Molasse nördlich der Alpen. Geol. Rundschau, **56**, 266–300 (1967).
Fujii, Koji: Petrography of the Upper Paleozoic sandstones from the Yatsushiro area, Kyusha. Mem. Fac. Sci., Kyusha Univ., Ser. D., 179–218 (1962).
Geike, A.: Textbook of geology, 971 p. London: Macmillan 1882.
Grabau, A. W.: On the classification of sedimentary rocks. Am. Geologist **33**, 228–247 (1904).
— Principles of stratigraphy, 1185 p. New York: A. G. Seiler and Co. Reprinted 1960 by Dover Publications, New York. The paperback reprint has the same paging as the 1913 edition.
— Textbook of geology, Part I., 864 p. Boston: Heath 1920.
Grout, F. F.: Petrography and petrology, 522 p. New York: McGraw Hill 1932.
Ham, W. E., Ed.: Classification of carbonate rocks – a symposium. Am. Assoc. Petroleum Geologists Mem. **1**, 279 p. (1962).
Holmes, A.: The nomenclature of petrology, 284 p. London: Murby and Co. 1920.
Hubert, John F.: Petrology of the Fountain and Lyons Formations, Front Range, Colorado: Colorado School Mines Quart. **55**, 242 p. (1960).
— Textural evidence for deposition of many western North Atlantic deep-sea sands by ocean-bottom currents rather than turbidity currents. Jour. Geology **72**, 757–785 (1964).
* Humble, William: Dictionary of geology and mineralogy, 279 p. London: H. Washbourne 1840.
Klein, G. deVries: Analysis and review of sandstone classifications in the North American geological literature, 1940–1960. Geol. Soc. America Bull. **74**, 555–576 (1963).
Konta, Jiri: Problem of the quantitative petrological classification in the rock series arkose-graywacke-quartz sandstone-clay shale. Contr. Mineral. and Petrol. **19**, 125–132 (1968).
Kossovskaia, A. G.: K voprosu o klassifikatsii peschanykh porod po mineralogischeskomu sostavu (On the question of classification of sandstones according to mineralogical composition): Uchen. Zap. Leningr. Gos. Un. V., No. 310, Seriia geolog. Nauk **12**, 201–211 (1962).
Krumbein, W. C., Sloss, L. L.: Stratigraphy and sedimentation, 491 p. San Francisco: Freeman 1951.
— Stratigraphy and sedimentation, 2nd Ed., 660 p. San Francisco: Freeman 1966.
Krynine, P. D.: Petrography and genesis of the Siwalik series: Am. Jour. Sci., ser. 5, **34**, 422–446 (1937).
— Petrology and genesis of the Third Bradford Sand. Pennsylvania State College Bull. **29**, 134 p. (1940).
— Paleogeographic and tectonic significance of sedimentary quartzites (abs.). Geol. Soc. America Bull. **52**, 1915 (1941).
— Sediments and the search for oil. Producers Monthly **9**, 12–22 (1945).

Krynine, P. D.: The megascopic study and field classification of sedimentary rocks. Jour. Geology **56**, 130–165 (1948).

— Petrology, stratigraphy and origin of the Triassic sedimentary rocks of Connecticut. Connecticut State Geol. Nat. Hist. Survey, Bull. **73**, 247 p. (1950).

* Lasius, G. S. Otto: Beobachtungen über das Harzgebirge mit Karte. Hannover 1789.

Lyell, Charles: Principles of geology, 3, appendix and glossary, 398 p. London: Murray 1833.

Marchese, H. G., Garrasino, C. A. F.: Clasificación descriptiva de areniscas. Rev. Assoc. Geol. Argentina **24**, 281–286 (1969).

McBride, E. F.: A classification of common sandstones. Jour. Sed. Petrology **33**, 664–669 (1963).

Michot, Paul: Classification et terminologie des roches lapidifiées de las série psammito-pélitique. Ann. Soc. géol. de Belgique **81**, 311–342 (1958).

Okada, Hakuya: Non-greywacke "turbidite" sandstones in the Welsh geosyncline. Sedimentology **7**, 211–232 (1966).

* Oldham, Thomas: Geological glossary, for the use of students, 62 p. London: Edward Stanford 1879.

Packham, G. H.: Sedimentary structures as an important feature in the classification of sandstones. Am. Jour. Sci. **252**, 466–476 (1954).

Page, David: Handbook of geological terms, geology and physical geography, 2nd Ed., 506 p. Edinburgh-London: Blackwood 1865.

Payne, T. G. and others: The arctic slope of Alaska: U.S. Geol. Survey, Oil and Gas Investig. Map OM 126, sheet 2, 1952.

Peikh, V.: Klassifikatsiya peschanikov po veshchestvennomu sostavu (Classification of sands according to their mineralogic composition). Vest. Moskov Univ. Geol. **24**, 87–98 (1969).

Pettijohn, F. J.: Sedimentary rocks (1st Ed.), 526 p. New York: Harper 1949.

— Classification of sandstones. Jour. Geology **62**, 360–365 (1954).

— Sedimentary rocks (2nd Ed.), 718 p. New York: Harper 1957.

Pettijohn, F. J., Potter, P. E.: Atlas and glossary of primary sedimentary structures, 370 p. Berlin-Heidelberg-New York: Springer 1964.

Ries, Heinrich: Economic geology, 589 p. New York: John Wiley and Sons 1910.

* Roberts, George: Etymological and explanatory dictionary of the terms and language of geology, 139 p. London: Longmans 1839.

Rodgers, John: The nomenclature and classification of sedimentary rocks. Am. Jour. Sci. **248**, 297–311 (1950).

Sahu, B. K.: Classification of common sandstones. Punjab Univ. Research Bull. **16**, N.S. 315–322 (1965).

Shutov, V. D.: Obzor i analiz mineralogicheskikh klassifikatsii peschanykh porod (Survey and analysis of mineralogical classifications of sandstones). Litol. i Polez. Iskop., I, 95–112 (1965).

— Klassifikatsiia peschanikov (Classification of sandstones). Litologiya i Poleznye Iskopalmye **5**, 86–103 (1967).

Tallman, S. F.: Sandstone types, their abundance and cementing agents. Jour. Geology **57**, 582–591 (1949).

Teodorovich, G. I.: Rasshirennaya klassifikatsiya peschanikov po veshchestvennomu sostavu (Comprehensive classification of sandstones based on their composition). Izvest. Akad. Nauk SSSR, ser. Geol., **6**, 75–95 (1967).

Thomas, H. H.: Refractory materials. Mem. Geol. Survey, Spec. Rept., Min. Resources Great Britain **6**, 241 p. (1918) also **16**, 159 p. (1920).

Tieje, A. J.: Description and naming of sedimentary rocks. Jour. Geology **29**, 650–666 (1921).

Travis, R. B.: Nomenclature for sedimentary rocks. Am. Assoc. Petroleum Geologists Bull. **54**, 1095–1107 (1970).

Twenhofel, W. H.: Treatise on sedimentation (2nd Ed.), 926 p. Baltimore: Williams and Wilkins 1932.

Tyrrell, G. W.: Some points in petrographic nomenclature. Geol. Mag. **58**, 501–502 (1921).

— The principles of petrology. 349 p. New York: Dutton 1926.

Wang, Chao-Siang: On the occurrence of quartz wacke and its bearing on the problems of sandstone classification. Geol. Soc. China Proc., **10**, 99–106 (1967).

Williams, H., Turner, F. J., Gilbert, C. M.: Petrology: 406 p. San Francisco: Freeman 1954.

Chapter 6. Petrography of Common Sands and Sandstones

Introduction

There comes a time in the study of a sandstone body or formation when it is necessary to look at the rocks of which it is made. It is not enough to measure a stratigraphic section, trace out the limits of a given sandstone, or even to study its structures in the field and map its paleocurrents. The rocks which constitute the formation cannot be ignored. A close look at these cannot be made without study of thin sections under the microscope. Such study is not to classify or name the rock but rather to understand it. It is necessary to know of what kinds and classes of minerals it is made and how they are put together. Petrographic studies can greatly expand our understanding of the geologic history – can contribute to the questions of provenance by supplementing or confirming paleocurrent studies, indicate probable source rocks as well as source areas, can assist, perhaps, in making environmental discriminations, provide insight into the nature of the rock fabric and pore system, and shed light on its diagenetic history – including grain alterations and cementation, changes that profoundly alter the porosity and permeability of the rock.

The objectives of this chapter are (1) to *describe* the principal families of sandstone and some of the more important species in each group and to point out some of the problems of origin of each of them, (2) to summarize what is known about their *relative abundance*, and (3) to summarize the principal theories of *sandstone petrogenesis.*

We have grouped sandstones into genetically meaningful classes as discussed in Chap. 5, namely those based on maturity as expressed by composition and sorting. These are (1) the immature sandstones, that is, the least modified residues – those closest in composition to the parent rock, (2) the mature sandstones, those nearest the theoretical end-product, and (3) various hybrid types such as tuffaceous sandstone, and other hybrids with a significant fraction of sand formed by chemical or biochemical processes, and (4) a few rare types of sand which do not fit readily in any of the above categories.

The petrography of a sandstone is determined in large measure by provenance (Fig. 6-1). The character of the immature sands, in particular, is controlled primarily by the nature of the source rock. The ultimate source of most sands is, of course, the quartz-bearing plutonic rocks – typically the granites and quartz monzonites, and the feldspar-rich metamorphic rocks such as the gneisses and coarse schists and granulites. A large class of highly feldspathic sands – the *arkoses* – are a direct product of the disintegration of these rocks. Those sands rich in rock particles – the *lithic arenites* – are derived from supracrustal rocks including pre-existing sediments (sandstones, shales and limestones), their

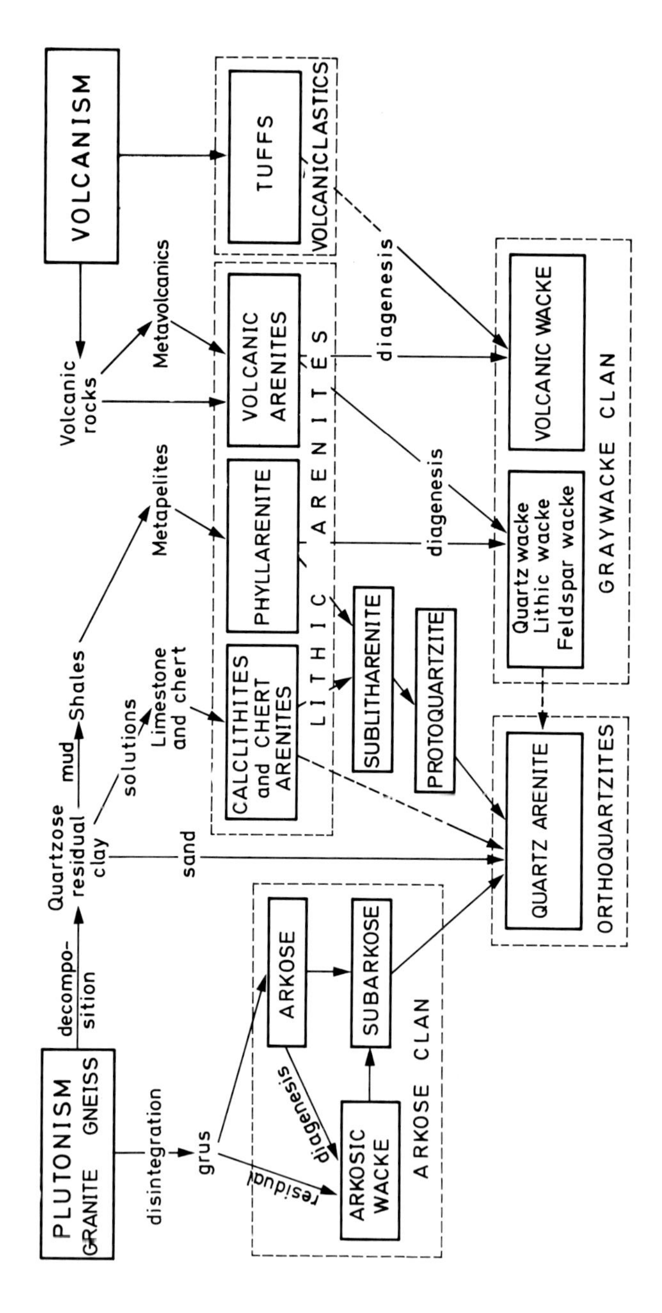

Fig. 6-1. Provenance and evolution of the noncarbonate sands

metamorphic equivalents, and the effusive volcanic rocks. The effect of provenance is greatly diminished in the mature sands – the *quartz arenites* (orthoquartzites) which tend to converge toward a common end-type. A great deal of skill is required to determine either the immediate or ultimate source of these sands.

Finally, all sands undergo diagenetic changes. These lead to a diversity of cementing agents. In some sands the changes lead to a breakdown of some of the framework elements and the production of a matrix. Matrix formation is accompanied by albitization of the feldspar and other changes profound enough to produce a separate class of rocks – *the graywackes.*

Feldspathic Sands and Arkose

Definitions

The term *arkose* was apparently first used by Brongniart who, in 1826, applied this term to some sandstones in the Auvergne district of France, the essential constituents being quartz and feldspar (Oriel, 1949). Incidental components include mica and clay, usually kaolinitic. The term has since been redefined as "sandstone containing 25 or more percent of feldspars usually derived from the disintegration of acid igneous rocks of granitoid texture" (Allen, 1936, p. 44). Most of the earlier definitions emphasized the coarseness of grain and the resemblance to granite from which these sands were presumably derived.

The above definitions are all in some degree unsatisfactory. Graywackes may likewise contain 25 or more percent of feldspar. Some authorities have applied the term arkose if the feldspar was conspicuous enough to be seen easily in hand specimen. Such rocks may contain 20 percent or even less feldspar (Krynine, 1950, p. 101). Furthermore, definitions based on inferred provenance are always difficult to apply but inasmuch as the feldspar of arkose, in many cases forming 40 to 50 percent of the rock, is characteristically potassium feldspar (microcline), the granitic derivation seems well established, although there are arkoses with substantial, and in rare cases dominant, plagioclase.

Although there is no general agreement on how little feldspar a sandstone can have and still be called an arkose, the 25 percent figure is most often cited. Krynine (1940, p. 50) suggested 30 percent but later (1948, Fig. 11) indicated 25 percent to be the *average* feldspar content and designated some rocks with less than 20 percent feldspar arkose. There is no natural discontinuity in the abundance of feldspar. But because sandstones with about 25 percent or more have a field appearance that is distinctive, this figure may be the best basis for the choice. The term *feldspathic sandstone* merely means any sandstone with appreciable feldspar – arkoses and graywackes included. It has also been used to designate a class of sandstones less feldspathic than arkose but more so than normal sandstones. For this group of sands (10 to 25 percent feldspar) generally lacking in rock fragments, the term *subarkose* has been suggested (Folk, 1954; Pettijohn, 1954, p. 364). Arkose has also been redefined as a sand containing 25 percent or more of labile constituents (rock fragments and feldspar) of which feldspar forms half or more (Pettijohn, 1957b, p. 322). By this definition arkose might contain as little as 12.5 percent feldspar. Subarkose similarly defined, containing 10 to 25 percent labile components of which feldspar is dominant, might have as little as 5 percent feldspar.

General Description

Typical arkose is a coarse-grained rock consisting of quartz and feldspar. The feldspar imparts a pink color to the rock. Though normal arkose is pinkish to reddish, some may be derived from granitic or gneissic rocks containing gray or white feldspar and such arkoses may themselves be gray or white – normally becoming lighter colored in outcrop. Some arkoses are associated with red beds but the two should not be confused as being necessarily the same.

In many cases the arkosic beds are massive and being of coarse grain and pink color, they resemble granite from which they were presumably derived. Some arkose is indeed no more than *in situ* disintegrated and weathered granite. Transported arkoses, on the other hand, usually display stratification, in some cases prominent crossbedding, somewhat better-rounded grains, and a higher proportion of quartz.

The dominant mineral of arkoses, as of most sandstones, is quartz though, exceptionally, feldspar may exceed quartz in volume (Table 6-1). Because of coarseness of grain, considerable polycrystalline quartz is present; also present may be composite granules consisting of both quartz and feldspar. The grains are generally irregular and poorly rounded.

Table 6-1. *Mineral composition of arkose and subarkose (Percent)*

	A	B[1]	C	D	E	F[1]	G	H	I	J
Quartz	60	57	57	71	60	35	37.7	57	51	53.1
Microcline	34	35[2]	27	25	13	59[2]	0.7	24	30	18.5
Plagioclase	—		1				45.4	6	11	0.4
Micas	—	—	—	—	T	—	4.2	3	1	6.9
Clay	—	—	—	—	5	—	12.0	9	7	17.0
Carbonate	—	P[3]	—	—	—	2	—	P[3]	P[3]	—
Other	6[4]	8[5]	14	4	8	4[5]	—	1	—	4.1

(1) Normative or calculated composition; (2) Modal feldspar, given by Mackie as 55 and 60, respectively; (3) Present in amounts under 1 percent; (4) Chlorite; (5) Iron oxide (hematite) and kaolin.

A. Sparagmite (Precambrian) Norway (Barth, 1938, p. 60).
B. Torridonian (Precambrian) Scotland (Mackie, 1905, p. 58).
C. Jotnian (Precambrian), Satakunta, Finland (Simonen and Kuovo, 1955, Table 2, No. 5).
D. Subarkose, Potsdam Sandstone (Cambrian), New York, U.S.A. (Wiesnet, 1961, p. 9). A subarkose.
E. Subarkose, Lamotte Sandstone (Cambrian), Missouri, U.S.A. (Ojakangas, 1963, p. 863). A subarkose.
F. Lower Old Red (Devonian) Scotland (Mackie, 1905, p. 58).
G. Arkose (Permian), Auvergne, France (Huckenholtz, 1963, p. 917).
H. Pale arkose (Triassic) Connecticut, U.S.A. (Krynine, 1950, p. 85).
I. Red arkose (Triassic) Connecticut, U.S.A. (Krynine, 1950, p. 85).
J. Arkose (Oligocene), Auvergne, France (Huckenholtz, 1963, p. 917).

The feldspar is, with few exceptions, dominantly K-feldspar, usually microline. It varies from extremely fresh to weathered (kaolinized) to a mixture of both fresh and weathered. In arkoses with a carbonate cement, the feldspar may show varying degrees of replacement – from grains with corroded borders to isolated but oriented residuals to completely replaced grains. Kaolinization, either before or after deposition, is common. In some porous arkoses the feldspars display limpid secondary overgrowths. If attached to clouded detrital cores, the alteration of the latter is clearly pre-depositional.

Table 6-2. *Chemical analyses of arkose and subarkose (from Pettijohn, 1963, Table 8, with additions)*

	A	B	C	D	E	F	G	H	I	J	K	L
SiO_2	79.30	75.80	80.89	87.02	92.60	73.32	59.24	92.13	85.74	69.94	72.21	76.6
Al_2O_3	9.94	11.74	7.57	2.86	3.52	11.31	6.65	4.42	6.84[1]	13.15	10.69	12.4
Fe_2O_3	1.00	0.59	2.90	0.49 }	0.44 (Fe₂O₃ + FeO)	{ 3.54	2.02	0.37	0.79 }	2.48 (Fe₂O₃ + FeO)	{ 0.80	0.7
FeO	0.72	1.31	1.30	0.28 }		{ 0.72	0.31	0.33	— }		{ 0.72	0.2
MgO	0.56	0.54	0.04	0.20	0.04	0.24	0.12	0.14	1.11	Trace	1.47	0.3
CaO	0.38	1.41	0.04	3.41	0.06	0.75	16.04	1.27	0.49	3.09	3.85	0.4
Na_2O	2.21	2.40	0.63	0.00 }	2.93 (Na_2O + K_2O)	{ 2.34	0.19	0.11	1.16	5.43	2.30	0.3
K_2O	4.32	4.51	4.75	1.98 }		{ 6.16	2.30	0.72	2.19	3.30	3.32	3.8
H_2O+	0.55	0.86 }	1.11 (H_2O+ and −)	{ — }	0.17 (H_2O+ and −)	0.30 (H_2O+ and −)	{ 1.26	—	—	—	1.46 }	2.7 (H_2O+ and −)
H_2O-	0.41	0.03 }		{ — }			{ —	—	—	—	0.08 }	
TiO_2	0.22	0.15	0.40	—	—	—	—	—	0.38[3]	—	0.22	0.6
P_2O_5	0.05	0.60	—	—	0.02	—	—	—	0.01	—	0.10	0.2
MnO	0.02	0.05	—	—	—	—	0.50[4]	0.24[4]	—	0.70	0.22	—
CO_2	—	Trace	—	—	0.06	0.92	12.16	None	—	—	2.66	—
Ign. loss	—	—	—	3.35	—	—	—	0.42	1.12	1.01	—	—
Total	99.68	99.99	99.63	99.65[5]	99.84	99.60	100.79	100.15	99.83	99.10	100.10[6]	100.6

(1) Contains MnO_2; (2) Total iron; (3) Contains ZrO_2 and V_2O_5; (4) Reported as MnO_2; (5) Includes 0.06 percent S; (6) Sum given in original as 99.90.

A. Jotnian (Precambrian) Köyliö, Muurunmäki, Finland. H B. Wiik, analyst (Simonen and Kouvo, 1955, p. 63). 44 percent normative feldspar.
B. Torridonian (Precambrian) Kinlock, Skye, M. H. Kerr, analyst (Kennedy, 1951, p. 258). 53 percent normative feldspar.
C. Sparagmite (Lower Cambrian), Engerdalen, Norway (Barth, 1938, p. 58). 33.5 percent normative feldspar.
D. Calcareous subarkose (Cambrian or Ordovician), Bastard Township, Ontario, Canada (Keith, 1949, p. 21). About 12 percent feldspar and 7 percent calcite.
E. Subarkose, Potsdam sandstone (Cambrian), New York, P. L. D. Elmore and K. E. White, analysts. 17 percent normative feldspar (Wiesnet, 1961, p. 9).
F. Lower Old Red Sandstone (Devonian), Foyers, Loch Ness, Scotland (Mackie, 1905, p. 58). 52 percent normative feldspar.
G. Calcareous arkose, Old Red Sandstone (Devonian), Red Crags, Fochabers-on-Spey, Scotland (Mackie, 1905, p. 58). 16 percent normative feldspar and 28 percent normative calcite.
H. Subarkose, Rosebrae Sandstone (Devonian), Rosebrae, Elgin, Scotland (Mackie, 1905, p. 59). About 12 percent normative feldspar.
I. Subarkose of Whitehorse Group (Permian), Kansas (Swineford, 1955, p. 122).
J. Portland Stone (Newark Group, Triassic), Portland, Conn. (Merrill, 1891, p. 420). 74 percent normative feldspar.
K. Molasse arkose (Oligocene, Zugertypus), Unterägeri, Kt. Zug, Switzerland, F. de Quervain, analyst (Niggli and others, 1930, p. 262).
L. Arkose (Oligocene), Auvergne, France (Huckenholtz, 1963, p. 917). 19 percent feldspar.

Large detrital micas characterize arkoses, both muscovite and biotite (and chloritized biotite) are common. The mica flakes, commonly considerably larger than the associated quartz and feldspar, tend to lie parallel to the bedding and hence to one another. The flakes may be bent or deformed by pressure of adjacent grains. The biotite may show chloritization or, more commonly alteration and oxidation. Mafic minerals, other than biotite, are absent implying that they were the chief loss by chemical weathering.

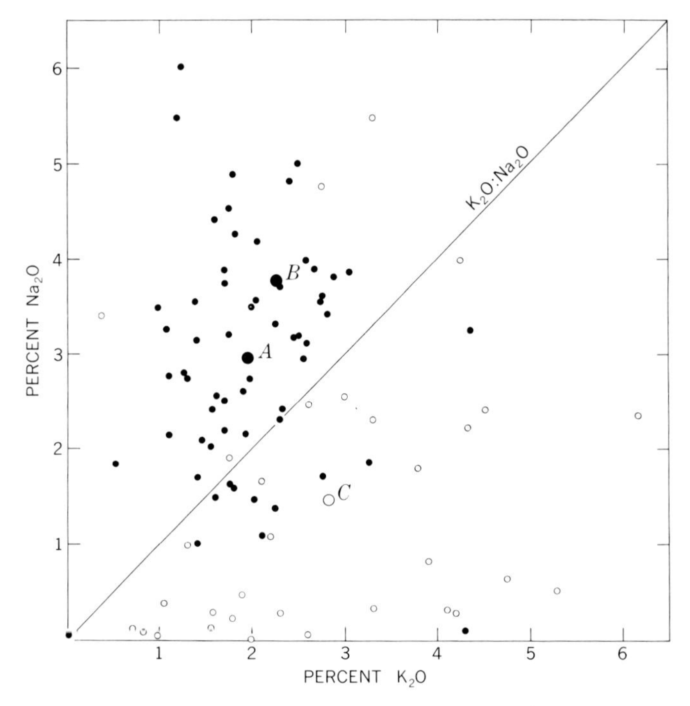

Fig. 6-2. Na_2O/K_2O ratio in arkoses and graywackes. Solid black circles, graywackes; open circles, arkoses. A, average graywacke; B, composite New Zealand graywacke (Reed, 1957, p. 16); C, average arkose (from Pettijohn, 1963, Fig. 2; data from same source)

Arkosic sandstones of mixed provenance may contain rock fragments and pass by degrees into coarse lithic arenites.

Some arkoses contain matrix clay – commonly kaolinitic and ironstained. Other varieties have but little such matrix and are generally carbonate cemented. The oldest arkoses show secondary overgrowth of the quartz and feldspar and if such enlargements are carried to completion, the resulting rock may so much resemble a granite or granitic gneiss as to be mistaken for such in small outcrops. This is especially true in some Precambrian terranes.

Because they are usually derived from K-rich granitoid rocks, arkoses form a chemically homogeneous group as representative analyses show (Table 6-2). Arkoses, like their source rocks, are rich in Al_2O_3 and K_2O, the latter, unlike in graywackes, generally exceeding Na_2O (Fig. 6-2). Most arkoses, perhaps because they are generally subaerial rather than submarine, have unlike graywackes, an excess of Fe_2O_3 over FeO. Those with a carbonate cement run high in CaO and CO_2.

Varieties and Types of Arkose

We have followed the usage suggested by Gilbert (Williams, Turner, and Gilbert, 1954, p. 310) and distinguished between those arkosic sands with little matrix – the *arkosic arenites* – and those with a significant clay content – the *arkosic wackes*. The latter might be confused with the feldspathic graywackes except that, in general, the latter have sodic rather than potassic feldspar and a chlorite-rich rather than clay-rich (or sericitic) matrix and hence are dark rather than light colored.

Some of the arkosic wackes are produced by the *in situ* disintegration of granite and related rocks. Such material has been termed "grus." Grus has a mineral composition similar to that of the parent rock. Although not sorted by a transporting agent, it has a characteristic Rosin's Law size-distribution (McEwen and others, 1959) – essentially the size-distribution which characterizes crushed materials. Such sedentary or *residual arkose* commonly marks an unconformity between a granitic basement and an overlying sedimentary sequence. The residual arkose may contain more quartz than normally found in granite; it grades upward to an arkose which is faintly stratified and which may contain scattered pebbles of granite and grades downward into fresh unweathered granite. When such a "graded unconformity" has been imprinted by metamorphism the relations become less clear and the relative age of the granite and the metasedimentary sequence may be misinterpreted. One of the best illustrations of this problem is the contact between the Archean Knife Lake sediments and the Saganaga Granite on Cache Bay on the Canadian-Minnesota boundary (Gruner, 1941, p. 1599).

Residual arkosic materials may be shifted downslope and deposited as fans or aprons of waste materials which extend into the basin and are intercalated with more normal, better stratified and sorted sediments. Such tongues of "granite wash" encountered in drill holes may be mistaken for "basement" and lead to abandonment of drilling. Many of these arkoses contain an abundant clay-rich matrix. They are the product of very limited transportation, perhaps by mass movement, with very imperfect or incomplete sorting. Rocks of this type having a deep red-stained matrix have been designated *redstones* by Krynine (1950, p. 103) from their occurrence on Redstone Hill in Connecticut. This term has also been applied to phases of the Fountain Formation in Colorado (Hubert, 1960, p. 65). These rocks have a minimum of 20 percent of matrix.

Although arkoses with a prominent matrix are probably sedentary or residual, or if "granite wash" a product of limited transport, others may be the result of post-depositional alteration that can be an important part of the diagenesis of continental beds. Arkosic sands so altered will show pseudomorphs of kaolin after feldspar.

Arkosic materials which have undergone considerable reworking by rivers, or by the sea, are fairly well sorted and matrix-free. These are the *arkosic arenites* of Gilbert (Williams, Turner, and Gilbert, 1954, p. 292). Older deposits of this sort may have a normal mineral cement precipitated in the pore system. Such cement is commonly a carbonate – generally calcite. In these arkoses the grains may show incipient to good rounding. The proportion of quartz rises as the badly weathered feldspars are eliminated by abrasion. Stratification becomes noticeable and crossbedding may be conspicuous. As is common in carbonate-cemented sands, the cement may marginally replace or embay the framework grains. Embayment and partial to complete replacement of the feldspar is especially noteworthy.

In addition to the textural variations which distinguish the arkosic wackes from the arkosic arenites, there are mineralogical variations in the composition of the framework components. Arkose normally is, and by tradition defined as, the product of disintegration of a granite. By "granite" is usually meant a coarse-grained plutonic rock, either igneous or metasomatic, of which K-feldspar is the dominant, or at least a major constituent. What about the comparable sandstones derived from rocks in which plagioclase is the dominant or even the sole feldspar? These have been called "plagioclase arkoses" and although they are, in general, not very common, some notable examples have been described. The Paleocene Swauk Formation in the State of Washington is predominantly a plagioclase arkose presumably derived from a quartz dioritic source.

Some volcanic sands resemble arkose in external appearance although they contain little feldspar. The role of feldspar is taken by light-colored, acid volcanic rock fragments – in many cases reddish rhyolitic or related materials. These rocks also contain quartz, some of volcanic origin, and feldspar, usually zoned and volcanic in origin. These sandstones are derived from volcanic terranes – the more acidic flows providing an undue share of the debris from such regions. These sands, like most arkoses, have had limited transport, are coarse, only moderately well sorted, and generally deposited in continental basins associated with coarser clastics. They are described more fully in Chapter 7.

Field Occurrence and Examples

Arkoses, as here defined, form no more than 15 percent of all sandstones (Pettijohn, 1963, p. 515) though some estimates are significantly higher.

Arkoses occur in all major geologic systems and range in age from Precambrian to Recent. Well-known Precambrian examples include the Sparagmites of Norway and Sweden (Hadding, 1929, p. 151; Barth, 1938, p. 60) and their presumed equivalent in the Northwest Highlands of Scotland, the Torridon sandstones (Peach and others, 1907, p. 278; Kennedy, 1951). The late Precambrian Jotnian sandstone of Finland (Simonen and Kuovo, 1955, p. 60) belongs here as does the lower part of the Huronian Lorrain Formation of the Canadian Shield which in places carries as much as 40 percent feldspar. All of these units form coarse, strongly crossbedded sequences and are associated with pebbly sands of conglomerates.

Well-known and well-described are the arkoses of the Old Red Sandstone of Scotland containing up to 60 percent feldspar (Mackie, 1899a), the arkoses of the Triassic Newark Group (Fig. 6-3) of Connecticut (Krynine, 1950) and the central and southern Appalachians (Fig. 6-4) and the arkosic beds of the Fountain and Lyons Formations (Pennsylvanian) of the Front Range of Colorado (Hubert, 1960). Another well-known arkose is that of the Swauk, over 5000 ft. thick, of Paleocene age in Washington. The feldspar of this formation, unlike that of most arkoses, is predominantly plagioclase (Foster, 1960, p. 105). Where the Tertiary Molasse of Switzerland had a granitic source, it is an arkose

Fig. 6-3. Sugarloaf arkose, Newark Series (Triassic), Mt. Tom, Massachusetts, U.S.A. Crossed nicols, × 20. A typical coarse red arkose consisting of a very poorly sorted mixture of angular quartz, feldspar, together with a little mica set in a red, ferruginous clayey matrix

commonly containing 50 to 60 percent of feldspar (Gasser, 1968, Table 10). For an extended tabulation of ancient arkoses, the reader is referred to Barton (1916).

The sandstones of the Precambrian Keweenawan Series of the Lake Superior region are in part arkosic (Irving, 1883, p. 128). For the most part, however, they are volcaniclastic arenites rather than true arkoses, rocks in which the quartz and feldspar is diluted by a large volume of acidic volcanic rock particles all cemented together by calcite.

Formations which themselves are not normally arkoses may have an arkosic basal facies at or near their contact with an underlying granitic basement. Such is the case with the Cambrian Lamotte Sandstone of Missouri (Ojakangas, 1963) and the Cambrian Potsdam Formation of New York (Wiesnet, 1961). These formations, which are primarily quartz arenites, locally contain as much as 30 percent and 52 percent feldspar respectively. Rocks such as the Lamotte show a commingling of little worn and partially weathered feldspar and quartz derived from the subjacent terrain and smaller but well-rounded quartz derived from more distant sources.

Fig. 6-4. Arkose from Newark Series (Triassic). Deep River Basin, North Carolina. Crossed nicols, × 20. A very coarse, immature sandstone, with abundant rock fragments as well as quartz and feldspar. Note composite character of some of the quartz, the alteration of the finely-twinned feldspar, and rock fragments

From a review of the known occurrences of arkose, it is clear that, leaving aside the basal arkoses related to marine transgression over granites, significant arkose accumulation can be expected only when sharp uplifts bring a granitic or gneissic basement up into the zone of erosion. If subsidence occurs adjacent to such elevated basement blocks, a local but thick arkose accumulation can be expected.

Significance and Origin

The question of presence or absence of feldspar in sandstones is at the heart of the arkose problem – its origin and significance. Obviously provenance is a major factor and arkose implies a richly feldspathic source area and one usually characterized by K-feldspar bearing plutonic granites and gneisses.

But beyond provenance the question is: under what conditions is feldspar released to the sediments rather than decomposed to clay in the source region? Normally, or certainly ultimately, the feldspar is so decomposed and the clay formed is separated from the quartz so that the resulting sand is quartz-rich and and feldspar-free. Years ago, Mackie (1899b, p. 444) concluded that under conditions of extreme aridity or extreme cold that the processes of weathering would

Fig. 6-5. Fountain Sandstone (Pennsylvanian). About 100 ft (30 m) above Precambrian basement, Red Mountain, 6 mi (10 km) south of Wyoming border, Laramie County, Colorado, U.S.A. Crossed nicols, × 55. An arkose marked by an abundance of detrital (?) dolomite grains, quartz, and feldspar cemented by calcite

be inhibited or retarded so that incompletely weathered materials would escape and become part of sediments produced. Arkoses thus became indicators of desert or glacial conditions. Many glacial sands are indeed highly feldspathic. But this concept has been challenged. It has been observed, for example, that many Eocene sandstones in California which contain up to 50 percent feldspar, yield a flora which could only have lived under warm, humid conditions (Reed, 1928). The Tertiary Catahoula Sandstone of Texas, also contains a similar quantity of feldspar (Goldman, 1915, p. 273) and likewise contains a flora characteristic of a tropical coastal region. Krynine (1935) observed modern arkose accumulation in tropical Mexico where the mean temperature is 80° F (27° C) and the annual precipitation 120 inches (305 cm).

These observations led Krynine to the view that the necessary condition for production of arkosic sands was high relief with consequent rapid erosion rather than adverse climatic conditions. Under conditions of high relief and rapid erosion producing deeply incised canyons, both fresh and partly weathered feldspars are incorporated in the sediments. If relief is low, slopes are more stable and covered by vegetation. Weathering is then more likely to go to completion and yield only a quartz sand. Detrital feldspar, therefore, is *the result of a balance between the rate of decomposition and the rate of erosion.* Arkose may be, therefore, a product of rigorous climate *if* decomposition is inhibited but it may be a product of high relief *if* erosion is accelerated. Distinguishing between these possibilities in an ancient arkose is difficult but on the whole, it seems more likely that high relief is more important than rigorous climate in arkose formation. The orogenic sediments, those related to orogenesis, are more feldspathic than those related to erosion and deposition in tectonically stable areas. Gibbs (1967) in his study of the Amazon River sediments has shown that high relief is the dominant factor. But Strakhov (1967, p. 48, 95) argues strongly for the effect of climate.

It has been presumed that feldspar is more subject to abrasion and destruction during transport than quartz. This conclusion, set forth by Mackie (1899a, p. 149) as a result of his studies of Scottish streams, seems to be confirmed in some measure by similar studies of the mineralogy of the steep-gradient streams draining the Black Hills of South Dakota (Plumley, 1948, p. 562). However, larger, low-gradient streams seem capable of transporting feldspar long distances without significant loss. The Mississippi sands, for example, near Cairo, Illinois, contain about 25 percent feldspar; at the delta, some 1100 miles (1760 km) downstream these sands still contain 20 percent feldspar (Russell, 1937, p. 1334). The presence of feldspar, therefore, does not imply short transport.

Lithic Arenites and Related Rocks

Definitions and Nomenclature

Sandstones which contain a substantial quantity of rock particles and which have little or no matrix materials but have instead an empty pore system or one filled with a precipitated mineral cement were early recognized and described but not specifically named. The term *subgraywacke* was applied to this group (Pettijohn, 1949, p. 255) because of the superficial resemblance of many of these sandstones to graywacke. In fact, the term "graywacke" or "low rank graywacke" itself has been applied to this class of rocks (Krynine, 1945, Table 1) but this usage is being gradually abandoned in favor of the term *lithic arenite* – a term proposed by Gilbert (Williams, Turner, and Gilbert, 1954, p. 304). The term has been shortened to *litharenite* (McBride, 1963, p. 669). Pettijohn (1954, Fig. 1) has used the term *lithic sandstones* to encompass both subgraywackes with 25 percent or more labile components, rock particles exceeding feldspar, and *proto-quartzites*, a term suggested by Krynine (Payne, 1952), with 5 to 25 percent labile constituents. The latter have also been designated *sublitharenites* (McBride, 1963, p. 667).

General Description

Lithic arenites are generally light gray, "salt and pepper" sands with an abundance of rock particles, especially of sedimentary and low-rank metamorphic rocks, subangular to rounded quartz, and with a chemical cement, either quartz or calcite. Mica flakes are common; feldspar is not. Mineral charcoal (fusain) and carbonaceous plant fragments are common in Devonian and younger lithic arenites. Shale pebbles may also be present. Detrital matrix is generally absent though a pseudomatrix (squashed shale particles) or authigenic precipitated clay may be present.

Of all sandstones, probably not even excepting graywackes, lithic arenites show the greatest diversity of both mineralogical and chemical composition. This variability reflects the importance and relative abundance of the diverse rock particles which these sands may contain. If the rock particle content is small, these sands pass over into the quartz arenites or orthoquartzites. On the other hand, rock particles may form over half of the framework and in a few rare cases, all of it.

The rock particles in lithic arenites are themselves diverse – their only common character is their fine grain. The number of rock types which have been described in these sandstones is formidable – even in a single sand as many twenty have been identified (Mattiat, 1960). In general, however, only a few kinds are prominent. They fall into three main classes: (1) *volcanic*, that is, particles of aphanitic flow rocks, (2) low grade *metamorphic* rock particles such as slate, phyllite and mica schist, and (3) *sedimentary* rock particles which include various kinds of shale, siltstone, argillite and related pelitic material but generally including also some detrital chert and, in some cases, micritic limestone and dolomite. There are sandstones in which these latter constituents, chert and micritic limestone, become dominant.

The arenites with volcanic rock particles are a special class and are a product of a restricted provenance. But because they constitute an important class of sands apart from the normal lithic sandstones, they have been treated in the following chapter. But as many lithic arenites, as well as graywackes, are of mixed provenance, volcanic rock particles are to be looked for as minor components in many of these sandstones.

The unique character of the lithic sandstones – the volcanic arenites excluded – is the abundance and variety of sand-sized particles of pelitic derivation: shale, siltstone, slate, phyllite and mica schist. These may form a substantial part of the sand as a whole – up to 50 percent or more. Their uniqueness lies in the fact that clay-sized materials have been built up – either through diagenesis or metamorphism – into rocks that yield sand-sized materials so that what were once two classes of material, sand and mud, sorted and deposited separately from one another, are now deposited in one and the same place as well-sorted sand. The lithic arenites with the pelitic materials of sedimentary and metasedimentary origin are the most common. Representative modal analyses and averages of these phyllarenites, together with a few abberrant or special types are given in Table 6-3.

Table 6-3. *Modal analyses of lithic sandstones (subgraywackes) and protoquartzites (sublitharenites) (from Pettijohn, 1963, Table 3, with additions)*

	A	B	C	D	E	F	G	H
Quartz	50	60	78 }	65.4	{ 32.0	71	30.9	27
Feldspar	3–5	3	3 }		{ 2.2	8	10.0	2
Mica	—	1	—	—	0.2	tr	0.5	—
Rock fragments	40	35	15	10.6	43.0[2]	22[4]	33.0[1]	46[5]
"Clay" or matrix	10	2	4	6.8	6.9	2	5.5	5
Silica cement	—[3]	—	—	11.9	trace	—	—	—
Calcite cement	—	present	—	8.5	13.0	—	19.2	20

(1) Includes 15.0 percent chert; (2) Includes 28.0 percent chert; (3) 5–10 percent, author's observation; (4) Includes 5.0 percent chert; (5) Includes 3.0 percent chert, 12.0 percent limestone, 27.0 percent dolomite.

A. Oswego Sandstone (Ordovician), Pennsylvania, U.S.A. (Krynine and Tuttle, 1941).
B. Bradford Sand (Devonian), Pennsylvania, U.S.A. (Krynine, 1940, C-1, Table 3).
C. Deese Formation (Pennsylvanian), Oklahoma, U.S.A. (Jacobsen, 1959, Table 4, Analysis D-112).
D. Salt Wash Member of Morrison Formation (Jurassic), Colorado Plateau, U.S.A. Mean of 25 thin sections (Griffiths, 1956, p. 25).
E. "Calcareous graywacke" (Cretaceous), Torok, Alaska. Average of 3 samples (Krynine in Payne and others, 1952).
F. Basal Claiborne Sand (Eocene), Texas, U.S.A. (Todd and Folk, 1957).
G. "Frio" Sandstone (Oligocene), Seeligson field, Jim Wells and Kleberg Counties, Texas, U.S.A. Average of 22 samples (Nanz, 1954, p. 112).
H. Molassesandstein (Tertiary), Germany (USM No. 186, Füchtbauer, 1964, p. 256).

Although the rock particles are the definitive component of lithic arenites, quartz is a prominent, and generally dominant detrital component. In those sands with rock particles of sedimentary origin (chert, limestone and dolomite, pelitic rock particles, etc.), it is probable much of the detrital quartz, perhaps most of it, is derived from sedimentary sources – mainly pre-existing sandstones. Quartz, released by disintegration of older sandstones, is apt to be better rounded than the quartz of most arkoses and graywackes. Inasmuch as many lithic arenites have a quartz cement, such original rounded detrital quartz grains may show good secondary overgrowths. In those sands rich in metamorphic rock particles, the quartz also is presumably largely of metamorphic origin, although this is difficult to demonstrate. Such arenites will, perhaps contain a higher proportion of undulatory and polycrystalline quartz than those of sedimentary or volcanic provenance. The quartz is apt to be angular to subangular.

As might be expected, sands derived principally from sedimentary and low-rank metamorphic sources will contain little feldspar. Such feldspar as is present is apt to be better rounded than the feldspar of arkose. Though one might expect a second-cycle arkose, the labile character of the feldspar makes this unlikely.

A notable constituent of many lithic arenites is detrital mica which tends to concentrate on certain bedding planes and impart a sheen to such surfaces. Both biotite and muscovite are present, the latter being the more common. The mica flakes are deposited parallel to the bedding and hence parallel to one another.

They may show deformation due to compaction and appear bent and wrapped around adjacent quartz grains.

The lithic arenites may be cemented either with carbonates or with silica or both. Little or no matrix material is present though some of the weaker argillaceous rock particles may, especially in the older, more compacted varieties, be deformed between adjacent more durable quartz grains in such a manner as to resemble a matrix-filled pore. Such squashed rock particles are recognized only

Table 6-4. *Chemical composition of lithic sandstones (subgraywackes) and protoquartzites (modified from Pettijohn, 1963, Table 4)*

	A	B	C	D	E	F	G	H
SiO_2	92.91	74.45	40.35	84.01	65.00	56.80	51.52	47.75
Al_2O_3	3.78	10.83	7.43	2.57	9.57	8.48	5.77	6.41
Fe_2O_3	trace	4.62	3.27	0.17	1.59	1.67	2.43	2.39
FeO	0.91	—	—	0.26	1.08	—	—	—
MgO	trace	1.30	10.28	0.67	0.40	1.24	0.95	4.48
CaO	0.31	0.35	12.00	5.41	10.10	15.25	16.96	18.75
Na_2O	0.34	1.07	0.54	0.17	2.14	1.31	1.32	1.20
K_2O	0.61	1.51	0.93	0.86	1.43	1.46	1.90	1.02
H_2O+	1.19	4.95	6.75	0.54	0.82	0.50	2.25	1.32
H_2O-		—	—	0.19	0.23	—	2.54	—
CO_2	—	trace	17.80	4.65	6.90	12.95	13.30	17.78
TiO_2	—	0.50	0.30	0.05	—	0.10	0.32	0.20
P_2O_5	—	trace	—	0.04	—	trace	0.10	0.10
SO_3	—	—	—	—	0.04	—	0.52	—
Cl	—	—	—	0.02	—	—	—	—
F	—	—	—	0.01	—	—	—	—
S	—	—	—	0.02	0.16	—	—	—
MnO	—	—	—	0.04	—	—	0.14	—
BaO	—	—	—	0.05	—	—	—	—
C	—	—	—	—	0.06	—	—	—
Ign. Loss	—	—	—	—	—	—	—	—
Total	100.05	99.58	99.65	99.73[1]	99.54[2]	99.76	100.06	101.40[3]

(1) Includes Cl, 0.02; F, 0.01, BaO, 0.05; (2) Includes C, 0.06; Cu, 0.002; V, 0.017; Zn, <0.03; Cr, 0.003; (3) Sum given as 99.40 in original.

A. Protoquartzite, Berea sandstone (Mississippian), Berea, Ohio, U.S.A., L.G. Eakins, analyst (Clarke, 1890, p. 159).

B. Coal measure sandstone (subgraywacke?) (Carboniferous) Westphalian coal basin, France-Belgium (Hornu and Wasmes), (Cayeux, 1929, p. 227).

C. Coal measure sandstone (calcareous subgraywacke?) (Carboniferous), Westphalian coal basin, France-Belgium (Hornu and Wasmes), (Cayeux, 1929, p. 227).

D. Protoquartzite, Salt Wash Member of Morrison Formation (Jurassic). Composite of 96 samples, Colorado Plateau, U.S.A. Unmineralized. V.C. Smith, analyst (Pettijohn, 1963, Table 4).

E. Calcareous subgraywacke (lithic arenite), (Oligocene, "Frio" Formation), Seeligson field, Jim Wells and Kleberg Counties, Texas, U.S.A. (Nanz, 1954, p. 114). Composite of 10 samples. Also included Cu 0.002, V 0.017, Zn less than 0.03, and Cr 0.003.

F. Calcareous subgraywacke (Tertiary Molasse Aquitanian), Lausanne, Switzerland (Cayeux, 1929, p. 161).

G. Calcareous subgraywacke (Tertiary Molasse), Grönchen, Burghalde, Kt. Aargan, Switzerland. J. Jakob, analyst (Niggli, and others, 1942, p. 263).

H. Calcareous subgraywacke (?) (Molasse Burdigalian), Voreppe (Isere), France (Cayeux, 1929, p. 163).

with difficulty (Allen, 1962, p. 669). They appear to fill some pores – not others. True matrix would be distributed among *all* pores. The pseudomatrix may also show relict bedding, albeit somewhat deformed, characteristic of shales and siltstones. Moreover, the fragments, though deformed by compaction, are not all alike either in color or in texture. True matrix materials should exhibit greater uniformity.

Representative chemical analyses of lithic arenites are given in Table 6-4. Most of the analyses in this table are of phyllarenites. These are characterized by the high Al_2O_3 and in some, particularly the older ones, relatively high K_2O contents, reflecting the pelitic nature of the rock particles. They are relatively low in Na_2O and MgO, unlike graywackes, and high in CaO and CO_2 in the carbonate-cemented varieties. As might be expected, those lithic arenites in which chert or micritic limestone particles are dominant, show an unusual quantity of SiO_2 or CaO and CO_2, respectively. A high $CaCO_3$ content, however, can denote a calcite cement rather than limestone detritus. The high MgO content of some of these sandstones is due to detrital dolomite.

The SiO_2 content is depressed by addition of carbonate cement, or by the abundance of detrital limestone and dolomite particles, and is augmented by added quartz cement or by the abundance of detrital chert.

Special Types

Folk (1968a, p. 131) apparently coined the term *phyllarenite* for those lithic sandstones in which the dominant rock particles are of metamorphosed pelites: slate, phyllite and mica schist. Those in which the rock particles are phyllite or schist had previously been termed *schist arenites* (Krynine, 1937, p. 427). The phyllarenites contain considerable mica and usually a little feldspar. With a diminishing proportion of such rock particles, these sandstones grade into a *subphyllarenite* (or protoquartzite) and ultimately into a *quartz arenite*. If a significant proportion of other rock particles is present, the term *polylitharenite* has been proposed (Folk, 1968a, p. 135).

Many sands contain detrital chert grains, in a few the detrital chert forms a considerable or even major part of the sand detritus. These are the *chert arenites*. Chert forms, for example, from 20 to 90 percent of the grains in the Lower Cretaceous Cut Bank Sandstone of Montana (Sloss and Feray, 1948, p. 6). Similarly some of the sands in the Jurassic Morrison Formation of Montana are extremely chert-rich (Suttner, 1969, Fig. 11). In these, as in other sandstones, there is difficulty distinguishing chert particles from particles of devitrified rhyolite materials. It is not uncommon for ordinary lithic sands to contain 5 to 15 percent detrital chert (Table 6-3).

Chert-rich sands are more prone to diagenetic change than are quartz sands inasmuch as chert is more soluble than the quartz. Consequently, microstylolitic contacts are apt to form between adjacent grains. Chert is probably more susceptible to carbonate replacement than is quartz.

The chert-rich sands probably have a very local provenance being the less soluble, sand-sized residue from a limestone terrane or being derived from a region with a significant amount of bedded chert.

Detrital carbonate grains are those derived from pre-existing carbonate rocks by the ordinary processes of weathering, erosion, and transportation. They are, in other words, the clastic products of the wastage of a land mass on which limestones and dolomites are exposed. Most such detrital grains appear as polycrystalline aggregates. The term "extraclast" has been suggested for these particles in contrast to "intraclast" (Chanda, 1969).

Detrital carbonates occur in modern sands; locally they form a significant part of the sand. They are comparatively rare in ancient sandstones but there are important exceptions. Sandstones in which detrital carbonate forms a large and significant part of the rock have been called *calclithite* to distinguish them from *calcarenites* in which the carbonate detritus is intrabasinal (Folk, 1968a, p. 141). Carbonate detritus may exceed 50 percent.

Detrital dolomite is reported in Upper Cretaceous sandstones in the Uinta Basin of Utah (Sabins, 1962, p. 1185). Sabins (1962, p. 1186) also recognized "primary dolomite" in sandstones – the dolomite being single crystals of primary origin modified somewhat by abrasion prior to inclusion in the framework of the sand. Dolomite of this kind is said to be common in Cretaceous sands of the Western Interior of the United States (Sabins, 1962, p. 1188). The rounded corners of some dolomite rhombohedra and the close correlation between the size of the dolomite grains and the associated clastic quartz are cited as evidence of the concurrent sedimentation of the dolomite and quartz. Because dolomite grains are confined to marine sandstones and absent in associated non-marine sands, they are considered "primary" rather than detrital. Dolomite grains of a similar nature, however, have been observed in the arkose of the Fountain Formation of Pennsylvanian age in Colorado. This occurrence in presumed fluvial sandstones suggests a detrital rather than a primary origin.

Although dolomite particles are common in some sandstones, calclithites as such are relatively rare. They are the analogue of and found in association with limestone conglomerates. The Oakville Sandstone (Miocene) of Texas consists of grains of Cretaceous limestones (Folk, 1968a, p. 141). Many of the Molasse sandstones north of the Alps are over half detrital limestone and dolomite (Füchtbauer, 1967, Fig. 3). The reduction of most limestone terranes of low relief in humid regions is by solution; the only residues moved by surface streams are red clays and cherts. Hence the calclithites probably record rapid erosion and therefore signify high relief whereas chert arenites indicate low relief and removal of the limestones by solution.

Calclithites are generally cemented with carbonate – probably self-cemented.

Field Occurrence and Examples

Lithic sandstones are very common, are widespread and of all ages.

Well-known and well-described examples include various Paleozoic sandstones of the central Appalachians: the Ordovician Juniata Formation (Yeakel, 1962), and Oswego Sandstone (Krynine and Tuttle, 1941), Fig. 6-6, the Devonian Third Bradford Sand (Krynine, 1940), the Mississippian Pocono Formation (Pelletier, 1958), and Mauch Chunk Formation (Meckel, 1967; Hoque, 1968), and the Pennsylvanian Pottsville Formation (Meckel, 1967), Fig. 6-7. All these are quartz-rich, feldspar-poor sandstones with a large component of rock particles of sedimentary and low-rank metamorphic provenances. Excepting the Third Bradford Sand, all are alluvial.

Many of the sandstones of the Lower Old Red Sandstone of England are lithic arenites (Allen, 1962, p. 671).

Most sandstones associated with coal measures throughout the world are lithic arenites; perhaps more are protoquartzites. The sandstones of the Illinois

Fig. 6-6. Oswego Sandstone (Ordovician). U.S. 322, Bald Eagle Mountain, near State College, Pennsylvania, U.S.A. Crossed nicols, × 20. The formation consists of 50 percent or more of quartz, and 30-40 percent of rock particles. Original outlines in quartz are difficult to see; the secondary quartz results in interlocking, sutured grain boundaries. Rock particles are mainly siltstone, fine-grained quartzite, and phyllitic low-grade metamorphic grains. Feldspar rare – up to 2 or 3 percent. A lithic arenite

Basin perhaps belong here, although their rock fragment content is low. They are, nevertheless characterized by a significant proportion – some 8 to 10 percent – of feldspar, mica, and rock particles (Siever, 1957, p. 240). The coal measure sandstones of the Westphalian coal basin of France and Belgium are lithic arenites (Cayeux, 1929, p. 227).

Many Jurassic and Cretaceous sandstones in the Western United States are excellent lithic arenites (Fig. 6-8 and 6-9). Included here are the chert-rich arenites of Montana, of Jurassic age (Suttner, 1969), the Cretaceous Cut Bank Sand (Sloss and Feray, 1948), the Cretaceous Belly River Sandstone from Alberta,

Canada (Lerbekmo, 1963), and the sandstones of the Cretaceous Chico Formation of California (Williams and others, 1954, p. 306). There are many other examples.

Many of the Tertiary sands of the Gulf Coast turn out to be lithic arenites. Some of the better-known examples include the Oligocene "Frio" (Nanz, 1954), and the Miocene Oakville Sandstone (Folk, 1968a, p. 141) of Texas and the Eocene Wilcox Formation of Louisiana (Williams and others, 1954, p. 305).

Fig. 6-7. Pottsville Formation (Pennsylvanian). U.S. 61 at Pottsville, Pennsylvania, U.S.A. Crossed nicols, × 20. A coarse, ill-sorted lithic arenite composed of quartz and rock particles, the latter being in part weak shale and siltstones, forms a pseudomatrix. The rock particles are pelitic, sedimentary and metamorphic. Quartz generally subangular

The Tertiary Molasse sandstones (Fig. 6-10) north of the Alps are mostly lithic arenites. They contain a large proportion of rock fragments and a relatively low content of quartz and feldspar (Füchtbauer, 1964, p. 256-7; Gasser, 1968). Quartz ranges from 17 to 75 percent, averages 20 to 30 percent. Feldspar ranges from 1 percent or less to a maximum of 27 percent and averages between 2 and 18 percent. Rock particles, on the other hand, run as high as 72 percent which in some sands are largely limestones and dolomite grains. The carbonate grains form 18 to 33 percent of these sands. The rock particles in the more feldspathic sands are of metamorphic rather than carbonate rocks. The carbonate-rich sands also carry considerable chert.

A large number of sandstones are best characterized as protoquartzites or sublitharenites. Good examples of these include the Precambrian Serpent Quartzite of the North Shore of Lake Huron, some parts of the Silurian Tuscarora Quartzite of the Appalachians and the Pennsylvanian Anvil Rock Sandstone of the Illinois Basin. These may be lithic arenites which have been somewhat "cleaned up" in the neritic zone; true lithic arenites are perhaps more characteristic of alluvial deposits.

Fig. 6-8. Frontier Formation (Cretaceous). SW$\frac{1}{4}$, SE$\frac{1}{4}$, Sec. 12, T. 42 N., R. 107 W., Fremont County, near Dubois, Wyoming, U.S.A. Crossed nicols, × 55. A moderately well-sorted lithic arenite consisting of about equal parts of subangular quartz and chert. The chert varies from dense isotropic, to well crystallized aggregates. See Fig. 6-9

Origin and Significance

It seems probable that most modern sands are lithic arenites or protoquartzites in composition. Although rock particles are commonly ignored in modern sands, perhaps largely because they cannot easily be identified from grain mounts but must be thin-sectioned, they generally exceed feldspar (Table 2-4). The sands of many large rivers contain an abundance of rock particles. Rock particles are present also on beaches but to a lesser degree than in the streams, perhaps, because they are destroyed by the more vigorous abrasion in the surf zone.

Ancient lithic arenites are very common. They are estimated to constitute some 20 (Pettijohn, 1963) to 26 percent ("low rank graywacke" Middleton, 1960,

p. 1021) of all sandstones and thus outrank arkose by a good margin. They are the most common sandstones in Cretaceous and Tertiary flysch sequences, playing the same role that graywacke does in the older flysch facies (Cummins, 1962a), but unlike graywacke, they occur outside the geosyncline as well as in it. The typical molasse sandstone probably is a lithic arenite. See Chap. 12 for definitions of "flysch" and "molasse."

Fig. 6-9. Viking or Cardium "B" sandstone (Cretaceous). Garrington Field, T. 35 N., R. 3 W., Alberta, Canada. Crossed nicols, ×55. A lithic arenite, similar to that shown in Fig. 6-8, consisting of a moderately well-sorted mixture of subangular to subrounded quartz and chert, the latter forming nearly one third of the framework of the sand

Lithic sandstones are, for the most part, immature sands and hence like other such sands require conditions favoring the production and deposition of relatively unstable materials. The mechanism of production of large volumes of sand from fine-grained rocks is not well understood. Thorough decomposition would yield silt and clay-sized materials. Disintegration without decomposition of coarse-grained rocks would yield sand, but the finer-grained rocks might be expected to yield either silt or block-sized pieces. Nonetheless, a large volume of sand of the finer-grained rocks is produced. The mechanically weak character of much of cleavable pelitic and metapelitic materials preclude prolonged transport or survival, particularly in the surf environment. For this reason, the lithic arenites may be indicators of relatively local provenance, perhaps even within uplifted parts of the same sedimentary basin.

The chemically unstable rock particles, limestone and volcanic rock fragments, require erosion of incompletely weathered materials. Such erosion is promoted by high relief and/or aridity. The chert arenites are an exception to this generalization. These, like the quartz arenites, are mature sandstones and imply prolonged and thorough weathering.

The lithic arenites reflect provenance. In the same way that the arkoses denote a plutonic provenance – granites and gneisses, lithic arenites denote a

Fig. 6-10. Altern molasse (Tertiary). Near Bregenz, Vorarlberg, Austria. Ordinary light × 100. An excellent example of a calclithite. Note relatively low content of quartz and great abundance of equigranular, microcrystalline rock particles – micritic limestone and dolomite. Some specimens of the sandstones contain as much as one-third to one-half detrital carbonate. Carbonate cement

supracrustal provenance – volcanic, low-grade metamorphic, and sedimentary. The composition of the lithic arenites reflects each of these major provenances. Most lithic arenites arise from the denudation of a mixed sedimentary and metasedimentary terrane. With deeper dissection, the plutonic and higher grade metamorphic sources should be unroofed with the result that the derived sands become increasingly feldspathic.

From the brief survey given above, lithic arenites seem to be either (1) alluvial sandstones deposited on the flanks of marked uplifts as thick accumulations in closely associated molasse basins, or (2) as alluvial sandstones deposited on cratons by large rivers that derived much of their detritus from marginal and distant uplifts, and (3) as marine turbidite sands both in Tertiary geosynclinal depressions and the *present deep sea*. The alluvial sands show the usual structures of such deposits: crossbedding, basal position in an upward-fining cycle and the like. The turbidite sands are part of a flysch-like sequence and have the usual graded-bedding and sole markings characteristic of this facies.

Graywackes and Related Rocks: The Wackes

Definitions and History of Term

The term "graywacke" has perhaps engendered more controversy than any other term in sedimentary petrography. There has been an outpouring of notes and papers dealing with this class of rocks and upon the proper definition of the term. Among the latter are those by Boswell (1960), Cummins (1962a), Krynine (1941a), McBride (1962a), McElroy (1954), Pettijohn (1960), Shiki and Mizutani (1965), Wieseneder (1961), and Dott (1964). Perhaps the most complete review of the nomenclatural problem is that of Dott.

Dott states that the term was originally a field term, attributed to Lasius (1789) who applied it to Upper Devonian-Lower Carboniferous Kulm strata of the Harz Mountains in Germany. It was, however, used by Werner in 1787 (Crook, 1970a). With the advent of microscopical methods of study, we now have a better understanding of these rocks and are capable of framing a sharper definition of the term. Recent studies by Helmbold and Van Houten (1958) and by Mattiat (1960) of the type graywackes of the Harz have provided us with petrological "bench marks" with which to compare graywackes elsewhere.

The authors are inclined to follow Cummins (1962a, p. 52) who accepts Jamieson's definition of 1808 in which graywacke is defined as a "... kind of sandstone ... composed of grains of sand, which are of various sizes ... connected together by a basis of clay-slate, and hence this rock derives its gray color and solidity." This definition clearly fits an early description of the original graywackes of the Harz Mountains by Naumann (1858, p. 663) as well as the recent descriptions of the same rocks by Helmbold and Mattiat. It is also consistent with the usage of the term graywacke by most geologists of the last one hundred and fifty years.

The most marked deviations from traditional usage are those definitions which broaden the term to encompass sands rich in rock fragments and, in some cases well-sorted, and exclude those rich in feldspar. The latter restriction has largely been abandoned inasmuch as the classic Harz graywackes are themselves notably feldspathic. The designation "graywacke" for the clean sands rich in rock particles has also now been generally replaced by the term "lithic sandstone" (lithic arenite or litharenite) or "subgraywacke."

Although difficulty has been encountered in giving a precise definition, there exists a fairly homogeneous group of rocks similar in their essential characteristics to the type Harz sandstone and which by tradition and usage have been called "graywackes." The dark, fine-grained matrix in which the sand grains are set is an essential characteristic of these rocks.

Although "matrix" is a relative term and different arbitrary upper size limits as well as differing percentages of matrix have been utilized in defining graywacke, these difficulties are second-order and do not affect the identity of the bulk of the graywackes that have been studied and described. Graywackes are not just muddy or silty sandstones nor arkoses with a clayey pore filling. Unlike graywackes, the clay content of the arkose is a separate fraction – a much finer-grained component. Arkoses seem to have a clear bimodal distribution whereas in graywacke there is a continuum from the coarsest to the finest particle (Huckenholtz, 1963). Moreover, the matrix of the arkoses is different from that of graywackes. It is apt to be kaolinitic and may also be red-stained rather than a dark chloritic "paste."

Muddy, quartz-rich sands are apt to be sands produced by organic burrowing, *bioturbates*, which are readily distinguished from graywackes. Neither they nor the clayey arkoses will resemble graywackes in color, in induration, or chemical composition. Moreover, their manner of occurrence and usual structure will be quite unlike that of the typical graywacke. It is with the water-laid tuffs and submarine ash flows with their zeolitic products of altered glass that the sedimentary petrographer will have the most difficulties.

General Description

Graywackes are, as noted by most investigators, dark gray or black, generally tough, well-indurated rocks. Many exhibit graded bedding, convolute and small-scale current laminations, and various sole markings such as flute, groove, and load casts. They are generally rhythmically interbedded with mudstones or slates. These features are neither universal nor restricted to graywackes and, although the occurrence of such features is of interest, they are not diagnostic of themselves and hence are not a part of the definition.

The sand fraction is generally rich in quartz, has a varying proportion of feldspar and rock particles and generally contains a little detrital mica. The quartz is varied in size and shape though it is generally quite angular and commonly shows pronounced undulatory extinction. It usually constitutes half or less of the sand fraction, the balance being feldspar and rock particles. The feldspar is largely plagioclase, showing the usual multiple twinning. It is more commonly sodic rather than calcic. The high Na_2O content of graywackes as a class suggests that the feldspar is nearly pure albite. K-feldspar may be wholly absent though it is present in small amounts in some (Bailey and Irwin, 1959). Although the feldspars are generally fresh, some contain inclusions of sericite, chlorite, and epidote – the last being, perhaps, a product of decalcification of the plagioclase. As with the quartz grains, the margins of the feldspars may be hazy as though encroached on by the matrix.

Rock particles are dominantly mudstone, shale, siltstone, slate and argillite, phyllite and mica schist. Chert and micritic limestone, especially the former, and polycrystalline quartz and fine-grained quartzite may be plentiful. Many graywackes, however, contain particles of fine-grained igneous rocks, some with microlites of feldspar. Particularly common are acid igneous flow rocks; less common is andesitic debris. In rare cases serpentine is present (Zimmerle, 1968).

Table 6-5. *Mineralogical composition of graywackes (based on modal analysis) (from Pettijohn, 1963, Table 5, with additions)*

	A	B	C	D	E	F	G	H	I	J
Quartz	33	22	37	4	24	27	33	56	9	trace
Feldspar	15	5	12	10	32	19	21	37	43	30
Rock fragments	3	26	15	50	19	30	7	7	10	13
"Matrix"	45	47	32	32	P[1]	21	33	P[2]	25	45
Mica and chlorite	—	—	—	—	16	—	6	—	4	—
Miscellaneous	—	—	3	2	8	3	—	—	4[3]	10[3]

(1) Not separately reported; 38 percent of rock is "clay and silt"; (2) Not separately reported; (3) Hornblende and pyroxene.

A. Feldspathic graywacke (Precambrian), Ontario, Canada; average of 3 analyses (Pettijohn, 1943, p. 946).
B. Lithic graywacke (Martinsburg Shale), (Ordovician) Pennsylvania, U.S.A. (McBride, 1962, p. 62).
C. Aberystwyth Grit (Silurian), Wales (Okada, 1967, Table 1, Analysis 70A).
D. Lithic graywacke (Devonian), Australia; average of 5 (Crook, 1955, p. 100).
E. Feldspathic graywacke (Devonian-Mississippian, Tanner), Harz Mountains, Germany (Helmbold, 1952, p. 256).
F. Graywacke (Kulm), Harz, Germany (Mattiat, 1960).
G. Graywacke (Lower Mesozoic), Porirua district, New Zealand (Webby, 1959, p. 472).
H. Feldspathic graywacke (Jurassic? Franciscan Formation), Calif., U.S.A.; average of 17 analyses (Taliaferro, 1943, p. 135).
I. Purari graywacke (Cretaceous), Papua; average of 4 (Edwards, 1950b, p. 164).
J. Tuffaceous Aure graywacke (Miocene), Papua; average of 2 (Edwards, 1950a, p. 129).

With an increasing proportion of volcanic rock and with the appearance of zoned plagioclase and broken crystal euhedra, the graywackes pass insensibly into water-laid tuffs and tuffaceous sandstone. The range in character and composition of graywackes is illustrated by Table 6-5.

Detrital micas, both biotite and chloritized biotite as well as muscovite, are common though not abundant constituents. Minor accessories include a carbonate, probably near ankerite, which occurs in irregular patches replacing matrix and also some rock particles and feldspar grains.

Other minor accessories include sulfides, probably mostly pyrite, which occur, like the carbonate, in irregular areas.

Unlike most other sandstones which are held together by a porefilling mineral cement, graywackes are bound by a fine-grained matrix consisting of an intimate intergrowth of chlorite, sericite, and minute, silt-sized particles of quartz and feldspar. The matrix content seems also to be something of a function of the size of the sand fraction – the finer the grain, the higher the matrix content (Okada, 1966). Huckenholtz (1963, p. 194) states that in a strict sense "matrix" requires two distinct maxima in the grain size distribution curve – one in the sand and the other in the clay-size region. By this view, most graywackes would not have matrix. This restrictive definition of matrix is generally not adhered to. The matrix is, to some extent, intergrown with the larger sand grains. In many cases one has great difficulty in distinguishing between the matrix as such and ill-defined, fine-grained pelitic rock particles. This difficulty may lead to an overestimation of the abundance of matrix (de Booy, 1966).

Table 6-6. *Representative chemical analyses of graywackes (modified from Pettijohn, 1963, Table 6, with modifications)*

	A	B	C	D	E	F	G	H	I	J
SiO_2	60.51	61.39	76.84	69.11	68.85	74.43	73.04	71.10	68.84	65.05
TiO_2	0.87	0.62	—	0.60	0.74	0.83	0.15	0.50	0.25	0.46
Al_2O_3	15.36	16.97	11.76	11.38	12.05	11.32	10.17	13.90	14.54	13.89
Fe_2O_3	0.76	0.39	0.55	1.41	2.72	0.81	0.56	trace	0.62	0.74
FeO	7.63	5.32	2.88	4.64	2.03	3.88	4.15	2.70	2.47	2.60
MnO	0.16	0.12	trace	0.17	0.05	0.04	0.18	0.05	nil	0.11
MgO	3.39	3.84	1.39	2.06	2.96	1.30	1.43	1.30	1.94	1.22
CaO	2.14	3.21	0.70	1.15	0.50	1.17	1.49	1.80	2.23	5.62
Na_2O	2.50	2.78	2.57	3.20	4.87	1.62	3.56	3.70	3.88	3.13
K_2O	1.69	1.25	1.62	1.76	1.81	1.74	1.37	2.30	2.68	1.41
H_2O+	3.38	2.44	1.87[1]	4.13	2.30	2.15	2.36[1]	1.90	1.60	2.30
H_2O-	0.15	0.06		0.05	0.77	0.20		0.26	0.35	0.28
P_2O_5	0.27	0.19	—	0.03	0.06	0.18	0.23	0.10	0.15	0.08
ZrO_2	—	0.07	—	—	—	—	—	—	0.05	—
CO_2	1.01	0.88	—	—	0.08	0.48	0.84	0.12	0.14	2.83
SO_3	—	—	—	—	—	—	—	—	0.15	—
	—	—	—	—	—	—	—	—	—	—
S	0.42	0.15	—	—	0.08	0.12	0.10	trace	—	0.05
Cr_2O_3	—	0.01	—	—	—	—	—	—	—	—
BaO	—	0.06	—	—	trace	—	—	—	0.04	—
C	—	—	—	—	0.07	0.17	0.17	0.09	—	—
	100.24	99.75	100.18	99.69	99.94	100.45	99.80	99.80	99.93	99.77

(1) Loss on ignition.

A. Graywacke (Archean), Manitou Lake, Ontario, Canada, B. Brunn, analyst (Pettijohn, 1957b, p. 306).

B. Graywacke (Archean), Knife Lake, Minnesota, U.S.A., F. F. Grout, analyst (Grout, 1933, p. 997).

C. Graywacke Tyler slate (Precambrian Animikian), Hurley, Wis., H. N. Stokes, analyst (Diller, 1898, p. 87).

D. Rensselaer Graywacke (Ordovician?), near Spencertown, N.Y., H. B. Wiik, analyst (Balk, 1953, p. 824).

E. Tanner Graywacke (Upper Devonian – Lower Carboniferous), Scharzfeld, Germany, R. Helmbold, analyst (Helmbold, 1952, p. 256).

F. Graywacke from Stanley Shale (Carboniferous), near Mena, Arkansas, B. Brunn, analyst (Pettijohn, 1957b, p. 306).

G. Kulm (Carboniferous), Steinback, Frankenwald, Germany (Eigenfeld, 1933, p. 58).

H. Composite sample (Lower Mesozoic), prepared by using equal parts of 20 graywackes exposed along shoreline between Palmer Head and Hue-Te-Taka, Wellington, N.Z., J. A. Richie, analyst (Reed, 1957, p. 16).

I. Franciscan Formation (Jurassic?), Quarry Oakland Paving Co., Piedmont, California, Jas. W. Howson, analyst (Davis, 1918, p. 22).

J. Graywacke (Cretaceous?), Olympic Mountains, Washington, near Solduc, B. Brunn, analyst (Pettijohn, 1957b, p. 306).

The graywackes are a surprisingly homogeneous group chemically (Table 6-6). Despite the uncertainty about the definition, the bulk chemical composition of graywackes is much alike the world over, regardless of geologic age. Moreover, their compositions are quite unlike those of arkoses (Table 6-2 and Fig. 6-2).

As can be seen by reference to the table, graywackes are rich in Al_2O_3, $FeO + Fe_2O_3$, MgO, and Na_2O. The high Na_2O content no doubt reflects the albitic nature of the feldspar; the MgO is related to the chloritic matrix as is the high FeO content. The distinguishing chemical attributes are dominance of FeO over Fe_2O_3, usually dominance of MgO over CaO and the dominance of Na_2O over K_2O. It is in these respects that graywackes differ notably from arkoses.

Many graywackes have a bulk composition closely resembling a particular source rock – commonly a granodiorite not, as some have thought, a mafic igneous rock. Sands are normally chemically differentiated from their source rocks by weathering and sorting. The chemically undifferentiated nature of graywacke suggests that one or both processes have been incomplete. The sands of the present Columbia River (Whetten, 1966) have a bulk chemical composition surprisingly similar to that of graywacke. Did ancient graywackes have a provenance similar to the Columbia River sands?

Varieties and Types of Graywacke

As noted elsewhere we distinguish between the matrix-free arkoses, that is, the arkosic arenites, and those with a kaolinitic clay matrix – the arkosic wackes. Is there a comparable distinction in the graywacke clan between matrix-free graywackes and matrix-rich graywacke? By our usage, matrix-free graywacke

Fig. 6-11. Recluse Formation, Epworth Group (Lower Proterozoic). Coronation Gulf, N.W.T., Canada. Crossed nicols, × 20. A feldspathic graywacke consisting of a poorly sorted mixture of angular debris in a fine-grained matrix. Mainly quartz and feldspar but includes also some rock particles

would be a contradiction of terms. A rock of graywacke composition without matrix would be, in most cases, a lithic arenite. Although the term "mature graywacke" was once applied to such sandstones, this usage has been replaced by "lithic arenite" even by Folk himself (1968a, p. 125).

It is possible, however, to recognize a number of varieties of graywacke. Those rich in rock fragments have been called *lithic graywackes*; those rich in feldspar are the *feldspathic graywackes*.

Fig. 6-12. Caradocian (Ordovician). Four hundred yards east by north of Garvald, Scotland. Crossed nicols, × 20. A coarse poorly-sorted graywacke consisting of subangular to subrounded quartz and feldspar with a few rock fragments, including aphanitic volcanic rocks

Some graywackes are very quartz-poor and contain evident volcanic contributions such as volcanic quartz, euhedra or broken euhedra of feldspar, and an appreciable content of ferromagnesian minerals. These are the volcanic graywackes. They pass by degrees into true subaqueous tuffs. There are other graywackes abnormally rich in quartz and lacking feldspar. These are the "quarzwackes" of Fischer (1933). In between the quartz-poor and quartz-rich varieties are the common and most abundant graywackes of the geologic record. Analyses D, I, and J, Table 6-5, are representative of the quartz-poor class; most of the other analyses in this table belong to the intermediate group.

Crook (1970b) has recognized these three major classes of graywacke and attributes their differences to provenance. The quartz-poor (under 15 percent quartz) being of volcanic provenance, the quartz-rich (over 65 percent and commonly 80 percent quartz) being of sedimentary provenance, and the intermediate class (15 to 65 percent quartz) being of a mixed provenance. Crook went so far as to attribute the quartz-poor types to the island arc environment, the quartz-rich class to tectonically inactive continental margins, and the intermediate type to tectonically active margins of continents or microcontinents.

Fig. 6-13. Kulm (Carboniferous). Quarry near Sösetalsperre, about 7 km from Osterode, Harz Mountains, Germany. Crossed nicols, × 20. The type or "classical" graywacke consisting mainly of angular, ill-sorted quartz (av. 26 percent), feldspar (av. 19 percent) and rock fragments (av. 38 percent) in fine-grained, chlorite-rich matrix

Field Occurrence and Examples

Graywackes typically occur in geosynclinal tracts, with or without associated volcanics. Certainly they are the characteristic sandstones of the eugeosynclinal belts where they are associated with pillow lavas or "greenstones" (Tyrrell, 1933; Bailey, 1936; Turner and Verhoogen, 1960, p. 270). Graywackes of this habit range widely in age from the oldest Precambrian to the Tertiary. Examples of the former are the many Archean sequences in the Canadian Shield (Pettijohn, 1943; Donaldson and Jackson, 1965), the Fennoscandian Shield, (Simonen and Kuovo, 1951), and South Africa (Anhaeusser and others, 1969, p. 2184). Included here also are the Precambrian Dalradian graywacke grits of Scotland (Sutton and

Watson, 1955). Similar rocks occur in the older Paleozoic regions of Wales (Woodland, 1938; Okada, 1967), the Southern Uplands of Scotland (Walton, 1955) Figs. 6-12, the Scandinavian Highlands and Spitzbergen (Tyrrell, 1933, p. 25). The Ordovician of Newfoundland is characterized by a similar assemblage (Espenshade, 1937). This association is found also in the younger Hercynian fold belt of Cornwall and Devon as well as in the corresponding areas of Europe,

Fig. 6-14. Franciscan Formation (Upper Jurassic?). U.S. 101, one quarter mile north of Golden Gate Bridge, near San Francisco, California, U.S.A. Crossed nicols, × 20. A typical feldspathic graywacke, chiefly quartz, feldspar, and rock fragments in a fine-grained chloritic and sericitic matrix. Note poor rounding and lack of assortment of the material

particularly in the Harz (Fischer, 1933; Helmbold, 1952; Mattiat, 1960), Fig. 6-13, and in the Rheinische Schiefergebirge (Henningsen, 1961). Well-described examples occur also in the Lower Mesozoic of the Wellington district of New Zealand (Reed, 1957) and in the Franciscan Formation (Jurassic), Fig. 6-14, of the Coast Ranges of California (Davis, 1918; Taliaferro, 1943). Graywackes with the greenstone and chert associations are found in both the Mesozoic and Paleozoic of Alaska, Fig. 6-15 (Cady and others, 1955; Loney, 1964), in the Olympic Mountains of Washington (Park, 1942), in the Paleozoic of New South Wales of Australia, Fig. 6-16 (Crook, 1955, 1960), and in the Cretaceous of the Northwest Borneo Geosyncline (Haile, 1963).

Graywackes also occur in geosynclinal tracts in which volcanism, if any, is minimal. The Animikian Tyler Slate, the Michigammee Slate, and the Thomson Slate near Duluth (Schwartz, 1942) of the Penokean fold belt south of Lake Superior contain good examples (Irving and Van Hise, 1892, p. 206f) as does the Chelmsford "sandstone," Fig. 6-17, of the Sudbury Basin. The sandstones of the Ordovician Martinsburg Shale (McBride, 1962b), Fig. 6-18, and the correlative

Fig. 6-15. Kuskokwim, (Cretaceous ?). Cribby Creek–George River divide, Kuskokwim region, Alaska. Crossed nicols, × 20. A graywacke marked by its low quartz content, and abundance of feldspar, some of which is subhedral, and many aphanitic flow rock fragments. Very poorly sorted. Clearly a graywacke with a considerable volcanic input

Rensselaer Graywacke (Balk, 1953) and the Normanskill Shale (Weber and Middleton, 1961) of the Appalachian Mountains belong in this category as do those of the Carboniferous Stanley Shale and Jackfork Sandstone of the Ouachita Mountains of Arkansas and Oklahoma (Bokman, 1953) and the correlative Haymond Formation of the Marathon district of Texas (McBride, 1966). Well-described examples of Cretaceous age occur in Japan (Shiki, 1962; Okada, 1960, 1961). Many of the graded sandstones of the Apennines are graywackes (Fig. 6-19). All of these examples of graywackes in miogeosynclinal tracts are apt to be characterized by their abundance of rock particles rather than by feldspar. Although a few tuffaceous beds are found in some of these geosynclinal sections, no extensive submarine volcanism is recorded.

Clearly graywackes are characteristically found in Alpine-type orogenic belts and are absent in sequences deposited on the undeformed stable or platform areas. They are generally marine, and, in the examples listed above, largely believed to be turbidite sands and part of what have called the "flysch" facies. Not all flysch sandstones, even though of turbidite origin, are graywackes. Neither are all sandstones in the orogenic belts graywackes; but graywackes do not seem to occur outside such belts.

Fig. 6-16. Graywacke, Neranleigh Group, probably Silurian. Arundale District, New South Wales, Australia. Ordinary light, × 20. A graywacke characterized by its low quartz content, abundance of feldspar and rock particles, including those of volcanic rocks

Graywacke has been estimated to constitute about one-fifth to one-fourth of all sandstones (Middleton, 1960; Pettijohn, 1963). It has been presumed that they were of greater importance in the earlier periods of earth history. The bases for these statements will be considered in Chapter 12.

The Matrix Problem

The matrix is the essential characteristic of graywacke, and is, as Cummins (1962a) noted, "the essence of the graywacke problem." By definition the matrix content exceeds 15 percent and is reported in some cases to exceed 50 percent.

These estimates, however, have been questioned by de Booy (1966) who has argued that immature detrital sediments, especially graywackes, are often wrongly analyzed. The detrital grain boundaries are in most cases camouflaged by recrystallization phenomena. As a result, the matrix content is greatly over estimated. De Booy contends that the matrix content of many typical graywackes is no more than a few percent. Certainly under crossed nicols one has great difficulty in distinguishing between somewhat altered rock fragments

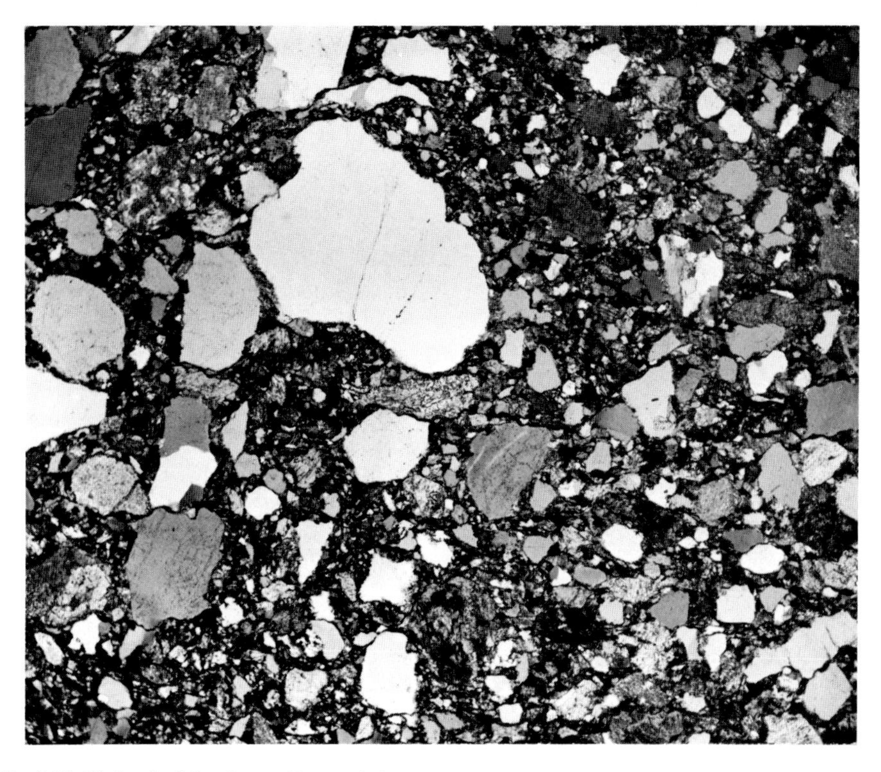

Fig. 6-17. Chelmsford Sandstone (Precambrian). Sudbury district, near Chelmsford, Ontario, Canada. Crossed nicols, × 20. A graywacke with the usual unassorted poorly-rounded quartz, feldspar and rock particles in a fine matrix

and the recrystallized sericitic and chloritic matrix. In many cases, too, it is difficult to distinguish between the softer pelitic rock fragments, which may become deformed and squeezed into pores between more durable quartz grains, and true matrix. Such fragments give rise to a false or pseudomatrix.

The matrix has been variously interpreted. It was thought to be the product of recrystallization of an original detrital component and having the composition of a slate, the original material was presumed to be mud. If so, the problem becomes one of explaining the simultaneous deposition of mud and sand in one and the same place. Normally, as a result of current action, the two part company and are separately accumulated. Even the sand bars in the lower reaches of the

Mississippi River – a dominantly mud-carrying stream – are clean. It has been suggested that flocculation by the electrolytes of sea water would have precipitated the clay-sized materials (Woodland, 1938). Evidence to support this view is generally lacking. Few, if any, modern near-shore, shallow-marine sands have the requisite character that is, interstitial mud, to become graywackes on lithification.

Fig. 6-18. Sandstone from Martinsburg Shale (Ordovician). Broadfording, Washington County, Maryland, U.S.A. Crossed nicols, × 20. A lithic graywacke consisting of poorly sorted, subangular sand-sized debris in a fine matrix. Martinsburg graywackes contain 20-60 percent quartz, 1-12 percent feldspar, and 3-44 (av. 24) percent rock fragments, mostly sedimentary in origin. Matrix forms 30-48 percent of the rock

The matrix of graywackes has more recently been attributed to their manner of transport and deposition. Many graywackes in the geologic record bear the earmarks of turbidity current action and are in fact turbidites. As noted by Kuenen and Migliorini (1950, p. 123) "... all the exceptional and puzzling features of typical series of graded graywackes can be readily accounted for by the activity of turbidity currents of high density." They noted that such currents of suspended fine muds are capable of transporting sands down submarine slopes and they reasoned that when ponded the current would cease and deposit both its load of sand and suspended mud producing a graded graywacke. The deposition in deep marine basins would explain our general failure to find the modern analogue of the ancient graywackes. But a consideration of settling velocities and analysis of flume studies do not substantiate the view that much mud matrix can be deposited by simple settling of clay together with sand (Chap. 9). In addition, graywackes, that is matrix-rich sands, do not appear to be common among present deep-sea sands of presumed turbidite origin (Hollister and Heezen, 1964). Kuenen (1966, p. 244) has more recently reviewed the problem and has concluded

Fig. 6-19. Pietraforte Sandstone (Cretaceous). Fiesole, near Florence, Tuscany, Italy. Crossed nicols, × 20. A good graywacke consisting of poorly sorted angular debris – mainly quartz, feldspar and rock particles. Some samples of this sandstone contain up to 17 percent detrital dolomite

on both theoretical and experimental grounds that the original matrix, if any, was well below ten percent in the coarser rocks.

Emery (1964) and Klein (1963, p. 571) believed that the clay might have been mechanically introduced – by movement of interstitial water from overlying or underlying beds and that initially clean and well-sorted turbidites were thus converted to matrix-rich graywackes. Kuenen rejected this suggestion and concluded that the matrix of the coarser graywackes is very largely of secondary origin and that there is no need to presume such graywackes were carried and deposited by some different kind of a turbidity current or by some different kind of mechanism than that assumed for Recent deep sea sands or Tertiary turbidites.

It is evident that the matrix is indeed recrystallized material. It consists of chlorite, sericite and quartz in the older graywackes and zeolites and montmorillonite in the younger ones. Irving and Van Hise (1892, p. 334) thought the matrix was produced by a "... micaceous or chloritic alteration of the feldspar." That some matrix material is truly authigenic was noted by Krynine (1940, p. 22) who observed sericite replacing detrital quartz, a relationship also observed between chlorite and quartz. The original water-worn boundaries of the quartz have wholly disappeared, the existing boundary being a kind of "chevaux-de-frise"

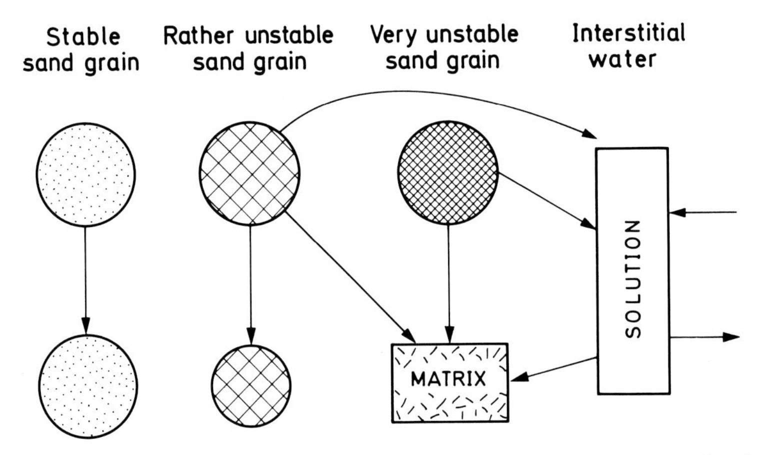

Fig. 6-20. Diagram showing presumed post-depositional origin of matrix of graywackes (Modified from Cummins, 1962, Fig. 4)

of green chlorite crystals projecting into clear quartz (Greenly, 1897, p. 256). More recently Cummins (1962a) has attributed all, or almost all, of the matrix to diagenesis. It has been shown by Brenchley (1969), that the matrix of some Ordovician volcanic graywackes, constituting 40 to 60 percent of the whole, is diagenetic. Where calcite cement is present, there is no matrix and the rock is a well-sorted sandstone. It is concluded that the deposition of an early calcite cement inhibited the formation of a matrix. Even adjacent parts of the same bed display these differences. Although Brenchley's observations are significant, it is not clear why, if carbonate cements are matrix inhibitors, the Paleozoic and older graywackes are so generally matrix-rich. Surely many more would have had carbonate cements.

Experiments with sands chemically similar to graywackes produced matrix minerals at 250° C and a water pressure of 1 kilobar. These observations support the notion that the matrix of graywacke may form diagenetically and need not be the recrystallized detrital fine fraction (Hawkins and Whetton, 1969).

Cummins noted that most Tertiary and Recent turbidites do not have the abundant matrix found in the comparable turbidites of the early flysch sequences. The latter have presumably undergone deep burial and incipient metamorphism involving pressure solution at grain contacts and especially alteration of the labile constituents – mainly unstable rock fragments but also including feldspar (Fig. 6-20). Although some deep-sea turbidites cored by the

JOIDES 1970 program disclosed lithification, much of that proved, on preliminary examination, to be due to silicification by dissolution and reprecipitation of biogenic silica. We may conclude, therefore, that the modern equivalents of the matrix-rich ancient graywackes are the immature sands with high amounts of argillaceous and/or volcanic rock fragments that were deposited in basins where they are destined for deep burial and low-grade metamorphism or high-grade diagenesis – termed "graywackisation" by Kuenen.

In conclusion, it can be said that matrix, the fine interstitial material characteristic of graywackes (and some arkoses), is most probably of more than one origin. The diverse types of matrix have been designated *protomatrix, orthomatrix, epimatrix*, and *pseudomatrix* by Dickinson (1970), the first being trapped detrital clay, the second being recrystallized material, the third is a diagenetic product of the alteration of sand-sized grains, and the fourth being deformed and squashed lithic fragments. Dickinson gives the criteria for the recognition of each type.

The Problem of Na_2O

Various explanations have been offered to account for the high Na_2O content of the graywackes (Engel and Engel, 1953, p. 1086-1091). The high Na_2O content seems to be due to the high content of albitic feldspar. The Tanner graywacke, for example, contains 3.5 percent Na_2O. This quantity of Na_2O would require 30 percent pure albitic feldspar, an estimate consistent with the observation that the feldspar actually forms from 30 to 40 percent of this rock, of which some 85 to 90 percent has a composition An_{3-10}.

Clearly the feldspar is a detrital component. But was the original material albitic or is its albitic nature a post-depositional feature? The close association of many (but not all) graywackes with spilitic rocks suggests that the problem is related to the origin of spilites (Turner and Verhoogen, 1960, p. 270). There is some evidence of albitization *in situ*. The irregular patches of carbonate, neither detrital nor a pore-filling cement and which replaces both detrital grains and matrix, is clearly a post-depositional product. Perhaps the albitization of the feldspar provided the CaO which was deposited elsewhere as a carbonate. One is reminded of the innumerable vugs and veinlets of calcite associated with the albitized greenstones. But if the alteration of the feldspar is post-depositional, why are not the associated and interbedded slates also Na_2O-rich? The pelitic strata interbedded with the graywackes of New Zealand, for example, have a normal Na_2O/K_2O ratio in contrast to that of the cogenetic graywackes (Reed, 1957, p. 28). See Fig. 6-21. Garrels, Mackenzie, and Siever (1971) suggest that the high K_2O/Na_2O content of shales and slates is also a diagenetic characteristic, and is a reflection of the growth of illite at the expense of montmorillonite, mixed layer, and kaolinite assemblage. Thus the Na_2O ends up in the feldspar and the K_2O in the mica. It is less likely that the original sands were so Na_2O-rich, being derived almost exclusively from albite-bearing source rocks while contemporaneously the finer materials, products of decomposition, being Na_2O-poor due to normal leaching during weathering. It has been suggested that the absence of K-feldspar is due to post-depositional solution and

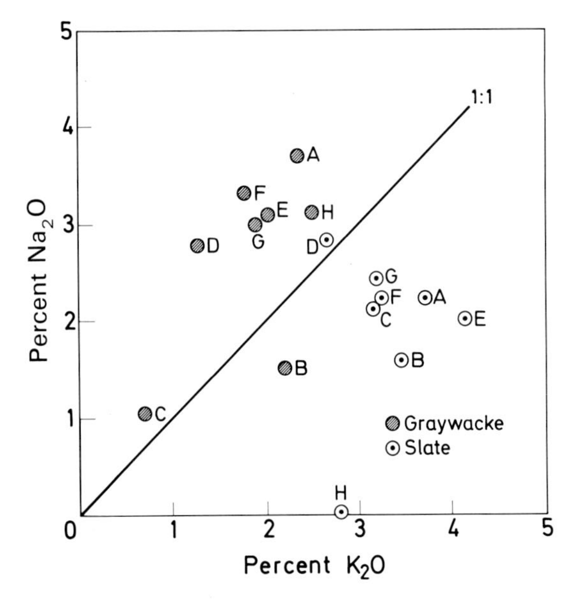

Fig. 6-21. Diagram showing Na_2O/K_2O content of graywackes and co-genetic shales. A) New Zealand Mesozoic, composite sample (Reed, 1957, Tables 1 and 3). B) New Zealand, Mesozoic, (Reed, 1957, Tables 1 and 3, Analyses 17 and 6). C) North Spirit Lake, Ontario, Archean (Donaldson and Jackson, 1965, Table 3, cols. 2 and 5). D) Knife Lake Minnesota, Archean (Grout, 1933, Table 1, cols. 1 and 2). E, F, and G) Yellowknife, Great Slave Lake, N. W. T., Archean (Henderson, pers. comm.). H) Casummit Lake, Ontario, Archean (Horwood, 1938)

is not a primary characteristic (Gluskoter, 1964, p. 341). This would fit the compatibility of albite and muscovite at low to moderate temperatures (see Chap. 10).

Significance of Graywackes

Many, if not most graywackes are redeposited sands. In general such redeposited sediments are not good guides to the climate either of the source area or the place of deposition. Insofar as the graywackes are marine, as most of them are, they tell us little about the climate of the depositional site. On the other hand, the detritus of which graywackes are composed must, like that of arkose require an environment in which erosion, transportation, and deposition are so rapid that complete chemical weathering of the materials does not take place. Inasmuch as arkosic sands can form under humid tropical conditions as well as under an arid or arctic climate, so also can the detritus of graywackes. Krynine (1937, p. 427 and Fig. 4) reports on graywackes of the Siwalik Series (late Tertiary) in northwestern India which he believed, on the basis of both lithologic and faunal evidence, were deposited under tropical conditions varying from savanna through prairie to steppe. Fischer (1933, p. 339) also concluded, from a study of graywackes in Germany, that they could form in several different climatic environments. Taliaferro (1943, p. 139) believed the graywackes of the Franciscan

Formation (Upper Jurassic) to have been derived "... from a high, rugged, recently uplifted landmass under rigorous climatic conditions, high rainfall, and possibly a cold climate in the highlands with well wooded lower slopes." This inference was based on the abundance of fresh feldspar and of carbonized fragments of wood.

Because graywacke is not confined to the Precambrian, or even the pre-Devonian, the absence of a vascular land plant cover does not seem to be a requirement for its formation as was suggested by Kaiser (1932).

Graywacke, however, does indicate a special tectonic environment. Fischer called it a "poured-in" type of sediment. To the extent that graywackes are turbidites, each bed is essentially an instantaneous event. Its restriction to geosynclinal tracts and its dominance in the flysch of the older systems suggests that much or most is a marine turbidite. Although graywackes are not by definition turbidites, the fact remains that most of those in the geologic record are turbidites. It may be, as Cummins suggests, that the turbidites of the younger orogenic flysch sequences are precursors of graywacke and will be converted to graywacke by diagenetic alteration or low-grade metamorphism. Such alteration accompanied by Na_2O-enrichment related to their marine environment could produce the distinctive characters of graywacke. By this interpretation, graywacke is a high-grade diagenetic or low-grade metamorphic product. Some authors do not, however, consider this a closed question (Shiki and Mizutani, 1965, p. 32).

It is uncertain how much, if any, graywacke is non-marine. As noted above, Krynine (1937) described some presumed non-marine graywackes (Cummins, 1962b) but an overwhelming proportion of the graywackes of the past seem to have been marine. Some have thought that graywackes reflect reducing conditions, perhaps even a reducing atmosphere (MacGregor, 1927), presumably because of the high FeO/Fe_2O_3 ratio. The existence of graywackes of Paleozoic and later age preclude a reducing atmosphere. The associated shales generally lack the abnormally high content of organic carbon or early diagenetic pyrite indicative of strongly reducing conditions.

To a certain extent graywacke and arkose do reflect a different provenance. Arkose records a granitic provenance. And although granite or related rocks are very common and widespread in some areas, they give rise to arkose only under relative restricted circumstances. On the other hand, graywacke seems to have had a less restricted provenance – the abundant quartz and feldspars of some, mixed with low rank metamorphic rock particles and even volcanic detritus, denote a more varied provenance than do arkoses. Is this because it is commonly a redeposited sediment, the detritus being supplied by large rivers, with an extensive drainage basin of varied lithology, and by long-shore drift? It may be, as Shiki (1962, p. 316) suggests, that there are at least two types of graywackes which differ in their provenance, the first being rich in feldspar and therefore of crystalline plutonic derivation, the second being a lithic graywacke and hence of a supracrustal source. Moreover, those characteristic of the eugeosynclines may have a significant volcanic input. These views are similar to those of Crook (1970b) who attributes the low-quartz graywackes to a volcanic provenance and the quartz-rich varieties to a sedimentary provenance. Certainly graywacke is *not* the ferromagnesian equivalent of an arkose (Twenhofel, 1932, p. 231).

Quartz Arenites (Orthoquartzites)

Definitions and Nomenclature

Almost all sands contain detrital quartz; most are quartz-rich, that is, have over 25 percent quartz, and a very large number consist almost wholly of detrital quartz. The term "quartzose sandstone" is therefore more or less redundant. But a name is required for those end-products of sand evolution which contain an overwhelming abundance of quartz. The term *orthoquartzite* has been used for this class of sands. Although apparently first used by Tieje (1921, p. 655), it was popularized by Krynine (1945) who applied this term to sands made up entirely of quartz grains cemented by silica. These are the sedimentary quartzites as opposed to the quartzites of metamorphic origin (metaquartzites). The term "quartzite" in the traditional sense applied to a rock so thoroughly cemented that the rock breaks across the grains rather than around them. However, as Krynine (1948, p. 152) pointed out, many contain some carbonate cement as an additional end member and as the proportion of this nonsiliceous material increases, the orthoquartzites begin to show less cohesiveness. Although the term "orthoquartzite" has come to be accepted and applied to these less cohesive and even to friable or modern quartz sands, it engenders some confusion because of the long-standing concept of quartzite as a durable, non-friable rock (see, for example, Mathur, 1958). Hence other terms have been suggested. Most acceptable is the term *quartz arenite* (or *quartzarenite*) proposed by Gilbert (Williams and others, 1954, p. 294) and McBride (1963). Although the authors have used the term "orthoquartzite" for many years, we now are inclined to consider "quartz arenite" the better term for this class of sandstones. The trend in the recent literature seems to indicate concurrence in this view.

General Description

The quartz arenites are very common – perhaps the best known type of sandstone. They may be defined simply as sands of which the detrital fraction is 95 or more percent quartz. In general, they are cemented with quartz deposited in crystallographic and optical continuity with the detrital quartz, and in many cases so well cemented that they are indeed quartzites.

Quartz arenites are generally white rocks, some are tinged with pink, a few are deeper red. The pink or reddish coloration is due to a film or coating of hematite on the grains. The iron oxide content, however, may be only a percent or less of the whole rock. Many quartz arenites are relatively thin widespread blanket sands with a minimum of interbedded shale. In other cases, the sand accumulation may be thick – several thousand feet (1000 m) or more as in the case of some Precambrian formations. Ripple-marking and/or crossbedding are characteristic structures.

As noted, many are indeed quartzites and hence durable and erosion-resistant and form high ridges or hills. Others, especially those cemented by carbonates, are less quartzitic and in some cases, perhaps due to leaching of the cement, are friable. In many of the friable sands, however, the quartz grains show silica

Table 6-7. *Chemical composition of representative orthoquartzites (Pettijohn, 1963, Table 2)*

	A	B	C	D	E	F	G	H	I	J
SiO_2	98.87	95.32	97.58	97.36	98.91	83.79	99.54	99.40	97.30	93.13
Al_2O_3	0.41	2.85	0.31	0.73	0.62	0.48	0.35	0.20	1.40	3.86
Fe_2O_3	0.08	0.05	1.20	0.63	0.09	0.063	0.09	0.01	0.30	0.11
FeO	0.11	—	—	0.14	—	—	—	—	—	0.54
MgO	0.04	0.04	0.10	0.01	0.02	0.05	0.06	0.01	0.03	0.25
CaO	—	trace	0.14	0.04		8.81	0.19	<0.01	<0.05	0.19
Na_2O	0.08	0.30	0.10	0.08	0.01	—	—	0.08	<0.05	—
K_2O	0.15		0.03	0.19	0.02	—	—	trace	0.20	—
H_2O+	0.17	—	—	0.54	—	—	0.25	0.04	—	—
H_2O-		—	—	0.14	—	—	—	0.01	—	—
TiO_2	—	—	—	0.05	0.05	—	0.03	0.02	0.28	—
P_2O_5	—	—	—	0.02	—	—	—	none	—	—
MnO	—	—	—	0.01	—	—	—	trace	0.003	—
ZrO_2	—	—	—	—	—	—	—	<0.01	0.06	—
CO_2	—	—	—	—	—	6.93[2]	—	—	—	—
Ign. loss	—	1.44	0.03	—	0.27	—	—	0.28	—	1.43
Total	99.91	100.00	99.62[1]	99.94	99.99	100.13[3]	100.51	100.05[4]	99.57	99.51

(1) Including SO_3 0.13; (2) Calculated; (3) Includes organic matter 0.006; (4) Includes Cr_2O_3 0.00008; BaO and SrO none; NiO less than 0.001; CuO less than 0.00027; CoO less than 0.0002.

A. Mesnard Quartzite (Precambrian), Marquette County, Michigan, U.S.A., R.D.Hall, analyst (Leith and Van Hise, 1911, p. 256).

B. Lorrain Quartzite (Precambrian), Plummer Township, Ontario, Canada, M.F.Connor, analyst (Collins, 1925, p. 68).

C. Sioux Quartzite (Precambrian), Sioux Falls, S. Dakota, U.S.A. (Rothrock, 1944, p. 151).

D. Lauhavouri Sandstone (Cambrian?), Tiiliharju, Finland, Pentti Ojanperä, analyst (Simonen and Kuovo, 1955, p. 79). Quartz 70–75; feldspar 0.1–1.4; rock fragments 0.1–5.6; silica cement 18–20.

E. St. Peter Sandstone (Ordovician), Mendota, Minnesota, U.S.A., A. William, analyst (Thiel, 1935, p. 601).

F. Simpson Sand (Ordovician), Cool Creek, Oklahoma, U.S.A. (Buttram, 1913, p. 50).

G. Tuscarora Quartzite (Silurian), Hyndman, Pennsylvania, U.S.A. (Fettke, 1918, p. 263).

H. Oriskany Sandstone (Devonian) Berkeley Springs quarry, Berkeley Springs, West Virginia, U.S.A., Pennsylvania Glass Sand Corp., Sharp-Schurtz Co., analysts. Analysis supplied courtesy of Pennsylvania Glass Sand Corp.

I. Mansfield Formation (basal Pennsylvanian), Crawford County, Indiana, U.S.A., M.E.Coller, R.K.Leininger, R.F.Blakely, analysts. Computed mineral composition: Quartz 95.3; orthoclase 1.2; kaolin 3.0; ilmenite 0.3 (Murray and Patton, 1953, p. 28).

J. Berea Sandstone (Mississippian), Berea, Ohio, U.S.A., N.W.Lord, analyst. A protoquartzite (Cushing, Leverett, and Van Horn, 1931, p. 110).

overgrowths leading to the reconstruction of the quartz crystal form with brilliant, sharp-edged facets that reflect light so that the sand has a prominent "sparkle" in bright sunlight.

These sands are, as noted above, those in which quartz forms 95 or more percent of the detrital fraction. In many, especially the quartz-cemented varieties, the bulk chemical analyses reflect this dominance of quartz, the silica content being 99 percent or even more (Table 6-7). Such deposits are the largest and purest concentrations of silica in the earth's crust and constitute a commercial source of silica for the manufacture of glass and other needs.

The quartz of these sands is largely monocrystalline quartz – the polycrystalline grains seem less stable and tend to be eliminated (Blatt, 1967, p. 422). The proportion of quartz with undulatory extinction tends also to be lower in these sands for similar reasons (Blatt and Christie, 1963, p. 571). The quartz grains of most orthoquartzites are highly rounded – some approaching perfect roundness. Many exhibit a frosted or "matte" surface. Inasmuch as sorting is excellent, these sands are both the most texturally and compositionally mature of all sands. Some approach the theoretical end point in sand evolution.

Other constituents, if any, are likely to include a few well-worn grains of chert or other equally durable rock particles. Though insignificant in volume, these few rock particles may be the clue to the provenance and "line of descent" of the particular quartz arenite in question. The heavy mineral fraction is generally highly restricted and consists usually only of very well-rounded tourmaline and zircon plus some ilmenite or leucoxene derived from the ilmenite.

The bulk chemical composition may be drastically altered by the addition of the cement. The silica-cemented sands, of course, are not so altered, but those with calcite or anhydrite will show a silica percentage depressed by the addition of CaO and CO_2 or SO_3.

Silica is the most common cementing agent and even in those sands in which another cement dominates, some silica cement is present. The silica is almost always quartz, deposited in optical and crystallographic continuity with the rounded detrital quartz. In sands incompletely cemented, the overgrowths exhibit well-formed crystal facets – which in thin section give the remaining pores straight-edged boundaries. In the more friable sands, the regenerated quartz euhedra show well-formed pyramidal terminations. Casual inspection shows that the crystal axis tends to coincide approximately with the long dimension of the detrital core. As cementation proceeds, the overgrowths meet along irregular boundaries so that the end-result of the enlargement process is an interlocking quartz mosaic within the elements of which, the rounded outlines of the detrital grains may be visible, being delineated by a thin ring of dust-like inclusions. In some quartz arenites this line of demarcation is faint or altogether absent.

Pressure-solution phenomena are well-known and especially notable and clearly displayed in many orthoquartzites. Clear and well-developed stylolitic seams are present in some cases (Heald, 1955). Microstylolites involving detrital chert grains are known in other cases. Pressure-solution along grain boundaries can presumably lead to transformation of a quartz sand into a quartzite. This conversion by pressure-solution at grain contacts has been described in detail by Skolnick (1965). In many cases, it is difficult to know whether or not the interlocking sutured boundaries of the quartz mosaic are the product of pressure-solution or whether they are nothing more than the result of mutual interference of the respective overgrowths. Examination by the method of luminescence microscopy (Sippel, 1968) suggests that many presumed cases of pressure-solution are nothing more than the end-product of quartz enlargement. Under cathodoluminescence the detrital grains are clearly visible and display uninterrupted detrital outlines surrounded by secondary quartz.

A few quartz arenites are cemented with other forms of silica – opal or chalcedony, some with both. Opal may form concretionary coatings on the grains which extend into the pores, partially or wholly filling them. That the opaline coatings are post rather than pre-depositional, is demonstrated by their absence at the points of contact of adjacent quartz grains. Chalcedony plays a similar role and differs from opal in that it may show a microfibrous fan-shaped structure and, unlike opal, is birefringent. In general, opaline cements are characteristic only of the younger sandstones. As in the cherts, the older rocks are chalcedonic or quartzose – presumed devitrification products of opal.

Carbonates, usually calcite, are a common cement in other quartz sands. They take several forms. In most cases each individual pore is filled by a single crystal of calcite; in others, the calcite crystallized in large "poikilitic" patches which enclose many sand grains. These are the so-called "luster-mottled" sandstones (Cayeux, 1929, p. 154). Such sandstones display striking reflections from the calcite cleavage of these coarse-textured cements. In others, the calcite forms drusy coatings on the individual quartz grains. Although calcite is the usual carbonate cement, a few sands contain other carbonate species. Siderite is present in some sands and gives rise, on oxidation, to iron oxide which then plays the role of cement. It is our impression, however, that the siderite-cemented quartz arenites are rare; siderite is more commonly found as cement in lithic sandstones.

Sands cemented with dolomite are common but not as abundant as those cemented with calcite (see, for example, Swineford, 1947, p. 79). Dolomite occurs as small rhombic crystals, in some cases about the same size as the sand grains.

Other cements are much less common. Sands cemented with anhydrite, barite and celestite have been described. The barite-cemented sands are restricted to small nodular bodies consisting of tabular barite crystals arranged in a rosette-like manner (Shead, 1923; Tarr, 1933). Celestite cementation is likewise a very local and restricted phenomenon (Swineford, 1947, p. 82). The quartz grains enclosed in the barite and celestite cements show markedly less corrosion by the cement than those cemented with carbonates. Unlike barite and celestite, anhydrite is a more common cement. It is generally associated with other cementing minerals and is usually the last to have been precipitated (Waldschmidt, 1941, p. 1886). Anhydrite is readily identified by its rectilinear cleavage.

Varieties

Because quartz arenites are 95 or more percent quartz, there is not much variation between one and the next except in the kind of cement which they contain. The most marked contrast is between the quartz-cemented and carbonate-cemented arenites.

On the other hand, it is important to make a distinction, if possible, between the quartz arenites derived directly from a plutonic source rock and those derived from pre-existing sandstones. It is very difficult, however, to make this distinction. First-cycle sands are apt to be less well rounded, to contain a greater proportion of polycrystalline and undulatory quartz, more likely to retain a little feldspar and to show a greater diversity of heavy minerals – or at least less well rounded zircons and tourmaline. Multicycle sands, on the other hand, might contain grains of older quartzites and detrital chert, especially in the coarser size grades. Worn overgrowths would be proof of second-cycle origin. Unfortunately, such features are uncommon.

Multi-cycle quartz arenites are commonly associated with transitional types – with subarkose and especially with protoquartzites (sublitharenites). They are presumed to be the better "cleaned-up" sands. Not uncommonly, the sands in a sedimentary sequence show a progressive decrease in feldspar or rock-particle content with decreasing age. The "evolutionary" succession that terminates in orthoquartzites suggests that the latter are indeed multicycle sands.

A peculiar feature of some quartz arenites is their bimodal texture. Attention has been called to this feature especially by Folk (1968b) who attributed it to the action of the wind in desert areas. It is presumed to be due to a selective removal of the fine sand fraction (0.1-0.3 mm) leaving behind both the coarser and finer sizes. Folk described present-day bimodal sands from the Simpson Desert of Australia. A similar bimodal texture is reported to be common in many Cambro-Ordovician and some Precambrian sands. These formations may be marine or fluvial as well as eolian but the final size distribution was created in the deflationary eolian environment. Folk (1968b, p. 12) tabulates some 21 North America examples of bimodal sands. A good Precambrian example is the Odjick Formation of the Coronation Gulf Geosyncline.

Field Occurrence and Examples

Quartz arenites have a world-wide distribution. Well-known North American examples include the thick, widespread Precambrian quartzites: the Sioux Quartzite (Fig. 6-22) of Minnesota, Iowa and South Dakota (Rothrock, 1944), the Baraboo and Waterloo Quartzites of Wisconsin (Brett, 1955), the Sturgeon Quartzite of northern Michigan and the correlative Mesnard Quartzite of the Marquette district on Lake Superior, the Palms Quartzite of the Gogebic district, and the uppermost Lorrain Quartzite (Fig. 6-23) on the north shore of Lake Huron in Ontario, and the Odjick Formation (Fig. 6-24) in the Northwest Territories. The Wishart Formation of the Labrador Trough contains excellent quartz arenites (Donaldson, 1966, p. 34). The Hinckley Sandstone of Minnesota (Thiel and Dutton, 1935) and the Sibley Sandstone of the Thunder Bay area of Lake Superior are orthoquartzites of Keweenawan age. The Athabaska Formation of northern Saskatchewan is fairly typical of many of the Precambrian quartz arenites (Fahrig, 1961). It appears to have covered a very large area and even today has an areal extent of 40,000 mi^2 (104,000 km^2), contains some quartz pebbles and granules, has a uniform transport direction, is largely free of shale and has a maximum thickness of about 5000 ft. (1510 m). The Thelon Formation of the Northwest Territories of Canada is largely sandstone and is similar in character and extent and perhaps in age, to the Athabaska (Donaldson, 1967). The Precambrian Uinta Mountain Group of Utah contains some quartz arenites said by Krynine to be first cycle (Krynine, 1941b).

The sandstones of the Cambro-Ordovician of the Upper Mississippi Valley are primarily quartz arenites. These include the Cambrian sandstones – the Dresbach Sandstone, Franconia Sandstone, and Jordan Sandstone (Graham, 1930) and the Ordovician New Richmond and St. Peter Sandstones, Fig. 6-25

(Dake, 1921; Thiel, 1935). Quartz arenites of equivalent age and similar character occur in Missouri, the Cambrian Lamotte Sandstone (Ojakangas, 1963) and the sandstones of the Ordovician Roubidoux Formation, and in Oklahoma, the sands of the Simpson Group (Ordovician) including, for example the sandstones of the Tulip Creek Formation of the Arbuckle region, and the Blakely and Crystal Mountain Sandstones of the Ouachitas. In the New York region, the Cambrian

Fig. 6-22. Sioux Quartzite (Precambrian). Dell Rapids, South Dakota, U.S.A. Crossed nicols, ×55. A typical orthoquartzite – a truly quartzitic rock produced by thorough cementation of a quartz arenite. Note well-preserved boundaries of highly rounded detrital quartz separating detrital core from secondary quartz overgrowth. Essentially 100 percent quartz

Potsdam Sandstone is in most places a quartz arenite (Fig. 6-26) as are also the sandstones in the Cambrian Gatesburg Formation and the Chickies Quartzite of Pennsylvania and the Antietam Sandstone of Maryland and West Virginia (Schwab, 1970). In the western United States the Cambrian is characterized by quartz arenites including the Flathead Quartzite of Wyoming and the Lower Cambrian quartzites of British Columbia. The Ordovician Eureka Quartzite of Nevada and its equivalent, the Swan Peak Quartzite of Idaho are high purity deposits consisting of over 99 percent highly rounded and frosted quartz (Ketner, 1966). The Pennsylvanian Tensleep Sandstone of Wyoming and its facies equivalents in the Casper Formation and the Weber Quartzite of Utah, are good quartz arenites, the latter being interpreted by Krynine (1941b) as second cycle. The Shawangunk Conglomerate and the Tuscarora Quartzite (Silurian) of Pennsylvania and New Jersey are similarly interpreted. The Devonian Oriskany Sandstone of Pennsylvania (Fig. 6-27), in places a glass sand, is a good example of a marine quartz arenite. The Dakota Sandstone (Cretaceous) of the Great Plains is in many places a quartz arenite (Fig. 6-28). Some of the Miocene sands of

Fig. 6-23. Lorrain Quartzite (Huronian, Precambrian). Near Bruce Mines, Ontario, Canada. Crossed nicols, × 55. An orthoquartzite (near 97 percent quartz), well-sorted and well-rounded quartz. Original grain boundaries well shown

New Jersey are very clean quartz sands, containing 97 to 99 percent quartz. These are locally well enough cemented with opal, chalcedony or quartz to deserve the name quartzite (Friedman, 1954). See Fig. 6-29.

In Europe there are a number of Cambrian quartz arenites. Among these are the Hardeberga Sandstone of Sweden (Hadding, 1929, p. 77), the sandstone of Lauhavuori, Finland (Simonen and Kuovo, 1955, p. 75), and the Cambrian Malvern Quartzite of England. The Cretaceous seems also to have been a time of quartz arenite development. The Lower Senonian Quadersandstein of the Harz of Germany is an excellent example (Rinne, 1923, p. 280) as is the Ashdown Sandstone (Lower Wealden) of Sussex (Allen, 1949). Other European quartz arenites include some sands from the Lias near Celle and the Dogger near Braunschweig (Füchtbauer, 1959, Table 2) and the Tertiary Fontainebleau Sandstone of France (Cayeux, 1929, p. 154). The Nubian Sandstone of Egypt is at least in part a good quartz arenite as is the Precambrian Scarp Sandstone of the Vindhyan System along the Son River of India.

Distribution in Space and Time

Because quartz arenites are by definition essentially quartz sands with less than five percent other constituents, they are exceptional sands. They are ex-

Fig. 6-24. Odjick Formation, Epworth Group, Lower Proterozoic. Coronation Gulf, Northwest Territories, Canada. Crossed nicols, × 20. A quartz arenite with a bimodal size distribution presumed to be indicative of an eolian origin. Note extraordinary rounding of the large quartz grains the interstices between which contain smaller (and less well-rounded) grains. Large microcrystalline grains are chert

ceptional, that is, in the sense that sands of this character seem not to be forming today. Modern sands of this type, if they exist at all, are small accumulations derived from or formed by redeposition of an older quartz arenite. Such seems to be the case with the pure quartz sands of the Libyan Desert in Egypt that contain 95 or more percent quartz (Mizutani and Suwa, 1966) which probably were derived from disintegration of the Nubian Sandstone. On the other hand, orthoquartzitic sands of the past were both common and widespread.

Although quartz arenites are estimated to constitute about one third of all sandstones (Pettijohn, 1963), they are not equally abundant everywhere or throughout geologic time. It is probably no accident that of the forty-four examples cited in the preceeding section all but five are Precambrian or Paleozoic and about one-third are Cambrian. There seem to be but few examples of Mesozoic and Cenozoic quartz arenites although the Nubian Sandstone of the Middle East approaches a true quartz arenite as do several Cretaceous Sandstones such as the Dakota sandstone of Kansas (Swineford, 1947), the Woodbine

Fig. 6-25. St. Peter Sandstone (Ordovician). Twin City Brick Yard, St. Paul, Minnesota, U.S.A. Crossed nicols, × 55. A famous quartz arenite consisting wholly of very well-sorted and highly rounded monocrystalline quartz. A friable sand with little or no cement in most places. Specimen photographed was impregnated with resin

sand of Texas, and the Quadersandstein of the Harz region of Germany. It would seem that the older sandstones were more largely derived from the craton whereas those of later times were derived from orogenic belts or lands and are, therefore, less mature. It has even been suggested that the abundance of Late Precambrian and Cambrian quartz arenites is related to more vigorous and effective tidal action due to a presumed lesser distance between the earth and moon (Merifield and Lamar, 1968).

Although quartz arenites seem to be the characteristic sandstone of the stable cratonic areas and of little importance in the geosynclinal or orogenic belts, there are some notable exceptions. Craton-derived quartz sands do extend into the miogeosynclinal belts being most prominant along the cratonic margin of these tracts. Ketner (1968), for example, has described an Ordovician quartzite, over 1000 ft. (300 m) thick, of high purity, in the Cordilleran miogeosyncline. It appears to have been derived by erosion of Cambrian sandstones in northern Alberta and transported over 1000 mi (1600 km) southward along the axis of the geosyncline. In general, eugeosynclines are without true quartz arenites – the typical arenite of these tracts being graywacke. In rare cases, however, clean

Fig. 6-26. Red Sandstone (Ordovician). Potsdam, New York, U.S.A. Crossed nicols, ×55. Unusually well-displayed example of secondary enlargement of well-rounded quartz. The boundaries between detrital core and overgrowth made very distinct by coating of iron oxide on the original detrital grains

quartz arenites are reported from eugeosynclines. The Silurian Clough Quartzite (Billings, 1956, p. 21), a metamorphic quartzite, clearly derived from a quartz arenite and containing marine fossils, lies in the midst of the eugeosynclinal sequence in New England. Ketner (1966) has described a high-purity Ordovician quartzite (Valmy Formation) in the eugeosynclinal tract in Nevada. This quartzite is associated with bedded cherts and volcanic greenstones.

Significance and Origin of Quartz Arenites

What is the geologic significance of these sands? The salient facts to be accounted for are: (1) their supermature petrography, (2) their sheet-like geometry, (3) their cratogenic derivation, and (4) their distribution, that is, primarily on the craton or its margins rather than in the geosynclines.

The quartz arenites exhibit the best sorting, the best rounding, the highest concentration of quartz, and the most restricted heavy mineral suit of any of the sands. On the basis of these characters most investigators have concluded that such sands could not have been derived directly from a weathered granite, that

Fig. 6-27. Oriskany Sandstone (Devonian). Near Berkeley Springs, West Virginia, U.S.A. Crossed nicols, × 55. A glass sand consisting solely of quartz. A well-sorted sand with a good marine fauna. Original grain outlines very indistinct

instead, they were derived from pre-existing sandstones. In short, they are multicycle. On the basis of experimental work, especially that of Kuenen (1959), and on field observation of some large modern streams (Russell, 1937, p. 1346), it seems improbable that a quartz arenite sand of the kind found in the geologic record could ever be produced by river transport no matter how prolonged.

According to Kuenen (1960, p. 109), only in the desert dune environment is sand effectively rounded. This conclusion would imply that most of the sand of the quartz arenites had an eolian episode at some time in its history – not necessarily the last stage represented by the present accumulation. The frosting observed on the grains of many of these sands has been taken as confirmatory evidence of eolian action, as has the bimodal texture described by Folk (1968b). The absence of shale has been thought to be related to the desert eolian episode in the history of the sand. Wind action effectively separates or winnows the clay-sized dust from the quartz sand. This deflation process is also almost always fatal to the micas (Krynine, 1942, p. 550).

Kuenen dismisses beach action also as too restricted to be quantitatively important even if it could be shown that such action is capable of concentrating and rounding quartz to the degree required. One cannot wholly dismiss the

Fig. 6-28. Dakota Sandstone (Cretaceous). Kansas, U.S.A. Crossed nicols. × 55. An excellent example of a calcite-cemented quartz sand showing local etching or embayments in quartz indicating solution and replacement. The cement is carbonate which has same crystallographic (and optical) orientation over entire field of view

possibility, however, that intense weathering followed by prolonged action of the surf might not produce an quartz arenite. Most modern beach sands are not quartz arenites although there are places, such as the Gulf coast of northwestern Florida, where the sand is nearly pure quartz, being 99.65 percent silica (Burchard, 1907, p. 382). These sands, however, may have been derived from coastal plain sediments of Cretaceous or Tertiary age. On the other hand, there is some support for the notion of a progressive "cleaning up" of sands by beach action. Not a few sedimentary sequences progress upward from lithic arenites or arkoses to quartz arenite. The lowest member of the 7000-foot (2100 m) Lorrain Quartzite (Precambrian) in the Bruce Mines area of Ontario is an arkose with 25-30 percent feldspar. The feldspar shows an upward decline so that the uppermost Lorrain is a clean orthoquartzite with less than one percent feldspar. The Tuscarora Quartzite (Silurian) of West Virginia is a further illustration (Folk, 1960). The lowest beds contain 50 to 70 percent quartz; the highest have over 95 percent quartz. This upward increase in quartz is accompanied by a compensatory decrease in the quantity of rock particles (15 to 35 percent) and chert (3 to 6 percent) to less than five percent. Similarly the lowest Pennsylvanian

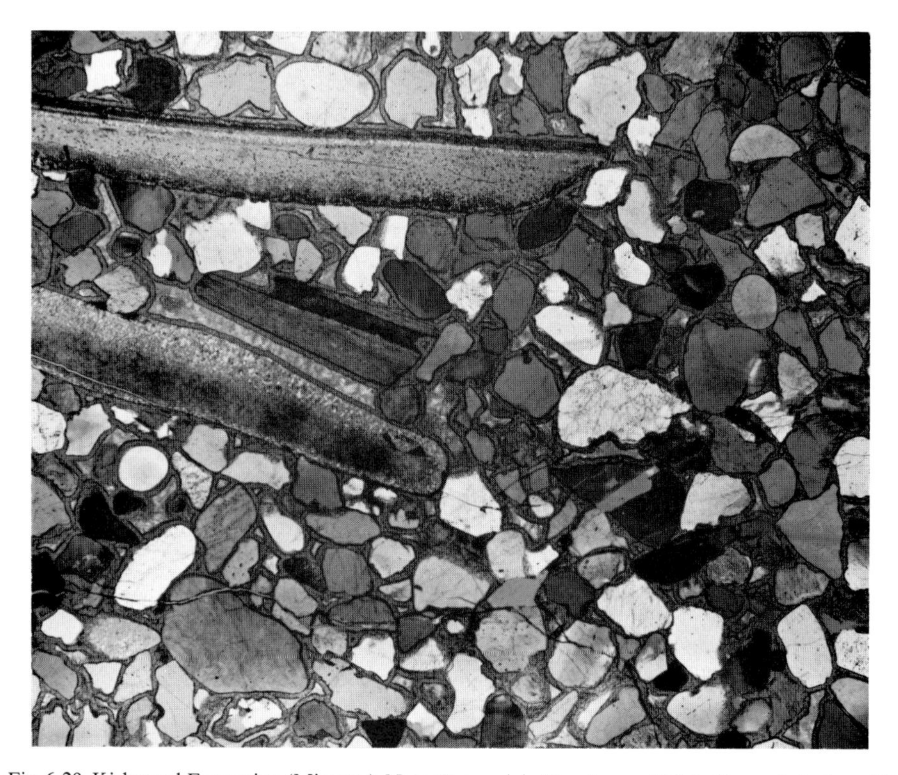

Fig. 6-29. Kirkwood Formation (Miocene). Near Greenwich, New Jersey, U.S.A. Crossed nicols, × 20. A chalcedony-cemented quartz arenite containing silicified shell fragments. Chalcedonic coatings, once opal, coat the quartz grains. Remaining interstices filled with fibrous chalcedonic material. Original quartz well-sorted and subangular to well-rounded

sandstones in the Anthracite Basin of Pennsylvania have 15 to 20 percent rock particles; the highest sands have less than 3 percent (Meckel, 1967). It seems unlikely in all the above examples that the provenance of the lower beds was appreciably different from that of the higher ones. If the change in character of the sands as one proceeds upward is not due to changing character of the sand supplied, it must be due to action in the environment of deposition. Folk interpreted the change as due to a shift from an essentially fluvial environment to that of the beach. He believed that it was to surf action that the highest sands owe their quartz enrichment.

Some of the quartz arenites are clearly marine – as the contained marine fauna testifies (Seilacher, 1968). Others are presumed to be marine even though fossils are absent, in part because these sands are interbedded with or pass laterally into marine limestones and dolomites. Because the shallow-water sands of today are mainly restricted to a narrow littoral zone, it has been supposed that these orthoquartzitic sand sheets record the advance of a transgressive shoreline. Thiel (1935), for example, thought the Ordovician St. Peter Sandstone of the Upper Mississippi Valley to be due to marine deposition related to several retreats and advances with intervals of wind action to account for the rounding and frosting. The size and character of the crossbedding and ripple-marking and, in some cases, association with limestones or dolomites seems to indicate a marine origin, or at least deposition in a standing body of water, for many of the Precambrian quartz arenites.

A problem which arises with these and other quartz arenites is the paucity of shale associated with them. Unlike the lithic and arkosic sandstones and the graywackes, little or no shale is interbedded with the quartz arenites, nor, indeed, are there significant shaly formations in the sequence of which the quartz arenites are a part. Although this absence of clays can be explained by eolian action, it seems more probable to us that because most quartz arenites of the past were deposited on shallow marine shelves, where, if the modern world can be taken as a guide, much if not most of the clay is in suspension and is carried over the shelf edge (see Chap. 11).

In summary, the necessary conditions for the formation of quartz arenites are extraordinary weathering to eliminate the feldspar coupled with the most effective removal of clay and rounding of the quartz. Whether the latter was achieved by desert eolian action or by action of the surf is uncertain.

The known quartz arenites, certainly the best North American examples, seem to have been deposited on the stable cratonic portions of the continents and such thin quartz arenites as occur in the marginal geosynclines are of cratonic or shield derivation. The best examples – the Cambro-Ordovician sands of the North American continental interior – are illustrative of the deposition on the craton. They are also of cratonic derivation (Potter and Pryor, 1961). Such sands as those of the Cambrian Chickies Quartzite and Antietam Sandstone which occur not on the craton but in a marginal geosyncline, are probably shield-derived. The great Precambrian orthoquartzites of the Lake Superior region seem also to have been derived from the interior of the Canadian Shield (Pettijohn, 1957a). All of the Paleozoic and some of the Precambrian quartz arenites seem to have been deposited by a marine transgression of this stable interior platform.

Hybrid Sands and Sandstones

Introduction

The sands we have so far dealt with are *epiclastic*, that is, sands in which the framework components, usually quartz, feldspar and rock particles, are the debris of a wasting landmass – are derived from pre-existing rocks or sediment. We shall now briefly discuss those sands in which a significant portion of the framework is of another origin – one formed *in* the basin or deposition by chemical or biochemical precipitation or produced elsewhere by volcanic action.

These sands are thus neither wholly epiclastic nor pyroclastic nor endogenetic but are instead a cross between several of these and are thus *hybrid* sands.

Examples of sands containing both intrabasinal and extrabasinal framework elements include the greensands in which glauconite is a significant constitutent, phosphatic sands in which the phosphate mineral is a framework component (rather than cement) and, most common of all, those sands consisting of a mixture of detrital quartz and shell or other calcareous debris indigeneous to the depositional basin.

Greensands

Glauconite occurs as granules which may be mixed in all proportions with ordinary sand. Some greensands contain over fifty percent glauconite. The glauconite may be concentrated in certain laminations or scattered throughout the sand. If abundant enough, it imparts a speckled appearance to the rock.

Under the binocular microscope the granules appear as very dark, almost black, grains with a smooth rounded surface, some being mammilated, and of diverse shapes, mostly ovoid. In thin section the glauconite varies from very pale bluish green to greenish yellow to a dark grass green. It may also be oxidized to a brownish yellow or reddish brown. The grains are rounded, are of about the same size as the associated quartz, exhibit a polylobate outline, are generally structureless, and may show some shrinkage cracks which taper inward. Some grains are deformed and are molded about the quartz or squeezed into a pore. Under crossed nicols, the glauconite appears as a microcrystalline aggregate. Birefringence is high so that the grain appears much the same under crossed nicols as in ordinary light. In the standard thin section it varies from a pale green to a deep grass green. Oxidized grains are yellowish or brown.

Glauconite is not confined to a particular type of sand. It occurs in some nearly pure quartz arenites; it also occurs in less mature feldspathic and micaceous sands. Commonly it is associated with shell materials, the glauconitic sand filling the shells, and, in the younger sands, the glauconite alone may fill foraminiferal tests. Very commonly the shell materials of the Cambrian sandstones are phosphatic. Clearly most glauconitic sandstones are marine. In some sands the glauconite is associated with pellets – many of which show partial conversion to glauconite – or with biotite flakes which likewise exhibit all degrees of glauconization. Some glauconite forms a coating on other mineral grains – even on heavy minerals, especially ilmenite (Grim, 1936, p. 201).

There is a large literature on the mineralogy of glauconite and hence many analyses of the mineral itself. There are few analyses of greensands. Glauconitic sands are represented by three analyses (A, B, and C) in Table 6-8. As might be expected, the composition of greensands is highly variable depending on the proportion of clastic material, interstitial calcite, and secondary siderite.

Greensands are characterized by their high total-iron content, principally Fe_2O_3, and by the high content of K_2O. Quite commonly also, as shown by the New Jersey greensands, they are high in phosphorus. As would be expected, they also contain a good deal of combined water. Some greensands are high enough

Table 6-8. *Chemical analyses of miscellaneous sandstones (after Pettijohn, 1963, Table 9, with modifications)*

	A	B	C	D	E	F
SiO_2	57.40	50.74	75.95	45.43	48.85	51.32
Al_2O_3	6.89	1.93	2.91	0.03	11.82	2.92
Fe_2O_3	11.98	17.36	10.29	2.92	1.83	0.72
FeO	3.04	3.34	—	—	1.22	—
MgO	2.41	3.76	1.37	0.61	0.45	0.58
CaO	1.78	2.86	0.10	26.21	12.85	24.70
Na_2O	1.11	1.53	0.35	0.34	0.47	—
K_2O	4.85	6.68	2.99	0.16	0.64	—
H_2O+	5.36	9.08	5.40	2.78	2.75	—
H_2O-	4.46	—	—	—	—	—
TiO_2	0.29	—	0.20	0.11	trace	0.30
P_2O_5	0.22	1.79	—	16.05	10.70	—
CO_2	—	0.88	—	3.12	3.40	20.00[4]
MnO	0.03	—	—	0.02	—	—
SO_3	0.45	—	—	0.86	—	trace
F	—	—	—	1.87	2.86	—
Total	100.29[1]	99.95	99.56	101.25[2]	97.84	100.54
Less 0				−0.79	−1.20	
				100.46[3]	96.64	

(1) Includes BaO, 0.02; (2) Includes C, 0.45, FeS_2, 0.29; (3) Given as 99.01 in original; (4) by calculation.

A. "Greensand" (Middle Eocene), Pahi Peninsula, New Zealand (Ferrar, 1934, p. 47).
B. "Greensand marl" (Upper Cretaceous), New Jersey, U.S.A., R.K. Bailey, analyst (Mansfield, 1920, p. 553).
C. Greensand, opal-cemented (Thanetien), Angre, Belgium (Cayeux, 1929, p. 130).
D. Phosphate sandstone, "Upper phosphorite stratum" (Cenomanian), Kursk, Schchigri, U.S.S.R. (Bushinsky, 1935, p. 90). About 38 percent quartz, 45 percent phosphorite, 5 percent glauconite.
E. Phosphatic sandstone, Saint Pôt, Boulannais, France (Cayeux, 1929, p. 191).
F. Calcarenaceous sandstone, Loyalhanna Formation (Mississippian), Pennsylvania, U.S.A. (Hickok and Moyer, 1940, p. 464).

in iron to be classed as an iron-bearing formation or ironstone (over 15 percent Fe).

Although glauconitic sandstones have a wide range in both space and time and perhaps occur in the rocks of every geologic system, one gains the impression that they are more common in some systems than others. They are particularly common, for example, in the rocks of the Cambrian. They occur in the sands of the Bright Angel Shale in the Grand Canyon, in the Reagan Sandstone of Oklahoma, in the Franconia Sandstone of Wisconsin, all of Cambrian age, and in the Lower and Middle Cambrian rocks of Wales and of Scandinavia (Hadding, 1932). Likewise the Cretaceous seems to have been a period of glauconite formation (Fig. 6-30). Greensands of this age occur in the east of England (Cambridge Greensand) and in Europe. This period of greensand formation continued in the Paleocene and Eocene of the Atlantic Coastal Plain (Aquia Formation, Drobnyk, 1965). Greensand is relatively rare in rocks of Precambrian age although glauconite sandstones are known from the Semri Series of the Vindhyan System of presumed Precambrian age in peninsular India (Auden, 1933, p. 160).

Glauconite has a wide distribution on the present sea floor. A good example of its occurrence and association with sand is that in Monterey Bay, California (Galliher, 1935).

There is still considerable uncertainty about the origin and significance of glauconite (Goldman, 1919; Hadding, 1932; Allen, 1937; Takihashi, 1939; Cloud, 1955; Burst, 1958). It apparently requires marine conditions for its forma-

Fig. 6-30. Greensand, Ft. Augustus Formation (Cretaceous). Ft. Augustus No. 1, T. 55 S., R. 21 W., west 4th Meridian, 2,883 feet, Alberta, Canada. Ordinary light, × 55. A sandstone rich in glauconite (large round pellets) together with subangular quartz cemented by carbonate. This formation also contains some detrital chert

tion though it may be redeposited, usually oxidized, in a non-marine sand. It seems to be more marine than chamosite inasmuch as the Clinton (Silurian) oolitic hematite-chamosite ironstone of the Appalachians grades to the east into semi-continental hematitic sandstone and to the west into fully marine glauconitic strata (Hunter, 1970, p. 118). Normal salinity, weakly reducing conditions, and a slow rate of deposition are required for its formation. The mineral appears to be a diagenetic product created by alteration of biotite (Galliher, 1935), or by reorganization and iron enrichment of mud pellets, replacement of shells and other debris (Takahashi, 1939), or even by precipation in cavities or as coatings on detrital grains (Grim, 1936, p. 201).

It does occur in the Precambrian (See, for example Auden, 1933, p. 160; Gulbrandsen and others, 1963) – despite some contrary opinions (Cloud, 1955): There is no good explanation for the apparently greater abundance in the Cambrian and Cretaceous-Eocene strata. Much yet remains to be learned about its origin and distribution.

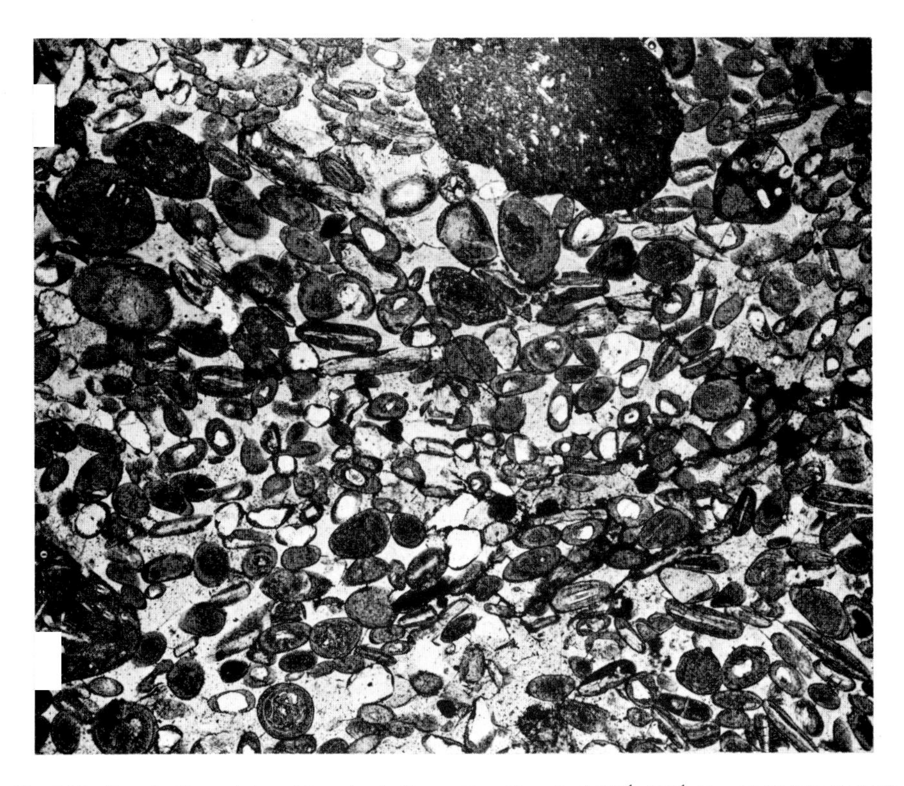

Fig. 6-31. Phosphatic sandstone Phosphoria Formation (Permian). SE$\frac{1}{4}$, NE$\frac{1}{4}$, Sec. 13, T. 1 N., R. 3 W., Jefferson County, Montana, U.S.A. Ordinary light, × 20. An oolitic phosphate rock. Many oolites have quartz sand grains as nuclei

Phosphatic Sandstones

Phosphatic sandstone is sand cemented with calcium phosphate (carbonate fluorapatite) or sandstone that contains an appreciable quantity of phosphatic debris or precipitated granules or oolites of phosphate.

Many sands contain a little phosphatic debris – largely that of phosphatic skeletal materials. The glauconitic sands are especially prone to be phosphatic. In most phosphatic sandstone, however, the phosphate forms a cement, or occurs as a drusy coating on the quartz grains or as a micro-crystalline pore-filling (Bushinsky, 1936). On the other hand, some phosphorites are themselves "sands" consisting of sand-sized phosphate granules or öoids which may be mixed in varying proportions with detrital quartz (Fig. 6-31).

Calcarenaceous Sandstones

By far the most common sands of a "mixed" origin are those which consist of a mixture of detrital quartz and sand-sized chemical or biochemical carbonate. These have been called *calcarenaceous sands* (Pettijohn, 1957b, p. 405) to distinguish them from a calcareous sand, the latter having carbonate as a cement rather than as a framework element. The calcarenaceous sands grade, with

Fig. 6-32. A well sorted skeletal limestone "Bedford limestone", Salem Limestone (Mississippian). Quarry near Bedford, Lawrence County, Indiana. Ordinary light, × 20. A biocalarenite cemented with clear crystalline calcite (biosparite). Skeletal debris includes foraminiferal tests, crinoidal debris, bryozoan, and brachiopod debris. Elongate fragments parallel to bedding. A good example of a skeletal carbonate sand

increasing proportions of the carbonate materials, into the calcarenite class of limestones (Figs. 6-32 and 33). Care should be taken to differentiate between calcarenaceous sandstone and calclithite, the latter being a lithic sandstone derived by the destruction of pre-existing limestones or dolomites.

Under the microscope, the calcarenaceous sandstones consisting of carbonate detritus such as foraminiferal tests, shell and other skeletal fragments, carbonate intraclasts and pellets, and carbonate colites, mingled in all proportions with quartz and other epiclastic debris. The usual cement is calcite. Inasmuch as all the sand-sized material is current-deposited, the structures of the rock are those of a sandstone. Many are strongly crossbedded. As would be expected, sandstones of this type are rich in Ca and CO_2 – richer than a calcareous sandstone – and relatively low in SiO_2 (analysis F of Table 6-8).

Modern examples of sands of this type are common, a notable one being the sands of the east coast of Florida where the siliceous components are carried by southward shore drift and mingled with shelly debris produced locally. In the northern section quartz forms most of the sand; further south the carbonate

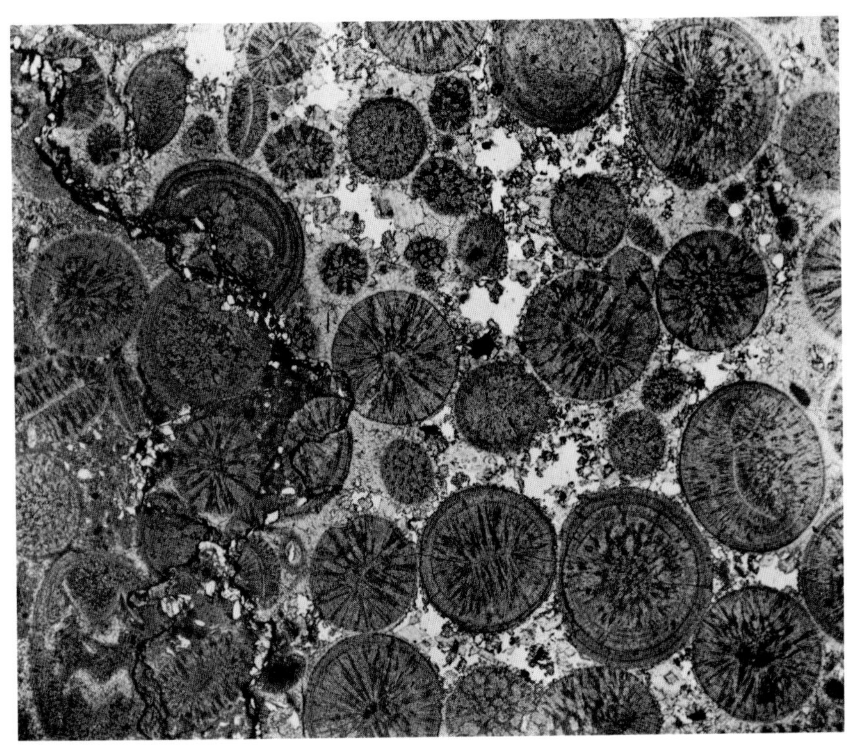

Fig. 6-33. Warrior Formation (Cambrian). Tyrone, Pennsylvania, U.S.A. Ordinary light, ×20, An oolitic calcarenite (oosparite) composed of spherical oolites with both radial and concentric structure embedded in a carbonate cement. Contains small scattered quartz silt particles. Note stylolite seam which transects oolites and along which is a residuum rich in quartz particles. A "concentration" sand – originally an oolitic sand

detritus increases until it constitutes more than half of the sand (Martens, 1931, p. 82).

Ancient examples are common, one of the best described being the Mississippian Loyalhanna Limestone of Pennsylvania and Maryland (Adams, 1970, p. 83). The Loyalhanna is, perhaps, the most crossbedded formation in the central Appalachians. A typical sample contains 38 percent monocrystalline quartz, 5 percent polycrystalline quartz, 1 percent chert, 26 percent carbonate detritus (intraclasts, ooids and skeletal grains) and 28 percent carbonate cement. The chemical composition of a rock of this type is given by analysis F in Table 6-8.

Portions of the Conococheague Limestone (Cambrian) and the Grove Limestone (Cambro-Ordovician) of Maryland contain calcarenaceous sandstone beds. The Cretaceous Cow Creek Formation (Fig. 6-34) of Burnet County, Texas, is likewise a calcarenaceous sand, probably a beach deposit. Almost any sequence of limestones interbedded with sandstones will contain some beds of mixed composition.

Fig. 6-34. Cow Creek Limestone (Cretaceous). Cow Creek locality, Edwards Plateau, near Fredericksburg, Texas, U.S.A. Crossed nicols, × 55. Consists of subrounded detrital quartz, skeletal debris (fibrous structure) and some micritic carbonate grains cemented by calcite

Tuffaceous Sandstones

A volcanic eruption very commonly generates a great volume of tuffaceous material which, on fall-out, overwhelms normal sedimentation and produces a tuff. Recognition of a fully developed tuff – especially one of recent origin – is no special problem. On the other hand, the detection of pyroclastic debris where such material is diluted by and mingled with sands of other origins is not easy. Such material is commonly overlooked. Moreover it is difficult to distinguish between such newly-formed pyroclastic materials and debris produced by disintegration of older volcanic rocks.

Reworking and redistribution of pyroclastic materials is common and such redeposited debris may be mixed with other sands in all proportions to produce a hybrid rock or tuffaceous sandstone (as opposed to a tuff). The alert petrographer will look for the tell-tale features which signify pyroclastic contamination. Criteria include euhedral feldspars, many of which are broken, are commonly zoned, generally oscillatory. Especially significant is volcanic quartz with its bipyramidal form which is commonly rounded or embayed due to magmatic resorption. Minerals rare in ordinary sands, such as olivine and pyroxene, usually denote tuffaceous origin. Glass in various forms – as envelops surrounding crystals, as collapsed pumice and as shards, all of which may be devitrified or altered but retain something of the pumice structure, is one of the best criteria of volcanic contamination. Highly tuffaceous sands have abnormally low quartz content (less than ten percent in some cases). Such a low quartz content is especially significant when associated sandstones contain a normal quantity of quartz. Likewise a high feldspar to quartz ratio is meaningful, particularly if the feldspar tends to be euhedral.

The discrimination between water-laid tuffs, tuffaceous sandstones, and graywackes is one of the most difficult tasks a sedimentary petrographer is called on to make. For an elaboration of this problem and for a more complete description of the volcaniclastic sands, the reader is referred to Chap. 7, which follows.

Relative Abundance of Sandstones and the "Average" Sandstone

Several estimates of the abundance of the common types of sandstones have been published (Krynine, 1948; Tallman, 1949; Pettijohn, 1957b; Middleton, 1960; Pettijohn, 1963). Because some disagreement exists on the defining parameters and on their limits, the several estimates are somewhat unlike (Table 6-9). The estimates of Middleton and Pettijohn are in reasonable agreement and because their estimates were derived by very different means – one from published chemical analyses and the other from a university rock collection – they can be accepted with fewest reservations.

Table 6-9. *Relative abundance of sandstone classes*

Class	Krynine[a] (1948)	Tallman[b] (1949)	Middleton[c] (1960)	Pettijohn[d] (1963)
Quartz arenite (orthoquartzite)	22.5	45	34	34
Arkose	32.5	17	16	15
Lithic arenite ("low rank" graywacke or subgraywacke)	35.0	17	24	26
Graywacke ("high rank" graywacke)	10.0	21	26	20
Miscellaneous	—	—	—	5

[a] Basis of estimate not stated.

[b] Based on sample of 275 sandstones, Cambrian to Tertiary in age, from all parts of the United States.

[c] Based on 167 sandstones for which chemical analyses appear in the published literature.

[d] Based on 121 sandstones in the Johns Hopkins University collection for which thin sections were available. Age and distribution of samples given in Pettijohn, 1963, Table 11.

If graywackes are indeed a diagenetic derivative of lithic arenites, we should include them with the lithic sands. If we do so, then the proportions are quartz-arenite sand 34 percent, arkosic sand 15 to 16 percent, and lithic sands 46 to 50 percent. Lithic sand is certainly dominant – nearly half of all sands belong in this category. On this point Middleton, Krynine, and Pettijohn all agree.

Elsewhere we have given the mean chemical composition of the principal sandstone types and also the various estimates of the bulk chemical composition of the average sandstone (Chap. 2). But what is the composition of this sandstone in mineralogical terms? Pettijohn's average sandstone (Pettijohn, 1963, Table 13) may be recast approximately as follows: quartz 59 percent, feldspar 22 percent, kaolin 6 percent, chlorite 4 percent, calcite 6 percent and iron oxide 2 percent. Such a calculated mineral composition does not distinguish between grains and cement, nor does it permit an estimate of the proportion of rock particles. Probably the latter are hidden in the figures for kaolin, chlorite and feldspar. If all of the kaolin and chlorite, and a third of the feldspar are considered to be present in the rock particles, the average sandstone would be basically 65 percent quartz, 15 percent feldspar, and 18 percent rock particles on a cement-free basis (carbonate and iron oxide omitted). Clarke (1924, p. 3) has given the calculated mineral composition of the average sandstone as 66.8 percent quartz, 11.5 percent feldspar, 11.1 percent carbonates and 10.6 percent "other."

How do these estimates based on calculations compare with actual modal analyses? The average sandstone of the Russian platform, according to Ronov and others (1966) is 69.7 percent quartz, 15.3 percent feldspar, 2.6 percent rock particles, 2.9 percent mica, and 1.8 percent carbonate.

Data are quite inadequate to get a very good estimate of what the average modern sand really is. Since an "educated guess" leads us to believe that alluvial sands are by far the most common types of sand in the geological record, especially in many thick geosynclinal sections, the sands of large rivers, therefore, should most closely resemble the average sandstone. The average river sand today consists of about 22 percent feldspar and about 20 percent rock particles and, by difference, 58 percent quartz. These figures agree reasonably well with the estimate made by recasting the chemical composition of the "average" sandstone considering the uncertainties in the recasting process and the inadequacies of the data on modern river sands.

These calculations and estimates, however crude they may be, point to several major conclusions. As noted in the Introduction (Chap. 1), the bulk of the sand of the world is on the continents and most of that occurs in the folded belts rather than on the cratonic platform. If the Appalachian geosyncline is typical of mio-geosynclinal belts, and we believe it is, most of its sands are alluvial. The average sandstone, as demonstrated above, most nearly resembles the sands of modern rivers. The differences may reflect the bias of the data on modern streams – the sample being overly weighted by glacial materials of Shield derivation and hence abnormally rich in feldspar and by the inclusion of nearshore marine sands in the average of ancient sands which tend to reduce the feldspar content and enhance the proportion of quartz.

Sandstone Petrogenesis

The Question

After our survey of the petrography of sandstones in this chapter, the question naturally arises: "How does one explain the diversity of sandstones?" How do we account for the marked contrast in composition and quantity of matrix between the white quartz arenites, the gray lithic arenites, the coarse reddish arkoses, and the dark graywackes? Are these diversities due to differing source materials or to differing environments of deposition, or, as some have supposed, to a differing tectonic milieu either in the source area or at the site of deposition or both? Or does climate play a significant role? Environment of deposition, source rocks, and tectonics are not fully independent of one another so that a combination of these three as well as climate is probably involved in any explanation. We briefly review the major hypotheses and then scrutinize the geologic record itself for critical evidence before drawing any conclusions.

The Hypotheses

Role of Source Materials. It has been assumed as self-evident that the composition of a sandstone is determined in a major way by the kind of source rocks from which it was derived. In other words, source materials control petrologic diversity of the framework components. The minerals available, however, are subject to modification by selective weathering so that some are destroyed and others are not. The selectivity of weathering is a function of the variation in stability and reactivity of the source minerals. This, in turn, is related to the climate – which controls the kind and intensity of weathering and is an important factor in the rate of erosion. A high rate of erosion will result in production of incompletely weathered materials – no matter what the climate.

According to the source materials hypothesis, arkoses and graywackes are thought to be the products of granitic and metamorphic terranes, respectively, volcanic arenites the product of volcanic regions, and so forth. The quartz arenites are attributed to recycling of earlier sands or are the end-products of decomposition of quartz-bearing plutonic rocks under the most severe climatic conditions.

Role of Environment. That the environment of deposition has a marked influence on the textural character of sands has long been taken for granted. It has been also presumed that it has an effect on the mineralogy. The softer and more cleavable minerals and rock particles are thought to be destroyed and eliminated from the sand fraction with concomitant improvement in sorting and rounding of the residue. Hence by prolonged abrasion an immature feldspathic or lithic sand could, theoretically, be converted into a clean quartz sand, the finer debris having been segregated from the framework by progressive sorting. These abrasion effects are believed to be minimal in streams and maximal on beaches or in dunes.

Role of Tectonism. Some writers consider the tectonic factor to be the overriding one in determining sandstone composition and equate quartz arenites with "stable shelves", arkoses with fault-bounded basins, and graywackes with eugeosynclines (Dapples, Krumbein, and Sloss, 1953). This view, with slight modification, was early expressed by Krynine (1942) who attributed first cycle orthoquartzites to peneplanation, graywackes (and second cycle orthoquartzites) to geosynclinal sedimentation, and arkose to a post-geosynclinal stage, commonly fault uplifts of a deformed and intruded geosynclinal tract (Fig. 6-35).

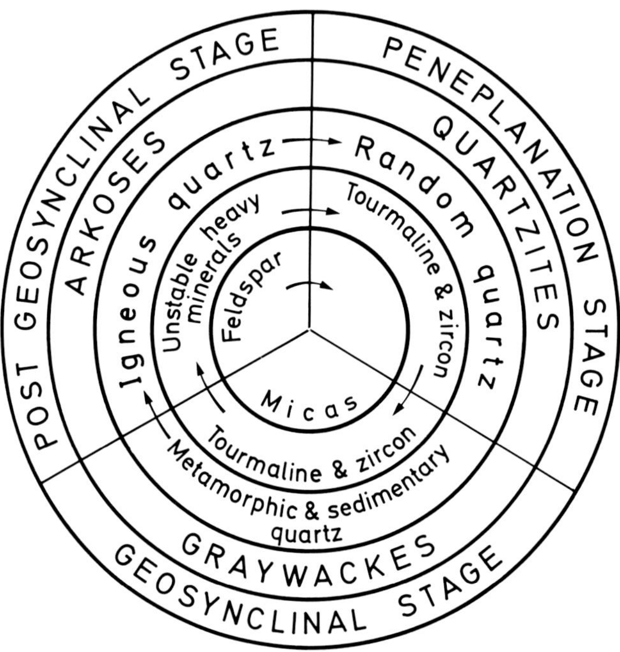

Fig. 6-35. Relations between tectonism, sandstone types, and mineralogy (Krynine, 1945, Fig. 2)

Role of Climate. Climate, which has received renewed attention in the past few years, has been suggested as an important factor controlling framework composition (Crook, 1968; Strakhov, 1967, p. 48, 95), especially for arkoses which traditionally were thought to be the products of arid or glacial climates where feldspar decomposition was minimal (Mackie, 1899b). Feldspathic sandstones and ferruginous cementation characterize many red beds which are commonly associated with evaporites and thus attributed to arid or semiarid climates. Climate has also been considered a prime factor controlling texture; the products of an arid climate being coarser than those of a humid region (Garner, 1959).

The Evidence

Other than plausibility what evidence is there to support or negate these theses of sandstone petrogenesis? What data from experiment or from a study of either modern or ancient sandstones relates to this question? Unfortunately critical observational data are none too plentiful.

Source Materials. That the composition of the source plays a major role in determining composition of a sand is confirmed by observations of modern sands, particularly stream sands, that can be related to their source rocks. Streams draining glaciated regions commonly have abundant feldspar as do streams draining granitic massifs. Plumley (1948, Fig. 24) reports as much as 35 percent feldspar in streams draining the Pre-Cambrian granites of the Black Hills of South Dakota. Volcanic terrains are rich in volcanic rock fragments, the streams draining a Mexican volcanic area contain, for example, 57 percent such particles. Moreover, streams eroding supermature quartz arenites could only have well-rounded quartz as framework grains.

Local control of detrital framework is undeniable. Along the Rhine, Koldewijn (1955, Fig. 4) clearly demonstrates how the composition of the light fraction reflects tributaries and bedrock: quartz and calcite come from Switzerland, quartz aggregates and reworked pink quartz from the Bunter-sandstone are added by the Neckar, much feldspar by the Main and rock fragments from the Rheinische Schiefergebirge become common farther downstream (Fig. 6-36). In ancient sandstones similar variations have been noted above unconformites. The basal Cambrian Lamotte Sandstone of the St. Francis Mountains of Missouri is arkosic where it rests on granite and it is a volcanic lithic arenite where it rests on rhyolite (Ojakangas, 1963, p. 863). But these differences are characteristic of only the lowest few meters of section. Its normal regional character is a mature sandstone – essentially a quartz arenite. Klein's observations (1962, p. 11-12) on the Triassic in the Maritime Provinces of Canada likewise demonstrate that basal sands may reflect the character of the subjacent bedrock. These observations do not explain why the sandstone of one formation, the Ordovician Martinsburg of the Appalachians, for example, is universally a graywacke whereas another, such as the Devonian Oriskany in the same region, is everywhere a supermature quartz arenite.

But is composition always a matter of source materials? Does the changing character of sands seemingly all derived from the same source region reflect changes in the climate of this area or unroofing of new sources? Or are environmental factors the cause of the variations?

Environment. Although much has been written on the relation between the textural parameters of modern sands and their environment, little is known about the relation of environment and framework composition. With the exception of graywackes (mostly marine turbidites), there is little direct correlation between rock type and depositional environment – a quartz arenite may have been deposited on a beach or marine shelf or in a desert dune. But is it possible, for example, for the beach or dune environment to convert a lithic sandstone into a quartz arenite? Inspection of Table 2-1 on the feldspar content of modern sands seems to show that beaches and dunes have less feldspar than streams in a given geographic area. The beaches and dunes along the Carolina coast, for example, contain only half as much feldspar as the streams of the same region (Giles and Pilkey, 1965, p. 906). Such evidence, which implies derivation of the beach and dune sands from those supplied by the rivers, is suspect as it has not generally been proved that the beach sands did not come, in part at least, from elsewhere, that the beaches may be "contaminated" by more mature sand lying offshore.

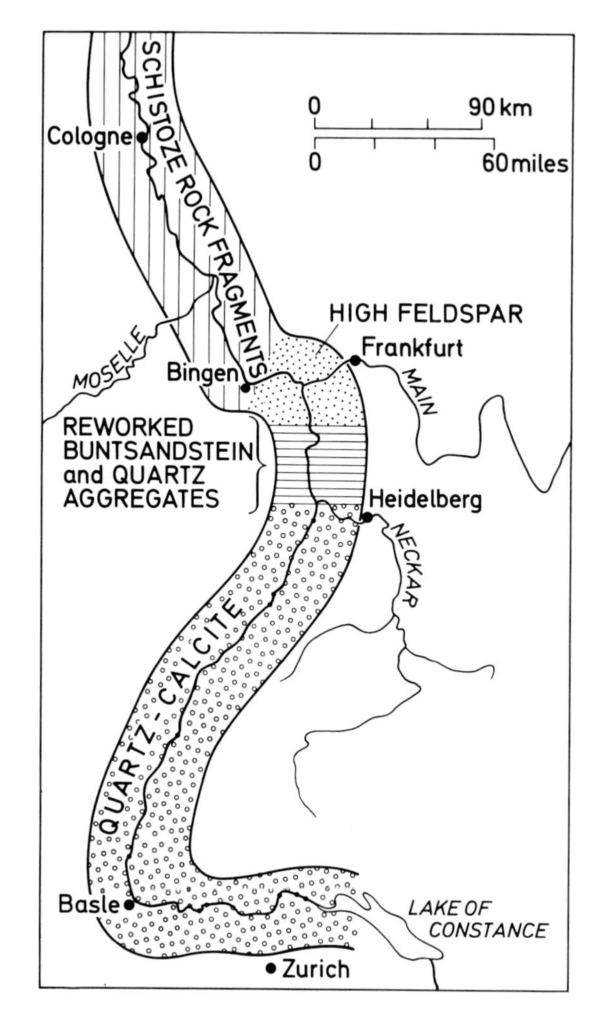

Fig. 6-36. Diagram showing distinctive light mineral content of sands of the Rhine below Basle (Modified from Koldwijn, 1955, Fig. 4)

Some evidence on environmental influence comes from the geologic record. Sandstones in the same sequence believed to be from the same source show marked differences in compositional and textural maturity which almost certainly is correlated with the environment of deposition. Folk (1960, p. 55), for example, has related the variations in character of the Silurian Tuscarora Quartzite of West Virginia to environmental factors – an estuarine environment for the lowest red Tuscarora with its angular quartz, poor sorting and high rock fragment content, and a "high-energy" beach environment for the highest white Tuscarora with its rounded and well-sorted quartz and virtual absence of rock particles. It could be argued that the differences noted could be related to tectonic or climatic changes in the source land. Rapid, meter by meter vertical alternations not only in mineralogy but also in roundness make this thesis rather unlikely, however. Nor does it seem to be result of a selective sorting process.

A good example is also afforded by the transition beds between the Ordovician Juniata Formation and the Silurian Tuscarora Quartzite in south central Pennsylvania. The red sandstones of the underlying Juniata, with subangular quartz and numerous pelitic rock particles are strongly crossbedded and are overlain by the white, very mature Tuscarora quartz arenite having ripple marks and only a little crossbedding. A pink crossbedded quartzite with less well-rounded quartz and with much polycrystalline quartz and a few rock particles follows. The regional paleocurrent patterns and other considerations indicate a common source for all these sandstones. The petrologic variation seems best accounted for by a "cleaning-up" and rounding of the sands supplied – the Juniata probably being fluvial, the lowest Tuscarora being a beach, and the higher Tuscarora reverting to a fluvial regime. When studying the role of environment on the framework composition of ancient sandstones, great care should be taken to establish that the *geographic position* and *composition* of the source region remained the same. Paleocurrent analysis, stratigraphic relations within the basin and knowledge of the regional geology beyond basin limits are vital elements in this judgement. Small scale meter by meter vertical petrographic variation is suggestive, as a general rule, of local depositional influence rather distal change in composition.

Experimental studies of sand abrasion have not convincingly demonstrated the effectiveness of the selective abrasion process, although some recent work suggests otherwise (Morris and Fan, 1962). Nor do we have any critical field studies of modern sands that unequivocally demonstrate it. Studies of the effects of transport in long rivers seem to suggest that little or no selective mineral loss occurs (Russell, 1937), although one should keep in mind that feldspar loss in high gradient mountain streams may be appreciably greater. Nor do rivers round sands greatly (Russell and Taylor, 1937). Comparable studies of beach and eolian action have yet to be made.

One should remember, however, that through its control of texture by varying current strength and sorting, the depositional environment can change sandstone type. For example, the Pennsylvanian Fountain Formation of Colorado is reported by Hubert (1960, Fig. 70) to be arkosic conglomerate and sandstones in ancient stream channels, micaceous arkose on flood plains, and quartzose arkose (feldspathic sandstone) in deltaic and lacustrine deposits.

Tectonism. The case for tectonic control of petrographic variation of sands rests largely on a study of the geologic record. The observation that the widespread blanket sandstones derived from and deposited on stable platforms or on the margins of a shield are highly mature whereas those deposited in the geosyncline during the end-stages of filling are immature graywackes and arkoses suggests a correlation between petrology and tectonic stability.

The geologic history of many geosynclines commonly shows a change in the character of the sandstones related to its structural evolution. In many cases the first orogenic activity is heralded by the appearance of immature detritus. This material becomes a flood and leads to a thick orogenic sedimentary pile –

the flysch. Following the paroxysmal stage, the sediments coarsen and commonly become feldspathic – an arkosic molasse.

The contrast between the petrologic character of sands of the "stable shelves" and those of the "mobile belts" led to the concept of tectonic control. This thesis has been challenged by Kay (1951, p. 85) who thinks that the "type of sediment is not an attribute of the geosyncline." Similarly van Andel (1958), in his study of the sandstones of western Venezuela, believes that source materials rather than tectonism are the basic controls over sandstone composition and that environment is the basic factor in determining only the texture and textural maturity. He states (p. 762) that "there is no systematic tectofacies control of texture and textural maturity" and moreover, "mineralogical maturity does not reflect the tectofacies of the depositional basin" and only insofar as the morphology in the source area is controlled by tectonics does the petrography of the sediments reflect the latter.

Source composition has also been called upon to explain Krynine's geosynclinal sequence of orthoquartzites-graywackes-arkoses, the quartz arenites largely having had a source external to the geosyncline whereas the later graywackes and arkoses largely had an internal source from cannabalism, volcanism, and syntectonic granitic plutons.

Others have noted that present-day sands in a given basin vary greatly in composition from place to place as is also the case in some ancient sedimentary basins, even in geosynclines (Klein, 1962 and 1966). This has led some to conclude that no particular type of sediment is characteristic of a geosyncline. Such observations, in some cases not well established, fail to explain the large-scale, striking vertical differences in sandstone petrology of a basin's fill – as in the Appalachians or the Marathon Basin of west Texas. It is not that specific sandstone types may not be found occasionally in all milieu, though no one has seen a graywacke on a cratonic platform, but that the dominant character of a geosynclinal sequence is different from that of a cratonic section. In other words, although one cannot identify a geosynclinal sandstone from a single hand specimen or thin section, one can recognize the geosynclinal suite from study of the whole geologic sequence.

Climate. What are the facts about the relation between sandstone petrography and climate? Krynine's observations (Krynine, 1935) on the present-day sands of tropical Mexico clearly show that feldspar is not necessarily a criterion of either arid or glacial climate. Sediments of a hot humid climate can also be feldspathic. Relief, rather than climate, controls feldspar content – a conclusion supported by Gibbs (1967) for the Amazon basin. Likewise it seems that the red color of some sandstones is no sure guide to deposition under arid conditions. Sands of present-day arid regions are only exceptionally red. Moreover the abundance of vertebrate remains found closely associated with red sandstones, as in the Triassic of Connecticut (Krynine, 1950, p. 171), and the observation that red soils characterize the warm humid regions have led to the rejection of red coloration as a certain criterion of aridity.

Garner (1959) as a result of a study of four regions in the Andes, concluded that climate rather than relief was the prime factor governing clast size. Arid weathering, a dominantly mechanical process, tends to produce coarse alluvium from which the fines are removed by deflation; humid weathering yields mainly silts and clays. On the other hand, Gibbs (1967) has shown that particle size is a function mainly of relief and that climate has a negligable influence in the Amazon River basin. Even if Garner were correct, it is not clear whether the climatic factor would be reflected in sandstone petrography.

Kuenen's experiments (Kuenen, 1960) led him to conclude that sands are effectively rounded only by wind action, such action being quantitatively important only in desert regions. Well-rounded sands are, therefore, a criterion of desert action. Rounding, unfortunately is an "inherited" character so that a sand may not have acquired its rounding in the environment in which it was last deposited.

In short, there seems to be no unequivocal imprint of climate on the petrography of sand.

The Verdict

What statement can we make after this brief survey of the evidence? Is the petrographic character of a given sandstone the result of the complex interplay of factors of which no single one plays a commanding role? We think not. What then is the relative importance of source materials, depositional environment, tectonism, and climate? Of the four we believe that tectonism is the primary control of sandstone petrography, partly because terrigenous sedimentation itself is very largely dependent on tectonism, for without earth movements such sedimentation would eventually come to a halt. Such a claim cannot be made for source materials, depositional environment, nor climate.

Moreover tectonism, to some degree, governs the other three. Tectonic uplift determines relief and hence affects the rate and depth of erosion and thus controls the volume and to some extent the kind of detritus delivered to the basin of sedimentation. In an indirect way tectonism controls the major rock types subject to erosion in that with shallow erosion only sediments, or perhaps extrusive volcanic rocks, are likely to be exposed whereas with deep erosion coarse-grained feldspathic igneous or metamorphic rocks are made available.

Tectonic uplift, if great enough, clearly modifies the climate in the source area. From a geologic point of view, however, relief is probably more important than climate in weathering for with high relief mechanical erosion rather than chemical weathering prevails.

Uplift creates a gradient, a slope down which detritus moves. Although far from being understood, transportation appears to be a minor factor in determining sand petrology although the action of high-gradient streams or movement on beaches may leave a recognizable effect on both the texture and composition of the sands.

Basin subsidence and tectonic uplift are not necessarily coupled together, especially where the source area is far removed from the depositional basin as is the case for most large river systems. But basin subsidence and uplift in the source area do nevertheless interact for it is the ratio of sediment input to basin capacity that broadly controls the depositional facies in the basin – whether the basin is mostly marine or nonmarine or even not filled at all. A large input with inadequate subsidence leads to marine regression and to a dominantly alluvial fill; with greater subsidence the level of the fill is carried below sealevel and the accumulation is largely marine. Minimal input produces a starved basin. The sands deposited in these several environmental realms are unlike in their structure and organization and to some extent unlike in their petrology.

In the end, the most convincing evidence of the dominance of tectonism as the basic control of sandstone petrology is, perhaps, the close correlation between the character of the sand and the tectonic or structural evolution of the geosyncline. Because this topic is reviewed more fully in Chapter 12, we will only note that such a correlation exists and that the changing character of the sands with time is related to the relative importance of supply of sand from the external or cratonic source and from the tectonic land or internal source. The character of the debris shed from these sources is different, reflecting probably the relief of each – a tectonically controlled factor – and the changing importance of these two source areas as the geosyncline evolves, the cratonic source being important in the early stages and generally overwhelmed in the end-stages.

We should note, in conclusion, that the petrology of a sandstone is the result of a complex interplay of many factors and even though we believe tectonism to be the overriding factor most of the time, a particular combination or interplay of such factors as climate, both temperature and precipitation, transportation, including its rigor and duration, supply and subsidence may yield a particular sandstone not greatly different than that produced by a somewhat different combination of factors. More than petrology may be needed, for example, to distinguish between a tectonic arkose – related to a sharp uplift – from a residual arkose formed on a stable old granitic massif and buried by a marine transgression.

Perhaps more than anything else we have been troubled by our inability to sort out and evaluate the effects of the several factors that control sandstone petrology. We have been unable to identify the crucial evidence for the several hypotheses. To put it more bluntly, geologists appear to have identified all the relevant hypotheses but at present cannot seem to formulate the critical experiments either in the field or laboratory that are needed to make the discriminations required. Perhaps one should proceed by studying modern sands, selecting what are believed to be the preferred paths of sand evolution. Careful petrographic studies of modern sands, collected from areas of known climate, relief, and source rock and known environment of deposition would seem to be a good starting point.

Annotated Bibliography of Comprehensive Petrographic Analyses of Sandstones

We have selected some fifty papers on sandstone petrography for annotation. Those chosen are in-depth studies of specific sandstones.

Most of these studies focus on the more conventional aspects of sandstone petrography: mineral composition, both light and heavy, and less commonly bulk chemical composition. We tend to favor those with tabulated modal and chemical analyses. We believe such published data will be of interest and value to the student of sandstones. Few are wholly descriptive and most contain much interpretative material. In some, the petrography is only a part of a larger approach in which all aspects, including stratigraphy, facies, and sedimentary structures, are investigated. We have also added a few early papers which were milestones in the development of the science and which are therefore of special historical significance. We also list a few larger compilations or reference works on the petrography of sandstones.

We have prepared also a short annotated bibliography of major petrographic studies of modern sands that contain good analytical data. It is noteworthy, however, that there are *no* studies of modern sands comparable to those of ancient sandstones. That is, none that cover all aspects: gross geometry of the sand body, its textural variations (including not only grain size and shape analyses but also pore pattern, packing and grain orientation), both its chemical and mineralogical composition, the latter based on point counts of thin sections, and its sedimentary structures and their orientation. No really adequate study has yet been made of modern sands – only partial studies which deal with one or another aspect of the deposit! We believe that significant advances will follow from integrated studies of modern sands, particularly those in which thin-section petrography is an important part.

Ancient Sandstones

Allen, J. R. L.: Petrology, origin and deposition of the highest Lower Old Red Sandstone of Shropshire, England. Jour. Sed. Petrology **32**, 657–697 (1962).

Penetrating observation and deduction combined with careful, quantitative measurement. A good example of how petrographic study of only a few geologic sections can expand geologic history.

van Andel, Tj. H.: Origin and classification of Cretaceous, Paleocene and Eocene sandstones of western Venezuela. Am. Assoc. Petroleum Geologists Bull. **42**, 734–763 (1958)

A critical analysis of the role of provenance, environment and tectonism in determining sandstone petrography; study based on quantitative data from about 400 samples and their relation to paleographic and structural elements in the depositional basin.

Assereto, Riccardo: The Jurassic Shemshak formation in central Elburz (Iran). Rivista Italiana di Paleontologie e die Stratigraphia **72**, 1133–1182 (1966).

An integrated, well organized study of a thick, widespread (over 1,275 km long) sandstone-shale unit that covers an area of approximately 1.5×10^6 km. Sandstones are mostly subgraywackes with some feldspathic subgraywackes and orthoquartzites.

Cadigan, R.A.: Petrology of Morrison Formation in the Colorado Plateau Region. U.S. Geol. Survey Prof. Paper **556**, 113 p. (1967).

Detailed quantitative petrography of 483 samples integrated with transport direction. Linear trend-surface analysis helps pinpoint sources of different detrital minerals. Excellent photomicrographs.

Casshyap, S. M.: Petrology of the Bruce and Gowganda formations and its bearing on the evolution of Huronian sedimentation in the Espanola-Willisville area, Ontario (Canada). Paleogeography, Paleoclimatol. Paleoecol. **6**, 5–36 (1969).

This paper is based on a thin-section and outcrop study of the Huronian sequence south of Espanola, Ontario, Canada. The author's goal is a synthesis of thin-section and outcrop data directed toward a paleogeographic and environmental reconstruction and an analysis of provenance.

Cayeux, Lucien: Structure et origine des grès du Tertiaire Parisien: Etudes des Gites Minéraux de la France, 131 p. Paris: Imprimeries Nationale 1906.

One of the classic works of Cayeux. A thorough petrographic study of seven Tertiary sandstones of the Paris Basin. Detailed thin-section descriptions without quantitative data. Ten plates of magnificant photomicrographs.

— Les roches sédimentaires de France: Roches siliceuses, 774 p. Paris: Impr. Nationale 1929.

A monograph by France's most famous sedimentary petrographer. A thorough petrographic description of the siliceous rocks – sandstones and cherts – of France. One-hundred eleven pages on sand and 131 pages on sandstones. Many chemical analyses of modern and ancient sands; superb photomicrographs.

Cipriani, Curzio, Malesani, Piergiorgio: Ricerche sulle arenarie – XIII. La Pietraforte. Soc. Geol. Italie Boll. **85**, 299–331 (1966) (Engl. Sum.).

A detailed textural and mineralogical analysis of the Pietraforte sandstone. A subgraywacke or feldspathic litharenite; 39 percent quartz, 15 percent feldspar, 37 percent carbonate rock fragments, 8 percent phyllosilicates; plagioclase exceeds orthoclase and dolomite exceeds limestone (as rock fragments). A descriptive rather than interpretative study.

Dake. C. L.: The problem of the St. Peter Sandstone. Missouri School Mines and Metall. Bull. **6**, 228 p. (1921).

One of the earliest comprehensive studies of all aspects of a sandstone formation. A classic and still perhaps the best publication on this well-known Ordovician quartz arenite of the Upper Mississippi Valley. A large section is devoted to the texture, composition, and structures of the sandstone.

Davis, E. F.: The Franciscan Sandstone. Univ. California Public. Bull. Dept. Geology **11**, 6–16 (1918).

A brief description of a Jurassic (?) graywacke sandstone from Coast Ranges of California. Two chemical analyses.

Edwards, A. B.: The petrology of the Miocene sediments of the Aure Trough, Papua. Royal Soc. Victoria Proc. (n. s.) **60**, 123–148 (1950).

Detailed thin section and heavy mineral analysis, physical properties, and chemical analyses of a 15,000-foot marine turbidite section in New Guinea. Volcaniclastic but not tuffaceous sands. Tabulated modal, size and chemical data.

Edwards, A. B.: The petrology of the Cretaceous greywackes of the Purari Valley, Papua. Royal Soc. Victoria, Proc. (N:S.) **60**, 163–171 (1950).

Modal, size, and chemical analyses are tabulated. The rock described is a graywacke with low quartz (4–15 percent) and high feldspar (33–55 percent) content.

Ehrenberg, H.: Sedimentpetrographische Untersuchungen an Nebengesteinen der Aachener Steinkohlenvorkommen. Preussischen Geol. Landesanstalt Jahrb. **49**, 33–58 (1928).

Petrographic study of the sands, particularly their cements. Eighteen photomicrographs. One of the earliest applications of petrography to stratigraphy.

Folk, R. L.: Petrography and origin of the Tuscarora, Rose Hill and Keefer formations, Lower and Middle Silurian of eastern West Virginia. Jour. Sed. Petrology **30**, 1–58 (1960).

A very thorough petrographic analysis of samples from a single vertical profile. A very perceptive study which illustrates the interpretative value of microscopical analysis. Lithic arenites, protoquartzites and quartz arenites. Twenty-nine photomicrographs.

Füchtbauer, Hans: Sedimentpetrographische Untersuchungen in der älteren Molasse nördlich der Alpen. Eclogae geol. Helvetiae **57**, 157–298 (1964).

Very detailed thin-section petrography, heavy mineral study, and size analyses plus some subsurface data. Special effort made to relate petrography to porosity and to distribution of porosity in Molasse trough of Bavaria. Good examples of what can be done by petrography when subsurface and outcrop data are limited. Mainly lithic arenites. Modal analyses.

Füchtbauer, Hans: Die Sandsteine in der Molasse nördlich der Alpen. Geol. Rundschau **56**, 266–300 (1967).

A very good summary of the petrography of the sands of the molasse trough in the south of Germany, showing quantitative evolution in composition of sands, light and heavy minerals, with time. The sands of the lowest strata are quartzose calcite-dolomite arenites (calclithites) derived from Alpine flysch and carbonate rocks; these are followed by either arkosic sands of granitic provenance or lithic sandstones (phyllarenites) of metamorphic derivation; the youngest sands are lithic arenites relatively poor in feldspar.

The sequence is related to the late tectonic history of the Alps, the phases of which produce sedimentary megacycles. A section normal to axis of molasse trough discloses dominance of conglomerate on proximal margin and general thinning away from source land.

Fujii, Koji: Petrography of the Upper Paleozoic sandstones from the Yatsushiro area, Kyushu. Kyushu Univ., Fac. Sci. Mem., Ser. D, **12**, 179–218 (1962).

Thoroughgoing petrography of a 5,000-meter section of clastics in the Chichibu geosyncline in Japan.

Gasser, Urs: Sedimentologische Untersuchungen in der äußeren Zone der subalpinen Molasse des Entlebuchs (Kt. Luzern). Eclogae geol. Helvetiae **59**, 724–772 (1966).

— Die innere Zone der subalpinen Molasse des Entlebuchs (Kt. Luzern). Eclogae geol. Helvetiae **61**, 229–319 (1968).

These two papers are parts of a single very detailed lithostratigraphic, sedimentologic study of the Swiss Molasse – a thick sequence derived from the uplifted Alps near Lucerne. The sediments are marine, brackish, and fresh-water origin. Much petrologic data especially for the conglomerates. Relates sedimentation to tectonics. The outer zone is Aquitanian; the inner Stampian.

Gilligan, A.: The petrography of the Millstone Grit of Yorkshire. Geol. Soc. London Quart. Jour. **75**, 251–294 (1920).

An early thin-section and heavy mineral study of a famous British formation; special emphasis on the petrography of the pebbles of the conglomerates. One of the earliest unified studies of an ancient sandstone and interpretative analysis.

Goldman, M. I.: Petrographic evidence on the origin of Catahoula Sandstone. Am. Jour. Sci., Ser. 4, **39**, 261–287 (1915).

A pioneer study, one of the first illustrating how much information and what conclusions can be extracted solely from limited sample material.

Gry, Helge: Petrology of the Paleocene sedimentary rocks of Denmark. Danmarks Geologiske Undersögelse II Raekke **61**, 171 p. (1935).

Mechanical and mineralogical analysis of loose sand and thin section studies. Limestones, greensands and phosphorites. Studies by geographic area. Tables of quantitative data. Two plates of photomicrographs.

Hadding, Assar: The pre-Quaternary sedimentary rocks of Sweden III. The Paleozoic and Mesozoic sandstones of Sweden. Lunds Univ. Arsskr., N.F., Adv. 2, **25**, nr. 3, 287 p. (1929).

A monographic survey of the Paleozoic and Mesozoic sandstones of Sweden based largely on microscopic examination of thin sections. One-hundred thirty-eight figures; many photomicrographs.

Hansen, Kaj: Die Gesteine des Unterkambriums von Bornholm. Danmarks Geologiske Undersögelse II. Raekke **62**, 194 p. (1936).

A descriptive petrography principally of arkoses and quartzites. Discussion by regions but also considers origin and practical significance. Eight plates of photomicrographs and one of sedimentary structures (German, Danish summ.).

Helmbold, Reinhard: Contribution to the petrography of the Tanner Graywacke (translated by F. B. Van Houten). Geol. Soc. America Bull. **69**, 301–314 (1958).

A paper based mainly on an intensive study of three specimens. Contains good chemical analyses of each of the three samples (coarse, medium, and fine-grained specimens), exhaustive modal analyses of each, and close observation of fabric (texture and structure) of each.

Hopkins, M. E.: Geology and petrology of the Anvil Rock Sandstone of southern Illinois. Illinois State Geol. Survey Circ. **256**, 48 p. (1958).

One of the first modern studies to integrate detailed subsurface data, petrology and sedimentary structures of a single sand unit throughout a basin. Thorough petrographic study including clay mineralogy, heavy minerals and thin-section analysis. A subgraywacke. Tabulated analytical data.

Hubert, J. F.: Petrology of the Fountain and Lyons formations, Front Range, Colorado. Colorado School of Mines Quart. **55**, no. 1, 242 p. (1960).

An example of extended petrographic analysis of Pennsylvanian arkosic sandstones based on outcrop and hand specimen study. Lengthy thin section analysis of the mineral composition and fabric. Microscopic study of the heavy mineral fraction. Strong emphasis on quantification of data.

Jacobsen, Lynn: Petrology of Pennsylvanian sandstones and conglomerates of the Ardmore Basin. Oklahoma Geol. Survey Bull. **79**, 144 p. (1959).

Size and mineralogical analyses, both light and heavy minerals, of four sandstones. Quantitative data, photomicrographs; interpretation of depositional history. Mainly lithic or sublithic sandstones.

Krynine, P. D.: Petrology and genesis of the Third Bradford Sand. Pennsylvania State Coll. Mineral Industries Expt. Sta. Bull. **29**, 134 p. (1950).

A good example of petrographic analysis of a lithic sandstone based almost wholly on microscopic analysis. Contains an extended section on mineralogy, including both major constituents and heavy minerals (26 p.), a section on petrography (25 p.), and a section on texture and pore pattern (14 p.).

— Petrology, stratigraphy and origin of the Triassic sedimentary rocks of Connecticut. Connecticut Geol. Survey Bull. **73**, 239 p. (1950).

An early exhaustive study of arkosic sands and associated sediments based largely on microscopic analysis. Contains a section on mineralogy, including both main rock-forming minerals and accessory heavy minerals (10 p.), stratigraphy (41 p.), and petrography (47 p.). The evidence marshalled from these sources is used to interpret the climate and paleogeography of the Connecticut Triassic.

Larsen, Gunnar: Rhaetic-Jurassic-Lower Cretaceous sediments in the Danish embayment. Geol. Survey Denmark. II. Series No. **91**, 127 p. (1966).

A heavy mineral study of the marine fill of the Danish embayment, based mostly on subsurface data. Discussion of diagenesis, weathering, transportation, source area, and basin development. Very well illustrated with many maps and excellent photographs of heavy minerals.

Laurent, J.: Recherches micrographiques sur quelques horizons greseux du terrain houiller de Belgique. L'Univ. Louvain, L'Inst. Geol., Mem. **7**, 43–123 (1933).

Qualitative description of coal measure sandstone plus 48 photomicrographs. One of the first studies of its kind.

Mattiat, Bernhard: Beitrag zur Petrographie der Oberharzer Kulmgrauwacke. Beiträge Min. Petrog. **7**, 242–280 (1960).

A very good "in-depth" study of a famous European graywacke formation. Contains ultra-detailed microscopic analysis with modal percentages of twenty-five samples of coarse, medium, and fine-grained classic graywackes. Chemical analysis included.

McBride, E. F.: Flysch and associated beds of the Martinsburg Formation (Ordovician), central Appalachians. Jour. Sed. Petrology **62**, 39–91 (1962).

A thorough description and interpretation of a flysch sequence based mainly on field and thin-section study of turbidite sandstones (lithic graywackes). Paleocurrent data. Modal analyses given.

Mellon, G. B.: Stratigraphy and petrology of the Lower Cretaceous Blairmore and Mannville Groups, Alberta foothills and plains. Research Council Alberta, Bull. **21**, 270 p. (1967).

Petrology and stratigraphy of a widespread, molassic nonmarine, clastic wedge 1000 to 4000 ft (330 to 1320 m) thick. Careful attention to diagenesis, mineral associations, and provenance. Much volcanic debris. Thirty plates, mostly photomicrographs.

Ojakangas, R. W.: Petrology and sedimentation of the Upper Cambrian Lamotte Sandstone in Missouri. Jour. Sed. Petrology **33**, 860–873 (1963).

A careful petrographic analysis of basal Cambrian sandstones that illustrate local control of the basal beds by the mineral composition of the underlying Precambrian. Locally arkose and lithic arenite; regionally quartz arenite. Modal analyses tabulated.

Okada, Hakuyu: Sandstones of the Cretaceous Mifune Group, Kyushu, Japan. Mem. Fac. Sci., Kyushu Univ., Ser. D., Geology **10**, 1–40 (1960).

Detailed modal and heavy mineral analyses and photomicrographs of graywackes and tuffaceous sandstones.

— Cretaceous sandstones of Goshonoura Island, Kyushu. Mem. Fac. Sci., Kyushu Univ., Ser. D., Geology **11**, 1–48 (1961).

A detailed petrographic analysis of Cretaceous sandstones, principally feldspathic and lithic graywacke plus some arkose. Discussion of mineral zonation and relation of provenance, tectonics and environment to petrography. Tabulated modal, chemical and heavy mineral analyses; photomicrographs.

— Composition and cementation of some Lower Paleozoic grits in Wales. Mem. Fac. Sci., Kyushu Univ., Ser. D., Geology **18**, 261–276 (1967).

Good quantitative description, with tabulated modal analyses and heavy mineral composition of the classic Silurian grits at Aberystwyth and other deposits. Mainly feldspathic graywackes and lithic arenites.

Potter, P. E., Pryor, W. A.: Dispersal centers of Paleozoic and later clastics of the upper Mississippi Valley and adjacent areas. Geol. Soc. America Bull. **72**, 1195–1250 (1961).

Systematic study of a few petrographic parameters – chiefly polycrystalline quartz and angularity of tourmaline – over a large area of the North American craton. Another example of specialized petrography, but with chief emphasis upon locating distant dispersal centers and determining the amount of sedimentary recycling.

Pye, W. D.: Petrology of the Bethel sandstone of south-central Illinois. Am. Assoc. Petroleum Geologists Bull. **28**, 63–122 (1944).

An early effort to treat a sandstone formation as a unit. Study based largely on petrography and the then-available subsurface data.

Ronov, A. B., Mikailovskaya, M. S., Solodkova, I. I.: 1963: Evolution of the chemical and mineralogical composition of arenaceous rocks. In: Chemistry of the Earth's Crust **1**, Acad. Sci. USSR (Trans. Israel Program Sci. Trans., **1966**, 212–262).

Tabulated summaries of 1818 chemical analyses and 918 granulometric and mineralogical (both light and heavy) analyses of sandstones of all ages from the Russian platform and adjacent geosynclinal tracts. Data analyzed for secular trends, for the effect of climate, distance of transport, depth of burial, tectonics and provenance on sandstone composition.

Siever, Raymond: Trivoli sandstone of Williamson County, Illinois. Jour. Geol. **57**, 614–617 (1949).

An excellent example of the amount of data obtainable from and the interpretative value of an intensive thin section analysis of a sandstone based on a few samples.

— Pennsylvanian sandstones of the Eastern Interior Coal Basin. Jour. Sed. Petrology **27**, 227–250 (1957).

Tabulated summary of stratigraphy, modal mineralogy, types of detrital quartz, and grain size data based on 176 samples; heavy mineral study of 35 samples. Orthoquartzites, protoquartzites and subgraywackes.

Shiki, Tsunemasa: Studies on sandstones in the Maizuru Zone, southwest Japan. Mem. College Sci., Univ. Kyoto, Ser. B. I. Importance of relations between mineral composition and grain size. **25**, 239–246 (1959).
— II. Graded bedding and mineral composition of sandstones of the Maizuru Group. **27**, 293–308 (1961).
— III. Graywacke and arkose sandstones in and out of the Maizuru Zone. **29**, 291–324 (1962).

A series of papers based on quantitative analyses encompassing relations between the proportion of quartz, feldspar, and rock fragments and the matrix and mean size, the change in the proportions of these constituents at intervals above the base of graded beds, and the mineralogy and provenance of several types of arkose and graywackes. The data are presented graphically.

Simonen, Ahti, Kuovo, Olavi: Sandstones in Finland. C.R. Soc. géol. Finlande, Bull. Comm. géol. Finlande **168**, 57–87 (1955).

Summary of the petrography of the non-metamorphic sandstones of Finland including the late-Precambrian Jotnian arkoses, the Muhos sandstones, and quartz arenites of probable Cambrian age. Tabulated modal and chemical analyses.

Stanley, D. J.: Etudes sédimentologique des Grès D'Annot et leur équivalents latéraux, 158 p. Paris: Societe des Editions Technip 1961.

One of the few integrated studies of a Tertiary turbidite sandstone formation. Based on stratigraphy, petrography, and paleocurrents.

Swineford, Ada: Cemented sandstones of the Dakota and Kiowa formations of Kansas. State Geol. Survey Kansas, Bull. **70**, pt. 4, 57–104 (1947).

Contains a section on petrography, with chemical analyses and photomicrographs of Cretaceous sandstones. Good discussion of cements.

Thiel, G. A.: Sedimentary and petrographic analysis of the St. Peter Sandstone. Geol. Soc. America Bull. **46**, 559–614 (1935).

Tabulated textural, heavy mineral, and chemical analyses of a widespread Ordovician quartz arenite of the Upper Mississippi Valley. Discussion of provenance.

Univ. Firenze (Florence), Gruppo di Ricera per la Mineral. dei Sedimenti, Inst. Mineral, Ricerche sulle arenaie.

Extended series of recent papers on well-known Italian sandstones characterized by an abundance of modal and textural analyses. Thorough modern descriptions; limited interpretation. See Cipriani and Malesani, this bibliography.

Valeton, Ida: Petrographie des süddeutschen Hauptbuntsandsteins. Heidelberger Beiträge zur Mineralogie und Petrographie **3**, 335–379 (1953).

Primarily aimed at the diagenetic aspects of the Bunter, this paper contains a great deal of information on the framework fraction. Many tables including one of diagenetic events. Sets the style for many subsequent petrographic studies of sandstones in Germany.

Walton, E. K.: Silurian greywackes in Peebleshire. Royal Soc. Edinburgh Proc., Ser. B, **65**, pt. 3, 327–356 (1955).

A good study of a particular class of turbidite sandstones derived from Ordovician volcanic and low-grade metamorphic rocks. Tabulated chemical and modal analyses.

Wieseneder, Hans: Zur Petrologie der Flyschgesteine des Wienerwaldes. Verh. Geol. Bundesanstalt **2**, 273–281 (1962).

Includes chemical and modal analyses of non-graywacke flysch sandstones of Cretaceous and Eocene age in the Vienna region.

Modern Sands

van Andel, Tj. H.: Provenance, transport, and deposition of Rhine sediments, 129 p. Wageningen: H. Veenman en Zonen 1950.

Heavy minerals determine source of Rhine sands. Much of general methodological value for the study of heavy minerals.

van Andel, Tj. H.: Sediments of the Rhone delta, pt. II, Sources and deposition of heavy minerals. Verh. Koninkl. Nederlandsch Geol. Mijnb. Genoot. **15**, 515–543 (1955).

A detailed study of heavy mineral varieties and their size distributions as a basis for mapping petrographic provinces of the delta and deducing an Alpine provenance for most of the sediment.

Baak, J. A.: Regional petrology of the southern North Sea, 127 p. Wageningen: H. Veenman and Zonen 1936.

Based on mineral analyses of about 400 samples of North Sea sands; the author defined and delineated five mineralogical associations and provinces; a very thorough study of provenance of modern marine sands. A pioneer study. About 90 references. Dutch summary.

Baker, George: Sand drift at Portland, Victoria. Royal Soc. Victoria Proc. **68**, 151–197 (1956).

Utilization of heavy mineral content to deduce directions of sand drift on beaches; utilization of artificially introduced minerals as marker species to measure rates and direction of sand migration.

Burri, Conrad: Sedimentpetrographische Untersuchungen an alpinen Flußsanden. I. Die Sande des Tessin. Schweiz. Min. Pet. Mitt. **9**, 205–241 (1929).

Mechanical and both light and heavy mineral analyses (but with emphasis on the heavy minerals) of river sands; discussion of down-stream changes in heavy mineral composition; provenance.

Christensen, Werner, Larsen, Gunnar: Tungsandsforekomster i Danmark. Danmarks Geol. Undersögelse III. Raekke **33**, 63 p. (1960).

An example of a modern study of placer sands. About 100,000 metric tons of placer sands estimated to occur mostly in northern Jutland. Chemical analyses.

Dickinson, W. R.: Singatoka dune sands, Vita Levu (Fiji). Sedimentary Geology **2**, 115–124 (1968).

An unusual paper in that it contains nine modal analyses obtained from thin sections of impregnated sands. The sands are feldspatholithic-volcaniclastic deposits. The average sand consists of quartz 34 percent, feldspar 25 percent, and rock fragments 42 percent. The rock fragments are mainly fresh and chloritized lavas of intermediate to basic character.

Du Rietz, Torsten: Composition of beach sand along the Swedish east coast. Geol. Fören. Förhandl. **75**, 381–395 (1953).

Size analysis, chemical analysis and petrology of light and heavy fraction *for each* of 11 samples, including one river sand. Finds that beach sands along the Baltic closely reflect the composition of the hinterland – microcline and plagioclase decrease sharply to the south because the Eocambrian sandstones cover part of the basement and are the principal source. Author compared average beach sand with average of 26 modal analysis of granites. Unusual. More studies of this type are needed.

Edelman, C. H.: Ergebnisse der sedimentpetrologischen Forschung in den Niederlanden and angrenzenden Gebieten. Geol. Rundschau **29**, 223–273 (1938).

Summarizes the work of the "Dutch" school of sedimentary petrographers primarily on the Tertiary and Quaternary sediments of the Netherlands and adjoining regions. The investigations reported on were made in the interval 1932 to 1937 and include mineralogical studies of the sands of the present-day streams.

Gasser, Urs, Nabholz, Walter: Zur Sedimentologie der Sandfraktion im Pleistozän des schweizerischen Mittellandes. Eclogae Geol. Helvetiae **62**, 467–517 (1969).

Comprehensive petrography (light and heavy minerals plus size analysis) of glacial Pleistocene sands in the Swiss "Mittelland" shows them to have been derived from the underlying Tertiary Molasse sands. Over 16 pages of quantitative data.

Hahn, Christoph: Mineralogisch-sedimentpetrographische Untersuchungen an den Flußbettsanden im Einzugsbereich des Alpenrheins. Eclogae Geol. Helvetiae **62**, 227–278 (1969).
Study of 226 samples of stream sand; size and both light and heavy minerals analyses. One of the most detailed studies of provenance of stream sands.

Hsu, K. J.: Texture and mineralogy of the Recent sands of the Gulf Coast. Jour. Sed. Petrology **30**, 380–403 (1960).
A thorough study of beach sands based on over 200 samples. Size, chemical and Rosiwal thin-section analyses of modern and Pleistocene sands. Good discussion of provenance and of effects of transport on sand petrology.

Hunter, R. E.: The petrography of some Illinois Pleistocene and Recent sands. Sedimentary Geology **1**, 57–75 (1967).
Mineral composition of 23 river, beach, and dune sands; tabulated analytical data.

Koldewijn, B. W.: Provenance, transport, and deposition of Rhine sediments. II. An examination of the light fraction. Geologie Mijnbouw (N. S.) **17**, 37–45 (1955).
Quantitative analyses of light mineral components of sands of Rhine and tributaries and assessment of provenance and downstream compositional variation.

— Sediments of the Paria-Trinidad shelf, Repts. Orinoco Shelf Exped., Vol. III, 109 p. The Hague: Mouton 1958.
Light and heavy minerals, composition and texture, of the sands (and other sediments) of the shallow marine shelf off the north shore of South America for purposes of provenance and environment.

Mackie, William: The sands and sandstones of eastern Moray. Trans. Edinburgh Geol. Soc. **7**, 148–172 (1899).
An oft-quoted classic study of the mineralogy of modern river sands, noteworthy for the recognition of the varietal types of detrital quartz and their relation to provenance.

Rittenhouse, Gordon: Sources of modern sands in the middle Rio Grande Valley. Jour. Geology **52**, 145–183 (1944).
Use of heavy minerals to determine proportion of total load furnished by each tributary. Use of hydraulic ratios and of absolute, rather than relative, mineral abundances.

Russell, R. D.: Mineral composition of Mississippi River sands. Geol. Soc. America, Bull. **48**, 1307–1348 (1937).
A significant study of the mineralogy of the sands of a large river with special reference to effects of transportation on mineral composition; tabulated quantitative data. The first of its kind and still not surpassed.

van Straaten, L. M. J. U.: Composition and structure of Recent marine sediments in the Netherlands. Leidse Geol. Mededeel. (Netherlands) **19**, 1–110 (1954).
A thorough description of mineralogy and texture of tidal flat sands, their sedimentary structures, and relation to other tidal flat deposits, as well as sediments of estuaries and the Zuider Zee.

Webb, W. M., Potter, P. E.: Petrology and chemical composition of modern detritus derived from a rhyolitic terrain, western Chihuahua. Bol. Soc. Geol. Mexicana, **32**, Num. 1 (1969), 45–61 (1969).
Modal, heavy mineral, and chemical analyses of seven sand samples. The average sand is 18 percent quartz, 18 percent feldspar (mainly K-spar) and 57 percent volcanic rock fragments. The bulk chemical composition is similar to that of the rhyolitic source rock. One of the few thin section studies of modern sands. See also Dickinson (1968).

References

Adams, R. W.: Loyalhanna Limestone – cross-bedding and provenance. In: Fisher, G. W., Pettijohn, F J., Reed, J. C., Jr., Weaver, K. N. (Eds.): Studies of Appalachian Geology – Central and Southern, p. 83–100. New York: Interscience 1970.

Allen, J. R. L.: Petrology, origin, and deposition of the highest Lower Old Red Sandstone of Shropshire, England. Jour. Sed. Petrology **32**, 657–697 (1962).

Allen, Percival: Wealdon petrology: The Top Ashdown Pebble Bed and the Top Ashdown Sandstone. Geol. Soc. London Quart. Jour. **104**, 257–321 (1949).

Allen, V.T.: Terminology of medium-grained sediments. Rept. Comm. Sedimentation 1935–1936, Natl. Research Council, p. 18–47 (1936).

— A study of Missouri glauconite. Am. Mineralogist **22**, 1180–1183 (1937).

van Andel, Tj.H.: Origin and classification of Cretaceous Paleocene and Eocene sandstones of western Venezuela. Am. Assoc. Petroleum Geologists Bull. **42**, 734–763 (1958).

Anhaeusser, C.R., Mason, Robert, Viljoen, M.J., Viljoen, R.P.: A reappraisal of some aspects of Precambrian shield geology. Geol. Soc. America Bull. **80**, 2175–2200 (1969).

Auden, J.B.: Vindhyan sedimentation in the Son Valley, Mirzapur District. India Geol. Survey Mem. **62**, pt. 2, 250 p. (1933).

Bailey, E.B.: Sedimentation in relation to tectonics. Geol. Soc. America Bull. **47**, 1716–1718 (1936).

Bailey, E.H., Irwin, W.P.: K-feldspar content of Jurassic and Cretaceous graywackes of northern Coast Ranges and Sacramento Valley, California. Am. Assoc. Petroleum Geologists Bull. **43**, 2797–2809 (1959).

Balk, R.: The structure of graywacke areas and Taconic Range, east of Troy, New York. Geol. Soc. America Bull. **64**, 811–864 (1953).

Barth, T.F.W.: Progressive metamorphism of sparagmite rocks of southern Norway. Norsk geol. tidsskr. **18**, 54–65 (1938).

Barton, D.C.: The geological significance and genetic classification of arkose deposits. Jour. Geology **24**, 417–449 (1916).

Billings, M.P.: The geology of New Hampshire. New Hampshire State Planning and Development Commission, 203 p. (1956).

Blatt, H.: Original characters of clastic quartz. Jour. Sed. Petrology **37**, 401–424 (1967).

— Christie, J.M.: Undulatory extinction in quartz of igneous and metamorphic rocks and its significance in provenance studies of sedimentary rocks. Jour. Sed. Petrology **33**, 559–579 (1963).

Bokman, John: Lithology and petrology of the Stanley and Jackfork formations. Jour. Geology **61**, 152–170 (1953).

de Booy, T.: Petrology of detritus in sediments, a valuable tool. Koninkl. Nederlandse Akad. Wetensch. Proc., Ser. B, **64**, 277–282 (1966).

Boswell, P.G.H.: The term graywacke. Jour. Sed. Petrology **30**, 154 (1960).

Brenchley, P.J.: Origin of matrix in Ordovician greywackes, Berwyn Hills, North Wales. Jour. Sed. Petrology **39**, 1297–1301 (1969).

Brett, G.W.: Cross-bedding in the Baraboo Quartzite of Wisconsin. Jour. Geology **63**, 143–148 (1955).

Brongniart, A.: De l'arkose, caractères minéralogiques et histoire géognostique de cette roche. Ann. sci. nat. **8**, 113–163 (1826).

Burchard, E.F.: Notes on various glass sands mainly undeveloped. U.S. Geol. Survey Bull. **315**, 377–382 (1907).

Burst, F.F.: "Glauconite" pellets; their mineral nature and applications to stratigraphic interpretations. Am. Assoc. Petroleum Geologists Bull. **42**, 310–327 (1958).

Bushinsky, G.I.: Structure and origin of the phosphorites of the U.S.S.R. Jour. Sed. Petrology **5**, 81–92 (1935).

Buttram, Frank: The glass sands of Oklahoma. Oklahoma Geol. Survey Bull. **10**, 91 p. (1913).

Cady, W.M., Wallace, R.E., Hoare, J.M., Webber, E.J.: The central Kuskokwim region, Alaska. U.S. Geol. Survey Prof. Paper **268**, 132 p. (1955).

Cayeux, Lucien: Les roches sédimentaires de France, Roches siliceuses, 774 p. Paris: Impr. Nationale 1929.

Chanda, S.K.: Calclithite fragments vs. extraclasts: A discussion. Jour. Sed. Petrology **39**, 1640–1641 (1969).

Clarke, F.W.: Report of work done in the division of chemistry and physics. U.S. Geol. Survey Bull. **60**, 174 p. (1890).

— Data of geochemistry. U.S. Geol. Survey Bull. **770**, 841 p. (1924).

Cloud, P.E., Jr.: Physical limits of glauconite formation. Am. Assoc. Petroleum Geologists Bull. **39**, 484–492 (1955).

Collins,W.H.: North shore of Lake Huron. Canada Geol. Survey Mem. **143**, 160 p. (1925).

Crook,K.A.W.: Petrology of graywacke suite sediments from the Turon River-Coolamigal Creek district, N.S.W. Jour. and Proc. Royal Soc. New South Wales **88**, 97–105 (1955).

— Petrology of Tamworth Group, Lower and Middle Devonian, Tamworth-Nundle District, New South Wales. Jour. Sed. Petrology **30**, 353–369 (1960).

— Weathering and roundness of quartz sand grains. Sedimentology **11**, 171–182 (1968).

— Graywackes. In: Encyclopaedia Brittanica **1970**a.

— Geotectonic significance of graywackes: Relevance of Recent sediments from Niugini. 42nd ANZAAS Congress, Port Moresby, Niugini, August, 1970b.

Cummins,W.A.: The greywacke problem. Liverpool Manchester Geol. Jour. **3**, 51–72 (1962a).

— Greywacke in Lower Siwaliks, Simla Hills. Nature **196**, 1085 (1962b).

Cushing,H.P., Leverett,Frank, Van Horn,F.R.: Geology and mineral resources of the Cleveland district, Ohio. U.S. Geol. Survey Bull. **818**, 138 p. (1931).

Dake,C.L.: The problem of the St. Peter sandstone. Missouri Univ. School Mines and Metal., Bull. (tech. Ser.) **6**, 158 p. (1921).

Dapples,E.C., Krumbein,W.C., Sloss,L.L.: Petrographic and lithologic attributes of sandstones. Jour. Geology **61**, 291–317 (1953).

Davis,E.F.: The Franciscan sandstone. California Univ. Publ., Bull. Dept. Geology **11**, 6–16 (1918).

Dickinson,W.R.: Singatoka dune sands, Viti Levu (Fiji). Sedimentary Geology **2**, 115–124 (1968).

— Interpreting detrital modes of graywacke and arkose. Jour. Sed. Petrology **40**, 695–707 (1970).

Diller,J.S.: The educational series of rock specimens, etc. U.S. Geol. Survey Bull. **150**, 84–87 (1898).

Donaldson,J.A.: Marion Lake Map-area, Quebec-Newfoundland. Canada Geol. Survey Mem. **338**, 85 p. (1966).

— Two Proterozoic clastic sequences. A sedimentological comparison. Geol. Assoc. Canada Proc. **18**, 33–54 (1967).

— Jackson,G.D.: Archean sedimentary rocks of North Spirit Lake Area, northwestern Ontario. Canadian Jour. Earth Sci. **2**, 622–647 (1965).

Dott,R.H.,Jr.: Wacke, graywacke and matrix – what approach to immature sandstone classification? Jour. Sed. Petrology **34**, 625–632 (1964).

Drobnyk,J.W.: Petrology of the Paleocene-Eocene Aquia formation of Virginia, Maryland and Delaware. Jour. Sed. Petrology **35**, 626–642 (1965).

Edwards,A.B.: The petrology of the Miocene sediments of the Aure Trough, Papua. Royal Soc. Victoria Proc. (N.S.) **60**, 123–148 (1950a).

— The petrology of the Cretaceous greywackes of the Purari Valley, Papua. Royal Soc. Victoria Proc. (N.S.) **60**, 163–171 (1950b).

Eigenfeld,R.: Die Kulmconglomerat von Teuschnitz im Frankenwalde. Sächs. Akad. Wiss. Abh., math. phys. Kl. **42**, 58 (1933).

Emery,K.O.: Turbidites – Precambrian to present. In: Studies on Oceanography 568 p. Tokyo: Univ. Tokyo Press. 1964.

Engel,A.E.J., Engel,C.G.: Grenville series in the northwest Adirondack Mountains, New York. Geol. Soc. America Bull. **64**, 1013–1097 (1953).

Espenshade,G.: Geology and mineral deposits of the Pilleys Island area. Dept. Natl. Resources Newfoundland, Geol. Survey Bull. **6**, 56 p. (1937).

Fahrig,W.F.: The geology of the Athabasca formation. Canada Geol. Survey Bull. **68**, 41 p. (1961).

Ferrar,H.T.: The geology of the Dargaville-Rodney Subdivision. New Zealand Geol. Survey Bull. **34**, 78 p. (1934).

Fettke,C.R.: Glass manufacture and the glass sand industry of Pennsylvania. Pennsylvania Topog. and Geol. Survey Rept. **12**, 278 p. (1918).

Fischer,Georg: Die Petrographie der Grauwacken. Jahrb. preuss. geol. Landesanstalt **54**, 320–343 (1933).

Folk,R.L.: The distinction between grain size and mineral composition in sedimentary rock nomenclature. Jour. Geology **62**, 344–359 (1954).

— Petrography and origin of the Tuscarora, Rose Hill, and Keefer formations, Lower and Middle Silurian of eastern West Virginia. Jour. Sed. Petrology **30**, 1–58 (1960).

— Petrology of sedimentary rocks, 170 p. Austin, Texas: Hemphill's Bookstore 1968a.

Folk, R. L.: Bimodal supermature sandstones. Product of the desert floor. XXIII Intern. Geol. Congress Proc. **8**, 9–32 (1968b).
Foster, R. J.: Tertiary geology of a portion of the central Cascade Mountains, Washington. Geol. Soc. America Bull. **71**, 99–125 (1960).
Friedman, Melvin: Miocene orthoquartzite from New Jersey. Jour. Sed. Petrology **24**, 235–241 (1954).
Füchtbauer, Hans: Zur Nomenklatur der Sedimentgesteine. Erdöl und Kohle **12**, 605–613 (1959).
— Sedimentpetrographische Untersuchungen in der älteren Molasse nördlich der Alpen. Eclogae geol. Helvetiae **57**, 157–298 (1964).
— Die Sandsteine in der Molasse nördlich der Alpen. Geol. Rundschau **56**, 266–300 (1967).
Galliher, E. W.: Glauconite genesis. Geol. Soc. America Bull. **46**, 1351–1366 (1935).
Garner, H. F.: Stratigraphic-sedimentary significance of contemporary climate and relief in four regions of the Andes Mountains. Geol. Soc. America Bull. **70**, 1327–1368 (1959).
Garrels, R. M., Mackenzie, F. T., Siever, Raymond: Sedimentary cycling in relation to the history of the continents and oceans. In: Robertson, E. C., (Ed.): The nature of the solid earth, p. 93–121. New York: McGraw-Hill 1971.
Gasser, Urs: Die innere Zone der subalpinen Molasse des Entlebuchs (Kt. Luzern). Geologie und Sedimentologie. Eclogae geol. Helvetiae **61**, 229–319 (1968).
Gibbs, R. J.: The geochemistry of the Amazon River System. Part I. The factors that control the salinity and the composition and concentration of the suspended solids. Geol. Soc. America Bull. **78**, 1203–1232 (1967).
Giles, R. T., Pilkey, O. H.: Atlantic beach and dune sediments of the southern United States. Jour. Sed. Petrology **35**, 900–910 (1965).
Gluskoter, H. J.: Orthoclase distribution and authigenesis in the Franciscan Formation of a portion of western Marin County, California. Jour. Sed. Petrology **34**, 335–343 (1964).
Goldman, M. I.: Petrographic evidence on the origin of the Catahoula Sandstone. Am. Jour. Sci., Ser. 4, **39**, 261–287 (1915).
— General character, mode of occurrence and origin of glauconite. Jour. Washington Acad. Sci. **9**, 501–502 (1919).
Graham, W. A. P.: A textural and petrographic study of the Cambrian sandstones of Minnesota. Jour. Geology **38**, 696–716 (1930).
Greenly, E.: Incipient metamorphism in the Harlech Grits. Edinburgh Geol. Soc. Trans. **7**, 254–258 (1899).
Griffiths, J. C.: Petrographical investigations of the Salt Wash sediments. U.S. Atomic Energy Comm., Tech. Rept. RME-3122 (Pts. I and II), 84 p., 1956.
Grim, Ralph: The Eocene sediments of Mississippi. Mississippi State Geol. Survey Bull. **30**, 240 p. (1936).
Grout, F. F.: Contact metamorphism of the slates of Minnesota by granite and gabbro magmas. Geol. Soc. America Bull. **44**, 989–1040 (1933).
Gruner, J. W.: Structural geology of the Knife Lake area of northeastern Minnesota. Geol. Soc. America Bull. **52**, 1577–1642 (1941).
Gulbrandsen, R. A., Goldich, S. S., Thomas, H. H.: Glauconite from the Precambrian Belt Series, Montana. Science **140**, 390–391 (1963).
Hadding, Assar: The pre-Quaternary sedimentary rocks of Sweden. III. The Paleozoic and Mesozoic sandstones of Sweden. Lunds Univ. Årsskr., N.F. Avd. 2, **25**, Nr. 3, 287 p. (1929).
— IV. Glauconite and glauconitic rocks. Medd. Lunds Geol. Min. Inst., no. 51, 175 p. (1932).
Haile, N. S.: The Cretaceous-Cenozoic Northwest Borneo Geosyncline. British Borneo Geol. Survey Bull. **4**, 1–18 (1963).
Hawkins, J. W., Whetten, J. T.: Graywacke matrix minerals. Hydrothermal reactions with Columbia River sediments. Science **166**, 868–870 (1969).
Heald, M. T.: Stylolites in sandstone. Jour. Geology **63**, 101–114 (1955).
Helmbold, Reinhard: Beitrag zur Petrographie der Tanner Grauwacken. Heidelberger Beiträge Min. Pet. **3**, 253–288 (1952).
— Van Houten, F. B.: Contribution to the petrography of the Tanner graywacke. Geol. Soc. America Bull. **69**, 301–314 (1958).
Henningsen, Dierk: Untersuchungen über Stoffbestand und Paläogeographie der Giessener Grauwacke. Geol. Rundschau **51**, 600–626 (1961).

Hickok IV, W. O., Moyer, F. T.: Geology and mineral resources of Fayette County. Pennsylvania Geol. Survey Bull. C-**26**, 530 p. (1940).
Hollister, C. D., Heezen, B. C.: Modern graywacke-type sands. Science **146**, 1573–1574 (1964).
Hoque, Momin ul: Sedimentologic and paleocurrent study of Mauch Chunk sandstones (Mississippian) of south-central and western Pennsylvania. Am. Assoc. Petroleum Geologists **52**, 246–263 (1968).
Hubert, J. F.: Petrology of the Fountain and Lyons formations, Front Range, Colorado. Colorado School Mines Quart. **55**, no. 1, 242 p. (1960).
Huckenholtz, H. G.: Mineral composition and texture in graywackes from the Harz Mountains (Germany) and in arkoses from the Auvergne (France). Jour. Sed. Petrology **33**, 914–918 (1963).
Hunter, R. E.: Facies of iron sedimentation in the Clinton Group. In: Fisher, G. W., and others, (Eds.): Studies of Appalachian geology, p. 101–121. Central and southern New York: Interscience 1970.
Irving, R. D.: The copper-bearing rocks of Lake Superior. U. S. Geol. Survey Mon. **5**, 464 (1883).
— Van Hise, C. R.: The Penokee iron-bearing series of Michigan and Wisconsin. U.S. Geol. Survey Mono. **19**, 534 p. (1892).
Jacobsen, Lynn: Petrology of Pennsylvanian sandstones and conglomerates of the Ardmore Basin. Oklahoma Geol. Survey Bull. **79**, 144 p. (1959).
Kaiser, E.: Der Grundsatz des Aktualismus in der Geologie. Zeitschr. deutsch. geol. Gesell. **83**, 401–402 (1932).
Kay, Marshall: North American Geosynclines. Geol. Soc. America Mem. **48**, 143 p. (1951).
Keith, M. L.: Sandstone as a source of silica sands in southeastern Ontario. Ontario Dept. Mines Ann. Rept. **55**, pt. 5, 36 p. (1949).
Kennedy, W. Q.: Sedimentary differentiation as a factor in the Moine-Torridonian correlation. Geol. Mag. **88**, 257–266 (1951).
Ketner, K. B.: Comparison of Ordovician eugeosynclinal and miogeosynclinal quartzites of the Cordilleran geosyncline. In: Geological Survey Research 1966. U.S. Geol. Survey Prof. Paper 550-C, C 54–C 60 (1966).
— Origin of Ordovician quartzite in the Cordilleran miogeosyncline. In: Geological Survey Research 1968. U.S. Geol. Survey Prof. Paper 600 B, B 169–B 177 (1968).
Klein, G. deVries: Triassic sedimentation, Maritime Provinces, Canada. Geol. Soc. America Bull. **73**, 1127–1146 (1962).
— Analysis and review of sandstone classification in the North American geological literature, 1940–1960. Geol. Soc. America Bull. **74**, 555–576 (1963).
— Dispersal and petrology of sandstones of Stanley-Jackfork boundary, Ouachita fold belt, Arkansas and Oklahoma. Am. Assoc. Petroleum Geologists Bull. **50**, 308–326 (1966).
Koldewijn, B. W.: Provenance, transport and deposition of Rhine sediments. II. An examination of the light fraction. Geol. Mijnb. (N. S.) **17**, 37–45 (1955).
Krynine, P. D.: Arkose deposits in the humid tropics, a study of sedimentation in southern Mexico. Am. Jour. Sci. **29**, 353–363 (1935).
— Petrography and genesis of the Siwalik Series. Am. Jour. Sci. **34**, 422–446 (1937).
— Petrology and genesis of the Third Bradford Sand. Pennsylvania State Coll. Bull. **29**, 134 p. (1940).
— Graywackes and the petrology of Bradford Oil Field, Pennsylvania. Am. Assoc. Petroleum Geologists Bull. **25**, 2071–2074 (1941a).
— Paleogeographic and tectonic significance of sedimentary quartzites (abstr.). Geol. Soc. America Bull. **52**, 1915 (1941b).
— Differential sedimentation and its products during one complete geosynclinal cycle. Anales 1st Congreso Panamerican Ingenieria de Minas y Geologia, Geology Part 1, **2**, 537–561 (1942).
— Sediments and the search for oil. Producers Monthly **9**, 12–22 (1945).
— The megascopic study and field classification of sedimentary rocks. Jour. Geology **56**, 130–165 (1948).
— Petrology, stratigraphy and origin of the Triassic sedimentary rocks of Connecticut. Connecticut State Geol. Nat. Hist. Survey Bull. **73**, 247 p. (1950).
— Tuttle, O. F.: Petrology of Ordovician-Silurian boundary in central Pennsylvania (abs.). Geol. Soc. America Bull. **52**, 1917–1918 (1941).
Kuenen, Ph. H.: Experimental abrasion, 3. Fluviatile action on sand. Am. Jour. Sci. **257**, 172–190 (1959).

Kuenen, Ph. H.: Sand. Sci. American **202**, no. 4, 94–110 (1960).
— Matrix of turbidites. Experimental approach. Sedimentology **7**, 267–297 (1966).
— Migliorini, C. I.: Turbidity currents as a cause of graded bedding. Jour. Geology **58**, 91–127 (1950).
Lasius, Georg: Beobachtungen im Harzgebirge, p. 132–152. Hannover: Helwing 1789.
Leith, C. K., Van Hise, C. R.: The geology of the Lake Superior region. U.S. Geol. Survey Mono. **52**, 641 p. (1911).
Lerbekmo, J. F.: Petrology of the Belly River Formation, southern Alberta Foothills. Sedimentology **2**, 54–86 (1963).
Loney, R. A.: Stratigraphy and petrography of the Pybus-Gambier area, Admiralty Island, Alaska. U.S. Geol. Survey Bull. **1178**, 103 p. (1964).
MacGregor, A. M.: The problem of the Precambrian atmosphere. South Africa Jour. Sci. **24**, 155–172 (1927).
Mackie, W.: The sands and sandstones of eastern Moray. Edinburgh Geol. Soc. Trans. **7**, 148–172 (1899a).
— The feldspars present in sedimentary rocks as indication of the condition of contemporaneous climate. Edinburgh Geol. Soc. Proc. **7**, 443–468 (1899b).
— Seventy chemical analyses of rocks. Edinburgh Geol. Soc. Proc. **8**, 33–60 (1905).
Mansfield, G. R.: The physical and chemical character of New Jersey Greensands. Econ. Geology **15**, 547–566 (1920).
Martens, J. H. C.: Beaches of Florida. Florida State Geol. Survey, 22nd Ann. Rept. **1931**, 67–119.
Mathur, S. M.: On the term "orthoquartzite". Eclogae geol. Helvetiae **51**, 695–696 (1958).
Mattiat, B.: Beitrag zur Petrographie der Oberharzer Kulmgrauwacke. Beitr. Mineralog. Petrogr. **7**, 242–280 (1960).
McBride, E. F.: The term graywacke (discussion). Jour. Sed. Petrology **32**, 614–615 (1962a).
— Flysch and associated beds of the Martinsburg Formation (Ordovician), central Appalachians. Jour. Sed. Petrology **32**, 39–91 (1962b).
— A classification of sandstones. Jour. Sed. Petrology **33**, 664–669 (1963).
— Sedimentary petrology and history of the Haymond Formation (Pennsylvanian), Marathon Basin, Texas. Univ. Texas, Bur. Econ. Geology Rept. Inv. No. **57**, 101 p. (1966).
McElroy, C. T.: The use of the term "greywacke" in rock nomenclature in New South Wales. Australian Jour. Sci. **16**, 150–151 (1954).
McEwen, M. C., Fessenden, F. W., Rogers, J. J. W.: Texture and composition of some weathered granites and slightly transported arkosic sands. Jour. Sed. Petrology **29**, 477–492 (1959).
Meckel, L. D.: Origin of Pottsville conglomerates (Pennsylvanian) in the central Appalachians. Geol. Soc. America Bull. **78**, 223–258 (1967).
Merifield, P. M., Lamar, D. L.: Sand waves and early earth-moon history. Jour. Geophys. Research **73**, 4767–4474 (1968).
Merrill, G. P.: Stones for building and decoration, 3rd Ed., 551 p. New York: John Wiley and Sons 1891.
Middleton, G. V.: Chemical composition of sandstones. Geol. Soc. America Bull. **71**, 1011–1026 (1960).
Mizutani, S., Suwa, K.: Orthoquartzitic sand from the Libyan Desert, Egypt. Nagoya Univ. Jour. Earth Sci. **14**, 137–150 (1966).
Morris, W. J., Fan, P.-F.: Abrasion effects on arkose mixtures. Jour. Sed. Petrology **32**, 231–239 (1962).
Murray, H. H., Patton, J. B.: Preliminary report on high-silica sand in Indiana. Indiana Dept. Conserv. Geol. Survey Rept. Prog. **5**, 35 p. (1953).
Nanz, R. H., Jr.: Genesis of Oligocene sandstone reservoir, Seeligson field, Jim Wells and Kleberg counties, Texas. Am. Assoc. Petroleum Geologists Bull. **38**, 96–117 (1954).
Naumann, C. F.: Lehrbuch der Geognosie, Vol. 1, 960 p. Leipzig: Engelman 1858.
Niggli, P., de Quervain, F., Winterhalter, R. U.: Chemismus schweizerischer Gesteine. Beitr. Geologie Schweiz, Geotechn. Ser. No. 14, 389 p. (1930).
Ojakangas, R. W.: Petrology and sedimentation of the Upper Cambrian Lamotte Sandstone in Missouri. Jour. Sed. Petrology **33**, 860–873 (1963).
Okada, H.: Sandstones of the Cretaceous Mifune Group, Kyushu, Japan. Kyushu Univ. Mem. Fac. Sci., Ser. D, Geology **10**, 1–40 (1960).

Okada, H.: Cretaceous sandstones of Goshonoura Island, Kyushu. Kyushu Univ. Mem. Fac. Sci., Ser. D, Geology **11**, 1–48 (1961).

— Non-graywacke "turbidite" sandstones in the Welsh geosynclines. Sedimentology **7**, 211–232 (1966).

— Composition and cementation of some Lower Paleozoic grits in Wales. Kyushu Univ. Mem. Fac. Sci., Ser. D, Geology **18**, 261–276 (1967).

Oriel, S. S.: Definitions of arkose. Am. Jour. Sci. **247**, 824–829 (1949).

Park, C. F., Jr.: Manganese resources of the Olympic Peninsula, Washington. U.S. Geol. Survey Bull. 931-R, 435–457 (1942).

Payne, T. G. and others: Geology of the Arctic slope of Alaska. U.S. Geol. Survey Oil Gas Invest., Map OM 126 (1952).

Peach, B. N., Horne, John, Gunn, W., Clough, C. T., Hinxman, L. W.: The geological structure of the North-West Highlands of Scotland. Mem. Geol. Survey Great Britain, 668 p. (1907).

Pelletier, B. R.: Pocono paleocurrents in Pennsylvania and Maryland. Geol. Soc. America Bull. **69**, 1033–1064 (1958).

Pettijohn, F. J.: Archean sedimentation. Geol. Soc. America Bull. **54**, 925–972 (1943).

— Sedimentary Rocks (1st Ed.), 526 p. New York: Harper and Bros. 1949.

— Classification of sandstones. Jour. Geology **62**, 360–365 (1954).

— Paleocurrents of Lake Superior Precambrian quartzites. Geol. Soc. America Bull. **68**, 469–480 (1957a).

— Sedimentary rocks (2nd Ed.), 718 p. New York: Harper and Bros. 1957b.

— The term graywacke. Jour. Sed. Petrology **30**, 627 (1960).

— Chemical composition of sandstones – excluding carbonate and volcanic sands. In: Data of Geochemistry (6th Ed.). U.S. Geol. Survey Prof. Paper 440S, 19 p. (1963).

Plumley, W. J.: Black Hills terrace gravels. A study in sediment transport. Jour. Geology **56**, 526–577 (1948).

Potter, P. E., Pryor, W. A.: Dispersal centers of Paleozoic and later clastics of the Upper Mississippi Valley and adjacent areas. Geol. Soc. America Bull. **72**, 1195–1250 (1961).

Reed, J. J.: Petrology of the Lower Mesozoic rocks of the Wellington District. New Zealand Geol. Survey Bull. (N.S.) **57**, 60 p. (1957).

Reed, R. D.: The occurrence of feldspar in California sandstones. Am. Assoc. Petroleum Geologists Bull. **12**, 1023–1024 (1928).

Rinne, Friedrich: Gesteinskunde, 374 p. Leipzig: Dr. Max Jänecke 1923.

Ronov, A. B., Mikhailovskaya, M. S., Solodkova, I. I.: Evolution of the chemical and mineralogical composition of arenaceous rocks. In: Chemistry of the Earth's Crust, v. 1 (Israel Program Sci. Trans., 1966), p. 212–262 (1963).

Rothrock, E. P.: A geology of South Dakota, part 3, Mineral resources. South Dakota Geol. Survey Bull. **15**, 255 p. (1944).

Russell, R. D.: Mineral composition of Mississippi River sands. Geol. Soc. America Bull. **48**, 1307–1348 (1937).

— Taylor, R. E.: Roundness and shape of Mississippi River sands. Jour. Geology **45**, 225–267 (1937).

Sabins, F. F., Jr.: Grains of detrital, secondary, and primary dolomite from Cretaceous strata of the Western Interior. Geol. Soc. America Bull. **73**, 1183–1196 (1962).

Schwab, F. L.: Origin of the Antietam Formation (Late Precambrian?-Lower Cambrian) central Virginia. Jour. Sed. Petrology **40**, 354–366 (1960).

Schwartz, G. M.: Correlation and metamorphism of the Thomson Formation, Minnesota. Geol. Soc. America Bull. **52**, 1001–1020 (1942).

Seilacher, A.: Origin and diagenesis of the Oriskany sandstone (Lower Devonian, Appalachians) as reflected in its shell fossils. In: Müller, G., Friedman, G. M., (Eds.): Recent Developments in Carbonate Sedimentology in Central Europe, p. 175–185. Berlin-Heidelberg-New York: Springer 1968.

Shead, A. C.: Notes on barite in Oklahoma with chemical analysis of sand barite rosettes. Oklahoma Acad. Sci. Proc. **3**, 102–106 (1923).

Shiki, T.: Studies on sandstone in the Maizuru Zone, southwest Japan. I. Importance of some relations between mineral composition and grain size. Kyoto Univ., Mem. Coll. Sci., Ser. B, **25**, 239–246 (1959).

Shiki, T.: Studies on sandstones in the Maizuri Zone, southwest Japan. III. Graywackes and arkose sandstone in and out of the Maizuri Zone. Kyoto Univ., Mem. Coll. Sci., Ser. B, **29**, 291–324 (1962).

— Mizutani, S.: On "graywacke". Geoscience **81**, 21–32 (1965) (in Japanese w. Engl. summ.).

Siever, Raymond: Trivoli sandstone of Williamson County, Illinois. Jour. Geology **57**, 614–617 (1949). (1949).

— Pennsylvanian sandstones of the Eastern Interior coal basin. Jour. Sed. Petrology **27**, 227–250 (1957).

Simonen, Ahti, Kuovo, Olavi: Archean varved schists north of Tampere in Finland. Soc. géol. Finlande Compte rendus, **24**, 93–117 (1951).

— Sandstones in Finland. Comm. géol. Finlande, Bull. **168**, 57–87 (1955).

Sippel, R. F.: Sandstone petrology, Evidence from luminescence petrography. Jour. Sed. Petrology **38**, 530–554 (1968).

Skolnick, Herbert: The quartzite problem. Jour. Sed. Petrology **35**, 12–21 (1965).

Sloss, L. L., Feray, D. E.: Microstylolites in sandstone. Jour. Sed. Petrology **18**, 3–13 (1948).

Strakhov, N. M.: Principles of lithogenesis, v. 2. First English ed. 1969. New York: Consultants Bureau, 609 p. (1967).

Suttner, L. J.: Stratigraphic and petrographic analysis of Upper Jurassic-Lower Cretaceous Morrison and Kootenai formations, southwest Montana. Am. Assoc. Petroleum Geologists Bull. **53**, 1391–1410 (1969).

Sutton, J., Watson, J.: The deposition of the upper Dalradian rocks of the Banffshire coast. Geol. Assoc. London Proc. **66**, 101–133 (1955).

Swineford, Ada: Cemented sandstones of the Dakota and Kiowa formations in Kansas. Kansas State Geol. Survey Bull. **70**, 57–104 (1947).

— Petrography of Upper Permian rocks in south-central Kansas. Kansas State Geol. Survey Bull. **111**, 179 p. (1955).

Takihashi, J.: Synopsis of glauconization. In: Trask, P. D., (Ed.): Recent Marine Sediments. Tulsa, Oklahoma, Am. Assoc. Petroleum Geologists 503–513 (1939).

Taliaferro, N. L.: Franciscan-Knoxville problem. Am. Assoc. Petroleum Geologists Bull. **27**, 109–219 (1943).

Tallman, S. L.: Sandstone types, their abundance and cementing agents. Jour. Geology **57**, 582–591 (1949).

Tarr, W. A.: The origin of the sand barites of the Lower Permian of Oklahoma. Am. Mineralogist **18**, 260–272 (1933).

Thiel, G. A.: Sedimentary and petrographic analysis of the St. Peter Sandstone. Geol. Soc. America Bull. **46**, 559–614 (1935).

— Dutton, C. E.: The architectural, structural and monumental stones of Minnesota. Minnesota Geol. Survey Bull. **25**, 160 p. (1935).

Tieje, A. J.: Suggestions as to the description and naming of sedimentary rocks. Jour. Geology **29**, 650–666 (1921).

Todd, T. W., Folk, R. L.: Basal Claiborne of Texas, record of Appalachian tectonism during Eocene. Am. Assoc. Petroleum Geologists Bull. **41**, 2545–2566 (1957).

Turner, F. J., Verhoogen, J.: Igneous and metamorphic petrology, 694 p. New York: McGraw-Hill 1960.

Twenhofel, W. H.: Treatise on sedimentation, 926 p. Baltimore: Williams and Wilkens 1932.

Tyrrell, G. W.: Greenstones and graywackes. Reunion intern. pour l'étude du Précambrien 1931 Comptes Rendus **1933**, 24–26.

Waldschmidt, W. A.: Cementing materials in sandstones and their influence on the migration of oil. Am. Assoc. Petroleum Geologists Bull. **25**, 1859–1879 (1941).

Walton, E. K.: Silurian greywackes in Peebleshire. Royal Soc. Edinburgh Proc. **65**, 327–357 (1955).

Webb, W. M., Potter, P. E.: Petrology and geochemistry of modern detritus derived from a rhyolitic terrain, western Chihuahua, Mexico. Bol. Soc. Geol. Mexicana, **32**, 45–61 (1969).

Webby, B. D.: Sedimentation of the alternating graywacke and argillite strata in the Porirua district. New Zealand Jour. Geol. and Geophys. **2**, 461–478 (1959).

Weber, J. N., Middleton, G. V.: Geochemistry of turbidites of the Normanskill and Charny formations. Geochim. et Cosmochim. Acta **22**, 200–288 (1961).

Whetten, J.T.: Sediments from the lower Columbia River and origin of graywacke. Science **152**, 1057–1058 (1966).
Wieseneder, H.: Über die Gesteinsbezeichnung Grauwacke. Tschermaks mineralog. petrog. Mitt. **7**, 451–454 (1961).
Wiesnet, D.R.: Composition, grain size, roundness and sphericity of the Potsdam Sandstone (Cambrian) in northeastern New York. Jour. Sed. Petrology **31**, 5–14 (1961).
Williams, Howel, Turner, F.J., Gilbert, C.M.: Petrography, 406 p. San Francisco: Freeman 1954.
Woodland, A.W.: Petrological studies in the Harlech Grit series of Merionethshire II. Geol. Mag. **74**, 440–454 (1938).
Yeakel, L.S., Jr.: Tuscarora, Juniata, and Bald Eagle paleocurrents and paleogeography in the central Appalachians. Geol. Soc. America Bull. **73**, 1515–1540 (1962).
Zimmerle, W.: Serpentine graywackes from the North Coast basin, Columbia, and their tectonic significance. Neues Jahrb. Mineral. Abh. **109**, 156–182 (1968).

Chapter 7. The Volcaniclastics

Introduction

Volcanism is a major source of detritus in island arcs and in the tectonic belts of continents. In ocean basins volcanism is potentially more important; Menard (1964, p. 95) has estimated that in the southwest Pacific in late Mesozoic and Cenozoic time the rate, duration, and effusion of outpouring of lava on the sea floor was comparable to that of plateau basalts on the continents but covered an area a hundred times greater. Verhoogen (1946, p. 746) has roughly estimated a total volume of 3×10^6 km^3 of plateau basalts in the geological record. In addition to lava, there are also vast quantities of airborne material, which on the continents appear to far exceed volume of lava. Obviously then, material derived from volcanism, either directly or indirectly, constitutes a most significant class of sandstones, one that cannot be ignored by the sedimentary petrologist.

Sands rich in volcanic debris that we call volcaniclastics are of two dominant types: those produced by explosive ejection from a volcanic vent, the pyroclastic sands, and those that are derived by normal erosion from a volcanic terrain, one composed of extrusives and/or pyroclastics. Except for their very immature mineralogy, the erosional volcaniclastics are comparable – in sedimentary structures, geometry, and thickness – to other terrigenous sandstones. But the pyroclastic sandstones stand somewhat apart; they originate as explosive, igneous rocks, but are deposited as sedimentary ones. Thus they lie partly outside the ken of both the igneous and the sedimentary petrologist. In part, their lack of study by sedimentary petrologists may be an historical accident in that sedimentary petrology had its beginnings in regions remote from active volcanism. In any case, they have not received their due.

Whether pyroclastic or not, volcaniclastic sands may be mixed in all proportions with other terrigenous sands, carbonates or pelitic sediments.

Some Characteristic Petrographic Features

Because they are either directly or indirectly the product of volcanism, volcaniclastic sands are more immature mineralogically than any other group of sandstones. Hence, detailed studies of their provenance are readily made. Special mineralogical earmarks of volcaniclastic sands include glass, pumice and scoria, euhedra, characteristic rock fragments, zoned feldspars, cristobalite, basaltic hornblende and much diagenetic mineralogical alteration.

Glass may occur as bubble-wall shards (Fig. 7-1; 7-2) as pumice shards (Fig. 7-3), or as a binding matrix where it plays the role of chemical cement.

Fig. 7-1. Undistorted bubble-wall glass shards (left) versus distorted and flattened (right)

Fig. 7-2. An ash fall consisting of an open framework of little abraded glass shards cemented by poikiloblastic calcite. Ellensburg Formation (Tertiary), Washington, U.S.A., × 100

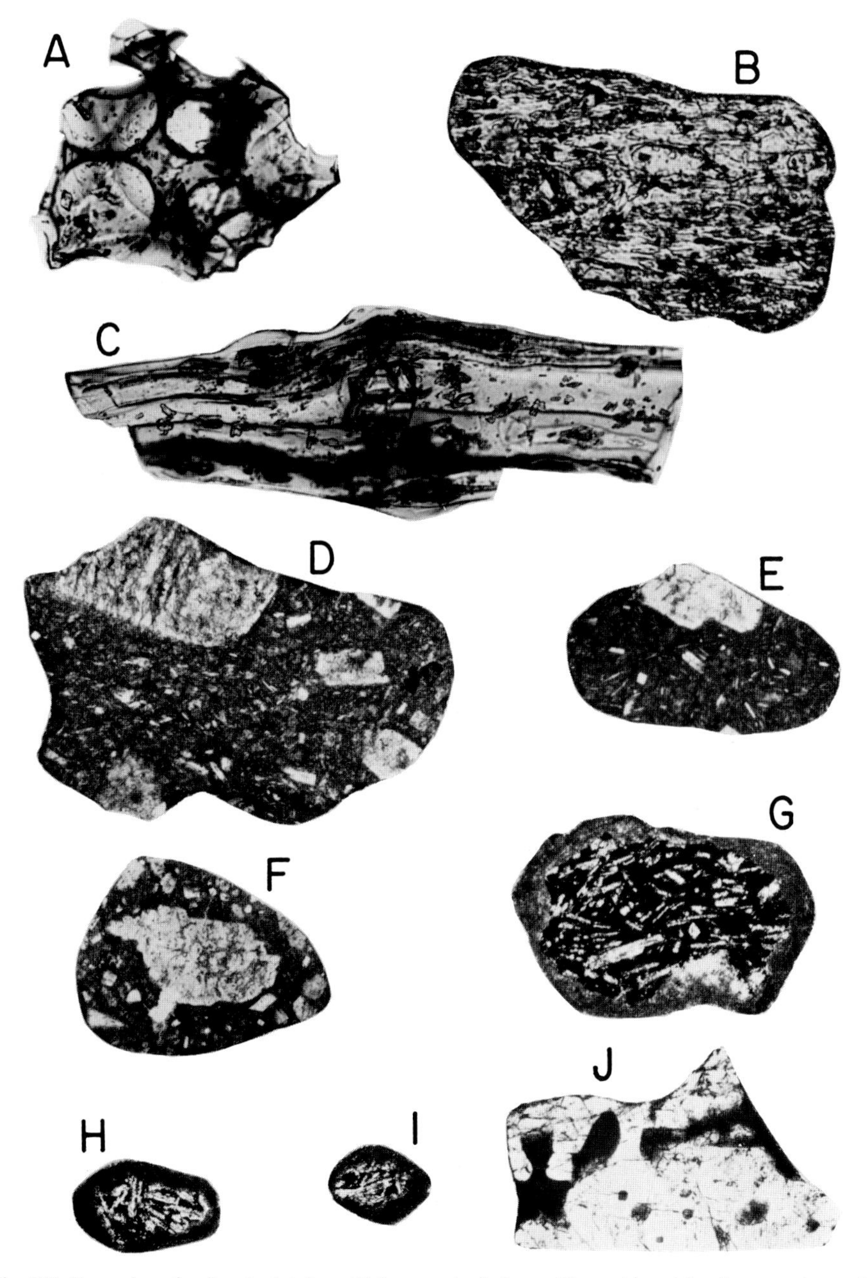

Fig. 7-3. Examples of volcanic detritus: A) fragment of glass with ovoid vesicles from modern ash, × 200; B) pumice fragment in Tertiary sandstone, Montana, × 80; C) pumice fragment with fluidal structures in modern ash, × 200; D), E) and F) large phenocrysts in fine-grained ground mass containing euhedral microlites from Judith Fancy Formation (Cretaceous), Virgin Islands, × 40; G), H) and I) microlites of feldspar in black glassy ground mass, fragments coated by carbonate, × 40; Pleistocene caliche; Dona Ana Co., New Mexico, U.S.A.: J) large phenocryst embayed by glass (black) from Currabubula Formation (Carboniferous) New South Wales, × 80. Photographs A and C courtesy of R. V. Fisher

The amount of glass seen in a thin-section of a volcaniclastic sandstone can be deceptive until one uses the upper nicol prism (Fig. 7-4). Glass may be colorless, red, yellow, brown, or black depending upon the oxidation state of its iron and on impurities. The approximate silica content of the glass can be determined by its index of refraction (Huber and Reinhart, 1966), although minor components, such as iron and water affect index too. Both glass and aphanitic microcrystalline material may also be attached as blebs to crystals

Fig. 7-4. Vitreous crystal tuff with crossed nicols (left) and unpolarized light (right). Note large percentage of glass. Currabubula Formation (Carboniferous), Werrie Basin, New South Wales

and grains (Fig. 7-5) or they may encrust a grain, defining bubble-wall texture (Fig. 7-6), a sure indication of volcanic origin.

Because they crystallize from a magma, the primary crystalline materials of volcaniclastics, both feldspars and mafic minerals, tend to be euhedra. In pyroclastics many such euhedra are commonly broken. Zoned feldspars (Fig. 7-7) are also abundant and reflect either high viscosity or fast cooling in the parent magma, both of which inhibit an early formed crystal from further reaction with the liquid phase. The high temperature feldspar, sanidine, is characteristic of quickly chilled volcanics and hence a good provenance indicator. Features worthy of note are embayments shown by quartz (Fig. 7-5) and feldspar. These form by resorption by the magma, perhaps shortly before eruption. Another notable feature of the quartz of volcaniclastic sands is that it consists almost entirely of strain-free, monocrystalline grains with unit extinction. Basaltic or oxyhornblende in the heavy mineral suite is also another telltale indicator of volcanic source rocks.

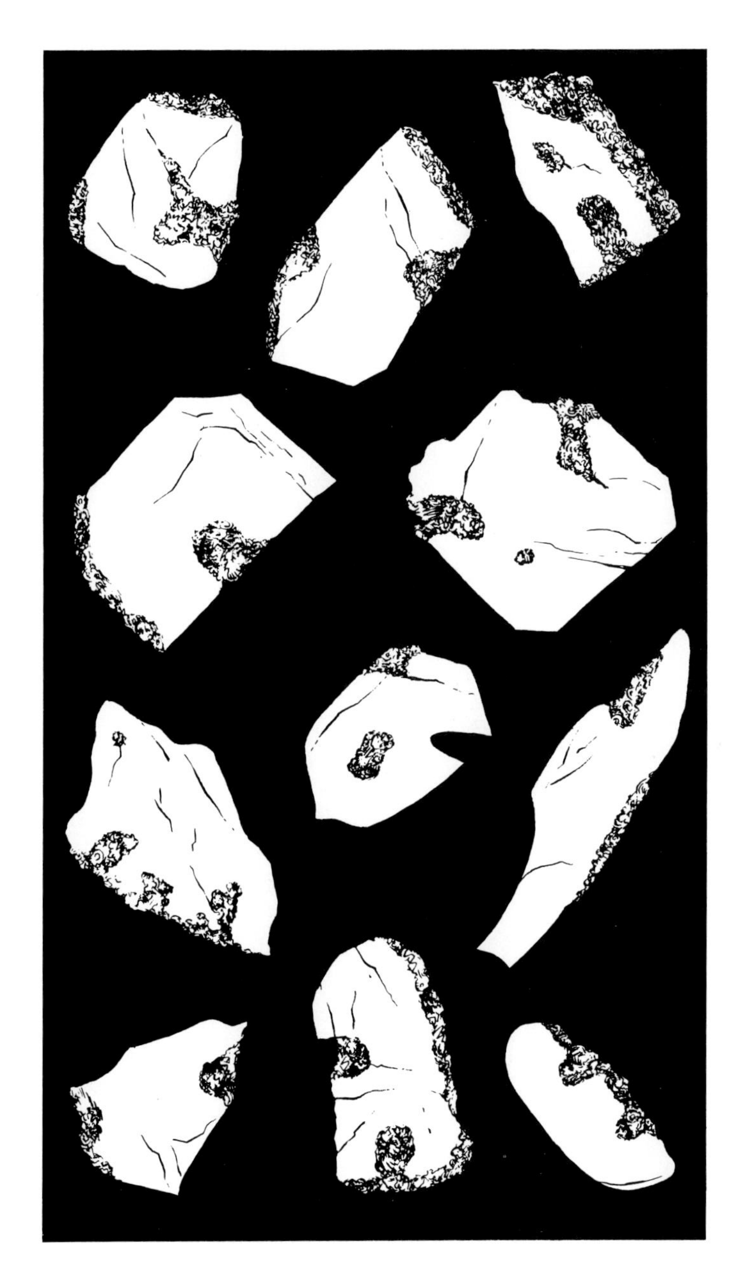

Fig. 7-5. Embayed quartz with attached aphanitic and glassy blebs (Webb and Potter, 1969, Fig. 3)

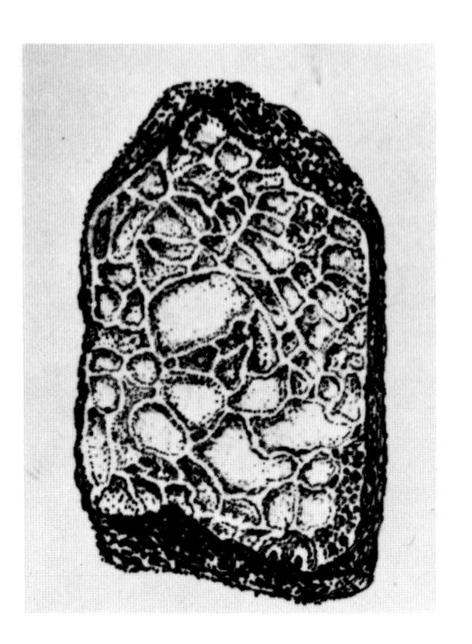

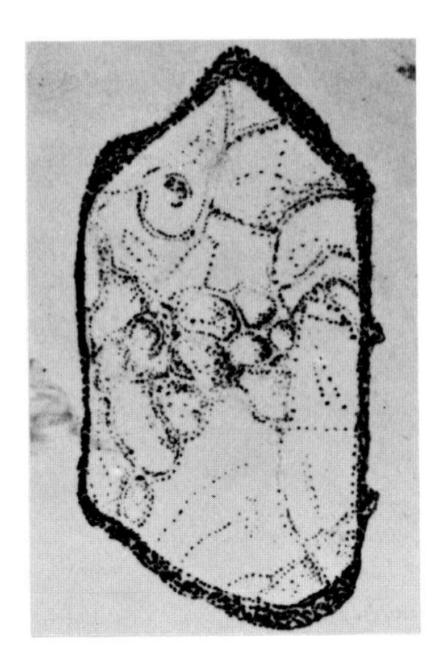

Fig. 7-6. Bubble-wall texture (Fisher, 1963, Fig. 4). Reproduced by permission of the Journal of Sedimentary Petrology

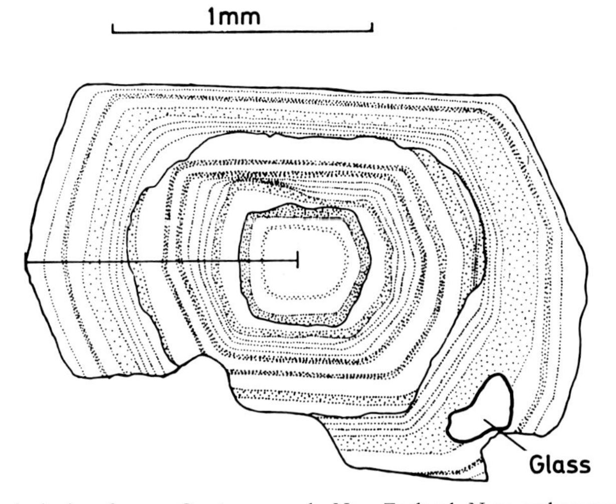

Fig. 7-7. Zoned plagioclase from a Quaternary ash, New Zealand. Note embayment (Ewart, 1963, Fig. 11). Reproduced by permission of the Journal of Petrology

Figure 7-3 shows some of the different types of rock fragments present in volcaniclastics. Randomly as well as preferentially oriented microlites are characteristic of such fragments, which may or may not contain phenocrysts. Glass is also commonly present as interstitial material between the microlites. The combination of fine grained volcanic rock fragments and detrital matrix makes for extremely difficult point counting – where does the rock fragment end and matrix begin?

Volcaniclastic sands are more susceptible to diagenesis than any other type of sand because of the chemical instability and reactivity of their framework grains: glass and pumice and the presence of fine grained rock fragments plus feldspathoids and mafic minerals. High porosity and surface area of the fine grained material accelerate diagenesis and promote devitrification of volcanic glass. Except where preserved in a "sealed environment", glass is commonly found only in mid-Tertiary and younger rocks. Glass alters to clay minerals,

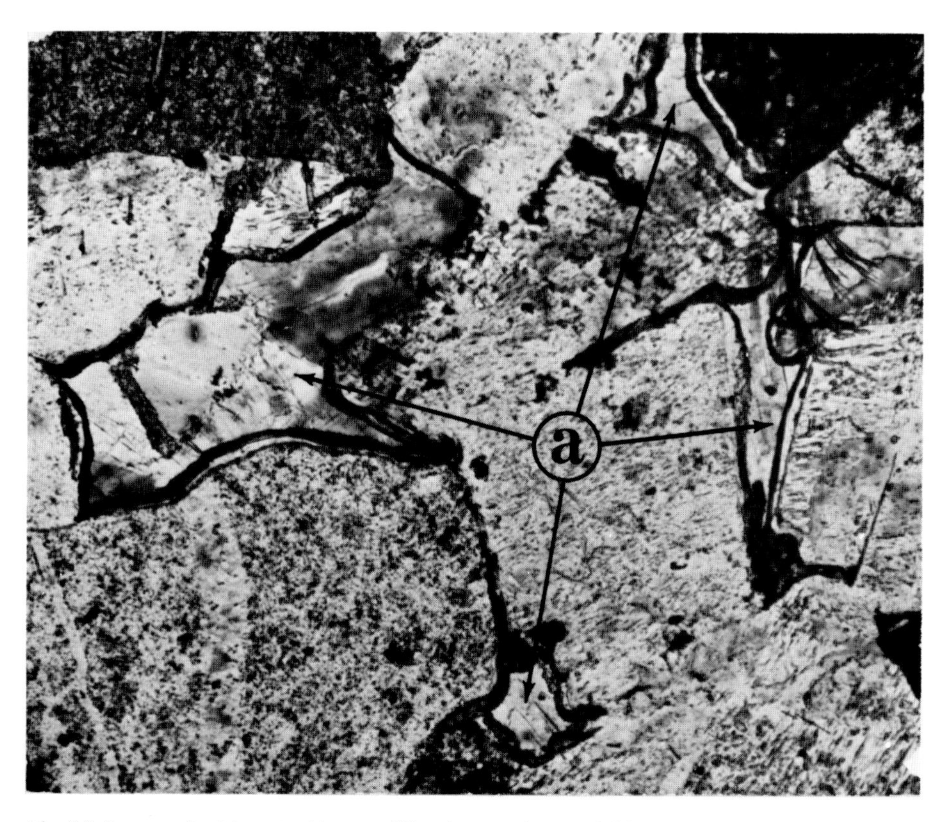

Fig. 7-8. Laumontite (a) as a void space filling in a sandstone of the Blairmore Group (Cretaceous), Alberta, × 100

zeolites, and silica. Montmorillonite and halloysite form colloform and vermicular masses which may precipitate in cavity fillings. Or glass may devitrify to a microcrystalline aggregate, which may resemble some chert; microlitic euhedra in such aggregates are a sure sign of their volcanic origin, however. Complete alteration of an ash may produce bentonite which consists almost entirely of montmorillonite. In outcrop this expandable clay has a very characteristic cracked, flaky weathering surface. Zeolites precipitated in void spaces (Fig. 7-8) and as coatings on framework grains are also a distinguishing earmark of many volcaniclastic sands. Zeolitic minerals may also replace glass. Silica, released from the alteration of volcanic glass, may be present as chalcedonic and opal cements or as discrete veins, pods, and nodules. Cristobalite is a common secondary feature of many ash flows; it can also be found in coarser grained rhyolitic rock fragments. High cristobalite may form during the cooling of an ash flow whereas a semi-amorphous variety of low-cristobalite forms later in the diagenesis of glass. Much of the alteration of subaerial volcaniclastic deposits may occur early in their history and be related to soil development.

Another distinctive feature of volcaniclastic sands, one that is more properly a fluid dynamic property rather than petrographic, stems from the density contrasts between pumice and scoria on the one hand and crystalline material on the other. Pumice, which may have densities as low as 0.22, may float long distances before becoming waterlogged and thus be finally deposited far from the crystalline material with which it was ejected. The wide dispersal pattern of scoriaceous materials comes from a combination of their pelagic nature and, following settling, their possible entrainment in turbidity currents. Because such materials have a low density, coarse grained pumice may be carried far by turbidity currents and sedimented with clay. Should the ash have been deposited in a desert, eolian transport is also an effective segregation mechanism: for a given size fraction the less dense pumice will travel farther and faster than the crystalline material. Within a single ash fall bed sorting in air currents can also produce marked changes in grain size and composition from bottom to top of the bed.

Classification

The classification of a volcaniclastic sand depends on whether it is of pyroclastic or erosional origin.

If of erosional origin one uses the standard terrigenous classification of Fig. 5-3, adding the prefix "tuffaceous." Because rock fragments commonly predominate in volcanic sands, such sands will be either tuffaceous lithic wackes or lithic arenites.

If the sand were eroded from a single source, which in turn received pyroclastic debris from a petrologically homogenous magma, feldspar composition can be determined (Heinrich, 1965, p. 334–370; Mizutani, 1959), so that one can add a specifying adjective such as andesitic, tuffaceous, lithic arenite. It is usually wise to study a number of samples before attempting to assign such a specific compositional name.

Pyroclastics are classified by size, composition, and origin (Table 7-1). The size classes used are after Fisher (1961, Table 3), who also provides a good review of the terminology of volcaniclastic rocks (1966a). Rocks composed of particles larger than 2 mm have been generally described as breccia. Ash and tuff both refer to material smaller than 2 mm, the former being unconsolidated, the latter consolidated.

Compositionally, one first determines the relative proportions of glass, rock fragments and crystals in a tuff and then the composition of the crystal fraction (Fig. 7-9). Having determined how much quartz, how much and what types of feldspar, what other crystalline minerals are present, the silica content of

Table 7-1. *Classification of pyroclastic debris*

Size (mm)		*Rock name*	*Origin*
Bombs and blocks —64— Lapilli —2— Ash or tuff[a]	Breccia	Determined by proportion of glass, crystals and rock fragments plus composition of crystals or via chemical analysis.	Three types: *juvenile* (derived from fresh magma); *accessory* (torn from volcanic neck); *accidental* (fragment of country rock).

[a] Termed ash, if uncompacted; tuff, if compacted.

any glass that may be present, and other compositional data, one uses a standard igneous classification scheme, such as those of Moorhouse (1959, Table 16), Rittmann (1962, p. 96–103), or Streckeisen (1967, p. 177–193) to assign a name. But because many pyroclastic rocks are very fine grained or glassy, a chemical analysis may be necessary so that one can calculate a probable mode from it (Jung and Brousse, 1959). X-ray diffraction or electron probe analysis may be needed to help identify the type of feldspar. If none of the above methods are available, a compositional name must be assigned on the basis of only the recognizable mineral content, perhaps only phenocrysts. If so, one prefixes "pheno" to the name. If one can specify composition, one can describe a rock, for example, as an andesitic, lithic tuff or a vitric, pheno-dacitic tuff. As with eroded materials, it is usually advisable to determine composition on several samples before adding a compositional, adjectival modifier. Williams and other (1954, p. 149) give a good brief introduction to the petrography of pyroclastics.

Tuffs may grade into volcaniclastic sandstones of erosional origin or into either non-volcanic quartzose terrigenous sandstones or into carbonate sand.

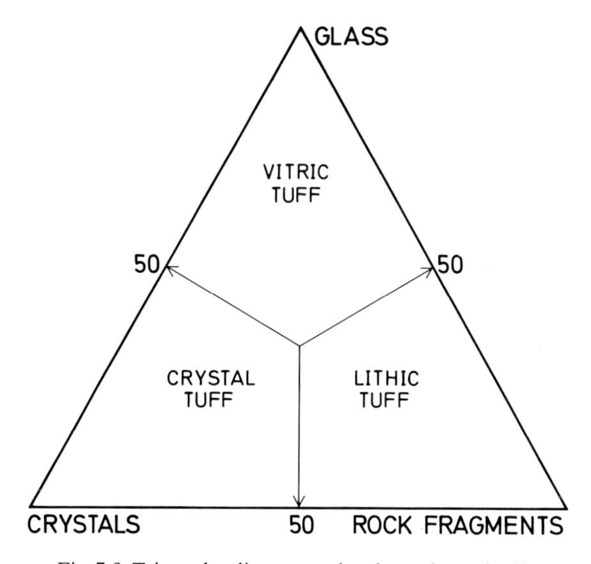

Fig. 7-9. Triangular diagram and end members of tuffs

If terrigenous material exceeds volcanic, the mixture is called sandy ash or sandy crystal tuff or perhaps, if appreciable skeletal material such as pelagic *Foraminifera* are present, a skeletal, crystal tuff.

Pyroclastic Sands

Pyroclastics constitute the most important product of volcanism in island arcs and along continental margins. In ancient rocks, they are principally restricted to former mountain systems. They are now recognized as an important concomitant of crustal resorption into the mantle in subduction zones associated with the convergence of crustal plates according to theories of global tectonics (Isacks and others, 1968; Dickinson, 1970).

Most pyroclastic materials are felsic in composition and tend to be rhyolitic, although dacitic and andesitic compositions also occur; magma rich in silica has greater viscosity and gas content than magma poor in silica, thus favoring explosive rather than the effusive volcanism that characterizes the outpourings of basic plateau basalts (Table 7-2). Pyroclastic material may form local agglomerates or cinder cones, or may be transported high in the atmosphere over much longer distances to form widespread beds of ash that serve as useful time-equivalent stratigraphic markers in many sedimentary basins. Volcanic debris that is blown high in the air may be deposited on either land or in water and is called ash fall; it contrasts with ash flow, material that accumulated from a hot, incandescent, turbulent mixture of debris and gas that explosively escaped from a fissure, cone, or vent and flowed downhill along the surface much in the same manner as a deep water, marine turbidity current. Both ash falls and ash flows may, of course, be interbedded with lava flows or other sediments. Although both flows and falls are commonly produced during a single major explosive eruption – their contrasting characteristics demand separate description.

Table 7-2. *Type of magma and volcanic activity (modified from Rittmann, 1962, Tables 4 and 5)*

Type of magma	*Quantity of magma*		*Type of activity*
	small	great	
Very hot, fluid, mafic ↓	Individual flows	Plateau basalts and shield volcanoes	Effusive ↓
Hot, moderately fluid, mafic	Pyroclastics and lava flows	Plateau basalts, strato-volcanic ridges and shield volcanoes plus cinder cones and scoriae	Mixed
Viscous, cooler intermediate			↓
Extremely viscous, with abundant crystals siliceous	Pumice and Scoria	Ash falls and ash flows from single volcanoes and fissures	Explosive

Ash Flows

Synonyms for ash flow include pyroclastic flow, ignimbrite, nuée ardente, and sand flow. A brief but informative discussion of ash flows is given by Aramaki (1957). Such deposits represent hot, "catastrophic" events – hot density currents that traveled swiftly downhill, perhaps at velocities approaching 100 mph, commonly

Fig. 7-10. Two cooling units of an ash flow overlying thin, white, evenly bedded ash fall which in turn overlies water deposited tuffaceous sandstones composed of pumice and ash. Pleistocene Bandelier Tuff (above) and Pliocene Peralta Tuff (below). Bland Canyon, Santo Domingo Pueblo Quadrangle, New Mexico, U.S.A. (Photograph courtesy of R. A. Bailey and U.S. Geol. Survey)

following the larger topographic lows. The temperatures of this mixture of gas and incandescent particles have been estimated to be between 550° and 950° C. Such units may extend as much as 20 to 60 miles from their source. Because ash flows cover the ground surface uniformly, they have flat tops, and variable thickness. In addition, they are generally internally zoned imparting a strong bedded character to the deposit (Fig. 7-10). Single cooling units, units representing a single pulse of sedimentation, can be as much as 100 m in thickness, but most vary between 15 to 30 m. Vertical zonation within a unit is defined by color, density and structures which reflect degrees of welding and crystallization. The internal zoning (Fig. 7-11) results from differential cooling between exterior and interior of the unit. Internally, one ash flow may differ from another by number of inclusions, grain size, color and welding characteristics as well as by mineralogical composition. Change in grain size at contacts between flows is particularly marked. Rapid accumulation of the incandescent tuff prevents rapid dissipation of heat so that the glass shards and pumice fragments

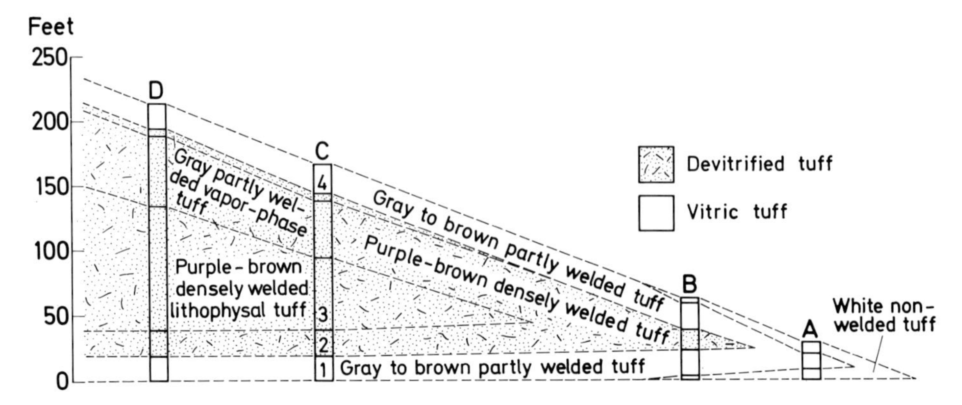

Fig. 7-11. Internal zoning of a welded tuff. Length of section about 5 miles (8 km). (Modified from Lipman and Christiansen, 1964, Fig. 2)

are either partly or totally fused or collapsed, forming a welded tuff (Fig. 7-12). The greater the degree of welding, the greater the density. Very characteristic field evidence for an ash flow are flattened, stretched pumice fragments which can show tension cracks, pull-a-parts, imbrication, and rotation, all of which can be used to infer transport direction (Fig. 9-12). In rare circumstances, movement during or after welding produces both deformation and laminar

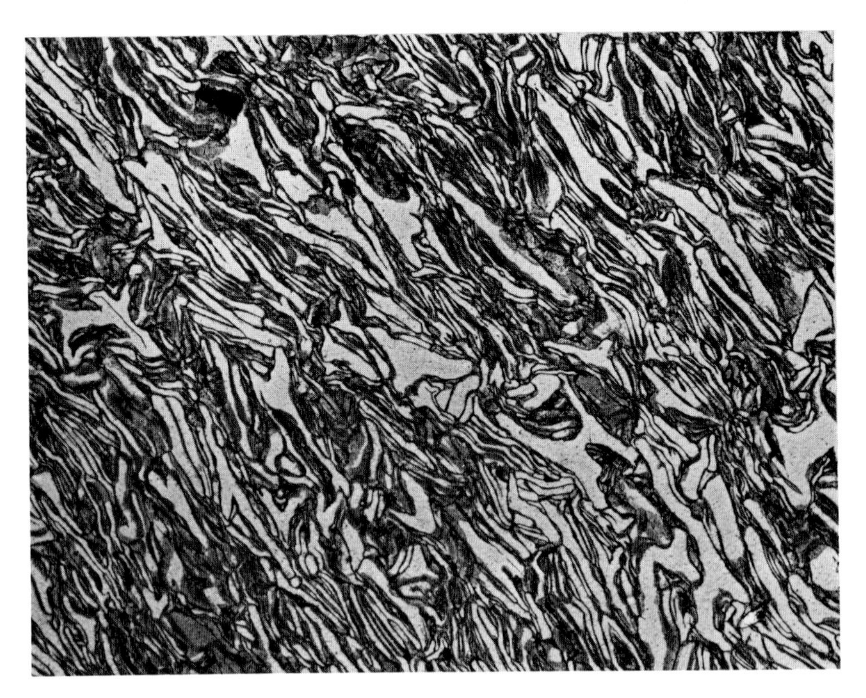

Fig. 7-12. Welded vitric tuff × 40. Walcott Tuff (Tertiary), Ferry Hollow, Sec. 6, T. 85., R. 31 E., Power Co., Idaho, U.S.A.

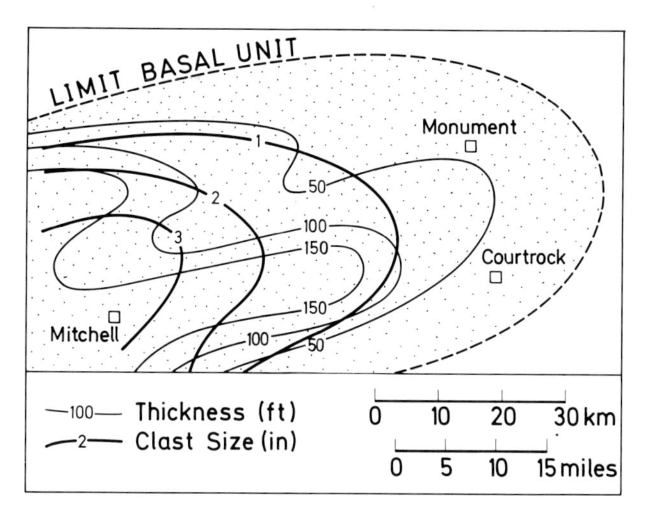

Fig. 7-13. Thickness and clast size in cooling units 1 and 2 of Picture Gorge ignimbrite (Oligocene-Miocene) in eastern Oregon. Clast size is from lower 20 feet of deposit. Transport from left to right (Modified from Fisher, 1966b, Figs. 1 and 4)

flowage features in the deposit. Features indicative of such movement are illustrated by Schmincke and Swanson (1966). Also useful is the downcurrent decline in flow thickness away from the source which may be accompanied by decrease in clast size (Fig. 7-13), although marked thickness decrease is not always conspicuous. Good exposures are commonly needed to demonstrate this, unfortunately. Sorting is extremely poor (Fig. 7-14).

Ash falls, stream deposits, mud flows, lava, or soil horizons may be found between ash flows. Crossbedding in associated stream deposits and the bent

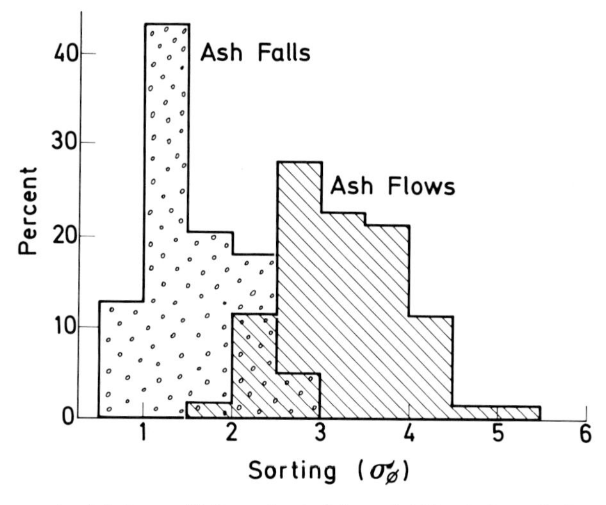

Fig. 7-14. Sorting contrast between 39 Recent ash falls and 192 ash flows in Japan (Musai, 1961, Tables 7 and 9)

Table 7-3. *Chemical analyses of volcaniclastics*

	Pyroclastics			Erosional sands	
	Rhyolitic tuff, Peru[a]	Andesitic pumice, Japan[b]	Ignimbrite Waihi district, New Zealand[d]	Modern Rivers Western Chihuahua, Mexico[d]	Columbia River, Oregon, USA[e]
SiO_2	70.81	63.53	75.5	73.50	61.69
Al_2O_3	13.12	16.01	13.5	13.3	13.89
Fe_2O_3	1.11	2.50	0.35	1.55	3.82
FeO	0.06	2.81	0.2	0.56	2.20
MgO	0.35	1.44	0.2	0.56	2.20
CaO	1.05	4.84	1.3	1.13	3.10
Na_2O	4.92	4.29	4.35	2.34	2.20
K_2O	4.20	0.84	3.5	4.01	1.88
H_2O+	2.34	2.49	0.45	} 1.80	} 1.81
H_2O-	0.10	0.75	0.45		
CO_2	0.02	—	—	0.12	—
SO_3	0.31	—	—	—	—
Cl	1.12	—	—	—	—
F	0.08	—	—	—	—
TiO_2	0.19	0.75	0.17	0.34	1.03
P_2O_5	0.00	0.16	0.06	0.02	0.25
MnO	0.07	0.07	0.01	0.04	0.11
BaO	0.09	—	—	—	—
	100.00	100.48		99.3	94.00[f]

[a] Jenks and Goldich (1956, Table 3, 1).
[b] Yagi (1962, Table 4, 6).
[c] Thompson and others (1965, Table B-L, 5).
[d] Webb and Potter (1969, Table 6, 5).
[e] Whetten (1966, Table 2, CC 29).
[f] Loss on ignition 5.99.

pipe amygdules of lavas (Waters, 1960) can help establish flow direction because all of the continental current laid deposits are gravity controlled and thus indicate paleoslope.

Fiske and Matsuda (1964) have given a good description of submarine ash flow deposits. When an explosion occurs in deep water, the pyroclastic debris is quickly chilled and settles to the bottom, where it forms dense turbidity flows. Such submarine ash flow deposits are much better sorted than their subaerial counterparts, because the greater density and viscosity of water reduce fall velocity, permitting better segregation of the different size and density. For example, fine grained ash may be carried long distances in suspension while pumice and lapilli will float to the surface and be dispersed by surface currents. In contrast, crystals and lithic fragments tend to be deposited first both laterally and vertically.

Chemical composition of volcaniclastics very largely reflects the composition of the parent magma (Table 7-3).

Ash Falls

Ash falls are the products of eolian differentiation wherein particles are segregated by their fall velocity as they are transported downwind. Size and density

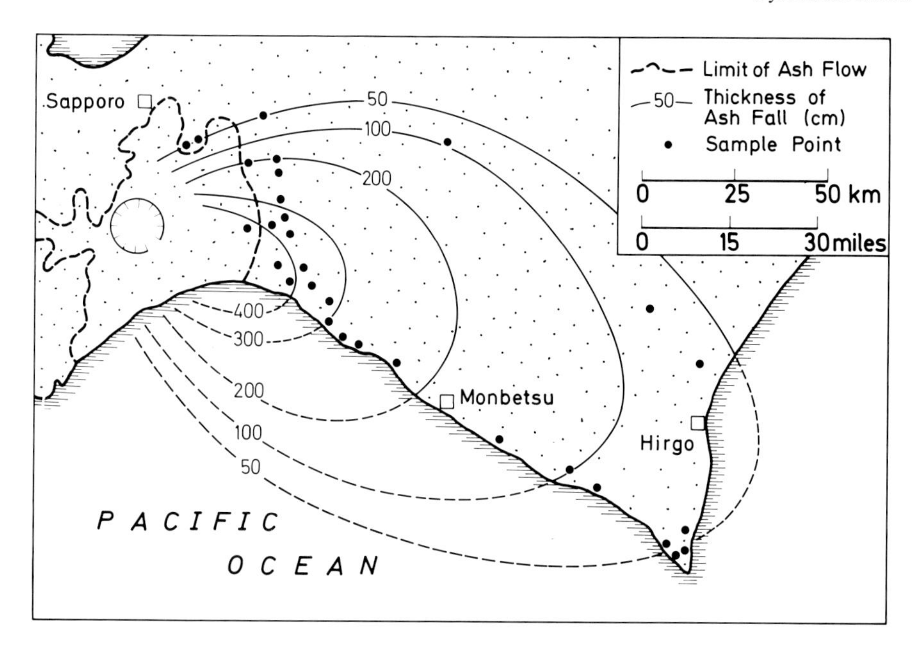

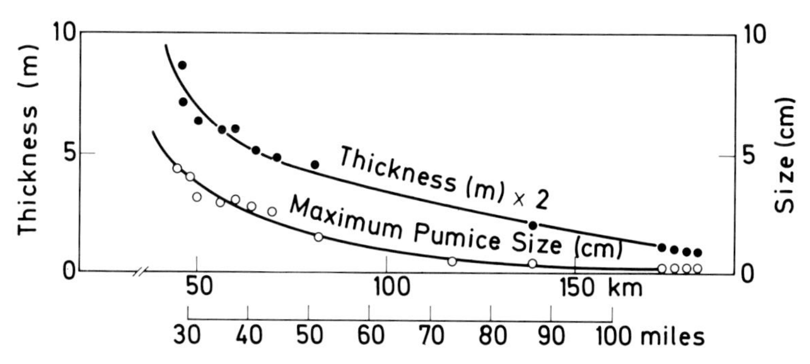

Fig. 7-15. Downwind thickness and grain size decline of an ash fall (Modified from Katsui, 1963, Figs. 4 and 5)

are primary factors that control fall velocity. Wind direction, velocity and turbulence as well as the height to which the particles are ejected control the resultant fallout pattern. This pattern may be roughly symmetrical or markedly asymmetrical and elongate as in Fig. 7-15. Ash falls tend to have an exponential decay of thickness downwind and a corresponding decrease in grain size (Fig. 7-16). In the ideal case, composition, grain size, and thickness are all interrelated and systematically change downwind. Scheidegger and Potter (1968) developed equations relating grain size and thickness to distance from source. Ash fall deposits are especially suitable for such study, because they are one of the most simple sedimentation systems in nature – progressive downwind decay of turbulence with essentially no complications due to reworking. As in ash flows, postdepositional alteration is usually extensive and is favored by high porosity and fineness of grain. Hay (1959) describes alteration of ash fall deposits in the West Indies. As a result of alteration, pre-Cenozoic ash falls have been almost always altered to bentonite and are composed primarily of montmorillonite. Silicification is common along the boundaries of bentonites. Typically, ash falls are marked by good to very good sorting (Fig. 7-14) and have well defined bedding (Figs. 7-10; 7-17). Unlike ash flows, which are restricted to

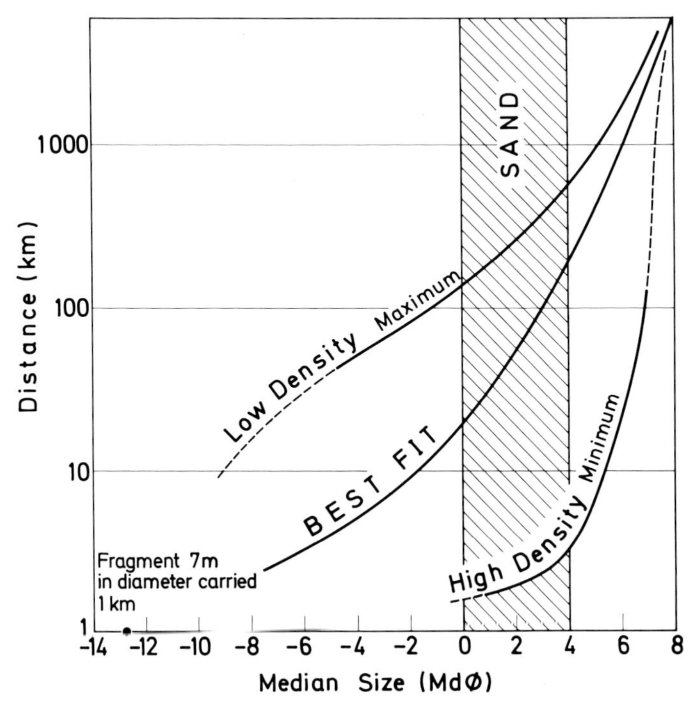

Fig. 7-16. Grain size and distance of air transport. Less dense particles travel farther than denser ones of the same size because of their smaller fall velocity (Modified from Fisher, 1964, Fig. 9)

mobile belts, large quantities of ash may be transported long distances in air and thus be deposited on stable platforms, either on land or in water. On the continents they have been noted in the stratigraphic record over areas as large as 300,000 to 400,000 square miles or 7.8×10^5 to 10.4×10^5 km^2 (Ross, 1955).

Table 7-4 summarizes the distinguishing features of ash flows and falls.

Intermediate between ash fall deposits and ignimbrites are *base surge deposits*, described by Moore (1967) and Fisher and Waters (1969). These transport mud, ash lapilli, and blocks at high speed; can leave well bedded deposits; and deposit debris well beyond the limit of throw-out trajectories. Antidunes may be present. The debris laden clouds are generated by large, shallow, phreatic volcanic explosions. Fisher and Waters suggest they may be significant in lunar sedimentation.

Fig. 7-17. Quaternary ash fall pumice in Tampo District, New Zealand. Observe perfection of bedding and its parallelism to topography, except below unconformity near base (Photograph courtesy of D. L. Homer and New Zealand Geological Survey)

Volcaniclastics of Erosional Origin

Pyroclastics and lavas are major sources of terrigenous debris in many mobile belts. This debris may come to rest on coastal plains, beaches, and shallow marine shelves or be carried into deep water by turbidity currents. Let us trace in some detail this seaward transport of debris and begin by considering what happens when ash inundates a landscape.

Vegetative cover is destroyed with the result that sheet, rill, and gulley erosion is greatly accelerated, thus destroying slope stability. Increased runoff can then greatly enhance channel erosion in the upper reaches of the drainage network (Fig. 7-18). Loss of vegetative cover usually is accompanied by landsliding, thus contributing more debris to valley floors. In such a setting a heavy rain can trigger the transport of an enormous quantity of debris. Waldron (1967, p. 16–29) gives an excellent recent account of the erosion and transport of volcanic debris following an ash fall and heavy rain in central Costa Rica. Debris flows, possibly containing 35 percent solids, developed in the headwaters and spread quickly downstream, covering lower valleys with a thick, poorly sorted, bouldery deposit (Fig. 7-18). Sorting in this debris flow averaged 3.1 ϕ and ranged from 2.6 to 4.0 ϕ. Ancient equivalents of such debris flows

Table 7-4. *Distinguishing characteristics of ash flows and falls*

Property	*Ash flow*	*Ash Fall*
Structures	Generally unbedded but may have "hot" flowage structures, welding, and fumaroles. Columnar structures in some zones.	Good horizontal bedding.
Size and size gradients	Clast size reduction commonly pronounced down dip. Most particles less than 4 mm.	Generally finer than ash flows. Strong lateral gradients and some vertical grading.
Sorting	Very poorly sorted with much variation within an outcrop. Marked tails on both ends of size curve.	Good sorting. Little variation within an outcrop. Tail of fines only.
Geometry	Elongate with shape controlled by major topographic lows. Flat top.	Wide spread, sheet-like and conformable, blanketing depositional surface.
Thickness and extent	May exceed several hundred feet near source and extend as much as 20 to 50 miles from it.	Commonly a few feet or less but exponentially thicker upwind; rarely more than 20 feet. May extend several hundred miles as very thin units.
Petrology	Welding common as is glass cement if not devitrified. Much welded and deformed pumice. Devitrified shards have axiolitic structure. Cristobalite.	Welding unusual.

have been identified (Anderson, 1944; Schmincke, 1967). Huge quantities of sand, silt, and clay were transported farther downstream to form coastal plain deposits and thence to the strandline and the shelf. Ash deposits and lava flows are common associated lithologies. Where lavas are interbedded, it may be possible to compare paleoslope inferred from both sands and lavas (Sandberg, 1938, p. 818–830). Volcanic detritus can be widely dispersed on marine shelves as in the Cretaceous of the high plains of North America.

Beyond the shelf, volcanic debris is carried into deep water by turbidity currents. Pillow lavas resulting from submarine eruptions may be interbedded with such turbidites. Volcanic glass in a Recent deep-sea ash turbidite in the northeastern Pacific Ocean has been correlated with continental ash deposits, all stemming from a cataclysmic eruption 6600 years ago, on the basis of stratigraphic position, refractive index, and radiocarbon dating (Nelson and others, 1968). The trenches associated with volcanic island arcs would seem to be prime sites to be filled by turbidites rich in volcaniclastics but modern trenches are not so filled; rather they are conspicuous for lack of any but relatively thin sediments. The explanation given in terms of global plate tectonics is that such trenches are sites of downward movement of crust into the mantle so that any sediments in the trench are continuously being destroyed (Isacks and others, 1968).

Fig. 7-18. Accelerated channel erosion in headwater stream (above) and resultant debris flow in valley (below) after an eruption and subsequent heavy rainfall in central Costa Rica (Waldron, 1967, Figs. 6 and 10)

Table 7-5. *Some modal analyses of volcaniclastics of erosional origin*

Unit	Quartz	Feldspar	Rock-fragments	Glass	Clay	$CaCO_3$	Cements Silica	*Remarks*	
Modern River Sand, Rio Tituaca, Western Chihuahua	19	14	35	17	4	—	—	11 percent heavies	Drains a mountain region underlain by rhyolitic tuffs
(Webb and Potter, 1969)	13	18	52	12	3	—	—	6 % heavies	
Gueydan Formation (Late Tertiary)	47	25	19	—	2	34	—	Caliche cement	Thin fluvial deposit
Southeast Texas (McBride and others, 1968)	41	23	35	—	—	—	37	Cement is opal	
Currabubula Formation (Carboniferous)	11	27	17	—	45	—	—	Coarse rhyolitic, vitric, crystal tuff.	
New South Wales (Whetten, 1965)	1	52	8	—	32	—	—	Andesitic crystal vitric tuff.	
Parry Group (Devonian-Carboniferous)	1	19	65	—	7	5	—	Parry group is 17,000 feet (5520 m) thick	
New South Wales (Crook, 1960)	—	35	47	—	14	—	—	4 % heavy minerals in framework.	
Umpqua Formation (Eocene), Oregon (Dott, 1966)	33	17	29	—	14	—	—	7 % accessories, mostly micas and heavy minerals. Some glass in matrix. Average of 10 analyses.	
Napere and Ouka Sandstones (Miocene)	—	31	7	—	51	—	—	11 and 9 % mafics in framework	
New Guinea (Edwards, 1947)	—	24	31	—	36	—	—	Thick marine turbidites.	

Because they are more likely to escape later subaerial erosion than continental deposits, volcanic-rich, deep water, marine turbidites are likely to be preserved. They have been reported fairly commonly in the Mesozoic and younger basins of the circumpacific region.

In terms of geometry, lithic associations and sedimentary structures, accumulations rich in volcanic detritus differ in no way from normal terrigenous water-laid deposits. Petrographically, however, volcaniclastic sands of erosional origin can be mineralogically very immature (Table 7-5) and are among the most variable: debris from different volcanic sources can be mixed together and combined with nonvolcanic debris. In addition, sorting and differential abrasion – the elimination of glass and aphanitic rock fragments either in streams or on beaches – can be most effective in producing remarkable placer sands (Fig. 7-19). But as a general rule, volcanic sands of erosional origin are marked by their abundance of rock fragments, diagenetic minerals and clay matrix, the latter primarily the result of diagenesis. Soil development on continental deposits can be an especially important factor that accelerates such diagenesis (McBride and others, 1968, p. 45–48). The diagenetic origin of the zeolites, which are conspicuous in many volcaniclastics, has been related in part to weathering (Chap. 10). In general, silica released from the conversion of glass to clays and zeolites forms opal and chalcedonic cements and concretions and silicifies organic debris such as buried wood. Calcium carbonate cements can also be conspicuous in sandstones rich in volcanic debris. Chemical composition of volcaniclastics depends on composition of the source rocks, the amount and kind of nonvolcanic contaminants, and nature of the cementing agents. As a group the chemical composition of the erosional volcaniclastics is probably more variable than that of the pyroclastics (Table 7-3).

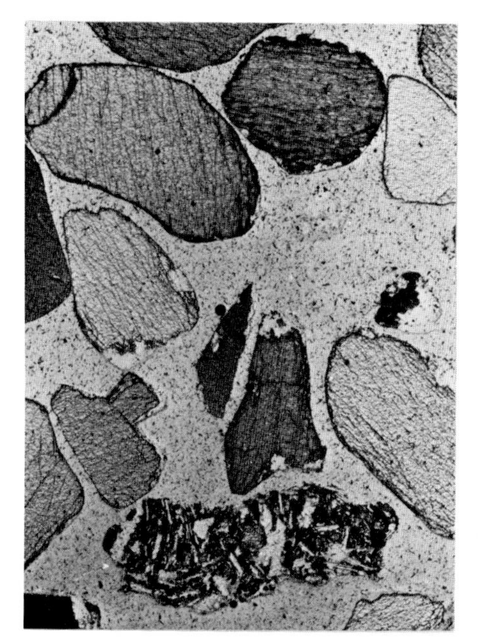
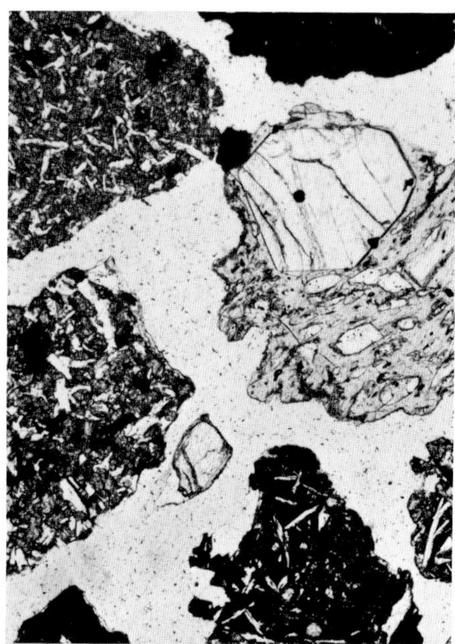

Fig. 7-19. Photomicrographs of olivine green sand from beach (left) and pumice sand of volcanic slopes (right). Pumice sand is source for olivine sand. Differential sorting on beach concentrates olivine (Hawaii, U.S.A.)

Distinguishing among the Volcaniclastic Sands

How successfully can one distinguish the various volcaniclastic sands from one another? For example, how can one distinguish a subaerial ashfall from one that fell in a standing body of water? There are many possibilities and the criteria for distinction have not been fully explored, in part, because sedimentological study is sparse.

Observations of sedimentary structures, bedding, lithologic associations and texture and mineralogy are needed just as they are for the recognition of the environment of terrigenous sandstones. In a slightly different form, the same general properties help to distinguish ash flow from falls (Table 7-4) and work just as well distinguishing ash falls from alluvial deposits derived from them. Thus an erosional basal contact, crossbedding, scour and fill, the probable presence of at least some nonvolcanic debris, and probably poorer sorting should be helpful in recognizing an alluvial deposit derived from an ashfall, the ash fall itself having even, laterally continuous graded bedding that largely parallels its flat or sloping basal contact. In ancient volcaniclastics, associated lithologies commonly play a critical role. For example, Fiske and Matsuda (1964) give such criteria to distinguish submarine ash flows from subaerial ones in a thick Miocene eugeosyncline in Japan. Interbedded volcaniclastic turbidites and a few marine fossils provide an essential part of their argument. The annotated references contain a variety of case histories that illustrate the role of lithologic associations, geometry, structures, basal contact, mineralogy and texture in environmental discrimination.

Provenance

The mineralogic immaturity of volcaniclastic sands makes possible detailed petrographic mapping of the fill of the eugeosynclines and their bordering platforms. Dispersal patterns of volcanic detritus commonly transcend the different depositional environments in a basin so that systematic petrographic mapping is the basis for reconstruction of the tectonic evolution of its source terrains.

Consider, for example, an early Mesozoic or older eugeosyncline whose source terrains can be discerned only incompletely, if at all, from regional geology. As has been emphasized by many, the volume of fill of a eugeosyncline is commonly so great as to suggest that large portions of its source terrain were destroyed during erosion and deposition. Thus even though a granodiorite batholith may now lie in the source region, in former times the same area may well have been mantled by volcanics venting from it. Hence if one can determine, perhaps by paleocurrent mapping and facies studies, that the source region remained geographically fixed, the record of much of its igneous history can be found in inverted order in the fill of the eugeosyncline. Because cycles of igneous activity may span several or more geologic periods, careful petrographic study of the clastic fill of eugeosynclines by sedimentary petrologists can thus contribute to some major geologic problems as, for example, the long term trend of magmatic evolution in island arcs. A good example of this approach via sedimentary petrology is Dickinson's (1969) summary of the evolution of the calcalkaline rocks of parts of the Mesozoic eugeosyncline of western North America. In such a sequence, for example, there can be a progression from early volcanic-rich sandstones to arkoses higher in the section as the volcanic superstructure of a batholith is removed by erosion and granite is exposed. Variations in amount and kind of rock fragments, changing ratios of type of plagioclase, ratio of plagioclase to total feldspar, and total quartz content – all framework elements – are the principal features to be relied on.

The classic view distinguishes three main phases of volcanic activity in orogenic belts. Initially, ophiolitic eruptions plus some mafic and ultramafic intrusions occur early in eugeosynclinal development leading to the widely occurring association of pillow lavas and turbidite graywackes. Palingenesis and uplift follow, producing granitic intrusions at depth plus andesitic and rhyolitic lavas and pyroclastics at the surface. The greater part of this volcanic material may be pyroclastic. The accompanying molasse sedimentation may spill over onto a stable craton as in the Cretaceous of the high plains of North America (Mellon, 1967, Fig. 48) or against a deep trough so that turbidity currents that flow into it may carry much volcaniclastic debris. A postorogenic phase that is the response of the crust to deep seated regional tension permits huge quantities of basaltic sima to be extruded as plateau basalts.

The above sequence is widely accepted, although not without major exceptions, as in western North America where early volcanism is andesitic rather than basic (Dickinson, 1962). And, of course, plateau basalts are not everywhere present to complete the cycle. One should recognize that agreement concerning volcanism and the tectonic cycle is far from complete so that summaries such as those by Aubouin (1965, Chs. 8 and 10) and Maleev (1963) should be considered as progress reports. The revolution in ideas of the genesis and evolution of volcanic and tectonic activity that has come with the ideas of global plate tectonics offers important alternative views to the classical ones given above. The relating of plate convergences and divergences to geosynclinal, tectonic, and igneous evolution is in a state of ferment at this time (Dickinson, 1970). Careful petrographic mapping by sedimentary petrologists can contribute importantly to a better understanding of trends in magmatic evolution in eugeosynclines for only the fill of the eugeosyncline contains the entire record.

Selected Petrographic Descriptions

Judith Fancy Formation (Cretaceous). North Star, north coast of St. Croix, U.S. Virgin Islands.

Geologic Setting: A rapidly subsiding, elongate geosynclinal trough with much volcanism. The formation is approximately 5,000 m thick and consists mostly of tuffaceous sandstone and tuff plus minor mudstone, conglomerate, limestone and lava flows. Graded bedding plus associated slumping and load casts.
Description: A medium-grained ($m_\phi = 1.1$), very poorly sorted ($s_\phi = 2.2$) tuffaceous wacke (Fig. 7-20). Mineralogically very complex and immature. Very poorly rounded, subequant volcanic rock fragments having a microcrystalline groundmass and containing phenocrysts dominantly of plagioclase; orthoclase, and some mafics form nearly half of the framework. A few of the rock fragments show weak flow structure; others appear to be recrystallized from glass. Phenocrysts are also present as individual crystals, some fragmental. Much of the plagioclase is andesine (Ab_{55-45}) with some possibly more sodic (Ab_{85-75}). The chief mafic minerals are several varieties of hornblende plus augite and diopside and together they from some 13 percent of the sandstone. A trace of quartz is present. Dark cryptocrystalline material, probably devitrified glass, forms part of the matrix along with very fine grained primary and secondary minerals. Matrix is conspicuous forming some 16 percent of the sandstone. Replacement of microcrystalline groundmass, matrix and mafic minerals by epidote, chlorite, and probably sericite and clay minerals is widespread. Many feldspars also show alteration. Some secondary magnetite is also present. Be-

cause it is hard to separate volcanic rock fragments from matrix and individual phenocrysts, point counting is difficult.

Interpretation: The ill-sorted, subequant, angular rock fragments, fragmental unabraded phenocrysts, and abundant matrix plus presence of graded bedding and interbedded limestone indicate deposition in a marine basin by turbidity currents. The general absence of pumice and the subrounded and equant rock fragments are strong evidence against deposition as ash fall on land. Instead,

Fig. 7-20. Judith Fancy Formation (Cretaceous), Virgin Islands. Note poor sorting and difficulty of distinguishing detrital matrix from volcanic rock fragments, × 35

turbidity currents carried ash, which may have been initially deposited on a narrow, unstable shelf, into a rapidly subsiding, deepwater basin. Much of the ash appears to have had an andesitic composition. Post-depositional alteration and replacement is extensive and was favored by the abundance of the fine grained unstable minerals of the matrix.

Reference: Whetten, J. T.: Geology of St. Croix, U.S. Virgin Islands. In: Hess, H. H., Ed.: Caribbean geological investigations. Geol. Soc. America Mem. **98**, 179–239 (1966).

Tokiwa Formation (Lower Miocene). Tunnel section, Fujigawa area, Honshu, Japan.

Geologic Setting: A thick (9,000 to 15,000 m), deep-water trough (Fossa Magna) filled with volcaniclastics plus some fossiliferous mudstones. Some of the sand bodies appear to be virtually single graded beds with thicknesses in excess of 30 m. The *Foraminifera* of the associated mudstones suggest water depths between 150 and 500 m, near the top of the continental slope. These mudstones enclose the sands of the sequence.

Description: A very coarse-grained ($m_\phi = 0.50$), very poorly sorted ($s_\phi = 2.4$), pumaceous, dacitic lapilli tuff with devitrified glass and matrix (Fig. 7-21). The euhedral and subhedral, little altered feldspar is mostly plagioclase – some

Fig. 7-21. Photomicrograph of Tokiwa Formation (Miocene), Japan, × 25: Crossed nicols (left) and partially crossed (right). Note large percentage of glass

18 percent. Regardless of size, the plagioclase appears to be unusually homogenous and has composition of An_{44-48}, the Ca-end of andesine. About 20 percent of the feldspar is zoned, some of it strikingly so. Quartz is present in minor amounts as fresh, angular monocrystalline grains. None of the feldspar and quartz shows evidence of rounding. Pumice fragments predominate and together with some devitrified glass form nearly 50 percent of the sand. Secondary minerals include minor amounts of serpentine and siderite, the latter possibly related to outcrop weathering. Chalcedony is also present as a minor cement.

Interpretation: As in many instances, field relations are necessary to maximize information gleaned from thin sections. The thick, massively graded sandstone bodies in a marine section are most simply interpreted as the result of an underwater marine explosion of dacitic lapilli and ash. Even the thickest, graded beds of normal, proximal turbidites do not have thicknesses much exceeding 5 to 8 m. The strongest single supporting thin-section evidencc for the interpretation is predominance of pumice, over 50 percent. The homogeneity of the plagioclase composition also indicates a single magmatic source of the detritus rather than a mixed source that would be characteristic of a turbidity current feeding on a shelf edge with volcanic detritus of mixed origin. Thus the deposit is interpreted as the submarine equivalent of an ash flow – a submarine eruption column that was transformed into a density current.

Reference: Fiske, R. S., Matsuda, Tokihiko: Submarine equivalents of ash flows in the Tokiwa Formation, Japan. Am. Jour. Sci. **262**, 76–106 (1964).

Recent Indurated Beach Sand. Playa del Medano, Tenerife, Canary Islands, Spain.

Regional Setting: Small cove on arid, lee side of 3,713 m strato-volcano, Pico Teide. Short, steep gradient, wet-weather streams, transport detritus to the strand.
Description: Fine-grained ($m_\phi = 2.05$), well-sorted ($s_\phi = 0.48$), weakly cemented, lithic arenitic (Fig. 7-22). Subrounded to rounded volcanic rock fragments form some 80 percent of the framework. Many of these grains have a well defined trachytic texture of elongate feldspars in a fine grained, largely glassy groundmass. Small opaque inclusions, probably magnetite, characterize many grains. About 15 percent of these fragments have an opaque groundmass of basaltic glass. Angular feldspar forms 9 percent of the framework and is about half plagioclase and half sanidine. Type of plagioclase is difficult to determine, although several grains have oligoclase-andesine composition. An-

Fig. 7-22. Photomicrograph of Recent beach sand, Tenerife, Canary Islands, × 100. Notice excellent sorting. Partially crossed nicols

gular quartz is fresh, entirely monocrystalline and has abundance of 5 percent. Although forming only a percent or so, hornblende, augite, and detrital carbonate are conspicuous, most of the carbonate being fossil debris. Cementation is incipient and consists of rims of radiating, birefringent scalenohedrons 4 to 10 microns long, probably calcite. Chalcedony fills some interstices and was deposited after calcite.

Interpretation: The well sorted framework and fossil debris indicate a marine sand and possibly a beach deposit, although the latter could not be inferred from thin section alone. Obviously, the source was almost entirely volcanic and one that included some basalts. Hornblende and augite are consistent with this view. Monocrystalline angular, strain free quartz also implies a felsic volcanic source. And what of the plagioclase composition? Because this sand is well sorted on a beach and contains both quartz and scoria, use of plagioclase composition to determine types of volcanic terrain seems unwise. In reality, rocks from the active volcanoes of Tenerife range from hyaloandesites and hyalotrachytes to olivine basalts.

Reference: Van Padang, Neumann, M. and others: Catalogue of the active volcanoes of the world and solfatara fields, Part 21, Atlantic Ocean: Naples and Rome, International Association of Volcanology, 128 p. (1967).

Glossary

Included here are a few of the many terms that are used to describe and interpret volcaniclastics and some of their associated rocks.

Agglomerate: A deposit composed of bombs (material larger than 64 mm) that largely solidified in flight.

Ash fall: Volcanic debris deposited from air on land or in water by gravity.

Ash flow: A turbulent mixture of gas and pyroclastic materials at high temperature that travels swiftly outward and downslope, having been explosively ejected from a crater or fissure. Same as a pyroclastic flow (after Ross and Smith, 1961, p. 3).

Ash flow tuff: Indurated deposit of ash flow.

Axiolitic structure: A fine, linear intergrowth of cristobalite and feldspar that develops from the devitrification of glass shards. Long axes of crystals perpendicular to shard boundary. According to Ross and Smith (1961, p. 37) axiolitic structure has been observed only in ash flows.

Bomb: Molten material larger than 64 mm ejected from a volcano.

Debris flow: Loosely speaking, a synonym for mud flow but restricted by some to describe rapid flowage of a high density sediment-water mixture. Generally, contains a large proportion of coarse fragments.

Fluidal texture: Flow structure in an aphanitic rock.

Hyalopilitic texture: A meshwork of randomly oriented microlites with interstitial glass.

Ignimbrite: A tuffaceous rock of acidic composition resulting from a nuée ardente type explosion – a "fiery rain cloud" (Marshall, 1935, p. 38).

Krakatoan caldera: A collapsed crater formed by a great and rapid outpouring of pumice-fall and especially pumice-flow deposits.

Lahar: Volcanic mud flow composed of volcanic debris.

Lapilli: Pyroclastic material between 2 and 64 mm.

Nuée ardente: Explosive ejection from volcano or rift produces overriding dust clouds plus a dense, basal gravity flow deposit of high temperature (Lacroix, 1903, p. 442–443).

Ophiolite: Includes a wide suite of dark greenish, mafic extrusive rocks. Generally same as greenstone. Most commonly formed by submarine eruption and associated with radiolarian chert.

Perlitic structure: Concentric, onion-like partings in glass produced by hydration and exfoliation of obsidian (Ross and Smith, 1955, p. 1.1, Fig. b).

Pilotaxitic texture: The groundmass of a volcanic rock with a meshwork of weakly oriented microlites. Glass may be present between the crystal meshwork.
Pumice: Highly vesicular volcanic glass. Porosity commonly exceeds 50 percent. Loosely speaking, a froth of volcanic glass.
Pumice-flow: A pyroclastic flow composed of pumice. Virtually the same as ash flow of some authors.
Pyroclastic flow: Includes ash and sand flows.
Scoria: Pumice of basic composition and thus having a very dark brown or black color.
Sideromelane: Basaltic glass.
Tephra: Volcanic material of any size transported from a crater through the air (Thorarinsson, 1955, p. 12).
Trachytic texture: Flow structure defined by elongate grains of alkali feldspar in a fine grained groundmass.
Tuff-breccia: A breccia of volcanic origin with abundant tuff matrix.
Vitroclastic texture: Pertains to typical fragmental structure of glass-rich volcanic rocks, those rich in glass shards.
Welded tuff: One that accumulated sufficiently rapidly so that its heat was not dissipated and as a result the debris became plastic enough to be welded together or in some cases to melt and exhibit various types of flow structure.

References

Anderson, C. A.: Tuscan formation California, with a discussion concerning the origin of volcanic breccias. Univ. California Pubs. Geol. Sci. **23**, 215–276 (1933).
Aramaki, Shiego: Classification of pyroclastic flows (in Japanese). Bull. Volcanol. Soc. Japan **1**, 47–57 (1957); Int. Geol. Rev. **3**, 518–524 (1961).
Aubouin, Jean: Geosynclines. In: Developments in tectonics, Vol. 1, 335 p. Amsterdam: Elsevier 1965.
Crook, K. A. W.: Petrology of Parry Group, Upper Devonian-Lower Carboniferous, Tamworth-Nundle District, New South Wales. Jour. Sed. Petrology **30**, 538–552 (1960).
Dickinson, W. R.: Petrology and diagenesis of Jurassic andesitic strata in central Oregon. Am. Jour. Sci. **200**, 481–500 (1962).
— Evolution of calc-alkaline rocks in the geosynclinal system of California and Oregon. In: McBirney, A. B., Ed.: Andesite Conf. Proc. Oregon Dept. Geol. Mineral Ind. Bull. **65**, 151–156 (1969).
— Global tectonics. Science **168**, 1250–1256 (1970).
Dott, R. H., Jr.: Eocene deltaic sedimentation at Coos Bay Oregon. Jour. Geology **74**, 373–420 (1966).
Edwards, A. B.: The petrology of the Miocene sediments of the Aure Trough, Papua. Royal Soc. Victoria Proc. **60**, 123–148 (1947).
Ewart, A.: Petrography and petrogenesis of the Quaternary pumice ash in the Taupo area, New Zealand. Jour. Petrology **4**, 392–431 (1963).
Fisher, R. V.: Proposed classification of volcaniclastic sediments and rocks. Geol. Soc. America Bull. **72**, 1409–1414 (1961).
— Bubble-wall texture and its significance. Jour. Sed. Petrology **33**, 224–227 (1963).
— Maximum size, median diameter and sorting of tephra. Jour. Geophys. Research **69**, 341–355 (1964).
— Rocks composed of volcanic fragments and their classification. Earth Sci. Reviews **1**, 287–298 (1966a).
— Geology of a Miocene ignimbrite layer, John Day Formation, eastern Oregon. Univ. California Pubs. Geol. Sci. **67**, 58 p. (1966b).
— Waters, A. C.: Bed forms in base-surge deposits: Lunar implications. Science **165**, 1349–1352 (1969).
Fiske, R. S., Matsuda, Tokohiko: Submarine equivalents of ash flows in the Tokiwa Formation, Japan. Am. Jour. Sci. **262**, 76–106 (1964).
Hay, R. L.: Origin and weathering of Late Pleistocene ash deposits on St. Vincent, B.W.I. Jour. Geology **67**, 65–87 (1959).
Heinrich, E. W.: Microscopic identification of minerals, 414 p. New York: McGraw-Hill 1965.
Huber, N. K., Reinhart, C. D.: Some relationships between refractive index of fused glass beads and the petrologic affinity of volcanic rock suites. Geol. Soc. America Bull. **77**, 101–110 (1966).

Isacks, B., Oliver, J., Sykes, L. R.: Seismology and the new global tectonics. Jour. Geophys. Research **73**, 5855–5899 (1968).
Jenks, W. F., Goldich, S. S.: Rhyolitic tuff flows in southern Peru. Jour. Geology **64**, 156–172 (1950).
Jung, J., Brousse, R.: Classification modale des roches eruptives, 122 p. Paris: Masson 1959.
Katsui, Yoshio: Evolution and magmatic history of some Krakatoan calderas in Hokkaido, Japan. Jour. Fac. Sci., Hokkaido Univ., Ser. 4, **11**, 631–650 (1963).
Lacroix, A.: L'eruption de la montagne Pelée en janvier 1903. Acad. Sci. (Paris) Comptes Rendus **136**, 442–445 (1903).
Lipman, P. W., Christiansen, R. L.: Zonal features of an ash-flow sheet in the Piapi Canyon Formation, Southern Nevada. U.S. Geol. Survey Prof. Paper **501**-B, 74–78 (1964).
Maleev, E. F.: Razvitiye tipov vulkanizma na primere Vostochnykh Karpat. Doklady Acad. Nauk. SSSR **148**, 1374–1377 (1963) [Evolution of volcanism in the Eastern Carpathians. USSR Acad. Sci. Proc., Earth Sci. Sec. **148**, 84–86 (1964)].
Marshall, P.: Acid rocks of Taupo-Rotorua volcanic district. Royal Soc. New Zealand Trans. **64**, 323–366 (1935).
McBride, E. F., Lindemaren, W. L., Freeman, P. S.: Lithology and petrology of the Gueydan (Catahoula) Formation in south Texas. Univ. Texas Bur. Econ. Geol. Rept. Inv. **63**, 122 p. (1968).
Mellon, G. B.: Stratigraphy and petrology of the lower Cretaceous Blairmore and Mannville Groups, Alberta foothills and plains. Research Council Alberta Bull. **21**, 270 p. (1967).
Menard, H. W.: Marine geology of the Pacific, 271 p. New York: McGraw-Hill 1964.
Mizutani, Shinjiro: Clastic plagioclase in Permian graywacke from the Mugi area, Gifu Prefecture, Central Japan. Nagoya Univ. Jour. Earth Sci. **7**, 108–136 (1959).
Moorhouse, W. W.: The study of rocks in thin section, 514 p. New York: Harper 1959.
Moore, J. G.: Base surge in recent volcanic eruptions. Bull. volcanol., Ser. 2, **30**, 337–363 (1967).
Musai, Isamu: A study of the textural characteristic of pyroclastic flow deposits in Japan. Earthquake Research Inst. Bull. **39**, 133–248 (1961).
Nelson, C. H., Kulm, L. D., Carlson, P. R., Duncan, J. R.: Mazama ash in the northeastern Pacific. Science **161**, 47–49 (1968).
Rittmann, A.: Volcanoes and their activity, 2nd Ed., 304 p. New York: Interscience 1962.
Ross, C. S.: Provenance of pyroclastic materials. Geol. Soc. America Bull. **66**, 427–434 (1955).
— Smith, R. L.: Water and other volatiles in volcanic glass. Am. Mineralogist **40**, 1071–1089 (1955).
— — Ash-flow tuffs – their origin, geologic relations and identification. U.S. Geol. Survey Prof. Paper **366**, 81 p. (1961).
Sandberg, A. E.: Section across Keweenawan lavas at Duluth, Minnesota. Geol. Soc. America Bull. **49**, 795–830 (1938).
Scheidegger, A. E., Potter, P. E.: Textural studies of graded bedding. Sedimentology **11**, 163–170 (1968).
Schmincke, Hans-Ulrich: Graded lahars in the type sections of the Ellensburg Formation, south-central Washington. Jour. Sed. Petrology **37**, 438–448 (1967).
— Swanson, D. A.: Laminar viscous flowage structures in ash-flow tuffs from Gran Canaria, Canary Islands. Jour. Geology **75**, 641–664 (1967).
Streckeisen, A. L.: Classification and nomenclature of igneous rocks (final report of an inquiry). Neues Jahrb. Mineral. Abh. **107**, 144–214 (1967).
Thompson, B. N., Kermode, L. O., Ewart, A.: New Zealand volcanology, central volcanic region. New Zealand Geol. Survey Handbook, Inf. Ser. 50, **1965**, 211 p.
Thorarinsson, S.: Discussions. Bull. volcanol., Sect. 2, **16**, 11–13 (1955).
Verhoogen, J.: Volcanic heat. Am. Jour. Sci. **244**, 745–771 (1946).
Waldron, H. H.: Debris flow and erosion control problems caused by the ash eruptions of Irazu Volcano, Costa Rica. U.S. Geol. Survey Bull. **1241-I**, 37 p. (1967).
Waters, A. C.: Determining direction of flow in basalts. Am. Jour. Sci. **258**, 350–366 (1960).
Whetten, J. T.: Carboniferous glacial rocks from Werrie Basin, New South Wales Australia. Geol. Soc. America Bull. **76**, 43–56 (1965).
— Sediments from the lower Columbia River and the origin of graywacke. Science **152**, 1057–1058 (1966).
Williams, Howel, Turner, F. J., Gilbert, C. M.: Petrography, 406 p. San Francisco: Freeman 1954.
Yagi, Kenzo: Welded tuffs and related pyroclastic deposits in northeastern Japan. Bull. Volcanol. Ser. 2, **24**, 109–128 (1962).

Annotated References

Aubouin, Jean: Geosynclines. In: Developments in tectonics, Vol. 1, 335 p. Amsterdam: Elsevier 1965.

Chapter 10, igneous activity and the geosynclinal concept, contains a good statement of European ideas. Compare with Dickinson (1962) for a differing role for volcanics in the tectonic cycle.

Cook, E. F.: Ignimbrite bibliography and review. Idaho Bur. Mines and Geology Inf. Circ. **13**, 63 p. (1962).

Historical outline and nomenclature plus stratigraphic and structural uses of ignimbrites. Over 500 references.

— (Ed.): Tufflavas and ignimbrites, 212 p. New York: Elsevier 1966.

Thirty Russian papers organized into three parts: general problems, tufflavas and ignimbrites in the Soviet Union, and practical uses.

Crook, K. A. W.: Petrology of Parry Group, Upper Devonian-Lower Carboniferous, Tamworth-Nundle District, New South Wales. Jour. Sed. Petrology **30**, 538–552 (1960).

Careful petrographic analyses and informative illustrations.

Dickinson, W. R.: Petrogenetic significance of geosynclinal andesitic volcanism along Pacific margin of North America. Geol. Soc. America Bull. **73**, 1241–1256 (1962).

Good statement of implications of geosynclinal andesites – the counter example to the Alps.

Dott, R. H., Jr.: Eocene deltaic sedimentation at Coos Bay, Oregon. Jour. Geology **74**, 373–420 (1966).

Stratigraphy, sedimentary structures and careful petrography of sediments in a volcanic milieu. Good photomicrographs.

Eaton, G. P.: Windborne volcanic ash. A possible index to polar wandering. Jour. Geology **72**, 1–35 (1964).

The best summary of thickness patterns of ash and the factors that control it.

Edwards, A. B.: The petrology of the Miocene sediments of the Aure Trough, Papua. Royal Soc. Victoria Proc. **60**, 123–148 (1947).

Careful petrography plus chemical analyses of the turbidite fill of a eugeosyncline.

Ewart, A.: Petrology and petrogenesis of the Quaternary pumice ash in the Taupo area, New Zealand. Jour. Petrology **4**, 392–431 (1963).

Detailed petrography of pumice plus excellent illustrations.

Green, Jack, Short, N. M., Eds.: Volcanic land-forms and surface features – a photographic atlas and glossary. 522 p. New York: Springer-Verlag 1971.

Nearly 200 multiple plates (about 420 separate photographs). Includes comprehensive glossary of terms dealing with volcanic processes and products.

International Association of Volcanology: 1951 to present, Catalogue of the active volcanoes of the world including solfatara fields. Naples and Rome.

Nineteen volumes to date. Describes topography, petrology and composition plus periods of activity. Good to use as "modern sediment" reference material to obtain an insight into a volcanic terrain of the past.

Ishii, Jiro: The weathering processes of volcanic ash and pumice deposits in Hokkaido. Jour. Fac. Sci. Hokkaido Univ. Ser. 4, **11**, 545–569 (1963).

Relates length of weathering to mineralogical changes using X-ray diffraction, differential thermal analysis, chemical composition, and electron microscopy.

Jenks, W. F., Goldich, S. S.: Rhyolitic tuff flows in southern Peru. Jour. Geology **64**, 156–172 (1956).

Unwelded Pleistocene ash flows of rhyolitic composition in the high Andes. Five chemical analyses plus a map showing flow direction. Flows up to 20 miles.

Katsui, Yoshio: Evolution and magmatic history of some Krakatoan calderas in Hokkaido, Japan. Jour. Fac. Sci. Hokkaido Univ. Ser. 4, **11**, 631–650 (1963).

Maps distribution of ash fall and pumice flow of Pleistocene calderas plus 12 chemical analyses. Excellent case history.

Losacco, U., Parea, G. C.: Saggio di un atlante di strutture sedimentaire e postsedimentaire osservate nelle piroclastiti de Lazio. Atti. Soc. Nat. Mat. Modena **99**, 30 p. (1969).

Structures of bedded tuffs organized into depositional (accumulation versus erosion) and postdepositional groups. Thirty references and 104 fine photographs.

McBride, E. F., Lindemaren, W. L., Freeman, P. S.: Lithology and petrology of the Gueydan (Catahoula) Formation in south Texas. Univ. Texas Bur. Econ. Geol. Rept. Inv. **63**, 122 p. (1968).

Directional structures, petrology and diagenesis of tuffaceous sands and clays. Well integrated with good photomicrographs.

Maleev, E. F.: Vulkanoklasticheskie gornye porody (Volcaniclastic rocks), 168 p. Moscow: Gosudar Nauch.-Tekh. Izd. Lit. Geol. i Okhrane Nedr. 1963a.

One of the few books devoted solely to volcaniclastic rocks: texture, structure, source and facies.

— Razvitiye tipov vulkanizma na primere Vostochnyk Karpat. Dokl. Akad. Nauk. SSSR. **148**, 1374–1377 (1963b) [Evolution of volcanism in the Eastern Carpathians. USSR Acad. Sci. Proc., Earth Sci. Sec. **148**, 84–86 (1963)].

Table 1 nicely summarizes the author's views, which he suggests may have some general application.

Pirrson, L. V.: The microscopical characters of volcanic tuffs – a study for students. Am. Jour. Sci., Ser. 4, **40**, 191–211 (1915).

One of the early papers and still worth reading.

Rittmann, A.: Volcanoes and their activity, 2nd Ed., 304 p. New York: Interscience 1962.

Ten chapters covering almost all aspects, but lacks a bibliography. Mostly European examples. Has extended treatment of volcanism and its relation to orogenesis and the mantle. Translated from the German.

Ross, C. S., Smith, R. L.: Ash-flow tuffs – their origin, geologic relations and identification. U.S. Geol. Survey Prof. Paper **366**, 81 p. (1961).

A good summary and helpful glossary plus 82 informative photomicrographs.

Semenenko, N. P., Boyko, V. L., Bordunov, I. N.: Geologiia osadochno-vulkano gennykh formatsii ukrainskogo shchita (The geology of the volcaniclastic formations of the Ukranian Shield), 378 p. Kiev: Izdatel'stvo "Naukova Dumka"

A case history, the third of 3 volumes, of a region containing some highly altered volcaniclastics.

Smith, R. L.: Ash flows. Geol. Soc. America Bull. **71**, 795–842 (1960).

Extended comprehensive review. A good place to start.

Snavely, P. D., Wagner, H. C.: Tertiary geologic history of western Washington and Oregon. Washington Div. Mines and Geology, Dept. Inv. **22**, 25 p. (1963).

Good paleogeographic maps and cross sections of an area with extensive volcanism.

Thompson, B. N., Kermode, L. D., Ewart, A.: New Zealand volcanology, central volcanic region. New Zealand Geol. Survey Handbook, Inf. Ser. **50**, 211 p. (1965).

Thirty-five short, informative articles containing numerous chemical and modal analyses with emphasis on pyroclastics. Well illustrated.

Waldron, H. H.: Debris flow and erosion control problems caused by the ash eruptions of Trazu volcano, Costa Rica. U.S. Geol. Survey Bull. **1241**-I, 37 p. (1967).

Excellent description of accelerated erosion and sedimentation resulting from ash falls. Compare with McBride and others (1968).

Williams, Howel, McBirney, A. R.: Volcanic history of Honduras. Univ. California Pubs. Geol. Sci. **85**, 101 p. (1969).

A systematic discussion of all rock units, starting with the metamorphic basement, and a particularly long section on ignimbrites and Quaternary volcanism. Six appendices on microscopic petrography. Good line drawings of photomicrographs.

Yagi, Kenzo: Welded tuffs and related pyroclastic deposits in north-eastern Japan. Bull. volcanol. Ser. 2, **24**, 109–128 (1962).

Concise description of petrography and 21 chemical analyses plus comments on origin. Excellent.

Part III

Processes that Form Sand and Sandstone

Having established a fundamental petrographic basis for the study of sand and sandstone, we now examine the subject largely from a process point of view – where does sand come from, what is known of the hydraulics of its transport, deposition and soft sediment deformation, and finally how is sand transformed into a lithified sandstone? Unlike Parts I and II, which are mostly geological, the three chapters of Part III draw upon fluid mechanics, engineering hydraulics and solution chemistry and thermodynamics. Perhaps better than any other portion of the book, Part III shows that the search for fundamental explanations in geology requires knowledge well beyond the domain of geology itself. We hope that Part III will interest sedimentologists to look deeper into neighboring fields in their search for ever more fundamental explanations. We should also be very pleased if Part III did, in fact, attract interested scientists in other fields to some of the problems associated with sand and sandstone as we see them.

Chapter 8. Production and Provenance of Sand

Introduction

In this chapter we consider two problems. One has to do with how sand is formed and the other deals with the source or place where sand is produced. In the first, we are concerned with the processes involved in the production of sand, how sand grains come into being. In the second, we wish to know what the source rocks were and where the source area was. It is presumed that a sand body or formation itself contains much of the information needed to answer these questions. We might ask about each and every sand grain – how did it form and where did it come from? This in a nutshell is the theme of this chapter.

The question of provenance is important to us for several reasons. A solution of the problem of provenance will immeasurably increase our understanding of the paleogeography of a region, enabling us to locate and identify possible source lands. We may also be able to trace the movement of materials and thus say something about paleocurrents and paleoslope.

Provenance has some practical aspects also. The study of the mineralogy of modern river and beach sands may disclose valuable minerals, such as chromite and gold, and if provenance is understood, we may know from whence the minerals came and so may prospect more intelligently. Minerals alien to a particular sand, or grains which have been dyed or made radioactive may be added to the sand as tracers in order to follow its movement (Baker, 1956, p. 182; Ingle, 1966, p. 15; Kennedy and Kouba, 1970). This is of interest in the study of shore drift, knowledge of which is of engineering importance. In this case the source is known and the transport pattern determined by systematic sampling. It is conceivable, also, that provenance and porosity of sands may be related. Sands derived from granitic sources may have a higher porosity than those rich in argillaceous rock fragments which tend to be squashed and to flow into pores forming a barrier to fluid movement and reducing storage capacity.

How Sand is Formed

In the broadest sense, sand is any non-coherent, granular material within some generally accepted size limits. Such materials are capable of being transported by currents, sorted and stratified, crossbedded and graded both laterally and vertically. Materials meeting this definition are of widely diverse origins (Fig. 8-1). Included here are detrital quartz and feldspar, the basaltic dune sands of Moses Lake, Washington, the gypsum dune sands of the Persian Gulf and of White Sands, New Mexico, the greensands rich in glauconite, the calcareous oolitic sands such as those in the Bahamas, calcarenites composed of fossil debris, and even the "clay dunes" of Corpus Christi, Texas. Although all these fit the size definition of sand and are mechanically transported and emplaced in the fabric of the rock as sand grains, they originate in radically different ways and are not all equally abundant in the geologic record.

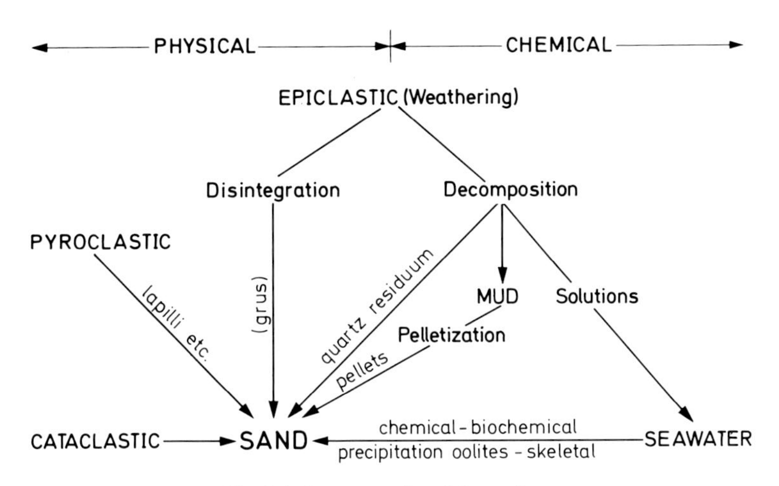

Fig. 8-1. Processes of sand formation

There are, perhaps, five basic processes that lead to the formation and release of sand-sized grains. These are (1) weathering, including both disintegration and decomposition, (2) explosive volcanism (pyroclastic), (3) crushing, both by rock movements (cataclastic) and impact, (4) pelletization, and (5) precipitation from solution, both chemical and biochemical. Those which relate to weathering have been termed *epiclastic*. One such process is primarily rock disintegration without much, if any decomposition. Under certain conditions there is a loosening of the grains in a rock, by changes in temperature, frost-action, or hydration so that the rock disintegrates into a loose granular mass. Such destruction of the coarser-grained plutonic rocks produces a residual sand or "grus."

Another process is rock decomposition which may also produce sand – or at least yield a significant sand fraction – as a product of this decomposition. This is probably the origin of the bulk of the quartz sands released by the decomposition of the quartz-bearing plutonic rocks. Most of the rock, mainly feldspar, is converted to fine-grained, clay-sized material from which the inert, undecomposed quartz grains, of sand size, are liberated and concentrated by the movement of fluids.

Sand-sized materials may be produced by crushing action but not by ordinary abrasion. As shown by experimental work, the abrasion of pebbles produces silt and not sand (Marshall, 1927). Moreover, as Kuenen (1959, p. 2) has pointed out, the average pebble is one million times larger than the average sand grain and a freighter load of gravel if worn down to sand size would yield only a tumbler full of sand. Crushing or the shattering of grains, however, may in some cases produce a significant volume of sand. The explosive action of volcanoes yields vast quantities of sand-sized debris: glass, crystal fragments, and lava particles. Likewise the explosion accompanying meteor impact will shatter rocks. Although the fall-back materials which fill an impact crater are often coarse, a considerable part is sand-sized. Earth movements may crush rocks and even though the sand yield is small and not likely to form a significant deposit, glacial crushing, a special case of cataclastic action, does produce a considerable body of sand-sized material. In short, although both pyroclastic and cataclastic action may be of local importance in the production of sand, such action is dwarfed in importance by the more common but less spectacular action of normal weathering.

Much sand-sized material is also produced by diagenetic action and by chemical and biochemical precipitation. Sands thus formed have been termed *endogenetic* (Grabau, 1904; 1913, p. 271). They are generated *within* the basin of sedimentation and, unlike the epiclastic sands, are not the products of the wastage of a land mass. Some of these intrabasinal sands are built up from finer materials – muds which are both pelitic and micritic. Included here are the sand-sized pellets produced by organisms and the pellets of other origins such as those blown into "clay dunes" (Huffman and Price, 1949).

Biochemical and chemical precipitation forms skeletal and oolitic sands that may form a significant deposit. Although intrabasinal carbonate sands, of whatever origin, may form sizeable deposits, they are usually included with the limestones rather than the sandstones. For this reason the problems of their generation are omitted. Likewise, the sands of volcanic origin form a special group, the problems of which are quite different from the common sands of epiclastic origin. We recognize, of course, that the epiclastic sands may be mingled in all proportions with these sands of a very different origin. In some cases, the problem becomes one of recognizing the carbonate or volcanic contributions as being intrabasinal or pyroclastic rather than being derived, as an epiclastic sand, by the breakdown of older carbonate or volcanic rocks.

What is the mechanical composition of the material fed into the sedimentary mill? There are few data which directly answer this question. On the other hand, some studies suggest that all possible sizes are not equally abundant. There seems to be a dearth of certain size grades and higher than normal concentrations of others. This has been interpreted as a reflection of initial abundance and, therefore, a question of provenance. The abundance by weight of grains in the 1 to 8 mm range (especially the 2 to 4 mm grade) seems to be relatively small (Pettijohn, 1940, Hough, 1942, p. 26; Krumbein, 1942, p. 9; Schlee, 1957, p. 1379). This deficiency has also been explained by differences in the competency of the several modes of transport (Wentworth, 1933; Sundborg, 1956, p. 191). Although sorting is effective in separating sizes, it can only result in concentration of particular sizes in different environments. Perhaps mechanical instability in the grains in the 1 to 4 mm range, in which the component mineral grains are large relative to the whole fragment, may be responsible for the destruction of these grades. A considerable quantity of quartz within these grades is produced by rock disintegration (Blatt, 1967a) but such quartz is commonly polycrystalline or is marked by many incipient fractures which likewise may lead to its early destruction during transport.

Russell (1968) rejected these concepts and thought, instead that the deficiency of very coarse sand and fine gravel in alluvial and in shallow-water deposits was due to a sorting process which concentrated such materials on beaches. He cited many examples of marked concentration of grains in the 1 to 6 mm range on beaches and pointed out that this observation may prove significant in the study of older sandstones and finer conglomerates.

There also appears to be a deficiency at the other end of the sand range. The coarser silt grades seem not to be as abundant as either sand or the finer silts and clays (Udden, 1914, Table p. 741; Pettijohn, 1940; Wolff, 1964). Although this gap has been attributed to the mode of transport, it is more likely that rock decomposition yields primarily two classes of material: (1) sand and (2) clay with little material in the coarse silt or fine sand grades. This concept has been designated the "Sorby Principle" (Folk and Robles, 1964, p. 287) inasmuch as Sorby was perhaps the first to recognize the relation between sediment grain size and the micro- and macrostructure of the source rocks, be they terrigenous or biogenetic.

The Problem of Provenance

Definitions and Concepts

The terms *provenance* and *provenience*, both derived from the French "provenir", meaning to originate or to come forth, have been used to encompass all the factors relating to the production or "birth" of the sediment. For a given sand, for example, we might ask from what kind of source rock (or rocks) was the sand derived? What was the relief and climate in the source area? How far and in what direction did the source area lie? What was its size? These are pertinent questions which must be answered if we are to achieve a complete paleogeographic reconstruction delineating not only the basin of deposition but also the source lands.

What kind of observations must we make to find an answer to our questions? The nature and character of the source land can be ascertained from a study of the composition of the deposit itself. Sands, more than any other sediment, lend themselves to this approach. The constituent detrital minerals and rock particles are determined by the nature of the source rocks. Some minerals are characteristic of a particular kind or class of source rocks – kyanite, for example, being indicative of a metamorphic source. Others, such as quartz, are ubiquitous but even these display varietal characteristics indicative of a particular source – volcanic quartz, for example. Hence individual mineral species or varieties, or better still, mineral suites, are important source-rock clues. Rock particles, present in many but not all sands, are invaluable aids in deciphering provenance.

Estimation of relief and climate is more difficult and is dependent on a knowledge of mineral stability – especially resistance to chemical change. Only the most stable species will survive the intense weathering promulgated by high rainfall and high temperatures. The problem is very complex as relief and climate may work in opposite directions – high relief leading to rapid erosion and removal of incompletely weathered materials even under conditions of intense chemical decay.

In addition to the "internal" evidence, that is, evidence furnished by the constituent grains of the sand, we may use other evidence in our analysis of provenance. We may utilize information gained from stratigraphy and facies studies to indicate or rule out possible source areas. A general knowledge of paleogeography and the tectonic elements of a region will provide clues to provenance. The question of provenance is inextricably bound up with tectonics. Positive or tectonically active areas are the source areas. These may shift from time to time or vary in importance with time. We need to know, therefore, not only what the source rocks were but where the potential source rocks lay. How much of the filling of a geosyncline, for example, is derived from the cratonic side? How much is furnished by an orogenic welt or tectonic land? Paleocurrent studies may also contribute in a significant way to provenance analysis. Paleocurrents may, for example, indicate the direction in which the source land lies. The situation is less complex if the sands are alluvial than if they are marine. It is clear, therefore, that we need to know a great deal more than just the mineralogy of a sand to determine its provenance.

The question of provenance is one of the most difficult the sedimentary petrographer is called on to solve. It is fraught with pitfalls due to the fact that sands are recycled – are derived from pre-existing sands which themselves had a provenance as well as a history of transport and rounding. Moreover, the source areas may be multiple, and may have changed with time. The record is further obscured by the complex interplay of relief and climate, the lack of simple relation between the recorded transport direction at the site of deposition and the overall transport path from source to depositional site, and by the effects of diagenesis – particularly intrastratal solution. Blatt (1967b) has written a thoughtful analysis of recycling of sediments and related problems.

It is clear that both the light and heavy minerals of a sand are important in the study of provenance. In fact, it is here, perhaps, that the heavy mineral studies make their principal contribution.

Evidence from Detrital Components

We will first consider the usefulness of individual minerals, especially the varietal characters of the "light" minerals in source rock determination and then discuss the role of mineral suites, especially of heavy minerals. It is with the sands of epiclastic origin – the products of rock disintegration and decomposition – that we are mainly concerned. Such sands are "washed" residues – the transported and sorted products of rock decomposition and disintegration. The character of the source rocks will limit or control what is put into the system. We need to know what kind of materials potential source rocks might be expected to furnish. Volcanic or plutonic quartz? Volume and types of polycrystalline quartz? Heavy mineral suites distinctive of source rock types? Let us review what is known about these and related topics as a guide to provenance.

Detrital Quartz as a Guide to Provenance: Quartz in sands and sandstones may be either mono- or polycrystalline and may be derived from either plutonic and volcanic igneous rocks, metamorphic rocks, or sedimentary rocks (Table 8-1). By *monocrystalline* or *unitary quartz* is meant a grain derived from

Table 8-1. *Genetic classification of monocrystalline and polycrystalline quartz*

Monocrystalline quartz

- Plutonic quartz (derived from plutonic igneous and metamorphic rocks and reworked from pre-existing sandstones)
- Volcanic quartz (derived from phenocrysts of acid volcanic rocks both flows and pyroclastics)

Polycrystalline quartz

- Metamorphic and igneous polycrystalline quartz
  - Polygonized quartz (product of "static annealling")
  - Sutured quartz (product of "cold-rolling")
- Quartzites (principally *in situ* cementation)
- Cherts (crystallized and recrystallized opaline or chalcedonic rocks)

a single crystal; by *polycrystalline* or *composite quartz* is meant a grain consisting of two or more quartz crystals. Because quartz sand is the principal product of rock disintegration and decomposition and is the dominant constituent of most sands, a number of attempts have been made to utilize quartz as a guide to provenance. Inclusions, grain shape and extinction pattern are the essential variables. Among the earliest were those of Sorby (1880) and Mackie (1899). Both thought that the inclusions in the quartz were a reliable guide to the source rock from which the quartz came. A more recent elaborate attempt is that of Krynine (1940, 1946a). Krynine's approach is based on grain shape, the character of the inclusions, and the extinction. Quartz was classified as (1) *igneous*, including plutonic, volcanic, and hydrothermal quartz, (2) *metamorphic*, including both pressure and injection quartz, and (3) *sedimentary*, including overgrowths, new crystals, and vein and vug fillings. Krynine's classification is largely deductive. Because of the difficulty of applying the criteria formulated by Krynine and because of our lack of knowledge about the quartz in potential source rocks, Krynine's classification is not satisfactory. Other studies on the quartz of both source rocks and sediments have been published by Mackie (1899), Dake (1921, p. 154–163), Feniak (1944), Keller and Littlefield (1950), Moss (1966), and Blatt (1967).

In general, attention was directed to discriminating between igneous (plutonic) and metamorphic origins of common monocrystalline quartz. It was presumed by Mackie and others that a discrimination could be made based on inclusions, shape, and extinction (undulatory or not). Many workers have found that these criteria are usually difficult to apply (for example, Bokman, 1952), in part because such attributes as shape and extinction show wide variations in the same rock and an assessment of them is very subjective, and in part also because a critical look at the quartz of source rocks shows that the presumed differences in inclusions, shape, and extinction either do not exist or that there is too wide a range of variations and overlap between the quartz of source rock types. Although there may be differences in the statistical average for the several source rock types, it is commonly impossible to assign specific grains in a sandstone to one or another source.

Even though one cannot distinguish with confidence the origins of specific grains of most monocrystalline plutonic quartz, one can frequently identify volcanic quartz. The quartz derived from volcanic rocks – notably the quartz porphyries – may find its way into sediments although only locally is it apt to be important. Volcanic quartz is essentially strain-free, commonly rounded or embayed, the embayments being filled in some cases with the groundmass of the porphyry from which the quartz came. Volcanic quartz also exhibits euhedral form, the grain displaying straight borders of well-developed hexagonal dipyramids without prism faces. Although volcanic quartz is relatively rare, its appearance suggests a volcanic provenance and is likely to be associated with felsic volcanic rock fragments. Modern sands of rhyolitic derivation may contain upwards of ten percent quartz (Webb and Potter, 1971). Volcanic quartz has been recognized in several Eocene formations in the coastal plain of Texas (Todd and Folk, 1957, p. 2550).

Many recent investigations have emphasized the usefulness of polycrystalline or composite quartz, that is, those grains composed of more than one crystal unit (Blatt and Christie, 1963; Blatt, 1967a; Conolly, 1965; Voll, 1960).

Blatt (1967), as a result of his study of a series of samples of weathered and disintegrated quartz-bearing igneous rocks, gneisses, and schists, attempted to set up criteria for assigning quartz grains of sandstones to one or another of these sources. He relied primarily on the ratio of monocrystalline to polycrystalline quartz and on the kinds of polycrystalline quartz present. He noted that for a given size grain, the polycrystalline quartz of gneisses is more finely crystalline than that of massive igneous rocks. Polycrystalline quartz showing two distinctly different sizes of crystals within a single grain is diagnostic of metamorphic quartz. A high ratio of polycrystalline quartz to total quartz also suggests a metamorphic source.

A somewhat different approach to the interpretation of polycrystalline quartz has been given by Voll (1960, p. 536), who noted that polycrystalline quartz of metamorphic origin is of two types: (1) *polygonized quartz*, in which the component grains form polygonal units, with straight boundaries, which tend to meet at 120 degree angles, and (2) *polycrystalline quartz*, the components of which exhibit sutured boundaries. The former is believed to have formed by "static annealing" of highly strained quartz whereas the latter is a product of "cold working." As Voll points out, these varieties may characterize differing source areas, as in the Scottish Highlands where polygonized quartz characterizes the Western Highlands whereas the quartz of the Moines south of the

Moine thrust and much of the eastern Dalradian has been affected by cold rolling and supplies sutured quartz.

As noted by Blatt (1967a), there is a great discrepancy between the proportion of polycrystalline quartz in the detritus from the source rocks and that in sandstones and also a significant difference between the immature and mature sandstones, being very high in the source rocks (30–70 percent) and low in the sandstones (about 7 percent) and lowest in the mature sands (2 percent). The most mature sandstones, the quartz arenites, also have, on the

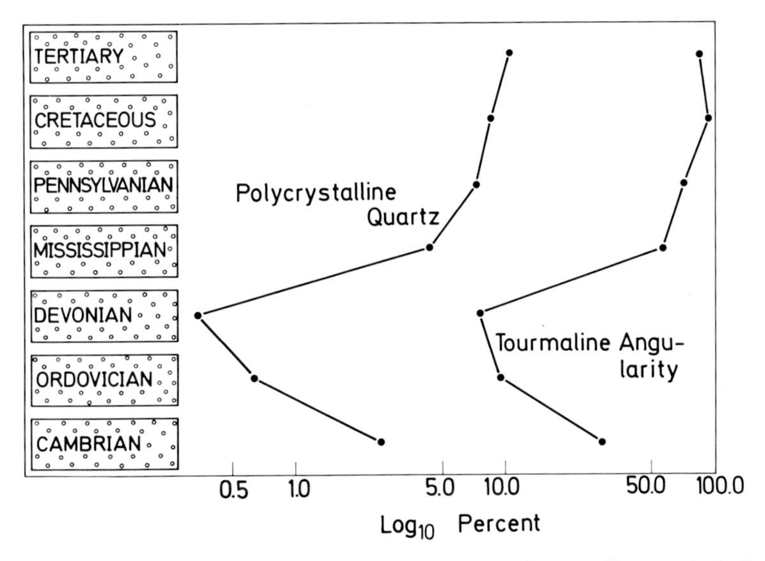

Fig. 8-2. Relation between percentage of polycrystalline quartz and tourmaline angularity in sandstones (by system) in Upper Mississippi Valley (Modified from Potter and Pryor, 1961, Fig. 6)

average, a lower proportion of undulatory quartz than either the immature sands or their source rocks (Blatt and Christie, 1963). Blatt (1967a) attributes these differences to the mechanical instability of polycrystalline and strained quartz. He believes most of such materials are eliminated during sand transport. There is a good correlation between abundance of polycrystalline quartz and percentage of angular grains in the Paleozoic sandstones of the North American Craton (Fig. 8-2). It is possible also that there is a selective loss by solution in some soils. Both rounding and embayments in quartz has been attributed to such action (Crook, 1968). As much of the polycrystalline quartz is in the 1 to 4 mm class, the more rapid destruction of this type of quartz may in part explain the dearth of these grades in most sediments.

It should be pointed out that the polycrystalline quartz of the preceding discussion is that found in igneous and metamorphic terranes – other varieties of polycrystalline quartz of major importance include fine-grained quartzites and chert – the latter being distinguished from acid felsic volcanic rock particles with great difficulty (Wolf, 1971).

It would seem, in conclusion, that meaningful observations can be made on detrital quartz. Even if it is not possible to determine with much certainty the provenance of common quartz, one can distinguish between polycrystalline quartz and common quartz and distinguish, perhaps, between sutured and polygonized quartz and between plutonic quartz and volcanic quartz. These distinctions have value in the analysis of provenance and can, in any case, be very useful in defining mineral associations in mature sandstones.

Detrital feldspar as a guide to provenance: Next to quartz, feldspar is the most common mineral of sand. Unlike quartz, little attention has been given to the initial characteristics of sedimentary feldspar though much has been written about its occurrence in sands. An exception is Rimsaite's 1967 study of the varietal features of feldspar in potential source rocks. Nine classes based on zoning, diverse intergrowth habits, twinning, fracturing and the like, could be defined.

The granular disintegration of the acid to intermediate plutonic igneous rocks and the gneisses releases feldspar. In the residual "grus," it may greatly exceed quartz and even in the sand or sandstone it may equal quartz in abundance. Several types of feldspar are usually present though the more alkali-rich varieties seem to be more abundant than the calcic feldspars.

The feldspars of many sediments display zoning. Oscillatory zoning, progressive zoning and lack of zoning in plagioclase may be clues to be provenance of the feldspar (Pittman, 1963). The plagioclase in volcanic and hypabysal rocks is characterized by oscillatory zoning whereas this type of zoning is rare in plutonic igneous and metamorphic rocks. Zoned plagioclase, regardless of type of zoning, is strongly indicative of an igneous rock.

The feldspar of the acid volcanics is likely to be sanidine; that of the acid plutonic rocks is either orthoclase or microcline. Perthitic feldspar is indicative of slow cooling and hence characteristic of the plutonic sources. The feldspars of pyroclastic origin tend to show euhedral forms, commonly broken, and in some cases display a thin envelope of glass, whereas the plutonic feldspars are anhedral.

Micas and provenance: The micas, though conspicuous in some sandstones, are never a major constituent. They are derived from schists and gneisses, from plutonic igneous rocks and from volcanic sources. Because of stability, muscovite is more common in the sediments than biotite. The latter, if it occurs as pseudohexagonal plates, is probably of volcanic origin. In general, abundant mica suggests a metamorphic provenance for the sand.

Rock particles as guides to provenance: Sandstones commonly contain rock particles in addition to the usual quartz and feldspar. By reason of the definition of sand, the rock particles will of necessity be fine-grained. These may be volcanic, mainly basaltic or felsitic, or be sedimentary, most notably pelitic particles, and may also be metamorphic, usually fine-grained metamorphics such as slate, phyllite and mica schists but including also metavolcanics and some rather exotic materials such as serpentine (Zimmerle, 1968). The mechanics which generate sand particles of these finer grained rocks and a discussion of their occurence in both modern and ancient sands is to be found in Chap. 2.

The rock particles themselves, perhaps more than any other type of grains, carry their own evidence of provenance. The volcanic assemblage includes glass, crystal fragments, corroded and embayed quartz with envelopes of glass or glass found in the embayed recesses, and volcanic rock particles. It may be difficult to distinguish between the immediate products of explosive volcanism and the reworked materials of this origin and the true epiclastic products of volcanic rock disintegration.

The sedimentary products include carbonate sands derived by the disintegration of the coarser dolomites and marbles, the calcareous debris so characteristic of the sands of the Alpine Molasse (Füchtbauer, 1967, p. 271) and the Miocene Oakville Sandstone of the Texas coastal plain (Folk, 1968, p. 141). More common are the abundant pelitic materials, the shales in particular, which tend to be deformed and molded about the more resistant quartz so as to appear more like pore-filling matrix rather than like a detrital grain (Allen, 1962, p. 669). Chert particles though never dominant are nearly ubiquitous in lithic arenites especially where a carbonate terrain of low to moderate relief is being eroded in a humid climate. In the United States the streams draining the cherty carbonates of the Ozark region of Arkansas and Missouri are particularly rich in chert detritus as are many Cretaceous sands in the northern Rocky Mountains area of the United States and Canada (Suttner, 1969, p. 1401).

Metamorphic materials are primarily schist particles but include also metavolcanic particles such as "greenstones," and not infrequently serpentine fragments which may locally be dominant (Zimmerle, 1968). Krynine (1937, p. 427) described "schist arenites" which contain 35 to 40 percent schist particles.

Rock particles are among the most informative of all detrital components and all efforts to determine provenance should include a close scrutiny of these materials.

Heavy minerals and provenance: The heavy minerals – the minor, high density, accessory detrital minerals of sandstone – have long been used as indices of provenance (Boswell, 1933, p. 47; Milner, 1926, p. 101). That certain species are characteristic of certain source rocks is well known. Sedimentary petrologists have defined certain detrital mineral associations each indicative of a major class of source rocks (Milner, 1926; Krumbein and Pettijohn, 1938, p. 463; Feo-Codecido, 1956; Baker, 1962, p. 90). These associations are summarized in Table 8-2.

Certain species appear in more than one of the major associations as does quartz, for example. Consequently attention has been given to the *varietal* characteristics of these species with the expectation that the several occurrences can be distinguished from one another. Minerals especially scrutinized include tourmaline and zircon. Krynine (1946) made a thorough study of the varieties of tourmaline and believed these could be related to their provenance. Krynine's approach to the problem was largely inductive whereas the work that has been done on zircons (Vitanage, 1957; Poldervaart, 1956) is based more largely on actual studies of zircon morphology in source rocks. All these efforts have been directed toward establishing generalizations applicable to the several mineral varieties wherever found. The size and predominant charac-

Table 8-2. *Heavy mineral associations and provenance (modified from Feo-Codecido, 1956, p. 997)*

Association	*Source*
Apatite, biotite, brookite, hornblende, monazite, muscovite, rutile, titanite, tourmaline (pink variety), zircon	Acid igneous rocks
Cassiterite, dumortierite, fluorite, garnet, monazite, muscovite, topaz, tourmaline (blue variety), wolframite, xenotime	Granite pegmatites
Augite, chromite, diopside, hypersthene, ilmenite, magnetite, olivine, picotite, pleonaste	Basic igneous rocks
Andalusite, chondrodite, corundum, garnet, phlogopite, staurolite, topaz, vesuvianite, wollastonite, zoisite	Contact metamorphic rocks
Andalusite, chloritoid, epidote, garnet, glaucophane, kyanite, sillimanite, staurolite, titanite, zoisite-clinozoisite	Dynamothermal metamorphic rocks
Barite, iron ores, leucoxene, rutile, tourmaline (rounded grains), zircon (rounded grains)	Reworked sediments

ters of zircons are apparently invariant or closely similar throughout a body of magmatic granite. There are marked differences between igneous bodies; the zircons of the extreme alkaline rocks, for example, display bipyramidal habit in contrast to the long prismatic forms of some granites. As first noted by Mackie (1923), the zircons of very old rocks, especially the Precambrian tend to be purple (hyacinth). Tomita's work (1954) on colored zircons suggests that such color is produced by prolonged radiation bombardment. A Precambrian source of such zircons is probable (Beveridge, 1960, p. 534).

Other studies have been directed toward a study of the minor accessory minerals of specific crystalline bedrock bodies which were or might have been the sources of heavy minerals found in younger sediments. Examples include the classic study of the Dartmoor granites by Brammall (1928) and that of Tyler and associates (1940) of the Precambrian granites and gneisses in the Lake Superior region. These and similar studies have made it possible to identify the sources of some of the heavy minerals in the younger sandstones. A notable early example of this type of investigation is that of Mackie (1923) on the purple zircon which is indigenous in the Precambrian Lewisian Gneiss of the north of Scotland but appears in sediments ranging from late Precambrian through Jurassic in age.

Our ability to make inferences about provenance depends a good bit on how much transportation and the environment of deposition may alter the detrital fraction. The extent of modification of the detrital minerals by the depositional environment is probably slight. There may be some segregation by hydraulic ratio and some rounding, but as we have discussed earlier (Chap. 2) there is no compelling evidence for disappearance of minerals likely to have been transported to a depositional environment with possible exception of carbonate rock fragments. Neither mechanical abrasion nor chemical solution have been established as major effects in such environments. Chemical precipitates may be added as pore fillings or as environmentally produced detrital grains but this in no way interferes with analysis of the truly clastic detritals. Diagenesis may play a more important role in selectively eliminating minerals and so some uncertainty is introduced. The role of intrastratal solution has been debated at length. This question is reviewed in Chap. 10. For heavy minerals the possibility of postdepositional intrastratal solution should always be considered.

Some work has been done which seems to show that the absolute ages of the potassium feldspars and the zircons of sands bear some relation to the ages of the rocks from which they were derived (Ledent, and others, 1964). Some knowledge of the latter might enable one to identify sands from such source areas. This approach to provenance is fraught with many difficulties due to mixing of sands from several sources giving mean ages in between those of the several separate source regions. This difficulty can be resolved in part if the age values contributed by the older and younger rocks can be sorted out, but in general the provenance questions can be more readily answered by other methods.

Mineralogy and Physical Geography

Although opinions vary, it is probably safe to state that only the most general conclusions can be drawn about relief and climate of the source area, particularly so for Paleozoic and older rocks. Such conclusions must be based on some knowledge of the stability of the various heavy mineral species. The stability of some of the more common species is given in Table 8-3. One should remember that the classification given in this table is a "statistical" one in that it is an over-all assessment of stability under "average" conditions. The stability of a mineral species in a specific case is a function of the composition of the mineral in question, of the composition of the formation waters, the pH of these waters and other variables. Inferences about climate and topography stem mainly from the ratio of chemically unstable to stable minerals, but, as discussed elsewhere, the effects of climate and topography on this ratio may be complementary and not individually determinable.

Source-area composition can be determined much more precisely, especially if weathering has not removed too many unstable components. Because it is likely that a few individual grains of even the most unstable minerals will have some probability of getting through the weathering barrier, careful study of the heavy mineral suite usually brings good results. It is clear that we need comparative studies on modern river sands from drainage basins differing from one another in relief, climate and some rock types.

In summary, we can state that the heavy minerals have proved to be the most useful indices of provenance. The usefulness of these minerals, however,

Table 8-3. *Stability of some detrital heavy minerals*

Ultrastable	Rutile, zircon, tourmaline, anatase
Stable	Apatite, garnet (iron-poor), staurolite, monazite, biotite, ilmenite, magnetite
Moderately stable	Epidote, kyanite, garnet (iron-rich), sillimanite, sphene, zoisite
Unstable	Hornblende, actinolite, augite, diopside, hypersthene, andalusite
Very unstable	Olivine

diminishes as one goes backward in time because of the degraded and somewhat impoverished suites in the older rocks. Heavy minerals have proved to be the most useful in the Tertiary basins of the world. The petrologist should not despair, however, as both tourmaline and zircon, ubiquitous in sandstones of all ages, may provide information on provenance for the alert observer.

We should add that conclusions relative to climate, based on sand mineralogy, apply to the climate of the source region and not to the depositional site. It is self-evident that if these two areas are far apart or differ greatly in elevation that their climates may be quite different. Our conclusions also often presume contemporaniety but it is conceivable that there might be a significant time lag between generation of the sand and its final deposition. The sands of the areas of North America formed during the Late Glacial epoch are now being reworked and deposited under non-glacial conditions.

Other Evidence Bearing on Provenance

As noted earlier, the sedimentary petrologist does not rely solely on a study of the detrital constituents to solve the problem of provenance. He utilizes other evidence – that provided by stratigraphy and paleocurrents.

Because much, if not most, sand is recycled, that is, derived from earlier formations which themselves may be derived from still older sandstone, regional stratigraphy will contribute to the analysis of provenance by establishing the relative ages of the strata. Only sands older that the formation in question are potential sources. Most significant is the recognition of unconformities which record episodes of erosion. A paleogeologic map will assist by indicating the formations exposed and subject to erosion. Such data as these indicate only *possible* sources. Proof that they did contribute to the formation in question must be established by other evidence. Stratigraphy may also show that large potential areas – such as a craton – were covered by carbonate rocks at the time of deposition of the sand in question precluding derivation from the crystalline rocks of this region.

Paleocurrent analyses have proved to be one of the most powerful approaches to the problems of paleogeography and provenance. They are especially helpful in the study of sandstones of alluvial origin. Water ran down hill in the past as it does now and the up-current direction of these alluvial sands, as shown by such features as crossbedding, is in the direction of the source. It may even be possible to estimate distance to the presumed source area if there are interbedded conglomerates (Pelletier, 1958, p. 1054).

Paleocurrent data may show that a geosyncline is filled by materials from both the craton and the fold belt itself. The character of the sands derived from the two sides may be strikingly different, as for example, in the Precambrian Coronation Geosyncline (Chap. 12). Striking changes in petrology correlated with marked differences in transport direction are good evidence of differing provenance. The source area can also often be located if one has some general knowledge of the petrography of these potential areas. This is especially true of modern Pleistocene and Tertiary sands.

It should be noted that, in general, shales contribute little to the question of provenance owing to the uncertainty about the precise role, if any, that the source region plays in determining the clay mineral assemblage found in a given shale. Nonetheless some have concluded that the clay minerals are primarily detrital in origin and reflect the composition of the source region. Biscaye (1965), for example, showed that the recent deep-sea clay minerals were not formed in place but were, instead, derived from the continents and were useful indicators of provenance. Weaver (1958, p. 302) applies this concept to a paleogeographic analysis of the Chester (Mississippian) and Springer (Pennsylvanian) strata of Oklahoma and Arkansas. Others have considered the clay mineral assemblage to be more largely a diagenetic product and hence of no value in provenance studies. Moreover shales, unlike sands, provide no paleocurrent data which are especially helpful in analysis of provenance.

Reading Provenance History

The ultimate goal of the sedimentary petrologist is to unravel the "line of descent" or "lineage" of the sandstone under investigation. As must now be evident from the preceding sections, reading the provenance history of a sand is a difficult task. Multiple sources and multiple cycles may be both involved. It may be difficult to distinguish between the immediate source (last cycle) and the ultimate source of the sand. There are several approaches to deciphering provenance. Even a single sand grain can shed some light on the problem. A tourmaline grain (Fig. 8-3), for example, may show a secondary overgrowth on a rounded detrital core. The overgrowth may be rounded also. Abraded overgrowths on an abraded core imply (1) formation of a tourmaline crystal in an igneous or metamorphic source rock, (2) weathering and release of this grain followed by transportation and abrasion, (3) deposition in a sand followed by (4) an authigenic or low-rank metamorphic regeneration of the crystal, (5) weathering and release of the tourmaline with transport and abrasion of the overgrowth, and, finally, (6) deposition in a new deposit of sand. It may even be possible, from the character of the original tourmaline to determine the nature of the source rock utilizing criteria formulated by Krynine (1946b).

A more complete analysis of provenance can be made from a sample than from a single grain. What could we infer, for example, from a single sample of Mississippi River sand about the nature and character of the rocks in the Mississippi Basin? A typical sand from the Delta (Russell, 1937, no. 1083-3/4) contains 64 percent quartz, 19 percent rock particles, and 15 percent feldspar. The quartz is subangular as is the feldspar. The latter is dominantly microcline with a lesser amount of oligoclase and andesine and a little orthoclase and sanidine. Principal rock particles are chert and fine-grained quartzite, but included are several varieties of volcanic rocks, some slates and schists. The sand contains a little detrital calcite and glauconite. The dominant heavy minerals are ilmenite, pyroxenes and amphiboles with lesser amounts of garnet, zircon, titanite, rutile and monazite. From this analysis we might conclude that the dominant source was a granite or granodiorite with minor sources from extrusive volcanics, metamorphic and sedimentary rocks. Clearly such an analysis fails to do justice to the geology of the Mississippi Basin, although it does reflect its glacial source which is characterized by a significant proportion of crystalline rocks. We may identify the source rocks and estimate their relative importance as contributors to the sand but we are unable to assess their areal extent in the drainage basin.

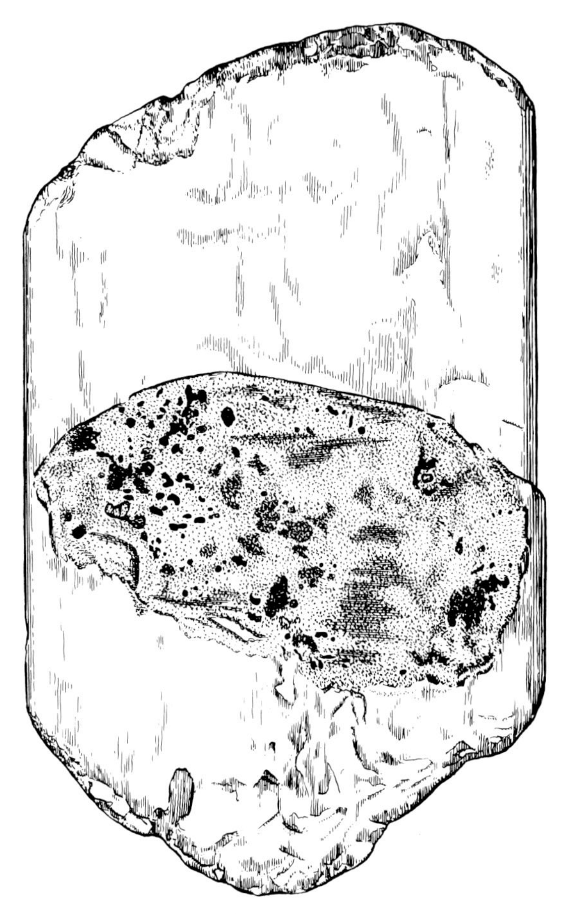

Fig. 8-3. Abraded tourmaline overgrowth on abraded detrital core, Cretaceous McNairy Sand, Henry County, Tennessee, U.S.A. (Redrawn from Potter and Pryor, 1961, Plate 2)

The use of more than one sample can improve our assessment. This is especially true if our samples are properly distributed throughout the basin or, in the case of ancient sandstones, from the whole area of sand accumulation. We can then delineate or *map* sedimentary petrologic provinces.

The investigator's conclusions are commonly presented as a kind of "flow sheet" or provenance diagram. These interpretative diagrams are of two kinds – the first is based solely on what can be seen in the rock itself – mainly from thin-section analysis. It records a judgment about the nature or kind of source rock for each of the detrital components of the sand.

The second type of provenance diagram is based not only on a study of the thin section but also upon a knowledge of the regional geology and stratigraphy. It depicts what is possible or, in the judgment of the petrologist, probable, in terms of the source rock contributions. To construct such a diagram, the investigator must piece together various lines of evidence – the distribution and kinds of heavy minerals, paleocurrent information, facies relations, both vertical and lateral, the distribution and kind of rock fragments both in the sands and in the associated conglomerates, knowledge of structure and the larger tectonic elements – in short, any and all manner of geologic data.

Good examples which illustrate this approach and in which there are diagrams summarizing the provenance of the sandstone have been published by Krynine (1940, Fig. 3), Payne (1942, p. 1754), Todd and Folk (1957, p. 2557), Potter and Pryor (1961, Fig. 14), Doty and Hubert (1962, Fig. 11), Hooper (1962, Fig. 8), ten Haaf (1964, p. 135), Suttner (1969, Fig. 15) and Walker and Pettijohn (1971). The diagrams of Krynine, (Fig. 8-4), Doty and Hubert, Todd and Folk (Fig. 8-10), and Walker and Pettijohn are diagrams of the first type, largely based on petrology – what can be seen in the sandstone itself – whereas those of Payne, Suttner, Potter and Pryor and ten Haaf (Fig. 8-5) belong to the second category and utilize regional geology, tectonics, and stratigraphy to help tell the story.

Evaluation and Summary

The major elements of a provenance methodology are shown in Fig. 8-6. The methodology relies primarily on petrography and directional structures. Such a combined attack on the question of provenance is especially useful where extrapolation beyond the present basin limits is required as it usually is in Paleozoic and Precambrian sedimentary basins. These basins are more generally separated from the sources of their sediments by subsequent deformation, erosion, and burial.

The methodology here presented makes possible better insight to the ultimate goals of regional tectonics and paleogeography, which are essential to understanding of the structural development of the continents. From the standpoint of continental evolution it is of the utmost importance, for example, to know that the Pennsylvanian sediments of the Illinois Basin were derived both from the interior of the continent as well as from bordering mobile belts. It may be possible that studies of the sediments in the geologic record will enable us to assess the importance of a continental interior as a major source of clastics throughout the geological history of a continent.

Geochemistry, in particular the determination of absolute ages, also offers promise of contributing to a better definition of mineral associations and more precise knowledge of the *age* of the original crystalline material. Ledent and others (1964) studied Sr-Rb in feldspar and Pb-U in zircons and Hurley and co-workers (1961) used K-Ar in illitic clays. Because isotopic analysis is expensive and requires bulk samples rather than single grains, geochemical studies of provenance are still very few in number. If, however, there were an analytical method that was both rapid and utilized single grains, benefits would be

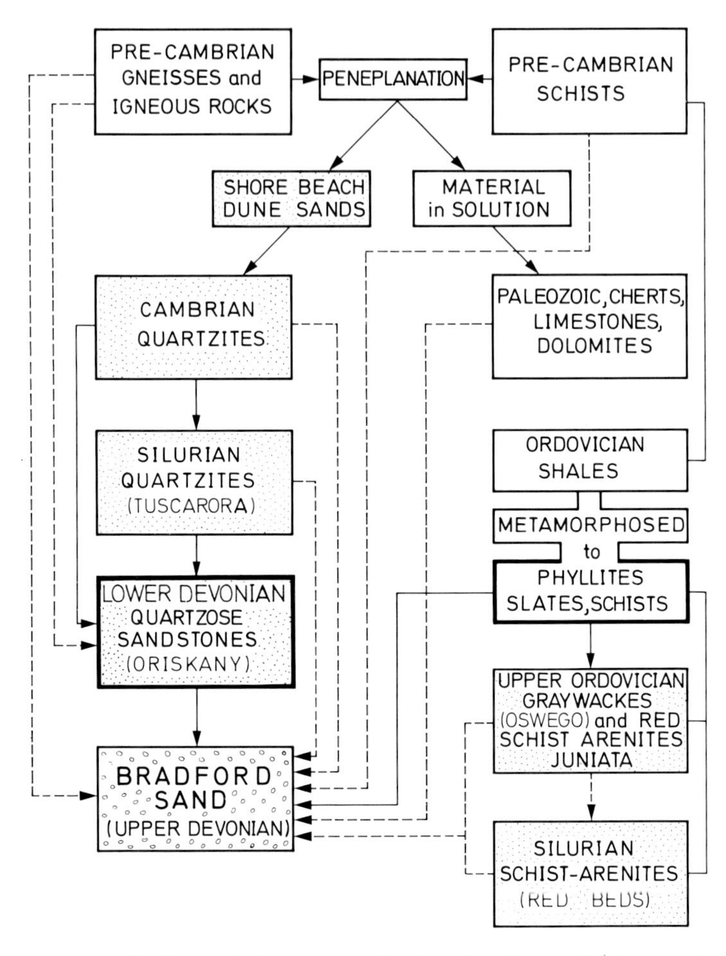

Fig. 8-4. Flow sheet indicating probable progress of Paleozoic sedimentation in central and western Pennsylvania. Solid lines indicate main paths of sedimentation; dashed lines are subordinate ones. Main source areas stippled. (Modified from Krynine, 1940, Fig. 3)

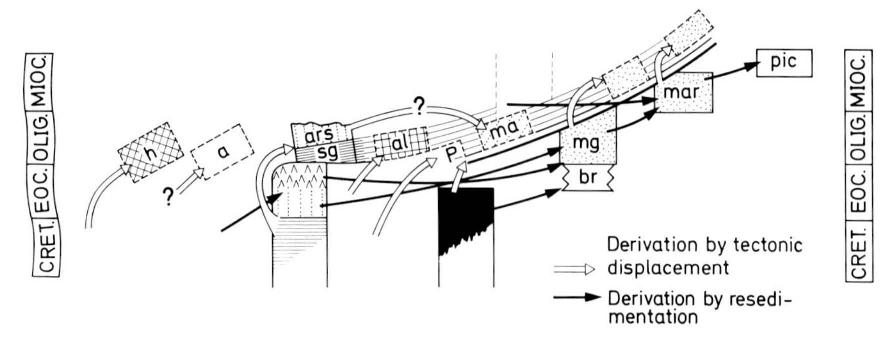

Fig. 8-5. Probable genetic relationships of northern Apennine flysches. Turbidite sandstones: p, ma, mg, mar, a (in part); argillaceous flysch; sg; chiefly calcareous flysch: h, al; brecciola: br; Picene flysch: pic. (ten Haaf, Fig. 2)

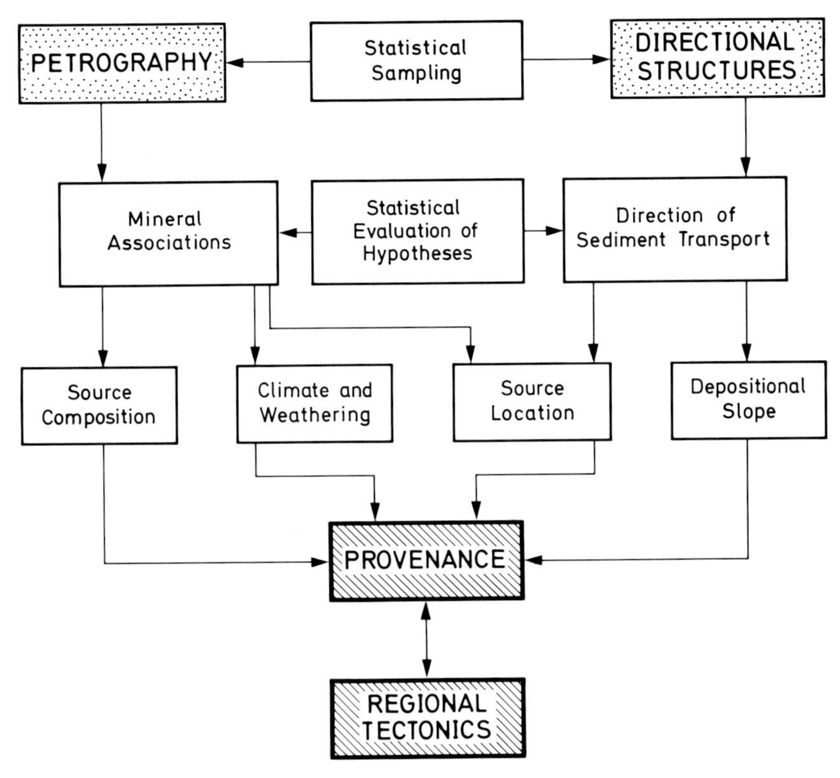

Fig. 8-6. Provenance methodology based on directional structures and petrography (Modified from Potter and Siever, 1956, Fig. 1)

substantial for, in essence, one could then tag or classify grains by their age – something that can be done only in exceptional cases today.

In its broadest aspect, the problem of provenance can be considered as a problem of accounting – making an inventory of the different types of grains contributed by different source rocks. To this, one should add the problem of *the same kinds of grains coming from different source rocks* so that, at present we are generally unable to distinguish between the quartz grains derived from a granite from those derived from a gneissic rock. More sophisticated accounting will be possible only when we can better petrographically "tag" grains and better evaluate the number of cycles of erosion in their abrasion history.

Mineral Associations and Petrologic Provinces

Definitions and Concepts

Because sediment sources are infinitely varied in mineral composition and any number of source rocks may contribute to the deposition of a given sand, it is possible for the mineral associations to be nearly infinite in number. Twenty mineral species, taken four at a time, for example, would yield 4,845 possible combinations. In fact, however, the number of mineral assemblages found is remarkably small. Doeglas (1940), for example, recognized five assemblages in East Java, designated (1) zircon-rich association, (2) zircon-staurolite association, (3) andalusite-rich association, (4) epidote-rich association, and (5) hornblende-rich association, each being named for the most prominent mineral components. Similar assemblages were recognized in the Tertiary of the Gulf Coast by Cogen (1940) who defined four zones on the basis of the appearance of a new index species – staurolite, kyanite, epidote and hornblende. Feo-Codecido (1956) likewise defined six major mineral suites in the Tertiary of Venezuela. It is noteworthy that the assemblages defined in these and other sedimentary basins, especially the Cretaceous-Tertiary basins, are quite similar to one another. Van Andel (1959, p. 157) in summarizing the problem, lists only six common associations – defined and named for their index species: hornblende-epidote, epidote, kyanite-zircon, staurolite-zircon, garnet-zircon-tourmaline, and zircon-tourmaline.

As noted by van Andel and also by Pettijohn (1957, p. 514), in many sedimentary basins these associations follow each other in a recurrent pattern. In general, the lower (and older) mineral assemblages are the more restricted in number of mineral species whereas the younger and uppermost contain more species and include many unstable minerals. This succession of mineral associations is particularly characteristic of the Tertiary basins throughout the world. It has been interpreted as due to intrastratal solution (Pettijohn, 1957, p. 514) and as due to progressive denudation and unroofing of new sources of supply (van Andel, 1959, p. 160).

In 1933 Edelman (quoted from Doeglas, 1940) defined the *sedimentary petrological province* as "... a complex of sediments which by their geographical distribution, age, and origin form a natural unit." This concept is generally attributed to Baturin (1931) as a result of his studies in the Caucasus. In a general way a province is defined by the geographic limits of a particular *mineral association* – usually but not always a heavy mineral assemblage. In a strict sense this is not wholly true. There may be subassemblages within the boundaries of a particular province. Within the Tertiary Molasse Basin of southern Germany, for example, there are subprovinces related to fans derived from the adjacent uplifted Alps (Füchtbauer, 1964). Although all of the material is of Alpine origin, each fan has its own subassemblage determined by the local drainage basin from which it was derived. Within the basin these assemblages merge and lose their identity and contribute to a mixed assemblage.

Systematic mapping of mineral associations is needed to delineate sedimentary petrologic provinces. This is one of the more useful applications of sedimentary petrography as it can help locate the source regions that lie beyond the present basin margins and define the paleocurrent systems and hence paleoslope where directional structures are absent or difficult to observe, as in subsurface cores or cuttings.

Provinces are perhaps best defined by a combination of light and heavy mineral data, although they may be defined by but a single characteristic such as roundness of tourmaline. In such cases, one needs to consider the relation between grain size and mineral composition. Because of hydraulic factors and differences in initial size, certain mineral species are concentrated in finer size grades than others. If the texture of the sands being compared are not all alike, wide fluctuations in the mineral proportions are likely to occur and reflect only these size-controlled differences rather than a significant difference in provenance or age. A plot of mineral frequency for several size grades in a given sand graphically illustrates the problem (Fig. 8-7). These variations have been studied in some detail by Rittenhouse (1943) and others.

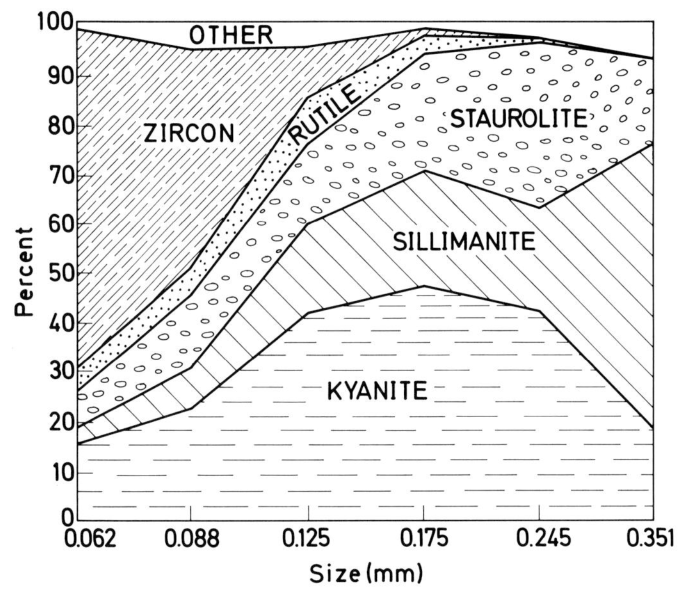

Fig. 8-7. Diagram showing relations between grain size and heavy mineral frequencies. Lafayette sand, western Kentucky (redrawn from Potter, 1955, Fig. 3)

By using varietal types of a single mineral, such as quartz or tourmaline, one can minimize the effects of grain size and possibly intrastratal solution as well. But commonly these size-controlled variations are not sufficient to obscure the differences between heavy mineral assemblages of differing provenance. From a detailed study of heavy minerals in the Rhine River, van Andel (1950) concluded that the study of the whole sand fraction was superior to that of single size grades, especially if many samples are studied.

Another type of complexity can result from sampling a unit which is not everywhere of the same age. For example, modern continental shelves may be covered by both Recent sands and relict sands of Pleistocene age related to an earlier lower sea level. Actual practice, however, has shown the persistance of source areas and paleocurrent patterns so that no significant changes in mineral associations are likely to be found over short stratigraphic intervals in ancient sediments.

As noted, mineral associations are geographically distinct and define petrologic provinces. Near the boundary between two provinces, however, there may be mixing of the assemblages of each. If extensive enough, this mixed assemblage will define another province or subprovince.

There is also, as noted above, a provincial succession with time. Change in character of the assemblage corresponds to unroofing of new source rocks or to drainage changes in the source area. It is upon this time-dependent factor that sedimentary petrologic correlation depends. Caution should be taken, however, to make sure the heavy mineral zones are truly controlled by provenance rather than by mineral stability and intrastratal solution.

Mapping Mineral Associations

Systematic petrographic mapping is the key to maximizing information about provenance. This is so because: 1) mapped petrographic provinces are commonly relatable to paleocurrents, facies patterns, and other major features of basin geology and 2) if two or more provinces are present, as is commonly the case, one can generally obtain a meaningful *comparative* rather than *absolute* explanation. Although many map patterns are possible, in practice there are only a few major recurring ones (Fig. 8-8). Regardless of type of map patterns, petrographic provinces will have maximum contrast when the source areas are

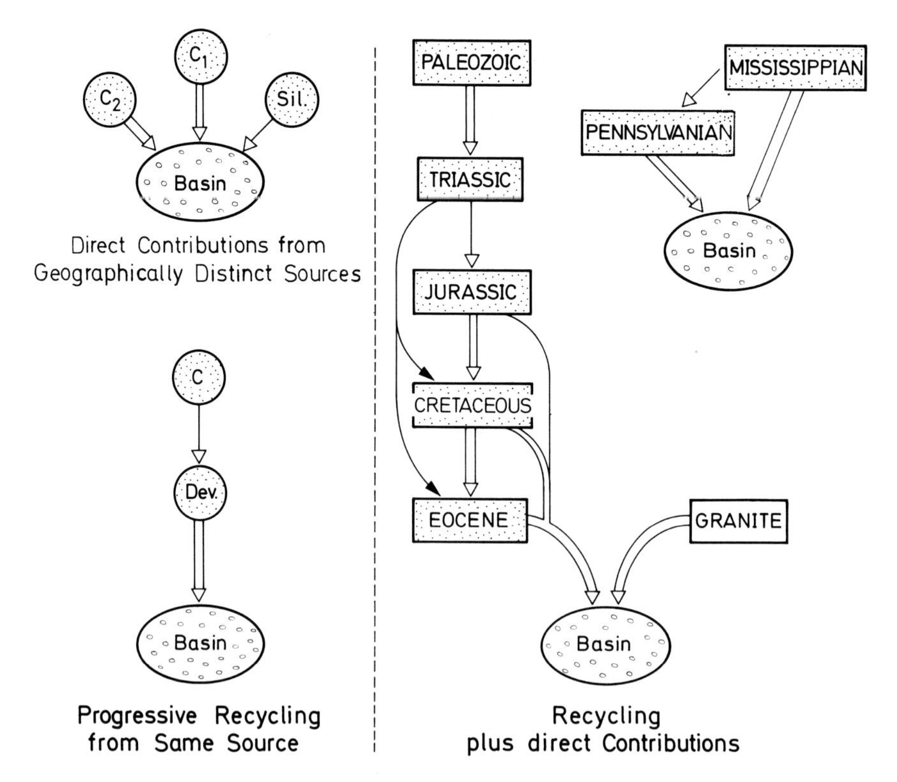

Fig. 8-8. Some common paths of mineral evolution

nearby and are compositionally distinct. Conversely, contrast will be minimal when source areas are distant and compositionally similar. It is here that many samples carefully studied, possibly using varietal types of minerals, are needed plus the supplementary information gleaned from paleocurrents and stratigraphy. Statistical analysis of the data will also play a greater role.

Krumbein and Graybill (1965) introduce the reader to the statistical background needed to more fully understand the statistical techniques, such as discriminant functions and the analysis of dispersion, that may be useful in analyzing mineral associations. A more elementary treatment will be found in Marsal (1967), Comité des Techniciens (1969), and the "Bibliography of Statistical Applications in Geology" by the American Geological Institute (1968). Also very helpful is Harbaugh and Merriam (1968) and Merriam (1969b), the latter containing many advanced studies.

From a statistical standpoint, petrographic analysis of a sample, be it an arkose or a fossiliferous calcarenite, yields a series of proportions of the petrographic variables, $x_1, x_2, x_3, x_4, \ldots\ldots,$ that have been counted. Almost always a petrographic province is defined by two or more variables such as feldspar abundance and tourmaline roundness and is thus an example of a *multicomponent* or *multivariate system*. In such a system each sample point is defined by a probability vector such as [0.20, 0.29, 0.31, 0.11, 0.06, 0.03] or more generally $[x_{11}, x_{12}, x_{13}, x_{14}, x_{15}, x_{16}]$, where the components of the vector are the proportions of the different minerals and thus sum to one. A vector may also be defined by combinations of different kinds of variables, for example, median size, percentage of round grains, and percentage of feldspar and polycrystalline quartz. In any case, one thinks of the components of a vector as defining a point in a multidimensional space. Collectively all the sample vectors define a *data array* or *data matrix*

$$\begin{bmatrix} x_{11} & x_{12} & x_{13} & \cdots\cdots & x_{1m} \\ x_{21} & x_{22} & x_{23} & \cdots\cdots & x_{2m} \\ - & - & - & - & - \\ x_{n1} & x_{n2} & x_{n3} & \cdots\cdots & x_{nm} \end{bmatrix}$$

consisting of n samples each with m variables. This data array is the table of modal analyses and is the starting point of all subsequent analysis of the data.

Sampling is generally best done by systematically covering a basin or the exposed outcrop so as to delimit as closely as possible its petrographic provinces. Paleocurrents and general stratigraphic knowledge may be a useful guide, if they are available. Commonly, it is best to first make a petrographic reconnaisance of the samples and then, if necessary, increase sample density at the boundaries of petrographic provinces where mixing is commonly greatest.

A number of statistical mapping techniques are now available (Table 8-4). Their general goal is to obtain more information from the map data than can be seen by direct inspection. Most of the techniques are multivariate ones and require a computer for their use. Fortunately, many computer programs

Table 8-4. *Major statistical mapping techniques*

UNIVARIATE

Linear combination

Combines two or more variables into a single one such as the ZTR index, which is the sum of zircon, tourmaline, and rutile and is thus a measure of the maturity of a heavy mineral suite. Many other combinations of sums of variables and/or ratios possible. Computer programs generally not needed.

Trend surface mapping

Reveals underlying systematic trends in a set map data. Fits successive mathematical surfaces to the areal variation of a single variable and computes the residual – the difference between the computed and observed value – at each point. Two principal models available: polynomial (linear, quadratic, cubic, and higher surfaced) and Fourier. Many computer programs.

MULTIVARIATE

Distance function

Shows the departure between a chosen sample and all others. Uses two or more variables and computes the distance between pairs of vectors. A related measure is the similarity function which defines the angle between vector pairs.

Entropy

Measures the degree of mixing of a sample. Entropy is maximal when mixing is maximal so that end-members, regardless of their composition, have minimum entropy. Samples with equal proportions of different components have like entropy.

Cluster analysis

Assigns samples into progressively larger hierarchical groups. Many different measures of similarity may be used. Groupings shown by a dendrogram.

Q-Mode factor or vector analysis

A method of data reduction that identifies the compositionally most distinct end members of a set of samples and expresses the rest in terms of these end members.

are available, the principal published source being the "Computer Contribution Series," of the Kansas Geological Survey, edited by D. F. Merriam.

The easiest way to simplify a map of a mineral association is to make a linear combination of some of its variables so that one obtains a new single variable. The Zircon-Tourmaline-Rutile (ZTR) index of Hubert (1962) is an example of this approach.

Another univariate technique is to make a trend surface analysis – a technique which was early popularized by Krumbein (1956) and has since been modified and extended by many others. Papers by Krumbein (1966), Merriam and Cocke (1967) and Miesch and Connor (1968) give computational details and discuss different methods. Simple graphical methods, by use of an areal moving average, may also be used to construct trend surface maps (see for example, Pelletier, 1958, p. 1036). Allen and Krumbein (1962) and Cadigan (1967) provide specific provenance applications. Trend surface mapping is also widely used in the analysis of facies maps. Because it is a univariate technique, one makes a trend analysis for each variable.

Distance or similarity maps are perhaps the simplest of all the multivariate mapping functions and are often the most informative, particularly if the reference sample is chosen correctly. Papers by Krumbein (1955) and Imbrie and Buchanan (1965, p. 256–258) describe their use. Entropy functions (Pelto, 1954) also play a useful role, particularly in providing a better delineation of the mixing of associations at their boundaries.

Cluster analysis (Parks, 1966; Wishart, 1969) is like factor analysis in that it is essentially a form of data reduction, but unlike factor analysis, is simpler to understand and compute. Cluster analysis permits one to graphically group a series of samples into progressively higher hierarchies – in other words, to first recognize the minor associations and then the progressively more important ones. If the samples have a geographic as well as compositional affinity, the areal distribution of the association is immediately apparent. Bonham-Carter (1967) provides a cluster analysis program for nonquantitative attributes such as the presence or absence of a mineral.

Q-mode factor analysis was introduced to geologists by Imbrie and van Andel (1964), who applied it to heavy mineral associations although carbonate petrographers have used it the most. There are now a number of variants of factor analysis and computer programs are available (Merriam, 1969a; Parks, 1970).

Examples of Provenance Studies

Modern Sands

There are instances in which it is desirable to know the source of a specific deposit of sand – the source of the sand on a particular beach or dune or area of shelf. Or we may wish to know from what areas the deposits of a particular present-day basin come and over what area they are deposited as, for example, the sands within the North Sea Basin or even within an particular estuary or harbour. We may also wish to know from what part of a drainage basin a particular mineral came – alluvial gold, for example, or from what part of a basin the sediment accumulating in a reservoir was derived. Similarly the mineralogy of glacial sands may provide clues to composition of bedrock formations now concealed by drift or by swamp or lake cover. Modern studies directed toward these ends generally depend on a study of the accessory, heavy mineral assemblages in the sand and a general knowledge of the petrology of potential or possible source areas. Several examples are briefly described below.

Northern Gulf of Mexico:

Goldstein, August, Jr.: Sedimentary petrologic provinces of the northern Gulf of Mexico. Jour. Sed. Petrology **12**, 77–84 (1942).

Samples from the shore and near-shore areas in the northern part of the Gulf of Mexico define four petrologic provinces, which are, from east to west, the Eastern Gulf Province, the Mississippi River Province, the Western Gulf Province and the Rio Grande Province. Each is defined by a heavy mineral assemblage. The Eastern Gulf Province is characterized by a high content of staurolite and kyanite. The Mississippi River Province is primarily an amphibole-epidote-pyroxene province, essentially that carried by the river today. The Western Gulf Province is similar to that of the Mississippi and may have been derived mainly from Mississippi sources at an earlier time. The Rio Grande Province is characterized by a basaltic hornblende-pyroxene assemblage. The distribution of these mineral assemblages is best explained by a slow westward drift carrying the Mississippi mineralogy westward but not eastward.

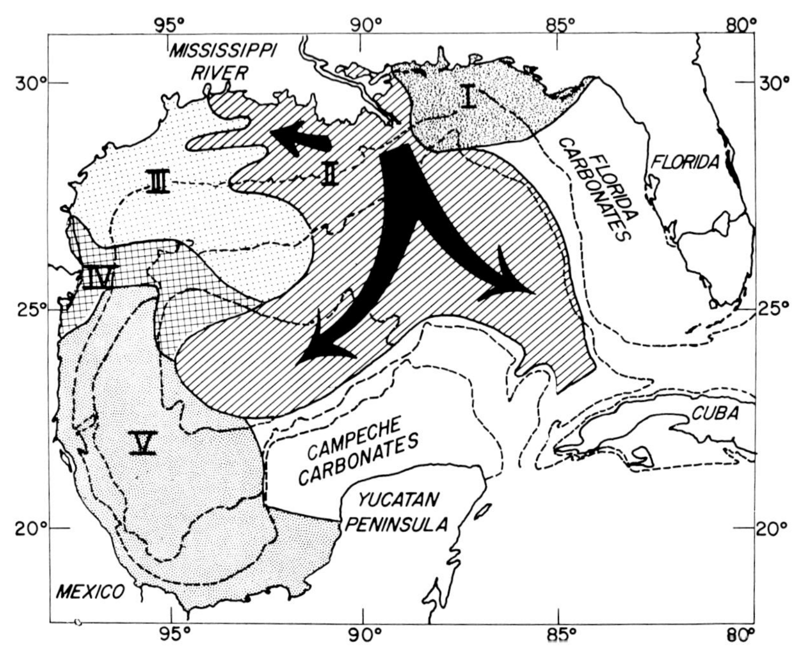

Fig. 8-9. Major heavy mineral provinces, inferred areal distribution of Mississippi sediment and principal sediment dispersal directions in the Gulf of Mexico. I – Eastern Gulf Province, II – Mississippi Province, III – Central Texas Province, IV – Rio Grande Province, V – Mexican Province. Data for map compiled from various sources by Davies and Moore (1970, Fig. 1)

In the western Gulf it is modified and diluted by materials brought to the shore by the smaller rivers of Texas (Brazos, Trinity, Colorado). The Rio Grande contributes a larger volume of sand and thus defines a province of distinctive character of its own.

For more recent studies of the provenance of northern Gulf sands, the reader is referred to van Andel and Poole (1960) and to Davies and Moore (1970). See Fig. 8-9.

Gulf of California:

van Andel, Tj. H.: Recent marine sediments of Gulf of California. Am. Assoc. Petroleum Geologists Memoir **3**, 216–310 (1964).

Van Andel describes the heavy-mineral provinces in the Gulf of California. The northern portion is dominated by the Colorado amphibole-epidote-pyroxene assemblages. There exist, in addition, narrow marginal zones of laterally-supplied sands locally derived. The central and southern portion of the Gulf is characterized by a pattern suggesting local sources – from both west and east – and with no evidence of significant longitudinal transport.

The North Sea:

Baak, J. A.: Regional petrology of the southern North Sea. Wageningen, 127 p. (1936).

This work is prefaced with a review of the philosophy and methods of the Edelman school of sedimentary petrology. The definition of a sedimentary petrologic province, the criteria for the delineation of a province, the several kinds of variations in mineral associations and their causes are briefly reviewed.

Following a review of all former studies of the petrology of the North Sea bottom and surrounding coasts, Baak presents his data based on the study of over 1,000 samples, in which over 30 mineral species and rock fragments are recognized. Five provinces or mineral association types are recognized, namely, an H-group marked by a garnet-epidote-saussurite-hornblende association, an A-group mainly garnet-epidote-hornblende, an E-group with a garnet-augite association, a North Hinder group, similar to recent Rhine sand and marked by an augite-hornblende-saussurite association, and lastly a bottom type of Tertiary derivation marked by a zircon-garnet-rutile assemblage. The regional distribution of the associations and the mixed border types are described and mapped. Special discussion is given of the Dutch and French coastal zones.

The geologic history of the North Sea basin and probable sources of the several mineral associations are outlined. Basically the North Sea bottom is a submerged landscape and not an area in which sedimentation is taking place on a large scale. Hence the bottom deposits do not relate to the later geologic history of this area. The A-group is mainly glacial from Scandinavia; the E-group is also glacial but of English and Scottish origin; the Tertiary group is locally derived from Tertiary subaqueous outcrops; the H-group is a mingling of Rhine-type sand with subordinate fluvioglacial material when the Riss ice blocked the North Sea and forced the drainage through the English Channel; the North Hinder group is a sand deposited by the Rhine during a "low terrace" stage.

Ancient Sandstones

Studies of the provenance of ancient sands are common as these are normally a part of the problem of analysis of a sandstone formation. They vary from a mere surmise to an extended effort to decipher the provenance. We present here several examples of the more extended analyses.

Eocene of the Coastal Plain of Texas:

Todd, T. W., Folk, R. L.: Basal Claiborne of Texas, record of Appalachian tectonism during Eocene. Am. Assoc. Petroleum Geologists Bull. **41**, 2545–2566 (1957).

The Texas Eocene formations, the Carrizo and Newby (sandstones) are quartzose subgraywackes with 5 to 10 percent feldspar and 5 to 25 percent rock particles; mainly chert, phyllite, slate and metaquartzite. The quartz, in part reworked (very well rounded) and partly volcanic (phenocrysts), is largely subangular ordinary quartz. The heavy minerals are several varieties of zircon and tourmaline plus staurolite, kyanite, rutile, and garnet.

The authors recognize five source rock classes; acid plutonic, high and low rank metamorphic rocks, volcanic rocks, and older sediments. The most important, the acid plutonic and metamorphics, supplied the bulk of the light components and the metamorphic heavy mineral suite. Although the underlying Wilcox Formation (beach) was the immediate source, the ultimate source was the southern Appalachians. Todd and Folk believe that only this area could supply the metamorphic suite. Local Texas sources contributed mainly to the coarse fraction. Included here was the volcanic quartz. See Fig. 8-10.

The appearance of the Appalachian materials in the basal Claiborne in Early Eocene is evidence of a sharp tectonic uplift in the southern Appalachians at this time.

Pennsylvanian of the Illinois Basin:

Siever, Raymond, Potter, P. E.: Sources of basal Pennsylvanian sediments in the Eastern Interior Basin. Part II. Jour. Geology **64**, 317–335 (1956).

Two regional mineral provinces are defined by tourmaline roundness and quartz varieties. The statistical significance of the observed differences was evaluated with the analysis of dispersion, a form of multivariate analysis.

The western Illinois province is characterized by no metamorphic quartz pebbles, high tourmaline roundness, insignificant feldspar and a low proportion of polycrystalline quartz. The eastern province has metamorphic quartz pebbles, low tourmaline roundness, a medium to high proportion of polycrystalline quartz and one to five percent of feldspar.

The orthoquartzitic nature of the sandstones, the absence or relatively small amount of feldspar, and the impoverished heavy mineral suite indicate derivation from a pre-existing sedimentary source in both provinces. The crossbedding pattern suggest derivation of the sediments of the western province from earlier Paleozoic sediments exposed on the Transcontinental Arch in Pennsylvanian time. The sediments in the eastern areas appear to have come from the eastern part of the Shield and especially from the Appalachian region.

This study demonstrates the value of utilizing paleocurrent data in conjunction with petrology and also how much can be done with petrology of older rocks with relatively impoverished mineral suites.

For a further discussion of these sands and their relation to the geologic history of the Illinois Basin, *see* Chap. 12.

Molasse of southern Germany:

Füchtbauer, H.: Sedimentpetrographische Untersuchungen in der älteren Molasse nördlich der Alpen. Eclogae geol. Helvetiae **57**, 157–298 (1964).

This monographic work is based on mineralogic analysis of over 1,500 samples collected in an area 450 km long lying north of the Alps, mainly in southern Germany. The Molasse sediments crop out along the tilted southern

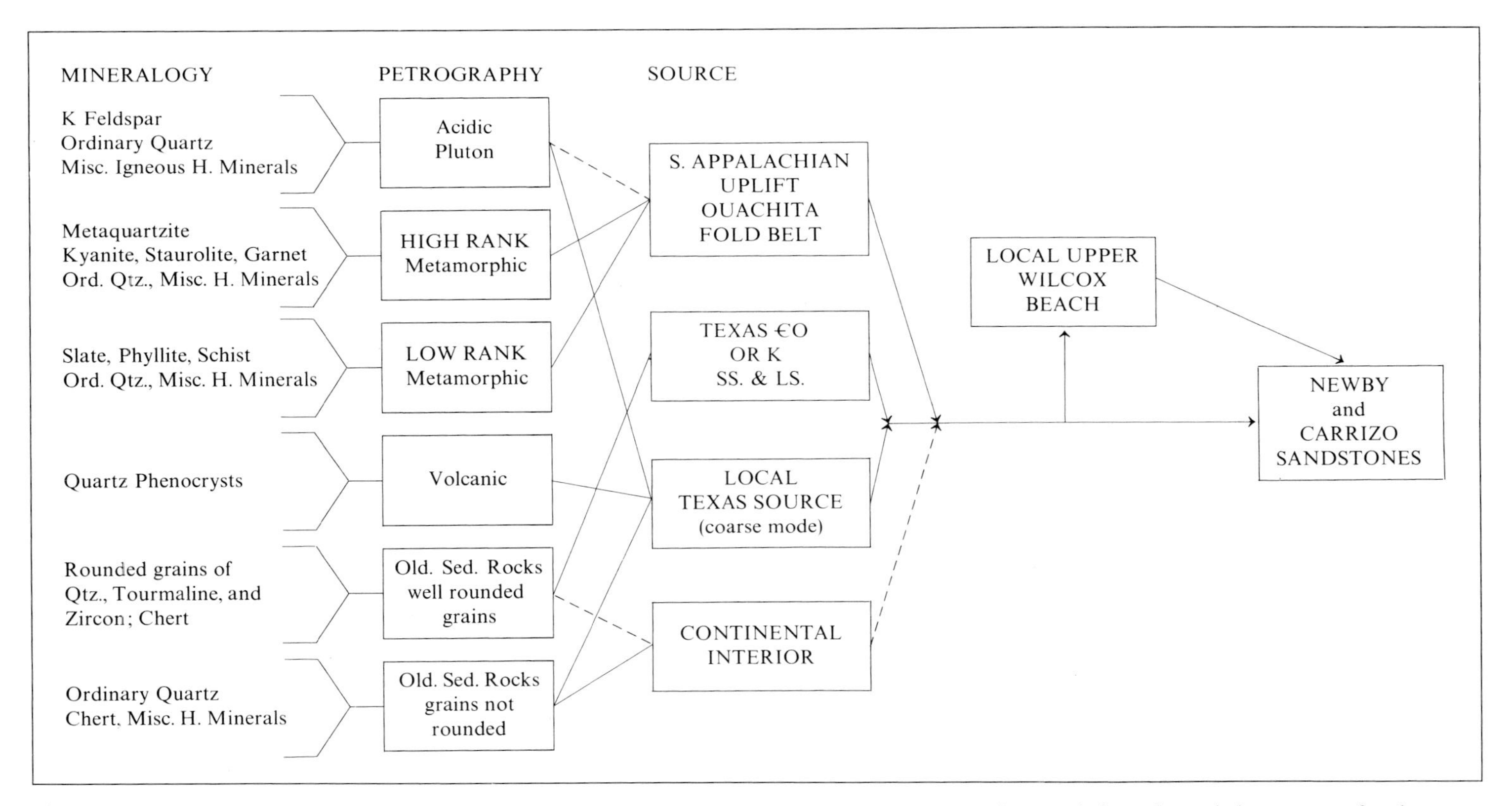

Fig. 8-10. Flow sheet derivation of Eocene Carrizo and Newby sandstones of Texas. Thickness of lines indicates relative volumetric importance of each source (Modified from Todd and Folk, 1957, Fig. 12)

margin of the basin. They are coarse conglomeratic fans which pass in the subsurface into fine-grained, horizontal deposits within the basin.

The sandstones, rich in feldspars and rock fragments, especially dolomitic rock fragments, have a varied heavy mineral assemblage which can be divided into subassemblages each correlated with a particular fan and which reflect a local Alpine source.

These assemblages are identified basinward and their mapping defines the paths of sediment transport. Near the margin transport is lateral, but in the basin, longitudinal transport prevails. There is a small input from the north side of the basin of an alien heavy mineral suite, mainly tourmaline, zircon and rutile, unlike the dominant Alpine assemblage, which consists principally of garnet, staurolite, apatite and zircon with an important epidote component in the southeast. The mineralogy of the sands also changes with time; the earlier sands are derived from the Flysch sediments including limestones and dolomites whereas the younger deposits contain significant contributions from the Alpine crystalline rocks.

Füchtbauer's study shows how the petrology of sands, coupled with stratigraphic data, can contribute to the understanding of the filling of a basin and the reconstruction of the paleogeography of the area. It is a nice integration of outcrop and subsurface data and shows what can be done when paleocurrent data are missing.

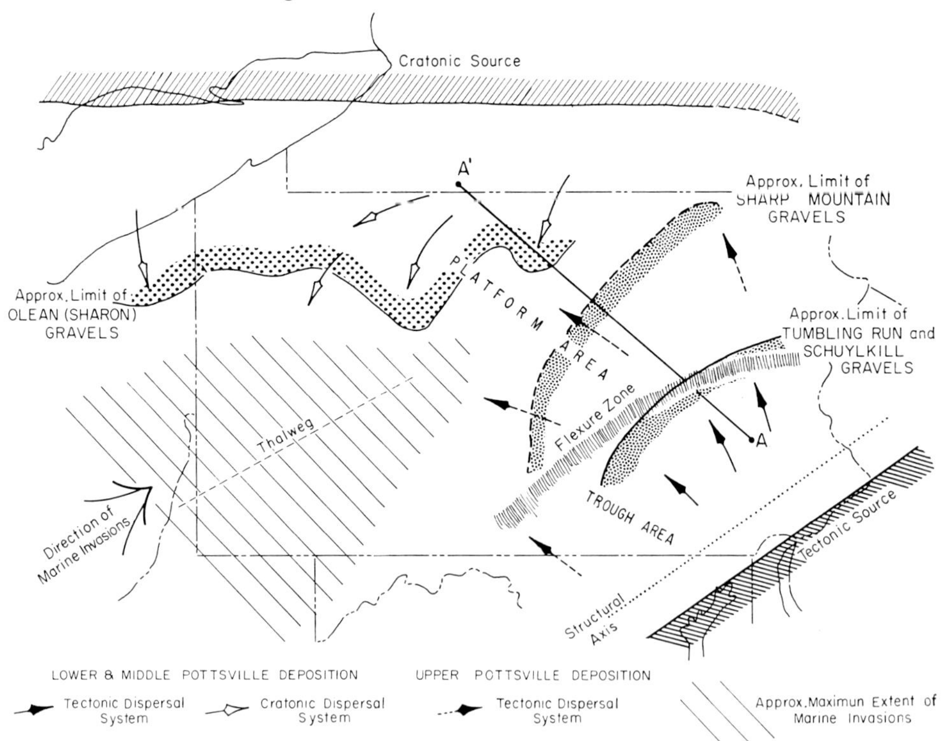

Fig. 8-11. Regional depositional pattern and paleogeography of the Pottsville (Meckel, 1967, Fig. 13). Reprinted by permission of the author and the Geological Society of America

Pottsville of Central Appalachians:

Meckel, L. D.: Origin of Pottsville Conglomerates (Pennsylvanian) in the central Appalachians. Geol. Soc. America Bull. **78**, 223–258 (1967).

The early and middle Pennsylvanian of the northern part of the central Appalachians was deposited in an elongate, northeasterly-trending basin flanked on its northern side by a stable cratonic platform and a tectonic source land on the southeast, the latter being composed largely of sedimentary and metamorphic rocks.

During Pottsville time, the basin was asymmetrically filled from the two sides (Fig. 8-11). This conclusion is supported by the paleocurrent pattern in the sandstones and pebble-size decline in the conglomerates. Although the sands contributed from the two source lands are similar in many respects, both being multicycle, they differ in some important respects, particularly the ratio of monocrystalline to polycrystalline quartz. In the cratonic sands this ratio averages 11.0; in those from the tectonic source the ratio is 3.0.

This study is a good example of the utility of paleocurrent data and facies changes (pebble size) in establishing provenance. The petrographic evidence alone is helpful, but is immeasurably strengthened by the paleocurrent and facies maps.

References

Allen, J. R. L.: Petrology, origin and deposition of the highest Lower Old Red Sandstone of Shropshire, England. Jour. Sed. Petrology **32**, 657–697 (1962).

Allen, Percival, Krumbein, W. C.: Secondary trend components in the Top Ashdown Pebble Bed. A case history. Jour. Geology **70**, 507–538 (1962).

van Andel, Tj. H.: Provenance, transport and deposition of Rhine sediments, 129 p. Wageningen: Veenman (Ph. D. Dissertation Univ. Wageningen) 1950.

— Reflections on the interpretation of heavy mineral analyses. Jour. Sed. Petrology **29**, 153–163 (1959).

— Poole, D. H.: Sources of Holocene sediments in the northern Gulf of Mexico. Jour. Sed. Petrology **30**, 91–122 (1960).

Baak, J.: Regional petrology of the southern North Sea, 127 p. Wageningen: Netherlands 1936.

Baker, George: Sand drift at Portland, Victoria. Royal Society Victoria, Proc. **68**, 151–197 (1956).

— Detrital heavy minerals in natural accumulates. Australian Inst. Mining Metall., Mono. **1**, 146 p. (1962).

Baturin, V. P.: Petrografia peskov i peschanikov produktivnoi tolshchi (Petrography of the sands and sandstones of the productive series). Trans. Azerbaidjan Neft. issled Inst., Bull. **1**, 1–96 (1931).

Beveridge, A. J.: Heavy minerals in lower Tertiary formations in the Santa Cruz Mountains, California. Jour. Sed. Petrology **30**, 513–537 (1960).

Biscaye, P. E.: Mineralogy and sedimentation of Recent deep-sea clay in the Atlantic Ocean and adjacent seas and oceans. Geol. Soc. America Bull. **76**, 803–832 (1965).

Blatt, Harvey: Original characters of clastic quartz. Jour. Sed. Petrology **37**, 401–424 (1967a).

— Provenance determinations and recycling of sediments. Jour. Sed. Petrology **37**, 1031–1044 (1967b).

— Christie, J. M.: Undulatory extinction in quartz of igneous and metamorphic rocks and its significance in provenance studies of sedimentary rocks. Jour. Sed. Petrology **33**, 559–579 (1963).

Bokman, John: Clastic quartz particles as indices of provenance. Jour. Sed. Petrology **22**, 17–24 (1952).

Bonham-Carter, G. F.: Fortran IV program for Q-mode cluster analysis of nonquantitative data using IBM 7090/7094 computers. Kansas Geol. Survey, Computer Contribution **17**, 28 p. (1967).

Boswell, P. G. H.: Mineralogy of sedimentary rocks, 393 p. London: Thos. Murby and Co. 1933.
Brammall, A.: Dartmoor detritals. A study in provenance. Proc. Geol. Assoc. **39**, 27–48 (1928).
Cadigan, Robert A.: Petrology of the Morrison Formation in the Colorado Plateau. U.S. Geol. Survey Prof. Paper **556**, 113 p. (1967).
Cogen, W. M.: Heavy-mineral zones of Louisiana and Texas Gulf Coast sediments. Am. Assoc. Petroleum Geologists Bull. **24**, 2069–2101 (1940).
Comité des Techniciens: Methodes modernes de traitement de l'information géologique sur ordinateur, 137 p. Paris: Editions Technip 1969.
Conolly, J. R.: The occurrence and polycrystallinity and undulatory extinction in quartz in sandstones. Jour. Sed. Petrology **35**, 116–135 (1965).
Crook, K. A. W.: Weathering and roundness of quartz sand grains. Sedimentology **11**, 171–182 (1968).
Dake, C. L.: The problem of the St. Peter Sandstone. Univ. Missouri School of Mines and Metall., Bull., Tech. Ser. **6**, 158 p. (1921).
Davies, D. K., Moore, W. R.: Dispersal of Mississippi sediment in the Gulf of Mexico. Jour. Sed. Petrology **40**, 339–353 (1970).
Doeglas, D. J.: The importance of heavy mineral analysis for regional sedimentary petrology. Rept. Comm. Sedimentation 1939–1940; Nat. Research Council, 102–121 (1940).
Doty, R. W., Hubert, J. F.: Petrology and paleogeography of the Warrenburg channel sandstone, western Missouri. Sedimentology **1**, 7–39 (1962).
Feniak, M. W.: Grain sizes and shapes of various minerals in igneous rocks. Am. Mineralogist **29**, 415–421 (1944).
Feo-Codecido, Gustavo: Heavy-mineral techniques and their application to Venezuelan stratigraphy. Am. Assoc. Petroleum Geologists, Bull. **40**, 984–1000 (1956).
Folk, R. L.: Petrology of sedimentary rocks, 170 p. Austin: Hemphill's 1968.
— Robles, R.: Carbonate sands of Isla Perez, Alacran Reef complex, Yucatan. Jour. Geology **72**, 255–292 (1964).
Füchtbauer, Hans: Sedimentpetrographische Untersuchungen in der älteren Molasse nördlich der Alpen. Eclogae geol. Helvetiae **57**, 158–298 (1964).
— Die Sandsteine in der Molasse nördlich der Alpen. Geol. Rundschau **56**, 266–300 (1967).
Grabau, A. W.: On the classification of sedimentary rocks. American Geologist **33**, 228–247 (1904).
— Principles of stratigraphy, 1185 p. New York: Seiler 1913.
ten Haaf, E.: Flysch formations of the northern Apennines. In: Bouma, A. H., Brouwer, A., Eds.: Turbidites: Developments in Sedimentology, Vol. 3, p. 127–136. Amsterdam: Elsevier 1964.
Harbaugh, J. W., Merriam, D. F.: Computer applications in stratigraphic analysis, 282 p. New York: John Wiley 1968.
Hooper, W. F.: Petrography of Lakota conglomerate, Casper Arch area, Wyoming. In: Symposium on Early Cretaceous rocks of Wyoming and adjacent areas – Wyoming: Wyoming Geol. Assoc., 17th Ann. Field Conf., Casper, Wyoming, p. 131–140 (1962).
Hough, J. L.: Sediments of Cape Cod Bay, Massachusetts. Jour. Sed. Petrology **12**, 10–30 (1942).
Hubert, J. F.: A zircon-tourmaline-rutile maturity index and the interdependence of the composition of heavy mineral assemblages with the gross composition and texture of sandstones. Jour. Sed. Petrology **32**, 440–450 (1962).
Huffman, G. G., Price, W. A.: Clay dune formation near Corpus Christi, Texas. Jour. Sed. Petrology **19**, 118–127 (1949).
Hurley, P. M., Brookins, D. G., Pinson, W. H., Hart, S. R., Fairbairn, H. W.: K-Ar age studies of Mississippi and other river sediments. Geol. Soc. America Bull. **72**, 1807–1816 (1961).
Imbrie, John, van Andel, Tj.: Vector analysis of heavy mineral data. Geol. Soc. America Bull. **75**, 1131–1156 (1964).
— Purdy, E. D.: Classification of modern Bahamian carbonate sediments. In: Ham, W. E., Ed.: Classification of Carbonate Rocks. Am. Assoc. Petroleum Geologists, Mem. **1**, 253–272 (1965).
Ingle, J. C., Jr.: The movement of beach sand. Developments in Sedimentology, Vol. 5, 221 p. Amsterdam: Elsevier 1966.
Keller, W. D., Littlefield, R. F.: Inclusions in the quartz of igneous and metamorphic rocks. Jour. Sed. Petrology **20**, 74–84 (1950).
Kennedy, V. C., Kouba, D. L.: Fluorescent sand as a tracer of fluvial sediment. U.S. Geol. Survey, Prof. Paper **562** E. p. E1-E13 (1970).

Krumbein, W. C.: Statistical summary of some alluvial gravels. Rept. Comm. Sedimentation for 1940–1941, Nat. Res. Council **1942**, 9–14.
— Composite end members in facies mapping. Jour. Sed. Petrology **25**, 115–122 (1955).
— Regional and local components in facies maps. Am. Assoc. Petroleum Geologists Bull. **40**, 2163–2194 (1956).
— A comparison of polynomial and Fourier models in map analysis. Northwestern Univ., Tech. Rept. No. 2, ONR Task No. 388-078, Contract Nonr-1228 (36), 45 p. (1966).
— Graybill, F. A.: An introduction to statistical models in geology, 475 p. New York: McGraw-Hill 1965.
— Pettijohn, F. J.: Manual of sedimentary petrography, 549 p. New York: Appleton-Century 1938.
Krynine, P. D.: Petrography and genesis of the Siwalik Series. Am. Jour. Sci., Ser. 5, **34**, 422–446 (1937).
— Petrology and genesis of the Third Bradford Sand. Pennsylvania State Coll., Mineral Ind. Expt. Sta., Bull. **27**, 134 p. (1940).
— Microscopic morphology of quartz types. Proc. 2nd Pan-Am. Cong. Mining Eng. and Geology **3**, 2nd Comm. 35–49 (1946a).
— The tourmaline group in sediments. Jour. Geology **54**, 65–87 (1946b).
Kuenen, Ph. H.: Sand – its origin, transportation, abrasion and accumulation. Geol. Soc. S. Africa, annexure to **62**, 1–33 (1959).
Ledent, D., Patterson, C., Tilton, G. R.: Ages of zircon and feldspar concentrates from North American beach and river sands. Jour. Geology **72**, 112–122 (1964).
Mackie, W.: The sands and sandstones of eastern Moray. Edinburgh Geol. Soc. Trans. **7**, 148–172 (1899).
— The source of the purple zircons in the sedimentary rocks of Scotland. Edinburgh Geol. Soc. Trans. **11**, 200–213 (1923).
Marsal, Dietrich: Statistische Methoden für Erdwissenschaftler, 152 p. Stuttgart: E. Schweizerbart'sche Verlagsbuchhandlung 1967.
Marshall, P. E.: The wearing of beach gravels. Trans. Proc. New Zealand Inst. **58**, 507–532 (1927).
Meckel, L. D.: Origin of Pottsville conglomerates (Pennsylvanian) in the central Appalachians. Geol. Soc. America Bull. **78**, 223–258 (1967).
Merriam, D. F., Ed.: Symposium on computer applications in petroleum exploration. Kansas Geol. Survey, Computer Contribution **40**, 41 p. (1969a).
— Computer applications in the earth sciences, 282 p. New York: Plenum 1969b.
— Cocke, N. C.: Computer applications in the earth sciences. Kansas Geol. Survey, Computer Contribution **12**, 62 p. (1967).
Miesch, A. T., Connor, J. J.: Stepwise regression and nonpolynomial models in trend analysis. Kansas Geol. Survey, Computer Contribution **27**, 40 p. (1968).
Milner, H. B.: Supplement to introduction to sedimentary petrography, 157 p. London: Murby 1926.
Moss, A. J.: Origin, shaping and significance of quartz sand grains. Jour. Geol. Soc. Australia **13**, 97–136 (1966).
Parks, J. M.: Cluster analysis applied to multivariate geologic problems. Jour. Geology **74**, 703–715 (1966).
— Fortran IV program for Q-mode cluster analysis on distance function with printed dendrogram. Kansas Geol. Survey, Computer Contribution **46**, 32 p. (1970).
Payne, T. G.: Stratigraphical analysis and environmental reconstruction. Am. Assoc. Petroleum Geologists, Bull. **26**, 1697–1770 (1942).
Pelletier, B. R.: Pocono paleocurrents in Pennsylvania and Maryland. Geol. Soc. America Bull. **69**, 1033–1064 (1958).
Pelto, C. R.: Mapping of multicomponent systems. Jour. Geology **62**, 501–511 (1954).
Pettijohn, F. J.: Relative abundance of size grades of clastic sediments (abstr.). Program Soc. Econ. Paleon. Mineralogists, Chicago, 1940.
— Sedimentary Rocks (2nd Ed.), 718 p. New York: Harper 1957.
Pittman, E. D.: Use of zoned plagioclase as an indicator of provenance. Jour. Sed. Petrology **33**, 380–386 (1963).
Poldervaart, Arie: Zircon in rocks. 2 Igneous rocks. Am. Jour. Sci. **254**, 521–554 (1956).
Potter, P. E., Pryor, W. A.: Dispersal centers of Paleozoic and later clastics of the upper Mississippi valley and adjacent areas. Geol. Soc. America Bull. **72**, 1195–1250 (1961).

Potter, P. E., Siever, Raymond: Sources of Basal Pennsylvanian sediments in the Eastern Interior Basin Part 3, Some methodological implications. Jour. Geology **64**, 447–455 (1956).

Rimsaite, J.: Optical heterogeneity of feldspars observed in diverse Canadian rocks. Schweiz. Min. Pet. Mitt. **47**, 61–76 (1967).

Rittenhouse, Gordon: Transportation and deposition of heavy minerals. Geol. Soc. America Bull. **54**, 1725–1780 (1943).

Russell, R. Dana: Mineral composition of Mississippi River sands. Geol. Soc. America Bull. **48**, 1307–1348 (1937).

Russell, R. J.: Where most grains of very coarse sand and fine gravel are deposited. Sedimentology **11**, 31–38 (1968).

Schlee, J.: Uplands gravels of southern Maryland. Geol. Soc. America Bull. **68**, 1371–1410 (1957).

Sorby, H. C.: On the structure and origin of non-calcareous stratified rocks. Geol. Soc. London Proc. **36**, 46–92 (1880).

Sundborg, Åke: The River Klarälven – A study of fluvial processes. Geografiska Annaler **38**, 127–316 (1956).

Suttner, L. J.: Stratigraphic and petrographic analysis of Upper Jurassic-Lower Cretaceous Morrison and Kootenai Formations, southwest Montana. Am. Assoc. Petroleum Geologists Bull. **53**, 1391–1410 (1969).

Todd, T. W., Folk, R. L.: Basal Claiborne of Texas, record of Appalachian tectonism during Eocene. Am. Assoc. Petroleum Geologists Bull. **41**, 2545–2566 (1957).

Tomita, T.: Geologic significance of the color of granite zircon and the discovery of the Pre-Cambrian in Japan. Kyushu Univ., Mem. Fac. Sci., Ser. D, Geol. **4**, 135–161 (1954).

Tyler, S. A., Marsden, R. W., Grout, F. F., Thiel, G. A.: Studies of the Lake Superior Pre-Cambrian by accessory-mineral methods. Geol. Soc. America Bull. **51**, 1429–1538 (1940).

Udden, J. A.: Mechanical composition of clastic sediments. Geol. Soc. America Bull. **25**, 655–744 (1914).

Vitanage, P. W.: Studies of zircon types in Ceylon Pre-Cambrian complex. Jour. Geology **65**, 117–138 (1957).

Voll, G.: New work on petrofabrics. Liverpool and Manchester Geol. Jour. **2**, pt. 3, 503–567 (1960).

Walker, R., Pettijohn, F. J.: Archean geosynclinal basin.: Analysis of the Minnitaki Basin, northwestern Ontario. Geol. Soc. America Bull. **82**, 2099–2129 (1971).

Weaver, C. E.: Geologic interpretation of argillaceous sediments. Am. Assoc. Petroleum Geologists Bull. **42**, 254–271, 272–309 (1958).

Webb, W. M., Potter, P. E.: Petrologia y geoquimica de detritos derivados de un terreno riolitico de la region occidental de Chihuahua Mexico. Bol. Soc. Geol. Mexicana, **32**, 45–61 (1971).

Wentworth, C. K.: Fundamental limits to the sizes of clastic grains. Science **77**, 633–634 (1933).

Wishart, Davis: Fortran II programs for 8 methods of cluster analysis (Clustan I). Kansas Geol. Survey, Computer Contribution **38**, 112 p. (1969).

Wolf, Karl H.: Textural and compositional transitional stages between various lithic grain types (with a comment on "Interpreting detrital modes of graywacke and arkose"). Jour. Sed. Petrology **41**, 328–332 (1971).

Wolff, R. G.: The dearth of certain size of materials in sediments. Jour. Sed. Petrology **34**, 320–327 (1964).

Zimmerle, Winfried: Serpentine graywackes from the North Coast basin, Columbia, and their geotectonic significance. Neues Jahrb. Mineral. abh. **109**, 156–182 (1968).

Chapter 9.
Transport, Deposition, and Deformation of Sand

Introduction

One of the major objectives of sedimentology is to infer from the properties of a bed how sand is transported and deposited by running water and wind on land and in the different realms of the ocean. This objective is an ambitious one and overlaps the objectives of those geomorphologists who study fluvial morphology and processes; of hydraulic engineers, who are concerned with the control of rivers, harbors, and beaches; of physical oceanographers, who study water movement in the ocean; and of fluid dynamicists, who study the fundamental laws of fluid motion with or without entrained particles. The problem is complex and bibliographies are long.

The flow of fluids alone, difficult as its study can be, is easier to cope with analytically than the behavior of a fluid-sediment mixture. With the latter one must study not only the behavior of fluid particles but also must correlate the behavior of an almost infinite number of sediment particles with flow characteristics which vary from point to point in both space and time. Even with the help of simplifying assumptions this is a most difficult task.

One can consider the problem of sand transport and deposition at many different levels. For example, much attention has been given in the literature to the forces acting on individual grains. Others have derived equations relating flow characteristics to transport expressed in weight per unit cross section per unit time. Although fundamental and practical to the engineer, neither are of direct interest and use to the student of ancient sandstones, for the rate of sedimentation of a specific body or bed of sand is very difficult, if not impossible, to obtain in ancient and even more modern sand accumulations. Though sedimentologists have devoted great energy to grain size analysis, hoping for a useful relation to fluid flow regimes, little of use has emerged from the fluid flow experimental literature. More relevant to sedimentologists are studies relating flow characteristics to type, magnitude, and migration of sand waves at the sediment-water interface or, more generally, the study of bedding and sedimentary structures. But what of larger morphological units such as point bars or sand accumulations in estuaries? Or the equilibrium profile of a beach? Clearly transport and deposition of sand at this level is closely linked to sand body morphology as treated in Chapter 11.

In our selection and treatment of topics we have been strongly influenced by the fact that as geologists our prime interest is an understanding of the past. For example, to make a careful study of the fluid dynamics of a cross section of a modern river reach, one measures discharge or average velocity, slope, determines bottom profile and median diameter of the sediment, and fluid properties such as viscosity and density. In a beach study one would include offshore bottom profile; breaker angle; period wavelength and amplitude of the waves; median diameter of the sand; and fluid properties. Many more and complex fluid dynamic parameters in use today have been used but almost all can be derived from these few. In both examples one specifies the process variables to understand the resultant phenomena – the size distribution, bed forms, and perhaps the morphology of the sand accumulation. In contrast, the geologist looking at an ancient river or beach deposit has only the resultant to work with, for the fluid phase has long since vanished and his job is to infer at least some of its characteristics. Thus when working with ancient sands and sandstones, the geologist sees formative processes through a "geologic filter", one that strongly conditions his view of transportation and deposition. Essentially only the size distribution, sedimentary structures, bedding, the vertical sequence, and information to be gleaned from the shape of the deposit are available to him. In other words, if the application of sediment transport mechanics to ancient sandstones is to have maximum value, it must be based on those variables that leave a tangible record.

Another question – and one that is not easy to answer – is what precisely are our goals of study? One might immediately respond by saying, "A better understanding of conditions at the time of deposition". In details, this might mean reconstruction of the velocity or discharge of the flow, an estimate of water depth, perhaps a map of tractive force across a sand body, or the making of a distinction between air and water or suspension and traction transport of sand. Or one might ask the questions, "Why do certain sedimentary structures commonly occur together in a given temporal sequence as in a graded bed? What conditions are needed to produce soft sediment deformation of sand?" In the broadest sense we are searching for physical explanations of bedding and its areal distribution in sand bodies. In so doing we realize, of course, that explanations are tied to the physical transport of sand and hence in the first instance largely independent of particular geographic environments. But we also have a hope that complex combinations of fluid regimes may be related to depositional environments, though not in a simple way. Finally we would like to stress that the fluid dynamic study of bedding and its distribution in sand bodies is only in its beginnings, much as was aerodynamic theory in the early part of the 20th century when J. C. Hunsaker wrote in 1915, "Inadequate theory, employed as a guide in a qualitative sense, is better than no theory at all".

The first part of this chapter, "Fluid Flow and Entrainment", provides the minimal background needed for an insight to the transport, emplacement, and deformation of sand. Here we begin by first considering, because it is relatively simple, fluid flow without particles, and subsequently the interaction of fluid and grains. To supplement "Fluid Flow and Entrainment" a reader who is not familiar with elementary fluid mechanics may benefit from consulting the annotated references, which contain a brief introduction to much of the hydraulic engineering and fluid dynamic literature and the glossary, which contains commonly recurring fluid dynamic and related terms. The subsequent headings of this chapter reflect some of the major realms of interest to sedimentologists. The fluid dynamic origin of sedimentary structures in these different realms is our main concern, although we also consider some of the broader aspects of sand transport and deposition, some of which are covered from a rather different point of view in Chapter 11. Because knowledge of fluid-particle behavior and the fluid dynamic origin of sedimentary structures in some of these realms is still incomplete, our treatment is not uniformly comprehensive.

Fluid Flow and Entrainment

Aspects of Fluid Flow

There are two types of flow, *laminar* and *turbulent*, based on the kinematic and dynamic properties of fluids. In the former, *streamlines* (imaginary curves connecting the tangents to the directions of motions of a fluid particle) are parallel and separate from one another, the flow moves in laminae parallel to the boundary and no mass transfer between layers takes place. Flow is relatively slow and velocity components other than those in the principal flow direction are negligible. In turbulent flow, on the other hand, the streamlines are complex and intertwined and continually changing with time in an unpredictable way so that rather than fluid moving in well defined laminae or sheets there are instead complex eddies which superimpose an irregular, random motion upon the larger general movement of the flow. In turbulent flow, masses of water move up, down, and laterally with respect to general flow direction, transferring mass and momentum. Although the irregular motions of such masses or lumps of fluid have velocities that deviate only a few percent from average velocity, they nonetheless exert a decisive effect upon the flow for it is turbulence which keeps particles in suspension, either constantly, as are the clays and silts of rivers and the sand of turbidity currents, or intermittently, as are most sand grains in rivers, beaches, and dunes.

Turbulence entrains particles in a combination of two ways: by added fluid force and by a related reduction of the local pressure as the turbulent eddy passes overhead. Both promote entrainment of sand along the bottom. Almost all the flows that transport sand in nature are fully turbulent ones. Turbulence is principally generated in rivers by shear due to gravity flow and may be increased significantly by bottom roughness; along shorelines and at sea it is produced by waves, surface wind stress, and shear between currents. In air it is turbulence that carries debris from a volcanic explosion as it is transported downwind. The magnitude of turbulent motion varies from micro to macro, the latter being easily seen in rivers as large complex eddies or boils that impinge against the surface, persist for some seconds and are then replaced by others. Matthes (1947) gives a useful classification of macroturbulence.

Whether laminar or turbulent, the effect that a flowing fluid exerts on its boundary at the sediment interface depends on fluid properties: *density*, ϱ, and *dynamic* or *absolute viscosity*, μ; some measure of fluid dynamic properties of *acceleration* or *velocity;* and on the geometry of the sediment – water boundary. From density and dynamic viscosity one obtains two other widely used properties: *specific weight*, $\lambda = \varrho g$, where g is the acceleration of gravity, and *kinematic viscosity*, $\nu = \mu/\varrho$. Commonly recurring symbols along with their dimensions are

Table 9-1. *Commonly recurring symbols and their dimensions*

Symbol	Meaning	Symbol	Meaning
d	grain diameter, L	γ	specific weight, $ML^{-2}T^{-2}$
D	water depth, L	μ	dynamic viscosity, $ML^{-1}T^{-1}$
F	Froude number	ν	kinematic viscosity, L^2T^{-1}
g	gravity, LT^{-2}	ϱ	density, ML^{-3}
L	length, L	ϱ_s	solid density, ML^{-3}
M	mass, M	ϱ_f	fluid density, ML^{-3}
P	pressure, $ML^{-1}T^{-2}$	σ_s	shear strength of a nonfluid substance, $ML^{-1}T^{-2}$
R	Reynolds number	τ	shear stress, $ML^{-1}T^{-2}$
R_*	boundary Reynolds number	τ_0	shear stress between a fluid and its boundary, $ML^{-1}T^{-2}$
T	time, T		
v	velocity, LT^{-1}	τ_i	shear stress between a turbidity current and its upper, *fluid* boundary, $ML^{-1}T^{-2}$
v_*	friction or shear velocity, LT^{-1}		
α	angle of inclination	τ_b	shear stress between two unconsolidated beds, $ML^{-1}T^{-2}$

given in Table 9-1. Knowledge of the dimensions of these and other hydraulic parameters provides helpful insight to the role they play in hydraulics.

Every real fluid such as water or air has an internal frictional resistance to flow called viscosity. The viscosity of a fluid is a measure of its resistance to shear – the extent to which a slower moving fluid mass retards a faster moving one so that a *shearing stress*, τ, is generated at the boundary of the two. Shearing stress is measured in force per unit area and always acts parallel to direction of movement (Fig. 9-1). The concept of a shear stress acting either between two fluid masses or between a fluid and a solid, such as the shear stress exerted by a stream on its boundary or by waves breaking on a beach, is fundamental to sediment transport mechanics. In laminar flow the shear stress or drag force per unit area between two such sliding masses is given by

$$\tau = \mu(dv/dy)\,, \tag{9-1}$$

where dv/dy is the velocity gradient normal to the boundary, y is the distance from the bottom, and v is the velocity (Fig. 9-2).

Dynamic viscosity is a fluid property that does not vary with the state of motion of fluid, but is very temperature dependent, decreasing with increase in temperature. The temperature dependence of viscosity is the result of the fundamental molecular forces of a fluid that are described by viscosity terms. This temperature effect is important to the sedimentologist, because viscosity plays a significant role in the settling or fall velocity of a particle; the colder and more viscous the water, the greater its resistance to deformation and consequently the slower fall velocity. Change in its fall velocity can thus significantly alter the capacity of a stream to carry sand and influence the microrelief of the sand-water interface. Another factor affecting dynamic viscosity is concentration of suspended fine clay. The viscosity of the clay-water mixture is called *apparent viscosity*; it is larger than that of clear water alone and has an important effect in retarding the fall velocity of suspended particles. This together with the ability of a muddy stream to exert a greater drag force, for a given discharge and temperature, accounts for the capability of a turbid stream transporting more sand than a comparable clear-water stream. It is vital to an understanding of the mechanics of turbidity currents.

In turbulent flow one must consider not only the viscosity generated by the frictional forces as described above but also that generated by the turbulent eddies as well so that the drag force is now defined by

$$\tau = (\mu + \eta)\,dv/dy \qquad (9\text{-}2)$$

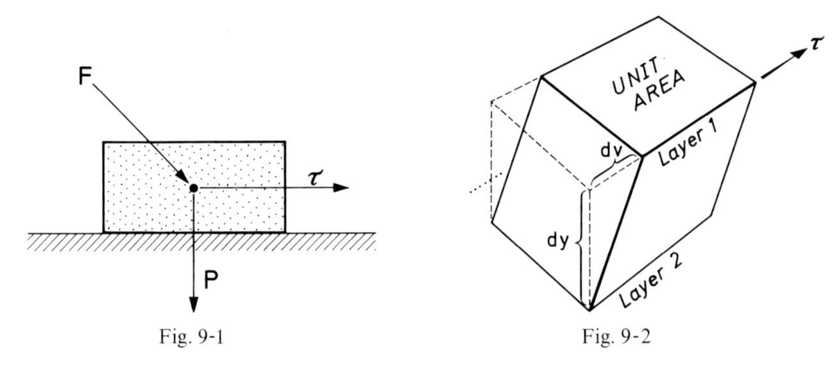

Fig. 9-1 Fig. 9-2

Fig. 9-1. Applied force, F, resolved into pressure P and shear stress, τ, both of which have dimensions of pressure ($ML^{-1}T^{-2}$)

Fig. 9-2. A geometrical representation of laminar shear stress between two sliding masses of fluid. Layer 1 slides over layer 2 and τ, the shear stress, may be thought of as the force that produces a change in velocity, dv, relative to height, dy

where η is called the *eddy viscosity* and is the rate of exchange of fluid mass between adjacent water masses, that is, the viscosity resulting from turbulent momentum transfer. Because of the complex random motion imparted to the fluid by the turbulent eddies, one uses the time-average velocity, $\bar{v}$, at any point to determine the gradient $d\bar{v}/dy$, rather than dv/dy as with laminar flow. The random interchange of material between masses not only transfers momentum, thus slowing faster particles and accelerating the slower ones, but also transfers sediment and diffuses salt concentrations. Eddy viscosity depends on the state of motion of the fluid and is much less temperature dependent than is dynamic viscosity. Numerically, eddy viscosity greatly exceeds dynamic viscosity in fully turbulent flows except very near the boundary.

To distinguish between turbulent and laminar flow the *Reynolds number*, R, is used and is defined by the equation

$$R = \frac{\varrho v L}{\mu}, \qquad (9\text{-}3)$$

where L is some measure of length, sometimes called a *hydraulic radius* (frequently taken as depth for rivers). Reynolds number is the ratio of the inertial force of a fluid to its viscous force and as such it represents the ratio between a driving and retarding force. Sir Osborne Reynolds, an English physicist, pointed out in 1883 the relationship between inertial and viscous forces that led to this dimensionless number. For a given flow geometry a large Reynolds number, certainly one in excess of 2000, indicates that the flow is turbulent and that inertial forces greatly exceed viscous forces. Thus the larger the Reynolds number, the less important the influence of dynamic viscosity upon flow pattern. Conversely, if R is small, much less than 500, viscous forces are dominant and the flow is laminar. For any given boundary, liquid, and temperature there will be a range of values that define a zone of transitional flow between the two types of flow, laminar and turbulent. The transition zone also depends on geometry of the flow and surface roughness of the boundary.

Another dimensionless number that plays an important role is the *Froude number*, F, the ratio of inertial force to gravity force. Froude number is defined as

$$F = \frac{v}{\sqrt{Dg}} \tag{9-4}$$

where v is velocity, D is depth, and g is gravity. Because it is the ratio of inertial to gravity force, Froude number is used in fluid flow problems when the flow has an unconfined or free surface as in a stream or in a shallow tidal estuary. The concept of free surface flow has been extended by some to apply to submarine turbidity currents, whose "free" surface is considered to be the interface between the denser turbid current below and the clear water above. In free surface flow gravity plays a significant role unlike, for example, pressure flow in a closed pipe where pressure and not gravity is the driving force. The Froude number also distinguishes between two types of flow, *shooting* or supercritical ($F > 1$) and *tranquil*, sometimes called streaming, or subcritical ($F < 1$). The transition between the two is called *critical flow* ($F = 1$). In shooting flow, a surface wave will be carried downstream by the current whereas in tranquil flow the wave front will move upstream against the current. Shooting flow occurs in rapids, constrictions, some floods, and where breakers rush up a beach. The type of flow – whether it is tranquil or shooting – plays an important role in molding the microrelief, the bed forms that produce many sedimentary structures on sandy bottoms. Two flows are dynamically similar if both Reynolds and Froude numbers are the same. Dynamic similarity is an important concept for model studies, for it allows the scaling down of natural large systems so that they can be studied in the laboratory.

A fluid flowing over a boundary exerts a shearing force on it and, conversely, the boundary exerts a retarding force on the flowing fluid. The zone where the fluid is appreciably retarded by the frictional drag of the boundary is called the *boundary layer* (Fig. 9-3). Depending upon the relief or roughness of the boundary and the velocity and viscosity of the flow, the thickness of the boundary layer may vary from a small fraction of a millimeter to several millimeters, or it can, under certain conditions, such as a shallow flow over a scour trough or a rough gravel bed, greatly expand to become an appreciable fraction or all of the total depth of the stream. Flow within the boundary layer may be either turbulent or laminar but, in any case, it is in the boundary layer where the flowing fluid loses much of its kinetic energy. If the flow is laminar in the boundary layer, the term *laminar sublayer* is used. A laminar sublayer in an otherwise turbulent boundary markedly reduces frictional resistance and thus may protect many small grains from entrainment by a turbulent flow.

The idea of the boundary layer allowed the conceptual separation of the fluid into two parts – one that is strongly affected by the boundary and a main flow that is largely isolated from it. The idea of separation of the total flow into these two parts that was proposed by Ludwig Prandtl in 1904 was a major achievement, for it permits one to apply the classical laws of fluid dynamics to the main body of flowing fluid and then combine this solution with a separate analysis of the flow in the boundary layer itself. For sedimentologists the boundary layer has a

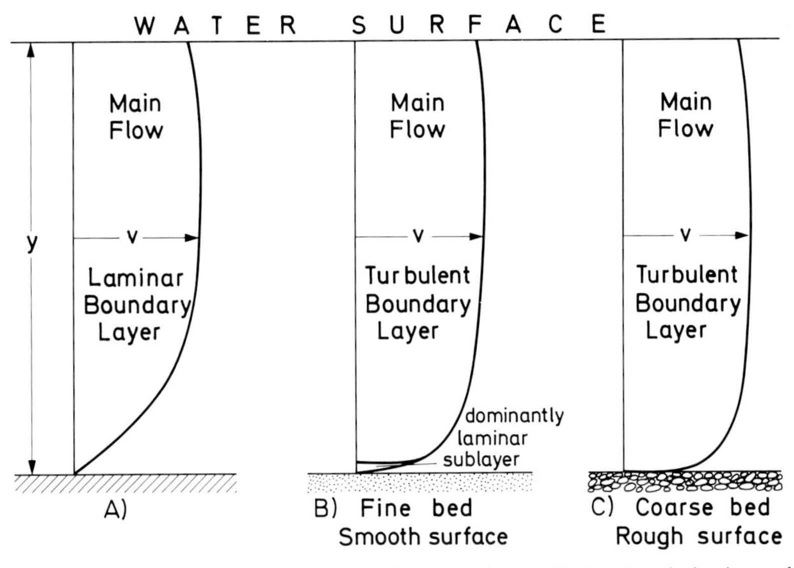

Fig. 9-3. Laminar boundary layer (A), turbulent boundary layer with dominantly laminar sublayer (B), and fully turbulent boundary layer (C)

threefold significance. It is in the boundary layer that the maximum exchange between the loose grains of the interface and those temporarily in suspension takes place, obviously a key zone for transportation and deposition. Secondly, the characteristics of the boundary layer have a marked influence on the shear stress exerted by the moving fluid on the bottom. Finally, the boundary layer may become expanded downstream from a sand wave or behind a boulder, where the main flow separates from the boundary. This phenomenon is called *flow separation* and is characteristic of the lee sides of obstacles such as sandwaves or pebbles and cobbles that project above the bottom (Fig. 9-4).

For fully turbulent flow over a rough surface such as a sand bed, the vertical velocity profile is given by an equation developed by von Kármán

$$\frac{\bar{v}}{v_*} = 8.5 + \frac{2.3}{k} \log \frac{y}{y_0}, \tag{9-5}$$

where $\bar{v}$ is the average velocity at a distance y above the bed, v_* is the shear or friction velocity, k is von Kármán's dimensionless constant, determined by experiment to be 0.4 for most fluids, and y_0 is a roughness length which depends upon the magnitude of the irregularities or roughness of the bottom. *Friction velocity* is related to shear stress by $v_* = \sqrt{\tau_0/\varrho}$ and has the dimensions of velocity. One should note that τ_0 is the shear stress between the fluid and the boundary and not the shear stress within the fluid. The above velocity law applies to both air and water.

The roughness or relief of the boundary is some fraction of grain diameter for a flat bed of sand but is much larger should the bed be rippled or duned.

Fig. 9-4. Side view of asymmetrical ripples in a flume (current from right to left). Aluminum powder defines main flow, flow separation, and its turbulent eddy in bed of ripples (Photo by A. V. Jopling)

Roughness has a pronounced effect on the velocity profile, the greater the roughness the greater the decrease in the velocity gradient or profile in the boundary layer because, as roughness increases, the more the frictional resistance of the boundary retards the flow near it (Fig. 9-5). Resistance resulting from the deformation of the boundary is called *form resistance* or *form drag*. Ripples and dunes are features that give rise to form drag.

The von Kármán equation is important to sedimentologists because it shows that the average velocity at any point in a flow depends upon the height of that point above the bottom, on the boundary roughness, and on friction velocity. The laminar sublayer plays a role in the von Kármán equation for it can alter $\bar{v}$ via boundary roughness and friction velocity. Shear stress, an impelling force acting at the boundary, is more commonly used than velocity when studying the entrainment of sand and the origin of bed forms, because velocity varies logarithmically as the height.

Settling Velocity

Four physical parameters dominate sediment transport mechanics: *settling velocity*, the *state of fluid flow* described by its turbulent structure and boundary shear stress, *gravitational sliding*, and *porewater pressure*. Irrespective of depositional environment or process these four commonly play an important role. In the sedimentological literature, settling velocity has received the most attention.

The properties of grains that are significant in relation to fluid flow are size, density, and shape. The combined effect of these variables as well as the density and viscosity of the fluid is summed up by settling or fall velocity, which is the terminal or steady state velocity of a particle falling in the fluid. Settling velocity is an essential feature of practically all theories of sedimentation, whether it is

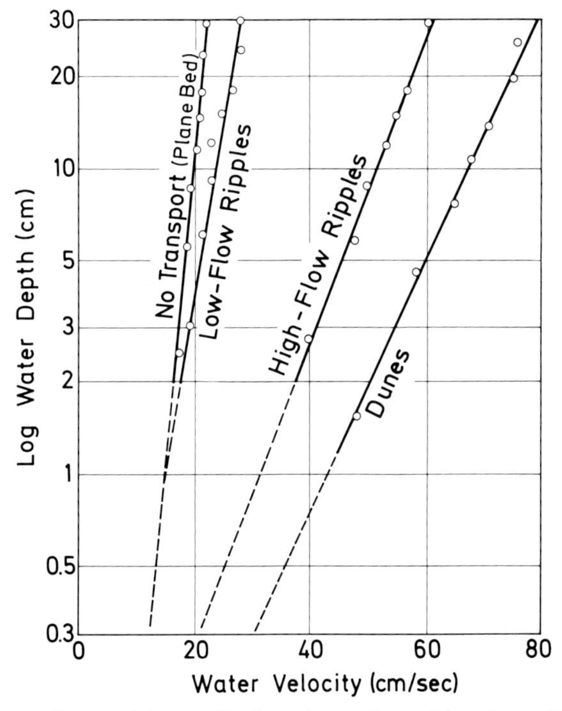

Fig. 9-5. Velocity gradients and forms of bed roughness observed in a flume. Note change in vertical velocity gradients as bed-form roughness increases (Modified from Harms, 1969, Fig. 10)

for bed load (sand and gravel moving along the bottom), saltation (intermittent suspension transport), or complete suspension. Simons and Richardson (1966, p. 16) suggest that settling velocity is also the primary variable that determines interaction between bed material and its transporting fluid. Certainly settling velocity is a convenient and meaningful way to sum up the interaction of a fluid and a single grain.

The equation for settling velocity, w, is given by the governing equation of motion: the algebraic sum of the downward force, the weight of the particle, and the upward or resisting forces of buoyancy and frictional resistance or drag are equated to $M \frac{dw}{dt}$ to obtain

$$M \frac{dw}{dt} = F_W - F_B - F_D \tag{9-6}$$

where M is particle mass, F_W is the weight of the particle, F_B is its buoyancy, and F_D is the drag force of the liquid on the falling particle. At terminal velocity there is no acceleration so $dw/dt = 0$ and thus $F_D = F_W - F_B$. Assuming that the drag force exerted by a fluid on a falling grain is proportional to the density of the fluid, cross-sectional area of the grain, and velocity, one obtains for spheres (Scheidegger, 1970, p. 176-179)

$$F_D = C_D \pi \frac{d^2}{4} \frac{\varrho_f w^2}{2}, \tag{9-7}$$

where C_D is the drag coefficient, which depends on Reynolds number and the shape of the particle, d is grain diameter, ϱ_f is the fluid density, and w is the settling velocity. C_D can be thought of as a measure of the relative resistance to the flow of differently shaped bodies with the same cross sectional area or length under identical flow conditions. F_W and F_B are given by

$$F_W = \frac{d^3}{6} \pi g \varrho_s \quad \text{and} \quad F_B = \frac{d^3}{6} \pi g \varrho_f, \tag{9-8}$$

where ϱ_f and ϱ_s are fluid and solid densities. Substituting Eqs. (9-7) and (9-8) into (9-6), for $\frac{dw}{dt} = 0$, we have:

$$w = \left(4/3 \frac{dg}{C_D} \frac{\Delta\varrho}{\varrho_f} \right)^{1/2} \tag{9-9}$$

where $\Delta\varrho = (\varrho_s - \varrho_f)$. In the common sedimentologic situation of quartz grains falling in water, for a given drag coefficient, Eq. (9-9) reduces to w varying as a constant times the square root of the diameter, where the constant term includes the drag coefficient, the densities, and g. A similar formula can also be obtained for non-spherical particles, but empirically determined shape factors must be used.

The drag coefficient, C_D, because of its dependence on R, varies with the type of flow. If the flow is fully turbulent (inertial), C_D is approximately one half for spheres. If it is slow laminar (viscous), C_D is found to be $24/R$ (Rouse, 1961, p. 244) from which we obtain Stokes' law by substituting Eq. (9-3) into (9-9),

$$w = \frac{1}{18} \frac{d^2 g}{\mu} \Delta\varrho = \frac{2}{9} r^2 g \frac{\Delta\varrho}{\mu} \tag{9-10}$$

and substituting for Reynolds number, where the length term of R is particle diameter, and converting diameter to radius, r. For Stokes' law (formulated by G. G. Stokes in 1845) to apply, the fluid must be isothermal, free of boundary effects, and completely non-turbulent. It works well for particles up to 0.18 mm that have viscous (laminar) settling but not for larger ones that have inertial (turbulent) settling. Rubey (1933) proposed an equation for larger particles. The comparison between Stokes' law, the settling behavior of spheres, natural quartz sand with a sphericity of 0.65, and Rubey's law is given in Fig. 9-6A. The behavior of large particles falling in quiescent liquids is described by Stringham and others (1969). A historical review of the different approaches to settling velocity including data based on natural sand is given by Graf and Acaroglu (1966).

As temperature decreases viscosity increases rapidly compared to a relatively constant density difference; as a result fall velocity decreases with decrease in temperature (Fig. 9-6B). As noted earlier, the temperature dependence of viscosity links stream carrying power and bed forms to the weather and climate – a rather disconcerting variable for the student of ancient sandstones to cope with, particu-

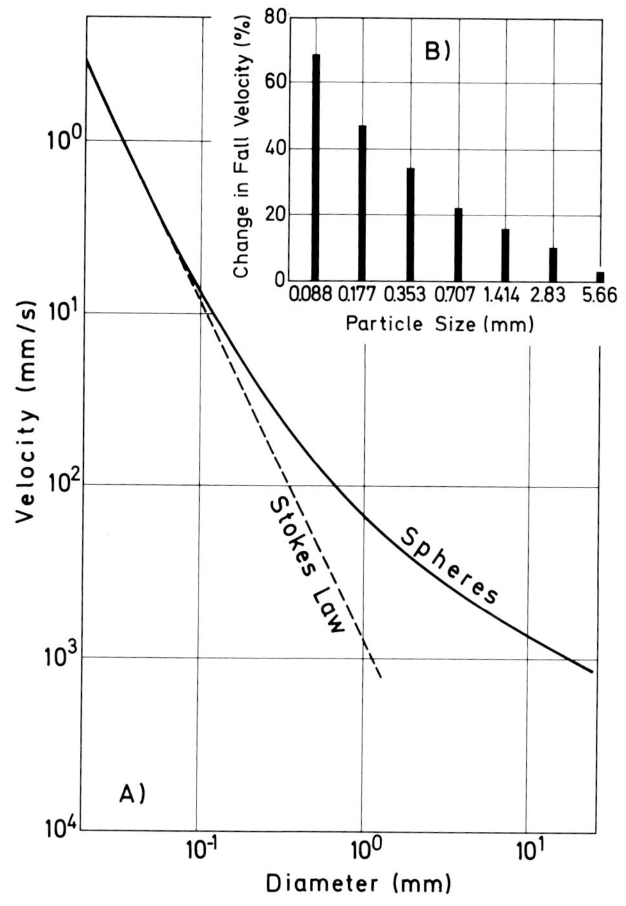

Fig. 9-6. (A) Fall velocity and diameter according to Stokes' Law and Rubey's formula for quartz spheres in pure water at 25° C. (B) Inset shows change in fall velocity of different sizes as temperature changes from 40° to 80° F (4.4 to 26.7° C) (Colby and Scott, 1965, Fig. 3)

larly if there are no independent paleoclimatic indicators. Settling velocity is also greatly reduced by high sediment concentration which has the effect of increasing the apparent viscosity and density of the fluid.

Fall velocity is the chief control on the vertical concentration gradient of suspended solid in a fluid as can be seen from data collected from the Mississippi River, in which finer sizes with their smaller fall velocities have the flatter (smaller) concentration gradients (Fig. 9-7). The governing equation is

$$-Q\,dc/dy = cw \tag{9-11}$$

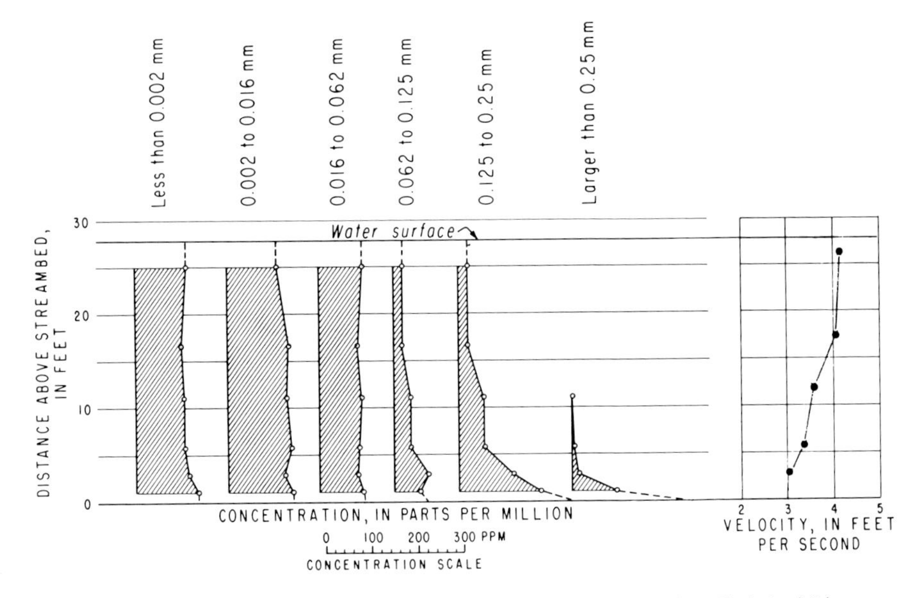

Fig. 9-7. Vertical distribution of suspended grains and velocity profile of the Mississippi River near St. Louis, Missouri. Note the steeper concentration gradients of the coarser grains (Colby, 1963, Fig. 7)

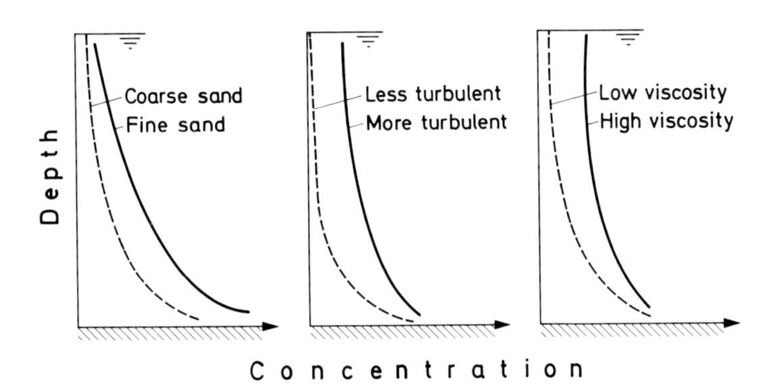

Fig. 9-8. Vertical distribution of suspended sediment as a function of grain size, turbulence, and viscosity in streams (Fig. 7-12 of Handbook of Applied Hydrology by V. T. Chow, ed. Copyright 1964. Used with permission of McGraw Hill Book Company)

where Q is the transfer coefficient for the particles and is proportional to the transfer of coefficient of momentum of the turbulent fluid, c is concentration of suspended grains, y is height above the bed, and w is fall velocity. The left side of Eq. (9-11) represents upward sediment diffusion due to turbulence and the right hand is the downward directed fall velocity. Eq. (9-11) shows that the larger w, the steeper must be the concentration gradient for any given value of c. Conversely, the greater Q (the more turbulent the flow), the larger w must be, which implies a larger grain size, if available. Qualitatively the effect of grain size, turbulence, and viscosity on depth-concentration curves are as in Figure 9-8. Eq. (9-11) can be integrated to obtain the relative concentrations of c_y and c_a

$$\frac{c_y}{c_a} = \left(\frac{(D-y)a}{(D-a)y}\right)^Z \tag{9-12}$$

where $Z = w/Rv_*$, D is depth of flow and y and a are depths, a being the depth having concentration c_a. Scheidegger (1970, p. 188-192) gives complete details.

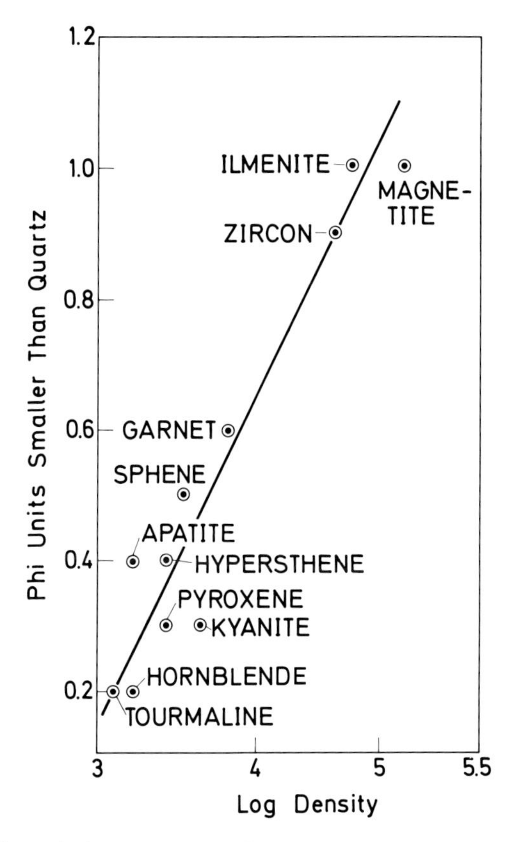

Fig. 9-9. Hydraulic equivalence of some different heavy minerals. Data from Rittenhouse (1953, Table 7)

Settling velocity also plays a role in the *hydraulic equivalence* of grains. Two different mineral grains are said to be hydraulically equivalent, if they have the same fall velocities (Rittenhouse, 1943, p. 1741). For the light minerals, size and shape are the principal factors affecting fall velocity. But with the heavy minerals density is all important. Because of their greater density, heavy minerals are finer than their associated light minerals (Fig. 9-9). More recently, Briggs and others (1962) found that shape has importance for heavy minerals too.

Although the sedimentological origin of placer concentrates and sands is poorly known, settling velocity and hydraulic equivalence no doubt play critical roles, as stressed by Ljunggren and Sundborg (1968, p. 133-134) and Tourtelot

(1968). Also needed for a placer is an abundant supply of heavy minerals in the source area. Immature, easily erodable, unconsolidated sands probably provide the best source. Repeated reworking under broadly similar hydraulic conditions also seems to be necessary. The origin of placers deserves much more attention than it has received.

Entrainment of Single Grains

One can obtain insight into the transport process and some of its problems by considering streams, where the process has been most thoroughly observed.

In streams the finest material, that less than 20 or 30 microns, is transported entirely in suspension, is supported by the upward components of turbulence, and moves approximately at the velocity of the water. Supply, ultimately from the weathering of the source area, controls concentration of suspended material, for even at low discharge the fall velocity of 20 to 30 micron particles is sufficiently small so that they can be supported by available upward components of turbulence.

Sand is transported by a different process. It moves mostly along the bottom as "bed load" going but briefly into suspension. Such short jumps are referred to as *saltation*. Instead of moving continuously, sand grains are likely to rest at the interface most of the time, the probability of entrainment being proportional to the ratio of fall velocity to turbulent uplift. The acting forces are those of gravity (weight, downward and buoyancy, upward), turbulent uplift (perpendicular to the current) and drag (parallel to it). Lift commonly does not appear explicitly in many analyses, but is represented in empirically determined constants.

Close observation of flumes clearly reveals the statistical, random nature of entrainment of grains. As the velocity at which a grain of given size and density will move, the critical velocity, is approached, a few of the smallest and lightest grains in different parts of the bed will move at irregular intervals. This irregular movement may result from random impingement of turbulent eddies on the bottom (Sutherland, 1967). The instantaneous velocity fluctuations associated with these eddies entrain the grains, starting with the smallest and lightest. As flow intensity increases, frequency and magnitude of the eddies increase and movement becomes more general; finally all of the grains, including the largest, are in motion everywhere on the bed and grain to grain collisions are the rule. General movement is accompanied by the progressive formation of ripples and dunes.

Different but interdependent measures of flow intensity in open channels have been used by different experimenters. Initially, velocity was used, but because it varies as the logarithm of the distance from the bottom, it is not commonly the most convenient measure, unless one can be satisfied with average velocity obtained from dividing discharge per unit time by cross sectional area. The most common measure is the *shear stress*, $\tau_0 = \gamma R_h S$, where γ is the specific gravity of water, R_h is the hydraulic radius (cross sectional area divided by the wetted perimeter), and S is the slope. Shear stress is also called *drag* or *tractive force*. As we have seen in equation 9-5, friction or shear velocity is another measure, for it is related to shear stress and has the dimensions of velocity. As friction velocity is a measure of shear stress, it describes conditions at the fluid-solid boundary.

It can be used together with shear stress as measures related to the mean velocity of the flow. Another measure of flow intensity is *flow* or *stream power* per unit of bed area, which is defined by Bagnold (1968, p. 46) as $\bar{v}\tau_0$ where $\bar{v}$ is average velocity. Flow power per unit area has the dimensions of power (work per unit time per unit area).

Of the many studies of the beginnings of grain movement in flumes, those by Shields (1935), an American working in Germany, are among the most

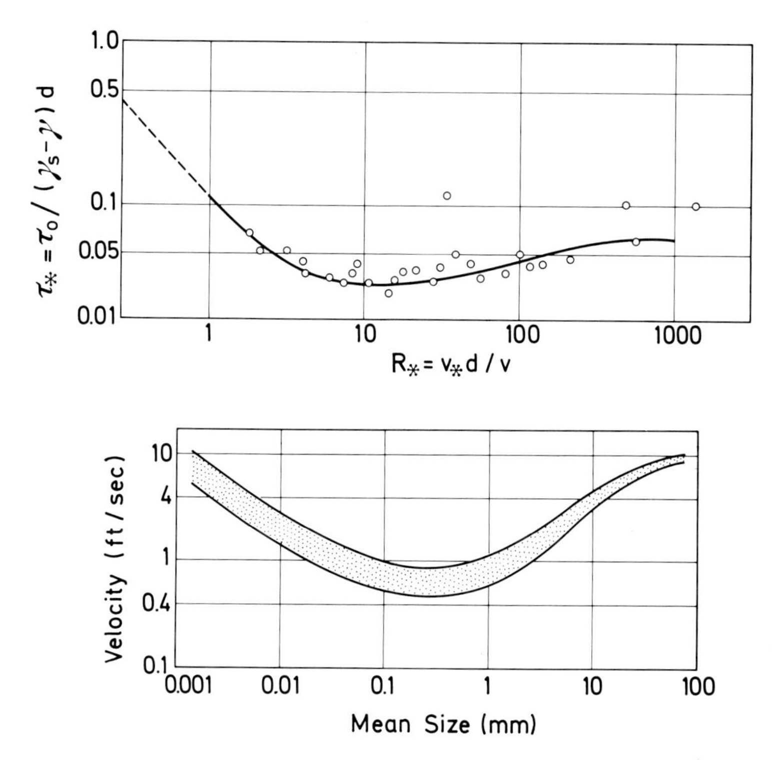

Fig. 9-10. Dimensionless entrainment function of Shields above and modified Hjulstrom curve below (Modified from Committee on Sedimentation, 1966, Fig. 2)

widely cited. He used fully turbulent flows and began each run with a flat bed. Repeated runs with different grain sizes and different water temperatures permitted him to establish the threshold of movement for a variety of conditions. Implicit to his approach was the idea that the beginning of motion is determined by τ_0 (shear stress), $\Delta\gamma$ (the difference of specific weight between fluid and grain), d, ϱ and μ. He combined these variables into two dimensionless quantities, τ_* and R_*, frequently called Shields' parameter (Fig. 9-10). τ_* is a dimensionless shear stress and R_* the boundary Reynolds number, a measure of turbulence at the boundary. Unlike ordinary Reynolds number, the velocity term of R_* is the shear velocity, v_*, and is a measure of turbulent eddying; the length term of R_* is particle diameter, d. One notes that both τ_* and R_* describe the hydraulic conditions at the boundary, which is the important site for initiation of grain movement.

The curve of Figure 9-10 separates the graph into two fields: if a point lies above the curve, the grains on the bed are fully in motion whereas a point below it indicates no movement. The curve itself indicates threshold or critical conditions, when movement is just beginning. From the standpoint of fluid flow, Shields obtained more generality by plotting τ_* against R_* then against d, as one might at first be tempted to do. By using R_* one can, for example, find the value of τ_*, the critical dimensionless shear stress, for different values of $\Delta\gamma$ or of $v = \varrho/\mu$, the latter a temperature dependent quantity. Thus, for example, Shields' curve can be used for wind as well as water. Hence a plot of τ_* against R_* conveys maximum information in a single graph. *When and only when ϱ, $\Delta\gamma$* and v *are either constant or assumed to be so*, as, for example, in fresh water streams in a given season during average flow, can one plot v versus d as shown in Figure 9-10. Figure 9-10 also shows, in striking manner, how τ_* varies with temperature because of the temperature dependence of v, especially for sand between 0.1 and 0.5 mm.

Shields' curve is concave upward with a poorly defined minimum between values of R_* of 8 and 15. For water at standard conditions this minimum corresponds to a size range of about 0.2 mm to 0.5 mm: larger grains require stronger currents for entrainment, as do smaller grains. The paradox of smaller particles requiring stronger dimensionless drag force for entrainment is explainable in terms of their shear strength according to Sundborg (1956, p. 177-180). If the clays and silts are unconsolidated they will require only a small shear velocity for entrainment, but if they are consolidated or semiconsolidated, a greater drag force is needed for entrainment. Hiding of small particles in the laminar sublayer may also be a factor. Cohesion of clay beds and thus shear resistance may be importantly affected by chemical variables, such as clay mineralogy and composition of the fluid, as well as the conditions under which the clay bed settled. However, the entrainment of silt and clays is not as yet fully understood and Sundborg's studies and those of the Committee on Sedimentation (1968) should be consulted for fuller details.

Slope of the bed and ground water conditions in natural streams also affect the entrainment of grains. A smaller drag force is needed to entrain a grain of given size on a surface sloping down current than up current, gravity making the difference. In natural streams one should also keep in mind that the drag force needed to entrain grains will vary depending on the inflow or outflow of water perpendicular to the bed. Inflow to the stream decreases the drag force needed whereas outflow from it increases it. The porewater pressure of the bed is the determining factor (see Soft Sand Deformation).

In its simplicity of variables, its generality, and its experimental reliability, the significance of Shield's work has not been greatly altered even though more recent experimental results are available. As an example of the latter, Meland and Norman (1966) studied the transport of single glass beads over a glass bead bed with rhombohedral packing. They found particle velocity to be a function of bed roughness, shear velocity, and diameter of bed materials. They thus confirmed some of the important variables included in Shields' two parameters τ_* and R_*.

Engineering Bed Load Formulas and Their Relevance

From the entrainment of single grains the next logical step would seem to be the detailed treatment of engineering bed load formulas most of which are concerned with discharge rates of either bed or suspended load or both. Though suspended load is easily measured, bed load is not and engineers have thus relied on calculations from these formulas. But because *transport rate* is almost impossible to estimate in ancient sandstones and even difficult to obtain in Recent sediments and modern streams, such formulas are not immediately useful to sedimentologists and here we only mention some of the major ones. Complete reviews are available in Raudkivi (1967, p. 39-95) and Scheidegger (1970, p. 176-204).

Because of their practical importance to waterways, engineers have proposed bed load formulas at least since the time of DuBoys in 1879. Of the numerous subsequent theories a formula developed by Einstein (1950) is widely used in America and one by Meyer-Peter and Müller (1948) in Europe. Almost all were initially designed for transport in streams. More recently, Bagnold has proposed a theory in a series of papers (1956, 1966, 1968) that has been accepted by many sedimentologists (see for example, Inman, 1963b, p. 132; Imbrie and Buchanan, 1965; Allen, 1969) and geomorphologists (Leopold and others, 1964, 1964, p. 173-184).

Bagnold's theory is based on the idea that the movement of solid grains by a fluid is accompanied by their shearing in the presence of an intergranular fluid, which also partakes in the shearing. The fluid drag force, τ_0, drives the grains forward and gravity pulls them downward. Bagnold also hypothesized that (1) shearing grains generate a dispersive pressure, P, which is directed upward and perpendicular to the bed; (2) there is a retarding tangential shear stress, τ_t, of the grains due to grain-to-grain-encounters; and (3) there is also a resisting shear stress τ_r of the intergranular shearing fluid. Thus at equilibrium on a flat bed, we have $\tau_0 = \tau_t + \tau_r$. A final step in his theory is that the normal dispersive pressure, P, is related to the tangential grain collision shear stress τ_t by $\tau_t/P = \tan\phi$, where ϕ is a friction angle corresponding to the equilibrium orientation of grains in the plane of transport, an angle of repose, so that $\tan\phi$ can be thought of as the dynamic analogue of the static friction coefficient. Bagnold (1956, p. 242) found that τ_r was very small in comparison to τ_t at high bed load concentrations in fluids such as air or water so that it could be ignored and he thus obtained

$$\tau_0 = P\tan\phi = [(\varrho_s - \varrho_f)/\varrho_s] Mg \tan\phi \qquad (9\text{-}13)$$

for a flat bed and

$$\tau_0 = P\tan\phi - t = [(\varrho_s - \varrho_f)/\varrho_s] Mg [\cos\alpha \tan\phi - \sin\alpha] \qquad (9\text{-}14)$$

for a sloping one, where t is the gravity-generated tangential stress of the sloping bed with an angle of inclination α and M is the mass of the moving grains. With the above equations as a starting point, Bagnold then calculates quantities such as the work done in transport, transport rate, and stream power all of which express his idea of fluid flow as a transporting machine. The appeal of Bagnold's theory is its deductive generality; its weakness is probably its experimental foundations and the difficulties of evaluating needed constants. Although accepted by many geologists, time and more testing are needed to evaluate its lasting significance.

Colby (1964) and Culbertson and Dawdy (1964) compared the application of some of the different bed load formulas to specific streams. Colby stresses the complexities of natural streams – in comparison to flumes – and points out that no precise relationships between discharge of sands on the one hand and flow and sediment characteristics on the other should be expected. Culbertson and Dawdy found the Einstein equation more suitable for field use than Bagnold's theory.

Qualitative descriptions of different types of bed load movement have been inferred from the study of the petrography and sedimentary structures of ancient sandstones using Bagnold's theory as a base. Dzulynski and Sanders (1962) have inferred the presence in some currents flowing over a mud bottom of a "traction carpet", a basal layer of saltating sand that allows tool marks to be made and excludes formation of scour marks. This result they ascribe to the collective grain collisions that they call the "Bagnold effect". Stauffer (1967, p. 500-502) concluded, on the basis of a study of some Tertiary sandstones of California, that there was a variety of structural flow of a granular bed that he called "grain flow", taking a term used by Bagnold (1956, p. 239) to describe the pseudo-laminar rectilinear flow of a sheared mass of fluid with a high concentration of dispersed grains in which grain collisions are the major effect.

Bed Forms in Alluvial Channels

Morphologically, the alluvial transport of sand creates forms ranging from the size of small ripples only a few centimeters high to meander belts many kilometers wide and 40 to 50 meters or more thick. A wide range of scales of dunes and point and lunate bars occur in between. Together all these forms form a hierarchy (Table 9-2), as has been emphasized by Allen (1966). Here we are primarily concerned with the fluid dynamic origin of ripples and dunes, the higher elements of the hierarchy being considered in Chapter 11.

Because of both the accessibility and economic importance of rivers to man, we know more about sand transport and bed forms in alluvial environments than in any other. Direct observation, experiment, and theory have all contributed. Flume studies have been particularly fruitful and in North America the United States Geological Survey has been foremost among these since the time of Gilbert (1914). Particularly notable are the many publications of the Survey's two series "Studies of Flow in Alluvial Channels" and "Sediment Transport in Alluvial Channels". But in spite of this and much other work, a generally accepted theory that can explain all of the bed forms in alluvial channels and relate them to sediment transport is lacking. Raudkivi (1967, p. 175-221) and Scheidegger (1970, p. 204-207) review most of the theories. Here we present only the principal experimental studies and relate them to field studies of modern sands.

An integral part of the transport of sand by a turbulent flow is the development of sand waves and other bed forms. When water of increasing velocity (discharge, tractive force, or flow power, all parameters of flow intensity) flows over a loose

Table 9-2. *Morphological hierarchy of alluvial deposits*

Meander belt

Formed by lateral migration of meandering channel. Mostly point bar deposition. Serrate edges. Thickness up to 40 m or more and widths up to 10 to 20 km or more in large rivers. Ancient equivalent: heterogeneous fluvial sand body.

Channel form

Straight or curved reaches. Chiefly point bars on insides of bends and large lunate bars in thalweg. On large sandy rivers point bars may be 1 to 5 km long. Depth of scour during high water controls thickness of point bar. Maximum morphologic change (scour and fill) at high discharge. Ancient equivalent: a channel fill either within a sand body or in a shale or limestone sequence.

Dunes

Occur in systems in thalweg and on upper reaches of point bar in high water. Height and spacing increase with flow depth. Preserved forms develop in lower flow regime. Slip slope down current, with small to moderate variance. Maximum movement at high discharge. Ancient equivalent: trough and planar crossbedding.

Ripples

Occur in systems and superimposed on dunes, bars or river reaches and form in lower flow regime. Spacing is proportional to grain size and independent of flow depth. Slip slope downcurrent with moderate to large variance. Variable patterns. Ancient equivalent: ripple marks.

Table 9-3. *Flow regime and bed forms*

Flow regime	*Bed form*	*Remarks*
Lower	Ripples Ripples on dunes Dunes	$F < 1$. Grains move in discrete steps and form roughness predominates. Water and bed surface are out of phase. Bed forms move downcurrent.
Transition	Washed out dunes	$F = 1$. Both grain and form roughness.
Upper	Plane bed Antidunes Chutes and pools	$F > 1$. Grains move continuously and grain roughness predominates. Water and bed surface are in phase. Antidunes move upcurrent, remain stationary, or go down current.

granular bed of sand and general movement begins, the sand bed is characteristically molded by the flow into a distinctive sequence of bed forms, each of which is recognizable in ancient sandstones. This movable boundary is both a response to flow and a conditioner of it, for as bed forms and microrelief change, they in turn influence – through their changed roughness – the resistance of the boundary to the flow (Fig. 9-5).

The flow in an alluvial channel has been divided by Simons and Richardson (1961) into lower, transition, and upper regimes based on bed morphology, sediment concentration, mode of transport, type of roughness, and phase relation between bed and water surface (Table 9-3 and Fig. 9-11). As a first approximation in flume studies, F is used to distinguish the three regimes: $F < 1$ (tranquil) for the lower regime, $F = F_{\text{critical}} = 1$ for the transitional regime, and $F > 1$ (shooting) the upper regime. As discharge or stream power increase, the following sequence of bed forms develops for sand finer than 0.6 mm: an initially flat bed → small asymmetrical ripples → ripples on dunes → dunes → washed out ripples and dunes (transition) → plane bed → antidunes → chutes and pools. For coarser sand, dunes may be the first to form.

Almost all the sand waves preserved in sand and sandstone formed under tranquil flow. In flumes and rivers almost all ripples and dunes are asymmetrical

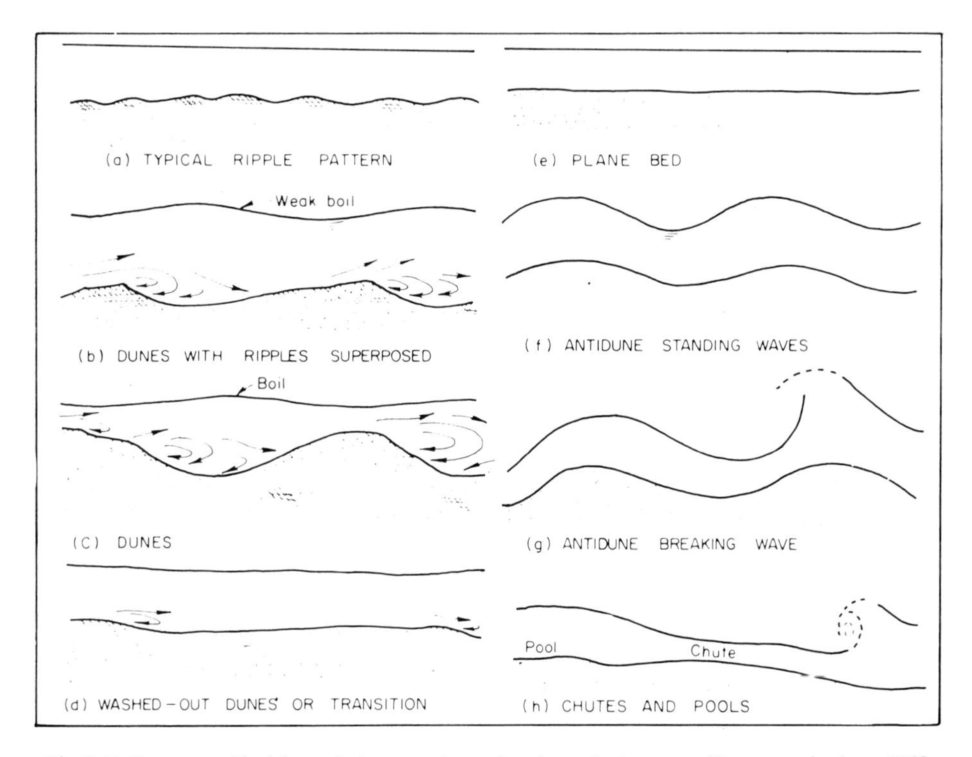

Fig. 9-11. Sequence of bed forms in loose grains as flow intensity increases (Simons and others, 1965, Fig. 3)

ridges and mounds transverse to flow. Slip slope is down current. Fig. 9-12 illustrates some of the many configurations. Ripples and dunes nearly always occur in fields or systems. There is no agreement on the precise separation of ripples from dunes but, as the terms are generally used, dunes differ from ripples in that they are larger (heights are rarely less than 7 cm) and commonly have ripples on their up current slopes whereas ripples are free of comparable ridges. The nomenclature of alluvial bed forms is summarized and briefly illustrated by the Committee on Sedimentation (1966b). They recommend the term ripple be restricted to sand waves less than 0.1 feet (3 cm) in height. There is some evidence of a statistical deficiency in abundance of sand waves having heights between 3 and 7 cm and this may reflect a fundamental difference in origin. The asymmetrical ripples and dunes of the lower flow regime have internal foreset bedding dipping downstream and are widely preserved in sandstones, although the extent to which maximum

Fig. 9-12. Asymmetrical alluvial sand waves of contrasting scale. Above: a large field of dunes in the Luusnan River northeast of Järvsö, Sweden (Lundquist, 1963, Fig. 64). River flows from right to left. Below: small asymmetrical sharp crested current ripples, current from left to right. Irrigation ditch, Mason Verte, Hassi Messaoud, Perfecture de Oasis, Algerie

thickness of crossbedding approximates maximum dune height is not absolutely certain because of later erosion of the top of the bed. Asymmetrical ripples and dunes have an angular, sharp, unrounded crestline. Many slipslopes are at the angle of repose in water, about 34° for sand sizes. Crests tend to be sharply curved and discontinuous. In the lower flow regime sediment transport is low to moderate and most of the resistance to flow is contributed by the bed forms. Dunes are out of phase with the water surface. Flow pattern, bed form, pressure and bed shear stress are all closely related in a sandwave field.

In spite of much and complicated theorizing, *why* the transport of granular bed load is almost always accompanied by ripples and dunes remains unexplained; one theory, for example, is that a chance piling up of grains on an initially smooth bed starts the process. Brush (1965, p. 20-22) and Raudkivi (1967, p. 175-219) review the various theories. The studies of Reynolds (1965) and Engelund and Hansen (1966) provide examples of some of the complex mathematics involved.

Many sedimentologists today equate the flat bed of the transitional flow regime of flumes with parting lineation (for example, Allen, 1964), but some field evidence suggests that not all was so formed. For example, sand beds that are transitional to shale commonly are very fine grained and current lineated and suggest deposition from low intensity flows. At issue here is the uniqueness of structure and flow condition. Is every resultant sedimentary structure the product of a unique fluid dynamic process with fixed parameters? Whatever the fluid regime, parting lineation is formed by grains washing along the bottom in ridges a few grain diameters thick. Grain orientation closely parallels ridge orientation, which is a good estimator of current direction. It has been suggested that possibly the distinctive ridges of parting lineation owe their origin to the transverse instability of a thin boundary layer (Allen, 1968a, p. 32).

Antidunes can move upstream, remain stationary, or move downstream and are in phase with the water surface (Fig. 9-13). They are rarely preserved in water-deposited sandstones, although some have been noted in base surge deposits of debris-laden density flows related to shallow phreatic, volcanic explosions (Fisher and Waters, 1970). Antidunes have a poorly defined internal bedding (Jopling and Richardson, 1966). In the upper flow regime sand transport is high and resistance to flow is low, coming mostly from grain roughness.

Simons and others (1965, p. 46) suggest that under both constant discharge and sediment supply the type of bed form is a function of

$$\phi\,[D, S, d, s_o, \varrho_m, g, w, S_c, P_w] \qquad (9\text{-}15)$$

where all the terms have been previously defined except S = slope of the energy grade line (a measure of the flow's energy loss), s_o = sediment sorting, S_c = shape of cross section, and P_w = pore water pressure. Both depth and slope are important determinants of shear stress, d and s_o determine grain roughness; d is the prime control on fall velocity; ϱ_m takes into account the concentration of the sand-water mixture in the bed load; g is the value of gravity; and w specifies the hydraulic behavior of a grain for a given liquid. S_c measures the shape of the boundary and plays a not fully understood role in determining overall boundary conditions; and P_w, the pore water pressure, can change the effective weight of the sand (inflow decreases it and outflow increases it) and thus alter the size of the material that can be moved by a given tractive force. With some simplification the essential variables in the above reduce to

$$\phi[D, S, d, s_o, \varrho_m, w] \tag{9-16}$$

Various combinations of hydraulic indices have been used to cast light on the origin of the different bed forms and are summarized by Allen (1968a, p. 130-149).

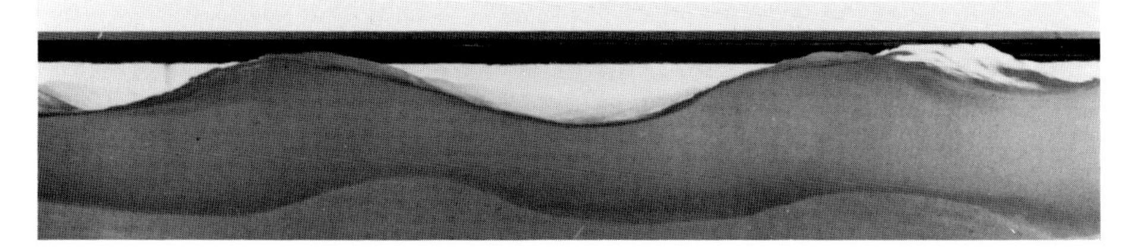

Fig. 9-13. Above: standing waves and antidunes in phase with the water surface in a small flume. Flow and antidune motion from left to right (Committee on Sedimentation, 1966, Fig. 8c). Below: standing waves in the Rio Puerco during a flood close to Bernardo, Socorro County, New Mexico. (Photo by J. P. Beverage, United States Geol. Survey)

Positions of bed forms on plots of two variables generally achieve only fair to moderate separation of fields of different forms indicating that more variables are needed. Part of the trouble may be the difficulty of precisely defining the appearance of different bed forms. The plots of Bogardi (1965) and Allen (1968) illustrate some of the parameters and the rationales that have been used to explain them. Bogardi plotted the reciprocal of a shear velocity Froude number, $F_* = \frac{v^2}{gd}$ against grain size, d (Fig. 9-14). Bogardi's Froude number differs from the con-

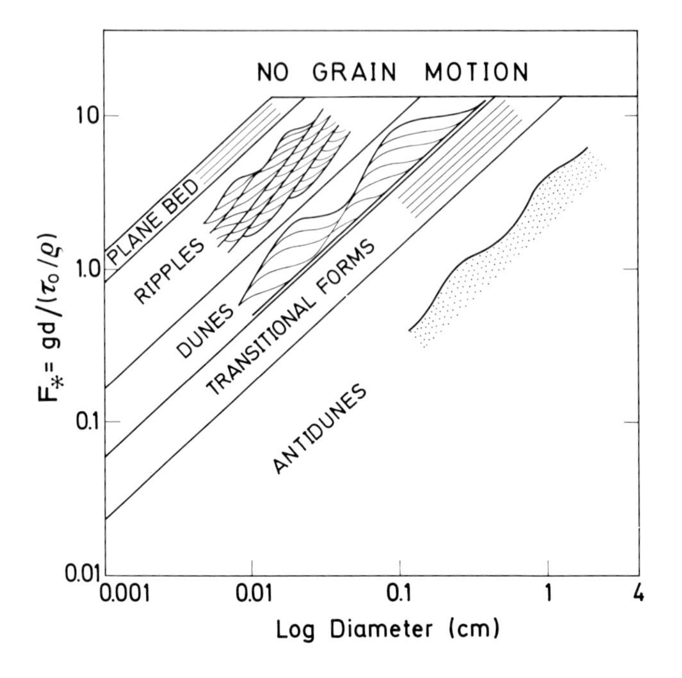

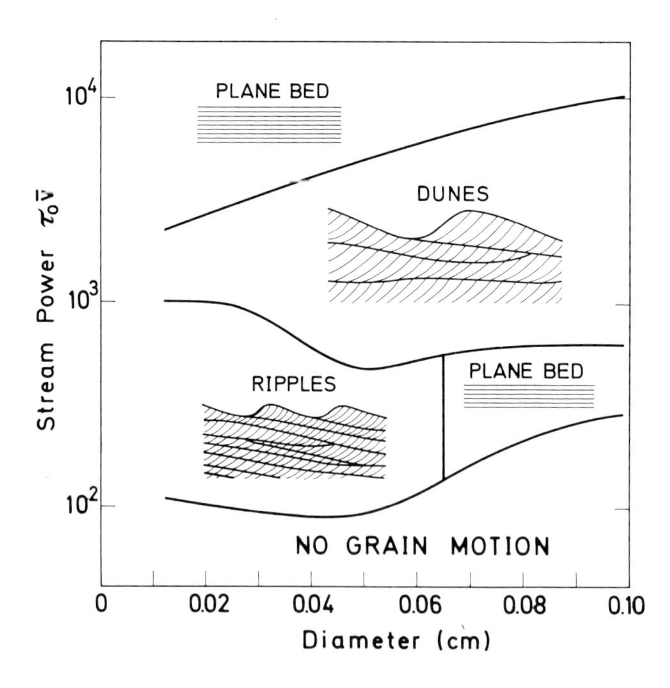

Fig. 9-14. Two views of bed forms and their relation to flow intensity and grain size. Above: a plot of F_*, the shear velocity Froude number, against d, grain size (Modified from Bogardi, 1961, Fig. 1). Below: a plot of stream power against median diameter (Modified from Allen, 1969, Fig. 1, based on data of Guy and others, 1966)

ventional one in that it uses the shear velocity at the interface rather than the average velocity of the flow; it is thus analogous to R_* of Shields. Another difference in F_* is the use of grain diameter, d, instead of water depth. By using shear velocity, Bogardi achieved a better measure of hydraulic conditions at the interface while his use of d incorporated the grain size of the bed load. As shear velocity increases, bed forms change from ripples to dunes to transitional flow for any given grain size. He suggested the existence of an initial plane bed as well as a second one transitional to antidunes. Much data has also been assembled by Guy and others

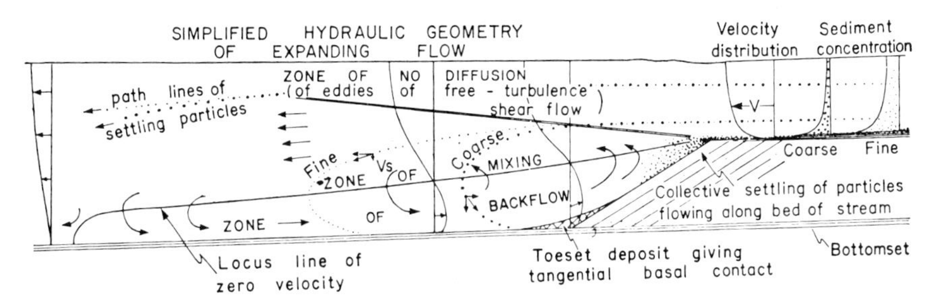

Fig. 9-15. Hydraulic conditions at the lip of a small delta produced in a flume. The flow structure developed over the slip slope is the same as that produced over the slip slope of a dune or ripple (Jopling, 1963, Fig. 2)

(1966) and plotted by Allen (1969, Fig. 1) to form the diagram of Fig. 9-14, a plot of stream power versus diameter which shows well the dependence of bed forms on stream power and, to a lesser extent, on grain size. Stream or flow power per unit area, $\bar{v}\tau_0$, was suggested by Bagnold (1968) and is the natural consequence of considering fluid flow as a transporting machine. In the fluviatile Devonian Wood Bay Series of Spitzbergen, Friend (1965, Fig. 4) found a good correlation between grain size and sedimentary structures which supports the conclusion from flume studies that bed form and grain size are interrelated because both depend on flow intensity. Harms (1969, Fig. 9) reports essentially the same conclusion. The dependence between grain size and sedimentary structures suggests that the use of grain size comparisons to fingerprint sedimentary environments might be best based on sampling similar sedimentary structures.

Jopling (1963, 1965) and Allen (1965, 1968b) have studied deposition on the slipslope of ripples and dunes. These studies have expanded our knowledge of the geometry of the internal foreset bedding of a crossbed and its relation to flow conditions. Tractive force and ratio of water depth at crest to water depth of a dune's frontal basin are the essential controls. Flow separation occurs at the crest of the slipslope producing a major turbulent eddy (Fig. 9-15). Regressive ripples in front of or on the foreset may form as the result of this eddy. If tractive force is small, most of the grains roll or slide down the foreset, the largest ones commonly travelling farthest, and angular crossbedding results; but as tractive force increases, more and more bed load grains are catapulted into the turbulent eddy and are deposited on either the lower slopes of the slip slope or in front of it so that angular foresets become tangential or concave upward. Finer grain size results in the same upward concavity. Shallowing of the basin in front of the dune (ratio of depth of crest to depth of basin increases) has a similar effect because, for a given tractive force, more of the suspension transported grains land beyond the bed load grains that roll over the crest. Thus both increasing suspension transport (greater tractive force) and a shallower basin in front of the slip slope

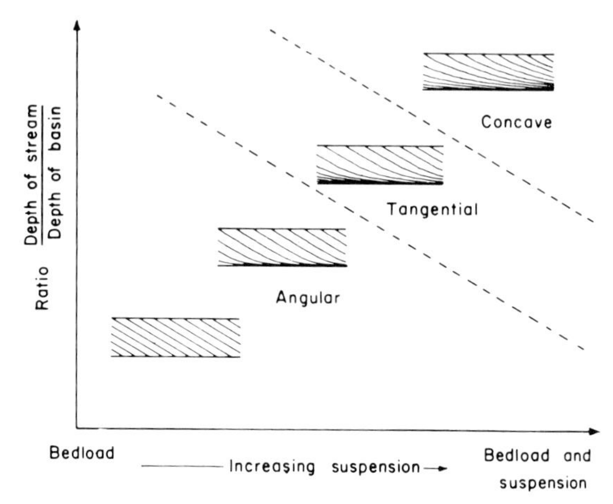

Fig. 9-16. Dependence of forset geometry on ratio of bed to suspended load and ratio of depth of stream to depth of basin (Allen, 1968a, Fig. 16.5). Reproduced by permission of North Holland Pub. Co.

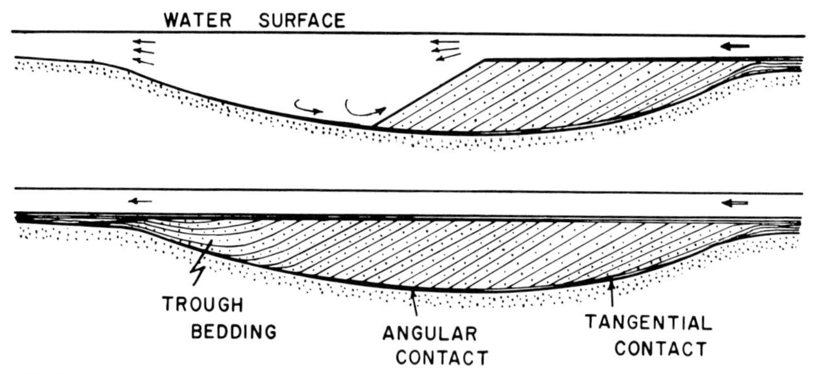

Fig. 9-17. Changing geometry of forset bedding and water depth. Trough axis parallel to page. Depth ratio of basin to water surface upstream of low separation controls geometry (Jopling, 1965, Fig. 9)

produce concave rather than angular foresets (Fig. 9-16). The infill of a scour pit illustrates these effects (Fig. 9-17).

In the lower flow regime all of the sand waves migrate downcurrent by traction transport on their back slope and deposition on their slip slope. An exception is ripple drift or climbing ripples where fallout from suspension transport is of prime importance. Climbing ripples occur when sedimentation is sufficiently rapid so there is accumulation on the back slope. They may be found in some flood deposits (McKee, 1965) as well as in the upper parts of turbidity cycles (Walker, 1963).

How do sand wave spacing, sand wave height, or average grain size in alluvial channels depend on fluid dynamic conditions? Although complete answers remain in the future, we do know that ripples and dunes behave differently. Yalin (1964) and Raudkivi (1967, p. 215-219) concluded that current ripples have a spacing proportional to particle size and are independent of depth of flow. Certainly observation of both modern and ancient sands and many experiments support the idea that ripple height and spacing are independent of water depth. Observation also shows that ripples can grade from one pattern to another

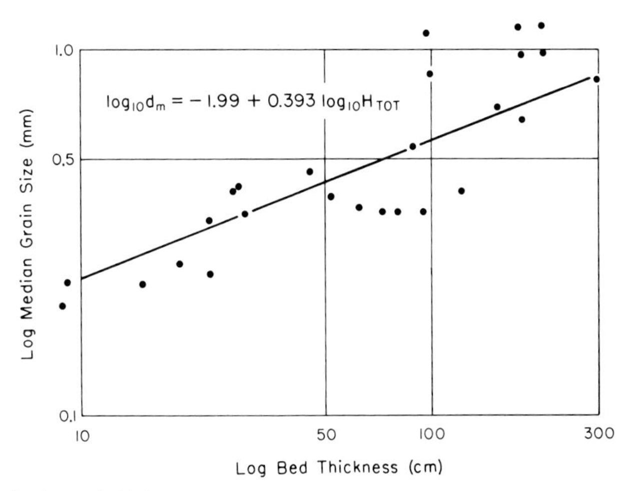

Fig. 9-18. Grain size and thickness of crossbedding (Scheidegger and Potter, 1967, Fig. 1). Reproduced by permission of Sedimentology

on the same surface in relatively short distances. Allen (1969, Fig. 4-61) suggested that relatively straight and continuous patterns change to curved and discontinuous patterns as depth and velocity increase in the shallow waters of streams and the strand.

Dune spacing is commonly difficult to determine in ancient sandstones, although Moore (1967, Fig. 12) mapped the spacing of antidunes that developed in a base surge deposit on Volcano Island, Luzon. He found spacing to decrease as distance from the crater increased, that is, as the intensity of the surge declined. Generally, dune height is closely approximated by maximum thickness of cross-bedding, a parameter that is as readily determinable as grain size. Some information on dune heights is available for rivers. Yalin (1964, p. 101), using both flume and some river data, postulated that the height, D_s, of a sand wave cannot exceed one sixth of the average depth of flow, D, of a stream, so that one has $D_s/D < \frac{1}{6}$ for limiting height. Field observations on some Russian rivers by Levi (1957, p. 101) yielded a comparable value. However, more testing of Yalin's hypothesis using data gathered from sonic depth surveys of modern rivers would be valuable. Yalin suggested that dune height depends upon shear stress and, because shear stress increases during high discharge, so does dune height. He minimized the role of grain size, although in nature thicker crossbeds usually correlate with coarser grain size (Fig. 9-18). Scheidegger and Potter (1967) suggested that greater dune height correlates with greater fluid turbulence and tractive force, which permit more and coarser grains to be transported should they be available. The general form of the dependence of grain size, d, on crossbedding thickness, t, is $d = \alpha t^{\beta}$, where α and β are constants. Little information is available on the relation between dune height and water depth in lakes and seas, although geological observation suggests that thickness of crossbedding may be appreciably greater than that in alluvial deposits.

In summary, what can be said of our knowledge of bed forms in alluvial channels? Clearly we know more about the origin of alluvial bed forms than those of any other environment. Flume experimentation and the relatively easy observation of bed forms at low water periods in natural streams have been the chief contributors to this greater knowledge. More data relating sand wave height and morphology to a wide range of flow conditions in rivers appears to be needed to further advance our knowledge, particularly with respect to the fluid dynamic significance of the scale of crossbedding and the difference between planar or trough morphology.

Beaches and Shelves

Sand dispersal on beaches and shelves is a complex process and depends on the interaction of basin geometry (shoreline configuration and underwater topography and hydrography) with the energy supplied by waves and currents, both of which condition each other and control circulation and turbulence, which in turn determine the nature of bottom sediment. Throughout the beach-shelf realm, it is nearly always a combination of waves and currents that transport sand and determine its sorting and bed forms. Inman (1963a) gives a good summary account of ocean waves and their associated currents as does Scheidegger (1970, p. 278-286 and 299-305). In addition to wind driven waves, which probably are most important in the near shore zone, there are also tidal and other oceanic currents to be considered. As in streams, sand is transported both as bed and suspended load, but mostly as bed load, except in the surf zone of beaches. We begin by considering first the shoreline and beaches. The annotated bibliography of Glen and Spooner (1968) gives a good idea of the scope and complexity of beach studies.

Beach geometry, wave pattern and fluid dynamic forces are given in Figure 9-19. As a deep-water wave moves shoreward toward a beach, it impinges on the rising bottom and is refracted so that its wave front tends to parallel shore. Waves impart an oscillatory circular motion to the water. As the water shallows, this motion impinges on the bottom and from here shoreward water and particle displacement, wave velocity, and wave steepness all increase and, if strong enough, cause loose grains to move back and forth on the bed. The process is called *wave surge transport*. Observation shows that the bed responds to increasing wave amplitude in the following way: oscillatory laminar boundary layer → appearance of turbulence → initial motion of some grains → general grain movement → formation of ripples and dunes (Abou-Seida, 1965, p. 7-8).

As a deep-water oscillatory wave moves shoreward and impinges on the rising bottom, the shear stress and pressure gradients associated with it increase so at

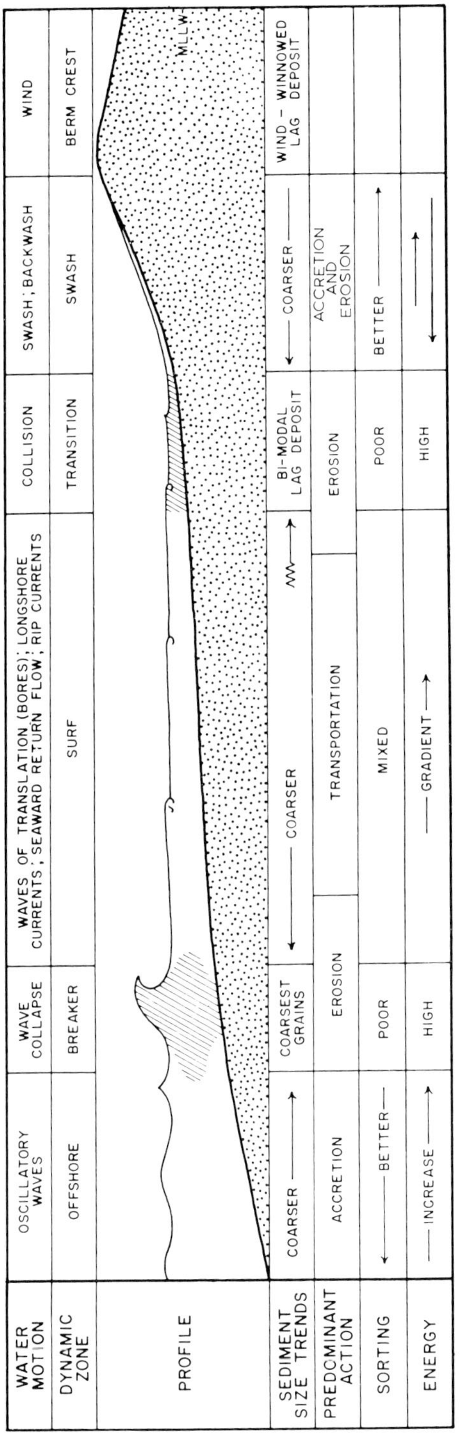

WATER MOTION	OSCILLATORY WAVES	WAVE COLLAPSE	WAVES OF TRANSLATION (BORES); LONGSHORE CURRENTS; SEAWARD RETURN FLOW; RIP CURRENTS	COLLISION	SWASH; BACKWASH	WIND
DYNAMIC ZONE	OFFSHORE	BREAKER	SURF	TRANSITION	SWASH	BERM CREST
PROFILE						MLLW
SEDIMENT SIZE TRENDS	COARSER →	COARSEST GRAINS	← COARSER →	BI-MODAL LAG DEPOSIT	← COARSER	WIND - WINNOWED LAG DEPOSIT
PREDOMINANT ACTION	ACCRETION	EROSION	TRANSPORTATION	EROSION	ACCRETION AND EROSION	
SORTING	← BETTER	POOR	MIXED	POOR	BETTER →	
ENERGY	INCREASE →	HIGH	GRADIENT →	HIGH	⇄	

Fig. 9-19. Summary of geometry, water motion, dynamic zones, and sedimentation characteristics on a beach. Hachured areas are zones of high suspended grain concentrations. MLLW = mean lower low water (Ingle, 1966, Fig. 116). Reproduced by permission of the author and Elsevier Publishing Co.

some point in the offshore region these forces will cause grains of a given settling velocity to move on a bottom having a slope α (Fig. 9-20). Shoreward of this point grains with this settling velocity will move to and fro with each wave cycle. Changes in pressure on the bottom induced by a wave as it passes may also change pore water pressure in the submerged sand and hence alter critical tractive force (see Soft Sand Deformation).

The direction and magnitude of the net movement of a grain with a given settling velocity on a beach having a seaward slope, α, depends on the relative magni-

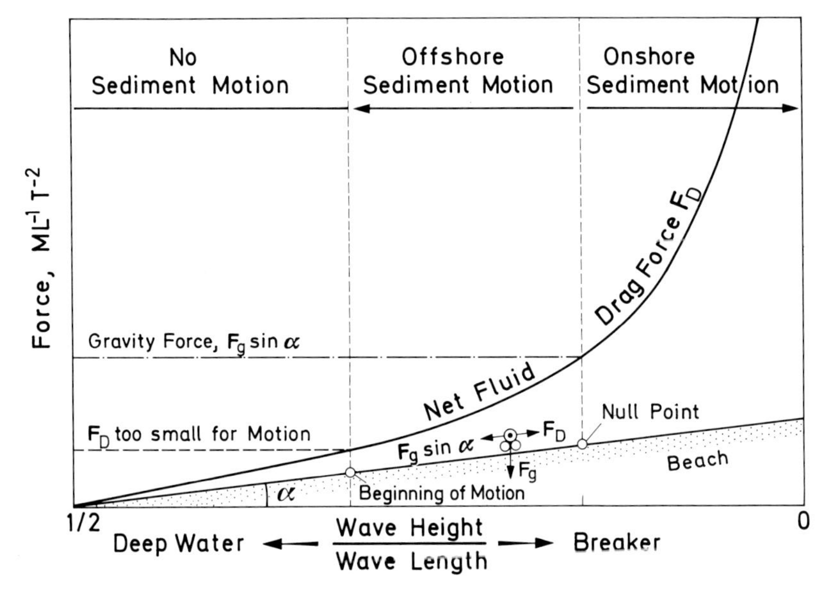

Fig. 9-20. Diagrammatic representation of forces acting on loose grains lying on a beach (Modified from Johnson and Eagleson, 1966, Fig. 9.12)

tude of three forces: the net drag force, F_D, averaged over one wave cycle; bottom resistance force, F_r, due to cohension between a grain and the bottom; and on gravity force, $F_g \sin\alpha$ (Fig. 9-20). The bottom friction force always retards movement. If $F_D > F_g \sin\alpha + F_r$, the grain moves shoreward; if $F_D + F_r < F_g$, the grain moves seaward; and if $F_D = F_g \sin\alpha + F_r$, the grain remains where it is and defines a line of no *net* movement, only oscillating back and forth.

This approach to sand movement on the beach is known as the *null line concept*. Johnson and Eagleson (1966) provide a full mathematical theory and Swift (1970, p. 10-11) reviews its applications. Although the theory has not been completely successful, it does give a unifying framework for the wave sediment dynamics of the beach zone for it relates settling velocity to slope angle and drag force, the latter a function of wave power. Raudkivi (1967, p. 253-306) reviews null point and other theories of sand movement and beach profiles at length.

Some field studies and laboratory studies have shown that, in addition to the above oscillatory movement of grains, short intense storm waves remove sand

Table 9-4. *Magnitude of some surface tidal and marine currents*

	knots		cm/sec
Tidal Currents			
Jade Estuary, North Sea	3.3		170
Bay of Fundy	2.2–8.8		132–453
San Francisco Bay	3.3	max. flood	170
	4.5	max. ebb	232
Cape Cod	3.3	max. flood	170
	2.3	max. ebb	118
Florida Current	1.7–3.5		88–180
Pacific Equatorial Countercurrent	0.6–1.2		31–62
Currents Induced by Wind	0.06–0.2		3–10
Longshore Currents			
Point Arguello, California	0.25–0.5		13–26
Nags Head,	0–2.6	range	0–132
South Carolina	0.2	mode	60
Ogoturuk Beach,	0–1.0	range	0–43
Alaska	0.2	mode	12
LaJolla,	0.01–0.5	range	0.001–26
California	0.2	mode	8

from the beach whereas long swells tend to add sand to it. Sand shuttles back and forth on the beach as a result of both continual wave action and variations in weather and, depending upon the magnitude of the longshore current, will move laterally along it. Repeated shuttling promotes good sorting and intense abrasion, the latter perhaps eliminating mechanically weak particles such as argillaceous rock fragments.

Waves play an important role in sediment transport along the beach, because they put sand into suspension and thus permit transport by a longshore or tidal current which by itself may be too weak to entrain sand. In other words, wave action supplies the energy to put sand in motion and a net current moves it. The greater the shear stress imparted by the wave, the more sand that goes into suspension and the greater the transport. Table 9-4 gives some representative values of marine currents. An additional result of waves impinging on a shore at an angle is the longshore drift that is a consequence of grains being thrown up by swash at an angle and returning back with backwash normal to the shore.

Just as with streams, engineers have attempted to formulate theories of bed load transport for beaches. Examples of this approach are provided by Inman and Bagnold (1963) and Abou-Seida (1965), the latter using a modification of Einstein's (1950) bed load theory first proposed for streams. For some California beaches Bowen and Inman (1966, p. 8) obtained a semi-empirical formula for longshore transport rate, S_t (measured in cu. ft./sec),

$$S_t = 1.13 \times 10^{-4} H \tag{9-17}$$

where H is the longshore component of wave power and depends on some nine variables including density of the water, gravity, wave height and period, angle the breaking waves make with the shore, etc. Bowen and Inman's approach not only brings out the very large number of variables involved, but emphasizes the added difficulty of predicting transport rate rather than simply analyzing forces on single grains as null line theory does. Changes in beach topography and the growth of shore features such as spits and bars are directly related to longshore currents and their intensity. Relevant theory is scant, although some work has been done. Longshore, transport is one of the most important processes by which sand may be carried long distances parallel to rather than down a paleoslope.

In response to passing waves, fluid particles follow more or less circular paths except near the bottom where orbits are flattened and the water particles move back and forth, roughly perpendicular to shoreline, above the bed. This oscillatory motion produces symmetrical rounded oscillation ripples having slopes that rarely exceed 23°, appreciably less than the angle of repose. Turbulent eddies formed by the passing waves are key factors in the formation of symmetrical ripples. As summarized by Harms (1969, p. 366) crests of oscillation ripples have uniform height and spacing, are very continuous, and tend to parallel the shoreline because wave fronts are refracted perpendicular to it. Newton (1968) studied oscillation ripples in the nearshore zone where he found that (1) they migrate, (2) they commonly consist of shoreward dipping laminations, and (3) ripple height and spacing are in part relatable to wave characteristics.

Ripples produced by a combination of currents and waves, which have characteristics intermediate between current and oscillation ripples, have been called combined flow ripples by Harms (1969, p. 387). Such ripples may be asymmetric but have crests that are more rounded and more continuous than normal asymmetric currents ripples and have slopes less than the angle of repose. Reineck and Wünderlich (1968) summarize most of the geometric properties of current and oscillation ripples.

The fluid dynamic interpretation of oscillation ripples is difficult, although Inman (1957) showed that ripple spacing is a function of both orbital wave diameter and velocity and of grain size as well, coarser sand forming ripples with longer spacing. Harms (1969, p. 383-387) reviewed the question more recently, but offers little encouragement. Moreover, as noted earlier for current ripples, there appears to be little dependence of oscillation ripple morphology on water depth.

Tidal currents can also play a significant role in the shallow waters of the shelf, in its associated estuaries, or where shoreline configuration markedly restricts flow as in a narrow inlet or strait. Such tidal currents and their corresponding bed forms perhaps have been studied best in the North Sea tidal flats of Holland and Germany; the Waddensymposium (1950) and more recent summaries by Reineck (1963) and Newton and Werner (1969) are instructive. Individual asymmetric sand waves and sand banks tend to exhibit distinctive bimodal crossbedding related to the ebb and flood tides (Fig. 9-21), although the orientation of asymmetric dunes can be complex (Newton and Werner, 1969, Figs. 8, 9, 10, and 11). Where the ebb and flood tides are about equally balanced, ebb and flood dune migration is about equal and crossbedding is bimodal, but where one predominates dune migration and crossbedding will be dominantly unimodal. A number of ancient marine limestones typically show bimodal crossbedding suggestive of tidal currents. Ball (1967) illustrates the complex sedimentation patterns of oolite sands in the Bahamas which must be comparable to many terrigenous sands in a similar setting. Stride (1963) has described the linear ribbons of sand in the North Sea that are a response to tidal currents. Shepard and Marshall (1969) recorded up and down canyon current velocities up to 34 cm/sec near the floors of submarine canyons off La Jolla, California, the high velocity down-canyon currents correlating with ebbing tides, though other factors, such as internal waves, are involved.

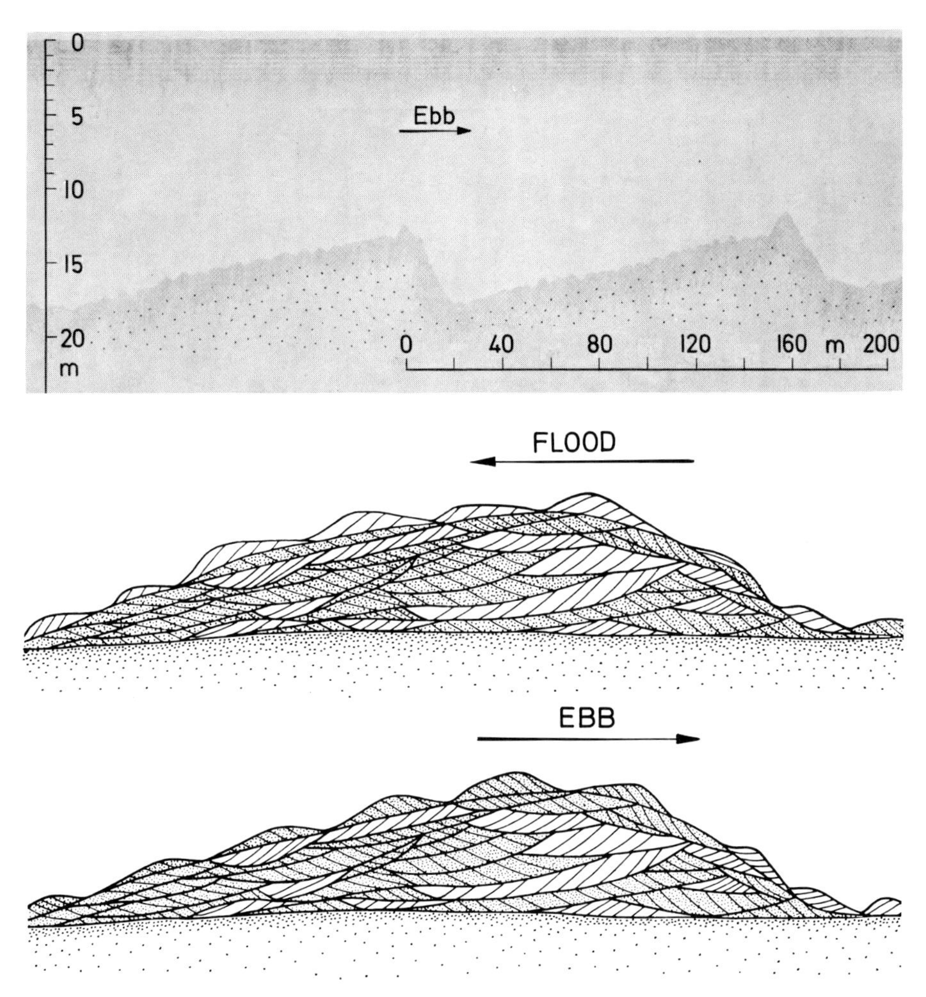

Fig. 9-21. Ebb and flood sand waves in the Jade estuary of the southern North Sea. Above: echo-sounding profiles. Below: internal structure as inferred from box-cores. Note bimodality of cross-bedding (Reineck, 1963, Figs. 16 and 19)

Seaward of the shore zone the processes active in modern shelf sedimentation are less well known for not only are most present day shelves a mixture of modern and relict sediments resulting from Pleistocene changes in sea level (Fig. 9-22), but, in addition, neither data nor theory are plentiful. A good summary of the present state of knowledge is provided by Swift (1970) who combines null-line theory with the concept of the graded or equilibrium shelf set forth by Johnson (1919, p. 199-271). As summarized by Swift (1970, p. 6), an equilibrium shelf profile has the shape of a concave upward, exponential curve whose steepest portion is at the shore. Seaward both slope and grain size become less. The

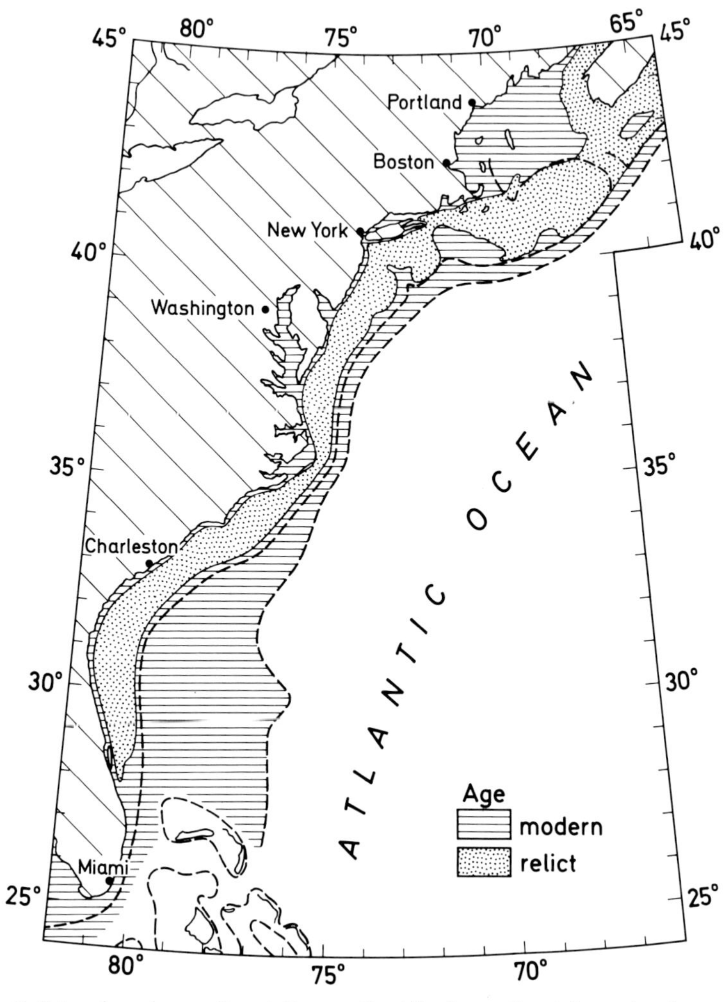

Fig. 9-22. Relict and modern sedimentation on the Atlantic continental margin of part of North America (Emery, 1965, Fig. 1)

profile shows a good correlation between grain size and slope just as does a graded stream. This is the classic idea largely inferred from studies of ancient sediments. If normal wind induced waves are to accomplish grading, water depth should probably not exceed 20 m, the depth at which wave refraction patterns are first commonly observed. For deeper shelves, Swift (p. 12) favors the idea that storm generated wave drift is a major factor in sand dispersal and that it acts in an essentially random manner, moving sediment relatively short distances, first one

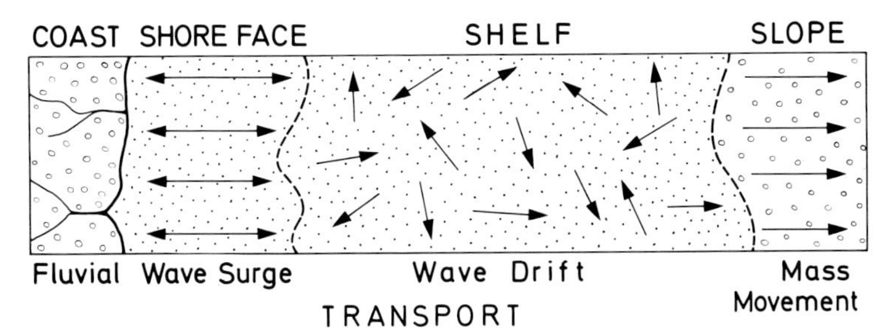

Fig. 9-23. Diagrammatic model of the different transport processes on shelves and their relation to the shoreline (Modified from Swift, 1970, Fig. 5)

way and then another, as it passes. Certainly crossbedding in many ancient marine sandstones is more variable in direction than most fluvial and deltaic crossbedding (Chap. 4) and may in fact partially be the response of storm induced sand waves. One should note, however, that in many marine sandstones the net crossbedding vector is commonly downslope, as inferred from paleogeographic evidence. Given a linear sediment source at the shoreline, a largely random movement of sand on the shelf, and a linear sink at the shelf break, as shown in Figure 9-23, a net transport of sand across the shelf by some type of weakly anisotropic diffusion process seems necessary (Swift, 1970, p. 12-13). Although details are not fully clear, some type of progressive sorting mechanism whereby smaller size fractions are eventually concentrated downslope in the direction of the shelf break seems needed. Donahue and others (1966, Fig. 2) found some evidence to support the

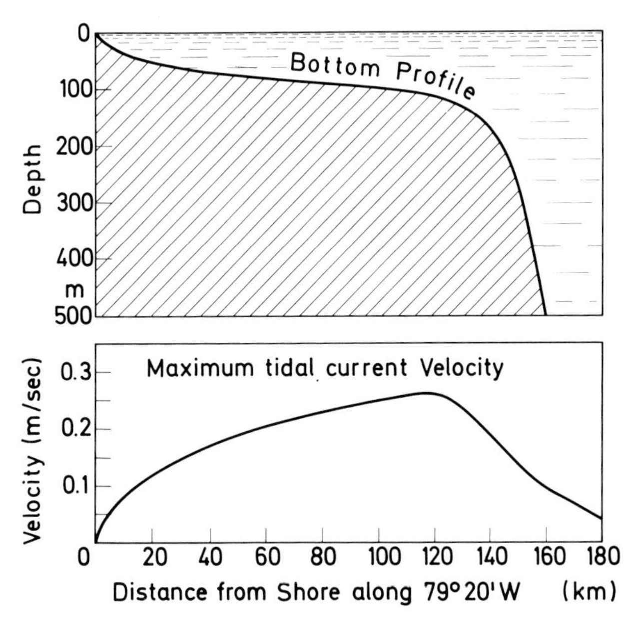

Fig. 9-24. Maximum tidal velocity and its relation to the profile of the continental shelf of Panama. Note maximum velocity near shelf edge (Modified from Fleming, 1938, Fig. 75)

idea of an incipient equilibrium shelf off the New Jersey coast, where a thin layer of sand, probably in part reworked from older relict sediments, becomes finer seaward.

Both theoretical and observational studies suggest that the area flanking the outer shelf break has more current action and consequently coarser sands than much of the central shelf described above (Swift, 1970, p. 13-14). Long period waves impinging against the shelf edge may be responsible as well as tidal currents (Fig. 9-24). Oceanic currents, such as the Gulf Stream where it impinges on the continental shelf of eastern North America, may also be factors in shelf sedimentation and, if strong enough at depth, would tend to "sweep it clean", although little data is yet available. Hulseman (1968) thinks that submarine slopes in deep water, such as the continental slopes, have erosional and constructional features, including crossbedding morphology, similar to those in shallower waters and quotes measurements of deep sea currents as high as 20 cm/sec, and even 1 m/sec, that are probably responsible. Such currents are ascribed to internal waves impinging on a slope.

Wind

Wind is a geologically important agent in sedimentation, for it transports vast quantities of sand in continental deserts, carries inland the terrigenous or carbonate sands of beaches that are supplied by longshore currents, and carries silt and clay as dust storms far from their source and supplies much of the pelagic sediment of the deep oceans. Hence wind transport is a major part of sedimentology even though it perhaps has received less attention than it deserves. General reviews are given by Raudkivi (1967, p. 28-38) and Scheidegger (1970, p. 383-401) and by the Committee on Sedimentation (1965). Sundborg's (1965) article is also very helpful. Much of the following is based on their work and thus indirectly on the classic study of Bagnold (1941).

Wind transports sand markedly differently than it does silt and clay. Sand travels close to the ground and mostly by successive jumps whereas silt and clay travel long distances in suspension (Table 9-5).

When wind reaches a critical velocity over loose, dry sand, grains begin to roll and accelerate and after a few centimeters may jump into the air, travel many times their diameter, and finally return to the surface in a long parabolic path.

Table 9-5. *Calculated flight time and height of particles*[a] *(Committee on Sedimentation, 1965, Table 2-I.3)*

Diameter mm	Fall velocity cm/sec	Flight time	Distance	Maximum height
0.001	0.00824	9–90 yr.	$4–40 \times 10^6$ km	6.1–61 km
0.01	0.824	8–80 yr.	$4–40 \times 10^2$ km	61–610 m
0.1	82.4	0.3–3 sec	46–460 m	0.61–6.1 m

[a] For a wind of 1500 cm/sec.

When it moves upward into the faster flowing air, the grain acquires additional energy and moves as a projectile at about the velocity of the wind. The impact angle, α, is commonly between 10° and 16° and is given by $\tan\alpha = v_g/v$, where v_g is the terminal velocity of the grain and v is wind velocity. Successive jumps of a grain are called saltation. The saltating sand of sand storms, in contrast to the silt and clay, has a clearly defined upper limit, usually about a meter. This upper limit will vary depending upon the surface; the harder the surface, the higher the grains bounce and the softer, the lower. As a rough approximation 13 mph (537 cm/sec), has been suggested as the minimum velocity needed to entrain dry sand. When a grain returns to the surface it may bounce back up, its impact may entrain another grain, or the grain may simply roll forward. Surface creep refers to the bed load transport of sand and has been estimated to account for about 25 percent of total sand transport. It is surface creep that permits grains too large to move by saltation to move down wind. Surface creep and saltation, like many other sedimentary processes, are gradational to one another.

Although saltation and surface creep occur in both air and water transport, saltation is much more pronounced in air. Kalinske (1943, p. 47) estimated the saltation jump in water to be only about 1/800 of that in air. It is the density contrast between air and water that is chiefly responsible for the greater role of saltation in air, which in turn alters some aspects of eolian sand wave morphology, for at 18° C the density of water is 869 times the density of dry air.

The drag force obtained from Eq. (9-7) permits us to obtain an estimate of the differing momentum and kinetic energies imparted to grains by air and water transport. For entrainment of the same grain in the two media, their drag forces must be equal so that

$$C_{D_a}[(\varrho_a v_a^2)/2] = C_{D_w}[(\varrho_w v_w^2)/2]\,. \tag{9-18}$$

But C_{D_a} and C_{D_w} are roughly the same because C_D is a function of Reynolds number and changes little for spheres for R values of 10^{-2} to 10^{-5} (Raudkivi, 1967, Fig. 2-1), so we can write

$$(\varrho_a v_a^2)/2 = (\varrho_w v_w^2)/2 \tag{9-19}$$

so that the velocities must be, as a first approximation, in the ratio of

$$v_a/v_w = \sqrt{\varrho_a/\varrho_w} = 29.3 \tag{9-20}$$

times greater in air than in water to obtain the same drag force. Thus the relative momentum of an airborne grain of a sand grain with mass M is 29.3 M times greater in air than in water. The corresponding kinetic energy is $(29.3)^2\,M/2$ or approximately 430 times greater in air than in water. This greater kinetic energy of wind blown sand explains the stronger abrasion by wind than by water transport that Kuenen (1960) and others have reported. From estimates based on experiments, Kuenen estimated sand to lose weight 100 to 1000 times faster by eolian rather than water transport (p. 442). He also suggested that the much smaller absolute viscosity of air – only 1.73×10^{-6} of that of water – means that cushioning in air prior to impact is also minimal. Presumably, the greater rounding that has been attributed by some to eolian sands (Kuenen and Perdok, 1962, p. 649) also stems from the higher kinetic energy of saltation transport by wind. The greater kinetic energy of wind driven sand is also responsible for the ventifacts cut by it. When saltating grains of high kinetic energy hit others on the ground, the added momentum may be enough to cause entrainment. This has been called the *impact threshold velocity*.

Von Engelhardt (1940, Figs. 3 and 4) suggested a difference in hydraulic equivalence between quartz and magnetite and quartz and garnet for wind versus water deposited sands. Recently Hand (1967) has modified this idea and tested it with some success using New Jersey beach and dune sands.

General analytical studies of sand movement by wind have been made by Bagnold (1941), Kawamura (1964) and Kadib (1966). Central to these theories is

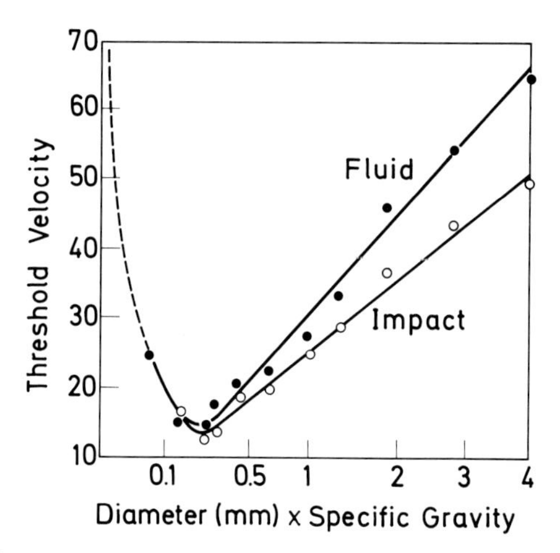

Fig. 9-25. The entrainment function (Modified from Chepil, 1945, Fig. 4), a plot of the variation of threshold velocity with grain size and specific gravity

an entrainment function for sand, which has been investigated in most detail by Bagnold (1941, Fig. 29), Chepil (1945) and Zingg (1953). This function is concave upward and has a minimum value for grains having diameters of about 0.1 mm (Fig. 9-25). The increase in shear velocity for smaller particles may be the result of their greater cohesion.

As in water transport, sand wave formation by wind is a complex and not as yet completely understood process, particularly for dunes, some of which can be as high as 200 m. Bagnold (1941), von Kármán (1947), and many geologists who have studied eolian sedimentation have speculated on sand wave formation. Sharp (1963) and Harms (1969) provide good geological summaries for ripples; their conclusions are condensed below.

Wind ripples tend to be long, uniform, and have asymmetric, straight crests. They also have low heights in the fine, well sorted sands that form most eolian deposits. Qualitatively, Sharp (p. 629) thought ripple index (ratio of wave length to height) to vary inversely with grain size and directly with velocity. Earlier Bagnold (1941, p. 64) had suggested that ripple spacing increases with wind velocity, which controls average length of saltation path. The straightness of ripple crests is the result of creep and saltation, for should a crest deflection occur, saltating grains promote lateral surface creep because they strike the deflection obliquely. According to Harms (p. 388), ripple index may range from 30 to 70 in the well sorted fine sands but may be as low as 10 to 15 in coarser and less well sorted ones. A distinctive feature that distinguishes wind from water ripples is their concentration of coarse grains at their crests whereas water ripples tend to have coarser grains in their troughs. Saltation drives and concentrates coarse grains to the crest in wind ripples whereas, in water, higher shear at the crest rolls the coarser grains into troughs.

Table 9-6. *Variables in transport and deposition by wind (modified from Chepil, 1945, p. 305)*

WIND	SURFACE
Speed	Topography
Direction	Flat
Constancy	Undulating
Temperature	Broken
Humidity	Small Scale Roughness
	Cover
	Materials
	Size distribution
	Moisture content
	Composition

Perhaps because eolian dunes are more readily accessible than subaqueous ones or perhaps because wind systems are more variable in direction and intensity than water currents, one has the *impression* that eolian dune forms are more complex than subaqueous ones. In any case, the relation between dune form, internal crossbedding, and wind direction is not fully known (see Eolian Sand Bodies, Chap. 11). Table 9-6 lists some of the major factors that condition the transport and deposition of sand.

Turbidity Currents

A turbidity current is a special variety of density current, which is defined as "the movement under gravity of a stream of fluids under, through, or over another fluid, the density of which differs by a small amount from that of the primary current" (Committee on Sedimentation, 1963, p. 78). Density currents commonly form because of density differences caused by differences in temperature or salinity. Turbidity currents owe their high density to suspended fine particles. Turbidity currents formed as muddy rivers enter fresh water lakes or reservoirs, in which the turbid water passes beneath the clear water above, have been known and observed for about 100 years; that kind of turbidity current has been described and analyzed most fully. One of the best described of such turbidity currents is that formed by the muddy Colorado River as it enters Lake Mead, first described by Grover and Howard (1938) and later given a more intensive study by Gould (1951). Experiments on turbidity currents have been made by

Kuenen (1937), Bell (1942), Middleton (1966a, 1966b, 1967) and Kuenen (1970). Menard and Ludwick (1951), the Committee on Sedimentation (1963) and Middleton (1966a, 1966b) have discussed the fluid dynamics of such currents, but our knowledge of the detailed behavior of such currents in nature is still fragmentary. Turbidity currents in the ocean are quantitatively most important; there they are the major process providing the sediments of the abyssal plains and the turbidites of the past geological record. Volcanic ash flows are another variety of turbidity current, the ash suspended in air giving the excess density.

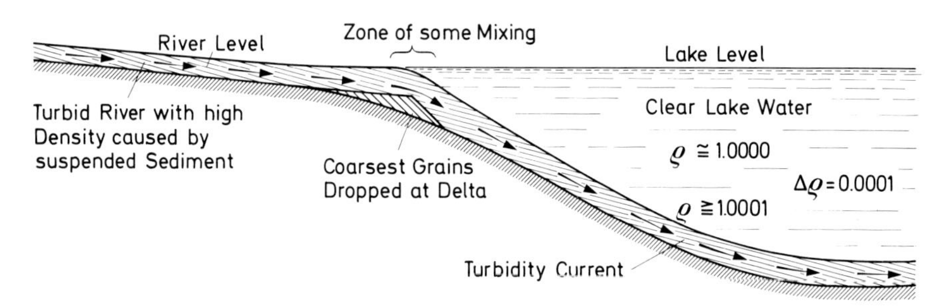

Fig. 9-26. The formation of a turbidity current by a stream with a high concentration of suspended sediment entering a lake. According to Bell (1942), the minimum effective density ($\Delta\varrho$) has to be 0.0001 for a density current to flow

Figure 9-26 shows diagrammatically how a turbidity current forms in fresh water lakes and reservoirs, the simplest case. The river current is sufficiently slowed in velocity as it enters the lake so that it does not mix completely with the surrounding lake water. Insofar as it does not mix it is maintained as a separate layer of water with a significantly higher density than the clear lake water. Under the force of gravity the turbid layer then moves downslope along the floor of the lake until it loses momentum, either by being stopped at the dam forming the reservoir, as in the case of Lake Mead, or by dissipation by frictional losses after a flat bottom is reached and the current spreads out in all directions. As the current loses velocity, its suspended material gradually settles out. The nature of the fluid flow is complex because it depends on the density of the suspension, the slope and roughness of the bottom, the amount of turbulence and mixing with the overlying layer, and the size distribution of the suspended material, the coarser grains of which settle out of the flow as its velocity decreases.

We can describe the conditions under which a two-dimensional turbidity current will flow as

$$\tau_0 + \tau_i = \Delta\varrho g h \alpha \tag{9-20a}$$

where τ_0 is the shear stress at the bottom, τ_i is the shear stress between the turbidity current and the overlying layer, $\Delta\varrho$ is the density difference between the two layers, called the *effective density*, h is the thickness of the flow, and α the slope of the bottom. The current will flow when the impelling gravitational force caused by the density difference operating over a slope exceeds the shear stresses on the fixed boundary below and the moving interface with the layer above. The depth of the turbidity current is assumed to be small with respect to the depth of the layer above. The interface of the two layers is assumed to be sharp and smooth without significant mixing.

Most workers who have developed theory for turbidity current flow have concentrated on the characteristics of a steady, uniform flow, though Middleton (1966a) has worked on the velocity of the head of the flow. For the steady uniform flow case, using a Chézy type of equation developed for flow in open channels (Leopold and others, 1964, p. 156-159), Middleton (1966b, p. 628) gives

$$v = C'\sqrt{\frac{\Delta\varrho}{\varrho} h\alpha} \tag{9-21}$$

where v is the velocity of uniform flow of the turbidity current, C' is a modified Chézy coefficient (a term combined from specific weight and flow resistance that relates mean velocity to hydraulic radius and slope), $\Delta\varrho$ is the effective density, ϱ is the density of the flow, h is flow thickness, and α is the bottom slope. Middleton discusses ways of estimating C' from bed roughness, Reynolds number, and the flow resistance offered by the fluid interface above the current. Estimation of the resistance of the fluid interface is complicated because the interface goes through stages from sharp and smooth, to wavy, to breaking waves, with turbulent mixing at the interface increasing through all these stages as the flow intensity increases. Keulegan (1949) proposed flow conditions for the formation of waves and mixing at the interface to be dependent on the *densimetric Froude number*, F_ϱ, where

$$F_\varrho = \frac{v}{\sqrt{\frac{\Delta\varrho}{\varrho} gh}} \tag{9-22}$$

and refers to gravitationally induced waves at the interface in an analogous way to the definition of F for free surface flow of water under air. The stability of the interface was measured by Keulegan in terms of F_ϱ and Reynolds number

$$\theta = \frac{1}{(F_\varrho{}^2 R)^{1/3}} \tag{9-23}$$

Keulegan experimentally determined an average critical value for θ of 0.18 for the turbulent range of flows. If θ exceeds this value, no mixing should occur.

It is now possible using equations (9-20) and (9-23) to see how a turbidity current might behave as the effective density and slope are varied. The greater the suspended load and thus the greater the effective density, the faster the current moves. As the slope increases, the densimetric Froude number, F_ϱ, increases and so controls the stability of the interface and, through its effect on flow resistance of the interface, the Chézy coeffizient, which affects the velocity. The faster and more turbulent the flow, the more mud, silt, and sand can be kept in suspension. But at the same time, the greater the density and viscosity of the suspension, the more turbulence is damped, so that there is an optimal Reynolds number for maximum current competence and capacity. As the velocity increases, so does F_ϱ and mixing at the upper surface of the flow; as the critical value of θ is exceeded, waves form at the interface, eventually breaking and mixing more turbulently with the overlying clear water. The mixing decreases the effective density, offers more flow resistance and so acts to dissipate the current. Thus a turbidity current traveling down a slope is poised between a lower bound of velocity, turbulence, and F_{ϱ} too low to keep much coarse material in suspension and an upper bound of the same parameters that would destroy the current by extensive mixing with the overlying water. When a turbidity current meets the flat floor of a lake or an abyssal plain of the ocean, it starts to decay as its inertial momentum is lowered by internal and boundary resistance, turbulence lessens, and suspended material gradually sediments as the laws of settling velocity become dominant.

Turbidity currents in lakes and reservoirs have been observed directly but their existence in the oceans of today and the past can only be inferred. The inference of turbidity currents in modern oceans has been based on the timing and distribution of cable breaks on the ocean floor, the presence of faunas displaced from shallow to deep water, the nature of the graded sediments of abyssal plains in contrast to the purely pelagic sediments of abyssal hills and seamounts, and the lack of any evidence of extensive deep non-density currents on the ocean floors that have the capability of transporting coarse grained material to the deep ocean. All of the evidence and discussion of the mechanism of setting turbidity currents in motion and the mechanism of the flows in the oceans have been given by Menard (1964, Chapter 9) and Kuenen (1967). One important factor to remember is that it is unlikely that turbidity currents can get started in shallow parts of the ocean, for there wave action mixes the water so thoroughly that no separate turbid layer can remain as such. Because the density of ocean water is about 1.02, a river entering the ocean would have to have an extraordinarily high density of suspended material to make its effective density great enough to form a flow. Even then the flow would be subject to extensive mixing by waves and tidal currents.

There are many other aspects of fluid flow of turbidity currents. Inman (1963b, p. 138) has discussed the relation of turbidity currents to the autosuspension concept, the idea that suspended grains add energy to a flow on a sloping bed such that it flows faster. Dzulynski and Simpson (1966) have experimentally produced sole markings by a plaster of Paris turbidity current over a mud bottom. Because calculations of settling velocities for turbidity flows of moderate density show that the amount of fine clay that would settle at the same time as sand is small, there appears to be not nearly enough clay settling to account for the matrix of graywackes. This has provoked a questioning of the existence of very high density turbidity currents in nature and whether they are capable of producing a sediment with much clay matrix. This has been a subject of discussion by Kuenen (1970, p. 113). The ability of turbidity currents to erode submarine canyons has been a source of lively controversy (Shepard and Dill, 1966, p. 320). Van Andel and Komar (1969) have discussed the role of turbidity currents and the fluid dynamics of the transport mechanism in the deposition of the graded sediments of a ponded valley of the flanks of the Mid-Atlantic Ridge and invoke repeated rebounds of the currents from the walls of the valley to explain the turbidite sequences. Many have speculated on the relationship between the various turbidite units proposed by Bouma (1962, p. 49) and the upper and lower flow regimes of a turbidity current, analogous to those of Simons and others (1965) for flow in alluvial channels (see, for example, Walker, 1967). The difficulty here is that the characterization of alluvial flow regimes in relation to bed forms in terms of fluid dynamic parameters such as Froude number are firmly based on many experiments, but so little is known of the relationship of turbidity current parameters, such as the densimetric Froude number, to bed forms that speculation is hazardous.

Finally, there are those in geology who are skeptical of both the existence of turbidity currents as such and the origin of flysch-type sediments from such currents; one of the most recent advocates of such doubt is van der Lingen (1969). Though there are many complexities and many unknowns in the theory of turbidity currents, it seems to us that we do not have available any unified theory that will explain as many interrelated phenomena of sedimentation nearly as well as the turbidity current concept.

Fabric

When sand is deposited, be it by surface creep, saltation or by suspension, it acquires a preferred orientation which, unless disturbed by animal activity or slumping, is preserved in consolidated sandstones. Theory explaining this fabric is limited and is reviewed by Rusnak (1957), who himself provided a qualitative discussion of the forces acting on single grains. More recently, Rees (1968) and Hamilton and others (1968) have considered fabric from the viewpoint of Bagnold's theory. Although more details of its application to fabric are needed, Bagnold's theory does seem particularly appropriate, because it deals with forces on aggregates of grains rather than on single ones. Below we briefly restate the salient points of Rees and Hamilton.

The fluid force impelling a shearing mass of sand can be resolved into a shear stress τ in the plane of shear (parallel to depositional surface) and a dispersive pressure P normal to it. The dispersive pressure depends on τ and the ratio of τ/P depends on the size and density of the grains, the viscosity of the intragranular fluid and the intensity of shearing of the grains. Observationally, we know that the longest axis of a grain lies in the plane of τ and its shortest in P. As a result the long axis that defines the imbrication angle of a sand grain can be considered as a function of the ratio of τ and P, its angle with the plane of shear being $\tan^{-1} \tau/P$. Bagnold (1956, p. 243) found the value of $\tan^{-1} \tau/P = 0.32$ for grain suspensions in a fully turbulent flow and $\tan^{-1} \tau/P = 0.75$ for laminar flow, both values being for low viscosity media such as water or air. The former corresponds to a value of 18° and the latter to 37°. An imbrication of 18° from the depositional surface closely approximates that observed in the grains of most sandstones, a striking agreement between theory and observation and thus lends support to Bagnold's theory of dispersive pressure generated by the flow of granular materials.

No one has yet formulated a quantitative theory on the complete orientation – in contrast to the imbrication – of aggregates of grains in unidirectional flow, but observational information is available on grain shape orientation in relation to current flow (Potter and Pettijohn, 1963, p. 44-48).

Soft Sand Deformation

Deformation of sand prior to lithification produces overturned and oversteepened crossbedding, ball and pillow structure, pull-a-parts, dikes, and slides. All involve flowage: if flowage is minimal, the original bedding and sedimentary structures remain easily recognizable whereas the extreme of flowage results in a quicksand – complete suspension of the grains by pore water pressure so that the original structure is completely destroyed. Although the mechanics of soft sand deformation have yet to receive much attention from geologists, it is possible to sketch some major outlines. Moore (1961), Artyushkov (1963), and Dott (1963) are geologists who have considered soft sediment deformation, the following discussion being largely drawn from them. More recently, Ramberg (1967, p. 133-144, and 179-200) has also dealt at length with unstable stratification. Two additional references are those by Morgenstern (1967) and Andresen and Bjerrum (1967) both of which contain many references of interest to sedimentologists. Kuenen (1958) has dramatically duplicated ball and pillow formation (Fig. 9-27).

Specific weight and shear between the layers are both involved. Two adjacent layers are unstable, if the upper has a larger specific weight than the lower ($\gamma_u > \gamma_l$), where γ is the weight per unit volume. Differences in specific weights may result from (1) original differences in packing (porosity), (2) degree of saturation (the more saturated the smaller γ), and (3) expansion of clays producing swelling. Because the contact surface of two adjacent layers rarely defines a perfect plane, there are always random deviations in elevation between high and low points at the interface so that one can relate specific weight, differential elevation, and shear between the layers as a second criteria for instability. Artyushkov (1963, p. 28) has postulated that one layer will intrude the other when $\Delta\gamma\Delta h > \tau_b$, where $\Delta\gamma$ is the difference between the specific weights of the two layers, Δh is the difference of elevation along the interface and τ_b is the shear *between* the layers, that is, tangent to their interface. The dimensions of $\Delta\gamma\Delta h$ are $ML^{-1}T^{-2}$, those of pressure or shear. Thus when $\Delta\gamma\Delta h$ exceeds τ_b, deformation results. This analysis assumes that the shear strengths of both beds, σ_u and σ_l, are much greater than $\Delta\gamma\Delta h$, the latter generally being small. The mechanics of soft sediment deformation deserves much more study than it has so far received. Its experimental study is attractive because problems of scaling appear to be minimal.

Shear strength, σ_s, of a sediment is given by

$$\sigma_s = C + (P_n - P_w) \tan\phi\,, \tag{9-24}$$

where C is cohesion resulting from physicochemical bonding between particles and is closely related to composition, P_n is the component of pressure normal to the interface, P_w the pore water pressure (given by $\varrho g h$, h being the height of water that would be measured if a manometer were inserted into the bed), and ϕ the angle of internal friction, which is a measure of the mechanical resistance to sliding and overturning of one particle with another. The internal friction angle depends on particle shape, sorting, and relative bulk density of the material; the greater the density the greater ϕ. Wetting has little effect on ϕ in sands but may be important in some clays, notably the "quick clays." Cohesion will generally increase with diagenesis.

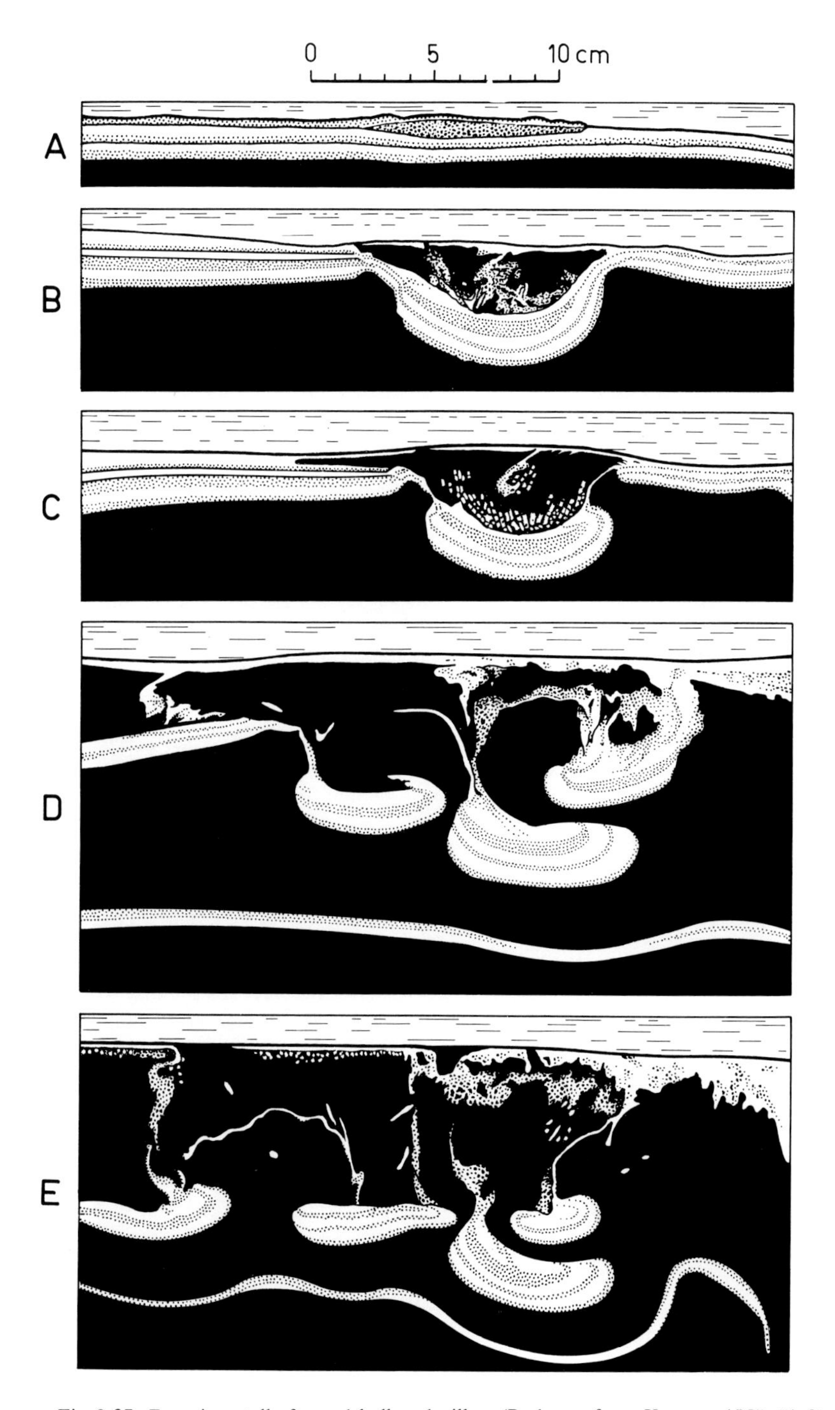

Fig. 9-27. Experimentally formed ball and pillow (Redrawn from Kuenen, 1958, Pl. 2)

The *effective pressure*, P, is defined by $P_n - P_w$ and is the major factor in most soft rock deformation through short term variation in pore water pressure. Increase in pore water pressure decreases shear strength and, should pore water pressure equal normal pressure, $\sigma_s = C$, the slightest disturbance will put the sand into suspension, forming a quicksand. Alternatively, if failure occurs, perhaps because of undercutting due to river bank erosion or tidal currents, there is an

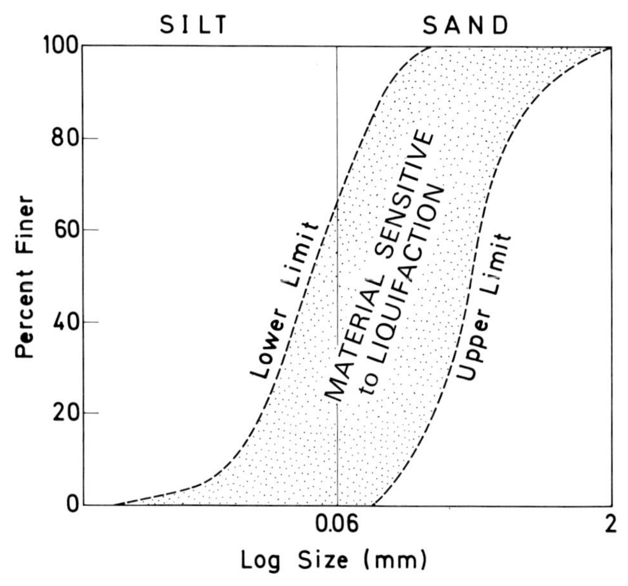

Fig. 9-28. Size limits of silt and sand most commonly found in slides (Modified from Andresen and Bjerrum, 1968, Fig. 8)

increase of pore-water pressure associated with strain beyond failure which thus causes liquifaction (Andresen and Bjerrum, 1967, p. 226). If the sand goes partially into suspension, shear strength is reduced and pressure from the overlying bed will exceed $\sigma_{s_{max}}$, promoting additional flowage. Should complete suspension occur on a slope, a viscous, high density flow such as a debris flow or a turbidity current will form.

Fine grained sand and siltstones are very susceptible to soft sediment deformation (Fig. 9-28) primarily because of their low permeability, that is, they cannot adjust quickly to exterior pore water pressure changes due perhaps to variation of water depth in the basin, to sudden change in groundwater flow or to sudden deposition of a layer of sand on top of an uncompacted bed. Via specific weight, gravity is the driving force wherever the overlying layer subsides into the underlying one, as in load cast formation. Pore water overpressure may well have been responsible for reduction of shear strength, however. Soft sediment deformation is favored where sedimentation is rapid (normal compaction does not have time to squeeze out pore water), where sands and muds are interstratified (contrasts in specific weight), where the sands have porosities in excess of 44 percent (Andresen and Bjerrum, 1967, p. 226), and where initial slopes are high.

Overturned crossbedding, crossbedding whose foresets are overturned in the downcurrent direction prior to deposition of the overlying unit, may also result from local upward groundwater seepage reducing τ_c, the critical drag force needed to entrain grains, so that the drag force of the current overturns the foresets

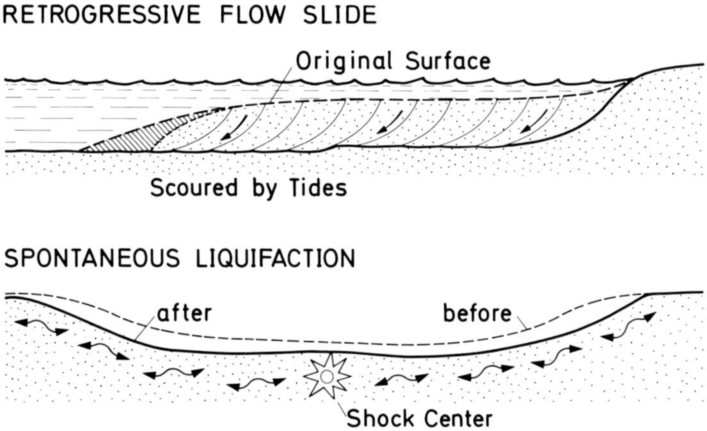

Fig. 9-29. Retrogressive flow and spontaneous liquifaction slides (Modified from Andresen and Bjerrum, 1967, Fig. 1)

downcurrent. Underseepage has also been offered as one explanation of different bed forms in stream reaches that otherwise appear to be fluid dynamically homogeneous: change in pore-water pressure changes τ_0 and thus may alter bed forms.

Slides in subaqueous slopes of loose sand and silt can occur on slopes as low as 1° and 2°, although the steeper the slope the more likely the failure. Morgenstern (1967, Table 1) cites some earthquake induced slumps with slopes of 3.5 to 10°. Andresen and Bjerrum (1967, Fig. 1) recognized two types of slides: retrogressive flow slides and spontaneous liquefaction (Fig. 9-29). The former starts in the lower part of a slope and expands, slip by slip, updip because of liquifaction associated with strain beyond failure. In this kind of liquifaction, the sand flows away, leaving an unsupported face and the processes continues until more resistant material is encountered. The potential of a sand for liquifaction *after* the initial slip is critical to the formation of a retrogressive flow slide. Spontaneous liquifaction slides, on the other hand, are caused by an initial shock so that liquifaction is believed to start at one point and spread in all directions uphill as well as down. Resedimentation follows the initial collapse.

Both types of slides may be the precursors of turbidity currents, although the actual mechanics are not yet fully known (Morgenstern, 1967). Even the operational distinction in ancient sandstones between a deposit formed by spontaneous liquifaction and some types of turbidity currents may not always be clear. Terzaghi (1956) proposed spontaneous liquefaction as an alternative to a turbidity current for the well known Grand Banks cable breaks of 1929.

Glossary

Antidune: Symmetrical sand and water surface waves that occur in trains and which are in phase and may move upstream, downstream or remain stationary (Simons, Richardson, and Nordin, 1965, p. 40–41). Characteristic of upper flow regime.

Apparent viscosity: The viscosity of a clay-water mixture that is greater than that of the clear water alone.

Boundary layer: Zone where the fluid is appreciably retarded by the frictional resistance of the boundary. May be turbulent throughout or may have a laminar sublayer next to boundary.

Cohesion: Roughly speaking, the bonding forces between particles. Appreciable for clays but virtually absent for sand and most silt.

Combined flow ripples: Ripples produced by a combination of currents and waves (Harms, 1969, p. 387).

Critical tractive force, τ_c: Shear stress exerted by a fluid on particles of a given size necessary to start movement. Same as critical shear stress.

Critical velocity: That velocity at some height above the bed necessary to initiate grain movement.

Densimetric Froude number, F_ϱ: Used for a flow with an interface with a fluid of slightly lower density, such as a turbidity flow, and defined by $F_\varrho = \dfrac{v}{\sqrt{\dfrac{\Delta\varrho}{\varrho} gd}}$.

Depth of flow, D: Roughly the same as average depth, but is commonly restricted to be the depth to the top of a dune or ripple field. Depending on the point of view one may have as many "depths of flow" in a stream as "velocities."

Dispersive pressure, P: Proposed by Bagnold as an important factor in transport of bed load material. Acts normal and upward to bed and is generated by shearing of moving, colliding grains.

Drag force: The force of frictional resistance offered by a fluid to a body moving through it. Also the frictional force exerted by a fluid moving over a fixed boundary.

Dune: A large sand wave formed by a fluid flow over a granular bed. Has gentle slope upcurrent and steep slip slope downcurrent and thus a roughly asymmetric triangular cross section. Related to but larger than ripples.

Dynamic viscosity, μ: Molecular internal resistance to flow, a kind of internal friction characteristic of the particular fluid. Defined by shear stress divided by the rate of shear or deformation.

Eddy viscosity, η: Internal resistance of a fluid to motion due to turbulence. Defined by shear stress divided by the average rate of turbulent shear and so a function of turbulence rather than the kind of fluid.

Effective density, $\Delta\varrho$: The difference between the density of a turbidity flow and that of the static body of fluid under which it flows.

Effective pressure, P': Defined as the difference between normal overburden pressure and pore water pressure. The principal variable determining shear strength in saturated sands.

Fall velocity, w: Same as settling velocity. Rate at which a particle falls through a fluid. Depends on density, diameter, and shape of particle plus density and viscosity of water.

Flat bed: See plane bed.

Flow separation: General flow separates from boundary as a result of abrupt change in geometry of boundary. In area of separation there are many turbulent eddies which make no net contribution to forward movement. Common downcurrent from bottom obstacles such as dunes.

Flow regime: Concept developed by engineers for streams and defined by a range of flows having similar bed forms, resistance to flow, and mode of transport.

Form drag: The resistance to a flow resulting from the deformation of the boundary layer, typically in natural situations by the formation of ripples and dunes. Also called form resistance.

Form roughness: Roughness of boundary generated by microrelief such as small ripples. Form roughness may be included as an important part of form drag.

Friction velocity, v_*: Defined as $\sqrt{\dfrac{\tau_0}{\varrho}}$, it appears in many fluid dynamic equations. Same as shear velocity. Applies to effects on boundary only.

Froude number, F: Dimensionless ratio of inertial to gravitational force, defined as $\frac{v}{\sqrt{Dg}}$ where v is velocity, D is commonly water depth, and g is the force of gravity. Also see densimetric Froude number.

Grain flow: A pseudolaminar structural flow of a sheared mass of water with a high concentration of dispersed grains. A variety of bed load movement inferred from ancient deposits (Stauffer, 1967, p. 500).

Hydraulic equivalence: Two grains are said to be hydraulically equivalent, if they have the same fall velocity. Commonly given with respect to a quartz sphere as a standard.

Hydraulic radius, R_h: Cross sectional area divided by wetted perimeter. Commonly taken as depth in rivers.

Internal angle of friction, ϕ: Depends on shape, sorting and density of material; the denser the packing the greater ϕ. Wetting to saturation apparently has little effect on ϕ.

Kinematic viscosity, ν: Dynamic viscosity divided by density.

Laminar flow: Fluid moves in parallel sheets or laminae with no mixing between layers.

Laminar sublayer: A basal zone of laminar flow in the boundary layer.

Plane bed: A smooth, firm sand bottom lacking ripples or dunes and thus having a low resistance to flow. Maximum grain size determines roughness.

Pore water pressure: Defined as ϱgh, where h is hydraulic head. Has a decisive bearing on shear strength. If pore water pressure equals or exceeds normal pressure, sand goes into suspension.

Resistance: Force exerted by the boundary on the fluid flow.

Reynolds number, R: Dimensionless ratio of inertial force to internal viscous retarding force. Depending on geometry and roughness, in most natural flows if R is large (>2000), flow is turbulent; if it is small (<500), flow is laminar.

Ripple: Small sand waves, transverse to flow, that may be symmetrical or asymmetrical depending on whether formed by currents or waves.

Roughness: Loosely speaking, the microtopographic "relief" on a fluid's boundary due either to grains, sand waves, or debris.

Saltation: The jumping behavior of grains that are temporarily entrained in a flow and then fall back to the bed.

Sand wave: A constructional ridge on the bottom formed by current movement of silt, sand, or gravel; ridge is commonly normal to flow. Includes ripples and dunes.

Scour: Net removal of sediment from interface.

Shear strength, σ_s: Component of force tangent to a boundary and defined as $\sigma_s = C + P' \tan\phi$ where C is cohesion, P' the effective pressure, and ϕ internal angle of friction. Shear strength is the critical value, which when exceeded, results in deformation.

Shear stress, τ: Acts parallel to bed to transport particles and is defined as the product of dynamic viscosity and rate of shear or deformation. It is also equal to γSD where γ is the specific weight of the fluid, S the slope of uniform flow, and D the depth of flow. The force per unit area that acts parallel to the bottom.

Shooting flow: A type of fluid flow in which gravity waves are swept downcurrent, and characterized by Froude numbers greater than 1.

Specific weight, γ: A ratio of weight to volume defined as product of density and gravity.

Streamline: Defined by tangent to path of moving fluid particle.

Surface creep: The generalized bed load movement of grains, many of which are too heavy to saltate.

Traction carpet: Coined by Dzulynski and Sanders (1962, p. 88) to describe the dense zone of saltation at the base of a turbulent flow.

Tractive force, τ_0: A shear stress exerted on the boundary of a particle by a moving fluid.

Tranquil flow: A type of fluid flow in which Froude numbers are less than 1. Gravity waves will be propagated upcurrent. Also called streaming flow.

Turbidity current: A density current flowing in consequence of the load of suspended sediment it is carrying, the load giving it the excess density.

Turbulent flow: Characterized by irregular flow streamlines with much lateral and vertical mixing.

Uniform flow: Velocity (or mean velocity) is the same over all sections through the flow normal to the boundary.

Viscosity: Internal resistance of a fluid to flow. See dynamic, eddy and kinematic viscosity.

Annotated References

Abou-Seida, M. M.: Bed load function due to wave action. Univ. California (Berkeley) Hydraulic Eng. Lab., Tech. Rept. Hel-2-11, 58 p., 1965.

Applies the Einstein formula to traction and transport by waves. Rate of bed load transport in any direction is product of number of oscillating particles put in suspension by waves times velocity of current.

Agriculture Research Service: Federal Interagency Sedimentation Conf. Proc., 1963, 933 p. Washington, D.C., U.S. Dept. Agriculture 1965.

Over 90 papers organized into four parts: land erosion and control; sediment in streams; sedimentaion in estuaries, harbors, and coastal areas; and sedimentation in reservoirs. An excellent compact source of material for the geologist.

Albertson, M. L., Simons, D. B.: Fluid mechanics. In: Chow, V. T. (Ed.): Handbook of applied hydrology, p. 7.1–7.48. New York: McGraw-Hill 1964.

Excellent condensation of a complicated subject. A must for both the beginner and those wishing to dig deeper.

Bagnold, R. A., Inman, D. L.: Beach and nearshore processes, Parts I and II. In: Hill, M. N. (Ed.): The Sea, Vol. 3, p. 507–554. London: Interscience 1963.

Bagnold's theory applied to beaches, shallow water and turbidity currents. Part II on littoral processes is one of the best, most concise treatments available.

Boswell, P. G. H.: Muddy sediments, 140 p. Cambridge: Heffer 1961.

Relevant soil mechanics as seen and applied by a famous sedimentary petrologist.

Colby, B. R.: Fluvial sediments – a summary of source, transportation, deposition, and measurement of sediment discharge. U.S. Geol. Sur. Bull. **1181**-A, 47 p. (1963).

Much of value written by an experienced hydraulic engineer. Section on transportation is perhaps most useful to the geologist.

Fairbridge, R. H. (Ed.): The encyclopedia of oceanography, 1021 p. New York: Rheinhold 1966.

Brief treatment of many topics relevant to fluid mechanics and sand transport. See also the "Encyclopedia of Geomorphology" in the same series.

Gilbert, G. K.: The transportation of debris by running water, 263 p. U.S. Geol. Sur., Prof. Paper **86** (1914).

The most widely cited set of experimental data, well ahead of its time. Carefully documents the three stages of flow that produce dune, smooth, and antidune bedforms and notes that this was observed in the Loire as early as 1871. A classic contribution by a lifelong student of streams.

Giles, R. V.: Fluid mechanics and hydraulics, 2nd Ed., 274 p. New York: Schaum 1962.

A good starting point, containing many worked examples.

Henderson, F. M.: Open channel flow, 522 p. New York: Macmillan 1966.

Eleven well-written chapters designed for engineering seniors. Six chapters are most useful to the quantitatively oriented geologist and especially Chapter 10, Sediment Transport.

Inman, D. L.: Sediments: Physical properties and mechanics of sedimentation. In: Shepard, F. P.: Submarine Geology, 2nd Ed., p. 101–151. New York: Harper and Row 1963.

Well written exposition by an expert. Complete acceptance of Bagnold's theory of transportation of granular material. A good place to begin.

Johnson, J. W., Eagleson, P. S.: Coastal processes. In: Ippen, A. T. (Ed.): Estuary and coastline hydrodynamics, p. 404–492. New York: McGraw-Hill 1966.

Well written, up-to-date discussion centering on beaches but including currents offshore plus comparison with estuaries. Moderately difficult mathematics. Well illustrated, 71 references.

Jopling, A.V.: Some principles and techniques used in reconstructing the hydraulic parameters of a paleoflow regime. Jour. Sed. Petrology **36**, 5–49 (1966).

A glacial sand and gravel pit provides an opportunity to estimate many hydraulic parameters. Good background required.

Kondrat'ev, N.E. (Ed.): River flow and river channel formation (Selected chapters from Ruslovi protsess, 1959). Jerusalem: Israel Program Sci. Translations for Nat. Sci. Foundation, Washington, D.C. 1962.

Three chapters of original are translated: theory of sediment movement, stream kinematics around curves and formation of bends, and morphometric characteristics of lowland river channels. Advanced.

Langhaar, H.L.: Dimensional analysis, 166 p. New York: Wiley 1951.

Theory and practicality nicely combined. Chapter 7 most relevant.

Leviavsky, Serge: An introduction to fluvial hydraulics, 257 p. London: Constable 1955.

Ten chapters reflecting both engineering and modern sediment aspects. Many references to the older literature. Good explanations of complex theories.

Middleton, G.V. (Ed.): Primary sedimentary structures and their hydrodynamic interpretation. Tulsa, Soc. Econ. Paleon. Mineral. Spec. Pub. **12**, 265 (1965).

Fourteen articles mostly about fluvial currents, and sand waves. Much stress on flow regime. Glossary.

Neumann, Gerhard, Pierson, W.J., Jr.: Principles of physical oceanography, 545 p. New York: Prentice-Hall 1966.

Chapters 5, 6, 7, and 10 through 13 give a clear and fairly advanced analysis of water movement in oceans. Chapter 6, the hydrodynamic equations, is particularly good. Designed as a text for seniors and first year graduate students in oceanography and as a reference text for other earth scientists.

Raudkivi, A.J.: Loose boundary hydraulics, 331 p. Oxford: Pergamon 1967.

Fifteen chapters written by an engineer but very well suited for the geologist.

Rouse, Hunter: Elementary mechanics of fluids, 376 p. New York: John Wiley 1946.

Clear, well illustrated exposition of fundamentals by a master American fluid dynamicist.

— Ince, Simon: History of hydraulics, 269 p. New York: Dover 1963.

Useful background for the teacher and serious student, young or old. Elementary explanations of many basic flow equations. Contains an interesting chronological table of leaders in hydraulics and related fields.

Scheidegger, A.E.: Theoretical geomorphology, 2nd Ed., 435 p. Berlin-Heidelberg-New York: Springer 1970.

Chapters 2 and 4, Physical Background and River Bed Processes, are especially relevant and good for the advanced reader. Many European references.

Schlichting, Hermann: Boundary layer theory, 4th Ed., 647 p. New York: McGraw-Hill 1960.

The bible for those who wish to delve deeply into the subject. Chapters 1 and 2 or part A, however, are not so technical and can be read with profit by all. Be sure and read the brief forward by H.L. Dryden. Chapter 18 on fundamentals of turbulence recommended.

Schmidt, M.: Gerinnehydraulik, 241 p. Wiesbaden: Bauverlag 1957.

Eleven chapters on open-channel flow written by and for engineers. Intermediate level, 139 references.

Streeter, V.L. (Ed.): Handbook of fluid dynamics, various paging. New York: McGraw-Hill 1961.

Twenty-seven chapters, seven of interest to the geologist: ideal fluid flow, turbulence, boundary layer theory, dimensional analysis, sedimentation and open-channel flow. Preparation needed.

Sundborg, Åke: The river Klarälven. Geografiska Annaler **38**, 127–316 (1956). Perhaps the best single concise statement of relevant theory and perceptive observation on fluvial transport and deposition of sand.

Wu, T. H.: Soil mechanics, 431 p. Boston: Allyn and Bacon 1966. Major topics are water movement, elastic deformation, and failure in soils. Thirteen chapters with minimal use of calculus. Chapters 4, 5, and 7 especially relevant.

References

Abou-Seida, M. M.: Bed load function due to wave action. Univ. California Hydraulic Eng. Lab. Tech. Rept. Hel-2-11, 58 p., 1965.

Albertson, M. L., Simons, D. B.: Fluid mechanics. In: Chow, V. T. (Ed.): Handbook of applied hydrology, p. 7.1–7.48. New York: McGraw-Hill 1964.

Allen, J. R. L.: Primary current lineation in the Lower Old Red Sandstone (Devonian), Anglo-Welsh basin. Sedimentology **3**, 89–108 (1963).

— Sedimentation in the lee of small underwater sand waves: an experimental study. Jour. Geology **73**, 95–116 (1965).

— On bed forms and paleocurrents. Sedimentology **6**, 153–190 (1966).

— Current ripples, 433 p. Amsterdam: North-Holland 1968a.

— The diffusion of grains in the lee of ripples, dunes, and sand deltas. Jour. Sed. Petrology **38**, 621–633 (1968b).

— Some recent advances in the physics of sedimentology. Geol. Assoc. Proc., Great Britain **80**, 1–42 (1969).

Andel, Tj. H. van, Komar, P. D.: Ponded sediments of the Mid-Atlantic Ridge between 22° and 23° North latitude. Geol. Soc. America Bull. **80**, 1163–1190 (1969).

Andresen, A., Bjerrum, L.: Slides in subaqueous slopes in loose sand and silt. In: Richards, A. F. (Ed.): Marine geotechnique, p. 221–229. Urbana: Univ. Illinois Press 1967.

Artyushkov, Ye. V.: O vozmozhnosti vozniknoveniya i obshchikh zakonomernostyakh razvitiya knvektivnoy neustoychivosti v osadochnykh porodakh. Doklady Akad. Nauk SSR **153**, 162–165 (1963) (Possibility of convective instability in sedimentary rocks and the general laws of its development. Earth Sci. Sec., p. 26–28).

Bagnold, R. A.: The physics of blown sand and desert dunes, 265 p. London: Methuen 1941.

— Flow of cohesionless grains in fluids. Royal Soc. Philos. Trans. **249**, 235–297 (1956).

— Beach and nearshore processes, Part I. Mechanics of marine sedimentation. In: Hill, M. N. (Ed.): The sea, Vol. 3, p. 507–528. London: Interscience Publishers 1963.

— An approach to the sediment transport problem from general physics. U.S. Geol. Survey, Prof. Paper **422-I**, 37 p. (1966).

— Deposition in the process of hydraulic transport. Sedimentology **10**, 45–56 (1968).

Ball, M. M.: Carbonate sand bodies of Florida and the Bahamas. Jour. Petrology **37**, 556–591 (1967).

Bell, H. S.: Density currents as agents for transporting sediments. Jour. Geology **50**, 512–547 (1942).

Bogardi, J. L.: European concepts of sediment transport. American Soc. Civil Eng. Proc., Jour. Hydraulics Div. **91**, No. HY1, Paper 4195, 29–54 (1965).

Bouma, A. H.: Sedimentology of some flysch deposits – a graphic approach to facies interpretation, 168 p. Amsterdam: Elsevier 1962.

Bowen, A. J., Inman, D. L.: Budget of littoral sands in the vicinity of Point Arguello, California. U.S. Army Coastal Eng. Research Center, Tech. Memo. **19**, 41 p. (1966).

Briggs, L. I., McCulloch, D. S., Moser, F.: The hydraulic shape of sand particles. Jour. Sed. Petrology **32**, 645–656 (1962).

Brush, L. M., Jr.: Experimental work on primary sedimentary structures. In: Middleton, G. V. (Ed.): Primary sedimentary structures and their hydrodynamic interpretation. Soc. Econ. Paleontologists and Mineralogists Spec. Pub. **12**, 17–24 (1965).

Chepil, W. S.: Dynamics of wind erosion, I. Nature of movement of soil by wind, II. Initiation of soil movement, III. The transport capacity of the wind. Soil Science **60**, 305–370, 397–474, 475–480 (1945).

Colby, B. R.: Discharge of sands and mean-velocity relationships in sand-bed streams. U.S. Geol. Survey, Prof. Paper **462-A**, 47 p. (1964).

Colby, B. R., Scott, C. H.: Effects of water temperature on the discharge of bed material. U.S. Geol. Survey Prof. Paper **462-G**, 25 p. (1965).
Committee on Sedimentation: Sediment transportation mechanics: density currents. American Soc. Civil Eng. Proc., Jour. Hydraulics Div. **89**, 77–87 (1963).
— Wind erosion and transportation. Proc. American Soc. Civil Eng. Proc., Jour. Hydraulics Div. **91**, Hy 2, Paper 4261, 267–287 (1965).
— Sediment transport mechanics: initiation of motion. American Soc. Civil Eng. Proc., Jour. Hydraulics Div. **92**, Paper 4738, 291–314 (1966a).
— Nomenclature for bed forms in alluvial channels. American Soc. Civil Eng. Proc., Jour. Hydraulics Div. **92**, Paper 4823, 51–64 (1966b).
— Erosion of cohesive sediments. American Soc. Civil Eng. Proc., Jour. Hydraulics Div. **94**, Paper 6044, 1017–1049 (1968).
Culbertson, J. K., Dawdy, D. R.: A study of fluvial characteristics and hydraulic variables, Middle Rio Grande, New Mexico. U.S. Geol. Survey, Water Supply Paper **1498-F**, 74 p. (1964).
Donahue, J. G., Allen, R. C., Heezen, B. C.: Sediment size distribution profile on the continental shelf off New Jersey. Sedimentology **7**, 155–159 (1966).
Dott, R. H., Jr.: Dynamics of subaqeous gravity depositional processes. Am. Assoc. Petroleum Geologists Bull. **47**, 104–128 (1963).
Dzulynski, Stanislaw, Sanders, J. E.: Current marks on firm mud bottoms. Connecticut Acad. Arts Sci. Trans. **42**, 57–96 (1962).
— Simpson, F.: Experiments on interfacial current markings. Geol. Romana **5**, 197–214 (1966).
Einstein, H. A.: The bed-load function of sediment transportation in open channel flows. U.S. Dept. Agriculture Tech. Bull. No. **1026**, 71 p. (1950).
Emery, K. O.: Relict sediments of the continental margin off eastern United States. In: Whittard, W. F., Bradshaw, R. (Eds.): Submarine geology, p. 1–20. London: Butterworth 1965.
Engelhardt, W. von: Die Unterscheidung wasser- und windsortierter Sande auf Grund der Korngrößenverteilung ihrer leichten und schweren Gemengteile. Chemie der Erde **12**, 445–465 (1940).
Engelund, Frank, Hansen, Eggert: Investigations of flow in alluvial streams. Acta Polytechnica Scandinavica, Civil Eng. Build. Const. Ser. No. **35**, p. 1–100 (1966).
Fisher, R. V., Waters, A. C.: Bed forms in base surge deposits: Lunar implications. Science **165**, 1349–1352 (1969).
— Base surge bed forms in Maar volcanoes. Am. Jour. Sci. **268**, 157–180 (1970).
Fleming, R. H.: Tides and tidal currents in the Gulf of Panama. Jour. Marine Research **1**, 192–206 (1938).
Friend, P. F.: Fluviatile sedimentary structures in the Wood Bay series (Devonian) of Spitsbergen. Sedimentology **5**, 39–68 (1965).
Gilbert, G. K.: The transportation of debris by running water. U.S. Geol. Survey Prof. Paper **86**, 263 p. (1914).
Glen, R. H., Spooner, E. L.: Annotated bibliography of BEB and CERC publications. U.S. Army Corps of Engineers Coastal Eng. Research Center, Misc. Paper No. **1–68**, 141 p. (1966).
Graf, W. H., Acaroglu, E. R.: Settling velocity of natural grains. Internat. Assoc. Sci. Hydraulics **6**, 27–43 (1966).
Grover, N. C., Howard, C. S.: The passage of turbid water through Lake Mead. American Soc. Civil Eng., Jour. Hydraulics Div. **103**, 720–732 (1938).
Gould, H. R.: Some quantitative aspects of Lake Mead turbidity currents. In: Turbidity currents and the transportation of coarse sediments to deep water – a symposium. Soc. Econ. Paleontologists and Mineralogists Spec. Pub. **2**, 34–52 (1951).
Guy, H. P., Simons, D. B., Richardson, E. V.: Summary of alluvial channel data from flume experiments, 1956–1961. U.S. Geol. Survey Prof. Paper **462-I**, 96 p. (1966).
Hamilton, N., Owens, W. H., Rees, A. J.: Laboratory experiments on the production of grain orientation in shearing sand. Jour. Geology **76**, 465–472 (1968).
Hand, Bryce M.: Differentiation of beach and dune sands, using settling velocities of light and heavy minerals. Jour. Sed. Petrology **37**, 514–520 (1967).
Harms, J. C.: Hydraulic significance of some sand ripples. Geol. Soc. America Bull. **80**, 363–396 (1969).
Hulseman, Jobst: Morphology and origins of sedimentary structures on submarine slopes. Science **161**, 45–47 (1968).

Imbrie, John, Buchanan, Hugh: Sedimentary structures in modern carbonate sands of the Bahamas. In: Middleton, G. W. (Ed.): Primary sedimentary structures and their hydrodynamic interpretation. Soc. Econ. Paleontologists and Mineralogists, Spec. Pub. **12**, 149–172 (1968).

Ingle, J. C., Jr.: The movement of beach sand, 232 p. New York: Elsevier 1966.

Inman, D. L.: Wave generated ripples in near shore sands. U.S. Beach Erosion Board Tech. Memo **82**, 30 p. (1957).

— Ocean waves and their associated currents. In: Shepard, F. P., Submarine geology, 2nd Ed., p. 49–100. New York: Harper and Row 1963a.

— Sediments: physical properties and mechanics of sedimentation. In: Shepard, F. P., Submarine geology, 2nd Ed., p. 101–151. New York: Harper and Row 1963b.

— Bagnold, R. A.: Littoral processes. Part II. In: Hill, M. N. (Ed.): The sea, Vol. 3, p. 529–543. New York: Interscience 1963.

Johnson, D. W.: Shore processes and shoreline development, 584 p. New York: Wiley 1919.

Johnson, J. W., Eagleson, P. S.: Coastal processes. In: Ippen, A. P., Ed.: Estuary and coastline hydrodynamics, p. 404–492. New York: McGraw-Hill 1966.

Jopling, A. V.: Hydraulic studies on the origin of bedding. Sedimentology **2**, 115–121 (1963).

— Hydraulic factors controlling the shape of laminae in laboratory deltas. Jour. Sed. Petrology **35**, 777–791 (1965).

— Richardson, E. V.: Backset bedding developed in shoaling flow in laboratory experiments. Jour. Sed. Petrology **36**, 821–825 (1966).

Kadib, A. A.: Mechanism of sand movement on coastal dunes. American Soc. Civil Eng. Proc., Jour. Waterways and Harbors Div., Paper **4817**, 27–44 (1966).

Kalinske, A. A.: Turbulence and the transport of sand and silt by wind. New York Acad. Sci. Annals **44**, 41–54 (1943).

Kawamura, Ryuma: Study of sand movement by wind. Univ. California Hydraulics Eng. Lab. **HEL-2-8**, 38 p. (Translated from the Japanese) (1964).

von Kármán, Theodore: Sand ripples in the desert. In: Collected works of Theodore von Karman, Vol. 4, p. 352–356. London: Butterworth 1947.

Keulegan, G. H.: Interfacial instability and mixing in stratified flows. National Bur. Standards Jour. Research **43**, 487–500 (1949).

Kuenen, Ph. H.: Experiments in connection with Daly's hypotheses on the formation of submarine canyons. Leidse Geol. Mededeel. **8**, 327–351 (1937).

— Experiments in geology. Geol. Soc. Glasgow Trans. **23**, 1–28 (1958).

— Experimental abrasion 4. Eolian action. Jour. Geology **68**, 427–449 (1960).

— Value of experiments in geology. Geol. Mijnbouw **44**, 22–36 (1965).

— Emplacement of flysch-type sand beds. Sedimentology **9**, 203–243 (1967).

— Experimental marine suspension currents, competency and capacity. Geol. Mijnbouw **49**, 89–118 (1970).

— Perdok, W. G.: Experimental abrasion, 5. Frosting and defrosting of quartz grains. Jour. Geology **70**, 648–658 (1962).

Leopold, L. B., Wolman, M. G., Miller, J. P.: Fluvial processes in geomorphology, 522 p. San Francisco: Freeman 1964.

Levi, I. I.: Dinamika ruslovykh potokov (Dynamics of alluvial streams, 2nd Ed.), 252 p. Moscow: Gosenergoizdat 1957.

van der Lingen, G. J.: The turbidite problem. New Zealand Jour. Geol. and Geophys. **12**, 7–50 (1969).

Ljunggren, Pontus, Sundborg, Åke: Some aspects on fluvial sediments and fluvial morphology. II. A study of some heavy mineral deposits in the valley of the River Lule Älv. Geog. Ann. **50**, Ser. A, 121–135 (1968).

Lundquist, G.: Beskrivning till Jourdartskarta över Gävleborgs län. Sveriges Geologiska Undersökning Ser. Ca, No. **42**, 181 p. (1963).

Matthes, G.: Macroturbulence in natural stream flow. Am. Geophys. Union Trans. **28**, 255–265 (1947).

McKee, E. D., Crosby, E. J., Berryhill, H. L., Jr.: Flood deposits, Bijou Creek, Colorado, June 1965. Jour. Sed. Petrology **37**, 829–851 (1967).

Meland, Nils, Norman, J. O.: Transport velocities of single particles in bedload motion. Geografiska Annaler **48**, Ser. A., 165–182 (1966).

Menard, H. W.: Marine Geology of the Pacific, 271 p. New York: McGraw-Hill 1964.

Menard, H. W., Ludwick, J. C.: Application of hydraulics to the study of marine turbidity currents. In: Turbidity currents and the transportation of coarse sediments to deep water – a symposium. Soc. Econ. Paleontologists and Mineralogists Spec. Pub. **2**, 2–13 (1951).

Meyer-Peter, E., Müller, R.: Formula for bed load transport. Stockholm, Internat. Assoc. Hydraulic Structures Research **3**, App. 2, p. 39–65 (1948).

Middleton, G. V.: Experiments on density and turbidity currents, I. Motion of the head. Canadian Jour. Earth Sci. **3**, 523–546 (1966a).

— Experiments on density currents, II. Uniform flow of density currents. Canadian Jour. Earth Sci. **3**, 627–637 (1966b).

— Experiments on density and turbidity currents, III. Deposition of sediment. Canadian Jour. Earth Sci. **4**, 475–505 (1967).

Moore, D. G.: Submarine slumps. Jour. Sed. Petrology **31**, 343–357 (1961).

Moore, J. G.: Base surge in recent volcanic eruptions. Bull. volcanol., Ser. 2, **30**, 337–363 (1967).

Morgenstern, N. R.: Submarine slumping and the initiation of turbidity currents. In: Richards, A. F., Ed.: Marine geotechnique, p. 189–220. Urbana: Univ. Illinois Press 1967.

Newton, R. S.: Internal structure of wave-formed ripple marks in the nearshore zone. Sedimentology **11**, 275–292 (1968).

— Werner, Friedrich: Luftbildanalyse und Sedimentgefüge als Hilfsmittel für das Sandtransportproblem im Wattgebiet vor Cuxhaven. Hamburger Küstenforschung **8**, 1–46 (1969).

Potter, Paul Edwin, Pettijohn, F. J.: Paleocurrents and basin analysis, 296 p. Berlin-Göttingen-Heidelberg: Springer 1963.

Ramberg, Hans: Gravity, deformation and the earth's crust, 214 p. London: Academic Press 1967.

Raudkivi, A. J.: Loose boundary hydraulics, 331 p. Oxford: Pergamon 1967.

Rusnak, G. A.: Orientation of sand grains under conditions of "unidirectional" fluid flow, 1. Theory and experiment. Jour. Geology **65**, 384–409 (1957).

Rees, A. I.: The production of preferred orientation in a concentrated dispersion of elongated and flattened grains. Jour. Geology **76**, 457–465 (1968).

Reineck, Hans-Erich: Sedimentgefüge im Bereich der südlichen Nordsee. Abh. Senckenberg naturf. Gesell. Nr. **505**, 1–138 (1963).

— Wunderlich, F.: Zur Unterscheidung von asymmetrischen Oszillationsrippeln und Strömungsrippeln. Senckenbergiana. Lethaea **49**, 321–345 (1968).

Reynolds, A. J.: Waves on the erodable bed of an open channel. Jour. Fluid Mechanics **22**, 113–133 (1965).

Rittenhouse, Gordon: The transportation and deposition of heavy minerals. Geol. Soc. America Bull. **54**, 1725–1780 (1943).

Rubey, W. W.: Settling velocity of gravel, sand, and silt particles. Am. Jour. Sci. 5th Ser. **25**, 325–338 (1933).

Scheidegger, A. E.: Theoretical geomorphology, 2nd Ed., 435 p. Berlin-Heidelberg-New York: Springer 1970.

— Potter, P. E.: Bed thickness and grain size: Crossbedding. Sedimentology **8**, 39–44 (1967).

Sharp, Robert P.: Wind ripples. Jour. Geology **71**, 617–636 (1963).

Shepard, F. P., Dill, R. F.: Submarine canyons and other sea valleys, 381 p. New York: Rand McNally 1966.

— Marshall, N. F.: Currents in the LaJolla and Scripps submarine canyons. Science **165**, 177–178 (1969).

Shields, A.: Anwendung der Ähnlichkeitsmechanik und der Turbulenzforschung auf die Geschiebebewegung. Mitt. Preuss. Versuchsanstalt für Wasserbau und Schiffbau **26**, 26 p. (1935).

Simons, D. B., Richardson, E. V.: Forms of bed roughness in alluvial channels. American Soc. Civil Eng. Proc., Jour. Hydraulics Div. **87**, (HY 3), p. 87–105 (1961).

— — Nordin, C. F., Jr.: Forms generated by flow in alluvial channels. In: Middleton, G. V., Ed.: Primary sedimentary structures and their hydrodynamic interpretation. Soc. Econ. Paleontologists and Mineralogists, Spec. Pub. **12**, 34–52 (1965).

Stauffer, P. H.: Grain-flow deposits and their implications, Santa Ynez Mountains, California. Jour. Sed. Petrology **37**, 487–508 (1967).

Stride, A. H.: Current-swept sea floors near the southern half of Great Britain [with discussion]. Geol. Soc. London Quart. Jour. **119**, 175–199 (1963).

Stringham, G. E., Simons, D. B., Guy, H. P.: The behavior of large particles falling in quiescent liquids. U.S. Geol. Survey Prof. Paper **562-C**, 36 p. (1969).

Sundborg, Åke: The River Klarälven, a study in fluvial processes. Geografiska Annaler. Ser. A, No. **115**, 127–316 (1956).

Sutherland, A. J.: Proposed mechanism for sediment entrainment by turbulent flows. Jour. Geophys. Research **72**, 6183–6194 (1967).

Swift, D. J. P.: Quaternary shelves and the return to grade. Marine Geology **8**, 5–30 (1970).

Terzaghi, Karl: Varieties of submarine slope failures. 8th Texas Conf. on soil mechanics and foundation Eng. Proc., Univ. Texas Bur. Eng. Research Spec. Pub. **29**, 41 p. (1956).

Tourtelot, H. A.: Hydraulic equivalence of grains of quartz and heavier minerals, and implications for the study of placer minerals. U.S. Geol. Survey Prof. Paper **594-F**, 13 p. (1968).

Waddensymposium: Fifteen papers on the geology, sediments, and faunas of the Waddensea. Tÿdschrift Koninkl., Nederlandsch Aardrÿkskundig Genootschap **67**, Groningen Stichting voor Marine Geologie, 148 p. (1950).

Walker, Roger G.: Distinctive types of ripple-drift cross-lamination. Sedimentology **2**, 173–188 (1963).

— Turbidite sedimentary structures and their relationship to proximal and distal depositional environments. Jour. Sed. Petrology **37**, 25–43 (1967).

Yalin, M. Selim: Geometrical properties of sand waves. American Soc. Civil Eng. Proc., Jour. Hydraulics Div. **90**, Paper 4055, 105–119 (1964).

Zingg, A. W.: Wind tunnel studies of movement of sedimentary material. 5th Ann. Hydraulics Conf. Proc., Iowa City, Iowa, p. 111–135 (1953).

Chapter 10. Diagenesis

Introduction

When one compares freshly deposited sands, for example, beach or river sands, with most ancient sandstones, the differences are obvious in terms of the rock's strength, coherence, porosity, and, upon closer inspection, perhaps its composition and textural details. The term *diagenesis* has been used for the many processes that are involved in post-depositional alteration of freshly deposited sediment.

Diagenetic processes, however, do not operate with complete uniformity and regularity. Thus we can find many sandstones hundreds of millions of years old which are relatively incoherent and can be easily disaggregated. On the other hand, we know there are relatively recently deposited sandstones which are already coherent, the grains being bound together in some way. So time or geologic age is not necessarily the crucial factor. The degree of coherence itself may not be the crucial factor either, for it may be seen upon comparison of chemical and mineralogical compositions of various sandstones that two sandstones with the same degree of coherence or lithification may have had drastically different post-depositional histories.

The possible combinations of ways in which diagenetic processes may affect the composition and texture of a sandstone are extraordinarily many but from the welter of possibilities it is possible to pick some general tendencies. The overall trend is towards a chemical equilibrium composition and texture. A sand as freshly deposited is a porous, non-equilibrium assemblage of detrital minerals; after diagenesis has had its effect the sandstone has become reduced in porosity through compaction and cementation, has lost many unstable detritals, and has gained stable authigenic precipitates. The end product of long continued deep burial diagenesis of a quartz arenite is a completely cemented quartzite; the end product of the same range of processes operating on a lithic arenite may be complete cementation by a combination of quartz, carbonate, and clay in which the clay is the illite-chlorite assemblage. Both are responses in composition and/or texture to the continued impress of pressure, temperature, and changing composition of pore waters. The character of the final product is dependent on the original composition and texture, the post-depositional geologic history of the formation, and the time span over which diagenetic processes operated. Because these factors are so extremely variable in sandstones, the many possibilities of final outcome arise.

We want to be able to do two things related to the historical geology of sandstones with an understanding of diagenesis: (1) to "look through" obscuring diagenetic features to better judge the provenance and environment of deposition of the original sand; and (2) to use the evidences of diagenesis to reconstruct a coherent post-depositional geologic history. A few practical questions point up the nature of the problems. In any particular sandstone, do the heavy minerals represent the original detrital assemblage and so reflect provenance or are they the residue of intrastratal solution? Does the clay matrix of a particular ancient graywacke represent an altered assemblage of argillaceous and volcanic rock fragments and, if so, can we use the clay composition to infer a provenance for the assumed precursor rock fragments? Or, does the replacement of authigenic quartz by calcite cement in a sandstone indicate that the sandstone was not buried deeply enough to have carbonate replaced by a second generation of silica? Does the presence of an authigenic kaolinite facies of a sandstone indicate that that facies was once exposed to recharge of fresh meteoric water?

To answer these kinds of questions the petrologist must start from a detailed mineralogic and microscopic assay of the composition and texture with the emphasis on the mineralogy of authigenic minerals and their paragenesis. To these facts must be added the general geochemical knowledge that allows the inference of the temperature, pressure, and composition of the pore solution that mediated all of the processes. Finally, the relation of the actual geological history, stratigraphic and structural, of the sedimentary basin to the diagenetic effects must be added for the complete post-depositional history of the sandstone to be read.

But these are not the only reasons we study diagenesis, for there is much of practical value too. An understanding of the ways in which porosity and permeability are affected by cementation and clay mineral alteration is vital both for the search for recoversable oil reserves in sandstone and for the reservoir engineering after the oil has been found. The same problems relate to the use of permeable sandstones as underground waste disposal reservoirs. But fluids are not the only economically useful products of sandstones; many important reserves of copper, uranium, and other important metals are diagenetic precipitates in sandstones and the economic geologist must use his knowledge of diagenesis as a guide to ore finding.

Finally, the amount of sediment of diagenetic origin in sandstones, when we include all of the cement and the altered clays, is significant in relation to the total bulk of sandstone in the earth. Anyone interested in the origin of sandstones must necessarily include the diagenetic component as a major concern when he considers the whole rock, for the original detritus is only part of it.

Definitions

There have been almost as many definitions of diagenesis and related words as there have been geologists who have written on the subject, starting with the first use of the term diagenesis in 1888 by von Gümbel (p. 334). It is probably no accident that the word has been used in so many different senses because it refers to a concept inherently difficult to delimit from other, related ideas. A complete review of the subject, its history, and its terminology has been given by Dunoyer de Segonzac (1968).

Diagenesis: Diagenesis is a word that has been used for the following concepts:

(1) All physical or chemical changes after deposition, for example, quartz disappears and calcite appears.

(2) A more restrictive use of the term diagenesis refers only to changes before lithification, demanding an explicit assessment when the lithification process was initiated or completed.

(3) Another restrictive use refers to changes before burial, using the term for reactions of sedimentary material with the environment.

(4) A usage contrary to that of (3) includes only those changes which have occurred in rocks *after* burial. Both (3) and (4) have the limitation that they are not easily applied, since we do not know except by inference, which of the changes occurred before or after burial, nor just what degree of burial separates (3) and (4).

(5) Still another definition restricts the usage of diagenesis to chemical rather than physical processes. Thus we would differentiate compaction, a physical process, from diagenesis.

(6) Next comes another restriction of the term for which one has to make a genetic assessment: diagenesis refers to processes that result in lithification or cementation, which are usually associated and may not be strictly separable.

Finally, almost all of the definitions exclude metamorphism as a stage of diagenesis. As clearly stated by Correns (1950), there is no simple and non-arbitrary dividing line between diagenesis and metamorphism in a pressure-temperature field. We inevitably find sandstones in which there have been diagenetic changes related to moderate pressure and temperature increases attendant on deep burial of the sedimentary sequence, transitional to rocks in which the pressures and temperatures have been sufficiently high that the geologist conventionally calls them metamorphic. In practice there is no great difficulty in distinguishing between metamorphic and sedimentary rocks where the degree of metamorphism is high or the degree of diagenesis is low. On the other hand, there are low-grade metamorphic rocks or high-grade diagenetic rocks in which there is no clear-cut separation. An outstanding example of this is the zeolite facies, rocks such as those described by Coombs and others (1959), which are of equal concern to metamorphic and sedimentary petrologists.

Lithification: The word lithification is another which has no clear-cut meaning, having been used synonymously with induration, consolidation, or compaction. But induration expresses a quality of hardness, whereas consolidation implies a quality of compactness or density. Compaction and consolidation are almost synonymous, those in soil mechanics and engineering tending to use the term consolidation, whereas geologists use the term compaction. Another important aspect of lithification of sandstones is the chemical binding of framework particles by introduced cements or by altered matrix. We consider lithification as an observable quantity to be an engineering property of sandstones, because the differentiation of a lithified sandstone from an incoherent sand is one which depends in some way on the strength of material, its hardness, and its coherence.

There are a great many other terms, some in common use, others used only rarely. In considering all of them and their usage, it seems clear that (1) there is no unanimity; (2) there is a great deal of connotation or implication in the use of certain words, even though they may not be formally used to state such implications; and (3) a variety of concepts are involved but many times not explicitly stated.

Usage followed: For simplicity we shall use the most general term, diagenesis, to include all those processes, chemical and physical, which affect the sediment after deposition and up to the lowest grade of metamorphism, the greenschist facies. In our use we will exclude the idea of mechanical reworking by currents after initial deposition. We recognize that we can only arbitrarily delimit the processes we call diagenetic from metamorphism on the one hand and weathering on the other. As a sandstone is buried deeper and deeper in a sedimentary basin, it begins to heat up and be subjected to increased pressure. At some stage there is a continuous transition into a metamorphic state, particularly if deeper regions are undergoing regional metamorphism. Certain rocks may be called either metamorphic or sedimentary, depending on the point of view of the particular geologist.

Though diagenesis and metamorphism represent a continuum in the geologic variables, pressure and temperature, there are important differences. The study of metamorphism is dominated by the interpretation of mineral assemblages of a given bulk composition as indicators of pressure and temperature, assuming equilibrium or a close approach to it (see, for example, Winkler, 1965, p. 8). The result is an order to the mineralogy of metamorphic rocks that is expressed either by isograds, metamorphic facies, or petrogenetic grids. But the mineral assemblages of many sandstones, even after diagenesis, tend to reflect the original, obviously non-equilibrium detrital mixtures more than the effects of pressure and temperature. This is so because the ranges of temperature, 0–200° C, and pressure, 1–1000 bars, involved in diagenesis are much narrower than those characteristic of metamorphism, and because the kinetics of mineral transformations are so slow at low temperatures. Some equilibrium assemblages of diagenetic minerals that have been recognized as such have more to do with a response to changing compositions of moving pore waters than to changing pressure and temperature.

We recognize also that some post-depositional changes in a sandstone may not easily be separated from those we conventionally call weathering processes. If a sandstone is uplifted and as a result of erosion is placed only some few or few tens of meters below the surface, such that fresh water flows through it, the chemical changes that are induced in the sandstone may be exactly the same as those that would be induced were the sandstone to be found at an outcrop. Thus it has been commonly observed that the intense weathering of a sandstone, with consequent removal or precipitation of minerals, for example, the precipitation of hematite cement, is restricted to relatively surficial parts of an outcrop, commonly not more than a few inches to a foot in thickness from the surface. Sandstones that lie within 20 meters of the surface may have much carbonate cement removed, as has been shown by comparison with the same stratigraphic horizon more deeply buried. Figure 10-1 shows the relation between stages in the geologic history of a sandstone and diagenesis. For different

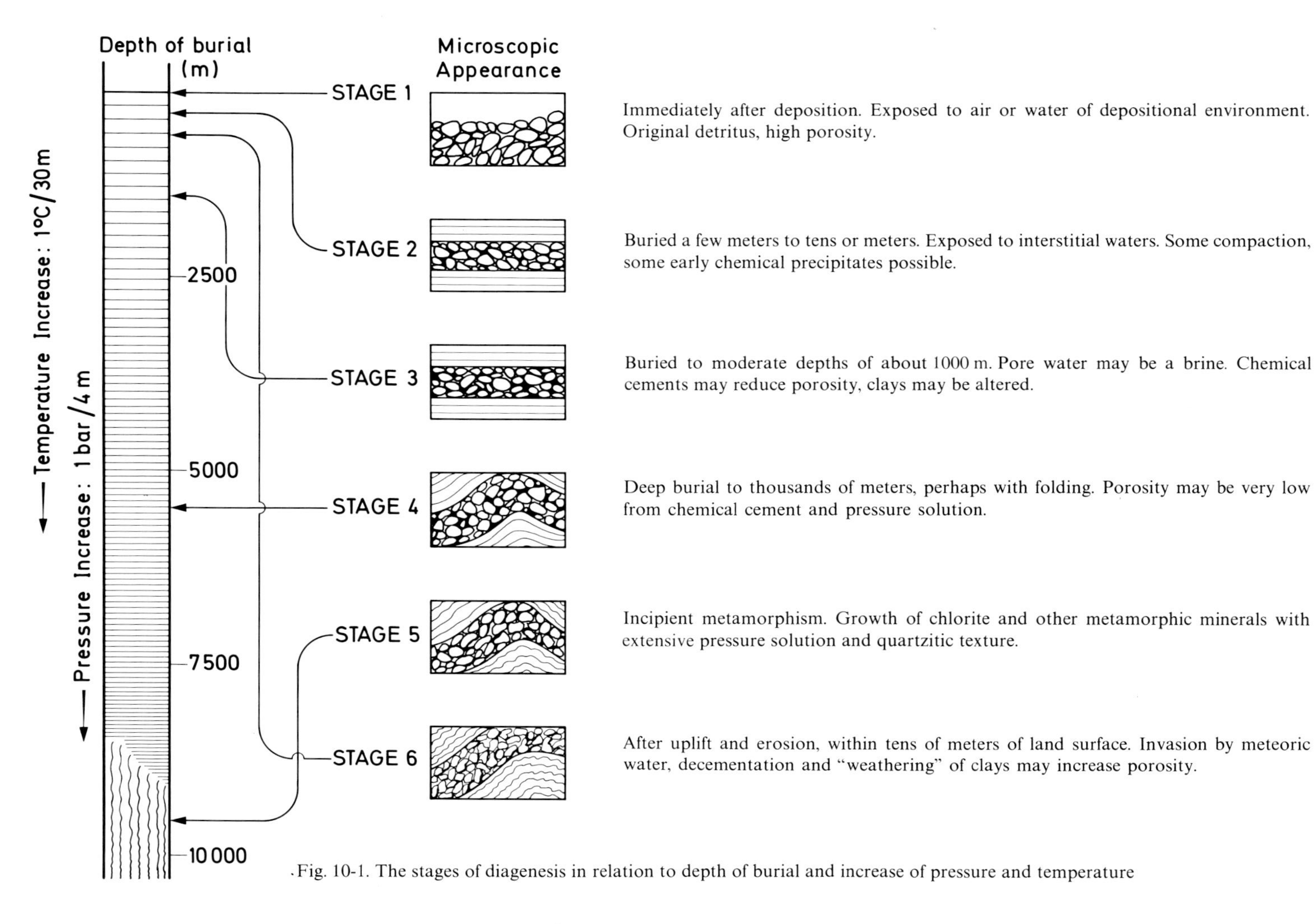

Fig. 10-1. The stages of diagenesis in relation to depth of burial and increase of pressure and temperature

versions of these stages and how terminology has been used see Dunoyer de Segonzac (1968, Tables 1-4).

It is convenient to use a time designation for diagenesis and diagenetic changes. We differentiate in a very simple way between "early" and "late" diagenesis. We recognize that we are arbitrary in this differentiation and furthermore that we cannot be precise about the dividing point. "Early" will be taken to mean those changes which take place within a few thousands to hundreds of thousands of years, presumably while the sediment is still buried at a rather shallow depth, less than 50 m, and may include indirect interaction with sea water and sea floor processes. "Late" refers to all later events. We imply, by this differentiation between early and late, differences between a sediment which is actively undergoing the initial processes of lithification and compaction and the much slower changes which come later and over a much longer time scale.

Aspects of Diagenesis

In this discussion we distinguish between several different aspects of diagenesis. One aspect is the objective evidence or data that is used to infer diagenesis. For example, a sharp, euhedral, tourmaline crystal projecting from a rounded grain of tourmaline that may be slightly different in composition, is a texture that can be objectively noted, photographed and described (Fig. 10-2). The geometry of it

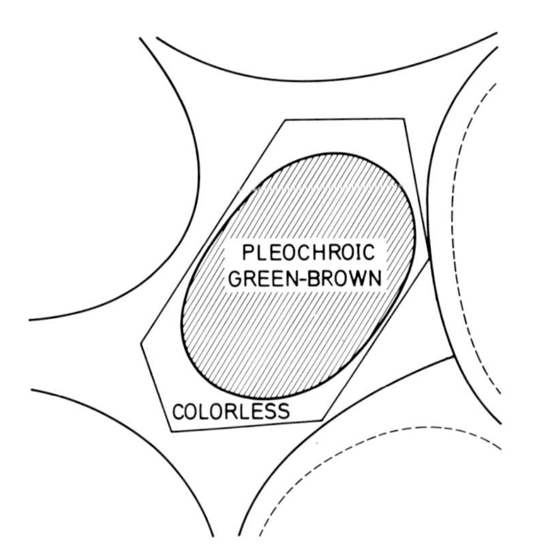

Fig. 10-2. A colorless euhedral overgrowth of tourmaline precipitated in optical continuity with a rounded detrital core of pleochroic green-brown tourmaline

alone plus our knowledge of the composition of tourmaline suggests that the sharp little projection was not part of an original detrital grain but was added to it in pore space surrounding the original detrital grain, sometime after the rounded grain was deposited.

Once the objective evidence is determined we may infer a process; in the preceding example we infer the precipitation of authigenic tourmaline from some aqueous solution which permeated the rock after deposition, a solution we will simply call formation water. Involved in this inference, of course, is the chemical composition of that fluid and how that fluid's composition evolved.

Next we must consider the time scale of the inferred process. Again with respect to the tourmaline crystal described above, we observe that we find most such overgrowths in older sandstones, none in very young ones, and so may hypothesize that, because of the extreme slowness of the precipitation of tourmaline and the relatively low temperature of the process, this process may have taken 10^7–10^8 years. When in the history of the sandstone that 10^7–10^8 years may have come is difficult to determine.

A final aspect is the ordering with respect to geologic events in the sandstone's history of the particular diagenetic event. Thus the tourmaline overgrowth that was precipitated from a formation water over a period of 10^7–10^8 years may be inferred to have taken place during the long period of time that this particular sandstone was deeply buried under a thick pile of sediment, perhaps 5,000 m.

One can conclude that though some limited answers to some of the questions of diagenesis may be given by inference from geochemical ideas, or by petrographic study of a limited suite of samples, all of the different aspects of diagenesis can be discussed only if one notes in some detail the general geologic history of the sedimentary basin in which the sandstone was deposited and how geologic events may be correlated with diagenetic episodes. It is this last, in fact, which is probably the ultimate justification for the study of diagenesis as an historical geological subject, for if we are able to make generalizations about the different kind of geological events which may be implied from diagenetic evidence, we may then be able to increase our knowledge of the historical geology of sandstone by studying those evidences that point to diagenetic change. Conversely, one may turn this process around so that we may use a thorough knowledge of the historical geology of sandstones, together with their diagenetic history, to learn a great deal more about geochemical processes in the sedimentary part of the earth's crust.

The Evidence of Diagenesis

Textural

The major direct evidence of diagenesis in sandstones is the nature of the textural relations between mineral grains and crystals. The most obvious are the pseudomorphic replacements of one mineral by another or of one type of crystalline aggregate by another. The replacement of a calcareous brachiopod shell by quartz is undoubted evidence of the dissolution of carbonate and precipitation of quartz after deposition. Another example is the replacement of a glassy shard by clay alteration products. The key point of these replacements is that the original, recognizable texture is preserved in a material that is known not to be original. True mineral pseudomorphs, in which the crystal habit of the replaced mineral is preserved by the replacing mineral, are uncommon in sandstones; one of this type that can be seen is a kaolinite mass that has the shape of a feldspar cleavage fragment.

The more abundant replacements are those which preserve faint outlines of the original, the phantom structures, in spite of the loss of grain boundaries. Silicified oolites are perhaps the most dramatic evidence of this kind. Outlines of fossil fragments will appear in a coarse mosaic of carbonate cement in some sandstones. The recognition of phantom structures depends on the existence of inclusions, most of which are so small as to be below the resolving power of the microscope.

Crosscutting relationships are conclusive evidence of replacement. The most frequently encountered texture of this kind in sandstones is the etching and embayment of detrital grains by a crystalline mosaic or single crystal. Such replacement can result in "floating" inclusions of the detrital grain, completely surrounded by the replacing material. Microstylolites that cut across detrital grains (Sloss and Feray, 1948; Heald, 1955) are other manifestations of the solution process that imply postdepositional chemical change.

A cursory examination of almost any ancient sandstone will show that the pores between detrital grains are filled with an aggregate or mosaic of minerals, most frequently carbonate, silica, or clay; that kind of pore filling is conspicuous by its absence in modern sands. If we neglect the clay matrix of graywackes, which may be primary in part, and consider only pore filling by carbonate, silica, and similar minerals, then the contrast between modern and ancient becomes even greater. The pore filling of ancient sandstones may be partial or complete. It may consist of only one mineral or several. The pore filling may or may not be associated with replacement of detrital grains, or one mineral of the pore filling may replace detrital grains and a second mineral of the same pore filling not. Recognition of pore filling as such may not be easy in some quartz-cemented quartz arenites or in clay matrix-rich graywackes with abundant argillaceous rock fragments.

Some sandstones display euhedral faces on detrital grains, typically quartz, that are sharp and show no signs of abrasion during transportation. On this basis alone they must be presumed to be authigenic overgrowths, even though the distinction between the detrital and authigenic part of the crystal cannot always be made. Studies of sands of modern sediments have shown that all detrital grains over 0.1 mm in diameter are affected at least slightly by the abrasion process and sharp euhedral faces are rarely found.

In some sandstones, euhedral overgrowths cannot be distinguished, and detrital grains cannot be separated from authigenic overgrowths; yet the pore space is almost completely filled. The observed interlocking crystalline mosaic texture can only result from chemical solution and/or precipitation rather than physical settling from a fluid medium.

Concretions and nodules in sandstones are textural evidences of diagenesis in sandstones that are a combination of pore-filling, with or without replacement, and any of the other textural elements that have been described. The gross external appearance of such structures is commonly related to cementation of grains being greater in the nodule than in the surrounding matrix or to cementation by a different kind of mineral. Thus we may find hematite cemented nodules in an otherwise calcareous cemented sandstone. If the concretions were formed early in diagenesis, the original mineralogy and texture may be preserved better than in the surrounding sandstone because the concretion was less permeable and so was not transmissible to fluids responsible for solution or precipitation later in diagenesis. Bramlette (1941) has discussed this effect on heavy minerals.

Mineralogical

The evidence for diagenesis that resides in the nature of the mineral substance rather than its shape, habit, or size is less conclusive than the textural evidence. A useful guide to diagenetic origin is purity of composition. Many authigenic overgrowths of quartz or other minerals are recognizable by the clarity and lack of crystalline, liquid, or gas inclusions so characteristic of the detrital contributions from igneous and metamorphic rocks. Authigenic tourmalines are colorless in contrast to colored, pleochroic detrital grains. Cathodoluminescence and electron probe studies show that authigenic minerals are much lower in trace element content and thus show lower luminescence than the detrital grains. Authigenic feldspars tend to be very pure Na- or K-forms and

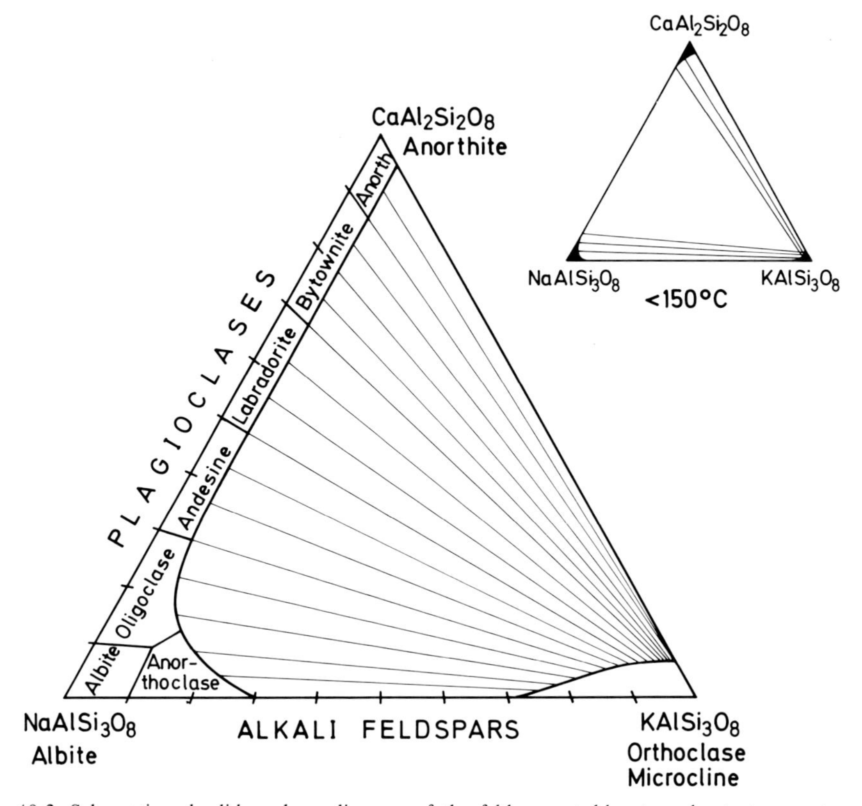

Fig. 10-3. Schematic subsolidus phase diagram of the feldspars stable at moderate temperatures (below about 500° C.) and low pressures, showing nomenclature in relation to composition. At temperatures below 150° C., characteristic of sedimentary environments, the schematic phase diagram would be as shown in inset, only the relatively pure end members being stable

thus distinct from the wide range of plagioclases and alkali feldspars that are represented in the detrital fraction (Kastner and Siever, 1968). Fig. 10-3 shows the phase relations of feldspars at low temperatures.

Some reliance may also be placed on the kind of mineral that is found. The more soluble minerals are unlikely to be found as detritals in most sandstones that were deposited by water. Though we know that sand dunes of gypsum and carbonate may be found, the first guess of the petrographer who finds gypsum or carbonate in sandstones is that they are diagenetic rather than detrital. Halite, if care is taken to preserve it by not using water during thin section grinding, is even more probably of diagenetic origin. But because there is always the possibility of detrital or primary chemical origin for such minerals, caution is called for and ultimate reliance should be placed on textural features. The fact that very insoluble minerals such as quartz and feldspar typically are found as detrital grains certainly does not predispose us to think that they can not be authigenic as well.

Mineralogy in combination with texture is an important criterion of diagenesis. We are able to recognize the diagenetic nature of a dolomitized fossil precisely because it is dolomite and we know of no organism that secretes dolomite for its shell though many deposit magnesian calcite. A zeolite mineral in the shape of a glass shard is diagnosed as a replacement because the shard shape of broken volcanic tephra is always associated with glass and zeolites do not form from the congealing of a melt.

Physical Properties

The differences in bulk physical properties between modern and ancient sands reflect the diagenetic modifications that are ultimately revealed in terms of the details of mineralogy and texture as seen in thin section. Thus bulk density of sandstones increases as the porosity decreases. Permeability is commonly correlated with porosity (Fig. 3-13). Permeability is also related to the fabric or packing of the grains and is reduced in more compact, more tightly packed ancient sandstones. As the porosity decreases, the velocity of elastic waves propagated through the rock increases, the basis for using sonic logging of drill holes to predict permeability. All of these changes in sandstones are attributable to the enlargement of detrital grains by overgrowths and the precipitation of mineral cements in pore space during diagenesis and only secondarily to compaction. Though porosity and permeability and related physical properties may be strongly dependent on the grain size and shape distribution of the sand, the prominent difference is between modern sands that are not cemented or compacted and ancient sandstones that are.

Little has been done by geologists on the hardness, coherence, and fracture of sandstones.

Chemical Properties

So few chemical analyses of modern sands have been made that there can be no statistically valid comparison of the composition of modern and ancient sands that would reveal the effects of diagenesis. We can only infer some changes on the basis of mineralogical differences, some of which are explored in Chap. 2. The general tendencies exhibited by ancient sandstones are to have more alkali metals, alkaline earths, and silica, and less water than modern sands; these are the expected consequences of carbonate and silica cementation and clay mineral alteration.

Though one does not ordinarily include the composition of the pore fluids of sandstones as part of its chemical analysis, there is every justification for doing so, for that fluid is part of the rock when it is buried in the subsurface, as the great bulk of sandstones are. The differences between pore water of modern and ancient sandstones is sufficient evidence by itself of the extensive changes wrought by diagenesis. Pore waters of modern sands are either fresh and similar in composition to river waters (alluvial sands) or are essentially sea water (beaches and offshore shelf sands, turbidites, deep sea sands). In contrast the pore waters of ancient sandstones may vary from potable water aquifers near the surface, not too dissimilar from the river water compositional range, to the concentrated brines of the deep subsurface. The brines are drastically different from sea water in composition not only because they may be much more concentrated in dissolved solids but because the ratios of various ions to each other may differ greatly from sea water. Many of the brines contain much more Ca than Mg; the reverse is true for sea water. Sulfate may be greatly depleted. Silica is usually increased. All of these changes are evidence of the chemical reactions that have taken place between pore fluids and rocks after deposition. Composition and origin of many subsurface fluids are covered in Young and Galley (1965).

Oil, gas, or other organic concentrations such as asphalts or pitch-like substances, do not appear in modern sands except for those places where such fluids are infiltering from external sources. Organic matter in any form is very low in abundance in modern sands, though there always is some. The contrast of this picture with the frequent occurrence of all kinds of oil and gas accumulations in ancient sandstones shows that some diagenetic processes have been active that are responsible for the production of the oil and the migration of it into the sandstone reservoir.

Summary

Our best evidence for diagenesis is the extent and nature of the differences between modern and ancient sands and the presence of textural elements, particularly the pseudomorphic replacements, indicating that post-depositional processes have modified the rock. We can illustrate the properties of the rock that give evidence of diagenetic change by giving two composite examples, one of a quartz arenite, the other of a graywacke.

A quartz arenite might show euhedral secondary enlargement of detrital quartz grains, partly intergrown to give a strongly cemented rock. The rock might have as much as 10–15 percent carbonate as pore filling, part of which has replaced both detrital and secondary enlargement quartz. The carbonate may be calcite with a few rhombohedra of dolomite replacing calcite. Some hematite cement may obscure both detrital grains and other diagenetic components.

Heavy minerals may include little besides tourmaline, zircon, and rutile, and the tourmaline may show secondary enlargement by a colorless, non-pleochroic variety. The rock may have a porosity of 15 percent but its permeability may be as low as a few millidarcys. It will be hard and resistant to breakage. The pore fluids may include oil and gas and a concentrated brine.

A different set of properties may be shown by a graywacke in which there is a clay matrix of about 30 percent, at least part of which has the appearance of squashed argillaceous rock fragments. Much of the matrix is of a chloritic composition and appears to replace parts of detrital silicate grains. Scattered patchy areas of calcite fill some pore space or replace both detrital grains and clay matrix and the carbonate areas may be associated with small rhombohedra of siderite or iron-rich dolomite. Shardlike fragments of zeolite, such as heulandite, are distributed through the rock, a few showing relics of an isotropic glass. The heavy minerals are abundant and varied, including many mafic minerals and show little or no evidence of corrosion or dissolution. The rock's porosity and permeability are very low. The rock may be hard and, if thinly bedded, may tend to fracture along bedding planes. There is no oil and gas, but the chemical analysis may show as much as 1 percent organic matter.

The genetic interpretation of such a catalogue of evidence is unlikely to be made in terms of a single, simple process but rather involves a group of varied physical and chemical processes that interact and so affect the rock in different ways and to different extents depending on the geologic milieu.

Physical Changes

Physical processes that affect the nature of a sand after deposition lead to an increase in bulk density, primarily by an adjustment to the force of gravitation. The major way in which this happens is compaction through a change in packing, which results in a loss of porosity. Sands as deposited may have a wide range of packing density, illustrated in the extreme by the difference between the windward and lee slope of a sand dune but easily observable also in various subenvironments of beaches and bars. So little has been done on packing that we cannot say whether such differences are ironed out as a result of compaction, but studies of imbrication such as that by Rusnak (1957) indicate that there is inheritance and that the effects of compaction are small. Brett (1955) found that the inclination of crossbeds was affected by folding, an effect attributed to grain rearrangement during shearing within the bed. The porosities of sandstones that are relatively uncemented but have been buried to sufficient depths to have been compacted, such as the Tertiary sandstones of the Gulf Coast of North America, are high, fairly close to the values expected for modern sands. Dickinson (1953) quotes a general average porosity of 30 percent for unconsolidated Miocene sands of the Gulf Coast region down to at least 10,000 feet (3226 m) below the surface and believes this confirms the very slight compressibility of loose sand reported by Athy (1930). We conclude that the amount of compaction of sand grains achieved by rearrangement under load is small and would account for a drop in porosity of only a few percent.

To the extent that graywacke sandstones with high amounts of matrix clay are rocks intermediate between sandstones and shales, they may show compaction effects much like those of shales. The decrease in porosity under load of clayey materials is large and in the natural case has been summarized most recently by Meade (1964). It is likely that the typical graywacke may have been physically compacted from an initial porosity of 50 to 60 percent to 30 to 40 percent by rearrangement of clay particles and squeezing out of water. Because there are few examples of modern graywacke sands with high amounts of matrix (Heezen and Hollister, 1964) it is not possible to report more definite conclusions. If many ancient graywackes present the appearence of clay matrix by virtue of the diagenetic modification and physical squashing of argillaceous rock fragments then the role of compaction would be doubly important, for it would then have been largely responsible for the matrix. The difference between the load required to squeeze water from clay and that required to squash clayey rock fragments is unknown. The ratio of identifiable rock fragments to matrix of an Eocene graywacke, the Tyee formation of Oregon, seems to decrease in the area of heaviest overburden, but stratigraphic control is insufficient to calculate the loading factor (Lovell, 1969). Experiments that lend support to the idea that the matrix originates by alteration of rock fragments were done by Hawkins and Whetten (1969). Samples of moderately well-sorted sands with abundant volcanic rock fragments from the Columbia River, sands that are similar in mineral and chemical composition to typical graywackes, were held at temperatures of 150–300° C and 1 kilobar water-pressure for 21 to 60 days, simulating burial to 3 or 4 km. Clay minerals and zeolites were produced from the volcanic fragments. Brenchley (1969) studied Ordovician graywackes from Wales and concluded that calcite cementation preserved detrital grains and inhibited matrix formation in some beds, but in non-calcareous beds, rock fragments were altered to matrix. Fractured or shattered grains in sandstones are uncommon and not completely restricted to strongly folded belts. The few that can be seen in undeformed sandstones are perhaps the result of alteration along lines of weakness inherited from the crystal in the source rock or induced during transport impact. Cathodo-luminescence studies have shown numbers of fractured and re-welded detrital grains that are not seen in ordinary microscopy (Sippel, 1968). In orogenic belts, strongly folded sandstones may show cataclastic textures, sheared grains, and other deformational structures that are clearly related to tectonic movements and so not properly in the province of diagenesis. The distribution and origin of fractured grains is a problem that has received practically no attention.

Ductile deformation of sand under moderate confining pressures, a few kilobars, at room temperature, followed by faulting at higher confining pressures has been reported by Byerlee and Brace (1969). Post-depositional physical changes such as slide brecciation, slump structures, and crumpling, are discussed in Chap. 4.

Chemical Changes

Chemical changes are far more important than physical changes in altering the character of sandstones after deposition. Cementation and lithification are mainly the result of chemical precipitation of a binding agent or the chemical welding of adjacent detrital grains. In all of the chemical processes one must consider the interaction between solid mineral grains and pore fluids. Because the bulk and mineral composition of the solid changes only slowly and slightly during diagenesis, most of the sequence of diagenetic events inferred from petrographic evidence must be related to evolutionary changes in pore water chemistry. It is convenient to consider the chemical changes in detail under the headings of precipitation, dissolution, recrystallization, alteration and composition of pore waters.

The chemistry of dissolution and precipitation is best analyzed by explicitly writing the balanced chemical equation and evaluating the change in *free energy* at constant pressure and temperature, accompanying the reaction. This free energy is properly termed *Gibbs free energy* or Gibbs function, and is here abbreviated as G; the finite change in G is designated by ΔG. From the equation and the value of ΔG we can quantitatively evaluate an *equilibrium constant*, which can then be used to predict the entire range of concentrations of reactants necessary for either dissolution or precipitation.

We model this approach by writing the reversible equation, at a given temperature and pressure, that describes the precipitation of a pure ionic crystalline solid compound, AY, by reaction of two completely ionized electrolytes, AX and BY, in a pure aqueous solution

$$A^{+} + X^{-} + B^{+} + Y^{-} \xleftrightarrow[25^\circ\text{C}]{1\text{ atm.}} AY_{\text{(crystalline)}} + B^{+} + X^{-}, \qquad (10\text{-}1)$$

where A and B are cations and X and Y anions.

We determine the standard free energy, ΔG^0_R, for this reaction by subtracting the sum of Gibbs free energies of formation, $\Delta G f^0$, of the reactants from the sum of the $\Delta G f^0$'s of products to obtain

$$\begin{aligned}\Delta G^0_{\text{Reaction}} &= [\Delta G f^0(AY) + \Delta G f^0(B^+) + \Delta G f^0(X^-)] \\ &\quad - [\Delta G f^0(A^+) + \Delta G f^0(X^-) + \Delta G f^0(B^+) + \Delta G f^0(Y^-)]\,.\end{aligned} \qquad (10\text{-}2)$$

Then, we use another relation to get the equilibrium constant, K_{eq},

$$\Delta G^0_R = -RT \ln K_{\text{eq}}, \qquad (10\text{-}3)$$

where R is the gas constant, T is absolute temperature (degrees Kelvin), and *ln* a natural base logarithm.

The equilibrium constant is formed in terms of the division of the product of the activities of the reaction products by the product of the activities of the reactants,

$$K_{\text{eq}} = \frac{a_{AY}\, a_{B^+}\, a_{X^-}}{a_{A^+}\, a_{B^+}\, a_{X^-}\, a_{Y^-}} \qquad (10\text{-}4)$$

which reduces in this case to

$$K_{\text{eq}} = \frac{a_{AY}}{a_{A^+}\, a_{Y^-}}\,. \qquad (10\text{-}5)$$

The *activity* of a substance, a_i, is frequently called *thermodynamic concentration* and is related to the true concentration of a substance, also referred to as *molarity* or *molality*. That relation is such that activity approaches concentration in very dilute solutions and deviates from it as total concentration of all substances in solution increases. By convention, the activity of any pure solid thermodynamically stable at the given temperature is taken as unity, and so equation 10-5 reduces to

$$K_{eq} = \frac{1}{a_{A^+} a_{Y^-}}. \qquad (10\text{-}6)$$

In the case of very dilute solutions, the activities may be replaced by concentrations.

Eq. (10-1) may be written from right to left to describe the dissolution of AY, in which case, B and X may be eliminated for dissolution in pure water and we get the simple result

$$K_{eq} = a_{A^+} a_{Y^-}. \qquad (10\text{-}7)$$

If AY is a relatively slightly soluble substance, the solution will be dilute and K_{eq} will approach the *solubility product constant*, $K_{S.P.}$, which is given in terms of concentration, m, and is

$$K_{S.P.} = m_{A^+} m_{Y^-}. \qquad (10\text{-}8)$$

Detailed expositions of this kind of calculation are given by Garrels and Christ (1965, Chap. 1) and Krauskopf (1967, Chap. 1). Both books cover a great many more geochemical calculations and many topics of interest to sedimentary petrologists. Only through such calculations can we deduce the chemical conditions needed for the various kinds of precipitation, dissolution, and other reactions discussed below.

Precipitation

Petrographic evidence is strong for direct precipitation of various kinds of cements from pore solutions onto surfaces of the same or different mineral grains. The most common precipitates are silica and carbonate. Others are gypsum and anhydrite, barite, hematite and limonite, feldspars, zeolites, and clays. It seems that a great variety of minerals, many of which are ordinarily dismissed as forming only from high temperature igneous or metamorphic processes, can be precipitated from low temperature solutions.

Strongly cemented sandstones in which the pore space is almost completely filled are evidence that the precipitation process may continue from a supersaturated pore solution until there is no open space remaining. Because the most concentrated brines known contain only a small amount of precipitable material, continued precipitation in such rocks could not have come from a closed system without replenishment of solution but must have come from a steady supply of supersaturated solution flowing through the rock, a fact recognized long ago by Van Hise (1904, p. 865-868). Then as the precipitation proceeded the pore spaces would have become smaller, the permeability steadily reduced, the flow rates diminished, and the rate of precipitation slowed (Fig. 10-4). The initial permeability and its rate of decrease is a function of grain size distribution, so that a fine-grained sandstone with lower initial permeability would become strongly cemented before a coarse-grained one (Adams, 1964, p. 1575-1577). A number of flow models, including those that are complicated by pressure solution, can be proposed for this kind of pore filling but they all have in common an exponential decrease in the amount of precipitate produced per unit time. To put it another way, the time required to completely cement a sandstone is very long compared to that needed for partial cementation.

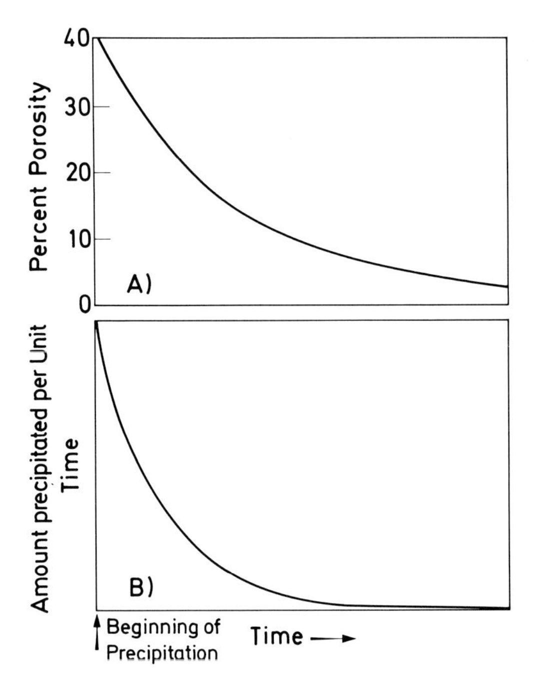

Fig. 10-4. Exponential decrease in (A), porosity and (B), amount of cement precipitated, as cementation of a sandstone proceeds. The precise shapes of such curves depend on the rate of fluid flow through the sand, the initial porosity and permeability, and the rate of precipitation per unit of fluid

What is not clear is the absolute length of time needed for complete cementation and when in the course of a sandstone's history such an episode is likely to have taken place. The only evidence available on this point is that few Recent or Tertiary sandstones are so cemented. Thus it seems likely that such an episode is linked to a long period of burial, of the order of magnitude of 10^8 years. The useful way in which to check this conclusion would be to develop statistical data on the degree of cementation of sandstones with geologic age. Unfortunately, little quantitative data is in the literature on the degree of cementation other than a few pilot studies (Gilbert, 1949; Heald, 1956; Siever, 1959). With current microscopy techniques such data is difficult to accumulate; distinguishing detrital quartz grains from quartz cement or overgrowths in many sandstones is difficult if not impossible. Cathodo-luminescence will be useful in making this distinction in the future (Fig. 10-5).

Partial filling of sandstone pores by chemical precipitates is most common. The form of the precipitate may be as overgrowths on detrital grains, as precipitated clay linings of detrital grains, and as isolated volumes of cement distributed in irregular patches. Pore fillings may be of the same mineralogy as the detrital grains. The resulting growths, called *syntaxial*, are frequently precipitates deposited in optical continuity with the original grain and can be looked on as a resumption of crystal growth, albeit under vastly different conditions from that of the original

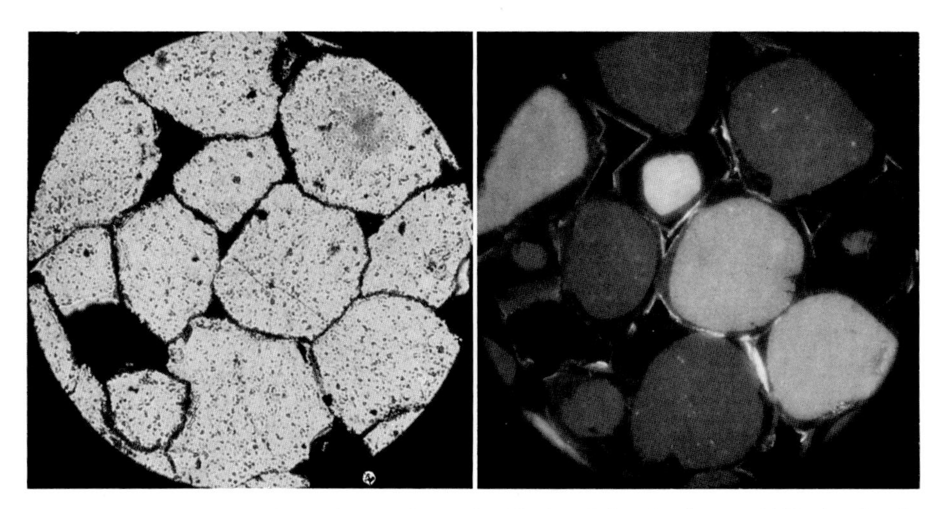

Fig. 10-5. Comparison of photomicrographs of the Silurian Hoing Sandstone of Illinois taken in plane polarized (left) light and by cathodo-luminescence (right). Detrital quartz grains appear to have sutured and concavo-convex borders indicating pressure solution in plane polarized light but cathodo-luminescence reveals that there is extensive secondary enlargement with little or no pressure solution (Photos by R. F. Sippel)

crystal and after a drastic interruption by erosion, transportation, and deposition. The mineral that shows this behavior most commonly is quartz; feldspar overgrowths are also common but less obvious; calcite may also show such overgrowths. Among the minor constituents, zircon and tourmaline show optically continuous overgrowths. (See Figs. 6-22, 6-26, and 10-2 for examples of overgrowths.)

Minerals precipitated on substrates of a different mineralogy form *epitaxial* overgrowths, such as carbonates on quartz surfaces or clays on quartz. The growth of chalcedony on quartz surfaces is probably this kind of growth too, for when originally precipitated the chalcedony was probably some form of amorphous silica or a form of disordered cristobalite that only much later was recrystallized into microcrystalline quartz.

The size of secondary precipitate crystals, syntaxial or epitaxial, is, if experience gained from artificial crystal growth technology is a guide, related to rapidity of crystal growth and to the nature of the substrate. In general, small crystals indicate relatively rapid growth and the most commonly noted large single crystals are the result of slow growth. It is frequently noted that an overgrowth starts as many tiny growths on the immediate substrate but as the crystals move away from the substrate they get progressively larger until only a single large crystal grows (Fig. 10-6). This can be ascribed to minor fluctuations in individual crystal growth rates whereby the faster growing crystals overtake and encompass the smaller, slower growing ones. It may also be the product of the slowing down of crystal growth as pore space becomes progressively filled, permeability is reduced, flow rate of solution is reduced, and crystal growth becomes slower: the consequence is larger crystals.

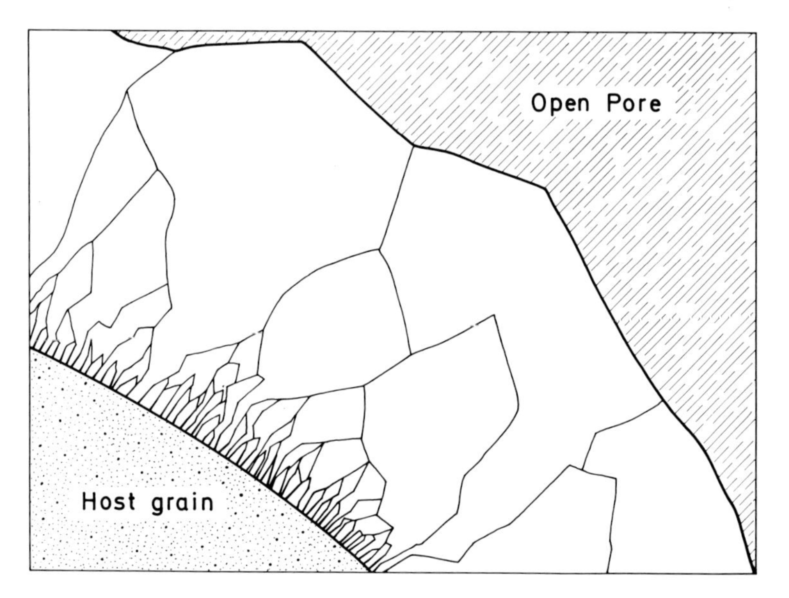

Fig. 10-6. Progressive growth in crystal size as a secondary precipitate grows outward from host grain into pore space. Faster growing crystals gradually take up more of the perimeter as they enlarge at the expense of smaller, slower growing ones

Another property of secondary precipitates that is related to the nature of crystal growth is the distribution of inclusions. Such inclusions may be randomly or regularly distributed. The inclusions are remnants of the rock that was replaced; secondary precipitates that have grown into voids are free of inclusions. The inclusion compositions and distribution can be a most useful guide to materials once present in the rock and now replaced. They also provide a guide to textures of those precursors preserved by the pattern of inclusions, such as those of oolites or fossils.

Some distinctive shapes of inclusions, such as the hourglass patterns seen on some authigenic feldspars (Kastner and Waldbaum, 1968), may be linked to the rapidity of crystal growth. Becke (1892) and Harker (1932) used hourglass patterns as evidence that inclusions were incorporated into rapidly growing crystal faces and pushed aside by slower growing faces. Experimental evidence to support this behavior comes from a study of rock and mineral particles incorporated into growing ice crystals (Corte, 1963) and a summary of the basic information on this kind of process is given by Buckley (1951, p. 445-446), who pointed out the ultimate in the incorporation of inclusions is the formation of large sand "crystals" in the Fontainebleau sandstone of France, in which growing calcite crystals have incorporated quartz grains up to two-thirds the weight of the whole crystal. Such "poikiloblastic" textures of sandstones have been noted for many years.

The most common silicates and oxides are precipitates of minerals present in the primary detrital assemblage. Quartz, feldspar, clays, zircon, tourmaline, anatase and other titanium oxide minerals are representative of this group. As noted elsewhere in this chapter (p. 391) diagenetically precipitated minerals are almost always purer than detrital phases. Diagenetic minerals may be of the same chemical composition but different crystal structure, as illustrated by the precipitation of opal on quartz surfaces and anatase overgrowths on brookite.

Other diagenetically precipitated minerals may be of phases not originally present in the detrital mix. Secondary carbonate typically is of this kind. Others are anhydrite, iron oxides, halite, pyrite, and other normally non-detrital minerals.

Precipitation may be integrally involved in recrystallization and alteration and it may not be clear from the morphology of the resulting crystal which it is, a recrystallization of a pre-existing crystal, the alteration of a pre-existing crystal by reaction with a pore solution, or a precipitate from a homogeneous solution. A homogeneous solution is one in which only one phase, typically a liquid solution, is present. Strictly speaking, when a solid precipitate forms, the system becomes one of several phases and thus heterogeneous. It is reasonable to assume that for some precipitation reactions the mineral surfaces bounding the pore space play no more of a role in the reaction than does the surface of a glass laboratory vessel. For example, quartz overgrowth crystals may be precipitated as the product of crystallization from a supersaturated pore solution passing over detrital quartz surfaces. The quartz surfaces do not play any part in the reaction other than to serve as nucleation sites. Another possibility is that the quartz overgrowth is the final precipitate in the recrystallization process by which an amorphous silica or chalcedony precipitate was locally dissolved and immediately reprecipitated as quartz. Finally, the quartz may be a precipitate of the excess silica produced in the local solution by the hydrolysis of nearby feldspar grains by water containing carbon dioxide according to the equation:

$$2\,KAlSi_3O_8 + 2H^+ + 2\,HCO_3^- + 9\,H_2O \longleftrightarrow Al_2Si_2O_5(OH)_4 + 4H_4SiO_4 + 2\,K^+ + 2HCO_3^- \qquad (10\text{-}9)$$

The distinction between these three origins is not easy but may be made possible in many sandstones by observation of relics of the former phases, the antecedents of either recrystallization or alteration.

Dissolution

We use dissolution here to include only *congruent dissolution*, a process by which all of the solid phase dissolves bit by bit, leaving behind fresh surfaces of as yet undissolved solid unaltered in composition. Such dissolution is characteristic of pure NaCl, SiO_2, and $CaCO_3$. This process is distinguished from *incongruent dissolution*, in which there is a kind of selective dissolution whereby the solid that is left undissolved is changed in composition, because only some components of the crystal are leached into solution or because the ratio of the components in the dissolved fraction may be different from that in the original solid. For example, a highly magnesian calcite will dissolve in such a way that the fraction dissolved has a much lower Ca/Mg ratio than the original solid, leaving undissolved a solid with a much higher Ca/Mg ratio (Chave and others, 1962). In this way, a magnesian calcite will gradually be converted to a purer calcite. This description of dissolution is not intended to imply a precise mechanism by which ions detach themselves from the crystal surface, solids become changed in composition, or become reprecipitated from solution; it only describes the overall result. Thus, Eq. (10-9) describing alteration – a variety of incongruent dissolution – of feldspar to kaolinite is schematic only; it does not pretend to give the exact reaction mechanism by which the alteration really takes place, a much more complex process (Wollast, 1967).

Dissolution is an integral part of the recrystallization process, for at the temperatures under which sediments are laid down or subjected to diagenesis, roughly from 0° – 200° C, "dry" recrystallization, that is, a restructuring of the mineral by solid diffusion without mediation by a surrounding fluid, would take so long that all of geologic time would not be sufficient for most transformations. So the recrystallization process in diagenesis is one during which the precursor is dissolved and the final crystal is precipitated from solution. The whole process may take place in a thin film of solution immediately adjacent to the solid surface in such a way that the dissolved ions are immediately reprecipitated without any wholesale communication with the pore water at large. In such a case the end result might show little observable difference from a dry recrystallization product.

The dissolution process is also a part of replacement of one mineral by another. A careful consideration of the geometric constraints on the nature of the replacement process indicates that in a large number of sandstones the substitution of one mineral for another takes place without volume change. Without necessarily specifying that the replacement has to be "atom by atom" or "ion by ion," the fact that the grains (and any cement) have to be in contact with each other to continue to support the rock without collapse demands a kind of formation during which all dissolution of the replaced and precipitation of the replacing mineral takes place in an exceedingly thin film between the bounding surfaces of the two phases (Fig. 10-7). Such a film is at least as thin as 0.1 mm and probably of the order of microns so that surface tension gives it enough strength to be mechanically supporting. The transport processes within this film must be adequate to move the dissolved material out and the precipitating material in from the open pore water reservoir. In the example shown in Figure 10-7, a typical case of calcite replacing detrital quartz in an arenite, some idea of the complexities involved can be seen from the following sequence of reactions:

1. SiO_4 tetrahedra dissolve from a quartz surface and hydrate to become H_4SiO_4 molecules in solution.

2. The concentration of H_4SiO_4 being higher in the film than in the pore fluid (necessary for dissolution to proceed), the H_4SiO_4 diffuses out into pores.

3. The concentration of Ca^{++} and HCO_3^- being higher in pore fluid than in thin film, the two kinds of ions diffuse into the film.

4. The ion activity product of Ca^{++} and HCO_3^- exceeding

$$K_{eq} = 10^{-8.35} = a_{Ca^{++}} a_{CO_3^{--}} \simeq m_{Ca^{++}} m_{CO_3^{--}} = K_{S.P.} \qquad (10\text{-}10)$$

the solubility product of calcite at the calcite surface, $CaCO_3$ precipitates according to the equation:

$$Ca^{++} + 2HCO_3^- = CaCO_3 + H^+ + HCO_3^- \qquad (10\text{-}11)$$

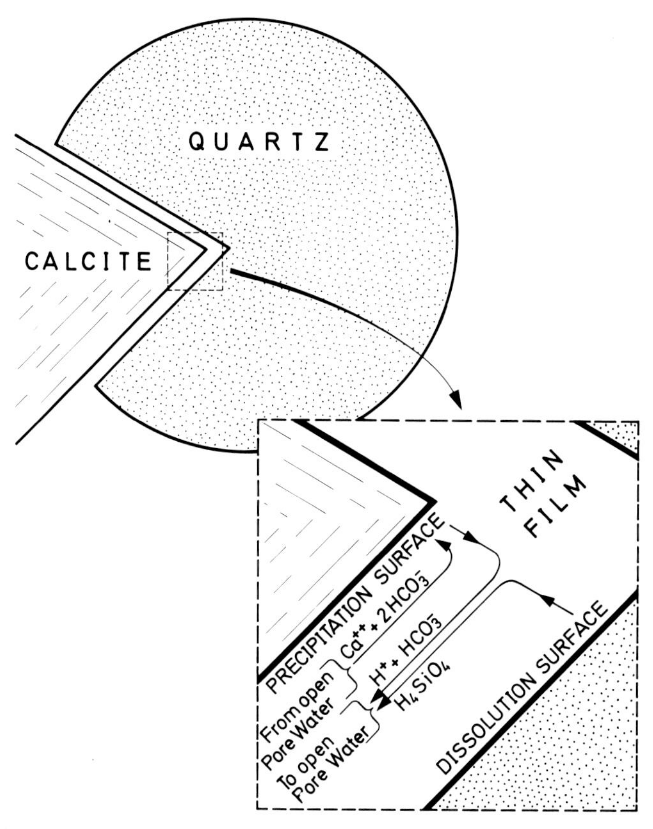

Fig. 10-7. Chemical transport in thin film at a front of replacement of quartz by calcite. Inset shows details of process

and, to maintain the concentration gradient that powers the diffusion, H^+ must diffuse out of the film into the pore space.

The critical matter in all of this transport is the *rate of diffusion in and out of the film*, which is a function of the thickness of the film, the temperature, and possibly other variables, for the behavior of such thin films is not well understood. In an analysis of a similar problem, the formation of stylolites, Weyl (1959) postulated a film including thin layers of clay such that the film included many subfilms or channels for diffusion; such a postulate seemed to him to be necessary to speed up the diffusion process sufficiently to accomplish much dissolution in reasonable geologic times and to be justified by the common presence of clay films along stylolites. Though clay films are by no means uncommon in arenites and have been used to explain pressure solution (Heald, 1955; Thomson, 1959), there are too many replacement interfaces or interpenetration interfaces without any visible clay to accept this as a general relation. It is likely that the replacement mechanism may have several explanations: (1) the diffusion rate may indeed be slow and the geometries observed formed only after very long geologic times; (2) the diffusion rate is speeded up by clay films in those arenites where there is interstitial clay; (3) the diffusion rate is speeded up by increase in temperature. That increase, to be effective, would be to the range 150 to 200° C. and would come about as a consequence of deep burial and a normal geothermal gradient or as a result of moderate burial in a region of abnormally high geothermal gradient related to tectonic deformation, regional metamorphism, or igneous activity, or the common combination of all three that is known from the historical geology of deformed geosynclinal belts.

The typical replacement of quartz by calcite, or the reverse, which is also common, implies necessary chemical characteristics of the pore waters; that is, the pore waters must, in the first case, be undersaturated with respect to quartz and supersaturated with respect to calcite. Because both processes are going on at the same time, there are severe constraints on the possible compositions. One must then propose reasonable mechanisms whereby a pore water will come to have the composition required for replacement of one mineral by another and flow through the sandstone for a long enough time to accomplish the dissolution and precipitation. Such mechanisms are discussed in detail in the sections on silica-carbonate cementation relations (p. 421).

One of the most important changes in the mineralogy of a sandstone may come about by the complete disappearance of accessory heavy minerals by dissolution, a process called *intrastratal solution* by Pettijohn (1941), who inferred such a process by a detailed analysis of the heavy minerals of sandstones as a function of geologic age. He noted the strong statistical tendency for some minerals to be absent in older rocks; for younger rocks consequently to have much more varied heavy mineral suites; and for certain authigenic minerals, in particular the titanium oxides, tourmalines, and zircons, to be much more common in ancient than in young sandstones. In general the minerals absent in older rocks are those, such as pyroxenes and amphiboles, which are chemically least stable in earth surface environments. The process, then, was one by which the chemically unstable minerals would gradually dissolve as a result of the passage of long geologic times; no trace of the dissolved minerals would be left, and some predominantly authigenic minerals, such as anatase, would form. This would obscure original differences in source areas, confuse attempts at stratigraphic correlation by heavy minerals, and make explicable the general tendency for Precambrian and early Paleozoic sandstones to have heavy mineral suites almost completely composed of zircon, tourmaline, and titanium and iron minerals. Support for the idea of diagenetic solution of heavy minerals comes from Bramlette's (1941) study that showed a much larger suite of heavy minerals in impermeable concretions in a sandstone than in their matrix of permeable sand.

It has been argued by van Andel (1959) that a more likely explanation for such heavy mineral sequences that go from an impoverished assemblage of zircon, tourmaline and anatase at the base to a varied suite including garnets, amphiboles, pyroxenes, and others at the top is the gradual unroofing of tectonically active source areas that expose to erosion deep seated igneous and metamorphic rocks containing unstable minerals in abundance. Such an unroofing, such as that first proposed in the classic paper on the unroofing of the Dartmoor granite (Groves, 1931), would be expected in the typical tectonic cyclical pattern of development of sedimentary basins linked to eroding source areas. Van Andel pointed to cyclical sequences of heavy minerals, each cycle going upwards from impoverished to varied, in some sedimentary sections. This repetition is strong evidence for the impossibility of intrastratal solution operating over long times as an explanation of the impoverished suites. A variant of this point of view is the interpretation of impoverished heavy mineral suites in terms of the relative amounts of zircon, tourmaline, and rutile – the ZTR index (Hubert, 1960, p. 208-221). Hubert believes this index correlates with general mineralogical maturity of a sandstone and hence is the result of the action of weathering at the source and abrasion during transport rather than diagenesis.

The dichotomy between these two points of view that emphasize source area control on the one hand and diagenetic processes on the other is more apparent than real. The general statistical tendency noted by Pettijohn continues to be supported in a general way as more and more petrographic data become available, though there are many exceptions. The time scale over which the statistical data used by Pettijohn show definite trends is one of 10^8 years and the mechanism is presumably a slow chemical equilibration with that order of magnitude time scale. The time scale of geosynclinal or basinal tectonic cycles that would correlate with cyclical changes in heavy mineral suites controlled by source area evolution is smaller by one order of magnitude, that is, 10^7 years. It is therefore entirely reasonable to suspect that both processes operate simultaneously. The resolution of this argument must come from the evaluation of much more regional petrographic mapping that involves not only the reconstruction of source area evolution but considers equally the diagenetic history of the deposit. The source area problem is usually solved by the application of techniques of petrographic province mapping, paleocurrent mapping, and general stratigraphic and historical geological considerations. The diagenetic problem will be attacked by systematizing, that is, mapping, observations on overburden, past and present; frequency of unconformities; and perhaps, most important, mapping of paleohydrologic regimes by which intrastratal solution took place. An example of the latter would be to infer the pattern of groundwater movement from the subcrop map pattern below an unconformity, such as that mapped by Siever (1951), and relate patterns of diagenesis to it. At present, provenance studies are much easier to do than diagenetic ones and more certain in their results.

A variety of intrastratal solution that is directly observable in thin section is *decementation*. Just as the name implies, this process is one by which a former precipitate, such as a pore filling calcite cement that had partially replaced detrital grains, dissolves, leaving behind its diagnostic relic texture of corroded grains and remnants of the pore space filling. The observation of this evidence leads directly to the interpretation of a reversal from precipitation to dissolution that required some geological change which made the composition of the pore fluids change from supersaturated to undersaturated. Such a change may come about, to give one example, by uplift of a sandstone from a deeply buried position in which the pore water was a slightly supersaturated or saturated brine to a new position close to the surface where it may be invaded by undersaturated surface waters of meteoric water derivation. Such a change in position can be inferred by ordinary historical geological reconstruction of episodes of burial, uplift, and unconformity. If a formation is subjected to a number of such episodes, as those thin sandstones normally found in cratonic interior platforms have been, it may show the evidence of repeated reversals of precipitation and dissolution, or to put it in terms of the evidence observed, cementation and decementation. This has been noted by Walker (1962) and elucidated in relation to geomorphic history for a Mississippian sandstone of Indiana and Kentucky by Hrabar and Potter (1969) and the Sedimentation Seminar (1969). It is conceivable that any given episode will obliterate the record of earlier events but it is more probable that some fragmentary record of the sequence will be left for careful petrographers to decipher.

Recrystallization

The energetics behind any recrystallization of a substance from one form to another more stable form is the tendency towards a minimum in the Gibbs Free energy of the chemical system. If we focus here on the equilibrium relations of a small system, that is, an assemblage of crystals only a few millimeters in size, and assume for the purposes of analysis that it is a more or less independent chemical system isolated from the rest of the rock, we can readily make explicit how such forces operate. Perhaps the simplest example is the recrystallization of a small group of intimately intergrown tiny crystals into a single large crystal. In this system:

$$G_{\text{total}} = G_{\text{mass}} + \sum_i G_{i_{\text{interface}}} \tag{10-12}$$

where G_{mass} is the contribution to the total made by the mass of all material in the system and $G_{i_{\text{interface}}}$ is the contribution of the interfacial ("surface") energy of each crystal. In growth by recrystallization the contribution to the total G of the assemblage that is made by the G_i term is reduced to a minimum as the total area of interface of crystals decreases to the point where it becomes the surface area of a single crystal. This is how we describe the energetics of recrystallization of tiny crystals of quartz in a microcrystalline quartz mosaic found in cherts to much larger crystals, or the formation of single large carbonate crystals from an aggregate of smaller ones. Another way of describing this process has been given by Smalley (1967), who calculated theoretical curves for the changes in topological properties and density as a result of cementation in a sandstone undergoing burial and diagenesis.

In the course of recrystallization, inclusions or impurities remain behind in their same relative positions so that relic structures may still be preserved and so give us the evidence needed to infer the crystallization. The same driving force behind recrystallization of small into big crystals could be conceived to cause the change of sandstones from assemblages of individual detrital grains to single crystals as large as the formation, but such a mechanism does not operate because the magnitude of $G_{\text{interface}}$ is negligible for most materials as grain size increases to magnitudes greater than 0.1 mm. This size is a general lower limit of preserved detrital grains in many metamorphic quartzites; finer particles have apparently recrystallized. Because the magnitude of $G_{\text{interface}}$ is a function of total surface area, it is clear how recrystallization will proceed in very fine grained materials but not in sand sized particles. An extended treatment of the thermodynamics of interfacial energy is given by Guggenheim (1967, p. 45-58) and a geological discussion by Verhoogen (1948).

A somewhat more complex kind of recrystallization is that by which one polymorph will invert to another without change in chemical composition. The most familiar change of this sort is the recrystallization of aragonite to calcite. The push toward chemical equilibrium is provided by the difference between the G's of the two polymorphs at any given temperature and pressure where $\Delta G_R^0 =$ the standard ΔG of the reaction: aragonite → calcite and ΔGf^0 = Gibbs free energy of formation in the standard state:

$$CaCO_{3(\text{aragonite})} \longrightarrow CaCO_{3(\text{calcite})} \quad (10\text{-}13)$$

$$\Delta Gf^0_{(\text{calcite})} - \Delta Gf^0_{(\text{aragonite})} = \Delta G_R^0 \quad (10\text{-}14)$$

$$(-269.78) - (-269.53) = -0.25 \text{ kilocalories} \quad (10\text{-}15)$$

At 25° C., 1 atm. pressure

Because it is only at one temperature at a given pressure that two such polymorphs can coexist at equilibrium, in general one is unstable with respect to the other and will be expected to recrystallize to it, if the system moves towards equilibrium. Aragonite is known from experimental studies to be the stable form of $CaCO_3$ only at high pressures (Simmons and Bell, 1963); under the temperature and pressure conditions of most sandstones, whether at the surface or deeply buried, calcite is the stable form. The observational confirmation of the tendency towards chemical equilibrium is given by the rarity of aragonite and the exclusivity of calcite in ancient rocks. In contrast, the lack of any strong tendency to equilibrium in the realm of modern sediments, where biological and non-biological processes both seem to ignore chemical equilibrium over short time scales, is attested to by the abundance of aragonite. Aragonite crystals or cements in sandstones are rare and found only in very young rocks. The inference of the former presence of aragonite is difficult if no relic structures are preserved. Where they are, as for example in the form of calcite oolites in calcarenitic sands, we can infer from the exclusively aragonitic composition of modern oolites that the calcite was once aragonite.

The recrystallization of original precipitates of amorphous silica or opaline silica (the original substance from which chalcedony forms) to microcrystalline quartz (the fine structure of chert) is a similarly controlled process, but one in which the thermodynamics are less well defined because of uncertainties in the precise value of ΔGf^0 of amorphous silica varieties (Siever, 1957). For vitreous silica glass we have

$$SiO_{2(glass)} \longrightarrow SiO_{2(quartz)} \quad (10\text{-}16)$$

$$\Delta Gf^0_{(quartz)} - \Delta Gf^0_{(glass)} = \Delta G^0_R \quad (10\text{-}17)$$

$$(-192.4) - (-190.9) = -1.5 \text{ kilocalories} \quad (10\text{-}18)$$

At 25° C. 1 atm. pressure.

The original precipitate of silica, usually called amorphous silica, opaline silica, or sometimes erroneously cristobalite, is a disordered form of silica that varies with its origin. Silica glasses and freshly precipitated gels or opals are truly amorphous silicas that cannot be properly characterized as crystalline, though on the level of unit cell dimensions they have a disordered cristobalite structure. Most fresh diatom and radiolaria tests and sponge spicules have this structure. There is another form of amorphous silica, including some of the opals, that appears to be more like a true cristobalite, though it is badly disordered. The path of recrystallization of amorphous silicas seems to go through an intermediate disordered cristobalite stage before ending as microcrystalline quartz. Each stage in this path is a reduction in the G for the silica as in the overall reaction given in Eqs. (10-16), (10-17), and (10-18), for at earth surface temperatures and pressures, low quartz is stable with respect to all other forms and cristobalite has a lower G (is "more stable") than amorphous silica. The decrease in G due to polymorphic inversion is accompanied by a decrease in the contribution of $G_{interface}$ as the precipitate becomes more coarsely crystalline.

Experimental evidence (Osborn, 1953) and thermodynamic data are compatible with the inference from geological occurrences that the observed authigenic anatase in sandstones represents the stable form of TiO_2 in earth surface temperatures and pressures and that rutile and brookite – both of which are also known to have authigenic occurrences – will with time invert to anatase. A complicating factor in the stability of and inversion to anatase is the amount of iron present; anatase has the lowest amount in solid solution and is most stable in low iron environments. This inversion is obviously such a slow one that it takes long geologic times, of the order of 10^8 years, to accomplish and may account in part for the increased abundance of anatase in older sandstones, a negative persistence related to diagenetic effects by Pettijohn (1941).

Incongruent Dissolution or Alteration

Alteration is the most general kind of chemical reaction and therefore the most complex. One of the more striking alterations, and one in which the controls on the reaction are fairly well understood, is the kaolinization of feldspar grains. The petrographic evidence for such an alteration is unequivocal; a grain that has all of the geometry of a detrital cleavage fragment of feldspar is seen to be composed entirely of small books and worms of kaolinite, a composition that would be highly unlikely to survive as an original detrital grain. Eq. (10-9) describes the reaction for the hydrolysis of a K-feldspar and indicates the relevant chemical variables: the activities of the dissolved aqueous species K^+, H_4SiO_4, and H^+. Activity of H^+ may be converted to pH$(-\log_{10} a_{H^+})$. It is important to note that aluminium does not appear as a species in solution but is conserved as a solid in the aluminosilicate structure. The equilibrium constant for Eq. (10-9) is simplified if the kaolinite and feldspar are pure and so have unit activities; and because the solutions are sufficiently dilute, we can consider $a_{H_2O} \simeq 1$ so that

$$K_{\text{feldspar} \rightarrow \text{kaolinite}} = \frac{(a_{K^+})^2 (a_{H_4SiO_4})^4}{(a_{H^+})^2}. \tag{10-19}$$

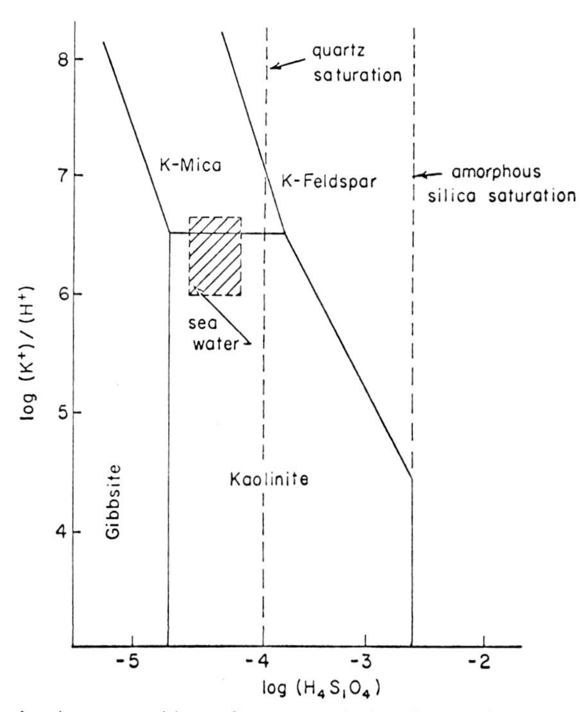

Fig. 10-8. Diagram showing compositions of aqueous solutions in equilibrium with K-feldspar, K-mica, kaolinite, and gibbsite. The compositions are given in terms of the logarithm of the activity of H_4SiO_4 and the logarithm of the ratio of the activities of K^+ and H^+ (Modified from Garrels and Christ, 1965, Fig. 10.6)

Because the equilibrium constants for the equations that relate transformations among K-feldspar, K-mica (muscovite, illite), kaolinite, and gibbsite all involve one or more of the same activities, and the ratio a_{K^+}/a_{H^+} recurs, equilibrium diagrams can be drawn, in terms of the logarithms of the activities, that show the composition of aqueous solutions in equilibrium with the solid minerals in the system (Garrels and Christ, 1965, p. 359-362; Hess, 1966). From this kind of diagram (Fig. 10-8) one can predict that, if log a_{K^+}/a_{H^+} values fall much below 5 or 6 and H_4SiO_4 concentrations drop much below 10^{-3} moles/liter, K-feldspar will kaolinize. This is a more precise way of stating the chemical environment in which kaolinization will occur, if thermodynamic equilibrium is reached, than saying qualitatively that acid conditions promote kaolinization or equivalent generalities. Once such variables are stated precisely, it becomes the task of the geologist to infer the geological environments, particularly the chemical environment of the subsurface, in which the variables assume the values specified. In considering the in-place kaolinization of feldspar, one can infer a groundwater that is low in dissolved solids, including being low in K^+ and H_4SiO_4, and has a relatively low pH that is most probably controlled by dissolved H_2CO_3. The most likely candidate is a groundwater of meteoric origin within a few hundred feet of the surface at most. Such a water fits the general character of the waters responsible for the formation of kaolinite described in a different manner by Millot (1963) as "aggressive".

Many of the clay minerals of diagenetic origin that are found in sandstones are derived from an alteration process that is more complex than simple precipitation or dissolution. The alteration of volcanic glass particles to montmorillonite takes place by reactions that involve incongruent dissolution according to schemes such as the following

$$\underset{\text{"volcanic glass"}}{Na_2KCaAl_5Si_{11}O_{32} + MgSiO_3} + H_2O + \underset{\text{carbonic acid}}{4H^+ + 4HCO_3^-} \longrightarrow$$

$$\underset{\text{montmorillonite}}{Na(Al_5Mg)Si_{12}O_{30}(OH)_6} + Na^+ + K^+ + Ca^{++} + 4HCO_3^- \qquad (10\text{-}20)$$

in which the composition of the glass is represented by a mixture of one part K-feldspar, one part anorthite, two parts albite, and one part enstatite; and the composition of the montmorillonite is somewhat simplified.

The formation of zeolites by devitrification of glass to heulandite-clinoptilolite, laumontite, phillipsite, or others can be represented by the same general kind of reaction scheme. Zen (1959) proposed another reaction, the formation of equilibrium assemblages of calcite and kaolinite by reaction of sea water with a volcanic glass.

All of these reactions by which many clays are thought to form involve the preservation in solid form of much of the aluminosilicate, though not necessarily in the same crystallographic structure. One can simply visualize the process as one in which cations are exchanged, lost, or gained, and silica is gained or lost relative to alumina, by reaction of a detrital solid with the surrounding solution. The petrographic character of clay mineral alterations and replacements fits well with this idea, for they occur typically in sandstones with either clay matrix or argillaceous rock fragments; in fact the idea of the origin of much clay matrix from alteration of rock fragments depends intrinsically on this mechanism. Table 10-1 gives some examples of the kinds of clay mineral transformations or alterations that have been inferred as diagenetic processes.

It is much less clear what are the origins of the pore filling clay minerals that have been deposited in sandstones with little or no detrital clay or aluminosilicate minerals; such pore fillings appear to the petrographer to be precipitates from a solution, for no relic of a different mineral as precursor is found. The most typical of these pore fillings are the relatively coarsely crystalline books and worms of kaolinite that are found in many quartz arenites. For such a precipitate to have formed by direct precipitation, the entering solution must have carried both dissolved silica and alumina. The form in which dissolved silica is carried is monomeric silicic acid, H_4SiO_4. Analyses of many groundwaters confirm the prediction based on the chemistry of H_4SiO_4 that silica can be carried in sufficient concentrations, from a few to a few tens of parts per million (ppm), to allow the precipitation of clay minerals (Siever, 1962, 1968a). The amount of dissolved alumina that can be carried in solution, on the other hand, is severely limited by the way in which its solubility is dependent on pH. In the pH region from about 4.5 to 8, the solubility of alumina is negligible for practical purposes. Below pH 4.5 it becomes very soluble, with the dominant species in solution being Al^{3+} ion; above pH 8 it becomes very soluble with the dominant species being AlO_2^- or some polymerized variant of it. But the difficulty is that the pH range in which alumina is insoluble is just that pH range in which most natural waters, and apparently the great bulk of subsurface waters, fall.

The apparent contradiction between the chemical behavior of alumina and the petrographic evidence of precipitation may be resolved if future study shows that the waters needed to precipitate kaolinite, which should have low cation and moderate silica concentrations, have a range of pH low enough, significantly below 5 (see kaolinite-solution boundary in Garrels and Christ, 1956, p. 352–358). Other possibilities are (1) that Al may form a soluble complex ion with other components of the solution, in some way as yet not known, and so be transported; (2) that enough groundwater containing the extremely small quantities of Al in solution at intermediate pH values, about 10^{-10} moles/liter, can be carried into pore space to precipitate over a long geologic time, probably of the order of 10^8 years at least; or (3) that the petrographic evidence is not conclusive and that there are in fact remnants of the precursor aluminosilicate left to be observed if a thorough search is made.

The incongruent dissolution of carbonates may play some role in diagenetic transformations in sandstones, though it is unlikely to be a major effect. The incongruent dissolution of dolomite or a magnesium-rich calcite will lead to a solution more enriched in Mg^{++} ion and a solid residue that is Ca-rich; from such a process we can get the formation of diagenetic calcite as a "replacement" of dolomite. Our knowledge of such diagenetic processes in carbonates in sandstones is small because we have few good chemical analyses of the precise compositions of carbonates. Optical determinations of the amount of Mg in calcite or excess Ca (or the amount of Fe) in dolomite are completely inadequate; X-ray diffraction or electron microprobe techniques must be used for this kind of information.

Pore Water Reactions

If we want to study the chemical reactions that characterize diagenetic processes, it is imperative that we consider the ways in which the pore fluids in the rock change their composition and participate in the reactions. The pore fluid, the formation water, is an important phase making up a porous rock in its natural condition in the ground. Only at or near surface outcrops, in the vadose zone above the groundwater table will sandstones be "dry." At the moderate to low temperatures and pressures under which all diagenetic reactions take place a fluid phase is always present, ranging from interstitial waters of modern sediments to concentrated brines of deeply buried ancient sandstones and including oil and gas. Because of the negligibility of purely solid-solid reactions under such conditions, the fluid is the important mediator in all reactions. A static pore fluid reacts with the rock just as does a solid with a fluid in a sealed laboratory vessel. The more common moving fluid reacts with the rock in such a way that one must consider the *rates* at which any dissolved reactants, the raw materials of reactions, are supplied and dissolved products removed from solid minerals, as well as the rate of the reaction itself. This leads to the necessity of evaluating possible reservoirs that supply the reactants and sinks that absorb the products, that is, the entire geological and geochemical system.

Naturally, geologists and petrologists working only with the solid phases of the rock will tend to ignore or insufficiently consider the chemical composition of the fluid phases. This tendency is increased by the ease of recovery of solid samples compared with the great difficulty of getting a representative sample of the fluid. Almost all of our information comes from formation water samples taken from oil wells. Information from springs and seepages is useful but is usually not traceable to any specific formation lying at any given depth. Even the information from oil well samples is subject to so much mixture from multiple sources, contamination from fluids in and out of the drillhole, and changes during sampling and storage that only a fraction of the analyses are reliable. The great bulk of the information on formation waters is not published although some is available (Young and Galley, 1965).

The composition of pore fluids is given by a chemical analysis, usually in terms of the major chemical species, measured in parts per million (ppm) assumed to be in solution rather than in terms of the oxides of the elements as in a rock analysis. The species usually included are the abundant alkali metal ions: Na^+ and K^+ (in older analyses these may be lumped); the abundant alkaline earth elements: Ca^{++} and Mg^{++}; the balancing anions: HCO_3^-, Cl^-, and SO_4^{--}; and sometimes dissolved SiO_2 and pH. Other elements, such as I^- and Br^-, may be included in specific studies. The analysis of organic fluids, almost always hydrocarbons, leads to compositions given in terms of the elements C, H, S, N, and O, or, more commonly, in terms of the relative proportions of the major types of organic compounds present: paraffinic, aromatic, asphaltic and naphthenic (Levorsen, 1967, p. 180). In addition, specific compounds may be identified, such as any of the normal paraffins, ranging from methane to those with carbon numbers greater than 20, or specific members of the aromatic (benzene) group. A further refinement of the chemical composition of pore fluids is an analysis giving stable isotope ratios of specific elements. The most common are O^{18}/O^{16}, C^{12}/C^{13}, S^{32}/S^{34}, and D/H (deuterium/hydrogen). Such isotope ratios are useful in determining whether the fluids are in equilibrium with the minerals of the sandstones and in deducing the processes involved in the formation of the waters, such as whether the water is of marine, meteoric, or magmatic origin, or whether reactions with carbonates or sulfides played a role. Degens (1965, p. 92–196) gives a useful summary and bibliography of some of this work. Studies of some near-surface waters include C^{14} or tritium analyses that can be used to give the "age" and therefore recharge history of the water.

Much of the information and ideas about the origins of formation waters comes from the study of fluids in sandstone formations. There is now a general consensus that, though many pore waters may have started as interstitial sea water in marine sediments, their present composition comes from reaction of water with the mineral components of the rock. The more soluble the minerals, evaporites being an extreme example, the more the effect on the composition. Meteoric waters migrate downward through soil zones and then downdip in permeable sandstones and mix with deeply buried waters that are partly inherited from trapped sea water. The sandstone petrologist needs to know the types and ranges of compositions of pore waters and how they relate to hydrodynamic movements of water through porous rocks, those movements

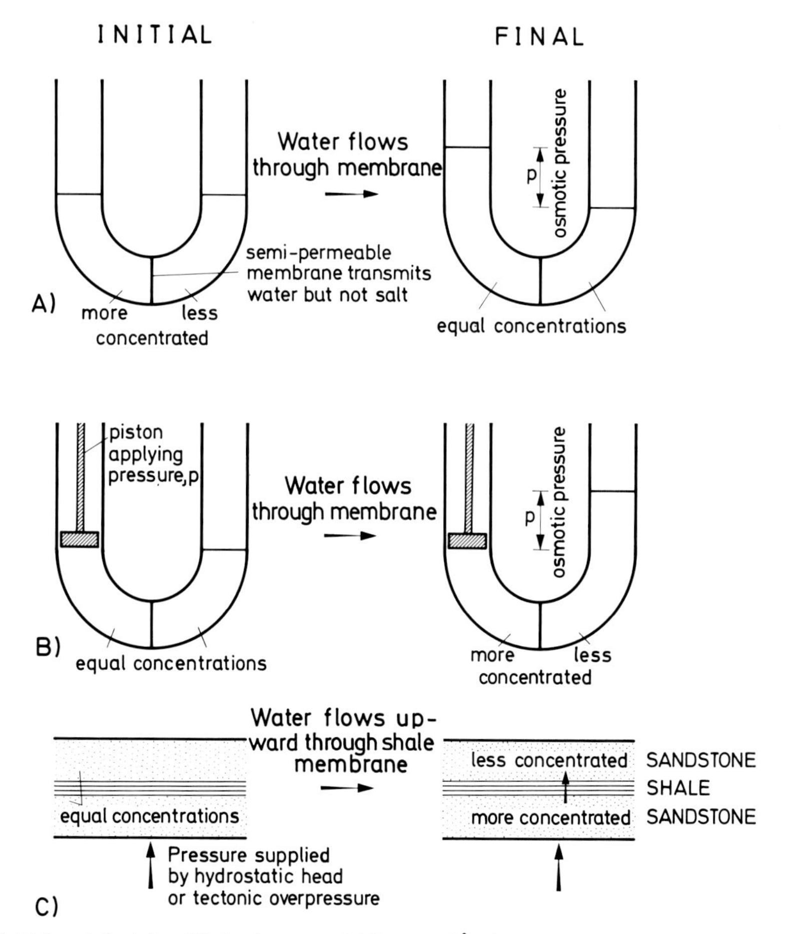

Fig. 10-9. Initial and final (equilibrium) states of (A) normal and (B) reverse osmosis in a U-tube and (C) reverse osmosis in two sandstones separated by a thin shale acting as a semi-permeable membrane

themselves functions of the structure, surface geomorphology, and climate of the sedimentary basin, before he can intelligently speculate on the origin and historical geology of the diagenesis of the sandstones he studies.

A special type of process which may alter the composition of pore waters is salt filtering by a semipermeable membrane, also called *reverse osmosis* (DeSitter, 1947; Bredehoeft and others, 1963). The osmotic process is dependent on the behavior of shales as semipermeable membranes and is powered by the pressure differential with depth. In any ordinary osmotic process there is a difference in concentration of salt between the two sides of a membrane which gives rise to a pressure differential, the osmotic pressure, which induces movement of the solvent from the low to the high concentration side (Fig. 10-9a). In reverse osmosis there is the same concentration on both sides of the membrane but a pressure differential is impressed between the two sides of the membrane and so the solvent flows from the high to the low pressure side and thereby gives rise to a higher salt concentration on the high pressure side and a lower one on the low pressure side (Fig. 10-9b). In a sedimentary column, the pressure differential is linked to varying depth of overburden so that a sandstone below a shale that acts as a semipermeable membrane will have a higher concentration of salts than a sandstone above the shale (Fig. 10-9c). This mechanism has been demonstrated in the laboratory (McKelvey and Milne, 1962; von Engelhardt and Gaida, 1963) but, though an attractive hypothesis, is difficult to prove as the major process causing concentration differences in pore waters in sedimentary basins, for it is necessary to demonstrate an association of pressure differences with differences in salt content that cannot be ascribed to other origins.

Oil, gas, and other organic deposits. These too are subsurface diagenetic fluids found in sandstones. It is now a commonplace to note that oil and gas are normal, not unusual, components of a great many sandstones. The composition, distribution, and origin of oil, gas, and asphalts of various kinds in sandstone deposits is a subject which has an immense literature, much of which pertains to deposits found in sandstones (Levorsen, 1967). It is important to consider oil and gas in sandstones not only for their economic significance but because the presence of oil may have an effect on other diagenetic processes, such as the inhibition of quartz diagenesis noted by Philipp and others (1963).

Sequence of Mineral Transformations

The ordering of a variety of diagenetic events with respect to time is one of the most important keys to the ultimate interpretation of the ways in which the geological and geochemical history of a sandstone affected its composition and texture. There is no doubt that replacement parageneses are the main guides to such an ordering. A good example would be the implied history of a pseudomorph of chert (microcrystalline quartz) after dolomite (Fig. 10-10): the original diagenetic precipitate was a rhombohedron of dolomite – which itself may have replaced an earlier precipitate of calcite – followed by a later replacement, either directly by chalcedonic (microcrystalline) quartz or first by an amorphous or opaline silica that later recrystallized to microcrystalline quartz.

In the absence of pseudomorphs a replacement series is usually subject to alternative interpretations, for a rational crystal face bounding the interface between one mineral and another may mean either that one mineral precipitated first with its rational face and that the other grew around it anhedrally, or that the anhedral mineral precipitated as an euhedron first, later to be replaced by the euhedral mineral. A first step in the study of such a boundary is to establish which mineral the euhedral face belongs to. An excellent example of how this is done is given by Gilbert (1949, p. 10) and Gilbert and Turner

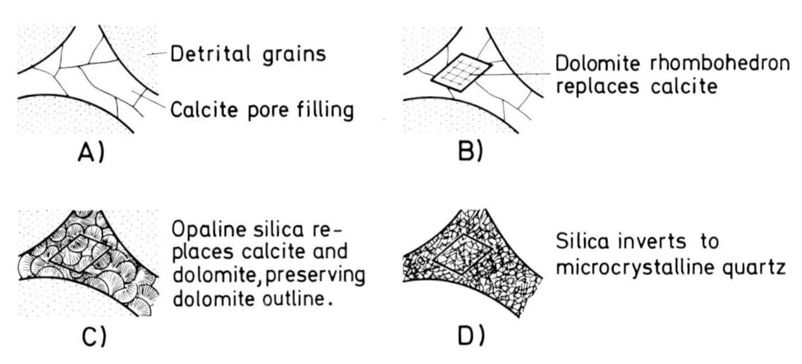

Fig. 10-10. Possible evolution of a chert pseudomorph after dolomite

(1948), who point out the necessity of using the universal stage for proving the correct assignment of the face to the appropriate crystal. When such work is done, it usually confirms the preliminary guess based on petrographic common sense.

It is not always necessarily true that the replacing mineral will have its own rational faces, for studies of etch figures of minerals show that dissolution, a part of the replacement process, may result in the formation of crystal faces of the dissolved substance; in such a case the replacing mineral simply precipitates anhedrally around the rational face of the replaced mineral. This caution does not seem to apply to either quartz or carbonate, the two most abundant participants in replacements in sandstones; both of these minerals assume euhedral faces only when precipitated and not when dissolved under the chemical conditions found in sediments.

It is possible to systematize the logic of analyzing the paragenesis of cements in sandstones in the way one programs a computer, though the variety of kinds of cement and geometries is too great to make such a program complete (Fig. 10-11). The analysis proceeds as follows.

1. The interface between two minerals is planar.

2. Establish whether one mineral is detrital by looking for a shape characteristic of abrasion rounding, combined with presence of typical inclusions, as in quartz of igneous derivation.

3. If the reasonably expected detrital rounded shape of the grain is interrupted by the interface then it has been replaced and in all likelihood the

plane face belongs to the other, diagenetic mineral. This is confirmed by examination with a universal stage.

4. See if neither mineral is detrital, as deduced from an absence of the kinds of inclusions characteristic of detrital minerals or from pore filling or replacement geometry. Such a diagenetic geometry is one that is entirely unexpected in a detrital grain and should satisfy the requirement that the framework of the sandstone must always be supported, that is, crystals cannot

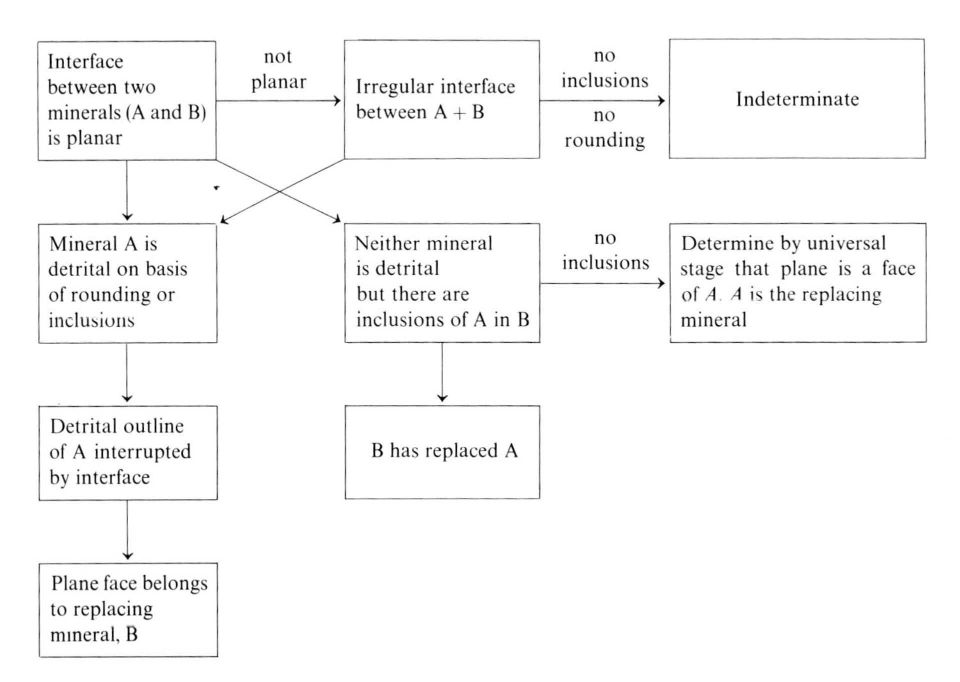

Fig. 10-11. Flow chart of logic establishing the sequence of replacement of two minerals, *A* and *B*

"float." Such a constraint, based on a "superposition" principle, would dictate that a crystal of diagenetic calcite in the center of a pore space lined with secondary quartz could not have precipitated before the secondary quartz, for then it would have been unsupported.

5. The possibility that one mineral has replaced the other is checked by noting whether there are any inclusions of one mineral in the other. If so, the mineral with the inclusions is the replacer.

6. If both minerals have grown in pore space and none of the above steps gives an unequivocal answer, then determine the mineral to whom the face belongs by universal stage techniques. On the assumption that the great bulk of minerals precipitate rather than dissolve to form rational crystal faces, that mineral to which the rational crystal face belongs was the first to precipitate and the other was the final anhedral filling of the pore.

Newer techniques of investigation may be helpful in deciding the order of events. Cathodo-luminescence will reveal the presence of growth (zone) lines in carbonates growing into pore space that are absent when the carbonate has replaced another mineral. The electron probe can be used for chemical analysis that may reveal the composition of inclusions too small to be resolved by ordinary microscopy. Electron probe chemical analysis may also show that a portion of a mineral is so pure an end member of a solid solution series that it almost certainly is of low temperature and therefore almost certainly of diagenetic origin.

The idea of a general "law" governing replacements has been explored by many who have studied sandstones under the microscope. Cayeux (1929, p. 234–251) and Waldschmidt (1941) are among those who thought that replacements are "one-way"; that is, mineral A always replaces mineral B but is never replaced by it. Waldschmidt, for example, stated that in the Rocky Mountain sandstones of various ages that he studied there was a definite order of precipitation, so that: if there were one cementing agent it was always quartz; if there were two, quartz was first and calcite second; if there were three, quartz was always first and dolomite second, followed by either anhydrite or calcite as the third; and so on. In contrast, other describers of cements have noted a great variety of sequences; most have found it difficult to see any overwhelming statistics for one kind of sequence or another. This is not to say that more careful work may not in the future reveal the general tendencies for such an order. For example, there does seem to be evidence that in the great majority of sandstones in which both quartz and calcite are diagenetic precipitates, the quartz was precipitated first and calcite second. The existence of such general tendencies should be powerful evidence for a generally applicable series of geologic events that determine the source of cementation of sandstones.

There is no doubt that a series of precipitates with a definite time order can be used to infer such mechanisms as fractional crystallization, which is characteristic of the differentiation of a basaltic magma or the precipitation of salts from sea water. No such sequence is known in the diagenetic precipitates of sandstones, however, and the two major minerals, quartz and calcite, do not interact in phase equilibria except at high temperatures. To the extent that the precipitation in sandstones of diagenetic gypsum, anhydrite, and carbonates of various kinds may be linked to the complex evaporation and groundwater movement on a sabkha flat, where aridity and supratidal flat environments coincide, there may be an ordered sequence that reveals the mechanism (for a discussion of this environment see Illing and others, 1965).

Though it is possible that one series of diagenetic events will obliterate evidence of previous episodes, the experience of petrographers shows that some relics are almost always left behind. The probability of a complete "wipe-out" is low and should be discounted. The sole possible exception to this may be the intrastratal solution of heavy minerals as discussed on p. 404. What the probability of some preservation does dictate is careful petrography in which a conscious hunt is made for the unusual that may prove to be a relic.

Age of Diagenetic Minerals

The "when" of diagenesis is the ultimate question that the historical geologist asks. Once mechanisms are proposed that account for the observed time-ordered sequence of diagenetic effects, it becomes important to deduce just *when* in the sandstone's history such events took place. Such dating can be with respect to either the stratigraphic or absolute time scale. The best direct evidence would be the radiometric dating of diagenetic precipitates, but such age analyses are fraught with difficulty, first because of the difficulty of extracting enough pure diagenetic precipitate to perform the analysis and second, because of inherent uncertainties linked to the ways in which the mineral was precipitated. Such uncertainties arise when one considers the possibility that a precipitate formed very slowly over long geologic times; such a precipitate would give an average age only. But it should be possible to date authigenic feldspars by potassium-argon (K^{40}–A^{40}) dating methods; the low temperature of diagenesis would not have allowed the extensive diffusion of argon that normally introduces inaccuracies in the data when higher temperature rocks are dated. Authigenic clays have been dated by implication by Hurley and others (1963), who show that the ages of some size fractions of shales were too young and therefore had to be authigenic. Dates based on uranium minerals are available for some sandstones of the Colorado Plateau. Dating of granitic zircon has been carried on for many years and it is possible that such methods can be applied to zircon overgrowths in sandstones. Any application of these methods is dependent on a method for separation and concentration of the overgrowth or authigenic mineral; once that is done, the dating techniques are available.

Remanent magnetism stratigraphy has been used to date the iron minerals in sandstones (Van Houten, 1968). The magnetic time scale is now most validated for the Cenozoic (Opdyke, 1968), where it is based primarily on magnetism of lavas, but it has been applied to more ancient sandstones as well (Picard, 1964). The method depends on reversals of magnetic polarity and the magnetic stratigraphy of the later Cenozoic is well established. The dating of sandstones rests on the hypothesis that the hematite-goethite of red sandstones is diagenetic but so early that for practical purposes it is contemporaneous with the deposition of the detritus. Because the iron minerals are precipitated in place they will have the magnetic orientation imposed by the polarity of the earth at that time. A major difficulty in establishing a world-wide time scale for magnetic reversals for older sandstones is that the time scale for reversals is of the order of 10^4 to 10^5 years, much too short for matching with the stratigraphic time scale, where the resolution is almost never much better than 10^6 years. But the strengthening of this hypothesis will reinforce the evidence for the very early diagenetic formation of secondary iron minerals in sandstones.

Still another way of dating diagenetic events comes from study of pebbles in intraformational conglomerates. For example, the presence of pebbles of well cemented, very late Mississippian sandstones in early Pennsylvanian conglomerates certainly indicates how early the cement was formed. But such occurrences of cemented sandstones pebbles are uncommon as compared with the variety of shale, limestone, and other lithologies that are found as pebbles in intraformational conglomerates. It may be that sandstones are not as a general rule cemented well enough, early enough in their history to form pebbles resistant to stream abrasion. Observation of modern streams draining even well cemented sandstone terrain rarely reveals significant quantities of small pebbles of sandstone; one finds either large blocks, boulders, and cobbles, or sand grains.

To the extent that a diagenetic event in a sandstone, such as the formation of a cement, can be related to an external geological event that can be dated, the age of the diagenetic event can be determined. If a thoroughly quartz cemented facies of a sandstone were to be genetically related to tectonic deformation then the cementation would be dated by reference to the structural event. Dating with respect to unconformities should also be possible.

The age of diagenetic minerals or textures in sandstones can be fully described only when one specifies the duration of the process, for if a process like intrastratal solution took place over a long time span, its "age" might vary from one era to another. Fine resolution of the evidence for diagenetic age such as radiometric dating would be the best way to specify the duration of the process. Another way of estimating the duration, more uncertain, comes from extrapolation of laboratory studies of the kinetics of chemical reaction, coupled with estimates of rates of flow of pore waters and rates of subsidence and uplift. For example, from laboratory experience we would predict the rapidity of the inversion of aragonite to calcite and the slowness of the inversion of amorphous silica to quartz. But we still do not know enough of the kinetics to extrapolate quantitatively with any great confidence. The best way to study the age effect on diagenesis still appears to be the study of similar sandstone types of different stratigraphic ages, such as that by Tallman (1949).

Major Diagenetic Effects

The most obvious diagenetic modification of sandstones is the introduction of cementing agents. Though there are a variety of minerals that have been observed, only two, carbonate and silica, account for the overwhelming bulk of cement of all sandstones. There are no good statistics on the relative abundance of carbonate and silica cement in all sandstones as a group; one can only quote subjective impressions of many petrographers that silica predominates. Tallman (1949) noted from his study what has been a matter of common experience, that there is a tendency for younger sandstones to be more cemented with carbonate and older ones to be dominated by silica. One might interpret this in terms of a secular change through geologic time but there seems to be no independent evidence for such a trend. If rather seems likely that the lesser importance of carbonate cement in older sandstones may be explained by a differential preservation similar in general nature to that based on solubility and ease of weathering which Garrels and Mackenzie (1969) proposed for the paucity of limestone in Precambrian sections. Because carbonate cement is much more soluble than silica it is much more likely to disappear as a result of the statistical tendency to be exposed to undersaturated groundwaters at various times during the geologic history of a sandstone. Silica, on the other hand, once precipitated, is much less likely – and less rapidly – to become dissolved, and so, persisting, becomes relatively more abundant.

Carbonate Cementation

Though calcite is the most common carbonate mineral cementing sandstone, many examples are known of dolomite (Swineford, 1947; Sabins, 1962), ferroan-dolomite (ankerite), and siderite (Siever, 1959). Rhodochrosite concretions are also known from sandstones. Aragonite cement is known only in Recent sands, apparently having inverted to calcite in older sandstones. The carbonate may be found as uniformly or patchily distributed pore fillings and replacements or segregated as concretions or in thin laminae. The nature of the well known sand crystals, with their "poikiloblastic" texture of calcite surrounding sand grains, have been described most recently by Fuhrmann (1968). Though every petrographic type of sandstone may be cemented in part by carbonate, the quartz arenites are the most typically carbonate cemented. Older graywackes, Paleozoic and Precambrian, most typically contain little or no carbonate but Mesozoic and Cenozoic graywackes frequently have large amounts of carbonate cement, probably related to the abundant pelagic foraminiferal remains found in many post-Cretaceous turbidites. Most arkoses contain some carbonate cement. Carbonate cemented sandstones may grade laterally into sandy limestones, an occurrence which clearly points to original environmental conditions as responsible for the cement. The absence of carbonate cement in surface or near surface exposures of some sandstones compared to its presence in the same horizon in the subsurface is clearly explainable in terms of weathering. The opposite pattern, the concentration of calcite cement in near-surface horizons of sandstones is analyzed as a type of caliche precipitation by Nagtegaal (1969) in his study of late Paleozoic and Triassic sandstones of the Pyrenees.

The idea that widespread carbonate cement is an early diagenetic precipitate related to a favorable (presumably marine) environment of sedimentation has been a popular one for a long time. According to this view, sand was transported into a carbonate precipitating environment and the cement was essentially a primary precipitate while the sand was exposed to the sea water. Support for this view has come from some modern sediment studies that have shown submarine lithification of carbonate and sandy sediments, especially the report of an aragonite-cemented sandstone from the Atlantic continental shelf (Allen and others, 1969). The early diagenetic cementation of sand in the delta of the Fraser River by low magnesian calcite has been discussed by Garrison and others (1969). They show that the composition of neither the Fraser River water nor the sea water of the delta is compatible with precipitation of carbonate and conclude that the responsible mechanism is the dissolution of carbonate shell material by pore water and reprecipitation higher in the sediment column by expressed water of compaction. Redistribution of shell material in other sandstones has been assumed on the basis of preservation of relic fossil structures in the carbonate cement.

Evidence that carbonate cement was the first diagenetic precipitate in many sandstones is abundant. Typical is the evidence given in an excellent summary by Glover (1963, p. 40): dolomite is molded directly against the rounded edge of a detrital quartz grain and is surrounded by a secondary quartz overgrowth. There is even more abundant evidence that carbonate cement was precipitated after secondary quartz overgrowths formed (Siever, 1959). Textures that show replacement of both authigenic and detrital quartz by carbonate that contains remnant inclusions of the replaced quartz is conclusive on this point. How late in the history of the sandstone this carbonate cement came can be deduced only from the genèral consideration of changes of diagenesis with depth of burial, age, and the geologic history of the bed in the sedimentary basin, as discussed on p. 432.

There is no well established sequence of precipitation among the carbonates: calcite may be first, followed by dolomite, but dolomite as the first precipitate is also common. Sequences of calcite followed by dolomite followed by a second generation of calcite are also known. Though calcite is most common as fairly large anhedral crystals molded around or replacing detrital grains or other cements, dolomite and siderite are much more common as rhombohedra and can be quickly recognized as such by that habit. The fabric of carbonate cements in sandstones may be complex and has many of the characters shown by the cements of carbonate rocks. A good guide to petrographic approaches to cements in relation to various aspects of limestone origin is given by Ham (1962), a discussion of diagenetic fabrics by Orme and Brown (1963), and a recent summary of cement fabrics in carbonate rocks by Friedman (1968).

The chemistry of carbonate precipitation is a guide to the possible mechanism by which carbonate cements are formed. The precipitation of aragonite or calcite, where it can be shown to be penecontemporaneous on geologic or petrographic grounds, comes as a result of prolonged exposure to supersaturated sea water, for few non-marine environments are carbonate supersaturated. But carbonate supersaturation, which in today's ocean is a function of temperature (and so latitude), may have been much more widespread and uniform in the oceans of the past. In any case, it is not easy from geologic or petrographic evidence to prove a primary or penecontemporaneous origin, and if modern sands are to be the basis for judgment, most carbonate cement is added during diagenesis, after at least slight burial.

Sea water that is trapped as pore water in marine sands may be carbonate supersaturated if the environmental conditions are favorable, such as now obtain over large areas of the Bahama Banks, Persian Gulf, and other well known areas. But if the pore water were static and precipitated all excess supersaturation as cement the amount would be negligible. For example, a volume of sand in a layer 1 cm thick and 50×50 cm in extent, that has 40 percent porosity, and contains sea water that is threefold supersaturated and precipitates all of the excess, could only precipitate a single spherical grain of authigenic calcite of approximately 1.25 mm diameter. Even if we invoke moving this kind of supersaturated pore water upward through the sand, the result of being squeezed out by compaction, there would be only slightly more calcite cement added, for sands do not compact greatly. Only if there are large thicknesses of muddy sediments being compacted in the section can there be any significant amount of cement added from this source.

It is more likely that carbonate shells dissolve in pore waters only to reprecipitate as cement. Aragonite and magnesian calcite would dissolve and reprecipitate as the stable calcite. Another mechanism is the dissolution of shells during the early period when organic matter was decomposing. A major product of such decay is CO_2, which increases the solubility of the carbonate. As the CO_2 diffuses upwards through the sediment column and is lost into the sea water above, the extra carbonate in solution that came from dissolved shells is reprecipitated in response to the drop in CO_2 pressure. We can describe the overall nature of this process as a conversion of organic matter to carbonate through the mediation of the biological world. Only recently has there been growing awareness of the importance of the tiny fauna found in the interstices of sands, typified by the description of a new metazoan phylum, the *Gnathostomulida*, a kind of "lower worm", found most commonly in fine-grained sands of coastal environments (Riedl, 1969). The population density of these worms is estimated to often exceed 6000 specimens per liter of sediment and they are associated with additional populations of nematodes and other small interstitial organisms. All of these organisms that consume organic matter produce CO_2 by respiration and so have an important effect on the carbonate system.

Dolomite cementation, where it is established as the first early cement, can be explained with reference to the modern environments where dolomitization is now proceeding, such as the supratidal flats of the Persian Gulf (Illing and others, 1965). The process involves the depletion of Ca relative to Mg by $CaCO_3$ and $CaCO_4$ precipitation from evaporating waters, followed by dolomite precipitation from the Mg-enriched solution. Much of the precipitation may take place in the subsurface as the result of sinking of the denser evaporite brine. The association of dolomite with anhydrite or gypsum cement, another product of this process, would strongly confirm the environmental inference made from cements. Sabins (1962) showed how one could distinguish among dolomite grains and crystals of detrital, primary, and diagenetic origin.

The calcite cement that is found replacing secondary quartz in so many sandstones must be a later diagenetic product and must be related to a redistribution of carbonate within the sedimentary rock pile. The solubility of carbonates decreases with increasing temperature and, with a much smaller effect, increases with increasing pressure. The net effect of burial is to decrease the solubility of carbonate and this may account for a small part of the cement but, unless large quantities of pore water are pumped through the sediment, relatively small amounts of carbonate would be produced. What does happen as a result of burial is the increase of pressure solution, which is usually hypothesized to explain silica cement (discussed in detail on p. 424). Pressure solution is also a factor in limestones and sandstones containing framework particles of carbonate which abut with small contact areas. Though the pressure solution effect is not known quantitatively its magnitude can be estimated from the hydrostatic pressure effect on solubility to be greater for carbonate than for quartz. If this is so, the likely explanation for much of the late carbonate cement lies in pressure solution of carbonate grains in the sandstone or in nearby formations, either limestones or sandstones, probably with pore water transport over limited distances.

Chemical mechanisms, such as pH variations, for appreciably changing carbonate solubility and so accounting for cementation would seem at first sight to be attractive but there is some question of the widespread geological applicability of such mechanisms. The relevant carbonate system equilibria are given by Garrels and Christ (1965, p. 74–92) and can be used to show dependence of carbonate solubility at a given temperature and pressure on the partial pressure of CO_2 and the ionic strength. Solubility in this system is dependent on pH *only* if the acid-base buffering is controlled by a chemical system other than the carbonate one. If not externally controlled, the pH is dependent on the CO_2 pressure. One can analyze the pH of pore waters of buried sediments and the mineral systems that buffer the pore water pH, using much the same considerations that led Sillén (1961) to the conclusion that the silicates control the acid balance of sea water, and keep it buffered around pH 8. Silicate buffering of pore waters might be much the same as that of sea water, for the same silicates, primarily clays, control the equilibria, and we would expect the pH to be about 8. No doubt there are many variations and local inhomogeneities. One cause of an increase in pH would be circulation of groundwaters through an actively altering basic volcanic section. A decrease might be caused by waters moving downward through acidic soils. These two are typical in that they are related to vadose water percolation through an active zone of weathering. But before such high or low pH waters will have traveled very far into the subsurface they will have reacted with the mixture of minerals that generally tends to keep the pH stabilized around or a little above neutrality. An example of how such waters evolve has been given by Garrels and Mackenzie (1967). Dapples, (1967, p. 118) suggests pH changes related to CO_2 changes near an outcrop may be important in the replacement of quartz by calcite in certain sandstones. In the analysis of a sedimentary series such groundwater movements may be related to unconformities. Sandstones that are deeply buried and remain so are not subject to such changing of infiltering waters except for volcanic or hydrothermal invasions. Changes in CO_2 content of pore waters follows much the same pattern. High CO_2 pressures may come from waters percolating through soils but sources of CO_2 variation at great depth, other than from volcanic emanations, are difficult to envision. Isotopic evidence based on C^{12}/C^{13} ratios has been used to infer an origin for some carbonate cement in which the CO_2 of the carbonate has been supplied by the oxidation of organic matter (Silverman, 1963; Spotts and Silverman, 1966).

The analysis above is highly deductive, for the sampling and analysis of subsurface waters is notoriously unreliable (Chave, 1960) and, though most measured pH values fall in the region around neutrality, we cannot be sure that there are not extensive variations. Analyses of dissolved gases such as CO_2 are rarely reported in any analysis. Until such data become available our geochemical deductions will remain uncertain.

Silica Cementation

The most common form of silica cementation is the secondary enlargement of quartz grains by optically continuous overgrowths that results in euhedral crystal faces or a mosaic of interlocking overgrowths. There is no evidence of any such cementation being formed in modern sands and so all of our clues must come from older sandstones. In a great many sandstones quartz is the first cement, though it is difficult to establish how early it started forming. In some it is the only cement. Quartz cement is most common in the arenites and much less conspicuous, if not entirely absent, in many wackes. The widespread and general distribution in arenites suggests a generally operating process that is common to most of those sandstones whose framework detrital grains completely support the rock in the gravitational field. That process, which we now call pressure solution, was inferred at least as long ago as the late 18th century by James Hutton and has a long history in the geological literature, recently summarized by Trurnit (1968). Heald's (1956) study of quartz cementation in relation to pressure solution was influential in calling renewed attention to the importance of the process as a major cause of cementation. Weyl (1959) provided a detailed analysis of how the process might work and proposed a functional relationship between time, depth of burial, and degree of pressure solution.

The basic idea of pressure solution is that the high effective pressures developed at point contacts of quartz grains increase the solubility at those points such that they preferentially dissolve. The solution process seems most advanced where clay films are present between grains. The dissolution liberates dissolved SiO_2 to the pore water, which then becomes supersaturated and reprecipitates the SiO_2 as quartz overgrowths as shown in Fig. 10-12. Pressure solution simultaneously provides a source for all of the dissolved silica that had to be precipitated to make the cement and gives an explanation for the interpenetration of one quartz grain by another. Such interpenetration is a prime cause of porosity reduction; the nature of the grain contacts, ranging in degrees of interpenetration from tangential through concavo-convex to sutured, has been related to porosity reduction and depth of burial (Taylor, 1950). A further evidence of pore space reduction that is related to pressure solution is the measurement of what Heald (1956) called "minus-cement porosity," the porosity which would be present if a specimen contained no chemical cement. Minus-cement porosity measurements on many sandstones indicate significant compaction that could not be accomplished by simple mechanical rearrangement and so must be due to solution.

The petrographic evidence for pressure solution is excellent where one can clearly see the boundaries between detrital grains and pore filling cement. In a great many sandstones, perhaps the majority, the boundaries cannot be seen, or such small portions can be seen that no quantitative measurement can be made. Some of these sand grains are deceptive, for though they may give every appearance of interpenetration and pressure solution under ordinary illumination, cathodo-luminescence shows that the detrital grains are not involved and that the authigenic overgrowths have grown together in a concavo-convex or sutured pattern as shown in Fig. 10-5 (Sippel, 1968). A less likely

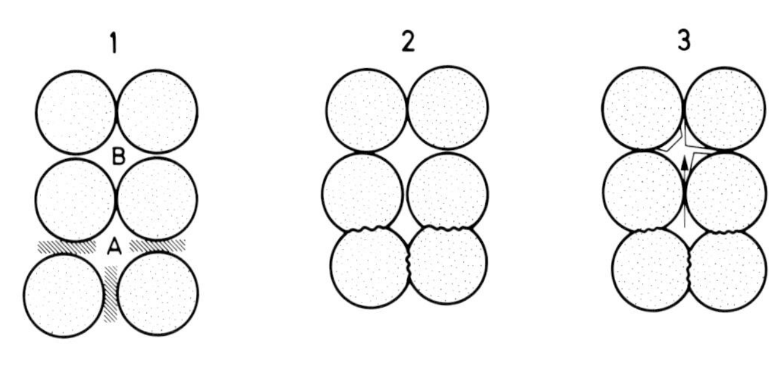

Fig. 10-12. Sequence of stages in the development of sand grains cemented by pressure solution. In stage 1 the two beds differ only in the presence of interstitial clay. In stage 2 pressure solution has begun, decreasing the volume of pore space *A*. In stage 3 some fluid from *A* escapes to *B*, reduces pressure, and precipitates quartz (Modified from Siever, 1962, Fig. 6)

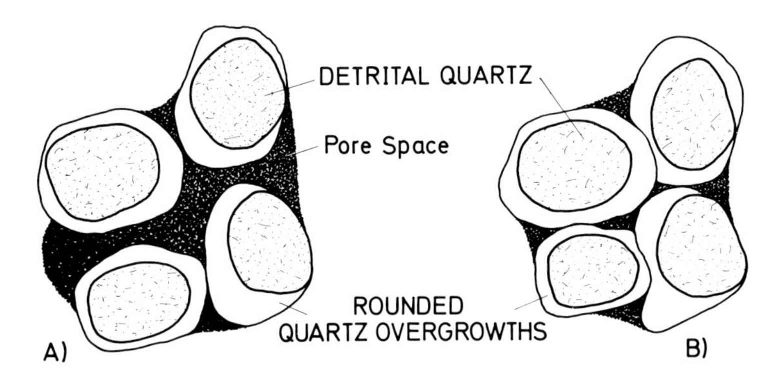

Fig. 10-13. Alternative possibility for origin of interpenetrating authigenic overgrowths shown in Fig. 10.5. Original detrital grains may have had rounded quartz overgrowths from a previous sedimentary cycle. (A) Original deposit; (B) After pressure solution

alternative is that such grains are really second-cycle grains and that the interpenetrated overgrowths are integral parts of the original detrital grains, having been added as cement in an earlier sedimentary cycle, and what we see is truly pressure solution (Fig. 10-13).

Though the evidence for pressure solution is firm, much work remains before we can relate the degree of pressure solution in a quantitative way to the fundamental parameter, which is the time duration of a given weight of overburden. That relation is now based only on theoretical models (Weyl, 1959; Füchtbauer, 1967) and to be more useful will have to depend on quantitative experimental data, on measurements of minus-cement porosity as a function of overburden history, or, preferably, both. Experiments that have produced quartz overgrowths did not involve pressure solution (Ernst and Blatt, 1964; Heald and Renton, 1966) but one pressure solution experiment did indicate increased quartz solubility (Siever, 1962). The relationship of pressure solution to overburden in Pennsylvanian sandstones was investigated with inconclusive results (Siever, 1959). There have been no detailed studies reported of pressure solution in relation to small or large tectonic features. The relation of pressure solution to chemical factors is still a subject for speculation though evidence suggests the importance of clay films in enhancing pressure solution (Heald, 1956).

Though pressure solution is important it is certainly not the whole story of silica cementation. Many authigenic quartz and chert rims are found in sandstones that show no pressure solution effects; this attests to the operation of other processes. As has been noted, authigenic quartz rims on detrital grains are most commonly the first diagenetic precipitate and, in the absence of pressure solution, must be derived from a silica supersaturated pore water. Such supersaturated pore waters have been shown to be widespread in modern sediments, where the concentrations of dissolved SiO_2 may reach 80 ppm (Siever and others, 1965), greatly exceeding the solubility of quartz at 25° C, which is about 10 ppm. Many river and groundwaters may also be supersaturated with respect to quartz. These natural waters are able to persist for long times in a supersaturated state because of the extreme slowness with which quartz precipitates; though amorphous silica precipitates quickly none of these waters is supersaturated with respect to its solubility, about 140 ppm (Siever, 1957).

In the case of quartz precipitation, just as in that of carbonate, there is not enough supersaturation, only a few tens of parts per million, to account for more than a tiny fraction of the authigenic quartz present if pore water does not circulate extensively through the sand bed. To produce a porosity reduction of even 10 percent, waters must circulate many times while constantly precipitating the supersaturation excess as overgrowths. Such extensive circulation is characteristic both of some alluvial sands, where shallow groundwaters of meteoric origin may be constantly moving through in the course of supplying water to perennial streams, and of shoreline or coastal plain sands in areas where groundwater recharge is great.

The dissolved silica in supersaturated pore waters comes from varied sources. The silica in marine sediment interstitial waters is mainly derived from the dissolution of the amorphous silica skeletons of diatoms, radiolaria, siliceous sponges, and other silica secreting organisms (Siever and others, 1965). Such organisms start dissolving immediately after sedimentation and as long as the waters continue to be undersaturated with respect to amorphous silica. Thus a marine sand may become emergent on a coastal plain some time after sedimentation and still contain siliceous skeletal material which will continue to dissolve in circulating artesian groundwaters. In this way biogenic silica ultimately gets converted to authigenic quartz overgrowths after reprecipitation.

Mineral transformations of silicates, including clays, can also account for much of the silica (Siever, 1957). Weathering of feldspars and other silicates by meteoric waters that move into the subsurface are a major source of dissolved silica that may get reprecipitated downdip in the same sandstone. In this way many alluvial sandstones may become depleted in feldspars by a form of weathering after deposition, the same process resulting in authigenic quartz production. Longer term diagenetic transformations of clay minerals, as, for example, in the conversion of a silica-rich montmorillonite derived from the alteration of volcanic glass to an illite, can supply dissolved silica to pore waters. The alteration of glass itself, to clays or zeolites, results in release of dissolved silica.

All of these sources of silica can be found in sandstones; they are even more abundant in finer grained sediments and thus the waters that may be expressed from muds into sands may serve to transport the dissolved silica. The association of augmented silica cement adjacent to clay beds, layers, or galls, is a common one. Füchtbauer (1967) showed how quartz cementation in a Dogger sandstone increases towards the shaly margin of the bed and ascribes it to an infiltration of water from the shale into the sandstone. Because the largest part of water of compaction leaves muddy sediments relatively early in their diagenetic history it is likely that this source of silica supersaturated water is most important in the early stages of sandstone diagenesis, before deep burial has taken place.

The chemical conditions that affect silica solubility are now well known and have been most recently summarized by Siever (1972). Increase of pH above 9 and increase of temperature are the two major factors that increase silica solubility. As previously noted, pore waters tend to have pH values around 8 and it is most unusual to find groundwaters with pH values greater than 9; these are usually associated with terrains of altering volcanics or of alkaline lakes in arid regions and the zeolites associated with such sediments should be diagnostic of such an origin. The temperature effect is of much more general importance, for in any section in which there are slowly rising groundwaters, or during periods of uplift, pore waters as they cool will become more and more supersaturated and may be expected to precipitate the excess as quartz. This may be one explanation for late diagenetic quartz that follows an earlier paragenesis of quartz and carbonate or other cements.

Opaline silica or chert authigenic rims and cement in sandstones are widespread and occur in a variety of sandstone types, most abundantly in volcanic rich sands, where the opal is clearly derived from the alteration of volcanic glass. Most of the chert is now microcrystalline quartz, but reasoning by analogy with the formation of chert in bedded cherts and chert nodules in limestones, it appears that the original precipitate of chert in sandstones was an amorphous silica, probably opal, which recrystallized and inverted to quartz with time. All of these occurrences imply the presence of pore waters that were supersaturated with respect to amorphous silica. It is doubtful if such pore waters would be stable for long times or could be transported over long distances without precipitating and so it is likely that the source was nearby. The source in volcaniclastics is indigenous but in the quartz arenites that show chert cement it is not so. The probable source for the latter is from soil waters in lateritic terrains, which are known to precipitate opal in the lower parts of the soil zone and in underlying permeable formations. Thus it is likely that chert-cemented sandstones were, at the time of cementation, placed not far beneath a lateritic soil, and thus below a future unconformity.

Alteration of Volcaniclastics

The diagenetic modification of volcaniclastic sandstones is covered in part in Chap. 7. There are a host of authigenic minerals, ranging from feldspars through many different kinds of zeolites to a variety of clay minerals, opal, and quartz. The major source of all of these minerals is the volcanic glass which as it devitrifies, provides to pore waters all of the components needed. The general nature of this kind of diagenesis has been ably reviewed by Hay (1966). The role of zeolites in the diagenesis and metamorphism of sediments and volcanics in eugosynclinal tracts has been emphasized by Coombs and others (1959), who have distinguished a zeolite facies as one of the lowest grades of metamorphism characteristic of deep burial of volcanic rich sediments. Packham and Crook (1960) have given the alternative designation of high rank diagenetic facies to these kinds of rocks on the basis that their essentially sedimentary texture and fabric is preserved. Their view of the relationships between metamorphic and diagenetic facies is given in Table 10-1. Regardless of whether we call them sedimentary or metamorphic, sandstones containing authigenic zeolites such as heulandite, clinoptilolite, erionite, chabazite, and analcime, and other authigenic silicates such as prehnite and pumpellyite, clearly show a progression of mineral facies that is a function of pressure and temperature and so can be used as an index of depth of burial. This facies owes its

Table 10-1. *Relations between metamorphic and diagenetic facies (modified from Packham and Crook, 1960, Table 4)*

CONTACT METAMORPHISM	EPIGENETIC DIAGENESIS	REGIONAL METAMORPHISM	ESKOLA-TYPE MINERAL FACIES
—	Heulandite-analcite (Southland, N.Z., Kuttung Beds, N.S.W., Australia)	—	Heulandite-analcite
— (Mt. Nelson, Tasmania)	Laumontite (Southland, N.Z., New England in Australia)	Zeolitic facies	Laumontite
—	Prehnite-pumpellyite (Southland, N.Z., New England in Australia)	— (Semischists of Auckland, N.Z.)	Prehnite-pumpellyite
Albite-epidote hornfels facies (in part) (Corsica)	Albite-epidote (New England in Australia)	Green-schist facies (in part) (Otago, N.Z.)	Albite-epidote

existence to the chemically reactive volcanics that were originally deposited. Thick, deeply buried sedimentary sections lacking volcanics do not contain these minerals. Furthermore, there is some tendency for zeolite facies rocks to be transformed with increasing age; most rocks of this type are of Mesozoic or Cenozoic age and the facies is conspicuous by its absence in Paleozoic and Precambrian rocks. Because of the tectonic position of eugeosynclinal facies, such rocks are highly susceptible to higher degrees of metamorphism that may have obliterated the early zeolite facies. The zeolitic rocks of those tracts which were not involved in further metamorphism may have undergone long-term mineral transformation to more stable assemblages of authigenic feldspars, quartz, and micas, presumably a thermodynamically more stable group of minerals.

Common Accessory Diagenetic Minerals

There are a host of diagenetically formed minerals that are quantitatively relatively unimportant but which may have a bearing on the post-depositional history of a sandstone. For each the textural relationships and mineral's place in the paragenesis of diagenetic minerals set the bounds for a genetic interpretation based on the known chemistry of the compounds in relation to geological processes, in the same way given in the detailed discussion of carbonate and silica cements.

Feldspars. Authigenic feldspars in sandstones have been known for many years and though most abundant in arkosic and volcaniclastic sandstones they are also present in many quartz arenites and graywackes. Because the feldspar overgrowths may go unnoticed in routine petrography because of mistaken identification as quartz or because they are such extremely small grains or overgrowths, they are systematically underestimated in abundance and occurrence. The feldspars, whether albite or K-feldspar are extremely pure end members of the alkali and plagioclase feldspar solid solution series as established by electron probe and x-ray diffraction studies (Kastner, 1971). They are most frequent as an early diagenetic mineral and predate any tectonic deformation. They are clearly not related to hydrothermal or igneous activity in most, though not all, sandstones and represent the general operation of diagenetic process in the same way that other authigenic minerals do. They are formed almost exclusively as overgrowths on detrital feldspar grains in sandstones. The chemical conditions for their formation, as noted earlier in this chapter, are a high enough concentration of dissolved silica and a high enough ratio of Na^+/H^+ or K^+/H^+ activities to make equations like 10-9 go to the left (see Fig. 10-8). The geologic conditions that seem to be necessary are moderately elevated temperatures, a source of silica, either from skeletal parts of organisms or from hydrolyzing silicates, and abundant Na^+ and/or K^+ ions that most likely come from pore waters of high salt concentrations. It may be that some of the detrital feldspars, which tend to be intermediates in either the plagioclase or alkali feldspar series and so less stable than the pure end members at low temperatures, first partially dissolve and then reprecipitate. This would be a form of "distillation" of an impure to a pure feldspar and would explain the general dependence on the presence of detrital feldspars.

Iron oxides. Authigenic iron oxides have been studied primarily in red bed sandstones though they have been known as a common outcrop weathering product in many other kinds of sandstones. Van Houten (1968) has summarized much of the data and conclusions on origin and relation to magnetism. The mineralogy of the common black oxide grains is complex and may include not only hematite but magnetite, ilmenite, and titaniferous hematites, maghematites, and magnetites. The red pigment consists solely of hematite. The diagenetic process includes the aging and dehydration of the brown amorphous ferric oxide found in modern sediments to goethite ("limonite") and hematite and the dehydration of goethite to hematite. Warm climates promote the alteration of brown (limonite) to red (hematite) pigments covering sand grains. Walker (1967) has argued for the in-place weathering of iron-rich minerals as a major source of ferric oxide pigment in desert red sands. The hematite that forms as an early diagenetic product acquires its chemical remanent magnetism as it grows through critical crystal sizes and continues to grow as long as the source, the preceding limonite, is available. Van Houten believes with Walker that most of the diagenetic hematite in sandstones, as opposed to muddy sediments, owes its origin to in-place alteration of iron-rich minerals in hot, dry regions.

Gypsum-anhydrite. The occurence of these minerals as cements is clearly related to evaporite conditions at the time of sedimentation or to movement of hypersaline pore waters from an overlying evaporite formation (Murray, 1964). The origin is much the same as for dolomite that is related to sabkha flat sedimentation. This kind of occurrence may not result in the scheme of order of cementation that Waldschmidt (1941) proposed as a general rule, of quartz first, dolomite second, and anhydrite third but may instead show anhydrite before or after dolomite and both before quartz. In any case the original mineral is an early diagenetic product related to environmental conditions, made complicated by the fact that the conversion of gypsum to anhydrite is reversible and dependent on the salinity of the water, the temperature, and pressure (see p. 51). Higher salinities, temperatures, and pressures favor anhydrite, which is by far the most common cement.

Clay minerals. The subject of diagenetic alteration of clay minerals is a huge one and much of the literature has been summarized by Grim (1968, p. 563–566). The major emphases on diagenesis of clays in sandstones has been on the formation of interstitial kaolinite or dickite that is most characteristically associated with quartz arenites and on the transformation of mixtures of originally deposited smectites and weathered micaceous clays to an illite-chlorite assemblage. The latter is characteristic of many sandstone types but is of prime importance in the problem of origin of the matrix of graywackes. The authigenic kaolinite that is so common in some sandstones has the appearance of a well crystallized, pure pore filling that was precipitated from a solution and shows no evidence of alteration from a precursor clay mineral (see p. 411). The presence of this kind of kaolinite is strong presumptive evidence for an invasion of relatively fresh groundwaters entering the sandstone from a recharge outcrop

Table 10-2. *Some clay mineral reactions during sandstone diagenesis*

Clay mineral formed	Precursor	Components added to (+) or subtracted from (−)	Stages of diagenesis (see Fig. 10-1)
Kaolinite	Feldspar	$-(K^+, SiO_2)$ $+H_2O$ (Eq. 10.9)	1, 2, 6
Kaolinite	Pore Space	$+(Al_2O_3, SiO_2, H_2O)$	2, 6
Illite	Kaolinite	$+(K^+, SiO_2)$ $-(Al_2O_3, H_2O)$	3, 4, 5
Muscovite	Kaolinite	$+K^+$ $-H_2O$	5
Illite	Montmorillonite	$+K^+$ $-(SiO_2, H_2O, Na^+, Ca^{++}, Mg^{++}, Fe^{++}$, etc.)	3, 4
Chlorite	Montmorillonite	$+(Fe^{++}, Mg^{++})$ $-(SiO_2, H_2O, Na^+, Ca^{++})$	3, 4, 5
Montmorillonite	Volcanic Glass	$+H_2O$ $-(Na^+, K^+, Ca^{++})$	1, 2, 3, 4
Glauconite	Illite	$+(Fe^{++}, Fe^{3+})$ $-(K^+, Al_2O_3)$	1, 2

belt in which there was a supply of dissolved silica from chemical weathering. Mapping clays in a sandstone that changes from kaolinitic to micaceous may be a good way of establishing the limits of fresh water infiltration either in this or a paleohydrologic cycle.

The formation of a stable clay mineral assemblage of illite and chlorite in sandstones, as in shales, is a result of long time and/or deep burial. This transformation is attested to by the illite-chlorite composition of both old – Paleozoic and Precambrian – and deeply buried rocks and is related to the long term substitution of K and Mg for Na and Ca in clay minerals. The substitution takes place in moderately concentrated pore waters and is part of a general geochemical "maturation" process that sediments go through (Siever, 1968b; Garrels and others, 1970). The relation of this transformation to depth of burial has been explored by Kossovskaya and Shutov (1958), Phillip and others (1963), and Füchtbauer (1967). Though there is substantial evidence that there is a relationship between depth of burial and illite-chlorite diagenesis, it is not certain that it is a case of simple dependence on temperature and pressure, although that must be a large part of it. Depth of burial is also linked, in most sedimentary basins, with increase in concentration and change in composition of pore waters; pore water chemistry undoubtedly has an important interaction with clay diagenesis, though the water composition may be more the result of clay diagenesis than the cause. Depth of burial is also related to time and the slow reactions of silicates may need long times for completion. All factors seem to work together and much more study of the relationships among the variables will be needed to assess the relative importance of pressure, temperature, age, and pore water chemistry in producing the illite-chlorite assemblage.

There is an important inverse relationship between the amount of clay and the amount of cement in sandstones. This generally observed relationship was shown explicitly for some Pennsylvanian sandstones by Siever (1959, Fig. 4). The explanation for the lack of cement lies partly in the decrease in permeability caused by clay matrix, either detrital or caused by alteration of rock fragments. It may also be related to lack of pressure solution, for in a clay rich sand there will be few grain to grain contacts between quartz and/or carbonate grains and many between quartz and carbonate and the easily deformable clay.

Miscellaneous diagenetic minerals. Few other minerals have been investigated in such detail as the preceding. Though there have been many studies of diagenesis of various minerals, few have been applied specifically to sandstones. Pyrite and its less common polymorph, marcasite, are known as diagenetic minerals in sandstones and imply reducing conditions, presumably reflecting a deoxygenated environment such as the sandy tidal flats in which black FeS forms just below the surface (Berner, 1964).

Carnotite and uranitite, economically important uranium minerals, occur as diagenetic cement in the Mesozoic sandstones of the Colorado Plateau. Their origin is linked to groundwater infiltration that transported uranium from a weathering primary granitic rock source into the permeable sandstones, many of which contain large accumulations of carbonaceous matter such as logs and branches. A discussion of the relation of the geochemistry of uranium in relation to these deposits is given by Garrels and Christ (1965, p. 388–394). Barite is another cementing mineral in sandstones but little is known of its geologic distribution and the kinds of geochemical regimes in which it forms.

Cementation in Relation to Depth of Burial

The relationship between diagenesis and depth of burial is a reflection of the general relationship between diagenesis and metamorphism in the pressure-temperature field of the earth (Fig. 10-14). As a sandstone is buried, its temperature and pressure increase along various paths related to rapidity of burial and igneous and tectonic behavior of the basin. The temperature and pressure increase strongly affect mineral equilibria and the rates of transformations, and so diagenesis. In this way diagenetic changes, particularly degrees and kind of cementation, have been assumed to be functions of depth by many but there are still relatively few detailed studies of sandstones such as that by Philipp and others (1963) and Füchtbauer (1967). These studies involve the combination of (1) sedimentation diagrams, in which the depth of burial of a given horizon is plotted in relation to geologic time after formation, (2) porosity of shales and sandstones vs. depth, and (3) the degree of quartz diagenesis as measured by the number of quartz grains that show varying degrees of secondary overgrowths. Quartz arenites are thought by Füchtbauer to decrease in porosity with depth chiefly by mechanical compaction down to about 1000 m; below that the decrease in porosity is primarily from pressure solution combined with secondary overgrowths. In this work Füchtbauer places heavy reliance on a relationship between porosity, permeability, and specific surface derived by von Engelhardt (1960, p. 83), which helps to distinguish between porosity decrease as a result of mechanical compaction, grain size variation and clay content, chemical addition to pore space, and pressure solution or replacement. It is hard to generalize from a few good studies. We need many more measurements of porosity and permeability in relation to quantitative measures of the degree as well as kind of cementation.

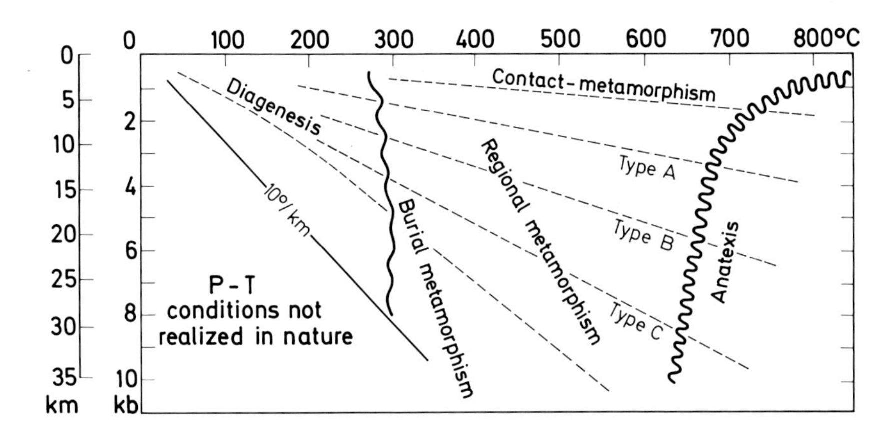

Fig. 10-14. Schematic pressure-temperature diagram for diagenesis and different types of metamorphism. The *P-T* region below the lowest possible geothermal gradient of about 10° C./km is not realized in nature (Modified from Winkler, 1965, Fig. 1)

Conclusion

Diagenesis is a complex web of physical and chemical processes related to the geologic history of a sandstone. Though our understanding of it depends on a knowledge of the chemistry and physics involved, and we have made some progress there, it also depends on more careful observation of the geological distribution of diagenetic textures and minerals. We have had many excellent descriptions of diagenetic features of particular sandstones but few have done what geologists profitably learned to do long ago – to map. Until we get maps – and cross sections – of diagenetic characters we will not be able to answer questions such as: How much stratigraphic control is there on cementation? How do different environmental facies control later diagenesis? What effect does structural deformation have on diagenesis? Can one delineate petrographic provinces based on diagenesis as well as the ones we now base on provenance and can the two be related? But maps have to be based on quantitative measurement and so we need more work too on how to do quantitative petrography in relation to diagenesis. When the combination of quantitative measurement of thin sections is coupled with mapping we will learn much more than we now know.

References

Adams, W. L.: Diagenetic aspects of Lower Morrowan, Pennsylvanian, sandstones, northwestern Oklahoma. Am. Assoc. Petroleum Geologists Bull. **48**, 1568–1580 (1964).

Allen, R. C., Gavish, E., Friedman, G. M., Sanders, J. E.: Aragonite-cemented sandstone from outer continental shelf off Delaware Bay: submarine lithification mechanism yields product resembling beach-rock. Jour. Sed. Petrology **39**, 136–149 (1969).

Andel, Tj. H., van: Reflections on the interpretation of heavy mineral analyses. Jour. Sed. Petrology **29**, 153–163 (1959).

Athy, L. F.: Density, porosity, and compaction of sedimentary rocks. Am. Assoc. Petroleum Geologists Bull. **14**, 1–24 (1939).

Becke, F.: Über chiastolith. Tschermak's mineralog. petrog. Mitt. **13**, 256–257 (1892).

Berner, R. A.: Iron sulfides formed from aqueous solution at low temperatures and atmospheric pressure. Jour. Geology **72**, 293–306 (1964).

Bramlette, M. N.: The stability of minerals in sandstone. Jour. Sed. Petrology **11**, 32–36 (1941).

Bredehoeft, J. D., Blyth, C., White, W. A., Maxey, G. B.: Possible mechanism for concentration of brines in subsurface formations. Am. Assoc. Petroleum Geologists Bull. **47**, 257–269 (1963).

Brenchley, P. J.: Origin of matrix in Ordovician greywackes, Berwyn Hills, North Wales. Jour. Sed. Petrology **39**, 1297–1301 (1969).

Brett, G. W.: Cross-bedding in the Baraboo quartzite of Wisconsin. Jour. Geology **63**, 143 148 (1955).

Buckley, H. E.: Crystal Growth, 571 pp. New York: Wiley 1951.

Byerlee, J. D., Brace, W. F.: High pressure mechanical instability in rocks. Science **164**, 713–715 (1969).

Cayeux, L.: Les roches sédimentaires de France, Roches siliceuses, 250 p. Paris: Masson 1929.

Chave, K. E.: Evidence of history of sea water from chemistry of deeper subsurface waters of ancient basins. Am. Assoc. Petroleum Geologists Bull. **44**, 357–370 (1960).

— Deffeyes, K. S., Weyl, P. K., Garrels, R. M., Thompson, M. E.: Observations on the solubility of skeletal carbonates in aqueous solutions. Science **137**, 33 (1962).

Coombs, D. S., Ellis, A. J., Fyfe, W. S., Taylor, A. M.: The zeolite facies, with comments on the interpretation of hydrothermal syntheses. Geochim. et Cosmochim. Acta **17**, 53–107 (1959).

Correns, C. W.: Zur Geochemie der Diagenese. Geochim. et Cosmochim. Acta **1**, 49–54 (1950).

Corte, A. E.: Vertical migration of particles in front of a moving freezing plane. U.S. Army Cold Regions Research Eng. Lab., Hannover, N.H., Report **105**, 9 p. (1963).

Dapples, E. C.: Diagenesis of sandstones. In: Larsen, G., Chilingar, G. V., Eds.: Diagenesis in sediments, Developments in sedimentology 8, p. 91–125. Amsterdam: Elsevier 1967.

Degens, E. T.: Geochemistry of sediments, 342 p. Englewood Cliffs, New Jersey: Prentice-Hall 1965.

DeSitter, L. U.: Diagenesis of oil-field brines. Am. Assoc. Petroleum Geologists Bull. **31**, 2030–2040 (1947).

Dickinson, G.: Geological aspects of abnormal reservoir pressures in Gulf Coast, Louisiana. Am. Assoc. Petroleum Geologists Bull. **37**, 410–432 (1953).

Dunoyer de Segonzac, G.: The birth and development of the concept of diagenesis (1866–1966). Earth-Science Reviews **4**, 153–201 (1968).

von Engelhardt, W.: Der Porenraum der Sedimente, 207 p. Berlin-Heidelberg-New York: Springer 1960.

— Gaida, K. H.: Concentration of pore solutions during the compaction of clay sediments. Jour. Sed. Petrology **33**, 919–930 (1963).

Ernst, W. G., Blatt, H.: Experimental study of quartz overgrowths and synthetic quartzites. Jour. Geology **72**, 461–470 (1964).

Friedman, G. M.: The fabric of carbonate cement and matrix and its dependence on the salinity of water. In: Müller, G., Friedman, G. M., Eds.: Carbonate sedimentology in Central Europe, p. 11–20. Berlin-Heidelberg-New York: Springer 1968.

Füchtbauer, H.: Influence of different types of diagenesis on sandstone porosity. 7th World Petroleum Cong. Proc., 353–369 (1967).

Fuhrmann, W.: „Sandkristalle" und Kugelsandstein. Ihre Rolle bei der Diagenese von Sanden. Der Aufschluss **5**, 105–111 (1968).

Garrels, R. M., Christ, C. L.: Solutions, minerals, and equilibria, 450 p. New York: Harper and Row 1965.
— Mackenzie, F. T.: Origin of the chemical composition of some springs and lakes. In: Stumm, W. Ed.: Equilibrium concepts in natural water systems. Advances in Chemistry Series 67, p. 222–242. New York: Reinhold 1967
— — Sedimentary rock types: relative proportions as a function of geological time. Science **163**, 560–571 (1969).
— — Siever, R.: Sedimentary cycling in relation to the history of the continents and oceans. In: Robertson, E. C., Ed.: The nature of the solid earth, 93–121. New York: McGraw-Hill 1971.
Garrison, R. E., Luternauer, J. L., Grill, E. V., MacDonald, R. D., Murray, J. W.: Early diagenetic cementation of Recent sands, Fraser River delta, British Columbia. Sedimentology **12**, 27–46 (1969).
Gilbert, C. M.: Cementation of some California Tertiary reservoir sands. Jour. Geology **57**, 1–17 (1949).
— Turner, F. J.: Use of the universal stage in sedimentary petrography. Am. Jour. Sci. **247**, 1–26 (1948).
Glover, J. E.: Studies in the diagenesis of some Western Australian sedimentary rocks. Royal Soc. Western Australia Jour. **46**, 33–56 (1963).
Grim, R. E.: Clay mineralogy, 2nd Ed., 596 p. New York: McGraw-Hill 1968.
Groves, A. W.: The unroofing of the Dartmoor granite and distribution of its detritus in southern England. Geol. Soc. London Quart. Jour. **87**, 62–96 (1931).
Guggenheim, E. A.: Thermodynamics, 5th, revised Ed., 390 p. Amsterdam: North-Holland 1967.
von Gümbel, C. W.: Grundzüge der Geologie, 1144 p. Kassel: Fischer 1888.
Ham, W. E. (Ed.): Classification of carbonate rocks. Am. Assoc. Petroleum Geologists Mem. **1**, 279 p. (1962).
Harker, A.: Metamorphism, 362 p. London: Methuen 1932.
Hawkins, J. W., Jr., Whetten, J. T.: Graywacke matrix minerals: hydrothermal reactions with Columbia River sediments. Science **166**, 868–870 (1969).
Hay, R. L.: Zeolites and zeolite reactions in sedimentary rocks. Geol. Soc. America Spec. Paper **85**, 130 p. (1966).
Heald, M. T.: Stylolites in sandstones. Jour. Geology **63**, 101–114 (1955).
— Cementation of Simpson and St. Peter sandstones in parts of Oklahoma, Arkansas, and Missouri. Jour. Geology **64**, 16–30 (1956).
— Renton, J. J.: Experimental study of sandstone cementation. Jour. Sed. Petrology **36**, 977–991 (1966).
Heezen, B. C., Hollister, C.: Deep-sea current evidence from abyssal sediments. Marine Research **1**, 141–174 (1964).
Hess, P. C.: Phase equilibria of some minerals in the K_2O–Na_2O–Al_2O_3–SiO_2–H_2O system at 25° C and 1 atmosphere. Am. Jour. Sci. **264**, 289–309 (1966).
Hrabar, S. V., Potter, P. E.: Lower West Baden (Late Mississippian) sandstone body of Owen and Greene Counties, Indiana. Am. Assoc. Petroleum Geologists Bull. **53**, 2150–2160 (1969).
Hubert, J. F.: Petrology of the Fountain and Lyons Formations, Front Range, Colorado. Colorado School Mines Quart. **55**, No. 1, 1–242 (1960).
Hurley, P. M., Hunt, J. M., Pinson, W. H., Jr., Fairbairn, H. W.: K-Ar age values on the clay fractions in dated shales. Geochim. et Cosmochim. Acta **27**, 279–284 (1963).
Illing, L. V., Wells, A. J., Taylor, J. C. M.: Penecontemporaneous dolomite in the Persian Gulf. In: Pray, L. C., Murray, R. C., Eds.: Dolomitization and limestone diagenesis. Soc. Economic Paleontologists and Mineralogists Spec. Pub. **13**, 89–111 (1965).
Kastner, Miriam: Authigenic feldspars in carbonate rocks. Am. Mineralogist **56**, 1403–1442 (1971).
— Siever, R.: Origin of authigenic feldspars in carbonate rocks. Geol. Soc. America Spec. Paper **121**, 155–156 (1968).
— Waldbaum, D. R.: Authigenic albite from Rhodes. Am. Mineralogist **53**, 1579–1602 (1968).
Kossovskaya, A. G., Shutov, V. D.: Zonality in the structure of terrigenous deposits in platform and geosynclinal regions. Eclog. Geol. Helv. **51**, 656–666 (1958).
Krauskopf, Konrad: Introduction to geochemistry, 721 p. New York: McGraw-Hill 1967.
Levorsen, A. I.: Geology of Petroleum, 2nd Ed., 724 p. San Francisco: Freeman 1967.

Lovell, J. P. B.: Tyee formation: a study of proximality in turbidites. Jour. Sed. Petrology **39**, 935–953 (1969).

McKelvey, J. G., Milne, I. H.: The flow of salt solutions through compacted clay. In: Clays and Clay Minerals. 9th National Clay Conf. Proc., p. 248–259. London: Pergamon 1962.

Meade, R. H.: Removal of water and rearrangement of particles during the compaction of clayey sediments – a review. U.S. Geol. Survey, Prof. Paper **497 B**, 23 p. (1964).

Millot, G.: Géologie des argiles, 499 p. Paris: Masson 1963.

Murray, R. C.: The origin and diagenesis of gypsum and anhydrite. Jour. Sed. Petrology **34**, 512–523 (1964).

Nagtegaal, P. J.: Microtextures in Recent and fossil caliche. Leidse Geol. Mededel. **42**, 131–142 (1969).

Opdyke, N. D.: The paleomagnetism of oceanic cores. In: Phinney, R. A., Ed.: The history of the earth's crust, p. 61–72. Princeton: Princeton Univ. Press 1968.

Orme, G. R., Brown, W. W. M.: Diagenetic fabrics in the Avonian limestones of Derbyshire and North Wales. Yorkshire Geol. Soc. Proc. **34**, 31–66 (1963).

Osborn, E. F.: Subsolidus reactions in oxide systems in the presence of water at high pressures. Jour. Am. Ceramic Soc. **36**, 147–151 (1953).

Packham, G. H., Crook, K. A. W.: The principle of diagenetic facies and some of its implications. Jour. Geology **68**, 392–407 (1960).

Pettijohn, F. J.: Persistence of minerals and geologic age. Jour. Geology **49**, 610–625 (1941).

Philipp, W., Drong, H. J., Füchtbauer, H., Haddenhorst, H.-G., Jankowsky, W.: The history of migration in the Gifhorn trough (NW Germany), 6th World Petroleum Congress, Frankfurt, Sec. I, Paper **19**, PD 2, 457–481 (1963).

Picard, M. D.: Paleomagnetic correlation of units within Chugwater (Triassic) Formation, west-central Wyoming. Am. Assoc. Petroleum Geologists Bull. **48**, 269–291 (1964).

Riedl, R. J.: Gnathostomulida from America. Science **163**, 445–452 (1969).

Rusnak, G. A.: A fabric and petrologic study of the Pleasantview sandstone. Jour. Sed. Petrology **27**, 41–55 (1957).

Sabins, F. F., Jr.: Grains of detrital, secondary, and primary dolomite from Cretaceous strata of the western interior. Geol. Soc. America Bull. **73**, 1183–1196 (1962).

Sedimentation Seminar: Bethel sandstone (Mississippian) of western Kentucky and south-central Indiana, a submarine-channel fill. Kentucky Geol. Survey Rept. Inv. **11**, Ser. X, 24 p. (1969).

Siever, Raymond: The Mississippian-Pennsylvanian Unconformity in southern Illinois. Am. Assoc. Petroleum Geologists Bull. **35**, 542–581 (1951).

— The silica budget in the sedimentary cycle. Am. Mineralogist **42**, 821–841 (1957).

— Petrology and geochemistry of silica cementation in some Pennsylvanian sandstones. In: Ireland, H. A. (Ed.): Silica in sediments. Soc. Econ. Paleontologists and Mineralogists Spec. Pub. **7**, 55–79 (1959).

— Silica solubility, 0° C–200° C, and the diagenesis of siliceous sediments. Jour. Geology **70**, 127–150 (1962).

— Establishment of equilibrium between clays and sea water. Earth Planetary Science Letters **5**, 106–110 (1968a).

— Sedimentological consequences of a steady-state ocean-atmosphere. Sedimentology **11**, 5–29 (1968b).

— The low temperature geochemistry of silicon. In: Wedepohl, K. H., Turekian, K. (Eds.): Handbook of geochemistry, Vol. II-14. Berlin-Heidelberg-New York: Springer 1972 (in press).

— Beck, K. C., Berner, R. A.: Composition of interstitial waters of modern sediments. Jour. Geology **73**, 39–73 (1965).

Sillen, L. G.: The physical chemistry of sea water. In: Sears, M., Ed.: Oceanography. Am. Assoc. Advancement Sci. Pub. **67**, 549–581 (1961).

Silverman, S. R.: Carbon isotope geochemistry of petroleum and other natural organic materials. 3rd International Wissenschaftlichen Konferenz für Geochemie, Mikrobiologie und Erdölchemie Vorträge, Budapest **2**, 328–341 (1963).

Simmons, G., Bell, P.: Calcite-aragonite equilibrium. Science **139**, 1197–1198 (1963).

Sippel, R. F.: Sandstone petrology, evidence from luminescence petrography. Jour. Sed. Petrology **38**, 530–554 (1968).

Sloss, L. L., Feray, D. E.: Microstylolites in sandstone. Jour. Sed. Petrology **18**, 3–14 (1948).

Smalley, I. J.: A simple model of a diagenetic system. Sedimentology **8**, 7–26 (1967).

Spotts, J. H., Silverman, S. R.: Organic dolomite from Point Fermin, California. Am. Mineralogist **51**, 1144–1155 (1966).

Swineford, A.: Cemented sandstones of the Dakota and Kiowa formations in Kansas. Kansas Geol. Survey Bull. **70**, Pt. 4, 53–104 (1947).

Tallman, S. L.: Sandstone types: their abundance and cementing agents. Jour. Geology **57**, 582–591 (1949).

Taylor, J. M.: Pore-space reduction in sandstones. Am. Assoc. Petroleum Geologists Bull. **34**, 701–716 (1950).

Thomson, A.: Pressure solution and porosity. In: Ireland, H. A., Ed.: Soc. Econ. Paleontologists and Mineralogists Spec. Pub. **7**, p. 92–111 (1959).

Trurnit, P.: Pressure solution phenomena in detrital rocks. Sedimentary Geol. **2**, 89–114 (1968).

Van Hise, C. R.: A treatise on metamorphism. U.S. Geological Survey Mon. **47**, 1286 p. (1904).

Van Houten, F. B.: Iron oxides in red beds. Geol. Soc. America Bull. **79**, 399–416 (1968).

Verhoogen, J.: Geological significance of surface tension. Jour. Geology **56**, 210–217 (1948).

Waldschmidt, W. A.: Cementing materials in sandstones and their probable influence on migration and accumulation of oil and gas. Am. Assoc. Petroleum Geologists Bull. **25**, 1839–1879 (1941).

Walker, T. R.: Reversible nature of chert-carbonate replacement in sedimentary rocks. Geol. Soc. America Bull. **73**, 237–242 (1962).

— Formation of red beds in modern and ancient deserts. Geol. Soc. America Bull. **78**, 353–368 (1967).

Weyl, P. K.: Pressure solution and the force of crystallization – a phenomenological theory. Jour. Geophys. Research **64**, 2001–2025 (1959).

Winkler, H. G. F.: Petrogenesis of metamorphic rocks, 220 p. Berlin-Heidelberg-New York: Springer 1965.

Wollast, R.: Kinetics of the alteration of K-feldspar in buffered solutions at low temperature. Geochim. et Cosmochim. Acta **31**, 635–648 (1967).

Young, A., Galley, J. E., Eds.: Fluids in Subsurface Environments. Am. Assoc. Petroleum Geologists Mem. **4**, 414 p. (1965).

Zen, E.-An: Clay mineral-carbonate relations in sedimentary rocks. Am. Jour. Sci. **257**, 29–43 (1959).

Part IV

Broader Aspects of Sand Deposition

The two chapters of Part IV are, like those of Parts I and II, mostly geological. These two chapters differ from earlier ones, however, in that they consider the biggest and broadest aspects of sand and sandstone. Chap. 11, "Sand Bodies and Environment," is a summary of the properties and origin of sand accumulations in seven major sedimentary environments; Chap. 12, "Sandstones, Sedimentary Basins, and Continental Evolution," is an inquiry into the historical geology of sandstone – what does sandstone tell us about the study of sedimentary basins and about the history of the earth? Together these two chapters draw upon all the preceding ones and constitute the "pay-off" of much previous labor by sedimentologists and many hundreds of geologists. We hope that Part IV will help convince all of you that the study of sand and sandstones is indeed needed for a better understanding of the history and evolution of the earth. We hope, in fact, it will convince you that it is the *essential key*.

Chapter 11. Sand Bodies and Environment

Introduction

A sand body can be defined as "a single interconnected mappable body of sand," the term *mappable* serving to distinguish a small sand body from most single beds.

Such a body is the response to processes that segregate and concentrate 10^{15} to 10^{18} grains in a restricted area. Such accumulations occur in many different physiographic settings: in a delta distributary, on a beach, in the seif dunes of a large continental desert, or on an alluvial apron bordering a graben, to name a few. A sand body is a natural and compelling object of study primarily because commonly it is the large-scale, unified response to some major geomorphic or sedimentologic process. Moreover, sand bodies are of economic importance: they are permeable conduits for groundwater or hydrocarbons in otherwise impermeable shaly sections; they may fill a channel cut into an underlying coal bed; they may harbor a placer of heavy minerals or be the sites of secondary copper or uranium mineralization. As we shall show, sand accumulations in the different major modern environments are distinctive enough so that the several kinds in ancient basins may be used to decipher an ancient physiography – a paleolandscape, if you will. As early as 1938 Rich stressed the role of physiography in the study of sand bodies. Once a sand body is identified as to origin, one can more readily interpret the associated sediments and, perhaps, better understand the basin itself. Knowledge of a sand's origin also improves prediction of both its spatial distribution and internal characteristics such as permeability and porosity. Obviously then, determination of the environment of deposition of a sand body is one of the more practical goals of sedimentary geology and an important key to delineating the history of many sedimentary basins.

Peterson and Osmond (1961) and Potter (1967) are two general sources for information on ancient sand bodies. Papers that are milestones in the study of sand bodies include those by Rich (1916, 1923, and 1938), Maksimov (1929), Bass (1934) and Bass and others (1937). "Stratigraphic Type Oil Fields," edited by A. I. Levorsen in 1941, contains a good résumé of the state of our knowledge prior to World War II.

Tools of the Trade

Determination of the environment of deposition of a sand body usually requires some knowledge of its shape and its internal characteristics including their spatial distribution within it. Although some notable studies have been made in areas of well-dissected, flat-lying beds, the all-important factor of shape is most effectively determined from subsurface data. Hence sand body morphology is largely studied best from well records; neither seismic nor gravity methods are routinely used to recognize and delineate sandstone bodies in ancient basins. On the other hand, geophysics has assisted in the study of modern marine sand bodies largely by acoustic reflection profiling, whereby a continuous cross-section of the sand body is obtained. Such cross sections may even disclose some of the sand's internal features. Talwani (1964, p. 43–45) describes briefly different kinds of continuous acoustic profiling and other marine geophysical methods.

Other important considerations are a sand's lateral and vertical lithologic associations.

Nomenclature and Geometry

Like any other topic, a certain amount of nomenclature is needed for successful study. The terminology applied to sand bodies, like that of sedimentary structures, is both descriptive and genetic. Descriptive terms are generally geometric ones whereas the genetic terms – reflecting the close connection between sand body origin and surface processes – tend to be geomorphic.

Quantitative description of sand bodies was proposed by Krynine in 1948 (p. 146–147), who regarded the ratio of width to thickness as a first approximation to an area-volume ratio. His types and measures appear to be inadequate to describe effectively the irregular and complex shapes that exist. As a result, qualitative terms with appropriate adjectival modifiers are most commonly used. Such terms may be descriptive or genetic. In any case, however, one conveys the essential information by map pattern, scale, thickness, relation to depositional strike and, should it be known, the appropriate genetic name. By *depositional strike* is meant a line parallel to the strand line (Chap. 4). One should also be aware that an appropriate name may depend on one's point of view. For example, what a reservoir geologist calls a blanket sand may be to the geologist viewing the basin as a whole but part of a large, elongate alluvial belt sand body.

There are at least four different basic recurring shapes and several derived ones (Fig. 11-1). Equidimensional sand bodies have length-width ratios of approximately 1 : 1 and may cover a few to thousands of square kilometers. These have been called *sheets* and *blankets*. Elongate sand bodies, on the other hand, are those with long dimension notably exceeding width and are one of three types: pods, ribbons, and dendroids (Potter, 1962, Fig. 3). *Pods* have length-width ratios of three or less where *ribbons* are much more elongate with length-width ratios of three or more and possibly as high as 20 to 1 or more. Rich (1923, p. 103) used the term *shoestring* for such bodies. *Dendroids* are commonly more sinuous and have branches, either tributaries or distributaries. By lateral migration, coalescent ribbons and dendroids may form *belts*, dendritic belts being the more common. Elongate sand bodies occur as beaches, barrier bars and on shallow marine shelves and also in river, deltaic, and turbidity current deposits. Other commonly used descriptive terms are dendritic, finger, anastomosing, bifurcating, branching, straight, *en echelon* or genetic terms such as washover or spillover sandbody (one transverse to a barrier bar), estuary sand, shelf-edge sand, etc. A good example of the use of genetic names is given by Ball (1967, Table 1), who recognized four types of modern carbonate sand bodies in the Bahamas: marine sand belts, tidal bar belts, eolian ridges, and platform interior sand blankets. Sand bodies may also be described by their relation to deposi-

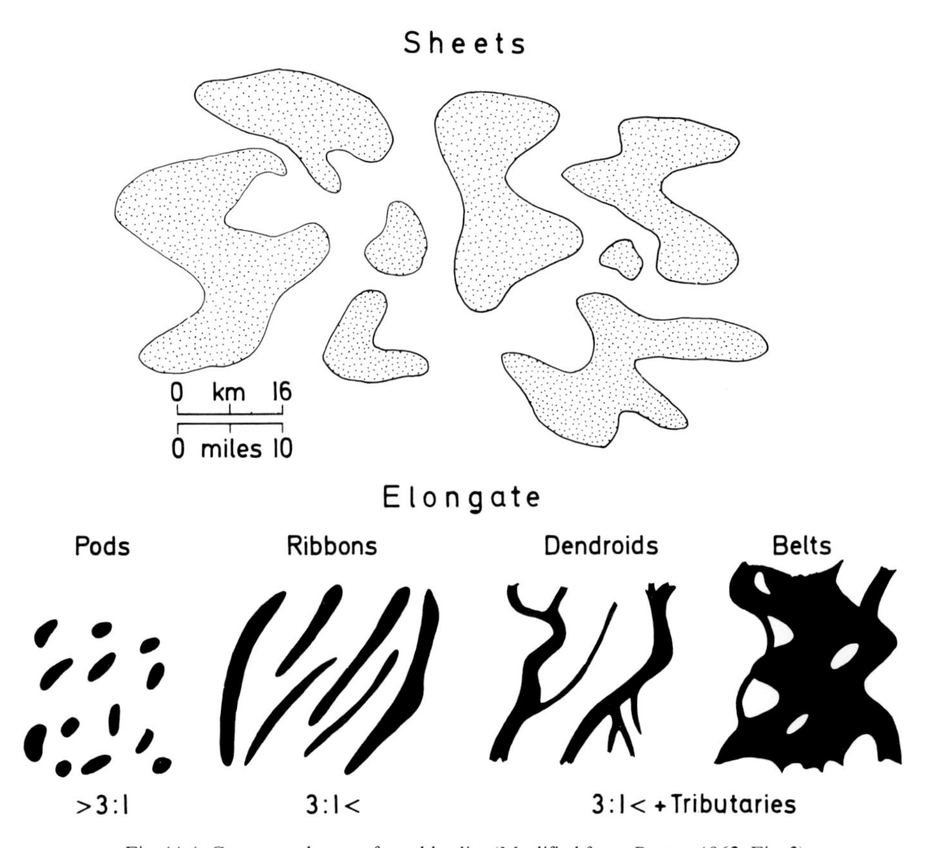

Fig. 11-1. Common shapes of sand bodies (Modified from Potter, 1962, Fig. 3)

tional strike: they have been called *strike sands*, if parallel to depositional strike and *dip sands*, if they extend down dip. If the sand body occurs above an unconformity, names such as *strike-valley* sand body (Busch, 1959, p. 2829) have been used. Sands associated with the shoreline may be described as transgressive or regressive. By *transgression* is meant the landward migration of the shoreline, by *regression* is meant its seaward migration. Because a physiographic name with appropriate scale and thickness modifiers conveys maximum information, it is the most useful if it can be assigned. It should be noted, however, that both scale and thickness are completely independent of either map pattern or origin.

An additional useful descriptive term is *multistory*, referring to the superposition of a sand body of one cycle on one or more earlier ones, the two or more forming an unusually thick sandstone section. Originally proposed for alluvial sand bodies the name is now used for all types of superposed sands. Sullwold (1961, Fig. 2) used "bundle" for multistory sand bodies. Failure to recognize multistory sands can lead to serious errors in the interpretation of sandy sections. The term *multilateral* refers to laterally coalescent sand bodies.

Representation

A thickness map and multiple cross sections are required to comprehend the shape of a sand body. The first presents the map pattern whereas the second shows relationship of the body to proximal beds. Both are essential.

Cross sections can be located effectively *only* after a thickness map has been made. Multiple cross sections are almost always desirable. Cross sections parallel or perpendicular to transport direction (or depositional strike) show maximum contrast in continuity of the body as well as the relationship of the body to proximal beds. Longitudinal cross sections of a valley-fill deposit, for example, may show downdip changes, that is, the separation of the base of the body from an underlying marker bed (Andresen, 1961, Fig. 9a). From our viewpoint, stratigraphic rather than structural cross sections are more informative. If possible, a marker bed just above or below the body should be used as a level line.

Differential compaction of mud around a sand body distorts its original shape thus complicating a subsequent interpretation of its origin. This is most apparent when cross sections of a channel or valley-fill body, one known to truncate underlying marker beds, are made. Because of compaction, choice of the first level line above the body may impart a synclinal appearance to lower marker beds. Representing the top of the body as level assumes that the upper surface of sand deposition was in fact so. As shown in Fig. 11-2, an underlying marker is probably the best, for in this case it shows several underlying coals (5A and 6) to be approximately horizontal, as in fact they are, and nicely displays the compactional bump or high on top of the body. An underlying marker has the great disadvantage, however, that it may be under the target of interest so that few wells reach it.

The effect of differential compaction also depends on type and consolidation of associated sediments. Unfortunately, our present knowledge of compaction and its effect on sand-body shape is far from adequate. Should a sand be deposited on consolidated mud or shale, subsidence of the sand into the mud will be minimal. If the underlying mud is but little consolidated, however, differential compaction under the body can appreciably modify the base (Fig. 11-3) and can produce unusually steep dips in the associated muds that can be detected by dipmeter and thus help establish position of the sand body (Reese, 1968). The key to correct interpretation of cross-section shape are associated marker beds – do they intersect or bend around the body? If they intersect the body, the sand is a valley or channel-fill deposit, but if they deform or bend beneath it, they indicate deformation and subsidence of the body so that present cross-sectional shape is not the original one. Rittenhouse (1961, p. 7-8)

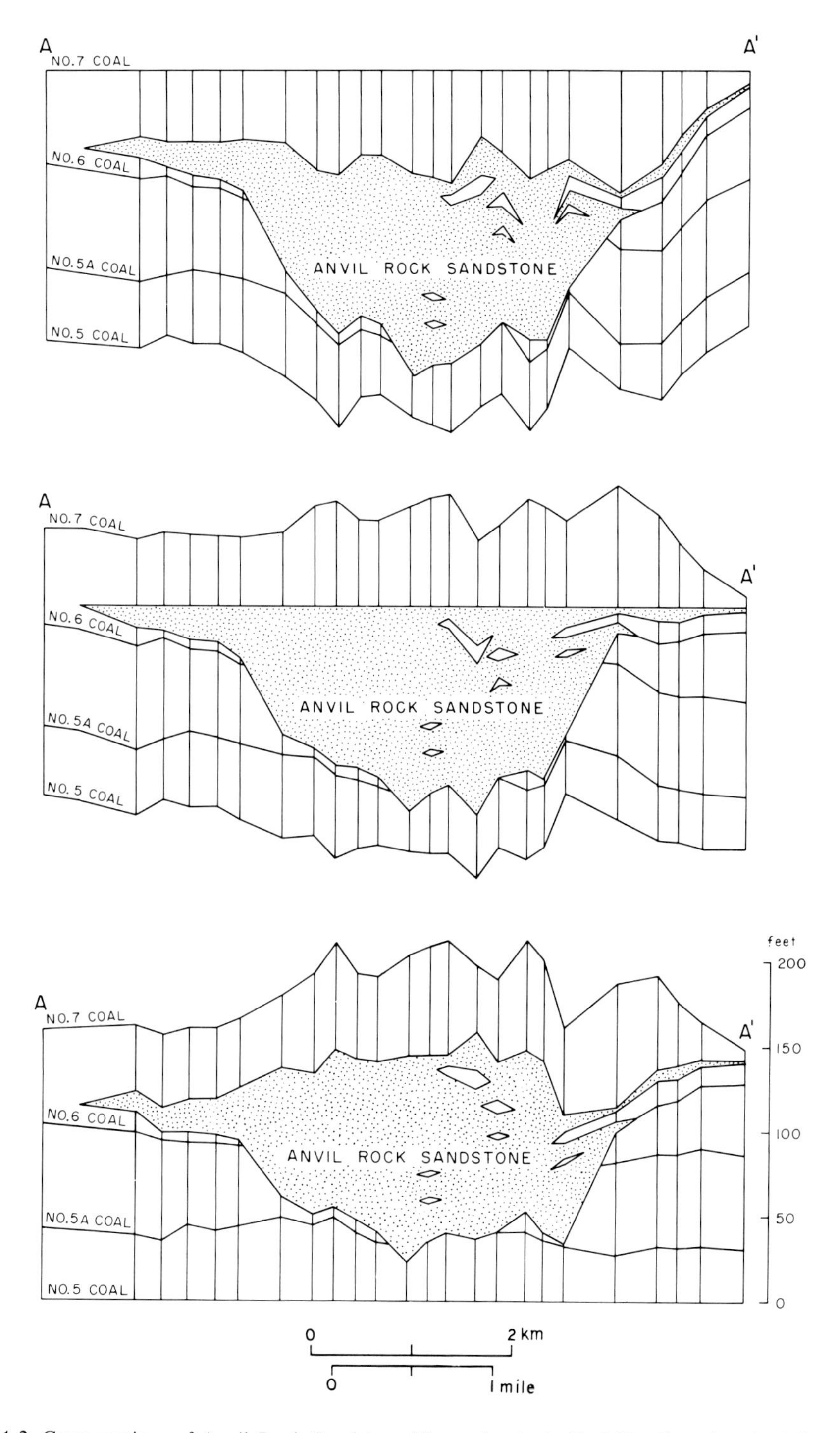

Fig. 11-2. Cross sections of Anvil Rock Sandstone (Pennsylvanian); Coal No. 7 used as level line (top), top of sandstone as level line (middle), and Coal No. 5 (below); from Potter, 1963, Fig. 37B

discussed some of the interrelations between compaction within a sand, topographic relief, and compaction of the enclosing shale. Conybeare (1967) published some graphs for estimation of compaction for associated clays and claystones and used them in decompaction calculations of some stratigraphic sections. Dickas and Payne (1967, Fig. 6) made similar calculations for compaction of mud in a deep channel. Such calculations make possible reconstruction of original

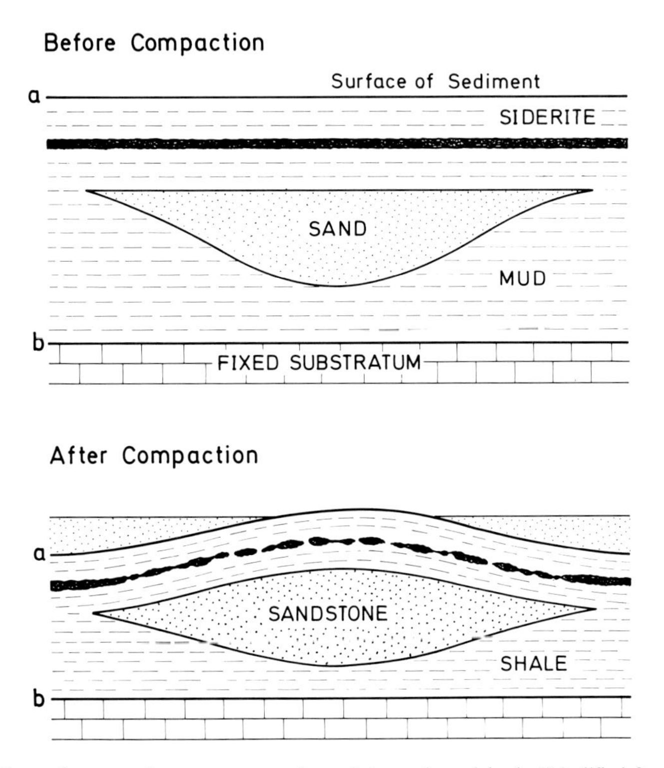

Fig. 11-3. Effects of compaction on cross section of channel sand body (Modified from Dupuy, and others, 1963, Fig. 16)

shape. Shelton (1962) approached this problem by studying small, contorted sandstone dikes in the Cretaceous shale of the western United States where he estimated the shale to have compacted approximately 2.6 times more than sandstone. Sand, whether quartz or carbonate, generally compacts but little. Knowledge of the original cross section of elongate sand bodies is most helpful in determining their environment of deposition.

Stratigraphic Models

The salient features of sand bodies are their internal fabric and structures, their external geometry, the nature of the basal contact and the bounding lithologies (Table 11-1). The spatial arrangement of the internal characteristics and their relation to sand-body geometry defines an internal organization which along with external geometry, nature of basal contact and bounding lithologies lead us to the *stratigraphic model*, that is, a concise, conceptual representation of these salient features. Each model correlates with a major sedimentary process. A model may be based on study of Recent sand accumulations and processes or in some cases, on the Tertiary, where paleontological control is commonly good, or even on older sand bodies. In any case, formulation of a model depends upon one's ability to critically observe salient geologic phenomena in depo-

Table 11-1. *Elements of stratigraphic models*

GEOMETRY	INTERNAL CHARACTERISTICS
Dimensions: length, width and thickness.	Sedimentary Structures: kinds, abundance and orientation. Biogenetic structures.
Orientation: parallel or perpendicular to depositional strike; random.	Bedding: type, thickness and homogeneity.
Position in basin: relation to major paleophysiographic features such as shelf edge, shoreline, etc.	Constituents: framework grains, fossils and cements.
	Texture: mean, sorting, and skewness.
BOUNDING LITHOLOGIES	ORGANIZATION
BASAL CONTACT	Putting the parts together, the essential key to it all. Based on modern and ancient analogues.

sits – both Recent and ancient. The elements of a stratigraphic model are commonly both qualitative (color, constituents, shape, and the like) and quantitative (size, sorting, current direction, bed thickness, faunal abundance, etc.) and may embody physical principles as well. The concept of a stratigraphic model has two main advantages: it focuses the geologist's attention on a relatively few relevant criteria and, because it integrates many variables, it brings an order, simplification, and unity to the study of sand bodies that is not otherwise attainable, for a varied array is now replaced by a few major types. Some models, of course, are better known than others, primarily because modern counterparts of some sand accumulations of ancient age have yet to be found or simply don't exist.

One should note that the model concept is essentially independent of scale so that the same model applies to both a large and small body of similar origin. For example, a delta may cover a few hundred or many thousands of square kilometers. If we did not assume independence of scale, then generalization would be almost impossible for no two products of the same environmental process – say barrier island formation – can ever be expected to have identical dimensions. Another feature of the model concept is that it not only requires a combination of criteria but their *spatial organization* as well. Commonly spatial organization is summarized in vertical profile. Visher (1965) and Shelton (1967) were among the first to emphasize the model concept for sand bodies, although the idea is implicit in an earlier study by Nanz (1954). The concept of a stratigraphic model applies equally well, of course, to carbonates, banded iron-formation, or any sedimentary facies or lithology. And it can, as we have seen, also be applied to the internal organization of single beds or, even be applied on a much larger scale, to sedimentary basins (Potter and Pettijohn, 1963, Chap. 9). Before we apply it to the environmental analysis of sand bodies, however, we need to examine in more detail what environmental information can be obtained from associated lithologies and vertical sequence.

Associated Lithologies, Vertical Sequence, and Memory

A sand body cannot be studied effectively without analysis of its relations to associated lithologies. It must be considered as part of a facies. For example, are lateral and overlying beds marine or nonmarine? Are the lateral equivalents on either side of an elongate sand body similar to each other or not? Is the body's basal contact conformable or erosional? Is a particular type of sand body commonly associated with other particular kinds of sediments? Relating the body to neighboring lithologies is best achieved by formulation of a *cycle*. The concept of a cycle, a recurring sequence of sedimentary lithologies, is one of the most fruitful in sedimentology, for a sequential pattern, if established, leads to a better understanding of the vertical and horizontal relations of the facies in which the sand body occurs and thus makes possible improved environmental analysis. The most up-to-date summary of sedimentary cycles is the "Symposium on Cyclic Sedimentation" published by the Kansas Geological Survey (Merriam, 1964).

The first step is the recognition and definition of the various lithologies in a given stratigraphic section. There may be few or many. Definition and recognition is obtained by careful, detailed analysis and description of drill cores, well-exposed sections, and possibly geophysical logs.

The second step involves a study of the relations of the lithologies to one another. Some follow one another with gradational contact, others have sharp contacts, and some overlie erosional surfaces. Recognition of erosional contacts is of particular importance for an erosional base means that the sand body has little or no dependence on the immediately underlying lithology. To see what sequential relations exist, one commonly studies *lithologic transitions* in vertical section. By lithologic transition is meant the sequential order of deposition – what follows what. Lithologic transitions may be based on the different lithologies of the facies in which the sand body occurs or on different bedding types within the sand body itself. One may be able to determine transitions qualitatively by visual inspection or it may be necessary to estimate their relative frequency by counting. Actual counting has the advantage of permitting one to estimate the *probability* of one type of sediment succeeding another. If counted, these probabilities are summarized in a *transition matrix*, wherein each element p_{ij} of the matrix gives the probability of lithology i being followed by lithology j (Fig. 11-4). In a three-component system of sandstone (S_1), siltstone (S_2), and shale (S_3), one simply determines the number of times sandstone follows itself (n_{11}), the number of times it follows siltstone (n_{21}), the number of times it follows shale (n_{31}) and similarly for other lithologies, where n_{ij} is the pair in question. To obtain probabilities, one divides each n_{ij} by the total number of beds of a particular lithology n_l so that $n_{11}/n_1 = p_{11}$, $n_{21}/n_2 = p_{21}$, $n_{23}/n_3 = p_{32}$, etc. Each row of the matrix sums to one, because *something* follows each lithology. A transition matrix is constructed either by recording lithologies at fixed intervals or by recording the frequency of the different types of transitions at lithologic contacts (Krumbein, 1967, p. 3–5 and Potter and Blakely, 1968, p. 155–156).

The third step involves examination of the empirically established relations to see whether or not there is a recurring sequence. A recurring sequence im-

A)

$$\begin{array}{ccccc} & S_1 & S_2 & S_3 & \\ S_1 & n_{11} & n_{12} & n_{13} & n_1 \\ S_2 & n_{21} & n_{22} & n_{23} & n_2 \\ S_3 & n_{31} & n_{32} & n_{33} & n_3 \end{array}$$

B)

$$\begin{array}{ccc} n_{11}/n_1 & n_{12}/n_1 & n_{12}/n_1 \\ n_{21}/n_2 & n_{22}/n_2 & n_{23}/n_2 \\ n_{31}/n_3 & n_{32}/n_3 & n_{33}/n_3 \end{array}$$

C)

$$\begin{bmatrix} p_{11} & p_{12} & p_{13} \\ p_{21} & p_{22} & p_{23} \\ p_{31} & p_{32} & p_{33} \end{bmatrix}$$

D)

$$\begin{bmatrix} 0.3 & 0.2 & 0.5 \\ 0.4 & 0.4 & 0.2 \\ 0.3 & 0.6 & 0.1 \end{bmatrix}$$

Fig. 11-4. Steps in calculating a transition matrix: (A) matrix of numbers of transitions, n_{ij}, (B) calculation of the p_{ij}'s, (C) 3×3 transition probability matrix, and (D) as in (C) but with specific values of p_{ij}'s. Three component sedimentation system of sandstone (S_1), siltstone (S_2), and shale (S_3)

plies a cycle and a "memory" – a dependence of one lithology on what was deposited previously. Memory may be long and extend back through a series of older lithologies or environments, it may depend only on the preceding one, or there may be no memory at all which means that a particular lithology occurs independently of previous sedimentary events. No memory whatsoever – be it in the progressive development of a beach or the sandy fill of a eugeosyncline – appears to be most unusual, however, for the vast majority of sedimentary sections without breaks are the result of a slow, orderly, lateral migration of related environments rather than the progression of unrelated, haphazard ones. Within the general framework of slow, progressive change with time, however, many smaller random events may occur – events to which we may not be able to assign to specific causes. Sequential variation in cross-bedding thickness in a fining-upward, fluvial sandstone body, one in which bed thickness as well as grain size gradually decreases upward, is such an example. Here we have a good grasp of the overall process, but cannot give a full deterministic explanation of successive bed-to-bed variation of cross-bedding thickness. It is in such a context that the recognition of *random events* and the probability of a transition from one sedimentary event to another is seen as one that is most appropriate for the study of sedimentation. If memory is short and extends back but one step, the sedimentation process is called Markov-1; if it extends back two steps, Markov-2, and so on. Krumbein (1968a, p. 15–17) published a brief exposition of the use of Markov models in sedimentation, and Allen (1970, p. 302–311) applied the concept to the facies transitions of fining-upward, alluvial cycles. Regardless of length of memory, however, analysis is made clearer by constructing a *tree diagram* (Fig. 11-5) from the transition

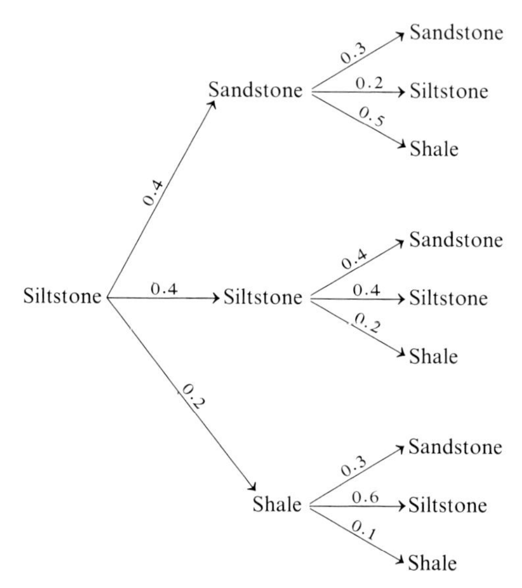

Fig. 11-5. Tree diagram based on transition matrix of Fig. 11-4. From such a diagram one can see and calculate probabilities for different paths. For example, the probability of the sequence siltstone → siltstone → sandstone is obtained by $p_{22} \times p_{21} = 0.4 \times 0.4 = 0.16$ whereas the probability of the sequence siltstone → shale → sandstone is $p_{23} \times p_{13} = 0.2 \times 0.3 = 0.06$

matrix or by formulating an ideal cycle in which one uses the transition matrix to estimate transition probabilities between the different steps of the cycle.

The vertical profile and cycle concept help us formulate a model, the unifying concept of which is the sedimentary process. Such a model, if well constructed makes possible a more adequate environmental reconstruction than would be possible without it. The environmental succession is restricted by the model for the possible choices for a given sand body are now sharply limited. Visher (1965) recognized six sand models and stresses their recognition through vertical profile.

The routine use of vertical profiles in the subsurface would be most effective, if the electrical and other geophysical logs could be *directly* linked to depositional environment, in other words, rather than simply determining lithology and thickness, one would determine directly the type of sand deposition be it barrier beach or distributory channel fill, from the log. Initial efforts in this direction have been made by Krueger (1968), who recognized nine different types of sand-shale log patterns based on type of sand-shale contact (sharp, gradational with no shale interbeds, and gradational with shale interbeds).

Krueger gives diagrammatic and actual examples of electric log patterns for some of the major sand types. Although we need much more study of the direct environmental interpretation of electrical logs, they offer promise – by systematic mapping of their different patterns – of maximizing the information to be gleaned from vertical profiles in the subsurface.

Sand Deposition and Sand Body Characteristics in Major Environments

In this section our objective is to describe concisely the processes that concentrate sand into distinct bodies in the different environments and to marshal the salient characteristics of the resultant accumulation. In so doing, we draw freely on all relevant previous material, for the problem of identifying a sand body's environment of deposition is one that can encompass just about all aspects of sedimentation (Fig. 11-6). But before we start, let us first state what we mean by the word "environment."

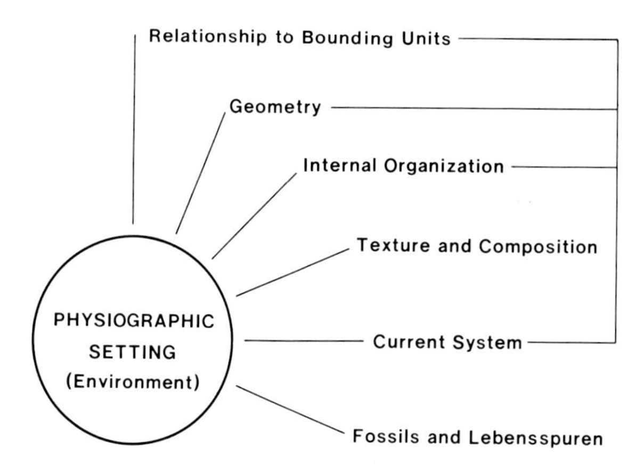

Fig. 11-6. Prinicipal contributors to environmental analysis

Concepts of Environment

The word environment is much used in sedimentary literature, although all too commonly its precise meaning is not defined. Here various uses are discussed and a definition offered.

Perhaps the most restricted use of the term occurs when one specifies the range of a physical or chemical variable, for example, pH greater than 7, O_{18}/O_{16} ratios less than some arbitrary quantity, or even "high" or "low" turbulence.

A completely different point of view is that in which one refers to "high" or "low" energy, usually meaning kinetic energy. At present, there is no unique way to evaluate this in an ancient sandstone. For example, in ancient sands grain size, sand-mud ratios, and bed thickness have all been used as qualitative measures of current competence, although their exact equivalence is yet to be fully explored.

The "process" point of view is well stated by Krumbein and Sloss (1963, p. 234) who define a sedimentary environment as "the complex of physical, chemical, and biological conditions under which a sediment accumulates." Shepard and Moore (1955, p. 1488) defined a sedimentary environment as "a spatial unit in which external physical, chemical, and biological conditions and influences affecting the development of a sediment are sufficiently constant to

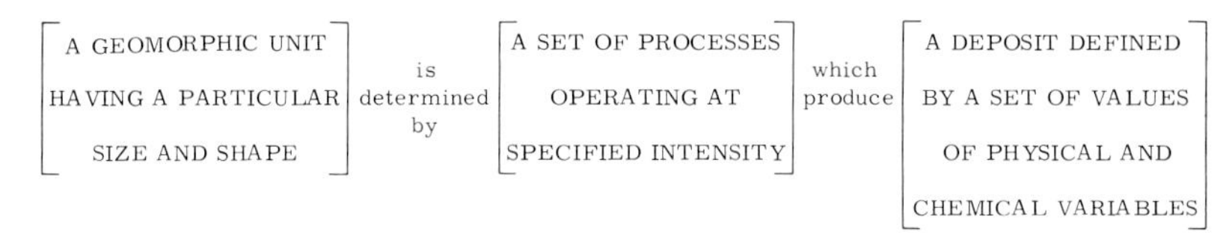

Fig. 11-7. Relations between geomorphic unit, processes, and resultant deposit (Potter, 1967, Fig. 2). Reproduced with the permission of the *Bulletin* of the American Association of Petroleum Geologists

form a characteristic deposit." The latter definition is reasonably close to the geomorphic concept of environment emphasized by Twenhofel (1950) who, in America, had much to do with shaping attitudes during the 1920's and 1930's. He emphasized a geomorphic unit described by a totality of variables.

The broadest and most meaningful concept of environment, especially for application in ancient sediments, appears to be the geomorphic one. A sedimentary environment is defined by a particular set of physical and chemical parameters that correspond to a geomorphic unit of stated size and shape. A beach, point bar, submarine sandy ridge in the North Sea, or a set of coastal dunes all satisfy this definition. Fig. 11-7 diagrams the links between a geomorphic unit, sedimentary processes, and resultant deposit and shows environmental analysis to be essentially paleogeomorphology. Hence in making an environmental reconstruction the sedimentologist becomes, in reality, a geomorphologist.

The optimistic view presumes that one can recognize an ancient environment by inferring a set of variables from their corresponding rock properties. A *set* or *combination* is required almost without exception because, if only one variable were used, it almost certainly would not be specific. Turbulence, for example, may be the same in kind and degree behind a sand wave either in a river or on a shallow-marine shelf. Thus a combination or set of variables is essential because it is difficult to define a geomorphic environment by one variable alone.

One's choice of variables is limited to those that leave a lasting impression on the sand. This is illustrated by the chemical variable pH. Although one can measure the pH of ground water in a sandstone, it is not necessarily the pH of an environment. On the other hand, one can, under favorable circumstances, say that certain minerals in a sandstone indicate very high pH values at the time of formation because the rock fabric indicates that these minerals were primary rather than diagenetic. Such a conclusion is important as high pH values are linked with certain restricted alkaline lake environments on the basis of chemistry, modern sediment study, and paleogeography. In general, the more variables the better, but in practice ordinarily not more than five to ten are commonly used. Table 11-2 gives a check list for sands.

We have assumed, optimistically perhaps, that a given set of processes operating at a specified intensity produces a unique deposit of stated size and shape. Although in general this is probably true, especially for the larger geomorphic units, it is possible that there may not be an exact one-to-one correspondence of geomorphic unit, processes, and resultant deposit.

Sand bodies of seven major origins are recognized: alluvial, deltaic, tidal, turbidite, beach and barrier island, shallow-water marine, and eolian. These types are described in terms of petrology, texture, sedimentary structures, associated fossils and lithologic features, internal organization, and external morphology. Existing descriptions of sand bodies in these different environments vary in detail, because of the widely differing kinds and amounts of information available.

The volume of sand preserved in these seven different environments differs markedly. We would hazard a guess that alluvial-deltaic sands are by far the most abundant in the geologic column, forming perhaps possibly 50 percent, and that turbidite sands are probably second – perhaps 30 percent. Far behind in abundance in the record are tidal, beach and barrier island, eolian, and marine shelf sands. Collectively, the latter may form only 20 percent of all ancient sands and possibly much less. Barrell in a classic 1906 paper, still very much worth reading, concluded (p. 441–449) that delta deposits should far surpass all others in volume. Volcaniclastics deposited in various environments, although not widely recognized in the pre-Mesozoic record, form a significant volume of Cenozoic and later sands (Chap. 7). Sand bodies of lacustrine origin are not recognized as a separate class, because they are neither volumetrically important in the geologic record nor do their physical properties differ greatly from those of shallow marine origin.

Like many classifications, the foregoing is a mixture of classifying criteria: agents (for alluvial, tidal, eolian, and turbidite), water depth (for example, marine shelves), and landforms (for example, barrier islands). Nevertheless, the sands in these seven different environments do seem to show – at least in our present state of knowledge – naturally distinct and recognizable associations of properties. Many of these environments can be subdivided as, for example, the barrier-island environment, which may include beaches and coastal-dune complexes. The tidal-flat environment provides another good example: the salt marsh, the higher mud flats, the inner and outer *Arenicola* sand flats, the lower mud flats, the lower sand flats, and creeks and bordering areas (Evans, 1965). Unfortunately, such detailed discrimination is not possible in the majority of ancient sediments. The seven groupings of the present study are a compromise between the present state of knowledge, on the one hand, and the distinctions that are deemed necessary for successful study of ancient sediments.

Finally, we offer a few generalizations that seem to be worthy of emphasis. One should remember that environmental boundaries may be – and commonly are – gradational, that the same sedimentary environments can vary greatly in dimensions as can the "life span" or persistence in geologic time of a partic-

Table 11-2. *Environmental check list for sand bodies*[a]

STRATIGRAPHIC

A. Shape
  1. Elongate
    a. curved
      (1) meandering
      (2) dendritic
      (3) bifurcating
    b. straight
    c. relation to depositional strike
  2. Blankets (sheets)
B. Lateral and vertical associations
  1. Continuous
    a. facies change
    b. oscillatory or cyclical
  2. Discontinuous
    a. unconformity
    b. structural cutoff
C. Lithologic associations
  1. Range and abundance of rock types
  2. Lithofacies maps

PRIMARY SEDIMENTARY STRUCTURES AND BEDDING

A. Inorganic structures
  1. Crossbedding
  2. Ripple marks
  3. Parting lineation
  4. Sole markings
  5. Deformational structures
  6. Small-scale channels
  7. Shale chips and clay galls
B. Burrows, tracks, and trails
C. Character and organization of bedding
D. Variability of directional structures
E. Vertical sequence of bedding and structures

PETROGRAPHIC

A. Mineralogy
  1. Primary
  2. Diagenetic
B. Texture
  1. Grain-size parameters
  2. Shape and roundness
  3. Surface texture
  4. Fabric
    a. grain-to-grain contacts
    b. orientation of framework fraction
  5. Vertical sequence of textures

PALEONTOLOGIC

A. Autochthonous fossils
B. Allochthonous fossils
C. Specialized indicators of marine and non-marine
D. Ecologic associations

[a] From Potter, 1967, Table I. Reproduced by *Bulletin* of the American Association of Petroleum Geologists.

Table 11-2 (continued)

GEOCHEMICAL
- A. Primary (Depositional)
  1. Minerals
  2. Isotopes
  3. Organic matter
  4. Trace elements
- B. Diagenetic (Secondary)
  1. Minerals
  2. Isotopes
  3. Organic matter, oil and gas
  4. Formation waters
  5. Structures: concretions, nodules, stylolites, etc.

ular environment. In any case, however, the *systematic mapping* of relevant variables is nearly always the key to effective identification for only mapping can provide the information needed to identify the correct stratigraphic model.

Alluvial Environment

Because of their accessibility, more has probably been written about modern alluvial deposits than any other environment of sand deposition. Moreover, there is a substantial, perhaps even a larger literature on their ancient equivalents, partly because of the worldwide association of alluvial sandstones and coal deposits. Only deltaic deposits rival alluvial deposits in terms of available literature.

We are here primarily concerned with alluvial sand deposits either in a broad alluvial valley on a coastal plain or on a sandy piedmont fan fed by a rising granitic fault block.

Flood plain morphology: The key to understanding most ancient, alluvial deposits is the origin and physiography of the modern alluvial flood plain. We summarize below the origin of those salient features of flood plains relevant to sand bodies by drawing freely on published accounts, especially those by Fisk (1944), Russell (1954), Allen (1965a, p. 115–127), and the U.S. Geological Survey's many physiographic and hydraulic studies of rivers.

The essential major features of an alluvial plain are its stream patterns, its point bars, the natural levees of the inner flood plain, abandoned channels and cutoff meanders (the latter called oxbow lakes) and the outer flood plain or backswamp (Fig. 11–8). Depending upon availability of sand and on climate, eolian deposits may also be present. The flood plain itself consists of channel sands and gravels plus overbank deposits of silt and mud and may include accumulations of plant material in swamps, accumulations that later may become peat and coal. Channel deposits, largely sand, generally far outweigh overbank deposits. In the ancient record, this appears to be especially true, perhaps because of the difficulty of distinguishing many overbank alluvial silts and muds from non-fossiliferous marine muds.

The river course on the flood plain may be straight, meandering, or braided, but is most commonly meandering. Meanders commonly define belts (Fig. 11–9) within which point bars on the inside of the meander accumulate sand in a distinctive fining-upward manner. Shifting of the meander belt takes place through

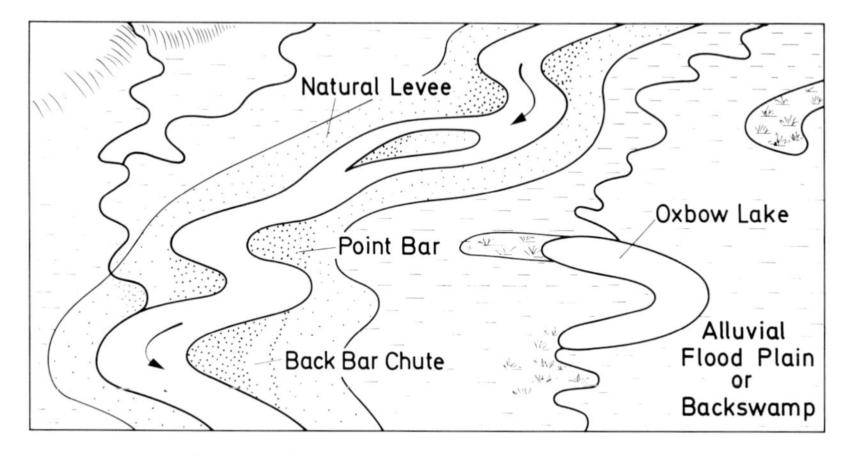

Fig. 11-8. Salient physiographic features of the flood plain

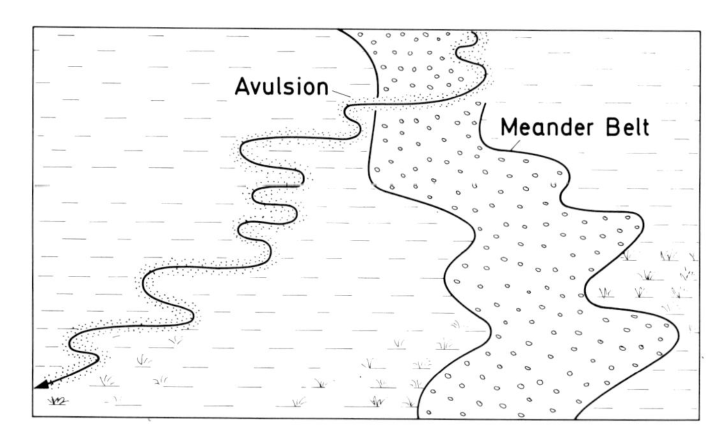

Fig. 11-9. Avulsion, meander belts and backswamp muds

avulsion – the sudden abandonment of a stream course by breaking through (crevassing) its natural levees. Crevassing occurs because vertical accretion deposits proximal to the channel raise the meander belt above the rest of the flood plain until, during flood, the stream breaks out of its natural levee into a lower part of the alluvial valley.

Stream discharge is the essential feature that determines the pattern and scale of most flood plain features. The type and quantity of detritus play a secondary role. Most rivers are subject to flooding at least once a year and during this high discharge erode their channel and deposit sand and gravel along the channel and mud and silt proximal to it, thus molding the flood-plain topography. One possible model for alluvial sand bodies is that of Figure 11–10 wherein discharge and detritus interact in a complicated, and as yet not fully understood manner, with the channel system: stream gradient, depth of thalweg, and channel sinuosity and its lateral stability. The amount and nature of detritus depend upon rock type, relief, discharge, and climate. Acting together, these energy and material factors determine the grain-size characteristics and stratification of an alluvial

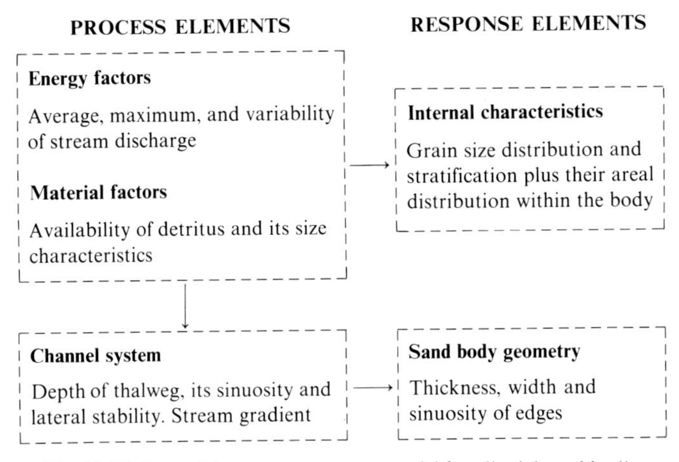

Fig. 11-10. A possible process-response model for alluvial sand bodies

sand body whereas stream behavior is most important in the control of sand body thickness, width and the sinuosity of its edges.

Some insight into flood plain morphology can be obtained by considering the behavior of single stream channels – the interrelations between their depth, discharge, meander wave length, bed load, and so forth or what has been termed their *hydraulic geometry* (Leopold, and others, 1964, Chap. 7). The key question is: how does variation of discharge and sediment load in a stream affect its channel width, depth and sinuosity? Schumm (1968a, p. 1579) finds, from observation of modern streams, that as discharge (Q) increases there are proportional increases in channel width (W), depth (D), meander wave length (l), and a decrease in channel gradient (s) so that $Q \alpha \frac{WDl}{s}$. He also emphasizes that type of sediment load must be considered. A significant increase in ratio of bed load to total sediment load (Q_s), leads to an increase in channel width, meander wave length, and slope but to a decrease in depth and sinuosity P (channel length/straight line distance between two points) so that $Q_s \alpha \frac{Wls}{DP}$. Here the width-depth ratio of the channel appears as an indicator of sediment load, that is, sand-carrying streams have greater W/D ratios than those carrying mostly silts and clays. For bed-load streams Schumm (1968b, Table 5) estimates W/D to commonly exceed 40, whereas in streams with 5 to 20 percent of silt and clay deposits W/D varies between

10 and 40. In dominantly suspended load streams the W/D ratio is less than 10. *Modern alluvial sands:* We emphasize here primarily those processes of stream deposition that control the well-established stratigraphic model for alluvial sand bodies (Table 11-3). Two major questions arise: what is responsible for the internal fining-upward structure of alluvial sand bodies and what controls their thickness and external geometry? Many of the problems posed by alluvial sands center about these two questions.

Table 11-3. *Characteristics of alluvial sands*[a] *(Alluvial-fan deposits excluded)*

PETROLOGY

Detrital. Abundant shale pebbles and shale-pebble conglomerates. Generally carbonaceous debris. Petrographically immature to moderately mature. Pebbles and cobbles, if present, may be both local and distal. Detrital plus chemical cements. Faunal content low to absent.

TEXTURE

Poor to moderate sorting and moderate to low grain-matrix ratio. Abundant silt in fine-end tail. Tendency to poor rounding. High variability.

SEDIMENTARY STRUCTURES

Asymmetrical ripple marks and abundant well-oriented crossbedding, commonly unimodal. Parting lineation and deformational structures are common minor accessories. Beds tend to be lenticular with erosional scour. Some tracks and trails.

INTERNAL ORGANIZATION

Strong asymmetry. Upward decrease in grain size and bed thickness, possibly with conglomerate near base. Larger channel-fill sandstone bodies tend to be coarser-grained than smaller ones.

SIZE, SHAPE, AND ORIENTATION

Commonly very elongate. Width ranges from a few tens of feet to composites of 30 miles. Dendritic as well as anastomosing and bifurcating patterns. Elongate downdip. Excellent correlation of internal directional structures and elongation.

ASSOCIATED LITHOLOGIC TYPES

Vertical: overlying silty shales, commonly of alluvial origin. Possible peat and coal. Basal contact commonly sharply disconformable. Multistory sandstone bodies. Marine units in mixed sections. Lateral: silty shale and siltstone commonly with abundant carbonaceous material as well as roots, leaves, and stems. Multilateral sandstone bodies. Correlation generally difficult.

[a] From Potter, 1967, Table II. Reproduced by *Bulletin* of the American Association of Petroleum Geologists.

The vertical sequence generated by the point bar (Fig. 11–11) is the key to understanding the fining-upward sequence of alluvial deposits. Ideally, there are four different facies: poorly-bedded, basal thalweg sands that may contain gravel and cobbles, crossbedded sands of the point bar, ripple- and horizontally-bedded fine sands and silts of the inner flood plain. The poorly-bedded, basal thalweg sands usually constitute only a small portion of the total section and may be indistinguishable from the point-bar sands which predominate. Parting lineation may be found in any of the four facies but it is most common in the ripples and horizontally-bedded facies. During flood these four facies are deposited simultaneously on different parts of the bar. Sand and gravel eroded from the opposite cut bank are transported downstream and probably are deposited largely on the next point bar. The back-bar slough also develops during flood by water seeking a shorter, more direct down-hill path. The size of point bars depends upon river size: they may vary from a few hundred feet (30–90 m) long on a small stream to more than four miles (7 km) on a river the size of the lower Mississippi. Grain size decreases downstream along the top of the bar. Point bars are present in both meandering and straight channels.

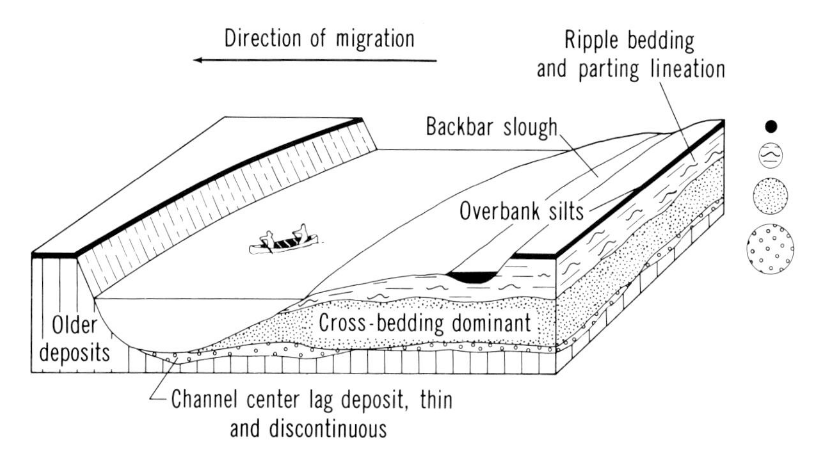

Fig. 11-11. Vertical sequence generated by a point bar on the inside of meander bend. Note fining upward. Current toward viewer (Potter and Blakely, 1967, Fig. 2). Published by permission of the Society of Petroleum Engineers Journal.

The velocity distribution within the channel at the stream bend sheds further light on the origin of the fining upward sequence. The highest velocity is commonly at the outside of the bend and there is an approximately linear decline in velocity shoreward that is accompanied by an upward decline of both sand-wave height (crossbedding thickness) and grain size as both correlate with each other. Hence, as one moves along the bottom from the thalweg to the inside of the meander bend, both sand-wave height and grain size decline in response to declining velocity and thus produce the observed upward decrease of bed thickness and grain size. Allen (1970, p. 314–320) presented a quantitative explanation for fining-upward alluvial cycles, based on a combination of hydraulics and empirical relations observed in river bends. These fining-upward cycles may contain subcycles, 3–20 ft (1–6 m) which have a sharp lower contract and an upward decrease in grain size. These are attributed by Meckel (1970, p. 64) to major floods.

Is the fining upward of alluvial sands solely attributable to the fining upward observed on point bars? We think not. Abandonment of the river channel by upstream avulsion (Fig. 11–9) also leads to fining upward. Like point bar formation itself, avulsion is a ubiquitous local feature of flood plains. Two other factors are rising base level and lowering relief of the source region.

What controls thickness of a single cycle alluvial sand body? It seems reasonable to believe that no point bar can be thicker than the depth of maximum scour (during flood) of its associated stream. Thus in a large river like the Mississippi in its lower reaches, point-bar deposits can be 200 or more feet (60 m) thick – the depth of scour of the river during flood. In using this criterion to estimate stream depth in ancient alluvial sands, one must be careful, of course, to make sure that the sand body is not multistory.

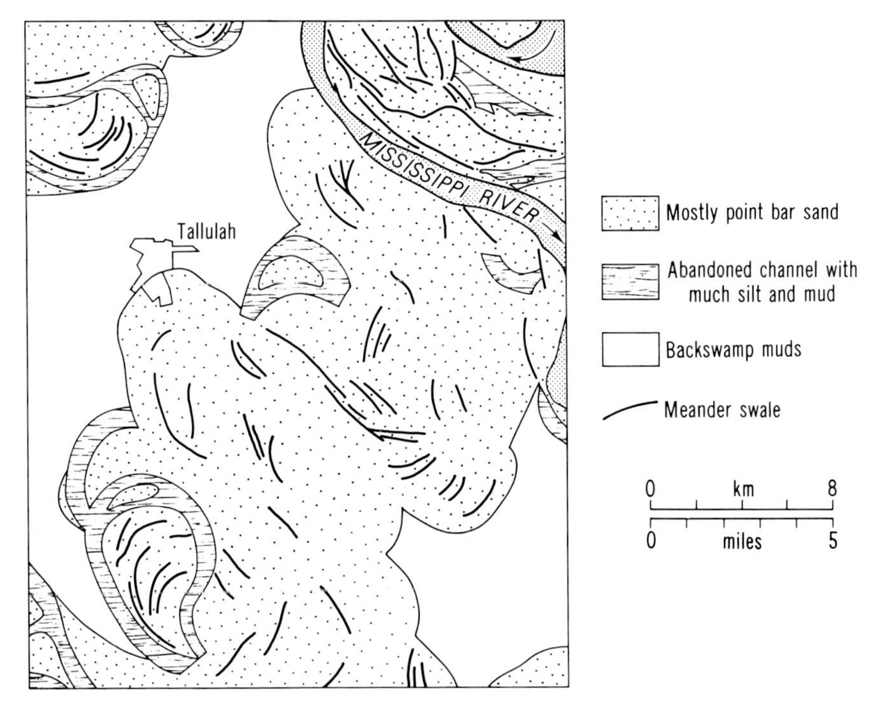

Fig. 11-12. Belt sand body deposited by former course of Mississippi River. Note marked serrate edges (Hrabar and Potter, 1969, Fig. 11). Reproduced with permission of the *Bulletin* of the American Association of Petroleum Geologists

And what can be said of the width of an alluvial sand body? Although increase in both stream discharge and ratio of bed load to total load are accompanied by increase in stream width, we cannot directly apply this relation to sand body width, which commonly reflects width of one or more meander belts rather than simple stream width. And in a broad alluvial valley, meander belt width may simply be proportional to time – the longer the time the greater the opportunity for the stream to shift back and forth and thus produce a wider meander belt sand body. Because time may be the critical factor, there appears to be little possibility of finding any analytical relationship between sand body width and thickness.

A characteristic of alluvial sand bodies are their serrate edges (Fig. 11–12) that are produced by meandering streams impinging upon the muds and silts of the outer flood plain or backswamp. The degree of serration depends upon the radius of curvature of the meanders. Leopold, and others (1964, p. 298), report that the ratio of meander curvature (r_c) to stream width (W) is fairly constant for most streams and is commonly between 2 and 3, averaging 2.7. In some alluvial sand bodies one can determine r_c from careful subsurface mapping along the edge of the body (Fig. 11–12) and thus possibly estimate stream width. Careful logging of cores would enable one to decide whether the sand body is multistory or not; then one might estimate stream depth (D) from thickness of single cycles within the body. Thus a W/D ratio for the stream might be estimated. If one could somehow estimate mean velocity from grain size and stratification, one could then very roughly estimate the paleodischarge.

Another feature of some alluvial sand bodies are the sand splays that form during high water by crevassing. In a big alluvial sand body, crevass deposits may form lobate delta-like appendages to the main body. Splays may, of course, also be associated with deltaic sand bodies.

And what about deposits from braided streams whose rivers are characterized by wide channels and frequent shifting of course? Despite much research, the factors that cause a stream to braid rather than meander are not fully understood, although a debris load too large for a single channel may be one important factor (Leopold, and others, 1964, p. 295). Whether or not one can distinguish an ancient sandstone deposited by a braided stream from one deposited by a meandering river is not certain. However, Berg and Cook (1968, Figs. 8–10) suggest that a meander belt sand will have a more uniform thickness than one formed by a braided stream.

The foregoing suggests that as we learn more about the deposits formed by different types of streams, we should be able to more closely determine the paleohydrology of ancient alluvial sands and thus better determine the paleoclimatic and tectonic conditions under which they were deposited. An ambitious attempt to do so was made by Schumm (1968 b) in his study of Murrumbidgee River and its paleochannels in New South Wales. Drawing on much previous work, especially soil mapping, Schumm compares the present river system, which is narrow, sinuous, deep and transports but little sand, with two sets of paleochannels: a younger set called "ancestral-river channels" and an older set called "prior stream channels". The ancestral river channels are similar to the modern, but larger, whereas the prior channels are relatively straight and filled with crossbedded sand. Because he can compare the paleochannels with discharge and sediment load of the present Murrumbidgee River, Schumm is able to utilize much of what has been learned from the hydraulic geometry of modern streams. Quaternary climate changes are considered the prime cause of change in characteristics of paleochannels of the Murrumbidgee system. Eicher (1969, p. 1087–1088) also attempts to estimate stream gradients in the alluvial sediments of Cretaceous age in western United States.

Ancient Alluvial Sandstones: The different major facies of alluvial sands are diagrammed in Figure 11–13.

Most alluvial sandstones have a disconformity at their base; many have basal gravel and fine upward (Fig. 11–14). Alluvial sandstone bodies range in width from 100 feet (32 m) to composites of more than 30 miles (48 km), the latter

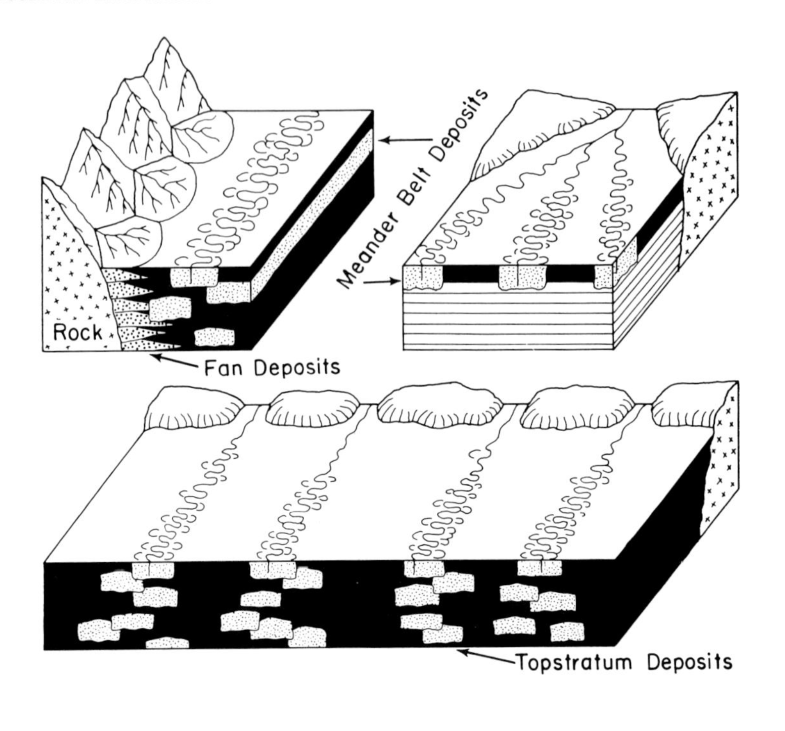

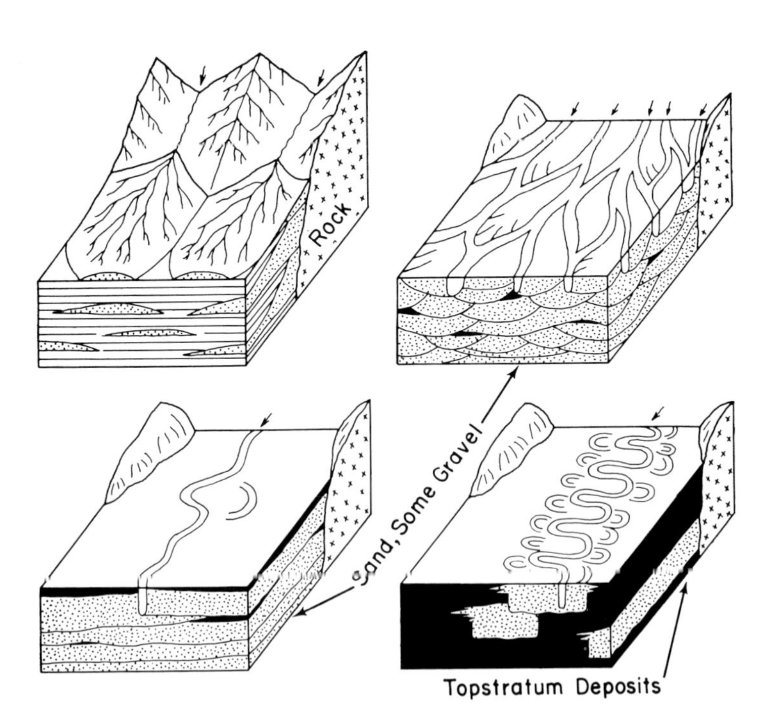

Fig. 11-13. Contrasting idealized alluvial sand patterns (Modified from Allen, 1965a, Figs. 35 and 36)

the product of meander-belt migration over the alluvial valley. Marginal boundaries may be weakly to markedly serrate (Fig. 11–15), the latter being a strong indication of alluvial origin. Splay deposits may also be present and, if oriented up structural dip, make excellent stratigraphic traps. Thickness ranges from a few feet to more than 100 feet (32 m), although probably few single cycle sands exceed 200 feet (64 m). Multistory sands are common. Along a major unconformity with a sharply entrenched valley system, one that is usually dendritic but may be trellis, the distribution pattern of basal alluvial sands is controlled almost completely by paleotopography.

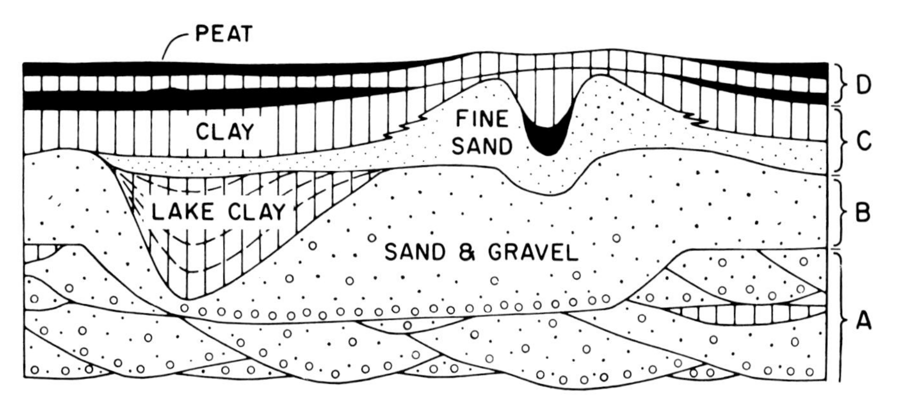

Fig. 11-14. Idealized sequence and structure of alluvial deposits in the Netherlands (Modified from Zagwijn, 1963, Fig. 4). Reproduced with the permission of the *Bulletin* of the American Association of Petroleum Geologists

Crossbedding and various types of current ripples are the dominant structures, although parting lineation and convoluted bedding are also known. Commonly over half of the beds are cross-laminated. In thickness crossbedded units rarely exceed three feet (100 cm) and are commonly one foot or less (30 cm). In any case, they thin upward and pass into ripple marked strata. In most alluvial sandstone bodies, crossbedding is a good predictor of elongation direction and correlates well with other directional structures (Fig. 4–27). Variability of dip direction is generally small but may be moderate. Small intrasand body channels are common. Lateral persistence of individual beds is limited. Hence, in terms of bedding and texture, alluvial sand bodies are among the most inhomogeneous. Multistory sands contribute to heterogeneity.

The mineral composition of alluvial sandstones is strongly dependent on source. But as a general rule, alluvial sands tend to be less mature than their more distal equivalents. In part, this is because the vast majority are the response to rejuvenation of the source region. Immaturity is also favored because, unlike the beach environment, alluvial transport produces negligible wear on sand-sized debris. Typically, an alluvial sandstone will be a lithic arenite or even a lithic wacke. Paine and Meyerhoff (1968, p. 109) list some ten thin-section criteria that collectively were found to distinguish fluvial from marine environments in the late Tertiary Catahoula Formation in Louisiana. Criteria of fluvial origin include textural and mineralogic immaturity, absence of marine fossils (presence of some abraded freshwater fossils), scarcity of glauconite (except for some abraded grains), and scarcity of mineral cements. In general, alluvial sands have abundant shale pebbles, may contain reworked clay-ironstone concretions, and, in some cases, distally derived gravels. Cementing agents are a mixture of detrital and

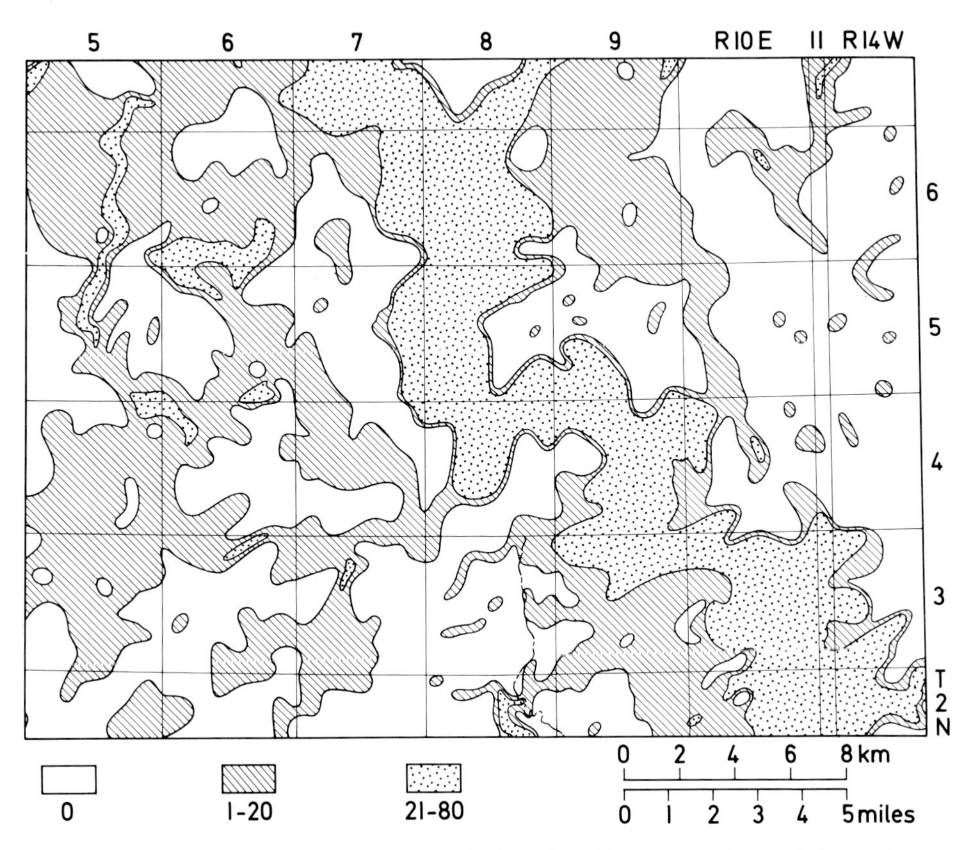

Fig. 11-15. Pennsylvanian sandstone body in Illinois Basin with pronounced meandering pattern (Modified from Potter, 1963, Fig. 51)

chemical materials. Modal diameter varies vertically with position in the body and also tends to be related to channel dimension which in turn is related to greater stream discharge. Alluvial sands tend not to be well sorted, the size distribution having a tail in the fines.

Fragmental plant remains, rootlets and some logs may be present, the latter commonly concentrated near the base. Alluvial sands are generally faunally barren containing only scattered fresh or brackish water shells, most commonly as channel-bottom concentrates. Acid ground water from associated peats may subsequently remove these. In the Mesozoic and later alluvial sands concentrates of vertebrate remains may be present.

Figure 11–16, a diagrammatic down-dip, longitudinal section, shows some of the lithologies that may be associated with fluvial-delta sands in cyclothemic sequences. These include the other sediments of the flood plain plus peat, lignite or coal, from which, if the sand body is part of a cycle that includes marine rocks, the sand body will be separated by basal unconformity. Silt and siltstone are nearly always present and tend to either cap the body or be lateral overbank equivalents. Because they are genetically related to the sands, the associated silts may thicken next to the sand body, an observation that may be useful in exploration.

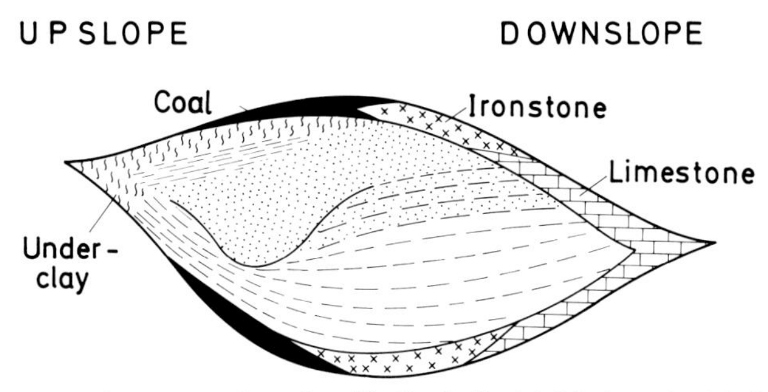

Fig. 11-16. Diagrammatic cross section of an idealized alluvial-deltaic cycle (Modified from Ferm and Williams, 1963, p. 356)

Hopkins (1968, pl. 1) shows a greater thickness of the overlying gray shale where it caps a sandstone that fills a narrow, low-sinuosity cutout in a major coal bed in the Illinois Basin. His study demonstrates how thickness of associated shale can be used as a guide to coal thickness and to proximity of a sandstone body. Should alluvial or deltaic sediments overlie an unconformity, differential compaction over large underlying reefs may be sufficient to affect the location of streams so that they follow the lower, more greatly compacted, interreef areas.

Correlation of sand bodies in an alluvial sequence is most easily made by associated coals or peats or by thin marine limestones, if these be present. If such associated markers are either not present or are not persistent, correlation of individual sandstone bodies may be virtually impossible. Elongation of alluvial sand bodies, truncated sections at their base, and differential compaction around them make alluvial sand bodies the most disruptive and "nonlayer-cake" elements of stratigraphic sequences.

Sand bodies of alluvial-fan origin are generally coarser grained than alluvial valley sands. Poorly-sorted gravel and even boulders are typical and sand may be locally subordinate, especially near the source. These materials are petrographically very immature; their composition is usually closed linked to bedrock. Size gradients are strong in the downcurrent direction, as determined from crossbedding dip, away from fan head. In many cases, size decline of pebbles and gravel can be approximated by Sternberg's law (Chap. 3). Lenticular bedding, cut-and-fill, and intrasand-body channels are common. Textural variation between samples is very high. In terms of geometry, alluvial fan sand bodies wedge out down dip and are probably composites (Fig. 11–13). By lateral coalescence they may form aprons many miles long parallel to a line source such as a rising fault block. Thickness of the resulting wedge, which may extend into the thousands of feet, is controlled by the subsidence of the down-thrown block. Bull (1968) published a good summary of the formative processes of modern alluvial fans. Stokes (1961, Fig. 14) interprets the Saltwash Sandstone Member of the Jurassic Morrison Formation in Arizona as a large alluvial fan (p. 173). Meckel (1967) similarly interprets the conglomerates of the Pottsville (Pennsylvanian) Formation of the central Appalachians.

Studies of ancient alluvial sands have been made by many, especially in Carboniferous Coal Measures throughout the world. Lee and others (1938, p. 12) early described alluvial sand bodies in the Pennsylvanian of north Texas. Zhemchuzhnikov and others (1959 and 1960) describe many examples from the Carboniferous of the Russian Donetz coal basin. A summary of the studies of the Paleozoic alluvial sands of Illinois Basin can be found in Friedman (1960) and Potter (1963). An excellent description of the geometry, sedimentary structures, and permeability of the Pennsylvanian Robinson Sandstone was published by Hewitt and Morgan (1965). They found maximum permeability to be parallel to the trend of the sand body. Doty and Hubert (1962) report on integrated studies of Pennsylvanian sandstones bodies in Missouri and Kansas. Schlee and Moench (1961) emphasize channel morphology, sedimentary structures and composition of the Jurassic "Jackpile" ore-bearing sandstone of New Mexico. Other helpful descriptions of sand body morphology by the U. S. Geological Survey in the Colorado Plateau can be found in Finch (1959, p. 134–145) and Witkind and Thaden (1963, p. 69–82). A comprehensive, outcrop-subsurface study by Palain (1966) describes a Triassic alluvial sandstone, the lateral equivalents of which include argillaceous and evaporitic sediments. Using both outcrop and subsurface data, he mapped a channel over 100 km long in northern France. In a classic early study, Rubey and Bass (1925, p. 54–62) outlined channel shape and related crossbedding direction to it in the Dakota Sandstone (Cretaceous) of eastern Kansas. They were the first to ever plot crossbedding orientation on a map. The alluvial Cut Bank Sandstone (Cretaceous) of Montana is explicitly described in terms of the model concept by Shelton (1967, p. 2443–2447). Riba and others (1967) mapped Tertiary channels in the Province of Zaragoza. They report good correlation between dip direction of crossbedding and channel trend. Although no reference is made to the model concept, Nanz's (1954) study of an Oligocene reservoir in the Texas subsurface contains all essential model elements and was well ahead of its time. Isolated pods of sand also can occur in subaerially cut channel systems at unconformities. Such bodies can be prolific reservoirs. Shiarella (1933) early reported such an occurrence.

Annotated References:

Allen, J. R. L.: A review of the origin and characteristics of Recent alluvial sediments. Sedimentology (Special Issue) **5**, 191 p. (1965).

Excellent integrated review emphasizing processes plus many references and informative illustrations. Table V lists ancient examples.

Bersier, A.: Séquences détritiques et divagations fluviales. Eclogae géol. Helvetiae **51**, 854–893 (1959).

A classic paper on cycles and alluvial sand bodies. Good diagrams of multistory sands.

Beerbower, J. R.: Cyclothems and cyclic depositional mechanisms in alluvial plain sedimentation. In: Merriam, D. F. (Ed.): Symposium on cyclic sedimentation, Vol. 1. Kansas Geol. Sur. **169**, 31–421 (1964).

Presents an alluvial-plain model with six subenvironments and relates them to cycles. Good summary tables.

Coleman, J. M.: Brahmaputra River: Channel processes and sedimentation. Sedimentary Geology (Special Issue), **3**, 122–239 (1969).

Hydrology, channel pattern, sedimentary structures and geologic history all very well done and written with ancient alluvial deposits in mind. Essential reading on river bed processes.

Fairbridge, R. W. (Ed.): The encyclopedia of geomorphology, 1295 p. New York: Rheinhold 1968.

See articles on alluvial fan, bajada, braided stream, flood plain, levee, and rivers.

Hewitt, C. H., Morgan, J. T.: The Fry *in situ* combustion test – reservoir characteristics. Jour. Petroleum Technology **17**, 337–353 (1965).

Short and full of relevant information. Good model to use when studying a sand body.

Hervieu, J.: Contribution à l'étude de l'alluvionnement en mileux tropical. Office Rech. Scientific Technique Outre-Mer, Mem. **24**, 465 p. (1968).

Modern alluvial sedimentation in Madagascar; in four parts and 16 chapters: part 1, geographic and environmental setting; part 2, climate, hydrology, and soils; part 3, sedimentation; part 4, budget of alluviation. Much data. Well illustrated.

Leopold, L. B., Maddock, Thomas, Jr.: The hydraulic geometry of stream channels and some physiographic implications. U.S. Geol. Survey Prof. Paper **252**, 57 p. (1953).

A widely cited analysis of the interrelationships between discharge and the four "dependent" variables of stream width, depth, mean velocity and suspended load. Implications for concept of the graded stream are discussed.

Levi, I. I.: Dinamika ruslovykh potokov (Dynamics of fluvial currents), 2nd Ed., 252 p. Moscow: State Pub. House 1957.

Ten chapters beginning with hydraulics and then focusing on the classification of rivers and river detritus, formation of stream channels, movement of detritus (both in suspension and by traction) and finally sedimentation in reservoirs and river reaches. Fifty-three references, all but eight in Russian.

McCowen, J. H., Garner, N. L.: Physiographic features and stratification types of coarse-grained point bars: modern and ancient examples. Sedimentology **14**, 77–111 (1970).

Authors compare point bars in some modern Louisiana and Texas rivers to some subsurface ones of the Texas Gulf coastal plain. Good illustrations.

Morisawa, Marie: Streams, their dynamics and morphology, 175 p. New York: McGraw-Hill 1968.

Moderately referenced, well illustrated, elementary treatment.

NEDECO: River studies and recommendations on improvement of Niger and Benue, 1000 p. Amsterdam: North-Holland 1959.

A six-part monograph covering all phases with Part III, the River (p. 169–666), of most relevance to geologists, especially the sections on physiography, transport of sediments, and channels. Essential reading for all those interested in integrated studies of large rivers, this one 2550 miles (4680 km) long.

Russell, R. J.: River plains and sea coasts, 173 p. Berkeley-Los Angeles: University of California Press 1967.

Four very readable and well-illustrated chapters written by a master physical geographer and a lifelong student of river and coastline morphology. Good for the beginner and useful to the advanced reader.

Rzhanitsyn, N. A.: Morphological and hydrological regularities of the structure of the river net (Translated from the 1960 Russian edition by Krimgold, D. B.), 380 p. Washington, U.S. Dept. Agriculture, Agriculture Res. Ser. 1969.

A Russian approach to river channel processes, morphology and stability. Clear treatment of stream orders. Many references, mostly Russian.

Schattner, Isaac: The lower Jordan valley. Scripta Hierosoloymitiana, Hebrew University **11**, 123 (1962).

Three parts devoted to geologic setting, physiography and valley floor of the Jordan River and the fourth part to its delta in the Dead Sea. Twenty high-quality photographs plus four plates of maps and longitudinal profiles. More geomorphology than sedimentology.

Shantser, Ye. V.: Alliuvii ravninnykh rek umerennogo poiasa i ego anachenie dlia poznaniia zakonomernostei stroeniia i formirovaniia alliuvial'nyk svit (Alluvium of the river plains of the temperate zone and its significance for knowledge of regularities of structure and formation of alluvial deposits). Akad. Nauk SSSR, Inst. Geol. Sci. **135**, Geol. Ser. **55**, 273 p. (1951).

Fourteen chapters covering most aspects of alluvium: its facies and normal cycle, geomorphic features of the flood plain, variants with climate and physiographic region, and its relation to terrace study and earth movements.

Sundborg, Åke: The River Klarälven – A study of fluvial processes. Geogr. Annaler **38**, 127–316 (1956).

A monographic treatment of the principles of motion and morphological activity of flowing water applied to a particular river by a member of the Uppsala school of fluvial geomorphologists. Good quantitative data on stratigraphy and composition of river sediments and on the river budget as a whole.

Zhemchuzhnikov, Ya. A.: Alliuvial'nge otlozheniia v uglenosnoi tolshche crednego Karbona Donbassa (Alluvial deposits in coal measures in the Middle Carboniferous of the Donetz Basin). Akad. Nauk SSSR, Trans. Inst. Geol. Sciences, No. 151 (Coal Ser. No. 5), 296 (1954).

Eight articles mostly on Carboniferous alluvial deposits in the Donetz Basin plus one on recent alluvium of the Don River and several others of general interest. Summary article by Zhemchuzhnikov gives general conclusions. Well illustrated.

Deltaic Environment

A delta is an area where deposition is dominant, proximal to and in a body of standing water at the place where a river system bifurcates and deposits its detritus. This bifurcating network of channels produces a distinctive pattern of sand deposition, one markedly different from that in the alluvial valley. Deltas range in size from a few square kilometers to giants that cover thousands of square kilometers. Some idea of the role that they can play in the ancient record can be obtained by noting that just the subaerial portions of the Ganges-Brahmaputra, the Niger, the Mississippi, the Lena, the Orinoco and the Nile, each exceed 20,000 km^2 (Coleman, 1968, Table 1). Moreover, many ancient basins have had a long history of delta filling. The major components of an ideal delta are shown in Fig. 11-17. Studies of modern deltas by Kolb and Van Lopik

(1958), Scruton (1960), Fisk (1961), Allen (1965b), Coleman and Gagliano (1964), Frazier (1967), and Coleman (1968) were sources for much of the following.

The processes of deposition at the mouth of the distributary produce a distinctive pattern of sand deposition – the *bar finger sands* (Fig. 11-18). These sands form the framework of the delta. They accumulate at the mouth of the distributary whereas finer silts and muds are carried in suspension and are

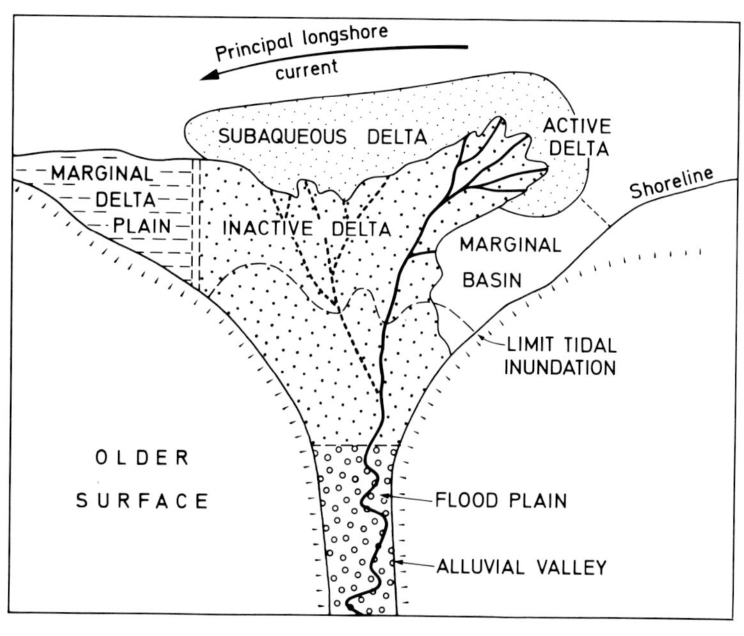

Fig. 11-17. Major physiographic components of a delta (Modified from Gagliano and McIntire, 1968, Fig. 1)

deposited farther offshore. Sand-silt deposition at a delta distributary mouth has been analyzed by a computer simulation technique (Bonham-Carter and Sutherland, 1968). The distributary acts as a straight flume or sluiceway discharging from the main stream. Bar-finger sands are elongate, have a gradational base, coarsen upward, have unidirectional, seaward-dipping cross-lamination in their upper part but multidirectional cross-lamination in their lower part, a product of the greater, directionally variable wave and current action of the open shelf. Bar-finger sands coarsen upward, because the maximum locus of deposition progrades forward. In the Mississippi delta, bar-finger sands are reported to reach 250 feet (75 m) in thickness (Fisk, 1961, p. 36), but are typically much less on the smaller deltas of stable cratons. Distinctive features of bar-finger sand bodies are their straightness and downdip bifurcation. Bifurcation occurs when the river crevasses through a natural levee to find a shorter course to the sea, usually during flood. Where a delta has a high proportion of mud, the cohesiveness of the muds into which the channel is cut may tend to keep the channel fairly straight even though the sand-fill is fluvial and fines upward. This may explain some ancient, low-sinuosity sand bodies that bifurcate downdip. Bar-finger sands may prograde downdip cutting into wave-transported marine shelf sands and in turn may be cut by channels filled with later and thicker alluvial sands. There is a regular progressive development of delta-finger sand bodies. Coleman and others (1964) have described the minor sedimentary structures in a prograding distributory.

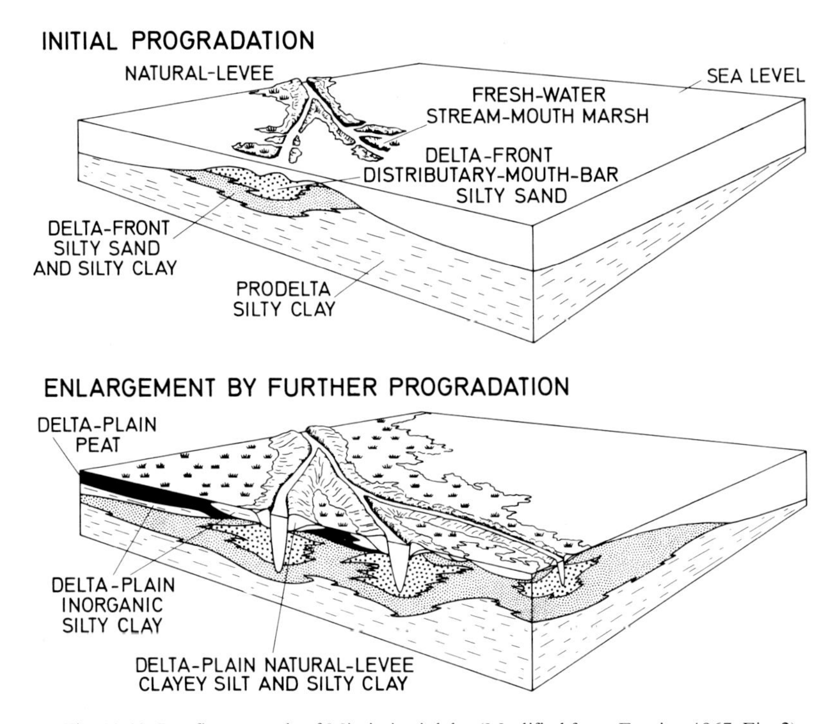

Fig. 11-18. Bar-finger sands of Mississippi delta (Modified from Frazier, 1967, Fig. 2)

In a large delta of a rapidly subsiding basin, abandonment of a distributary is followed by peat formation and subsequent encroachment of the sea. Associated lithologies are prodelta muds, peats, natural levee deposits, marine-shelf deposits (both sands and limestones) and alluvial deposits. Most deltas consist of successive lobes such as have been described in the Mississippi delta (Fig. 11-19). When a lobe is abandoned, it is subject to wave attack which will redistribute its sand and mud along the shelf. Much of the sand of nearby marine shelf and barrier islands is thought to be obtained from wave attack on abandoned delta lobes. These sands, unlike the bar-finger sands from which they were derived, will show greater maturity and perhaps marine fossils and glauconite. New heavy minerals introduced by long-shore currents may also be present. Thin marine bands of limestone, glauconite zones, and intensely bioturbated zones intercalated with delta deposits have been interpreted as marking destructional phases. Such bands may have value as local markers. Transgressive marine deposits tend to follow abandonment of a deltaic lobe so that the cyclicity of some deltaic sequences, wherein marine units are intercalated with deltaic and fluvial ones has been attributed to abandonment of successive delta lobes. With rising base level, younger deltaic lobes may regress landward in contrast to the majority of modern deltas which commonly transgress seaward. In any case, rate of forward growth is determined by sediment supply, depth of water, and rate of removal by longshore currents. Where a delta is marginal to a deep basin, turbidity currents may be generated on its lower subaqueous slopes – slopes that tend to be

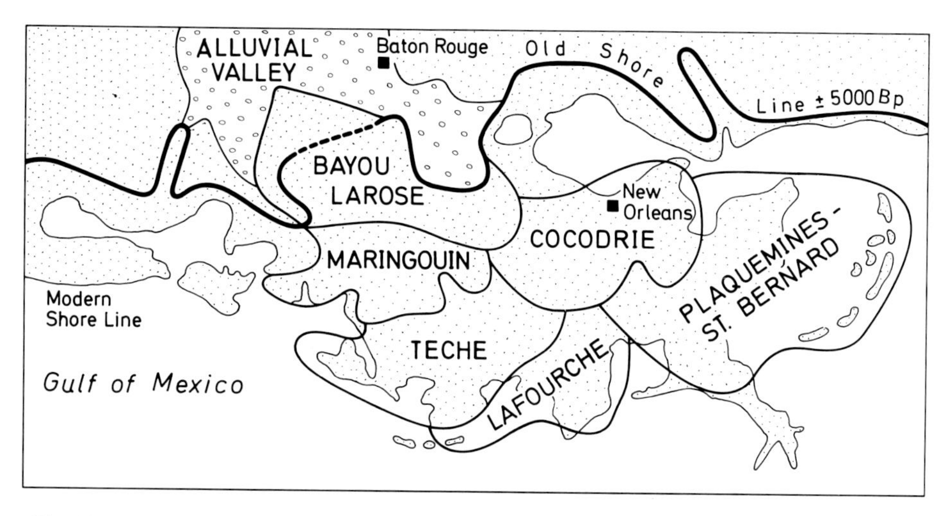

Fig. 11-19. Successive lobes in the step-by-step development of the Mississippi delta: 1) Bayou Larose, 2) Maringouin, 3) Cocodrie, 4) Teche, 5) LaFourche, 6) Plaquemines-St. Bernard, and 7) Modern (Modified from Scruton, 1960, Pl. 1)

mechanically unstable and thus subject to sliding because of the rapid sedimentation of fine sands, silts, and clays beyond the delta front. In some ancient basins, only the subaqueous portion of the delta may now be preserved.

The marine fauna associated with the modern Mississippi delta has been studied, the larger invertebrates by Parker (1956), the ostracods by Curtis (1960), and the foraminifers by Phleger (1955). Reyment (1969) investigated the ecology of the Niger delta.

Although deltaic deposits are mentioned rather frequently in studies of ancient basins, detailed descriptions emphasizing the pattern of sand distribution and the sand body geometry are not numerous. Shallow-water deltas of the craton are best known, probably because of greater availability of subsurface data. Busch (1959, Fig. 12) was one of the earliest to describe a cratonic delta in detail. He studied the Booch delta (Pennsylvanian) over a 2000 square mile (5200 km^2) area in the McAlester Basin of Oklahoma. Unusually clear delta patterns of sand distribution are displayed in the Chester (Mississippian) Palestine and Hardinsburg Sandstones (Fig. 11-20) of the Illinois Basin. The preserved parts of these two deltas cover some 7500 and 18,000 square miles (19,500 and 46,800 km^2) and are a product of a southwest-flowing river system, the Michigan River System, that supplied sands and muds to the Illinois Basin in late Paleozoic time. Both alluvial and deltaic sands are conspicuous products of this system. In the cyclic sedimentation of the Chester, intervening marine units become thicker down the paleoslope to the southwest. Brown and Wermund

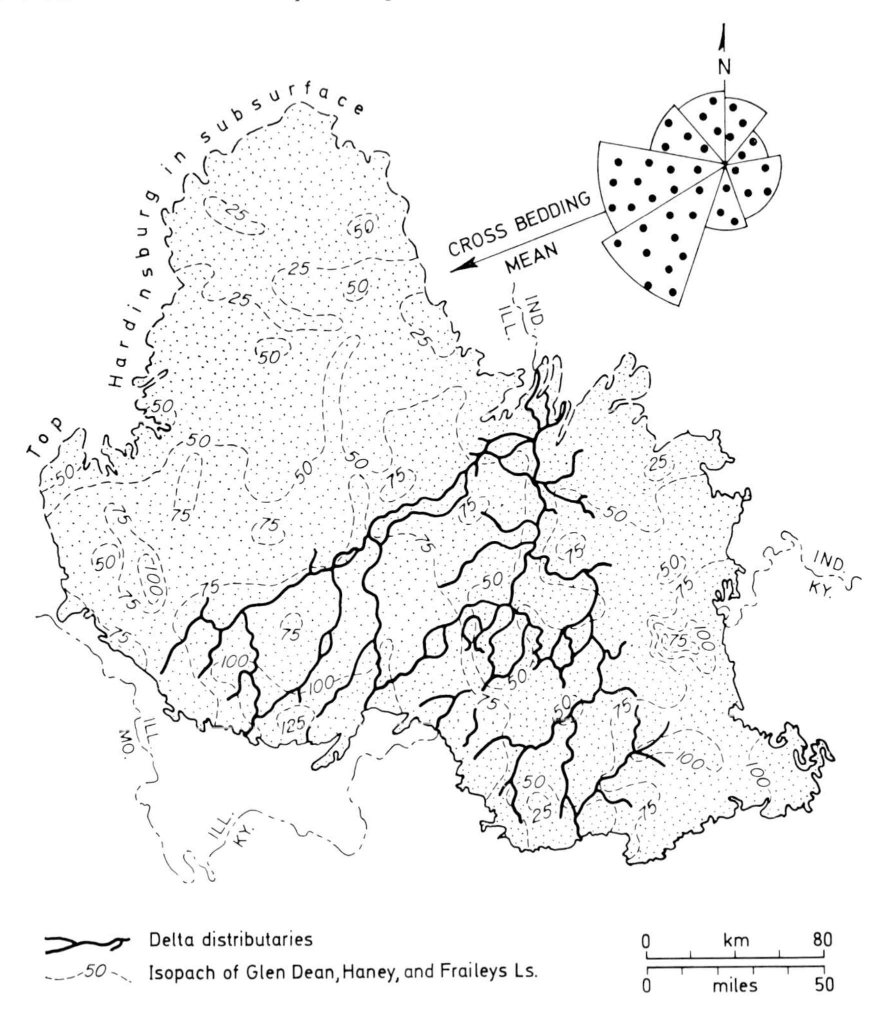

Fig. 11-20. Deltaic pattern of Mississippian sandstone, the Hardinsburg, in the Illinois Basin (Potter, 1963, Fig. 54)

(1969), utilizing both subsurface and outcrop data, have described the late Pennsylvanian shelf sediments of north central Texas. They make a detailed comparison of the Pennsylvanian deltaic deposits with the modern deltas of the Mississippi. Ferm and Williams (1965) collected outcrop and subsurface data on sand bodies and associated sediments in portions of the coal measure of the Appalachian Basin. In Germany, Wurster (1964) has taken advantage of dissection along more than 250 km of outcrop to describe what he interpreted as the deltaic pattern of the Schilfsandstein (Triassic). His many detailed maps show a strong correlation between crossbedding direction and sand-body elongation. Montadert and others (1965) have mapped outcrops of Devonian sand bodies in south Algeria that disclose a small-scale distributary pattern of delta fingers some of which are of the order of only 50 to 60 meters wide with thicknesses of 5 to 10 meters. Such detail is probably close to the lower limit of what would be recognizable from subsurface data. Rather different is the work

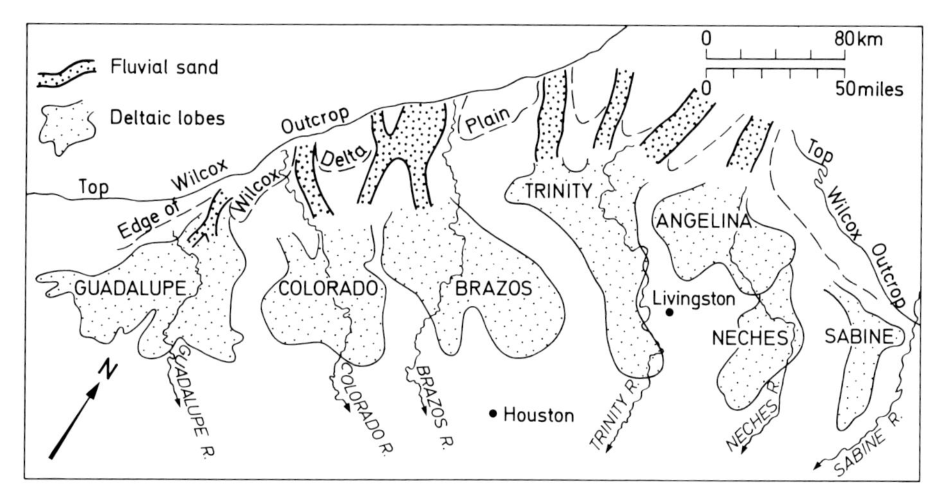

Fig. 11-21. Six Wilcox (Tertiary) deltas of the Gulf Coastal Plain in Texas and present-day rivers that occupy the same positions. A remarkable demonstration of continuity between present and ancient dispersal systems (Modified from Fisher and McGowan, 1967, Fig. 5)

of Dondanville (1963) who discriminated petrographically between marine shelf and deltaic sandstones in the Cretaceous Fall River Formation of Wyoming, where delta-finger sand bodies are as much as 80 feet (245 m) thick. The detrital mineralogy of the delta sands differed from that of the marine shelf sands. Obviously such a contrast, detectable only if the heavy mineral suite is sufficiently varied, might be useful in exploration. In its outline and grain size distribution, the glacial delta described by Baker (1967) is probably typical of small deltas that may form at the margins of shallow seas and lakes if coarse detritus be available.

Delta systems that were formed in unstable basins marginal to deep water have not generally been described, the study of the Wilcox Formation in a portion of Texas by Fisher and McGowen (1967) being a notable exception. Their study covered an area of some 40,000 square miles (104,000 km^2), and was based mainly on subsurface information. These authors infer an almost one-to-one correspondence between the subenvironments of the modern Mississippi and Wilcox delta system. As in modern deltas, fluvial and marine shelf-sand bodies are here intimately interbedded with deltaic ones. They recognized six delta centers each with two or more lobes, the position of the Wilcox lobes closely corresponding to position of present day southward flowing streams (Fig. 11-21). They attribute the stability of delta systems to the fact that they were deposited marginal to an unstable basin in which subsidence occurred in response to sediment load. This led them to the important generalization (p. 122) that on stable cratons delta systems prograde extensively so that vertical sections display many different lithologic units, whereas delta systems in rapidly subsiding basins do not prograde appreciably and thus have facies that tend to persist vertically.

Perhaps one of the most famous examples of a presumed ancient delta complex is that of the Devonian Catskill Formation of eastern New York State, originally described in great detail by Barrell (1913 and 1914). For many years the Catskill-type of delta was used as a universal model but more recently, as a result of restudy (Frakes, 1967; Allen and Friend, 1968), it has come to represent only one important type, that of the "tectonic delta complex" (Friedman and Johnson, 1966, p. 185–186). Such a complex forms as a thick sequence marginal to a mountain front where the delta develops at the edge of piedmont alluvial fans rather than on broad coastal plains. The tectonic delta complex commonly includes coarse sandstones, conglomerates, and much higher sand-shale ratios than its coastal plain, large river counterpart. Both the tectonic delta complex and the alluvial fan can be thought of as clastic wedges that, unlike other deltas, thicken and coarsen towards their source.

The typical continental facies of the Catskill Formation includes much coarse sandstones and conglomerate, the composition of which indicates an eastward derivation from the Taconic tectonic highlands. The sandstones everywhere show trough crossbedding, troughs ranging in width from a foot (30 cm) to 30 feet (9 m). The sandstones, lithic arenites, are parts of cycles including gray to green silty shale, red shale, and sandstone. Over a relatively narrow lateral transitional zone, this facies gives way to a marine sequence much more dominated by fine clastics; here the sandstones, the so-called Portage facies, are turbidites. The transitional zone is characterized by a cyclical alternation of marine and nonmarine sediments, principally shales. These sediments deposited along a prograding muddy shoreline are described by Walker and Harms (1970). These shoreline deposits are unusual in that sand is minimal. In the Eocene Tyee Formation of Oregon, the transition from delta-fluvial or shallow shelf to well-defined turbidite sandstones has been mapped and studied by Dott (1966) and Lovell (1969). There seems to be no definite zone of shallow marine shelf intervening between the fluvial deltaic and the slope-turbidite environments.

Annotated References:

Axelsson, Valter: The Laiture delta, a study of delatic morphology and processes. Geografiska Annaler **49** A, 127 p. (1967).

The first 39 pages are devoted to general information about deltas followed by two chapters on experimental studies mostly concerning bifurcation. The remainder describes the delta itself in great detail. Good illustrations and many references.

Barrell, Joseph: Criteria for the recognition of ancient delta deposits. Geol. Soc. America Bull. **23**, 377–446 (1912).

Classic illustration of deductive logic in sedimentation.

Bates, C. C.: Rational theory of delta formation. Am. Assoc. Petroleum Geologists Bull. **37**, 2119–2162 (1953).

Applies jet theory to different types of deltas and especially the formation of elongate bay-mouth bars. A good starting point for a complicated subject.

Bernard, H. A., Leblanc, R. F.: Resumé of the Quaternary geology of the northwestern Gulf of Mexico province. In: Wright, H. E., Jr., Frey, D. G. (Eds.): The Quaternary of the United States, Part I, Geology, p. 157–185. Princeton: Princeton Univ. Press 1965.

Well-illustrated, concise description of 34 geomorphic units of the coastal plain emphasizing their importance for sedimentation. Many subsurface cross sections.

Fisher, W. L., Brown, L. F., Jr., Scott, A. J., McGowen, J. H.: Delta systems in the exploration for oil and gas: A Research Colloquium: Univ. Texas (Austin), Bur. Econ. Geol., 78 p., 178 figs. (1969).

An outstanding process-oriented summary that is particularly commendable for its scope, illustrations and references (some 240). Highly recommended.

Kolb, C. R., Van Lopik, J. R.: Geology of the Mississippi River deltaic plain southeastern Louisiana: U.S. Army Eng. Waterway Expt. Sta., Tech. Rept. No. **3–483**, 1 and 2, 120 p. (1958).

Contains much of value and is very well illustrated. Section on subsidence in a large delta particularly good. Many references.

Mackay, J. R.: The Mackenzie delta area, N.W.T. Canadian Dept. Mines Tech. Surveys, Geog. Branch, Mem. **8**, 202 p. (1963).

Excellent description of physical geography of an arctic delta.

Morgan, J. P.: Deltas – a resumé. Jour. Geol. Education **18**, 107–117 (1970).

Short summary with most of the relevant references. Table 1 nicely summarizes the role of rivers, coastal processes, climatic factors and structural behavior of basin on deltaic sedimentation.

— Ed.: Deltaic sedimentation, modern and ancient. Soc. Econ. Paleon. Mineralogists, Sp. Pub. **15**, 312 p. (1970).

Nine papers on modern deltas and five on ancient. Well illustrated and informative.

Research Committee: Delta symposium: Am. Assoc. Petroleum Geologists Bull. **55**, pp. 1135–1280 (1971).

Eight papers and a forward emphasize the processes, paleogeography, petroleum content and how to find it.

Shirley, M. L., Ragsdale, J. A., Eds.: Deltas. Houston: Houston Geol. Soc. 251 p. (1966).

Eleven papers, all but three of which are on old and new deltas in the United States. Appendix has 24 maps of modern deltas from around the world, all drawn to same scale.

Tricart, J.: Notice explicative de la carte géomorphologique du delta du Sénégal. Mém. Bur. Recherches Géol. Minières **8**, 136 p. (1958).

Describes Quaternary geomorphology mapped on three 1 : 100,000 maps of the Senegal delta where six stratigraphic units (each with six environmental-lithologic units) are distinguished. Some petrologic data. Good bibliography.

UNESCO: Scientific problems of the humid tropical zone deltas and their implications, Proceedings of the Dacca Symposium, 1964, jointly organized by the Goverment of Pakistan: Paris: UNESCO, 422 p. (1966).

Fifty-three short papers arranged into six parts: Parts I (Geomorphology, sedimentation and petrology), II (Hydrography and hydrology), VI (Descriptive studies) and VII (Classification of deltas) are of most interest to geologists. A good source of additional references.

Welder, F. A.: Processes of deltaic sedimentation in the Lower Mississippi River: Louisiana State Univ., Coastal Studies Inst. Tech. Rept. **12**, 90 p. (1959).

Good general description with emphasis on crevassing and bifurcation of delta distributaries. Well illustrated and referenced.

Zhemchuzhnikov, Yu., A., and others: Stroenie i usloviia nakopleniia asnovnykh uglenosykh svit i ugolnykh plastov arednego karvona Donetskogo Basseina chast'pervaia (Structure and conditions of accumulation of the principal coal-bearing formations and coal beds of the Middle Carboniferous of the Donetz Basin). Akad. Nauk SSSR, Trans. Geol. Inst., No. 15, Pt. 1, 332 p. and Pt. 2, 347 p. (1959 and 1960).

Probably one of the most comprehensive studies of a coal-bearing interval ever made. Many tables and illustrations plus 12 plates including interesting maps showing recurring deltas. Well referenced.

Estuaries and Tidal Flats

Estuaries: An estuary may be defined as that portion of a river mouth which is subject to tides and where river water is significantly diluted with seawater. Narrow marine shelves and coastal areas with moderate to high relief favor formation of estuaries in drowned valleys, which may be small or have dimensions comparable to those of the Congo and Amazon Rivers. Ratio of sand to mud in an estuary is controlled by the relative strength of wave-generated, tidal, and river currents and availability of mud versus sand. But as a rule, mud is more prevalent than sand. Principal sources of information for our review of estuaries include van Straaten (1954a), Robinson (1956), Reineck (1963), and Postma (1967).

A special earmark of estuaries where tidal activity is small is the invasion by denser seawater, the so-called salt-water wedge (Fig. 11-22). The inshore limit of the salt wedge is commonly marked by a bar, which defines the inshore limit of marine sand and mud invasion. Salt-water invasion also makes possible a predominantly marine fauna, one that becomes progressively less diverse up estuary as salinity decreases. With a few exceptions, all major groups of marine organisms can be found at least in the more saline portions of estuaries.

Depending on where one is in the estuary, the relative magnitudes of the flood (landward) or ebb (seaward) tide, transport in the estuary may be inshore and landward, may be reversing, or may be seaward. The ebb tide is generally the stronger of the two reversing currents. Inshore transport of marine sand is assisted by tides and wave-generated currents in the estuary. Tidal currents rather commonly attain velocities of 1 knot (51.5 cm/sec) and may appreciably exceed that (Table 9-4) so that tidal velocities are more than adequate to transport sand, partly because wave action initially puts the sand into suspension. An unusually wide range of sand waves from small to giant ripples, the latter with wave lengths of 10 to 30 m or more, form in response to such tidal currents. Consequently, crossbedding and many types of ripple mark are the predominant sedimentary structures of tidal sand bodies, although virtually all structures have been

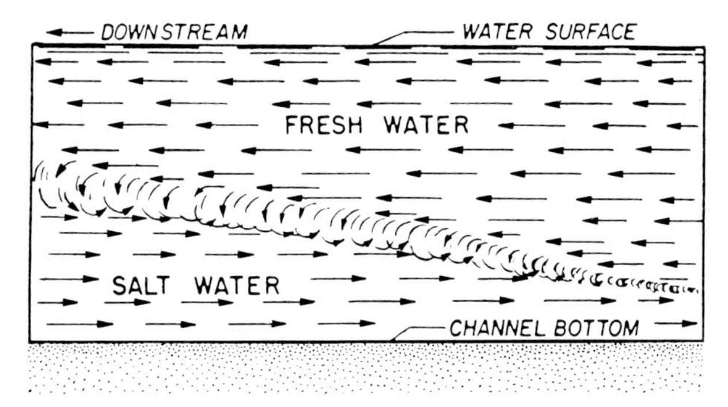

Fig. 11-22. Schematic drawing of well-defined salt-water wedge (Wiegel, 1964, Fig. 13-18). Published by permission of Prentice-Hall, Inc., Englewood Cliffs, New Jersey

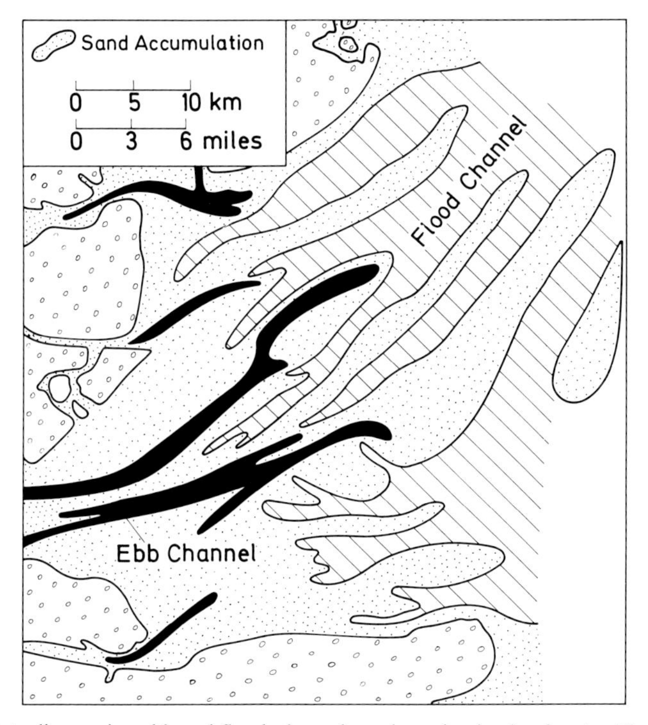

Fig. 11-23. Mutually evasive ebb-and-flood channels and sandy shoals of outer Thames Estuary, England (Modified from Robinson, 1956, Fig. 4)

reported. Zhemchuzhnikov (1926, p. 60–63) was one of the first to suggest that bimodal crossbedding characterized tidal sand bodies, the two modal directions being 180 degrees apart and formed by the reversing tidal flow (Fig. 9-21).

Where sand prevails in an estuary, a well-defined system of mutually evasive, shifting, ebb-and-flow channels is present (Fig. 11-23). This system is sometimes referred to as a *tidal delta*. Submerged point-bars are believed to be an essential feature of these channels so that tidal sands fine upward (Fig. 11-24). Like fluvial sands, tidal ones have a pronounced internal organization. In sandy estuaries, the sand bodies can be expected to be closely parallel to the estuarine axis as are the internal directional properties. Little information is available about sand body dimensions, although the thickness of single-cycle sands probably does not exceed depth of scour of the ebb-and-flow channel system. Frequent shifting of these channels in the estuary probably produces many multistory sand bodies. In response to scour, their basal contact is sharp and disconformable and may have a lag concentrate of skeletal debris and/or gravel. Studies of modern

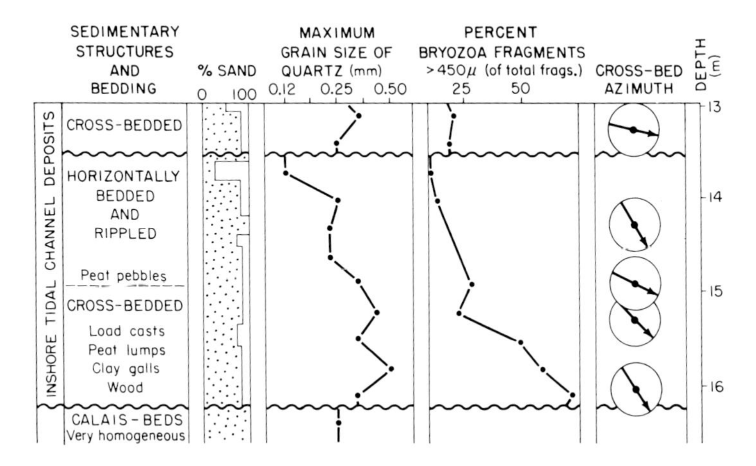

Fig. 11-24. Vertical profile of inshore tidal and body in Rhine-Maas estuary (Modified from Noorthoorn van der Kruijff and Lagaaij, 1960, Figs. 3 and 5). Reproduced by permission of the *Bulletin* of the American Association of Petroleum Geologists

estuarine sand bodies, unlike those of alluvial and deltaic sands, are scant, so that much of what we believe is inferred – not seen. However, use of continuous acoustic profiling should greatly enhance our knowledge of tidal sand body form and structure. Using such methods, Swift and McMullen (1968) mapped two tidal sand bodies each up to 30 m thick and 30 km long in the Minas Basin of Nova Scotia. Tidal sand bodies, either as lunate bars, tidal deltas and channel deposits may also be present between barrier islands. Hoyt and Henry (1965) describe sand deposition in tidal inlets between barrier islands along the Georgia coast. As in alluvial deposits, depth of tidal scour between barrier islands controls sand body thickness, which approaches 100 feet (30 m). Longshore drift introduces sand to the channel so that lateral accretion takes place in the direction of longshore drift (Fig. 11-25). Crossbedding, perpendicular to strandline, is related to sand waves that move up and down the channel. Because of the depth of sedimentation in such inlets, Hoyt and Henry suggest (p. 193) that the tidal deposit is much more likely to be preserved than the barrier island itself.

The Dutch have made some of the best observations of modern tidal sand bodies (Noorthhoorn van der Kruijff and Lagaaij, 1960; Oomkens and Terwindt, 1960; Terwindt and others, 1963). Point bars in a meandering estuary have also been described by Land and Hoyt (1966). Table 11-4 summarizes essential characteristics.

With stable base level, the estuary will be gradually filled so that fluvial deposits may cap it. In contrast, further submergence will lead to an overlying littoral and marine sequence possibly including barrier islands. Correlation in such a complex is not easy. Provided mineralogy is not too restricted in kind, careful petrographic mapping may delineate better sorted inshore marine and

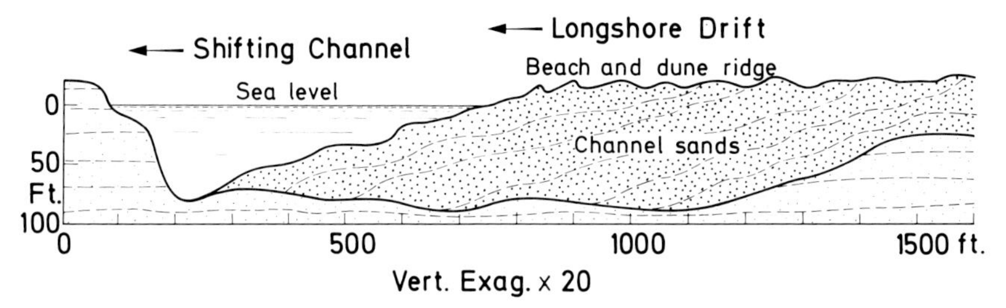

Fig. 11-25. Idealized cross section parallel to shoreline showing portion of barrier island and inclined, lateral accretion bedding in lee of shifting inlet (Modified from Hoyt and Henry, 1965, Fig. 4)

Table 11-4. *Characteristics of tidal sands*[a]

PETROLOGY

Detrital. Argillaceous rock fragments, skeletal debris and collophane plus argillaceous material. Possibly some glauconite and authigenic feldspar. Detrital and chemical cements.

TEXTURE

Fair sorting, moderate to high grain-matrix ratio. Possibly shell conglomerate at base. Very similar to texture of alluvial sandstone. Peat, clay galls, and wood common.

SEDIMENTARY STRUCTURES

Asymmetrical and symmetrical ripple marks common. Abundant crossbedding of variable thickness; commonly bimodal but may dip seaward or landward. Lenticular bedding dominant. Tracks, trails, and burrows abundant. Channels and washouts.

INTERNAL ORGANIZATION

Good asymmetry. Strong vertical decrease in grain size and bedding thickness, possibly with some conglomerate at base.

SIZE, SHAPE, AND ORIENTATION

A few tens of feet to more than 1,000 feet wide, mostly very elongate. Long axis at right angles to shoreline or parallel with estuarine axis. Straight to moderately meandering, dendritic patterns, the latter as tidal inlets. Also lunate bars in passes between barrier islands. Crossbedding parallel with elongation; principal mode may point seaward as well as landward in estuaries.

ASSOCIATED LITHOLOGIC TYPES

Vertical: variable according to regressive or transgressive origin. Associated with interlaminated shale and sandstone of tidal origin and marine sediments. Disconformable basal contact. Lateral: interbedded siltstone and shale of tidal-flat origin, commonly with mollusks, worms, crustaceans, and possibly algae as well as marine sediments.

[a] From Potter, 1967, Table III. Reproduced by *Bulletin* of the American Association of Petroleum Geologists.

tidal sands of coastal origin from river sands (Kulm and Borne, 1966, Fig. 9). Marine fauna, skeletal debris, glauconite, and bimodal crossbedding and ripple mark are other features distinguishing tidal from alluvial sands.

Tidal flats: On low-lying coasts, tidal flats may be conspicuous behind barrier islands. On the majority of such tidal flats, mud is dominant with sand, if present at all, being restricted to creeks and inlets of the intertidal zone and salt marshes. The thickness of these sand accumulations is but a small fraction of that of the

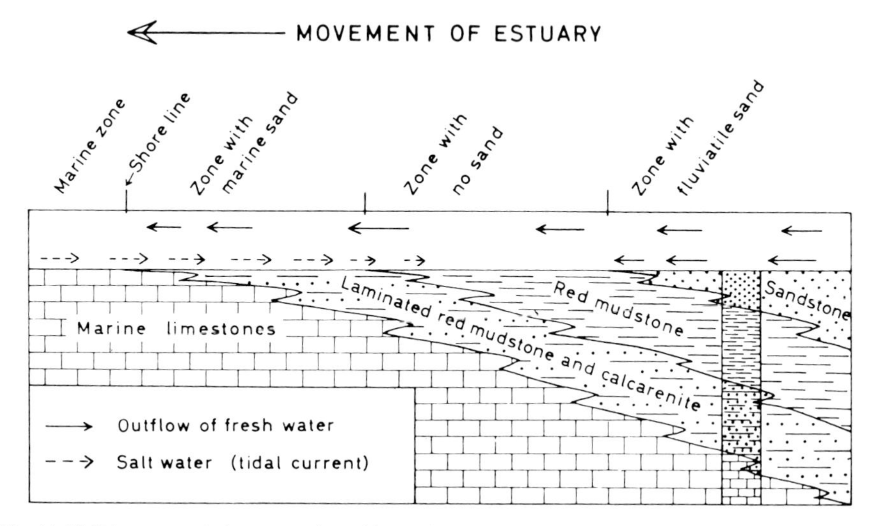

Fig. 11-26. Diagrammatic interpretation of formation of estuarine beds at base of Ringerike Formation at Kolsas, Norway (Spjeldnaes, 1966, Fig. 5)

sand bodies of large estuaries. Thin but abundant crossbedding and ripple mark are their dominant structures. Lateral accretion bedding in small scour channels is also reported (van Straaten, 1954a, Fig. 4). Gravel, mud pebbles, together with cobbles and boulders of peat, may be present in the tidal channels. In the interchannel and marsh areas of the higher portions of the tidal flat, sand is much less prevalent and, if present at all, is interlaminated with mud. Disruption of stratification by burrowing and roots is appreciable in such areas. Characteristics of tidal channel, tidal flat, and marsh environments of the Netherlands are well summarized by van Straaten (1954a, p. 88–93). But where tidal range is very large and velocities are strong, sand may be the dominant sediment of the tidal flat. Along arid coasts, gypsum, dolomite, and Mg-calcite are conspicuous in supratidal flat deposits, especially behind coastal barriers.

Ancient analogues: Detailed studies of ancient estuarine and tidal flat sand bodies are scant. An exception is Selley's (1967) work on beds of Miocene age in Libya, where fluvial and deltaic sands can be traced into marine shallow shelf carbonates. Earlier, van Straaten (1954b) compared a Devonian sand in Belgium with that of the modern Wadden Sea. Spjeldnaes (1966) using sedimentary structures, facies relations and fossil data, believed that the Ringerike Formation (Silurian) of Norway contained estuarine deposits (Fig. 11-26).

Annotated References:

Bajard, J.: Figures et structures sédimentaires dans la zone intertidale de la partie orientale de la Baie du Mont-Saint-Michel. Rev. Géog. Phys. Géol. Dynamique **8**, 39–111 (1966).	One hundred and sixty illustrations, mostly high-quality photographs, of the physical and biogenic structures of a famous tidal flat.
Brunn, P., Gerritsen, F.: Stability of coastal inlets, 123 p. Amsterdam: North Holland 1960.	Six chapters: introduction, natural inlet regimen, previous theories, discussion of pertinent factors affecting stability, actual data, and design of tidal inlets. Moderate mathematics. Good starting point for digging deeper.
Coastal Research Group. Coastal environments of northeastern Massachusetts and New Hampshire, Fieldtrip Guide Book, May 9–11 (1969). Univ. Massachusetts, Geol. Dept., Contr. No. **1**, 462 p.	Fifteen well-illustrated articles on estuaries, beaches, offshore bars and nearshore hydrography. Many high quality photographs and maps. Two glossaries. Twenty field trip stops described. Good source of information on modern tidal sand bodies and their directional structures.
Dillo, H. G.: Sandwanderung in Tideflussen. Mitt. Franzus-Inst. für Grund- und Wasserbau der Tech. Hochschule Hannover **17**, 136–253 (1960).	Observation of sand and suspension transport in three German estuaries plus some experimental work. Uses bedload functions and regime theory. Discusses sandwaves.
Evans, Graham: Intertidal flat sediments and their environments of deposition in the Wash. Geol. Soc. London, Quart. Jour. **121**, 209–245 (1965).	Good description and integration with emphasis on organisms.
Gripp, K.: Das Watt; Begriff, Begrenzung und fossile Vorkommen. Senckenbergiana Leth. **37**, 149–181 (1956).	Defines and describes briefly the German North Sea Watt, attempting to distinguish it from delta, estuarine and shallow shelf deposits.
Guilcher, A.: Estuaries, deltas, shelf, slope. In: Hill, M. N. (Ed.): The Sea, Vol. 3, p. 620–654. New York: Interscience 1963.	Well-illustrated and integrated treatment. Many European references. Recommended for a concise overview and good bibliography.
Lankford, R., Rogers, J. J. W. (Eds.): Holocene geology of Galveston Bay area. Houston: Houston Geol. Soc., 141 p. (1969).	Seven chapters and a summary cover the bay: sedimentology of estuarine deposits, delta of Trinity River and its suspended load, comparison with Trinity River terraces, and biological aspects.
Lauff, G. H. (Ed.): Estuaries. Washington: Amer. Assoc. Adv. Sci., Pub. **83**, 757 p. (1967).	Ten sections and 68 short papers. Among the more relevant parts are Physical Factors (II), Geomorphology (III), and Sediments and Sedimentation (IV), the last containing 13 papers. Part IV is probably the best single place to find a modern summary of estuarine sedimentary processes. J. W. Hedgpeth's concluding paper is outstanding.
Pimienta, Jean: Le cycle Pliocène-Actuel dans les bassins paraliques de Tunis. Mém. Soc. Géol. France **85**, 176 p. (1959).	Description, in six chapters, of fluvial, deltaic, tidal flat, lagoonal and marine sedimentation Pliocene to Recent. Cycles, diagenesis, and Quaternary tectonics. Large bibliography.
Reineck, Hans-Erich: Sedimentgefüge im Bereich der südlichen Nordsee. Abh. Senckenberg naturf. Gesell., No. **505**, 138 p. (1963).	An outstanding, well illustrated (52 figures and 15 maps) study of sand in estuaries. Much of relevance for understanding ancient deposits.

van Straaten, L. M. J. U.: Composition and structure of Recent marine sediments in the Netherlands. Leidse geol. Mededeel. **19**, 1–110 (1954).
A well-illustrated, informative classic stressing lithology, sedimentary structures, and biota but also including some petrology. Chapter 10 summarizes results in outline form.

Thompson, R. W.: Tidal flat sedimentation on the Colorado River delta, northwestern Gulf of California. Geol. Soc. America Mem. **107**, 133 p. (1968).
Describes tidal sedimentation in an arid zone delta; many illustrations.

Beaches and Barriers

Where preserved as distinct deposits, beaches and barrier islands are strikingly elongate and have distinctive characteristics. They may be sources of high-silica sand or sites of economically important heavy mineral placers. If buried, they may be excellent reservoirs for fluids. They are also worthy of study because it seems to us that most ancient marine shelf sands, and probably many fluvial and deltaic sands as well, consist of grains that have, in earlier cycles, passed through the beach or barrier island environment. We suggest, moreover, that many of the ancient, widespread sheet sands that occur on cratons may be composites formed by migrating barrier islands and beaches. Thus even though beach and barrier island deposits – shoreline sandstones as they are sometimes called – are not widely reported in the geologic record, they fully deserve our attention. We emphasize here the beach as a linear sand generator and as the key formative feature of barrier islands.

Modern beaches: Sandy beaches may range from small deposits of a few hundred feet (30–100 m) in length in protected coves along rocky mountainous coasts to straight or gently curving bodies that may have lengths of 10 to more than 100 miles (16–160 km) along the shores of low lying coastal plains with gently sloping (less than 0.01) shallow shelves. Along an embayed, drowned coastline, beaches form between headlands and are concave seaward. Beach deposits may range from a few to a few tens of feet (1–10 m) in thickness. Along the Gulf Coast in Louisiana, Hoyt (1969, p. 300) reports that abandoned raised beaches, or *cheniers* as they are called locally, rarely exceed 20 feet (7 m) in thickness and have widths of 150 to 1500 feet (47–475 m). Such dimensions appear to be fairly typical. Byrne and others (1959) and Gould and McFarlan (1959) describe, with cross sections, cheniers and their lithologic associations on the Gulf Coast. Abandoned beaches are fairly common on the low lying coastal plains of the world where they form systems of beach ridges or cheniers. The foreshore or beach face (Fig. 9-23) dips seaward at a low angle generally less than 10 degrees; many are less than 5 degrees. For a given grain size, the stronger the wave action the steeper the seaward slope (Fig. 11-27). Such long, regular, gently sloping, seaward-dipping bedding is a distinctive earmark of beach deposits and has been termed *beach accretion bedding* (Ball, 1967, Fig. 36). Thin laminations of heavy minerals help define these seaward dipping beds, the laminations being finer, better sorted, and darker than the ordinary sand. In beach laminations the long axes of the grains tends to be perpendicular to the shoreline. The seaward-dipping bedding planes may be well rippled or be smoothly laminated by the swash. Varieties of ripples are many. Other sedimentary structures include crossbedding formed by small sand waves migrating parallel and or obliquely to the strand line (van Straaten, 1953, Fig. 7) and small, seaward-dipping channels. Thompson (1937, p. 723–745) and McKee (1957, p. 1706–1718) give additional information on the structures of beaches and Werner (1963) and Nachtigal (1968) describe the morphology and sedimentary structures of submerged longshore bars closely associated with the beach. Werner (Pls. 1 and 2) reports low angle bedding, dipping both shore- and seaward, as well as crossbedding in these bars.

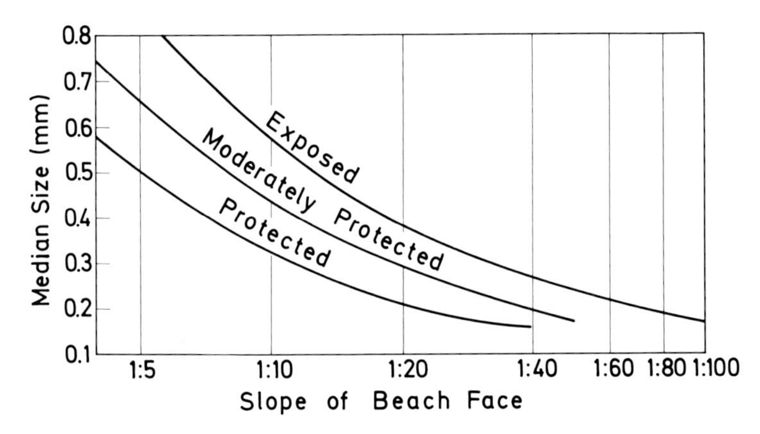

Fig. 11-27. Size-slope relationships of beaches. Data collected at midpoint between mean lower low water and mean higher high water (Modified from Wiegel, 1964, Fig. 14-17)

The chief source of energy for erosion, transportation, and deposition of sand along the strand are wind-generated waves and the secondary currents that they induce (Chap. 9). Direction of approaching waves, wave height and length, and configuration of the sea floor are critical in determining how wave action affects the shoreline. Sand may be transported both inward and parallel to the beach by wave action; during storms sand may be carried to deeper water, later to gradually return to the beach. Where the surf forms an angle (usually less than 20 degrees) with the beach, the momentum of the breaking wave has a component parallel to the shore and generates a longshore current, which imparts the striking linearity to beaches and their associated deposits. Where water depth is 20 meters or less, shoaling waves can put much sand into suspension and thus permit its transport by relatively weak longshore currents. Zig-zag movement in the swash zone also contributes to sand transport along the beach. McKee and Sterrett (1961) used a 46-foot (14 m) wave tank to simulate form and internal structure of beaches and associated near-shore topography.

Sand may be supplied to the beach by rivers, by wave erosion of unconsolidated sediments along the shore, by eolian transport from the interior and, in some cases, by shoreward transport of sand reworked from the shelf. Sand is exported from the beach either landward by dunes, or longshore drift or to the deeper part of the shelf. If supply locally exceeds export, the beach progrades seaward; if supply is less than export, the beach regrades. Climatic or other changes that reduce sand supply from contributing rivers will ultimately trigger beach attrition.

Calcareous debris is present on most low-latitude, modern beaches and locally may predominate. The preservation of calcareous skeletal debris on a beach is largely a function of the organic productivity of shoal waters and the degree of supersaturation of those waters with respect to calcite or aragonite. Supersaturation of both is strongly temperature dependent.

Some petrographers, working almost entirely with ancient rather than modern sands, have concluded that repeated back and forth movement of sand in the surf zone is the prime aqueous environment for rounding of grains, elimination of argillaceous and other soft rock fragments, and enhancement of framework sorting. More studies of modern beach sands and of sand in associated contributing environments are needed to test this idea. Folk and Robles (1964, p. 290) suggested that $\sigma = 0.3$ to $0.6\,\phi$ is the limiting value of sorting for beach sands. Owing to the uniformity of hydraulic processes on the beach, shore sands are remarkably homogeneous. Because clay and silt are largely removed, they can form no matrix to bind the framework so that later induration of the sands depends on chemical cements.

Modern barrier islands: The term "barrier-island complex" (Hayes and Scott, 1964, p. 238) includes sand bodies that formed as submerged or offshore bars, as beaches and spits, and as dunes. The latter are commonly nourished by sand blown from an associated beach. These complexes are what Shepard (1960) called "Gulf Coast barriers". In essence, they are elongate sand bodies parallel to depositional strike that lie in the zone that separates marine from nonmarine deposits.

The dune and water-transported sands of the barrier differ in their sedimentary structures. The beach sands have both conspicuous lamination and lineation and typically display weakly-inclined, seaward-dipping bedding. Crossbedding occurs in the water-laid sands and dips either parallel or oblique to strandline. Ripples are conspicuous and of many varieties. By contrast, crossbedding of a complex variety is the dominant structure in the associated dunes and is likely to be of a much larger scale than its waterlaid equivalents. Variability of dip direction may be appreciable depending upon summer and winter wind regimes. Any general relation between mean dip direction of eolian crossbedding and direction of elongation of the barrier island seems most unlikely. Unusually steep crossbed dip angles have been reported in some coastal dunes (McBride and Hayes, 1962, Fig. 7 and Mackenzie, 1964, p. 55). Both Land (1964) and Logvinenko and Remizov (1964) also report on the orientation of eolian and aqueous crossbedding in barrier islands. Ball (1967) shows many informative block diagrams of the sedimentary structures of coastal barriers composed of skeletal carbonate and oolites in Florida and the Bahamas. Horn (1965) gave a good description of a Baltic barrier. Very probably many quartz sand bodies of the littoral zone have similar characteristics. Burrows and trails are abundant in the aqueous part of the barrier as are marine organisms, which tend to be the more robust forms. Weimer and Hoyt (1964) suggest one burrow type in particular, *Callianassa major* Say, as an indicator of littoral and shallow neritic environments. In summary, the best general means of distinguishing dune from subaqueous sands in barriers appears to be a combination of sedimentary and biogenic structures and fossil evidence. Petrologic and textural discrimination has been attempted but with little success.

Barrier islands commonly have a complex sequential development (Fig. 11-28) which implies either an equilibrium between land and sea or, more commonly, at least a temporary regression of the sea. Vertical increase in grain size is the result of progressive migration of the surf zone seaward so that coarse sands accumulate over finer, deeper water sands and silts (Fig. 11-29). The basal sand passes downward into mud. Modern barriers may be up to 200 feet (62 m) thick, up to 6 miles (10 km) wide, and many miles long. Their elongation is parallel to the strand line and they are fairly straight to gently curving. In a longitudinal direction their thickness is fairly uniform, most of the variation probably stemming from variable dune height. After formation and prior to burial, however, marine planation may equalize thickness along the barrier. Laterally, barriers may coalesce just as do cheniers. The dip direction of crossbedding foresets, formed by longshore drift, is parallel or subparallel to the elongation; beach accretion bedding dips gently seaward perpendicular to long dimension of the barrier. Washover or spillover lobes of sand, formed during storms that break the barrier, may extend into the lagoon behind it. Later these breaks may become enlarged tidal inlets. The landward side of the barrier may be more irregular – as the result of spillover or washover lobes (Fig. 11-30) – than the seaward side. Marshes, tidal flats, or lagoons, whose salinity may vary from brackish to hypersaline depending on ratio of evaporation to fresh water supply, may lie landward of the barrier. Hence one should expect maximum biotic contrast on opposite sides of the barrier – marine fossils and possibly glauconite seaward in contrast to a brackish or continental biota landward. Swamp deposits, subsequently altered to coals, may occur behind the barrier whereas in arid and semiarid climates, a hypersaline lagoon may exist and produce salt deposits. Rusnak (1960) and Nichols (1964) summarize features of the lagoonal deposits behind the barriers. Eventually fluvial sediments may fill in the lagoon or overlap the barrier. Or the barrier may be largely destroyed by rising sea level so that only portions of it remain or perhaps only tidal sands deposited in the deep tidal channels between barriers (Hoyt and Henry, 1965) will be preserved. Barriers form best on gently shoaling coasts with low to moderate tidal range. Too high a tidal range inhibits their formation. Theories of their origin recently have been summarized by Hoyt (1967). Table 11-5 gives the essential characteristics.

Corbeille (1962) describes the slightly buried New Orleans barrier island, consisting of some 35 feet (11 m) of coarsening-upward, clean, well-sorted sand. It is 3 miles (5 km) wide and at least 10 miles (16 km) long. Shepard (1960) has written a good descriptive account of barriers. Hayes and Scott (1964) have also written a short but informative summary of the environments and processes that produce barrier islands. Seibold (1963) summarizes the results of granulometric, petrologic, and sedimentary structure study of the bars and barrier along parts of the Baltic and North Sea Coasts. Horn (1965) gives an excellent description and cross sections for a "Baltic" barrier of late Holocene age. Noteworthy is the work of van Straaten (1965) on coastal barrier deposits in the Netherlands. Hoyt (1969) suggests criteria to distinguish cheniers from barrier islands.

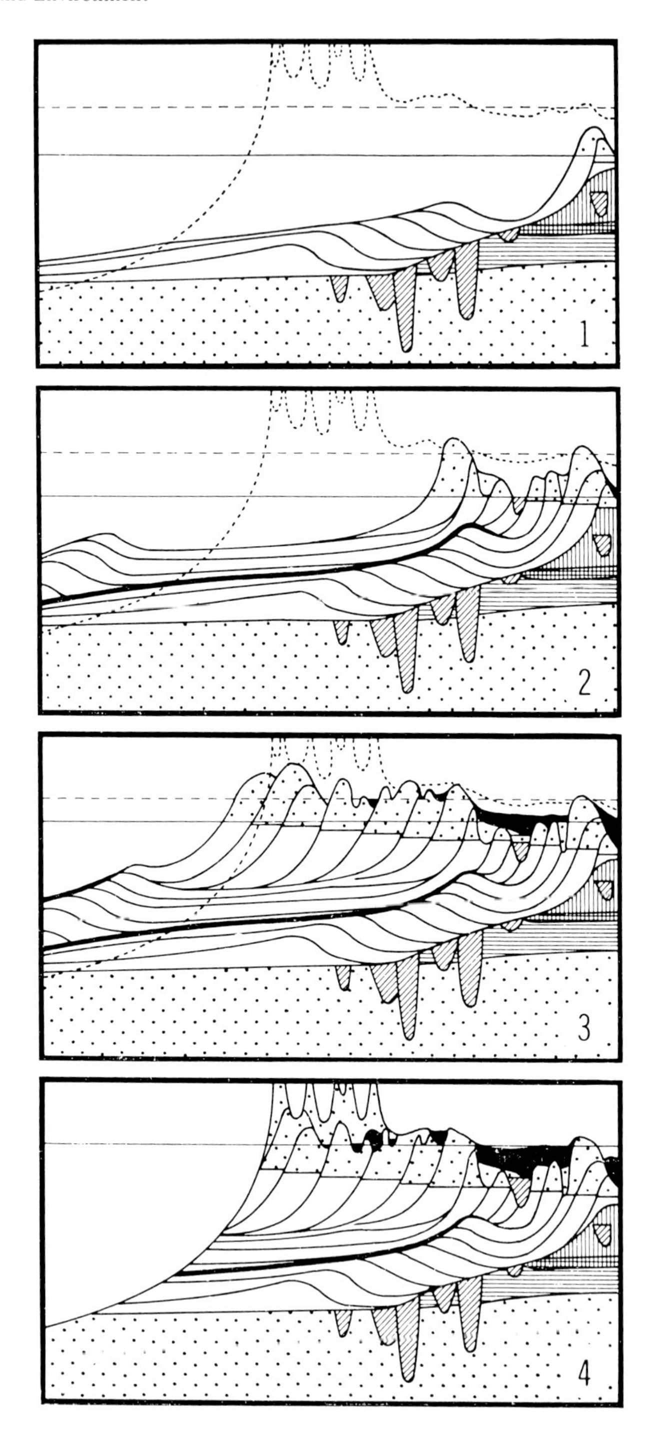

Fig. 11-28. Part of the complex sequential development inferred by van Straaten (1965, Fig. 26) in a barrier island along the North Sea

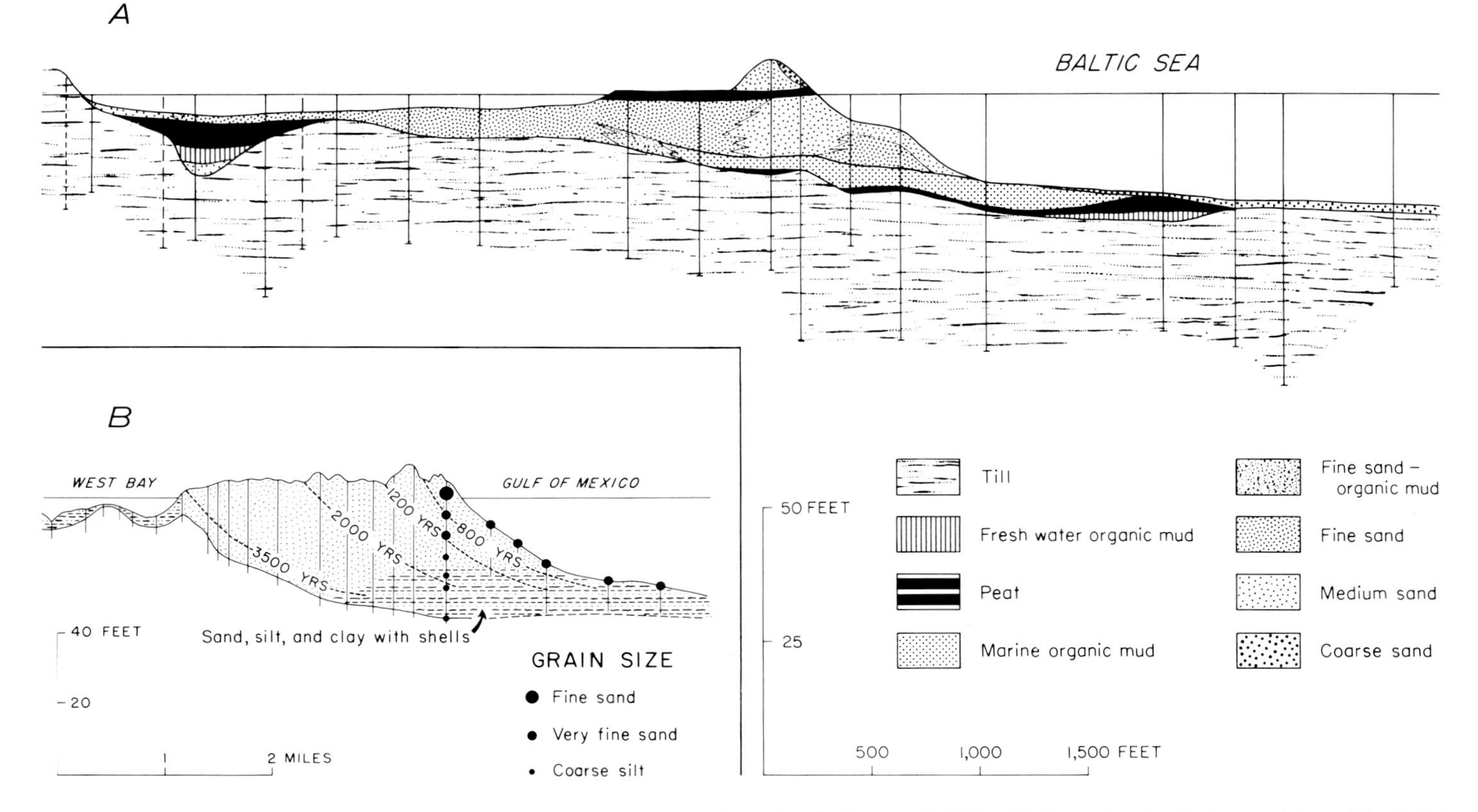

Fig. 11-29. Two cross sections of barrier islands: A) a Baltic barrier and B) Gulf Coast barrier (Potter, 1967, Fig. 10). Reproduced with the permission of the *Bulletin* of the American Association of Petroleum Geologists

Ancient beach sandstones: Well-established examples of ancient beaches or barriers are rare in the geologic literature. This is partly because beaches and barriers are not easily distinguishable as distinct bodies in many ancient shelf sands. In addition, probability of destruction prior to burial is high. Below we briefly summarize most of the published examples.

Long subparallel quartzitic ridges of Stampien (Tertiary) age in the Paris Basin have a pattern strikingly similar to cheniers, although Alimen (1936)

Fig. 11-30. Moriches Inlet, Long Island, New York in September 1947. Tidal or washover delta developed after hurricane breached barrier island. Photograph by U.S. Dept. of Agriculture. Courtsey M.M. Nichols

interpreted them as seif dunes. Fettke (1941) mapped a barrier island in the Silverville Sandstone (Devonian) of western Pennsylvania. Hollenshead and Pritchard (1961) describe examples of barrier sandstone bodies in the Cretaceous Mesaverde Group of the San Juan Basin of Colorado and New Mexico. Sabins (1963a and 1965), integrating petrology, grain size, and electric log cross sections, gives a good description and interpretation of the Gallup Sandstone (Cretaceous) of New Mexico. He distinguished forebar, bar, back bar and beach sand bodies. Behrn (1965, p. 19-30) interprets the heavy mineral placers in the Ecca Stage of the Karoo System in South Africa as regressive beach deposits formed during isostatic adjustment following cessation of Dwyka glaciation. He summarizes most of the relevant English-language literature of modern and ancient beach placers and also has a brief but effective summary of beach sedimentation (p. 26-28). He suggests that the beach placers were localized in coves formed in valleys cut in the pre-Karoo landscape. Certainly the beach environment, as part of the marine transgression of a dissected massif, provides an effective concentrating mechanism for the heavy minerals eroded from the massif.

Generalized descriptions of two large composite Gulf Coast barrier island systems, the Terryville Sandstone (Jurassic) and the Frio bar system (Oligocene), have been published by Thomas and Mann (1966) and Boyd and Dyer (1966). Both bodies are thick (600 to 1400 ft: 183 to 427 m), regressive, multistory composites 80 to 160 miles (128 to 256 km) long and 10 to 30 miles (16 to 48 km) wide.

Table 11-5. *Characteristics of barrier-island sands*[a]

PETROLOGY

Detrital. Heavy-mineral concentrates common. Skeletal debris, collophane, and minor glauconite. Conglomerates, if present, generally locally derived and may be mostly shells. Fauna are generally the more robust species. Mostly chemical cements.

TEXTURE

Commonly excellent sorting and very high grain-matrix ratio. Low variability of textural parameters between samples. Commonly good rounding.

SEDIMENTARY STRUCTURES

Asymmetrical ripple marks. Abundant gently inclined bedding. Lamination and lineation conspicuous on beach. Crossbedding moderately abundant and may be eolian as well as water-laid. Variability of crossbedding orientation, moderate to large. Bimodal distributions may occur on the beach. Burrows and channeling common.

INTERNAL ORGANIZATION

Sparse data indicate vertical increase in grain size and bed thickness, especially in regressive sequences.

SIZE, SHAPE, AND ORIENTATION

Widths from a few hundreds of feet to more than several miles. Thickness 20–60 feet. Very elongate, parallel with strand line. Sandstone bodies generally straight to gently curved. Grain fabrics and crossbedding can be variable, especially if eolian transport important.

ASSOCIATED LITHOLOGIC TYPES

Vertical: variable according to regressive or transgressive origin. Basal contact generally fairly even and may be transitional. Lateral: separates marine from lagoonal or terrestrial deposits giving maximum lithologic contrast. Multilateral sandstone bodies common.

[a] From Potter, 1967, Table V. Reproduced by *Bulletin* of the American Association of Petroleum Geologists.

Their great dimensions may be the result of deposition in a rapidly subsiding basin where vertical persistence of environments is the rule rather than long lateral migration as on stable cratons (see Fisher and McGowen, 1967, p. 122). Shelton (1965) presents convincing evidence for the barrier island origin of the Cretaceous Eagle Sandstone of Montana: accretion beach bedding, mottled structure, upward increase in grain size, transitional basal contact and long dimensional parallel to depositional strike. Young (1966, Fig. 2) published cross sections of shoreline sands separating lagoonal-paludal from marine facies in some 7000 feet (2134 m) of Upper Cretaceous sediments in the vicinity of the Book Cliffs of Utah. Weimer (1965) and McCubbin and Brady (1969) also describe two

Cretaceous shoreline sand bodies. Baars and Seager (1970) describe a 10 mile (16 km) long bar with 200 feet (62 m) of relief in the White Rim Sandstone Member of the Cutler Formation (Permian) of Utah.

Finally, what can be said about conditions needed to incorporate beaches and barrier islands into the geologic record? Unfortunately, a satisfactory answer is elusive. Perhaps rapid regional subsidence is necessary for burial of the barrier to prevent its destruction by later marine transgression or erosion from prograding streams. On stable cratons and intracratonic basins, a rapid eustatic rise in sea level may incorporate barriers into the geologic record. Slow transgression on cratons favors destruction. Barrell (1906, p. 442-446), who made one of the best geological analyses of the conditions necessary for preservation of littoral deposits, suggests that they are more likely to be found in association with major deltas (rapid subsidence) than with unconformities.

Annotated References:

Allen, J. R. L.: Coastal geomorphology of eastern Nigeria. Beach ridge barrier islands and vegetated tidal flats. Geol. Mijnb. Jahrg. **44**, 1–21 (1965).
Many line drawings and air photos plus brief comparison with the Netherlands.

Bird, E. C. F.: Coastal landforms, and introduction to coastal geomorphology with Australian examples, 193 p. Canberra: Australian Natl. Univ. (1964).
Ten chapters with many unusually informative diagrammatic line drawings. Elementary and an excellent starting point to dig deeper. Glossary.

Coastal Studies Institute: International Geographical Union Commission on coastal sedimentation, Bibliography 1955–1958. Louisiana State Univ., Coastal Studies Inst., Contr. **60-2**., 148 p. (1960).
References arranged by 16 different countries, some annotated and some not.

Dolan, Robert, McCoy, James: Selected bibliography on beach features and related nearshore processes. Louisiana State Univ. Studies, Coastal Studies Inst. Ser. No. **11**, 59 p. (1965).
Five valuable parts: morphology, techniques and properties; processes; coastal engineering; quantitative analysis and collected source materials.

Fairbridge, R. W. (Ed.): The encyclopedia of geomorphology, 1295 p. New York: Rheinhold 1968.
See articles on barriers, bars, beaches, coastlines – theoretical shapes, littoral processes, and sediment transport.

Guilcher, A.: Coastal and submarine morphology, 294 p. New York: John Wiley 1968.
Eight chapters in two major parts (coastal geomorphology and submarine geomorphology). Good classified bibliographies.

Johnson, D. W.: Shoreline processes and shoreline development, 584 p. New York-London: Hafner 1965.
A facsimile of the 1919 classic with with strong overtones of Davisian geomorphology and style.

King, C. A. M.: Beaches and coasts, 403 p. London: Arnold 1959.
An up-to-date book with 12 chapters similar to that of D. W. Johnson. Cover both processes and deposits. References few. Most examples cited are English.

Inman, D. L., Bagnold, R. A.: Beach and nearshore processes, Part II. Littoral processes. In: Hill, M. N., Ed.: The sea, Vol. 3, p. 529–553. New York: Interscience 1963.
Physical processes explained largely with Bagnold's theory. Integrates field and laboratory experiments on longshore drift.

Larras, J.: Plages et côtes de sable, 117 p. Paris: Collection du Lab. Natl. d'Hydraulique Eyrolles 1957.
Thirteen short chapters covering all major topics. Mostly European examples. Some mathematical treatment. Good bibliography.

Longinov, V. V., Ed.: Dynamics and morphology of sea coasts. Jerusalem, Israel Program for Scientific Translations, 372 p. (1969). (Translated from the Russian and available as TT68-50355, U. S. Dept. Commerce, Springfield, Va., 22151.)
Sixteen articles by Russian authors, some of whom have a mathematical flair. More geomorphology than engineering.

McGill, J. T.: Selected bibliography of coastal geomorphology of the world, 50 p. Los Angeles: Univ. California, Los Angeles 1960.
Primarily an index to coastal land forms. Useful subdivisions of 933 items.

Ottmann, Francais: Géologie marine et littorale, 259 p. Paris: Masson 1965.
Six chapters, the first (marine erosion) and the fourth (sandy coasts) most relevant to the geologist.

Philipponneau, Michel: Contribution a l'étude du Golfe Normano-Breton et de la Baie du Mont-Saint-Michel. Mém. Soc. Géol. Mineral. Bretagne **11**, 215 p. (1956).
Morphology of a famous tidal flat in three parts: morphologic evolution (3 chapters, earliest beginnings to present); beach and shoreline (3 chapters); good regional aspects and their evolution (3 chapters). Well-illustrated and 245 references. See also Pelhate's study in the same volume.

Putman, W. C., Axelrod, D. I., Bailey, H. P., McGill, J. T.: Natural coastal environments of the world. Los Angeles, Geog. Branch, Office of Naval Research and Univ. of California Contract Nonr - 233 (06), NR 388 - 013, 140 p. (1960).
Consists of three major parts (coastal landforms, vegetation and climates) plus 61 aerial photographs of coastlines from all over the world and their interpretation.

Rich, J. L.: Shorelines and lenticular sands as factors in oil accumulation, Vol. 1, p. 230–239. In: Dunstan, A. E. Ed.: The science of petroleum. London: Oxford University Press 1938.
Deductive physiography plus the then available modern sediment information are combined to produce a paper far ahead of its time. Outstanding.

Wiegel, R. L.: Oceanographical engineering, 532 p. Englewood Cliffs: Prentice-Hall 1964.
Nineteen chapters in all. Chapter 14, Shores and Shore Processes, gives a nonmathematical, well-illustrated summary.

Zenkovich, V. P.: Processes of coastal development, 738 p. Edinburgh-London: Oliver and Boyd 1967 (Translated from 1962 Russian edition).
Fourteen, well-illustrated chapters covering the ocean's edge. Over 1200 references almost half of which are Russian.

The Marine Shelf

In the study of ancient cratonic sediments geologists have long noted the presence of relatively thin sandstones in association with fossiliferous carbonates and shales. Typically, the principal structures of these sands are crossbedding and ripple mark. These sands may form either widespread sheets or small, discrete and discontinuous bodies. From this association came the idea of a shallow submerged marine shelf as a distinctive site of sand deposition, one that can be found on cratons throughout the world. Although "shallow" is rarely defined, most would agree that these sand bodies formed in water depths probably no greater than that of the outer edge of present day continental shelf whose world average is 132 m (Shepard, 1963, p. 257) and that probably most formed in only a very small fraction of this depth. As estimated from the position at which waves begin to refract and transmit energy to the bottom, wave base is about 20 m (Emery, 1966, p. 9). Much of our knowledge about the formative processes of marine shelf sand bodies, if that they be, comes from the ancient rather than the modern. This is in part due to contrast in scale, for the submerged portions of ancient cratons were far larger than the modern continental shelves. For example, the continental shelf off eastern North America between Cape Cod and Florida generally varies between 75 and 150 km in width, whereas throughout most of the Phanerozoic history of the North African or North American cratons, shorelines could migrate hundreds of kilometers. Study has also shown that most of the sands on modern continental shelves are predominantly relict deposits formed at lower sea levels (Fig. 9-22) so that it is sometimes difficult to relate present-day processes to the shelf sands.

Modern shelves: We emphasize below the distribution and characteristics of modern (in contrast to relict) sand on the Atlantic and Gulf Coast continental shelf of the United States and from the northwestern European shelf, two areas that have been studied in more detail than most. Much, of course, remains to be learned about each. Informative summaries have been given by Curray (1960 and 1965), Emery (1966), Uchupi (1968) for the United States and Stride (1963) for portions of the northwestern European shelf.

Upon entering the sea, sand and mud are segregated. Sand, moving by traction transport and propelled by longshore and tidal currents, forms barrier islands, beaches, offshore bars and sheets and ribbons. Off the eastern United States such modern sands – in contrast to widespread relict ones – extend to depths of but 10 to 20 meters and, in response to diminishing wave and tidal currents, show a seaward diminution in grain size. Depending upon water depth, they extend but a few kilometers seaward. Emery (1966, p. 12) thinks that they do not extend farther largely because insufficient time has elapsed since the ocean reached its present level about 5000 years ago. Sands beyond this point are generally relict, coarser grained, may be iron-stained, and may contain "misplaced" fauna that lived in shallower water during lower Pleistocene sea levels. Burrowing is appreciable in these sands. Some parts of the shelf are subject to turbulence from waves and currents sufficient to keep mud in suspension and disperse it into deeper water beyond the shelf edge. But on a protected shelf, mud should come to rest seaward of the sand limit.

Between Cape Cod and Florida, the Atlantic Continental Shelf is probably most similar to ancient cratonic shelves in terms of relief and sedimentary processes. It has many minor irregularities, most having relief less than 10 meters, some of which are of erosional origin, such as low terraces and shallow channels, whereas others are constructional, such as sand swells and the coral and algal banks and reefs. The sand swells (Fig. 11-31) are linear bodies that are typically about 4 km wide, symmetrical and broad in cross section with sides that slope one or two degrees, and are tens of kilometers long (Uchupi, 1968, p. 17). Whether these are relict barrier islands or modern features that may form in response to occasional intense storms is uncertain. The latter possibility is supported by calculations made by Hadley (1964, p. 164), that

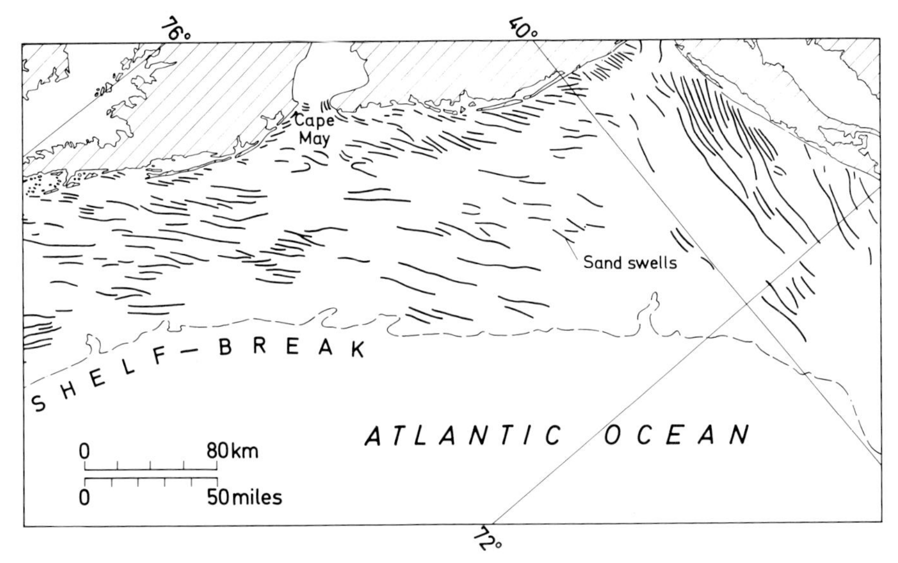

Fig. 11-31. Sand swells on continental shelf of eastern United States. (Modified from Uchupi, 1968, Fig. 14)

indicate that winds of 90 km/hr may produce wave induced currents of as much as 4 km/hr (104 cm/sec) at depths of 180 m, capable of moving sand (Fig. 9-10). Sandy shoals with complex, dune-like topography and sand waves appear to be largely the result of flood and ebb tidal currents that rework relict sands (Uchupi, 1968, Fig. 4). Some sand waves are perpendicular to and others parallel to flood and ebb tides.

Northwestern Europe presents a rather different view of sand dispersal on modern shelves primarily because stronger tidal currents transport sand into deeper water. In the Celtic Sea between England and Ireland, Belderson and Stride (1966) report tidal currents with surface velocities up to 3 knots (154.5 cm/sec). These currents erode near-shore areas and deposit gravel, sand, and mud seaward (Fig. 11-32). Belderson and Stride report a transgressive underlying conglomerate and suggest that its time span is of the order of 10,000 years, for its formation probably began with the lowest Wisconsin stage sea level and continues today near-shore.

Individual sand bodies have also been carefully studied in the southern Bight of the North Sea by Houboult (1968). These bodies consist of straight and gently to strongly curved ridges of reworked Holocene sand that rest on a flat bedrock floor (Fig. 11-33). The ridges are asymmetric in cross section and their surfaces are covered by megaripples. Spiral tidal currents are believed to segregate the sand into ridges and may be slowly moving them to the northeast. Off (1963) called attention to tidal currents as possible producers of linear sand bodies on shallow shelves. Evans (1970) described imbricate linear sandstone bodies in the Viking (Cretaceous) of Saskatchewan.

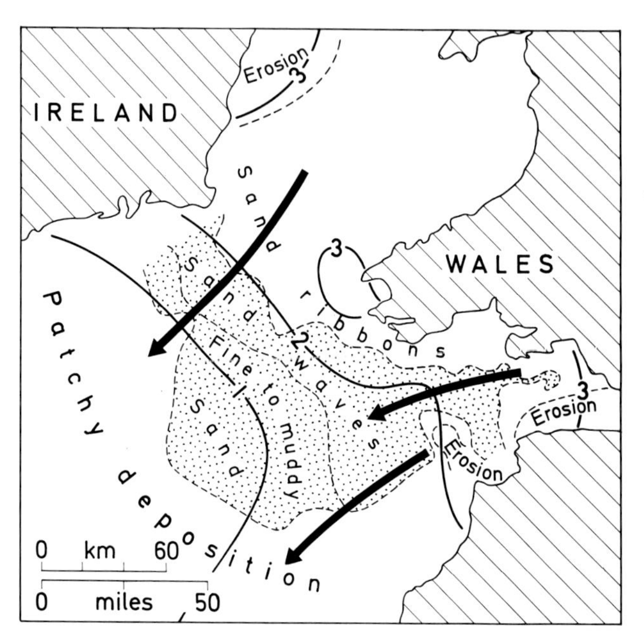

Fig. 11-32. Average maximum speed (in knots) and nature of sea bottom in northeastern part of Celtic Sea (Modified from Belderson and Stride, 1966, Figs. 2 and 4)

What generalizations pertaining to terrigenous sand dispersal and concentration can be obtained from the foregoing? Foremost among these is the segregation of sand from mud. This segregation is no doubt largely responsible for the deficiency of shale in the limestone-quartz arenite association of ancient shelves (see Chap. 12). Turbulent currents may separate the mud completely from the sand by taking it over the shelf edge as off eastern North America or, if water depth increases progressively rather than abruptly, the result will be a sand to mud transition downslope.

Tidal currents play a significant role on shallow shelves. They may concentrate sand into "windrows" or ribbons parallel to current direction as in the southern part of the North Sea, where sediment transport is largely parallel to the strandline (Stride, 1963, p. 10). Tidal currents are also responsible for the diverse and variable orientation of the sand-wave systems of shoaling shelves. Although one can but speculate, tidal currents may have been high in inland seas connected to the ocean. Relation of crossbedding direction to shape of marine sheet sands is not well established but is probably complex.

Phleger (1960) and Parker (1960) summarize the modern micro- and macrofauna for shelf and littoral areas of the northern Gulf of Mexico.

Ancient shelves: In the absence of plentiful, good descriptions of ancient marine shelf sand bodies, we have attempted to summarize and synthesize their properties from few and scattered data (Table 11-6). Much remains to be learned.

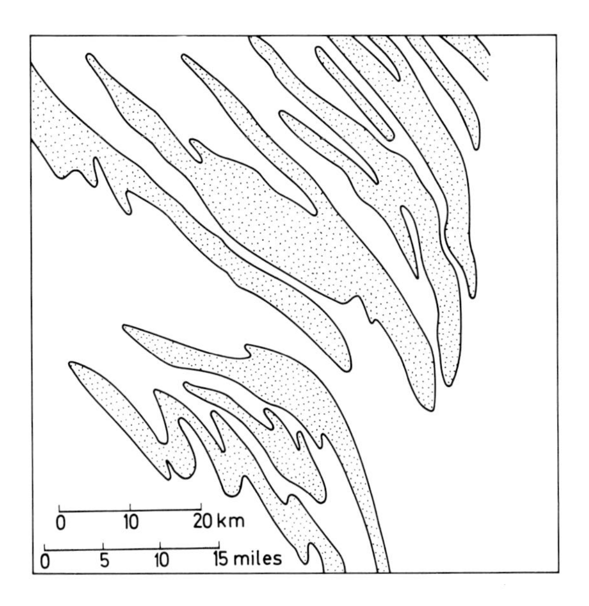

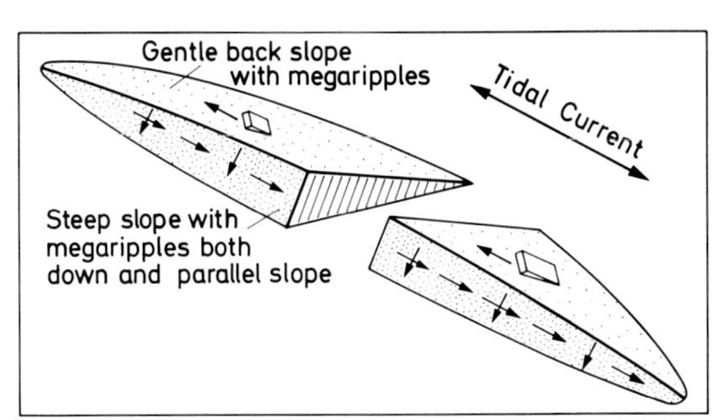

Fig. 11-33. Elongate sand bodies in portions of the North Sea (above) and schematic representation of their internal structure and surface sandwaves (Modified from Houboult, 1968, Figs. 9 and 17)

The sand of many ancient marine shelves tends to be mineralogically mature, especially on cratons, presumably because it has passed through the shoreline complex. Thus argillaceous rock fragments and unstable minerals generally are not very abundant. Glauconite, detrital carbonate skeletal debris, marine fossils, and collophane are commonly present. A relict fauna and anomalously coarse grains are common in the upper few feet of some shelf sands. Gravel or conglomerate may be derived locally. Cementing agents are mostly chemical, but may include appreciable detrital material proximal to clay

Table 11-6. *Characteristics of marine-shelf sands*[a]

PETROLOGY

Detrital. Commonly few argillaceous fragments and micas. Conglomerates, if present, are locally derived. Authigenic feldspar and glauconite and some detrital carbonate may be present. Mostly chemical cements. Relict faunas are common at tops of sand.

TEXTURE

Good to excellent sorting and high grain-matrix ratio. Variability of textural parameters between samples is very low. Tendency toward good rounding.

SEDIMENTARY STRUCTURES

Symmetrical and asymmetrical ripple marks. Crossbeds may be abundant; orientation commonly variable and may be bimodal. Trails and burrows. Minor channels and washouts.

INTERNAL ORGANIZATION

Data sparse, but vertical trends in grain size and bedding thickness either may be irregular or may increase or decrease, according to possible transgressive or regressive origin.

SIZE, SHAPE, AND ORIENTATION

Size and shape, highly variable, ranging from irregular, small pods through elongate ribbons to widespread sheets of many miles. Bifurcating and dendritic patterns absent. Elongate bodies have variable orientation with respect to depositional strike: parallel, perpendicular, and random. Relation of crossbedding orientation to elongation not well-known, but probably variable.

ASSOCIATED LITHOLOGIC TYPES

Vertical: variable according to regressive or transgressive origin, but mostly marine shale and (or) carbonates. Basal contact may be disconformable, but generally not of great magnitude. Lateral: generally relatively uniform silt-free shales or carbonates, commonly with rich and varied marine fauna.

[a] From Potter, 1967, Table VI. Reproduced by *Bulletin* of the American Association of Petroleum Geologists.

or shale transitions. Sorting generally is good to excellent, and there is perhaps a tendency for better-than-average rounding. The variability of textural parameters between samples is generally very low.

Both symmetrical and asymmetrical ripple marks generally are abundant, although varietal types are probably less diverse than in the barrier-island complex. Crossbedding is commonly conspicuous and as a rule much more variable in orientation than in alluvial and turbidite sands. Typically, the variance of crossbedding in a marine shelf sandstone is 6000 or more whereas in fluvial sandstones variances of 6000 or less are the rule (Potter and Pettijohn, 1963, Table 4-2). Zhemchuzhnikov (1926, p. 60–63) early suggested that bimodal crossbedding distribution was due to tidal currents (see Fig. 9-21). Moreover, essentially random orientation of crossbedding in some marine shelf sandstones is not uncommon. Changeable wind-driven waves also probably contribute to variability. Marine-shelf arenites appear to have a closer affinity to calcarenites of shallow shelf origin than to many shoreline sands. Thus a shelf calcarenite such as the Salem Limestone (Mississippian) of Indiana has a bimodal and variable crossbedding (Fig. 11-34) and may well have a pattern of sand bodies closely comparable to that of many marine-shelf arenites. Trails and burrows are abundant, especially where sedimentation was slow. A few minor channels and washouts – commonly of smaller magnitude than in alluvial, turbidite, and tidal sand bodies – also occur.

Vertical trends in grain size of marine-shelf sand bodies are known best from the widespread sand sheets of the past, especially in the Cretaceous of western North America, where grain-size declines upward in transgressive sandstones and increases upward in regressive ones. Similar vertical grain-size

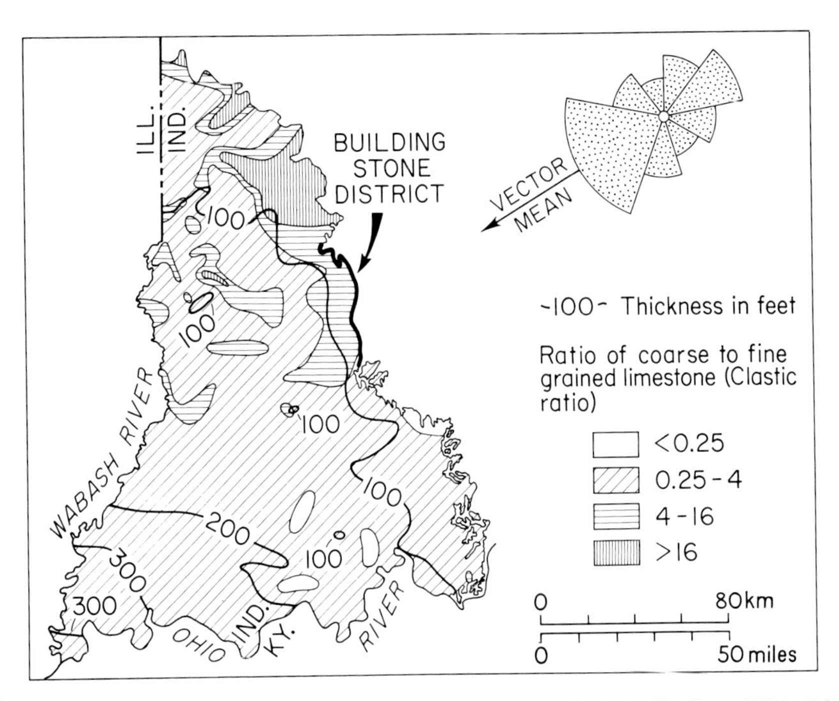

Fig. 11-34. Crossbedding direction in the Salem Limestone (Mississippian) of Indiana, U.S.A. (Modified from Sedimentation Seminar, 1966, Fig. 6)

gradients may prevail for smaller isolated marine-shelf sand bodies, if they are part of a regressive or transgressive sequence. On the other hand, vertical trends in grain size may be weak if present at all. Unfortunately, little information is available.

Size, shape, and orientation of marine-shelf sand bodies pose challenging problems, principally because of the diversity of sand-body types that can occur. Both thick and thin widespread sheets have been interpreted to have had a shallow marine-shelf origin. Undoubtedly such widespread sheets are the complex record of transgressive-regressive strand line deposits. Better evidence is generally available for the thin rather than thick sheets which may be, in reality, complex composites and thus may include some shoreline sand bodies. Most of these sheets have a relatively simple geometry forming either relatively uniform blankets or wedges. Textural gradients tend to be slight. Near their margins these sheets commonly pass into discrete sand bodies of diverse shapes which, if elongate, may have variable orientation with respect to depositional strike. Relation of elongation to crossbedding orientation is not well known, but it is probably variable. Marine shale, usually relatively sand free and having a rich and varied marine fauna, is commonly the lateral equivalent of discrete, isolated sandstone bodies. But many carbonates are closely associated with marine-shelf sands. In ancient basins, laterally-persistent, marine marker beds, usually thin limestones, calcareous shales, or possibly bentonites, greatly facilitate correlation in these mixed sequences. Less is known about isolated, elongate, shallow-water marine sand bodies than probably any other type.

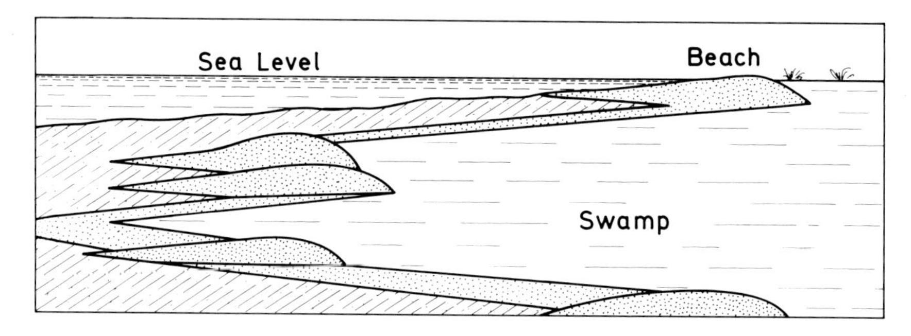

Fig. 11-35 Barrier and marine sand bodies combine to form a complex sheet as the result of strand-line migration (Modified from Hollenshead and Pritchard, 1961, Fig. 6)

Figure 11-35 illustrates how a marine sheet sand may form as the strand line migrates to and fro. As judged by modern continental shelves, migration of the strand line seems a more likely mechanism for spreading an extensive, thin, marine sand sheet than does simultaneous deposition. Although distinct shoreline sand bodies are seldom recognized, careful study near suspected strand lines may reveal them.

By use of a computer, Krumbein (1968b) has been able to generate cross sections similar to Figure 11-35, using continuous-time Markov models. These models assume a sedimentary memory or linkage of environments, use Markovian transition matrices and incorporate a random element in the lateral migration of environments. Krumbein's model might be particularly useful in projecting marine sheet and strand line deposits from outcrop downdip into the subsurface, especially where the outcrops are perhaps 20 to 50 miles (32-80 km) from the area of immediate interest. Correct assessment of the environment of deposition and knowledge of depositional strike is essential for such projection.

Emrich (1966) has described two wide-spread Cambrian sandstones, the Ironton and Galesville, in the upper Mississippi Valley (Fig. 11-36). Although control is too sparse to delineate details of sandbody morphology, broad outlines are clearly discernable. These sands have a combined maximum thickness of 220 feet (71 m), are sparingly fossiliferous, contain some oolites and glauconite, and are mature, well-sorted quartz arenites. Crossbedding orientation is variable in the Galesville Sandstone. However, its modal value is to the southwest approximately at right angles to the shoreline whereas in the overlying Ironton crossbedding is essentially random and only locally is there an indication of long-shore transport. Both sands become thinner and finer to the south and pass into dolomite. If the behavior of other marine sheet sandstone bodies can be taken as a guide, detailed mapping near the margins of these two sandstones would probably reveal a series of isolated, irregular sand bodies possibly with an essentially haphazard map pattern. The St. Peter Sandstone

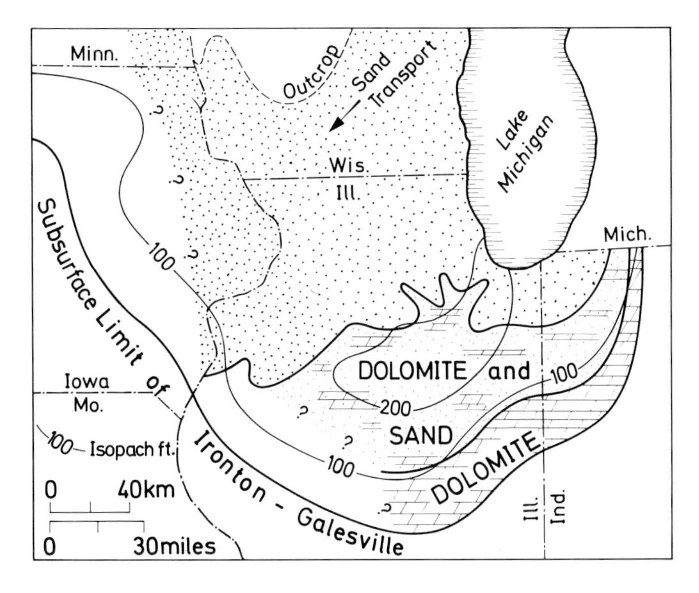

Fig. 11-36. Simplified paleogeography of Ironton-Galesville deposition on a portion of North American Paleozoic craton (Modified from Emrich, 1966, Fig. 21)

(Ordovician) of the Upper Mississippi Valley shares a number of characteristics in common with the Ironton-Galesville sandstones. Dake's (1921) classic study of the St. Peter is a summary still worth reading. There seem to be no published studies of Recent marine-sands similar to these Paleozoic examples.

Very thin sheets (less than 2 m) as well as thin, isolated mature sands may occur in many shelf carbonates. The Devonian Dutch Creek, Beauvais, and Hoing Sandstones (Collinson, 1967, p. 951–952, 955–956, and 965) of the Midcontinent region of the United States are examples of this type. Little is known about the processes that concentrate arenites on carbonate shelves or in and around evaporite basins. Horn (1964), however, provides unusual detail about texture and sedimentary structures of the Dogger-β (Jurassic), marine, near-shore sand that forms a reservoir proximal to a rising salt dome in Germany.

Thin, straight, and very elongate Cardium sand bodies considered to be offshore bars by Bervin (1966) occur in the late Cretaceous seaway near Calgary Alberta. They are flat bottomed, commonly are only 5 to 10 feet (1.5–3.0 m) thick, have gradational base, and are capped by a chert-pebble conglomerate. Linear sand bodies of the Blackleaf Formation (Cretaceous), possibly in part shore line sands, have been described by Cannon (1966) in Montana.

Elongate, marine, channel-fill sandstones also occur in marine shelf sections. They are, however most unusual and quite possibly require a proximal shelf edge to provide the potential energy to cut the channel into the shelf. Cutting may either be subaerial, during a lowering of sea level, or perhaps by some as

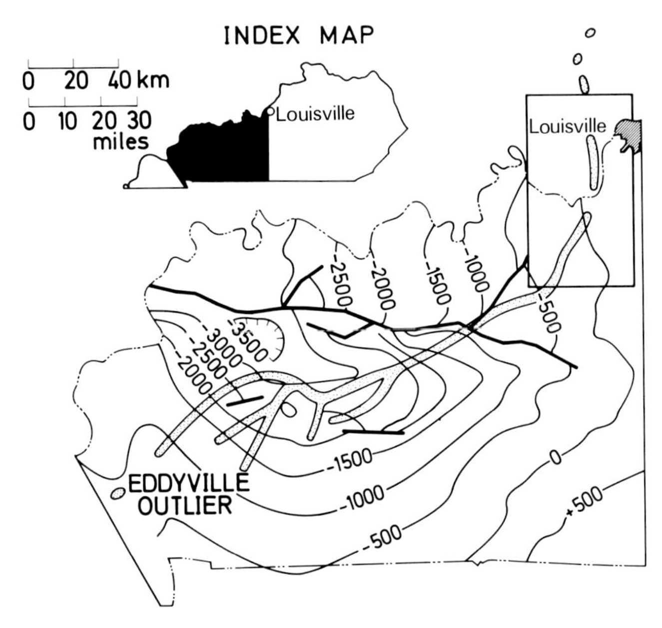

Fig. 11 37. Bethel channel and structure on top of New Albany Shale in western Kentucky and southern Indiana. Distributaries are from top of channel. Channel was cut into carbonate and believed to have been filled mostly by marine sand (Sedimentation Seminar, 1969, Fig. 1)

yet poorly understood type of shallow water turbidity current or shelf edge, tidal currents. Reynolds and Vincent (1967) and the Sedimentation Seminar (1969) describe a sand body in the Bethel Sandstone (Mississippian) that is over 170 miles (272 km) long in the Illinois Basin (Fig. 11-37). The weakly sinuous channel is cut some 250 feet (81 m) deep into carbonates and appears to have been the product of submarine processes. Its fill is also submarine and consists chiefly of fine to medium sand having abundant unidirectional crossbedding. Weakly entrenched distributary fingers occur near the top of the body and imply deposition in standing water. Fossiliferous limestone occurs in an irregular manner near the top of the channel. Sand may completely or partially fill such shelf channels, the remainder being shale and/or carbonate. A much wider, and deeper shale-filled shelf channel occurs in the Wilcox Formation (Eocene) of Texas (Hoyt, 1959). Isolated sand bodies in such channels are possible petroleum reservoirs.

Annotated References:

Curray, J. R.: Transgressions and regressions. In: Miller, R., Ed.: Papers in marine geology, p. 175–203. New York: Macmillan 1964.

Defines transgression and regression, discusses causal factors, analyzes nine modern shelves and uses the modern to evaluate the ancient record. Article concludes with a classification of transgressions and regressions.

— Moore, D. G.: Pleistocene deltaic progradation of continental terrace, Costa de Nayaret, Mexico. In: van Andel, Tj. H., Shor, G. G., Jr., Eds.: Marine geology of the Gulf of California. Tulsa: Amer. Assoc. Petroleum Geologists Mem. **3**, 193–215 (1964).

Geophysical records plus coring reveal structure and facies in sediments deposited since early Wisconsin time on a 150 mile (390 km) segment of continental terrace. Both deltaic and slope deposits played an important role in progradation of the terrace.

Irwin, R.: General theory of epeiric clear water sedimentation. Am. Assoc. Petrol. Geologists Bull. **49**, 445–459 (1965).

A limestone paper very relevant to marine shelf sandstones. Clear exposition of regression, transgression, cycles, and rhythms.

Kuenen, Ph. H.: Marine geology, 568 p. New York: John Wiley 1950.

An elementary, very well written and well illustrated classic rich in its discussion of major ideas.

Nota, D. J. G.: Sediments of the western Guiana shelf. Reports of the Orinoco Shelf expedition, Vol. 2. Mededeel. Landbouhogeschool Wageningen, Nederland **58**, 1–98 (1958).

Lithology, mineralogy, grain size, and depositional history of the western Guiana shelf between the Essequibo and Orinoco rivers. Mostly relict sediments.

Shepard, F. P.: Submarine geology, 2nd Ed., 557 p. New York: Harper and Row 1963.

Seventeen chapters contain a well-written overview of submarine geology including instrumentation, current systems and mechanics of sedimentation (both by D. L. Inman) plus description of continental shelves and beyond. Many references and a chart of the world.

— Phleger, F. B., van Andel, Tj., H. Eds.: Recent sediments, northwest Gulf of Mexico; 1951–1958, 394 p. Tulsa, Oklahoma: Am. Assoc. Petroleum Geologists 1960.

Fourteen articles plus a preface by F. P. Shepard. General review articles as well as new research reports. An ambitious attempt to put the parts together – including stratigraphy, petrology, and fauna – of the Holocene deltaic, shelf, lagoonal, and barrier island sediments of a large part of the Gulf of Mexico.

Sears, J. D., Hunt, C. B., Hendricks, T. A.: Transgressive and regressive Cretaceous deposits in southern San Juan Basin. New Mexico. U.S. Geol. Survey Prof. Paper **193-F**, 101–121 (1941).

Clear analysis of possible causes of migrating strandlines. A classic.

Turbidite Basins

Thick sequences of preparoxysmal marine, terrigenous sediments consisting mostly of rhythmically-interbedded shale and argillaceous, immature sandstone occur throughout the geologic column in geosynclines. Most of the sandstones show at least some graded bedding and commonly conspicuous sole markings. The term "flysch" is loosely applied to such deposits (Fig. 11-38). Although we see little of their vertical dimension, modern sands and muds with broadly comparable lithologic and bedding characteristics also occur on submarine fans on the continental rise near the base of the continental slope and in the deep-sea on abyssal plains. Density or turbidity currents, currents caused by mud suspensions that periodically travel downslope along the bottom, are considered by most sedimentologists to be the principal mechanism for suspension transport of silt and sand into deep-water basins, modern and ancient. Evidence for turbidity currents and discussion of possible alternatives is given by Kuenen (1967). Sources

Fig. 11-38. Medial turbidites with long even beds of sand interbedded with shale. Marnoso-arenacea (Miocene), valley of Santerno River near Castel del Rio, Emilia Romagna, Italy

of information about transport of sand into the deep-sea include Menard and others (1969), Shepard (1965), and Shepard and others (1969).

There is a prolific literature on the distinctive bedding and structures of turbidite sandstones (see Chap. 4). Only slightly smaller is the petrographic literature on graywackes, most of which are turbidites. But unfortunately, comparable studies of the geometry of turbidite sand bodies are scant. Lack of good markers in most turbidite sequences restricts mapping of individual sand bodies. In addition most turbidites are formed in the preorogenic stage of the tectonic cycle so that later complex structural deformation is also a limiting factor. Lack of good subsurface data is still another. We have drawn upon a combination of evidence from ancient and modern basins in our treatment of turbidite sand bodies (Table 11–7).

Table 11-7. *Characteristics of turbidite sands*[a]

PETROLOGY

Detrital. Rock fragments and immature minerals abundant as well as marine skeletals of both deep-water and shallow-water origin. Carbonaceous material generally present. Shale-pebble conglomerate. Cements are very largely detrital.

TEXTURE

Very poor to fair sorting and low grain-matrix ratio. Rhythmic alternation of beds produces abrupt juxtaposition of shale and sandstone. Rounding almost all inherited.

SEDIMENTARY STRUCTURES

Graded beds rhythmically interbedded with shale. Absence of large-scale crossbeds. Beds notable for their long lateral persistence. Sole marks, asymmetrical ripple marks, and laminated and convoluted beds are common. Trails and tracks are generally present.

INTERNAL ORGANIZATION

Few data. Grain size may be related to sand-body dimensions.

SIZE, SHAPE, AND ORIENTATION

Elongate sandstone bodies up to several miles; fairly straight but dendritic and bifurcating possible. Extend downdip into basin. Excellent correlation of directional structure and shape. But sheet and blanket-like deposits probably predominate. Large olistostromes not uncommon.

ASSOCIATED LITHOLOGIC TYPES

Vertical: commonly other marine shale and turbidite sandstone. Multistory sandstone bodies possible. Lateral: except for lower sandstone-shale ratio, notably little lithologic contrast. Mixed benthonic and pelagic faunas in shale and possibly reworked shelf faunas in sandstone.

[a] From Potter, 1967, Table IV. Reproduced by *Bulletin* of the American Association of Petroleum Geologists.

Most ancient turbidite sands are immature; they are rich in clay matrix and have a framework of poorly rounded and sorted unstable mineral grains (Figs. 6-12 to 6-19). Rock fragments, terrigenous and skeletal carbonate, and volcanic debris may be conspicious, even predominant components of the framework. Skeletal debris commonly is a mixture of both shallow and deep-water origin. Glauconite may be present and carbonaceous material is common. Shale clasts are common and even conglomerate is present in some ancient turbidite sections. Modern turbidites are relatively immature but they are generally much finer grained than many ancient ones – gravels being very rare. *Foraminifera* are present where depths are not too great (below the $CaCO_3$ compensation depth). Clay matrix is the principal binding agent, but it is notably absent in most modern and many Cenozoic turbidites probably because in most ancient turbidites the matrix is largely diagenetic – not original. Thus grain-matrix ratio is much lower in ancient than modern turbidites.

Graded bedding, abundant sole marks, convoluted lamination, asymmetrical ripple mark, small-scale crossbedding and long, even beds of sandstone interlaminated with shale are the most striking sedimentary structures of ancient turbidites. This combination in thick stratigraphic sections is unique to turbidites. Ksiazkiewiez (1961) and Kuenen (1968) summarize Lebenspuren and macrofossils in flysch basins.

Studies of the topography and sediments of the continental slope and deeper areas show that deposition of sand occurs on the gently sloping broad fans at the base of submarine valleys or canyons that cut into the continental slope and shelf. Some of these valleys and canyons may extend virtually to the strand line where they can intercept sand carried by longshore drift along the shelf. Or sand may be introduced into a valley or gulley directly by a stream as, for example,

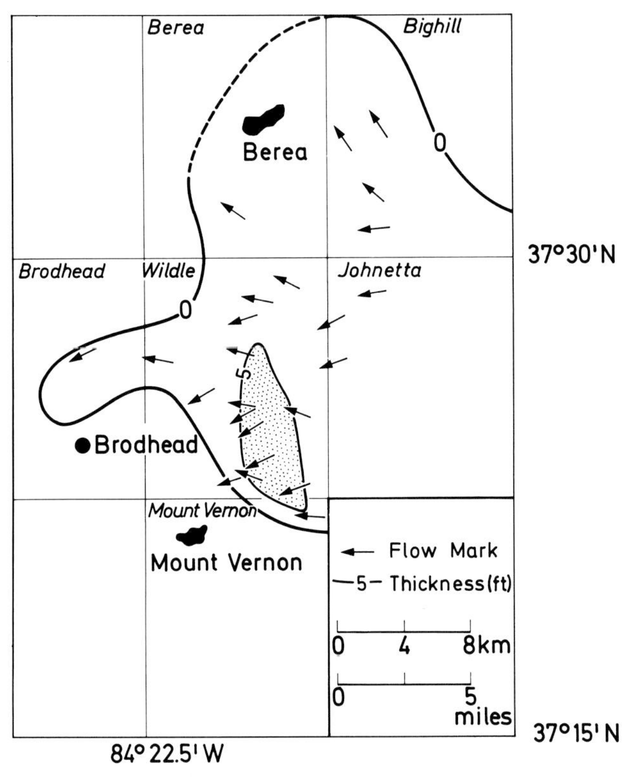

Fig. 11-39. Thickness and flow system (furrows and flute casts) in the "Rockcastle Freestone" a Mississippian delta-front turbidite (Modified from Weir, 1969, Fig. 13). Note relation between flow system and isopach

off a delta growing into deep water. Unstable accumulations may be temporarily stored and then carried down valley or down canyon by slumping, sand flow, or turbidity currents onto a submarine fan and finally onto a basin floor. Channels with natural levees have been mapped in detail across the upper parts of submarine fans (Grim, 1969) and are part of an integrated system of bifurcating distributaries (Hand and Emery, 1964, Figs. 3 and 8). As with alluvial fans, there is a direct correlation between slope of the fan and grain size. Turbidity currents, whatever the initiating mechanism, are periodic, each flow depositing an individual sandy bed as the flow moves out of its valley or canyon onto the fan and finally dissipates on the basin floor. Fig. 11-39 shows the pattern of deposition of an individual unit, the "Rockcastle Freestone" (Mississippian), which is interpreted as a delta-front turbidite by Weir and others (1969, Fig. 13). Some ponded-sediment basins in the rugged topography of the Mid-Atlantic Ridge, possibly akin to intermontane basins in some ways, have turbidite fills of locally-derived materials and may have successive beds from the same flow – because of rebound from the opposite walls of the basin (van Andel and Komar, 1969).

By combining the above observations with, as yet incomplete, knowledge of sand distribution in ancient flysch basins, several types of turbidite sand bodies can be inferred. Unfortunately, our present knowledge leaves very much to be desired. One should recognize, moreover, that turbidite sands may be rhythmically interbedded with much or little shale. Because thin, uniform sandstone beds interbedded with shale are one of the characteristics of the flysch facies, a turbidite sand body – unlike most other types – is somewhat difficult to define and delimit. Most such sand bodies will consist of more than one turbidite bed and may include appreciable shale.

Turbidite sand bodies may be elongate as well as sheet-like. Some linear turbidites fill the lower reaches of submarine valleys and canyons. These bodies will extend downslope, be perpendicular to the strand line, coarse-grained, contain minimal shale and have almost perfect correlation between internal directional properties and elongation direction. These form multistory bundles. Near the apex of a submarine fan, a relatively fixed channel system is also probably responsible for superposition of one sand turbidite upon another. Fan-head bodies extend down fan and downslope, probably with slight radial pattern, show a strong correlation between directional properties and shape and are coarse to medium grained. They probably have a distinctly erosional channel at their base. Internal slump, within individual beds in the body, may show transport direction in addition to the usual flutes and groves. Away from the apex of the fan, the current spreads out depositing a single thinning and fining and widening bed that terminates at the distal margins of the basin floor as a very thin, silty layer. A radial pattern of directional properties prevails (Fig. 11-39). On basin plains, silt is typical and is interbedded with much mud so that sand-shale ratio will be less than one, whereas near the fan apex the sand-shale ratio will exceed one and possibly appreciably so. In the down-current direction beds become finer grained, thinner, and have a different suite of sedimentary structures (Walker, 1967) so that thick graded beds with large sole marks become thin siltstones that are weakly graded, if at all, and have but a few delicate sole marks. These observations led to the division into proximal, intermediate, and distal facies. Should the fan debouch at right angles into a linear trough, however, the current will seek the lowermost part of the trough and follow it. Thus the resulting sand bed may be confined to a fairly narrow L-shaped or T-shaped pattern. The Repetto Formation (Pliocene) of California contains a marginal fan that spreads laterally in a narrow trough-like basin (Fig. 11-40). Water depth is estimated at 6000 to 8000 feet (1741–2580 m) by Shelton (1967, p. 2459). According to him, single sand bodies in this sequence rarely exceed 150 to 200 feet (48–64 m) in thickness. California also contains examples of Miocene turbidite sands in channels 1200 feet (372 m) deep (Martin, 1963, Figs. 5 and 6), where individual sand bodies are as much as 200 to 400 feet (62–124 m) thick. Dickas and Payne (1967) describe a similar channel in the Paleocene of California. Such channel-fill deposits are differentiated from surrounding shale by their higher sand-shale ratio and truncation of any marker beds that may be present. Grain size may be related to channel dimension and may, together with bed thickness, decrease upward as in alluvial and tidal sand bodies, although confirmatory data on this point are lacking. As with subaerial fans, prolonged turbidite sedimentation may completely bury initial bottom topography. Continuous acoustic profiling commonly shows this to be true in the oceanic abyssal plains.

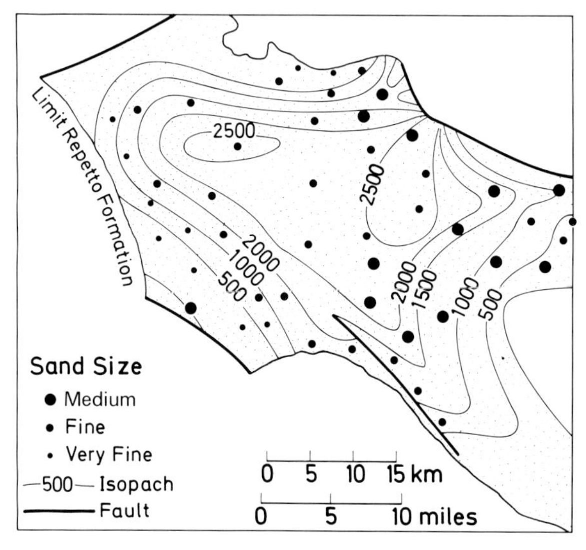

Fig. 11-40. Thickness of sandstone and grain size in Lower Pliocene Repetto Formation, Los Angeles Basin, California (Modified from Shelton, 1967, Fig. 9)

Because turbidites are deposited during short-lived pulses and not appreciably reworked by later currents, they form, perhaps, the most continuous individual beds known – ash falls and some ash flows excepted. Hesse (1965, p. 13) correlated individual turbidites over 50 km in the Lower Cretaceous Bavarian flysch trough. Presumably single beds of tuffaceous turbidites formed by underwater volcanic explosions may be correlated over even longer distances.

Where a delta terminates in deep water, deposition of turbidites may occur on the lower reaches of its front and on the basin floor in deep water beyond it as in the Gulf of Mexico, where the Mississippi delta extends to the edge of the continental shelf (Davies and Moore, 1970). Lineback (1968) considers the sands of the Carpers B, of the Borden Formation (Mississippian), in the Illinois Basin (Fig. 11-41) to be a turbidite deposit. This unit is an evenly bedded siltstone with fine grained sand beds that are weakly graded and have some sole marks. The silt and fine sand beds are interbedded with shale. He regards the sand as having been carried to a nearby delta front by a distributary channel, which was a part of the Michigan River System (see p. 470). Submarine deposition on a delta front or shelf slope produces sand bodies with high initial dip, a dip that defines the foreset slope of the delta or of a shelf edge where it is prograding into deeper water.

Rich (1951) proposed the terms "undaform", "clinoform", and "fondoform" for shelf, slope and deep basin. Sands presumed to be deposited on a delta front possibly by turbidity currents, such as those of the Borden Formation described above, would be clinoform (also called "falloff") deposits. Fig. 11-42 shows how falloff can drastically change stratigraphic correlation of slope deposits. Knowl-

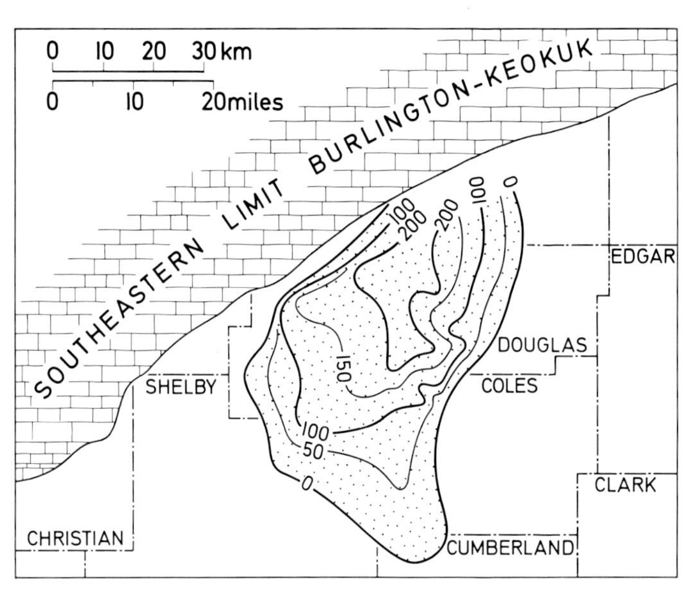

Fig. 11-41. Thickness of the Carper B, a turbidite siltstone in the Mississippian Borden Formation of the Illinois Basin (Modified from Lineback, 1968, Fig. 9). Southeastern limit of Burlington-Keokuk limestones marks a temporary basin edge

edge of an originally horizontal reference bed is vital when such falloff deposits are suspected. The true situation can best be established if it can be shown that level marker beds lie both above and below the inclined strata. Clinoform structure results from the basinward and upward growth of a shelf edge. The sands and silts of slope deposits can be turbidites or share many of their features. Carbonate as well as terrigenous sand may be deposited on the slope and interbedded with mud, if available. Typically, the sands are inclined sheets (Fig. 11-39). Long dimension may parallel strike of the slope. Asquith (1970) illustrates many of the above points in the marine Cretaceous of Wyoming.

Carbonate turbidites are not widely described but do constitute a significant part of turbidite deposition. Rusnak and Nesteroff (1964) describe modern biogenic turbidites. Carbonate turbidites are also reported from the abyssal plain of the Gulf of Mexico (Davies, 1968). Sestini (1970) summarizes knowledge of ancient turbidite limestones in the northern Apennines.

Another feature of turbidite sand bodies that should be mentioned are large mass slump sheets – olistostromes – which not only disrupt the normal strati-

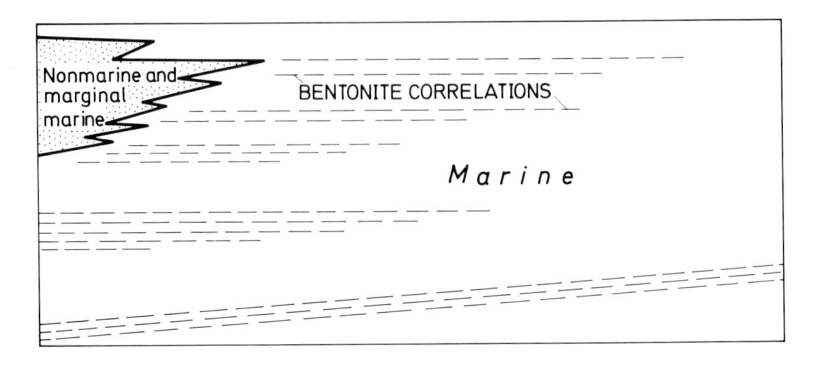

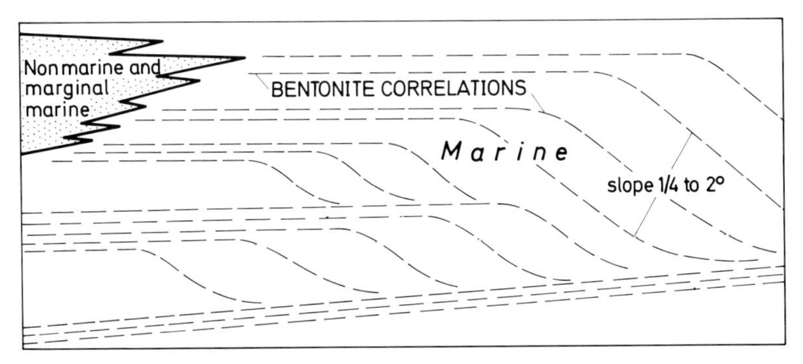

Fig. 11-42. Downslope cross sections before and after recognition of "falloff" in the Cretaceous of Wyoming (Modified from Asquith, 1970, Figs. 2 and 3). Recognition of falloff can significantly alter correlations and history of basin filling

graphic section but through rotation and crumpling during sliding can vastly complicate the interpretation of sandstone bodies in turbidite sequences. Descriptions by Flores (1959) and Lucchi and D'Onofrio (1967) are informative.

Annotated References:

Bouma, A. H., Brouwer, A., Eds.: Turbidites, 264 p. Amsterdam: Elsevier 1964. — Fifteen papers, mostly in English, on turbidite deposition.

Crowell, J. C., Hope, R. A., Kahle, J. E., Ovenshine, A. T., Sams, R. H.: Deep-water sedimentary structures Pliocene Pico Formation Santa Paula Creek, Ventura Basin California. California Div. Mines Geol., Spec. Rept. **89**, 40 p. (1966). — A teacher and his graduate class record the bedding and sedimentary structures of an outcrop of turbidites. Many quantitative data.

Emery, K. O.: The sea off southern California, 366 p. New York: Wiley 1960. — Integrated analysis of all aspects of the sea, its sediments and tectonic setting in a geosynclinal area.

Fairbridge, R. W., Ed.: The encyclopedia of oceanography, 1021 p. New York: Reinhold 1966. — A wide range of topics briefly treated in marine geology and oceanography. A good place to start.

Hill, M. N., Ed.: The sea, Vol. 3, 963 p. New York: Interscience 1963.

Section II, Topography and structure, and Section III, Sedimentation, are most relevant. Particularly recommended are chapters 12 (Topography of the deep-sea floor), 13 (Continental shelf and slope), 14 (Abyssal plains), 17 (Trenches), 20 (Submarine canyons), 27 (Turbidity currents), and 33 (The preserved record: Paleontology of pelagic sediments).

Holtedahl, Hans: Recent turbidites in the Hardangerfjord, Norway. In: Whittard, W. F., Bradshaw, R., Eds.: Submarine geology and geophysics, p. 107–142. London: Butterworths 1965.

Grain size, sedimentary structures and *Foraminifera* help trace turbidity currents down a deep fjord.

Kingma, J. T.: The Tongaporutuan sedimentation in central Hawke's Bay. New Zealand Jour. Geol. Geophysics **1**, 1–30 (1958).

Maps of grain size, sand-shale ratio, and thickness plus paleocurrent system.

Lajoie, J., Ed.: Flysch sedimentology of North America. Geol. Assoc. Canada, Special Paper No. 7, 272 p. (1970).

Fifteen papers by North American geologists. Turbidites galore. Mostly ancient examples. Many illustrations.

Rad, Ulrich, von: Comparison of sedimentation in the Bavarian Flysch (Cretaceous) and Recent San Diego Trough (California). Jour. Sed. Petrology **38**, 1120–1154 (1968).

Effective description of modern and ancient sediments and a detailed comparison of the two. Author suggests an open mind rather than assuming turbidity currents as the only agent capable of depositing flysch. See also Scott (1966), Klein (1967), and Kuenen (1967).

Scott, K. M.: Sedimentology and dispersal pattern of a Cretaceous flysch sequence, Patagonian Andes, southern Chile. Am. Assoc. Petroleum Geologists Bull. **50**, 72–107 (1966).

Description of a typical flysch sequence plus discussion of whether or not gravity currents were the exclusive depositional mechanism. Also discusses dispersal pattern of flysch in longitudinal basins. Compare with Kuenen (1967).

Shepard, F. P., Dill, R. F.: Submarine canyons and other sea valleys, 367 p. Chicago: Rand McNally 1966.

Twenty-two well organized and well illustrated chapters plus many references. Much on turbidites and their causes.

Eolian Environment

Wherever an abundant supply of unconsolidated sand is exposed to the wind, it will be transported and deposited in a great variety of dune forms. Eolian sands accumulate in several areas: in large deserts and semi-arid regions and along many shorelines and flood plains, the last two forming in response to a linear source of supply in both humid as well as arid climates. Judged by the present, the volume of arid zone eolian sands far exceeds that of dunes associated with beaches and dunes for approximately one-third of the world's present land surface has been classified as semi-arid or arid (Peel, 1966, p. 3). Thus, if the present is any key to the past, deserts should have been fairly common in the past history of the earth for regardless of whether the continents have been relatively fixed or mobile on the earth's surface, the world pattern of atmospheric circulation must have been as it is today with subtropical high-pressure and arid zones parallel to a humid equatorial belt. Moreover in the early Paleozoic and Precambrian times, vascular plants were absent or less abundant than now so that eolian activity probably played an even larger role than it does today. Could it be that eolian environments had their peak in earliest Precambrian time?

Modern eolian sands: The best insight into the character and origin of ancient eolianites is through knowledge of sand accumulations in modern deserts. How much of modern deserts is covered by sand? What kinds of dunes predominate? What is the relation between dune form and wind regime? Between external dune form and internal structure? As we have noted earlier in this chapter, critical information is commonly lacking so that our answers are all too often incomplete and far from satisfactory.

In most modern deserts bedrock predominates at the surface. For example in the Arabian Peninsula of 1,060,000 square miles, (2.75×10^6 km^2), sand covers only some 300,000 square miles (0.78×10^6 km^2) or about one third (Holm, 1960, p. 1369) and in the Sahara only about 10 percent (Peel, 1966, p. 17). But in large deserts, as in Arabia and the Sahara, such sand deposits, even though they cover but a fraction of total desert area, are of appreciable size: the Grand Erg Occidental of Algeria is larger than England (Peel, 1966, p. 11) and in the Arabian Peninsula the Rub'al Khali occupies about 230,000 square miles or 598,000 km^2 (Holm, 1960, p. 1369). Thus, if the world's present deserts are comparable to those of the past, extensive dune sand deposits must have existed.

The larger sand deposits of deserts occupy topographic lows on either narrow plains or broad basins. This is primarily because such large quantities of sand can only be concentrated by the centripetal drainage of wadis or by former shore lines along either inland lakes or marine embayments. In either case, water initially concentrated the sand and subsequently wind removed silt and clay leaving sand and gravel as a lag deposit. And still later, wind separated the sand from the gravel. In short, although direct supply from underlying sand bedrock is one source, concentration by intermittent streams or former shorelines appears to be essential to produce the substantial volumes found in the larger deserts. Because an eolianite is a subaerial deposit, there should be an unconformity bordering the area covered by the sand accumulation as well as one at its top. Associated gravel plains, fluvial sands and gravels, and salt deposits, the last either forming in inland playa lakes or lagoons bordering a former sea, should be associated with the eolian sands.

Dunes have many and varied forms, so many in fact, that it has been difficult to formulate a universal classification even though simple forms such as the barchan, longitudinal, transverse, and sand shadows have long been recognized (Table 11-8). Gradation from one form to another has undoubtedly contributed to difficulty of classification. Holm (1960) and Smith (1963) illustrate many of the complex forms revealed by aerial photography.

The conditions under which these forms develop are not well known, although in principle any specific dune type probably represents the most efficient transport form for a given wind pattern, sand supply and vegetative cover. Judging by the literature, there appears to be more variety of subaerial than of subaqueous dune forms. One possible explanation may be that wind currents at the earth's surface, because they are 0.00012 times less dense than water and because they are very responsive to diurnal heating and cooling, are much more variable both in magnitude and direction than water currents. Shifting winds appear to be

Table 11-8. *Some major dune types and characteristics*

TYPE	MORPHOLOGY	RELATION TO WIND	INTERNAL STRUCTURE
Longitudinal (Seif, parallel and whaleback).	As much as 100 m high, 1 to 2 km wide and 20 to 200 km long. Linear, sinuous. Cross winds increase height and width. Form spectacular "windrows" in some continental deserts. May be straight or slightly sinuous. Whalebacks are flat topped.	Extended downwind roughly parallel to average wind vector.	Poorly known but probably has bimodal crossbedding the two modes pointed downwind and separated by 100 to 150 degrees.
Crescentic (Barchan, sigmoidal transverse and parabolic).	Heights up 100 m.	All perpendicular to wind: barchan has horns downwind; parabolic ("blowout") has horns upwind. Transverse dunes have relatively straight crest.	Slip slopes downwind with variance comparable to many alluvial sands – unimodal.
Complex (Pyramidal, hooked, dune "massifs", heaps, star, and reversing).	Many and varied but pyramidal dunes to 200 m or more.	Variable, shifting winds plus merging dune forms.	Virtually no data but undoubtedly complex reflecting external morphology.
Sheets	Flat to gently undulating. Can cover wide areas.	Accretion probably predominates over avalanching.	Little data but probably mostly ripples and dips less than 15°

particularly effective in producing complicated dune forms. In analyzing wind data, prime attention is given to direction and frequency of winds that are capable of moving sand, usually in excess of 15 to 25 mi/hr (24 to 40 km/hr) rather than average conditions. In attempting to relate subaerial dune form to wind system there is, just as in river bed morphology, the time lag problem – ascertaining that the bed forms are really in equilibrium with present current system and not the response to some earlier one.

Longitudinal dunes (Fig. 11-43) prevail over vast areas in modern deserts. Some are as long as 200 km in the Rub'al Khali (Holm, 1960, p. 1373). Madigan (1946) reports comparable longitudinal dune systems of comparable dimension in Australia. Cross winds are believed to increase longitudinal dune height and width, although elongation is presumed to be parallel to prevailing wind. One type of crescentic dune, the barchan, is particularly common where a limited sand supply is transported across a bedrock surface. These can grade into transverse dunes (Fig. 11-44). Complex dunes come in many forms, may be composite and some, such as the pyramidal dunes, can reach heights of over 200 meters (Fig. 11-44). Complex and longitudinal dunes are more important in modern deserts than barchans.

Information on external dune form and internal crossbedding is not abundant, the only notable exception being the data obtained by McKee (1966) for the gypsum sands of White Sands National Monument, New Mexico. In the barchan,

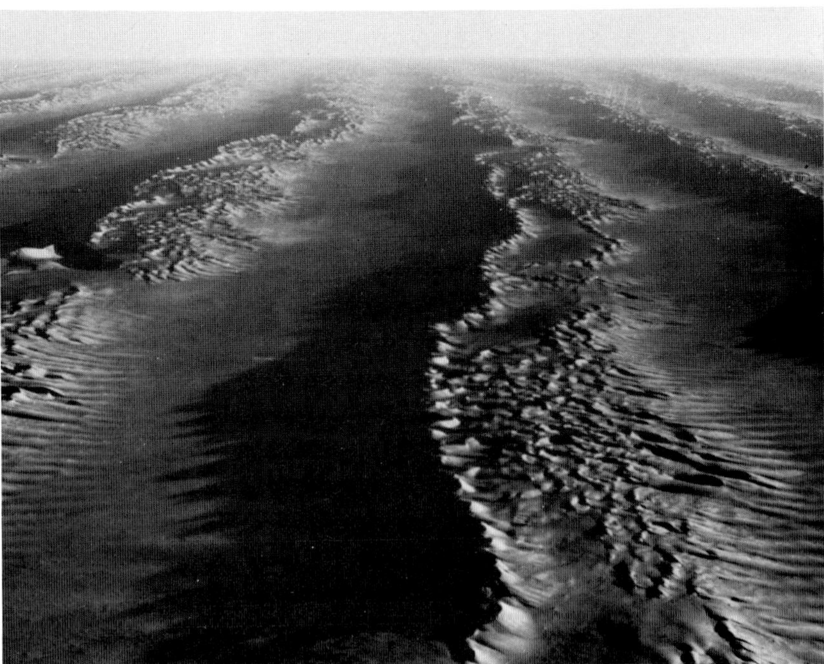

Fig. 11-43. Longitudinal dunes of the Rub'al Khali in Arabian Peninsula as seen from Gemini IV (NASA S-65-3476), and from airplane (photo courtesy Arabian American Oil Co.). These longitudinal dunes are about 1 to 2 km wide, 100 meters high and up to 200 km long

Fig. 11-44. Transverse dunes (above) and "sand mountains" (below) in the Rub'al Khali of the Arabian Peninsula (photographs courtsey of the Arabian American Oil Co.). Sand mountains may reach heights of as much as 250 m. Large areas of desert are swept bare of sand to produce concentrations as shown in Figures 11-43 and 11-44 so that there are also vast areas of exposed bedrock and gravel plains in most large continental deserts

domal, parabolic and transverse dunes that form down-wind from a playa lake, McKee (p. 58–59) found 1) crossbedded thickness to vary from one to five feet (0.3 to 1.6 m) with sets thinning upward within dunes, 2) unimodal crossbedding with very small variance (2051), 3) steeply dipping (30–34°) foresets, 4) good agreement between crossbedding direction and wind system and, 5) mostly tabular crossbedding with its bounding upper surface becoming steeper down-wind. Also present are a little contorted bedding, ripple laminae and local scour-and-fill. Because McKee studied mostly crescentic dunes, the applicability of his results to reconstruction of wind pattern in ancient eolianites, probably largely formed by longitudinal and complex dunes, is uncertain.

Dunes associated with the strand line or with alluviating flood plains are common features of humid and desert climates alike for both strand line and flood plain provide a renewable, linear source of sand that can be transported down-wind. Varieties of crescentic dunes appear to be the common types, although pyramidal dunes may develop in response to seasonal wind changes. The size of the dune field depends on sand supply, effectiveness of onshore or prevailing winds, and vegetation. On the shorelines of carbonate basins, oolites and skeletal debris may form most of the dune sands. And, should such a coastline have an arid climate, gypsum supplied by salt flats bordering saline lagoons may also contribute to dune sand. Among the many studies of coastal dunes are those of Cooper (1958 and 1967) along the Oregon and California coasts and Gay (1962), who describes spectacular dune migration from the strand line high up into the mountains along the arid Pacific coast of Peru. Crossbedding in dunes on a barrier island was studied by McBride and Hayes (1962), who found it to have 1) a large percentage of dips steeper than 30 degrees and 2) bimodal cross-bedding resulting from seasonal wind changes.

Problem of ancient eolianites: Judging by the present with its semi-arid and desert regions that cover almost one third of the world's land surface, much of the sand of the geological past must have had some history of wind transport, and it would seem reasonable that some of it is preserved as distinct eolian deposits. But recognition of these deposits is perhaps the most difficult problem in the environmental analysis of sandstones. Primarily this is so because we have so few integrated studies of the sedimentary structures, texture, petrology and geometry of sand deposits in modern continental deserts, which presumably are the most likely candidates for ultimate preservation as widespread ancient eolianites. In short, the stratigraphic model is still to be formulated and as a consequence organizing principles are lacking.

Three questions are pertinent. Is a large sand deposit of a continental desert likely to be incorporated into the geologic record? How can we recognize an ancient eolianite? Once recognized, can its crossbedding be used to infer paleo-wind direction?

Barrell (1906, p. 442), following an earlier proposal by Passarge, suggests that the probability of preservation in the geological record of a widespread eolian sand is unlikely because it will be largely destroyed by subsequent marine transgressions. A climatic change, with increased rainfall and greatly increased fluvial erosion, could also remove it. Thus, although throughout most of its geological history the earth has had substantial deserts containing large volumes of eolian sand, most of this sand probably has been reworked into marine shoreline and fluvial-deltaic deposits. Only very rapid local subsidence could preserve the sand before removal by erosion and thus incorporate it into the rock record.

Nonetheless, some widespread, ancient sandstones have been interpreted as eolianites (McKee, 1933; Shotten, 1937; Reiche, 1938; Bigarella and Salamuni, 1961; and Poole, 1964). Stokes (1961, p. 154–161) gives perhaps the best rationale for why desert or semi-arid conditions are believed to have existed repeatedly in the Colorado Plateau from Permian to the present day. From Permian to Jurassic he interprets five sandstones (Cedar Mesa, Permian; Wingate, Triassic?; Navajo, Entrada, and Bluff, Jurassic) as eolian, one of which (Cedar Mesa) is 1200 feet (372 m) thick. Some of these formations are believed to have covered from 10,000 to possibly as much as 200,000 square miles (26,000 to 520,000 km^2). There is not complete agreement about the eolian origin of all these sandstones, however.

Among the criteria for recognition of ancient eolianites, comparable in size to large modern deposits in deserts, are: 1) association with evaporite deposits and fluvial sands, 2) probably unconformities either above, below or lateral to the deposit, 3) absence of widespread key beds, 4) predominance of crossbedded sandstone, 5) good sorting (due to clay and silt removal so that chemical rather than detrital cements will later prevail), 6) paucity of micaceous minerals, 7) a dull opaque or mat surface on quartz grains (Kuenen and Perdok, 1962, p. 648), and 8) absence of marine invertebrates and coals. Ventifacts, formed by sand-blasting of associated fluvially-transported gravel, may also be present. Depending on the source, the framework fraction may vary widely from feldspathic to orthoquartzitic. It is interesting to note that to the best of our knowledge no ancient eolianites have ever been convincingly identified by size analysis alone – in spite of the many analyses of modern dune sands.

Critical for the eolian interpretation of an ancient sandstone is the character of its crossbedding. How thick are single crossbedded units and what is the variability of their dip direction? Judging from heights of modern dunes, thick crossbeds – perhaps in excess of 20 or more meters – with steep dips of 30 to 34 degrees might be expected and preserved in the ancient record. And indeed thick sets of crossbedding have been taken by some to imply an eolian origin. But do we really know enough about crossbedding in modern desert sands and on modern continental shelves to be certain of this? To explain relatively thin sets of crossbedding in sandstones believed to eolian, Stokes (1968) suggests that eolian erosion of loose sand is limited by the upper surface of the water table and that this surface is recorded by long, level erosion surfaces resulting in what he termed multiple, parallel, truncation bedding planes. Shotten (1937, Fig. 3), however, thought these surfaces to simply be the back slopes of barchans.

Variability of crossbedding poses another, perhaps more troublesome question, for if longitudinal and complex dune forms were as prevalent in ancient eolianites as they are in modern deserts, it seems most unlikely that well-oriented crossbedding would exist. An eolianite formed by longitudinal dunes (Fig. 11-43) would have a distinctive bimodal pattern of crossbedding (Table 11-8) and one consisting of complex dunes would have a very disperse, probably random pattern. Yet inspection of crossbedding maps of all the widespread ancient sandstones believed to be eolianites, shows variability of dip direction comparable to that of most fluvial and deltaic sands. Only coastal eolianites where parabolic, transverse, and possibly barchan dunes prevail have low variability of crossbedding and even then only if wind transport of sand were largely onshore. Could it be that most of the sands regarded as ancient eolianites are in fact water-laid?

The ripple index (wave length/wave height) has long been considered a means of distinguishing wind from water ripples. Tanner (1967) evaluated a number of ripple mark indices and concluded that wind-formed ripples could be distinguished from water-formed ones, where the index exceeds 18. Sharp (1963, p. 629) found ripple index to increase with grain size and wind velocity and that the average index in the Kelso Dunes of the Mojave Desert to be about 18.

Coastal dunes, especially those capping barrier-island sand bodies, may be somewhat more common in the geologic record, although they too have yet to be recognized in ancient terrigenous sandstone bodies. However, Mackenzie (1964) has studied Pleistocene eolianites (skeletal limestones) on Bermuda. He found 66 percent of the foresets to dip between 30 and 35 degrees and that the dunes migrated shoreward at times of high sea level so that the prevailing crossbedding dip direction is largely perpendicular to shoreline. He considered transverse dunes to be the most common type.

Finally, what can be inferred about the pre-Pleistocene paleowind systems from crossbedded sandstones? To make an inference, we need first strong evidence for eolian origin of the sand in question and second, a simple relation between atmospheric circulation and crossbedding dip direction. We regard identification of the eolian environment of a sand as an exceedingly difficult problem, one that requires a very careful, thorough analysis of all aspects of the deposit and its associated lithologies. Critically needed are more studies of the internal crossbedding of dunes in modern deserts such as McKee has made of the White Sands dunes. Better knowledge of crossbedding and relations of its dip direction to atmospheric circulations is also vital. Until the above requirements are met, it seems to us that paleowind inferences can be much more securely obtained from down-wind decline of grain size and thickness of ash falls (Eaton, 1964) than from the study of crossbedding in sandstones.

Annotated References:

Alimen, Marie-Henriette: Sables Quaternaires du Sahara Nord-Occidental (Saoura-Ougarta). Publ. Serv. Carte géol. L'Algerie (N.S.), Bull. **15**, 207 p. (1957).

Divided into two major parts: water-worked and eolian sands. Dune profiles, some size analyses, plus color and surface textures mostly of the dunes of the Grand Erg. Considers most of Grand Erg dunes as relict features. Modest bibliography, mostly local.

Bagnold, R. A.: The physics of blown sand and desert dunes, 265 p. London: Methuen 1941.

The original "process" book on sand dunes.

Glennie, K. W.: Desert sedimentary environments. In: Developments in sedimentology, **14**, 222 p. Amsterdam: Elsevier 1970.

Nine chapters and a glossory cover the main aspects of desert sedimentology and physiography. Useful discussion of how to recognize an ancient eolianite (about which the author is most confident).

Holm, D. A.: Sand dunes. In: Fairbridge, R. W., Ed.: The encyclopedia of geomorphology, p. 973–979. New York: Rheinhold 1968.

Short, informative, illustrated account with a rather different viewpoint. Includes a map of dune types in the Rub'al Khali. See also deflation, deserts and desert land forms, and eolian transport in the same volume.

Lowman, P. D., Jr.: Space panorama. Feldmeilen Zürich: Weltflugbild R. A. Miller 1968.

Numerous and superb color photographs of deserts and desert dunes from earth satellites of the Gemini flights.

Lustig, L. K.: Inventory of research of geomorphology and surface hydrology. In: An inventory of geographical research on desert environments. Univ. Arizona, Office of Arid Lands Research, Chap. IV, 189 p. (1967).

Many references, some of which are annotated, mostly arranged by geographic areas plus three chapters: background for modern geomorphic concepts, availability of information, and evaluation and recommendations.

Mabbutt, J. A., and others: Land systems of the Wiluna-Meekatharra area. In: General report on the lands of the Wiluna-Meekatharra area, Western Australia, 1958. Melbourne: Commonwealth Sci. Indus. Research Organization, Land Research Ser. No. 7, Part II, p. 24–72 (1963).

Desert area subdivided into many local regions each described by land form, soil type and vegetation plus block diagrams.

McGinnies, W. G., Goldman, B. J., Paylore, Patricia: Deserts of the world, 788 p. Tuscon: Univ. Arizona Press 1968.

Composed of seven parts: weather and climate, geomorphology, surface materials, vegetation, fauna, coastal desert areas, and ground-water hydrology. Each section organized into state of knowledge and pertinent publications. Many references. Pages 8 to 16 give a very brief description of major deserts.

McKee, E. D.: Structures of dunes at White Sands National Monument, New Mexico. Sedimentology (Special Issue) **7**, 69 p. (1966).

The most detailed study of the geometry and internal stratification of dunes available.

N.A.S.A.: Earth photographs from Gemini III, IV, and V, 266 p. Washington, D.C.: Natl. Aeron. Space Adm. 1967.

A large number of colored plates are of desert regions and show proportion of sand to bedrock and the many dune types. Also see Lowman (1968).

Peel, R. F.: The landscape in aridity. Inst. British Geographers, Trans. and Papers, Pub. **38**, 23 p. (1966).

A well-illustrated, general review stressing both erosional and depositional aspects using mostly North African examples. Over sixty references. Section on modern sand formations (p. 16–20) notable.

Smith, H. T. U.: Eolian geomorphology, wind direction and climatic change in North Africa. Bedford, Mass.: Air Force Cambridge Research Lab., No. **63–443**, 48 p. (1963).

Recognizes three types of active dunes (simple, compound and complex) and discusses stabilized dunes. Twenty photographs, 19 of which are of dune forms.

— Dune morphology and chronology in central and western Nebraska. Jour. Geology **73**, 557–578 (1965).

A geomorphologist analyzes the geomorphic history of a 22,000 square mile (57,200 km^2) area recognizing three major dune types and infers several paleowind systems.

Stone, R. O.: A desert glossary. Earth-Sci. Rev. **3**, 211–268 (1967).

Desert terminology organized by geographic area plus 37 pictures.

Problems of Sand-Body Prediction

Prediction has two major aspects – where to look for a new sandstone body and how to outline efficiently its areal extent with a minimum of data once it is found. Both problems are among the most challenging that a geologist can be asked to solve. The location problem is generally much more difficult than the extension problem.

As a rule, four kinds of knowledge are required: 1) depositional environment, 2) regional distribution of facies in the basin, 3) paleoslope, and 4) prior experience with similar bodies of the same facies, preferably but not necessarily in the same basin.

Perceptive contouring requires a proper model, one that indicates the probable shape of a sandstone body of a particular environment. To identify the correct model, a combination of internal and external characteristics are used. As we have seen, the principal external characteristics are the associated lithologies and character of the basal contact of the sand. Paleoslope is vital to proper interpretation.

Another source of information that can be most helpful may be gleaned from the vertical sequence of lithologies that preceed and follow the sandstone body in question, for ideally, lithologic transitions in vertical section reflect laterally migrating environments, as was early recognized by Walther (1894, p. 62). In the west this observation is known as Walther's Law. Shaw (1964, p. 50) and Visher (1965, p. 41–42) have recently re-emphasized Walther's Law as an effective means of predicting lateral occurrence from a vertical sequence. This procedure assumes that there are no erosional breaks in the section and that a vertical dependence or memory does in fact exist in the sequence.

Whether a vertical section displays a memory (see p. 447), and whether memory is short or long (does the presence or absence of a lithology or bed depend only on the immediately preceding or on several preceding lithologies?), is primarily a consequence of the spatial distribution of environments at the time of deposition (Fig. 11-45). Suppose, for example, that a series of depositional environments is related to water depth which increases systematically offshore and that each environment produces a distinctive sediment type. Ideally, slow transgression will result in a stratigraphic sequence wherein all or nearly all of the environments will be represented in a stratigraphic section *in the same order as they occur offshore*. As the different environments transgress shoreward, they leave behind a strongly linked sequence, one with a memory extending back one or more units or steps. Suppose on the other hand, that the different depositional environments occur more or less at random in a "crazy quilt" or haphazard pattern rather than having systematic relations to one another or to the strand line (Fig. 11-45). Shoreward transgression will now produce a stratigraphic section without strong vertical dependence – essentially a "memoryless" section.

These two hypothetical and rather extreme examples, assuming stratigraphic sections free of depositional breaks, suggest the following: sections with strong vertical memory accumulated in basins which had well-defined, strong systematic environmental zonation and, conversely, sections with little or no memory accumulated in basins with little or no systematic zonation of environments. Krumbein (1967) and also Potter and Blakely (1968, p. 159–160) discuss problems of determining the existence of memory and its length.

Effective prediction of a sand body or any other lithology with Walther's Law in a new borehole requires 1) that memory does in fact exist in the sequence, 2) knowledge of depositional strike (paleoslope), 3) keeping in the same cycle, and 4) that map pattern remain the same during migration. Successful prediction is enhanced if boundaries between environmental realms are fairly straight.

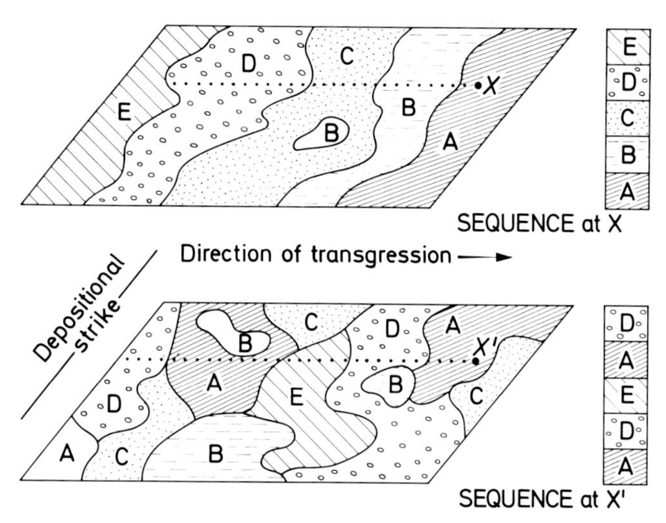

Fig. 11-45. Regular (above) *versus* "crazy quilt" (below) distribution of environments and resultant depositional sequences at X and X^1 produced by transgressions from left to right. See text for assumptions (Modified from Potter and Blakely, 1968, Fig. 2)

Once experience with a sandy facies in a basin has been acquired, it may be possible to develop a relatively small number of criteria – perhaps largely empirical – by which one can distinguish not only major but also subenvironments. For example, Sabins (1963b) showed how a combination of fossils, mineralogy, and information on median grain size could be analyzed in terms of a "yes" or "no" response in a computer-like flow diagram to distinguish seven subfacies of barrier-island and marine-shelf sand bodies (Fig. 11-46). With present knowledge it is generally not possible to utilize a particular set of criteria derived from a sand facies in one basin to identify a like facies in a different basin.

Extension Problem

Once the environment has been identified, how successfully can one predict the probable direction of elongation and the lateral extent of a sand body?

The direction of elongation can be attacked in one of several ways. Probably the simplest is by analogy with the orientation of similar sand bodies elsewhere in the basin. From what is known, this approach yields good results simply because paleoslope in many basins was stable through appreciable periods of time and,

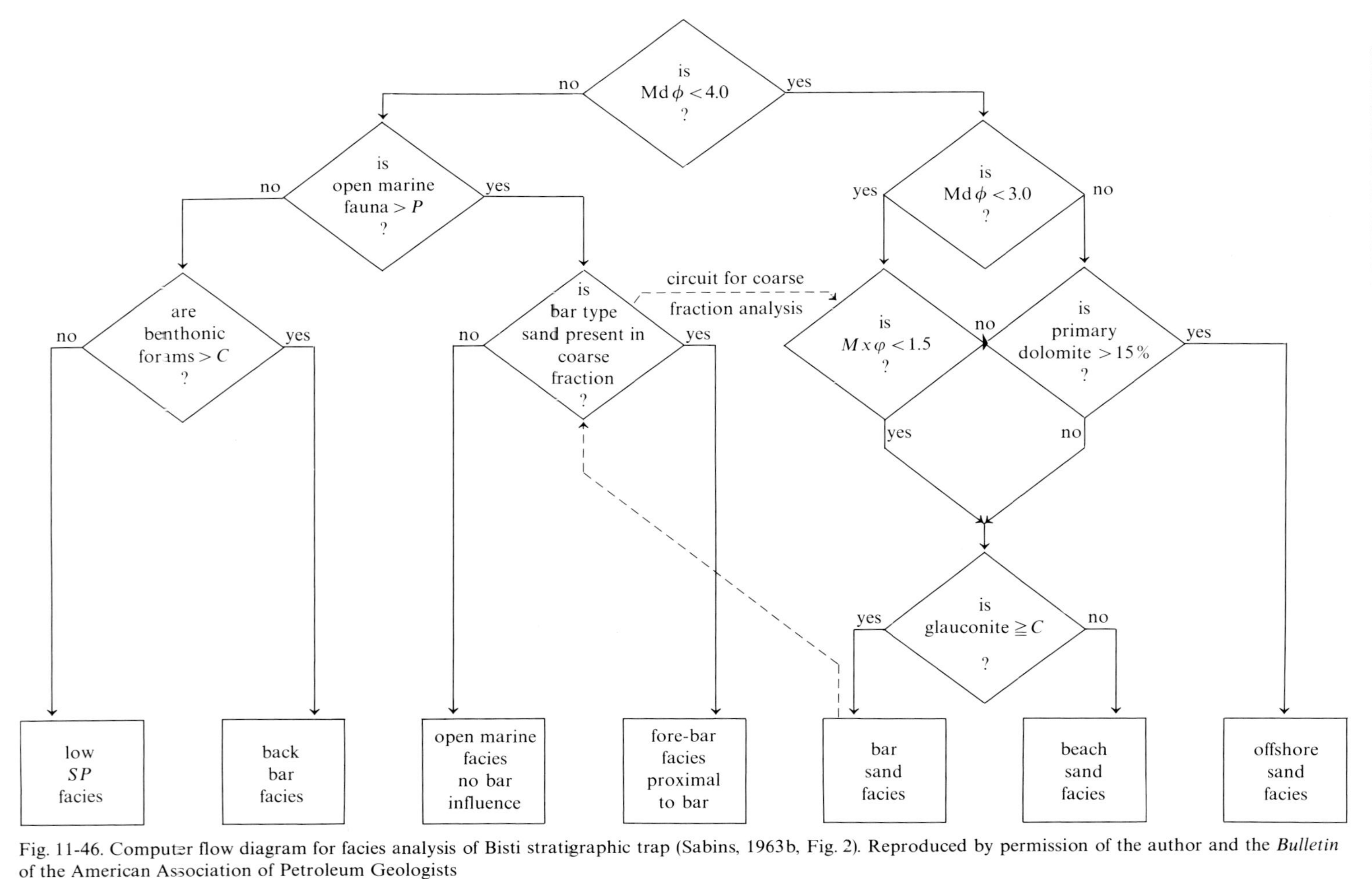

Fig. 11-46. Computer flow diagram for facies analysis of Bisti stratigraphic trap (Sabins, 1963b, Fig. 2). Reproduced by permission of the author and the *Bulletin* of the American Association of Petroleum Geologists

as a result, sandbody orientation changes very little in successive cycles. The foregoing empirical approach presumes a close correlation between orientation of a sand body and the paleoslope. Almost all tidal, barrier-island, turbidite, deltaic, and alluvial sand bodies have such a close relation to paleoslope: alluvial, deltaic, turbidite, and many tidal sand bodies are essentially perpendicular to it whereas barrier island and beaches are parallel to it. Marine-shelf sand bodies also may have a close relation, but so little is known about them that generalization is difficult. Orientation of internal directional structures also aids in prediction of the elongation of alluvial, tidal, and turbidite sand bodies. Crossbedding orientation is most helpful for alluvial, deltaic and tidal deposits, because it correlates well with long dimension. It is also the best structure to measure in marine-shelf sands, but its relation to long dimension in these sands is not well established. Nor do barrier-island sandstone bodies generally have a simple relation between their cross-bedding and elongation. It is possible in some cases to determine crossbedding orientation with a dipmeter as shown by Reese (1968) for some of the sands of the Gulf Coast. However, calibration with oriented cores or with known trends is recommended. Jizba, and others (1964) discuss some of the problems involved. Insofar as sand fabric correlates with sand body elongation, it can be used to predict direction.

The paleotopography of an unconformity also can be a valuable guide to to elongation of the sand bodies that lie just above it. It appears to be useful because – for a particular amount of well control – one can usually outline paleo-drainage and paleovalleys much more accurately than one can outline the sand bodies that later may have partly buried pre-existing topography. Paleotopography probably is most helpful when one is working with a youthfully dissected surface – an unconformity with entrenched second-cycle valleys – of either dendritic or trellis origin for under these conditions there is good correlation between sand body shape and valley orientation. Topographic control on basal sedimentation is likely to be less strong where an unconformity is in maturity. Busch (1959) stresses paleogeomorphology in prospecting for stratigraphic traps above unconformities. Rukhin (1962) discusses paleotopography in the search for fossil fuels (p. 584–619). Martin (1966) summarizes and illustrates applications of paleogeomorphology in oil and gas exploration in western Canada.

Where location of elongate sand bodies is believed to have been influenced by structure, structure maps also have been used to predict sand-body orientation.

More difficult than the direction of elongation of a sandy body is prediction of its lateral extent. In a sandstone-shale sequence a dipmeter may indicate effectively an edge of a sandstone body by the steeper dip of either its top or basal contact as it thins near an edge. Although published information is scant, there is probably a relation between thickness, grain size and bedding for most sandstone bodies that might be helpful, especially if cores are available. For example, alluvial sands are commonly finer grained and thinner bedded near their margins. In addition, crossbedding predominates in the axial region and is weak or absent in the marginal areas of many alluvial sands.

Geophysical techniques also, under certain conditions, may be rewarding, although few published case histories are available. For example, it may be possible to use some recent developments in common-depth-point techniques (Marr and Zagst, 1967) to resolve the sonic energy from small sand bodies. As with most geophysical techniques, it would be most helpful to first test this or other methods where sand body geometry is well established.

Proper contouring also can contribute to location of the margins of a sandstone body. Contouring sand thickness on the basis of arithmetic spacing between control points generally yields poor estimates of sand-body width. More realistic is the determination of the width of a sand body and of the rate of change of thickness along its margins in areas of good well control and extrapolation of these values into areas of less dense control.

Location Problem

The techniques of locating a new sandstone body are not as good as those for outlining its areal extent after discovery.

The pessimistic view of this problem is that elongate sandstone bodies – at least some types of them – occur more or less randomly within a particular facies in a basin. Much evidence in some well-drilled basins tends to support this view. Certainly, not much is known about locating small, isolated or apparently isolated marine shelf sandstone bodies from subsurface data.

Geologic solutions to the location problem may be either regional or local. A regional solution attempts at an early stage to outline areas – mostly groups of townships (squares with area of 91 km^2) – of greater and lesser probability of sand-body occurrence. One way to do this is to outline roughly the sand dispersal system of the basin by determining sand occurrence and directional structures in outcrop and supplementing this with the available subsurface data. Both structure and paleotopography have been used in some basins in the belief that these features, which have substantial continuity, exercise a control on sand occurrence. Because narrow elongate sandstone bodies in a shale section usually produce a compactional distortion in the adjacent shales, small density contrasts between the shale and sandstone may be sufficient to permit identification at shallow depths of the position of a sandstone body in much the same manner gravity has been successfully used in exploration for buried valleys of Pleistocene age (Hall and Hajnal, 1962). Buried sand and gravel deposits in till plains have been identified routinely by electrical resistivity (Buhle, 1953). More recently, temperature gradients have also been used (Cartwright, 1968).

Geologically, the most difficult prediction problem is the newly discovered basin with isolated marine-shelf sand bodies and no outcrop. Comparatively little can be done at an early stage except to drill structural highs. In contrast, best success can be expected in basins where an outcrop is nearby and the sand bodies are principally of alluvial origin. With such conditions it may be possible to make downdip predictions by using directional structures and sandstone thickness measured in outcrop or, if a structure has been successfully drilled, to use paleoslope and the downdip orientation of alluvial sand bodies to locate the first step-out well. Probably the biggest effort at downdip prediction with directional structures in an alluvial sequence has been made in the Triassic and Jurassic uraniferous

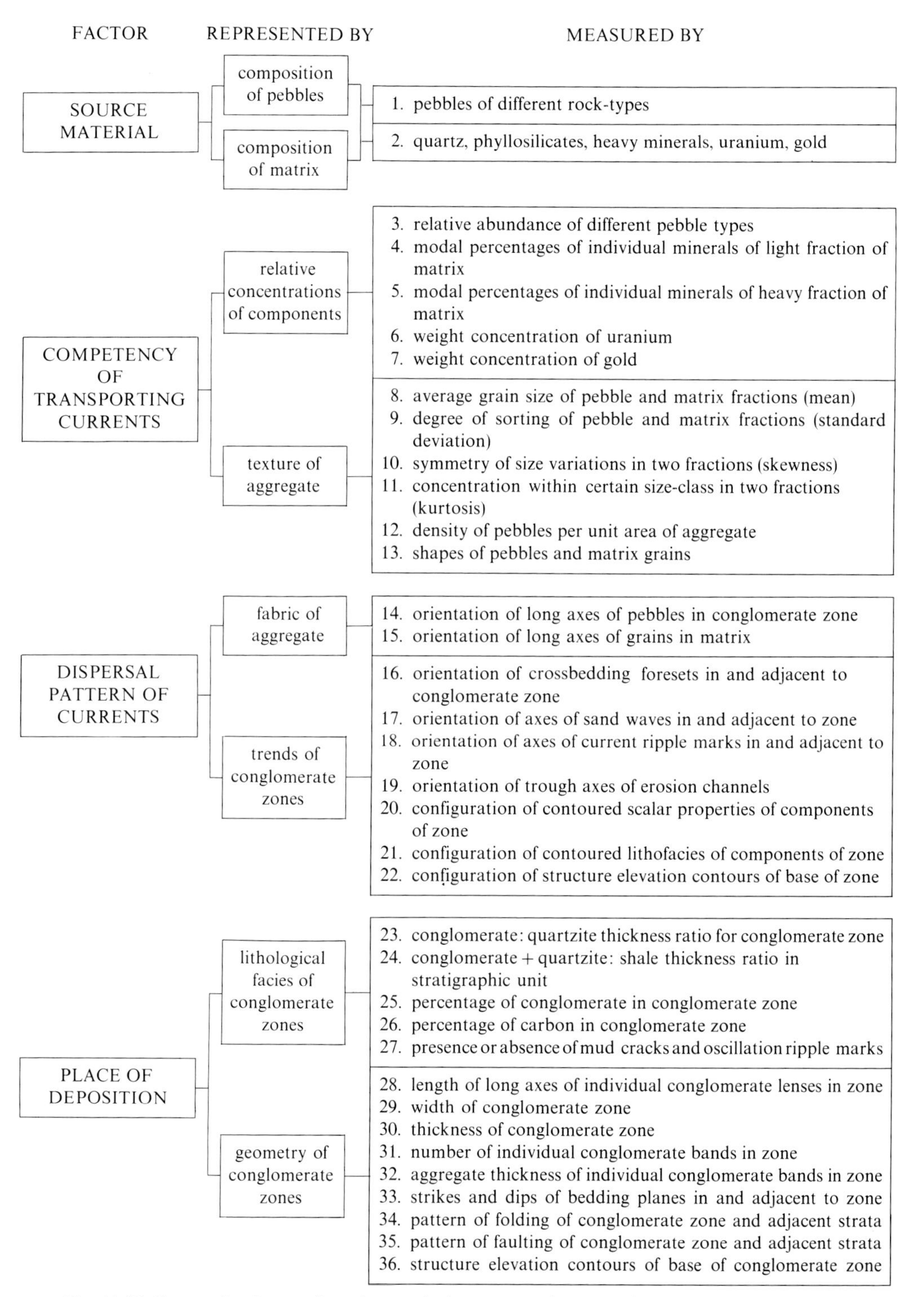

Fig. 11-47. Four major factors that play a role in concentrating "pay shoots" of detrital gold and how they are measured in the Precambrian conglomerates and quartzites of the Witwatersrand Basin of South Africa (Pretorius, 1966, Fig. 7)

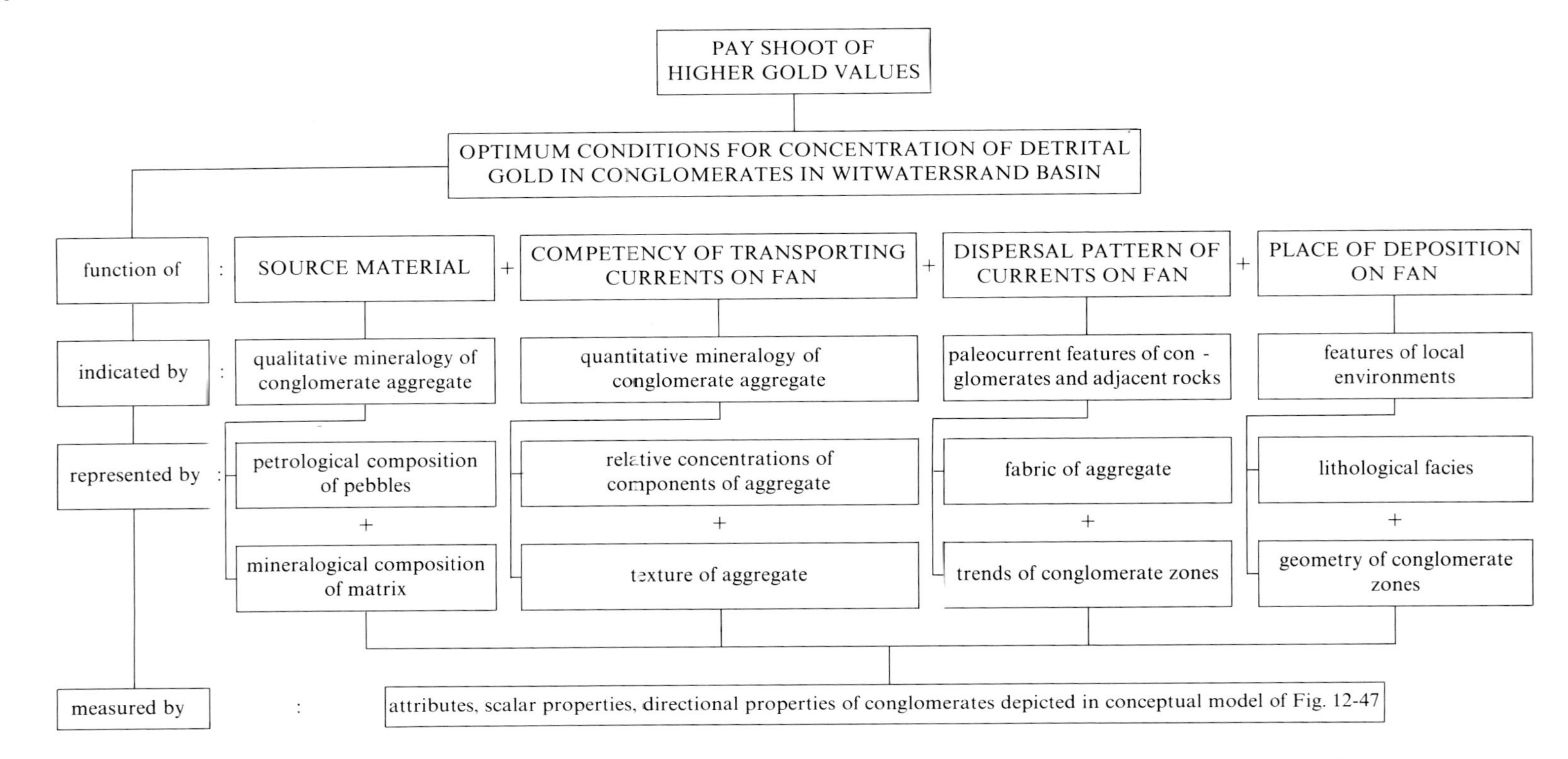

Fig. 11-48. Flow diagram showing complex interrelationship of some of the parameters listed in Fig. 11-47 (Pretorius, 1966, Fig. 6)

sandstone bodies of the Colorado Plateau and for gold in the Precambrian Witwatersrand System of South Africa.

Gold in the "reefs" of the Witwatersrand System is believed to be detrital and associated with alluvial fans that debouched into a large lake. Pretorius (1966), drawing upon a vast amount of previous work, has presented conceptual models for gold exploration in these deposits. His models analyze and identify major causal factors and show their interrelations (Figs. 11-47 and 11-48). The many other publications of the University of Witwatersrand's Economic Geology Unit are rich in sedimentary exploration methodology – both geological and statistical.

The foregoing assumes that sandstone bodies of a basin have a rational arrangement related to sand-dispersal system, basin geomorphology, and distribution of sources of supply. This rational arrangement has been termed a *facies model* (Potter, 1959). The concept of sand-body distribution in a basin has been best formulated for alluvial sand bodies but, as knowledge increases, more models for other environments will be worked out. While it may not be possible, by use of such a model, to locate a specific sand body, it may be possible to predict the areas or subregions where sand bodies of a predictable type are most likely to occur. It might have been possible, for example, to have anticipated the occurrence, the kind, and orientation of the oil-bearing sands of the late Devonian and early Mississippian in Pennsylvania from a sedimentologic analysis of the exposed lowest Mississippian (Pocono) of that region (Pelletier, 1958). Computer simulation of sandstone distribution in a basin appears to be the next logical step in the development of the facies-model concept for sands.

The facies-model concept with its emphasis on the existence of relatively few recurring models represents cause-and-effect "deterministic geology" – an approach that attempts to relate distribution and orientation of sand bodies in a basin to measurable, causal factors. Future research will undoubtedly incorporate a random, probabilistic element in such models as has been done by Krumbein (1968b) in his simulation of regressive and transgressive strand line sand bodies. Incorporation of such a probabilistic element is recognition of the fact that, for a stated amount of information, wholly deterministic prediction may be impossible.

Another approach is search theory, developed from operations research. Here one outlines a drilling campaign on the assumption that sand bodies of assumed dimensions and orientation are essentially randomly distributed either within an entire basin, or more likely, within a small part of it, perhaps (an area of 200 to 300 km^2). Applied first to ore bodies, application to sand-bodies would seem to be the next logical step (Celasum, 1964 and Marshall, 1964). Dowd (1964 and 1969) has suggested that some aspects of information theory and Bayesian probability analysis can be helpful in delineating the "statistical geometry" of sandstone oil-bearing reservoirs.

Sand Bodies as Permeable Conduits: A Sedimentological View

One of the most important economic aspects of sand bodies is their capacity to store and transmit water, oil, gas and mineralizing solutions. We are primarily concerned here with transmissive characteristics.

Variation in permeability is enormous and can range from hundreds of darcys in well-washed, unconsolidated terrace and river sands to virtually zero in a sandstone fully cemented by quartz, the quartz-cemented sandstone having a permeability comparable to that of some silts and clays (Fig. 11-49). Thus, the range of variation can be of the order of 100,000 times. Moreover, small-scale spatial variation is also great, especially in cemented sandstones, so that the permeability of a sandstone is one of its most variable properties, far exceeding that of porosity, for example. This variability has made permeability difficult to predict.

Specific Permeability, K.(Darcys)	10^5 10^4 10^3	10^3 10^2 10 1	1 10^{-1} 10^{-2} 10^{-3} 10^{-4}	10^{-4} 10^{-5}
Material	Clean gravel	Clean sands; mixtures of clean sands and gravels	Very fine sands; silts; mixtures of sand, silt, and clay; glacial till; stratified clays; etc.	Unweathered clays
Flow characteristics	Good aquifers		Poor aquifers	Impervious

Fig. 11-49. Magnitude of specific permeability, k, of gravel, sand, silt, and clay (Modified from Todd, 1960, Fig. 3-4)

But because of its economic importance, there is a very large engineering literature pertaining to flow theory and various measurement techniques. There is a much smaller literature pertaining to the analysis of the *geological factors* that control the size and shape of permeability contrasts within a reservoir. These contrasts are commonly called *reservoir inhomogeneities*. To gain insight to some of the problems of this aspect of sand bodies consider the following questions. In either the initial development or later water flooding of an oil field in an elongate sand body, how should wells be located or spaced? What ratios of maximum to minimum horizontal permeabilities might be anticipated – if any – in such a sand? And what relation is there between maximum horizontal permeability and elongation direction of the sand? Would it be the same for an alluvial channel fill and a Gulf Coast barrier?

These questions call attention to some problems which we cannot solve today as well as we should. We stress here the geological variables that affect permeability contrasts, our main interest being not so much in the details of the flow through the pore system but in the large scale features of a sand body, especially sedimentological ones that affect permeability (Table 11-9).

Flow Systems and Their Geologic Controls

Transmission of fluid in a sandstone depends on a combination of features formed by both primary and secondary processes. Fabric and orientation of the framework grains, sedimentary structures, and the bedding facies of a sand, and some extent its cementation, are all linked to primary deposition. The controls over permeability form a hierarchy (Table 11-10): texture and fabric are the fundamental building blocks of sedimentary structures and individual beds, which together define a bedding facies. One can, in turn, consider a sand body as nothing more

Table 11-9. *Rock properties and flow response*

ROCK PROPERTY	EFFECTS ON PERMEABILITY AND POROSITY
TEXTURE	
Grain Size	Permeability decreases with grain size; porosity unchanged.
Sorting	Permeability and porosity decrease as sorting becomes poorer.
Packing	Although little data is available, tighter packing favors both lesser permeability and porosity.
Fabric	In absence of lamination, controls anisotropy of permeability; permeability is a maximum parallel to mean shape fabric.
Cement	The more cement, the less permeability and porosity.
SEDIMENTARY STRUCTURES	
Parting lineation	Maximum permeability most probably parallels fabric in plane of bedding.
Crossbedding	Scant available data suggests that horizontal permeability parallels direction of inclination and that the steeper the dip of the foreset, the weaker the horizontal vector of permeability.
Ripple mark	Little data, but fine grain and more laminations combine to cause low permeability and hence ripple zones are commonly barriers to flow.
Grooves and flutes	As judged by fabric, permeability should parallel long dimension.
Slump structures	No data, but probably always greatly reduce horizontal permeability.
Biogenic structures	Destroy depositional fabric and bedding and thus drastically reduce permeability and cause minimal, if any, horizontal anisotropy of permeability. Effect on porosity is unknown, but may be negligible.
LITHOLOGY	
Sandstone	Thicker beds tend to be coarser grained and thus more permeable, if cement is not a factor. If weakly or uncemented, ratio of maximum to minimum permeability is perhaps less than 5 to 1; if cement controlled, ratio may reach 100 to 1 or more.
Shale	The prime barrier to flow that outshadows all others by far. Thus it is the *arrangement* of sand and shale much more than permeability variation *within* the sand that controls flow in most reservoirs.

than a particular arrangement of different bedding facies. Superimposed upon these are secondary patterns of cementation, fracture systems and faulting, all of which may locally transcend the importance of the primary controls on fluid flow. For example, Hutchinson and others (1961, p. 64) report that in an unconsolidated, well-sorted sand with regular bedding the ratio of maximum to minimum horizontal permeability is commonly less than 5 to 1; but if permeability is cement-controlled, ratios of perhaps 100 to 1 in the same depositional unit can be expected. If later cementation is associated with faulting or related fracture systems, a once-continuous reservoir may be effectively compartmentalized into isolated segments. Continuity may also, of course, be interrupted by faulting or because the reservoir consists, in reality, of discontinuous sands as in some alluvial sequences. Where fully cemented by silica, and thus more brittle than usual, later fracture systems, either related to faulting or folding, may constitute the principal paths of flow.

The most readily recognized barriers to fluid flow within a sandstone body are the siltstone and shale partings and shale beds within it. As emphasized by Polesek and Hutchinson (1967, p. 397) it is the *arrangement* of such barriers within a sand

Table 11-10. *Hierarchical sequence of primary controls on permeability*

CONTROL	REMARKS
TEXTURE AND FABRIC	
Defined by grain size, sorting, packing, and shape orientation of framework grains. Scale: 1 to a few cm^3.	Fundamental "building blocks" that define the primary pore system. Depositional fabric may be completely destroyed by burrowing organisms.
SEDIMENTARY STRUCTURES	
Crossbedding, ripple mark, and parting lineation are most common and nearly always have preferred orientation and anisotropic fabrics. Scale: 1 to 10^2 m^3.	Directional structures consist of anisotropic fabrics so that individual structures should behave as "flow packets."
BEDDING FACIES	
Defined by bed thickness, types and abundances of sedimentary structures and frequency of shale beds. Scale: 10^2 to 10^5 m^3.	Probably the most important primary control on permeability distribution in a sandstone body. Shale beds act as impermeable barriers to flow and are one of the more continuous lithologies.
COMPOSITE SAND BODIES	
Superposition of one "cycle" of sand upon another, cycles commonly separated by unconformities. Scale: 10^6 to 10^{10} m^3.	Characteristic of many alluvial and deltaic sands. Multilateral as well as multistory bodies possible.

body and not the permeability variations within the sand, that are of critical first-order concern in reservoir analysis. The distribution of impermeable shaly beds within a sand may determine whether it will behave as a single or as a stratified reservoir, the latter being one consisting of several flow units. Figure 11-50 illustrates the importance of siltstone beds to the smoothed permeability profile of a well-cemented, crossbedded sandstone. Shale partings have a similar effect as do thin laminations of other materials such as siderite. The lateral extent of such barriers is critical: if lateral extent is little greater than the diameter of a vertical core plug, such laminations may have little importance for vertical flow; but if the horizontal barriers are widespread, they may almost totally inhibit it. Commonly, shale beds and zones within a sand body, unless locally removed by erosion, are its most widespread lithologic units. Continuity of shale beds is maximal in turbidite sands (Fig. 11-38) and may also be high in marine shelf sands. The shale beds of alluvial sands, on the other hand, probably have the lowest continuity, largely because of their disruption by intrasand-body channels. Another important factor affecting continuity in alluvial and tidal sands, are the relatively impermeable ripple-bedded zones (part of their fining upward cycle) that can act as barriers to flow. Shale laminations of a millimeter or less commonly separate individual ripples in such zones and, along with fine grain and clay matrix, are probably responsible for the general low permeability of this facies. Zeito (1965) gives some quantitative data on lateral lithologic continuity in marine, deltaic, and fluvial sands. Polasek and Hutchinson (1967, p. 405–406) propose a quantitative heterogeneity factor, *HF*, that may be useful in assigning a numerical value to the lateral continuity or stratigraphic variability of a reservoir. Papers by Semin (1962) and

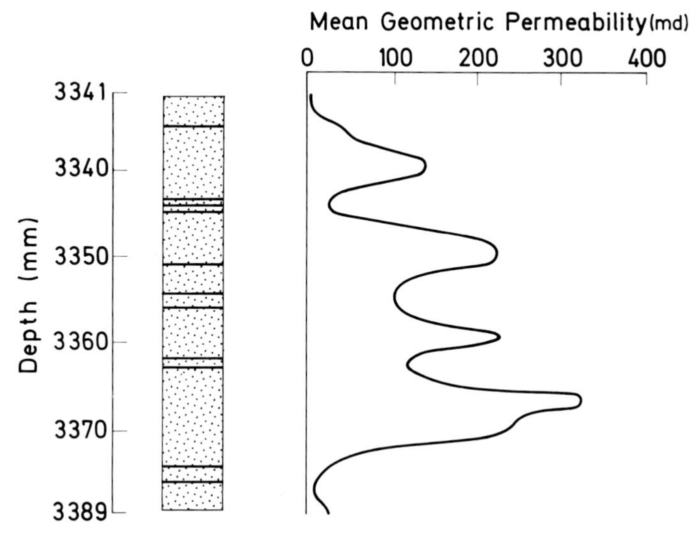

Fig. 11-50. Silicified siltstone beds (dark lines) and their affect on the smoothed permeability curve of the Hassi Messaoud Sandstone (Cambro-Ordovician) in well OMP 73 of the Hassi Messaoud oil field, Algeria (courtesy Compagnie Française des Pétroles d'Algerie)

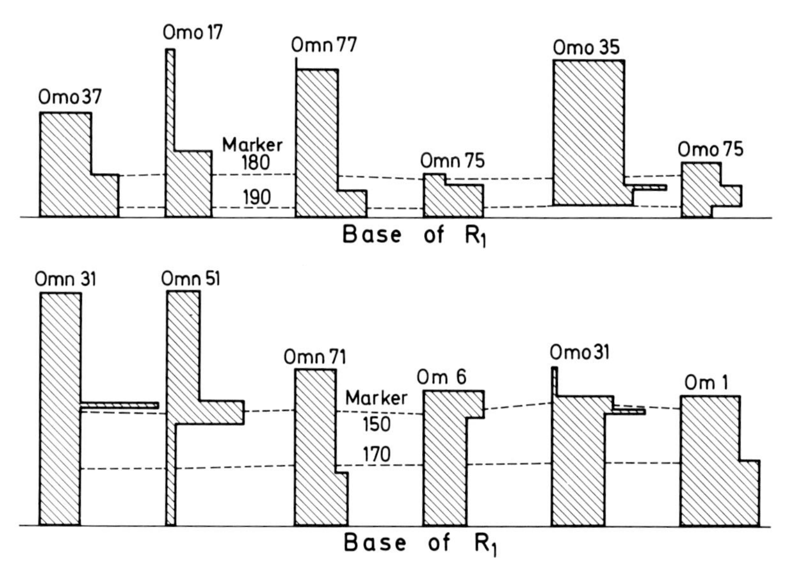

Fig. 11-51. Comparison of stratigraphic units (base of R_1 and Markers 170 and 150) and reservoir units as defined by an analysis of variance technique (Testerman, 1962) in the Hassi Messaoud oil field of Algeria (courtesy Compagnie Française des Pétroles Algerie). Note general lack of correspondence

Voinov (1967) contain the Russian approach to assessment of nonuniformity of oil-bearing strata.

In the analysis of a reservoir, it is essential to determine if its reservoir units correspond to stratigraphic units or not (Fig. 11-51). By *reservoir unit* is meant some segment of the sand body that has a characteristic permeability and porosity distribution, one different from its neighbors. These have also been called *hydrostratigraphic units* (Maxey, 1964). Table 11-11 summarizes the consequences of the correspondence – or lack of it – between reservoir and stratigraphic units. Predicting or modeling reservoir behavior is almost always simpler if reservoir units correspond to stratigraphic units than if they do not; if there is correspondence, the reservoir can be considered as a regular, ordered flow system with appreciable lateral continuity. When visual inspection of cross sections leaves doubt as to the correlation of reservoir and stratigraphic units, statistical

Table 11-11. *Correspondence between reservoir and stratigraphic units*

If Stratigraphic Units *Are* Reservoir Units,

1. stratigraphic units can be used with confidence to extrapolate reservoir units between wells recognizing, of course, the possibility of interruption by faults and unconformities;
2. reservoir units depend chiefly on primary sedimentation; and
3. the reservoir is vertically stratified and is thus vertically inhomogeneous.

If Stratigraphic Units *Are Not* Reservoir Units,

1. stratigraphic units cannot be used to extrapolate reservoir units between wells;
2. reservoir units are probably related to postdepositional factors, which are often unsystematic unless they are related to, for example, alteration along an unconformity or cementation proximal to fault;
3. the reservoir may or may not be vertically stratified; and
4. statistical zoning of permeability may be useful.

testing with an analysis of variance technique proposed by Testerman (1962) or time series analysis (Bennion, 1968) may be helpful. Well spacing is all important in this decision for a reservoir judged to have no correspondence between reservoir and stratigraphic units with wide well spacing might be judged otherwise with close spacing.

A few data are available for horizontal variation of permeability in sandy oil reservoirs. Study of oriented cores by Johnson and Breston (1951) showed that the permeability of the Third Bradford Sand (Mississippian) of Pennsylvania was anisotropic. Greenkorn and others (1964) also reported anisotropic permeability in a reservoir. They used tensor methods to determine major and minor axes of permeability based on measurements on a set of eight horizontal plug samples spaced 45 degrees apart. With present incomplete knowledge we can but speculate on the likelihood and magnitude of horizontal anisotropy in sandstones and on its origin as well.

Although horizontal anisotropy may be linked to primary deposition, the really critical supporting field evidence is yet to be obtained. We can, therefore, only surmise and speculate. The starting point is grain shape, fabric, bedding and their relation to permeability. There is surprisingly little published on the relations between these three and permeability even though some of the first observations were made in the 19th century.

Newell (cited in King, 1899, p. 126) noted in 1885 that in sandstones permeability was greater parallel to bedding than perpendicular to it. Fettke (1938) measured permeability in different directions on one-inch cubes and found vertical permeability (perpendicular to bedding) to be commonly 20 to 35 percent less than horizontal permeability (parallel to bedding) and that permeability varied in the horizontal direction also. Johnson and Hughes (1948), using a radial flow system, found that the average ratio of maximum to minimum permeability was about 1.4 to 1.0. Rühl and Schmidt (1958) report that the ratio of vertical to horizontal permeability varied from 0 to 1 and that the greater the permeability and porosity the more nearly a ratio of unity was approached. In the plane of the bedding they noted that ratios of maximum to minimum permeability rarely exceed 1.5. Wyllie and Spangler (1952, p. 394) also have commented on the anisotropy of electrical resistance in sandstones.

From a sedimentological viewpoint, fabric and bedding, the latter perhaps only visible to the X-ray (Fig. 4-3), control permeability variations. Griffiths (1948), using three samples, found best shape-orientation to be parallel to maximum permeability. He noted (p. 161) that the maximum permeability is parallel to the angle of imbrication of the quartz grains. Permeability variations may be shown by imbibition patterns of water or other liquids. The direction of maximum imbibition is parallel to both direction of maximum permeability and mean fabric direction (Mast and Potter, 1963, p. 549–558). Fig. 11-52 shows idealized imbibition patterns for laminated and nonlaminated sandstones with anisotropic shape fabrics. The correlation in sandstone is best shown in vertical sections where grain fabric is strongest. The imbibition ellipse is elongated parallel to the fabric axis. Contributing to a stronger anisotropy in vertical sections is lamination itself with its attendant reduction of permeability. In the plane of the bedding, fabric is much weaker and imbibition patterns disclose minimal anisotropy so that correlation with fabric is also weak. Rutledge (1966) found that *cryptostructures* – obscure inhomogeneities due to sedimentary structures and revealed only by X-rays – were a major factor causing irregular imbibition patterns in sandstones with an otherwise good effective porosity.

Data collected by Fondeur (1964) convincingly demonstrates the relative importance of lamination and fabric in sandstones. He studied permeability in the silicified Hassi Messaoud Sandstone (Cambrian) of Algeria (Fig. 11-53). There is little difference between K_B and K_C, the permeabilities in the B and C directions, in the plane of the foresets, even though grain fabric on foresets is known to be parallel to dip direction (Potter and Mast, 1963, p. 454–458 and Hamilton and others, 1968). There is appreciable difference between K_A and $(K_B + K_C)/2$, however, for K_A ranges from only one-half to one-fourth as much as average permeability in the plane of the foresets. Thus from the scant data available, small-scale lamination in a sandstone affects permeability anisotropy more than does grain fabric. In the ripple-bedded facies, it is the frequency of such laminations, as well as fineness of grain, that cause low permeability.

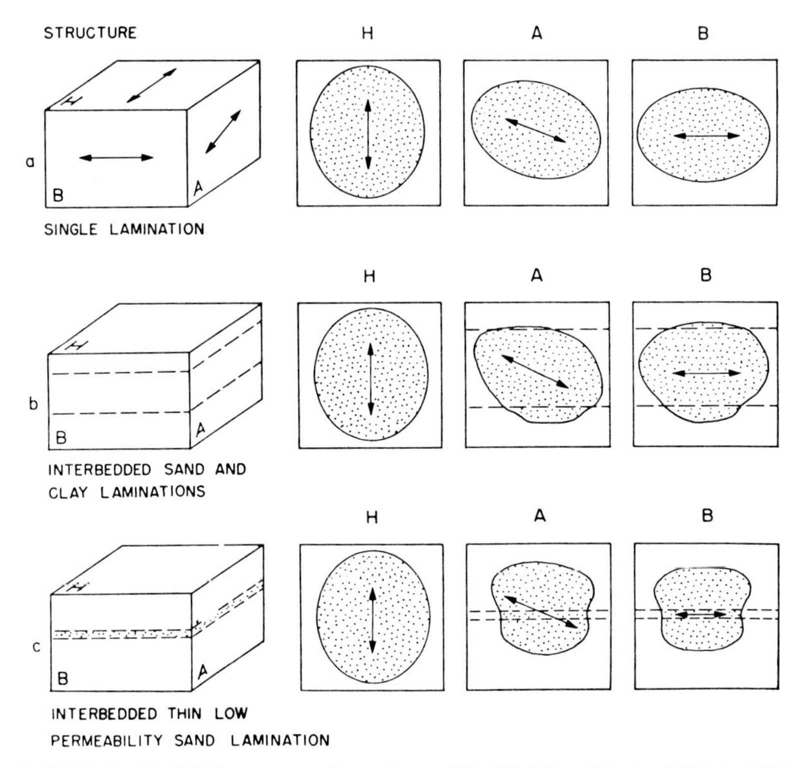

Fig. 11-52. Idealized imbibition patterns in sandstone (Modified from Mast and Potter, 1963, Fig. 6)

A number of investigators have shown that directional structures such as crossbedding, ripple marks, parting lineation and others have shape fabrics related to shape and orientation of the sedimentary structure and the current flow vector that formed it. Hence where fluid flow follows fabric or internal bedding, it seems reasonable to consider all directional structures as anisotropic "flow packets" within the sand body. It is such packets that probably control the primary anisotropic horizontal permeability of elongate sand bodies. Be-

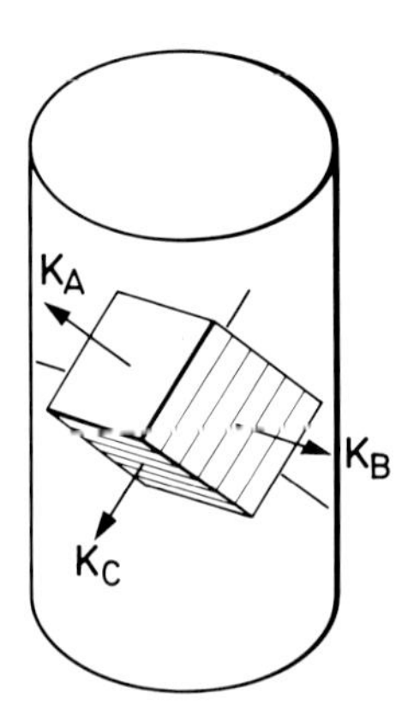

Fig. 11-53. Cube oriented with respect to foresets of a crossbedded sandstone and definition of three mutually perpendicular permeabilities, K_A, K_B, and K_C (Modified from Fondeur, 1964, Fig. 2)

cause crossbedding with preferred orientation is the dominant structure of many, if not most, sand reservoirs, it is the prime suspect for depositional control of anisotropic horizontal permeability in a reservoir. Hewitt and Morgan (1965, p. 342) reported horizontal permeability to be greatest parallel to crossbedding in the Pennsylvanian Robinson sand of the Illinois Basin. Minimum permeability, at right angles, was about 90 to 95 percent of maximum permeability. To establish this in other sandstones, crossbedding orientation should be checked in at least a few wells and compared with horizontal permeability. Primary sedimentary structures such as crossbedding have greatest significance for reservoir behavior when permeability and porosity are low rather than they are high.

Case Histories

Careful mapping of bedding facies and shale zones within a sand body is commonly the key to understanding primary controls of its reservoir characteristics. Mapping cements may also be essential, although it is a much slower and more laborious task.

Jobin (1962) contrasted the transmissive characteristics of eolian and marine sandstones with sandstones of fluvial origin in the Colorado Plateau region. His object was to relate regional transmissivity of groundwater to uranium mineralization. He noted that eolian and marine sands occurred as blankets and wedges with relatively simple internal geometry. These sands have moderate to high permeability with small variance and slight local and regional gradients. By contrast, fluvial sands are multistory and internally more complex and in-

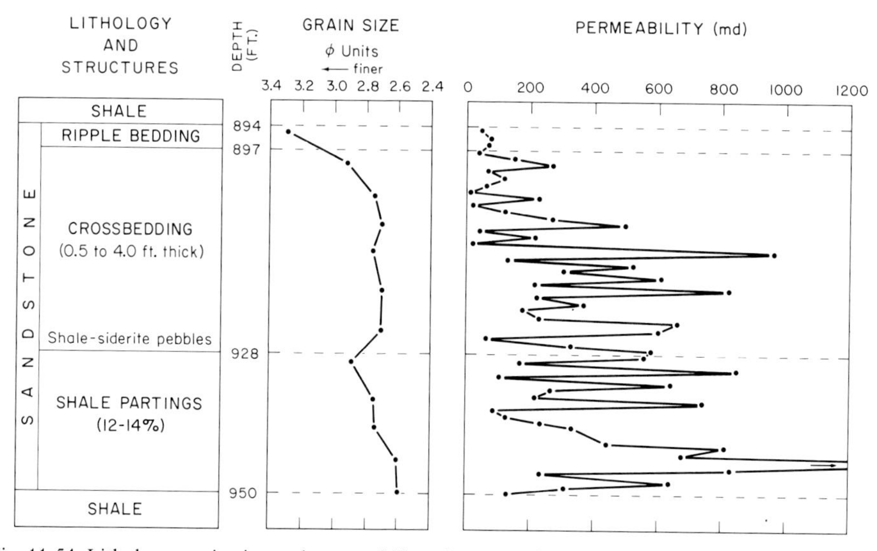

Fig. 11-54. Lithology, grain size and permeability of a Pennsylvanian sandstone in the Illinois Basin (Modified from Hewett and Morgan, 1965, Fig. 7). Reproduced by permission of American Association Petroleum Geologists

homogeneous lithologically. They have low to moderate permeability with large variance and high local but uniform regional gradients. Primary sedimentation, in terms of gross geometry and bedding facies, thus controls hydrologic behavior in all three environments. Jobin's study is regional in scope and encompasses many rather than single sand bodies.

Very different is the work of Hewitt and Morgan (1965) on a single Pennsylvanian sandstone body in the Illinois Basin. They studied oriented diamond drill cores from a multistory alluvial sand and found good correlation between bedding facies and reservoir characteristics (Fig. 11-54). Their study

Table 11-12. *Rock properties and flow characteristics of a sand body*[a] *(Hrabar and Potter, 1969, Table 1)*

	LITHOLOGY and TEXTURE	PETROLOGY		POROSITY and PERMEABILITY
CROSS-BEDDED	Dominantly crossbedded, fine to medium, well sorted (0.3 to 0.5 ϕ) sand. Crossbeds from 4 to 187 cm thick. Some massive beds.	Mostly quartz (95%+) with traces of chert and feldspar	Cement (12 to 17%) is mostly authigenic quartz, chalcedony and iron oxide plus minor clays.	Porosity ranges from 22 to 27%; permeability ranges from 137 to 3300 md and averages 600 md.
RIPPLE-BEDDED	Ripple bedding with shale laminations. Poorly to moderately sorted (0.7 to 1.5 ϕ) silt and fine sand.		Sparry calcite and less abundant clay minerals form 30 to 55% of sand.	Porosity ranges from 7 to 22%; permeability averages only 10 md.

[a] Reproduction with permission of the *Bulletin* of the American Association of Petroleum Geologists.

preceded ignition of the reservoir to determine its effect on tertiary oil recovery. They found horizontal permeability to be parallel to long dimension of the sand body. During later combustion, the fire front was influenced by the sand's bedding facies.

Hrabar and Potter (1969) made a reservoir analysis from outcrop data on a delta finger sand body of Bethel Sandstone (Mississippian) age in the Illinois Basin and found that the low-permeability, ripple-bedded facies within the reservoir was the principal barrier to flow (Table 11-12). The ripple-bedded facies is finer grained, more poorly sorted, and contains much more cement (30 to 55 percent) than does the crossbedded facies (12 to 17 percent). As a result, its permeability averages 10 millidarcys whereas that of the crossbedded facies averages 600. Although outcrops are insufficient to fully demonstrate it, the ripple-bedded facies probably lacks appreciable continuity because the delta finger is multistory so that there is local erosion at the base of the crossbedded facies. Movement of groundwater downward into underlying carbonates has greatly increased the permeability of this outcrop sand body. If not later recemented, sand bodies immediately underlying buried unconformities may also display higher than average permeabilities.

Papers that are noteworthy for their emphasis on the methodology of description of the lithology of sand reservoirs, descriptions having maximum relevance for fluid flow, are those by Verrien and others (1967) and Polasek and Hutchinson (1967). More relevant description by sedimentologists is essential if maximum use is to be made of engineering theory.

Annotated References:

Dupuy, M., Lefebvre du Prey, E.: L'anisotropie d'écoulement en milieu poreaux presentant des intercalations horizontales discontinues. IIIe Colloque A.R.T.F.P. (Pau 23/26 Sept. 1968), Common. 34, Inst. Francais du Pétrole 15982, 48 p. (1968).

This report summarizes earlier studies by the authors. Their starting point is the distribution of impermeable silt layers as mapped in a cliff of crossbedded Cambrian sandstone exposed in south Algeria. They simulated fluid flow in this system by electrical flow in a paper model and then compared theory with experiment. Unusual integration of geology and engineering.

Ferradon, Jean: Les lois de l'écoulement de filtration. Le Genie Civil **75**, 24–28 (1948).

Brief, clear mathematical exposition includes triaxial flow, both isotropic and anisotropic. Theory.

Groult, J., Reiso, L. H., Montadert, L.: Reservoir inhomogeneities deduced from outcrop observations and production logging. Jour. Petroleum Tech. **18**, 883–891 (1966).

Summary in English of most of the work of the Institute Francais du Pétroles on the geological aspects of reservoir inhomogeneities.

Hutchinson, C. A., Jr.: A frontier in reservoir technology. Reservoir inhomogeneity assessment and control. The Petroleum Engineer **31**, 19–26 (1959).

A good general review stressing fundamentals and emphasizing effects of stratification (permeability zones) and directional permeability on flow.

Marcus, H., Evenson, D. E.: Directional permeability in anisotropic porous media. Berkeley, Univ. California Hydraulic Laboratory, Water Research Center, Contr. No. **31**, 105 p. (1961).

Theory and model results compared for continuous and discontinuous layered porous media. Equations presented in graphical form when possible.

Matheron, G.: Eléments pour une théorie des milieux poreux, 166 p. Paris: Masson 1966.

Background for those who wish to see the theoretical as viewed by the foremost French geomathematician. Two parts. Part I is the geometry of the pore system (4 chapters) and part II is the hydrodynamics of flow (3 chapters). Free use of set theory, Boolean algebra and Markov processes in Part I, to describe the geometry of the pore system. Theory.

Scheidegger, A. E.: The physics of flow through porous media, 236 p. New York: MacMillan Co. 1957.

Eight chapters covering almost all aspects. Mathematical theory.

Warren, J. E., Price, H. S.: Flow in heterogeneous porous media. Am. Inst. Mining Metall. Petroleum Engineers Trans. **222**, 153–169 (1961).

A mathematical attempt to infer the heterogeneity of a reservoir's physical properties using single phase flow and based on data obtained from conventional well tests and cores. Numerical methods used to solve equations. Engineering background required. Theory.

World Petroleum Congress, Panel on Reservoir Geology: Reservoir geology, in Proceedings 7th World Petroleum Congress, v. 2, Origin of oil, geology and geophysics, p. 301–458. Barking: Elsevier Publ. Co., Ltd. 1967.

Eight high quality papers on different aspects of flow and barriers to it in sandstones and limestones. In broad terms, these papers relate both primary and diagenetic rock properties to flow barriers.

References Cited

Alimen, Marie-Henriette: Étude sur le Stampien du bassin de Paris. Soc. géol. France Mem. **14**, 304 p. (1936).

Allen, J. R. L.: A review of the origin and characteristics of Recent alluvial sediments. Sedimentology (Special Issue) **5**, 191 p. (1965a).

— Late Quaternary Niger delta, and adjacent areas: Sedimentary environments and lithofacies. Am. Assoc. Petroleum Geologists Bull. **49**, 547–600 (1965b).

— Studies in fluviatile sedimentation: a comparison of fining upwards cyclothems, with special reference to coarse member composition and interpretation. Jour. Sed. Petrology **40**, 298–323 (1970).

— Friend, P. F.: Deposition of the Catskill facies, Appalachian region: With notes on some other Old Red Sandstone basins. In: Klein, G., de Vries, Ed.: Late Paleozoic and Mesozoic continental sedimentation, northeastern North America. Geol. Soc. America Spec. Paper **106**, 21–74 (1968).

van Andel, Tj. H., Komar, P. D.: Ponded sediments of the mid-Atlantic Ridge between 22 degrees and 23 degrees north latitude. Geol. Soc. America Bull. **80**, 1163–1190 (1969).

Andresen, M. J.: Geology and petrology of the Trivoli sandstone in the Illinois Basin. Illinois Geol. Survey, Circ. **316**, 31 p. (1961).

Asquith, D. O.: Depositional topography and major marine environments, Late Cretaceous, Wyoming. Am. Assoc. Petroleum Geologists Bull. **54**, 1184–1224 (1970).

Baker, C. H., Jr.: New observations on the Sheyenne delta of glacial Lake Agassiz. U.S. Geol. Survey Prof. Paper **575** B, 62–68 (1967).

Ball, M. M.: Carbonate sand bodies of Florida and the Bahamas. Jour. Sed. Petrology **37**, 556–591 (1967).

Baars, D. L., Seager, W. R.: Stratigraphic control of petroleum in White Rim Sandstone (Permian) in and near Canyonlands National Park, Utah. Am. Assoc. Petroleum Geologists Bull. **54**, 709–718 (1970).

Barrell, Joseph: Relative geological importance of continental littoral and marine sedimentation. Jour. Geology **14**, pt. 1, 316–356; pt. 2, 430–457, and pt. 3, 524–568 (1906).

— The upper Devonian delta of the Appalachian geosyncline. Am. Jour. Sci. 4th ser. **36**, 436–472 (pt. 1); **37**, 87–107 (pt. 2); and **37**, 225–253 (pt. 3) (1913 and 1914).

Bass, N. W.: Origin of Bartlesville shoestring sands, Greenwood and Butler Counties, Kansas. Am. Assoc. Petroleum Geologists Bull. **18**, 1313–1345 (1934).

Leatherock, Constance, Dillard, W. R., Kennedy, L. E.: Origin and distribution of Bartlesville and Burbank shoestring oil sands in parts of Oklahoma and Kansas. Am. Assoc. Petroleum Geologists Bull. **21**, 30–66 (1937).

Behrn, S. H.: Heavy-mineral beach deposits in the Karoo System. Geol. Survey South Africa, Mem. **56**, 116 p. (1965).

Belderson, R. H., Stride, A. H.: Tidal current fashioning of a basal bed. Marine Geology **4**, 237–257 (1966).

Bennion, D. W.: Time series analysis for reservoir zonation. Jour. Petroleum Technology **20**, 913–914 (1968).

Berg, R. M., Cook, B. C.: Petrography and origin of Lower Tuscaloosa sandstones, Mallalieu Field, Lincoln County, Mississippi. Gulf Coast Geol. Soc. Trans. **18**, 242–255 (1968).

Bervin, R. J.: Cardium sandstone bodies, Crossfield-Farrington area, Alberta. Bull. Canadian Petroleum Geology **14**, 208–240 (1966).

Bigarella, J. J., Salamuni, R.: Early Mesozoic wind patterns as suggested by dune bedding in the Botucatú Sandstone of Brazil and Uruguay. Geol. Soc. America Bull. **72**, 1089–1106 (1961).

Bonham-Carter, G. F., Sutherland, A. J.: Mathematical model and Fortran IV program for computer simulation of deltaic sedimentation. Kansas Geol. Survey, Computer Contr. **24**, 56 p. (1968).

Boyd, D. R., Dyer, B. F.: Frio barrier bar system of south Texas. Am. Assoc. Petroleum Geologists Bull. **50**, 170–178 (1966).

Brown, L. F., Jr., Wermund, E. G., Eds.: A guidebook to the Late Pennsylvanian shelf sediments north-central Texas, 69 p. Dallas: Dallas Geological Soc. 1969.

Buhle, M. B.: Earth resistivity in ground-water studies in Illinois. Am Inst. Mining Metall. Engineers Trans. **34**, 395–399 (1953).

Bull, W. B.: Alluvial Fans. Jour. Geol. Education **16**, 101–106 (1968).

Busch, D. A.: Prospecting for stratigraphic traps. Am. Assoc. Petroleum Geologists Bull. **43**, 2829–2843 (1959).
Byrne, J. V. and others: The chenier plain and its stratigraphy, southwestern Louisiana. Gulf Coast Assoc. Geol. Soc. Trans. **9**, 1–23 (1959).
Cannon, J. L.: Outcrop examination and interpretation of paleocurrent patterns of the Blackleaf Formation near Great Falls, Montana. Billings Geol. Soc. 17th Ann. Field Conf. p. 71–111, 1966.
Cartwright, Keros: Temperature prospecting for shallow glacial and alluvial aquifers in Illinois. Illinois Geol. Survey, Circ. **433**, 41 p. (1968).
Celasun, Merith: The allocation of funds to reconnaissance drilling projects. Colorado School Mines Quart. **59**, pt. A, 169–185 (1964).
Coleman, J. M.: Deltaic evolution. In: Fairbridge, R., Ed.: Encyclopedia of geomorphology, Vol. 3, p. 255–261. New York: Rheinhold 1968.
— Gagliano, S. M.:: Cyclic sedimentation in the Mississippi River deltaic plain. Gulf Coast Assoc. Geol. Soc. Trans. **14**, 67–80 (1964).
— — Webb, J. E.: Minor sedimentary structures in a prograding distributary. Marine Geology **1**, 240–258 (1964).
Collinson, C. W.: Devonian of the north-central region United States. In: Oswald, D. H., Ed.: International Symposium on the Devonian System, Calgary, 1967, Vol. 1, p. 933–973. Calgary: Alberta Soc. Petroleum Geologists 1967.
Conybeare, C. E. B.: Influence of compaction on stratigraphic analysis. Bull. Canadian Petroleum Geology **15**, 331–345 (1967).
Cooper, W. S.: Coastal sand dunes of Oregon and Washington. Geol. Soc. America Mem. **72**, 169 p. (1958).
— Coastal dunes of California. Geol. Soc. America Mem. **104**, 131 p. (1967).
Corbeille, R. L.: New Orleans barrier-island. Gulf Coast Assoc. Geol. Soc. Trans. **12**, 223–230 (1962).
Curray, J. R.: Sediments and history of Holocene transgression, continental shelf, northwest Gulf of Mexico. In: Shepard, F. P., Ed.: Recent sediments northwest Gulf of Mexico, p. 221–266. Tulsa: Am. Assoc. Petroleum Geologists 1960.
— Late Quaternary history, continental shelves of the United States. In: Wright, H. E., Jr., Frey, David G., Eds.: The Quaternary of the United States, p. 723–735. Princeton: Princeton University Press 1965.
Curtis, D. M.: Relation of environmental energy levels and ostracod biofacies in east Mississippi delta area. Am. Assoc. Petroleum Geologists Bull. **44**, 471–494 (1960).
Dake, C. L.: The problem of the St. Peter Sandstone. Missouri Univ. School Mines and Metallurgy, Bull., Tech. Ser. **6**, 228 p. (1921).
Davies, D. K.: Carbonate turbidites, Gulf of Mexico. Jour. Sed. Petrology **38**, 1100–1109 (1968).
— Moore, W. R.: Dispersal of Mississippi sediment in the Gulf of Mexico. Jour. Sed. Petrology **40**, 339–353 (1970).
Dickas, A. B., Payne, J. L.: Upper Paleocene buried channel in Sacramento Valley California. Am. Assoc. Petroleum Geologists Bull. **51**, 873–882 (1967).
Dondanville, R. F.: The Fall River Formation, northwestern Black Hills: lithology and geologic history. Guidebook, First Joint Field Conf., Wyoming Geol. Assoc. and Billings Geol. Soc., Aug. 8–10, 1963, p. 87–99 (1963).
Dott, R. H., Jr.: Eocene deltaic sedimentation at Coos Bay, Oregon. Jour. Geology **74**, 373–419 (1966).
Doty, R. W., Hubert, J. F.: Petrology and paleogeography of the Warrensburg channel sandstone, western Missouri. Sedimentology **1**, 7–39 (1962).
Dowd, J. P.: Application of information theory in establishing oil field trends. In: Parks, G. A., Ed.: Computers in the mineral industries, p. 577–610. Stanford: The School of Earth Science, Stanford Univ. 1964.
— Statistical geometry of petroleum reservoirs in exploration and exploitation. Jour. Petroleum Technology **21**, 841–852 (1969).
Dupuy, J. P., Oswaldt, G., Sens, J.: Champ de Cazaux-géologie et production. In: 6th World Petroleum Cong. Proc., Frankfurt am Main, Sec. II, p. 199–212 (1963).
Eaton, G. P.: Windborne volcanic ash. A possible index to polar wandering. Jour. Geology **72**, 1–35 (1964).
Eicher, D. L.: Paleobathymetry of Cretaceous Greenhorn sea in eastern Colorado. Am. Assoc. Petroleum Geologists Bull **53**, 1075–1090 (1969).

Emery, K. O.: Atlantic continental shelf and slope of the United States. U.S. Geol. Survey Prof. Paper **529-A**, 23 p. (1966).
Emrich, G. H.: Ironton and Galesville (Cambrian) sandstones in Illinois and adjacent areas. Illinois Geol. Survey Circ. **403**, 55 p. (1966).
Evans, Graham: Inter-tidal flat sediments and their environments of deposition in the Wash. Geol. Soc. London Quart. Jour. **121**, 209–245 (1965).
Evans, W. E.: Imbricate linear sandstone bodies of Viking Formation in Dodsland-Hoosier area of southwestern Saskatchewan, Canada. Am. Assoc. Petroleum Geologists Bull. **54**, 469–486 (1970).
Ferm, J. C., Williams, E. G.: Characteristics of a Carboniferous marine invasion in western Pennsylvania. Jour. Sed. Petrology **35**, 319–330 (1965).
Fettke, C. R.: The Bradford Oil Field, Pennsylvania and New York. Pennsylvania Geol. Survey, 4th ser., Bull. **M-21**, 1938.
— Music Mountain oil pool, McKean County, Pennsylvania. In: Levorsen, A. I., Ed.: Stratigraphic type oil fields, p. 492–506. Tulsa: Am. Assoc. Petroleum Geologists 1941.
Finch, W. I.: Geology of uranium deposits in Triassic rocks of the Colorado Plateau region. U.S. Geol. Survey Bull. **1074-D**, 164 p. (1959).
Fisher, W. L., McGowen, J. H.: Depositional systems in the Wilcox Group of Texas and their relationship to occurence of oil and gas. Gulf Coast Assoc. Geol. Soc. Trans. **17**, 105–125 (1967).
Fisk, H. N.: Geological investigations of the alluvial valley of the Lower Mississippi River, 82 p. Vicksburg, Mississippi: Mississippi River Comm. 1944.
— Bar-finger sands of Mississippi delta. In: Peterson, J. A., Osmond, J. C., Eds.: Geometry of sandstone bodies, p. 29–52. Tulsa: Am. Assoc. Petroleum Geologists 1961.
Flores, G.: Evidence of slump phenomena (olistostromes) in areas of hydrocarbon exploration in Sicily. 5th World Petroleum Cong., Proc., New York, Sec. 1, Paper 13, 259–275 (1959).
Folk, R. L., Robles, R.: Carbonate sands of Isla Perez, Alacran reef complex, Yucatán. Jour. Geology **72**, 255–292 (1964).
Fondeur, C.: Étude pétrographic détaillée d'un grès à structure en feuillets. Rev. Inst. Francais Pétrole **19**, 901–920 (1964).
Frakes, L. A.: Stratigraphy of the Devonian Trimmer's Rock in eastern Pennsylvania. Pennsylvania Topog. Geol. Survey, Gen. Geol. Rept. **G-51**, 208 p. (1967).
Frazier, D. E.: Recent deltaic deposits of the Mississippi River: their development and chronology. Gulf Coast Assoc. Geol. Soc. Trans. **17**, 287–311 (1967).
Friedman, G. M., Johnson, K. G.: The Devonian Catskill deltaic complex of New York, type example of a tectonic delta complex. In: Shirley, M. L., Raysdale, J. A., Eds.: Deltas, p. 171–188. Houston: Houston Geol. Soc. 1966.
Friedman, S. A.: Channel-fill sandstones in the Middle Pennsylvanian rocks of Indiana. Indiana Geol. Survey, Rept. Prog. **23**, 59 p. (1960).
Gagliano, S. M., McIntire, W. G.: Delta components in Recent and ancient deltaic sedimentation; A comparison. Louisiana State Univ., Coastal Studies Inst., various paging 1968.
Gay, Parker, Jr.: Origen, distribucion y movimiento de las arenas eolicas en el area de Yauca a Palpa. Bol. Soc. Geol. Peru **37**, 37–58 (1962).
Gould, H. D., McFarlan, E., Jr.: Geologic history of the chenier plain southwestern Louisiana. Gulf Coast Assoc. Geol. Soc. Trans. **9**, 1–10 (1959).
Greenkorn, R. A., Johnson, C. R., Shallenberger, L. K.: Directional permeability of heterogeneous anisotropic porous media. Soc. Petroleum Engineers Jour., June, 124–132 (1964).
Griffiths, J. C.: Directional permeability and dimensional orientation in the Bradford Sand. Pennsylvania State Coll. Mineral Industries Expt. Sta. Bull. **54**, 138–163 (1949).
Grim, P. J.: Seamap deep-sea channel. U.S. Dept. Commerce, ESSA Tech. Rept. **ERL 93-Pol 2** 27 p. (1969).
Hadley, M. L.: Wave-induced bottom currents in the Celtic Sea. Marine Geology **2**, 164–167 (1964).
Hall, D. H., Hajnal, Z.: The gravimeter in studies of buried valleys. Geophysics **27**, 939–951 (1962).
Hamilton, N., Owens, W. H., Rees, A. I.: Laboratory experiments on the production of grain orientation in shearing sand. Jour. Geology **76**, 465–472 (1968).
Hand, B. M., Emery, K. O.: Turbidites and topography of north end of San Diego trough, California. Jour. Geology **72**, 726–742 (1964).
Hayes, M. O., Scott, A. J.: Environmental complexes south Texas Coast. Gulf Coast Assoc. Geol. Soc. Trans. **14**, 237–240 (1964).

Hesse, R.: Herkunft und Transport der Sedimente im bayerischen Flyschtrog. Zeitschr. deutsch. geol. Gesell. Jahrg. 1964, **116**, 403–426 (1965).

Hewett, C. H., Morgan, J. T.: The Fry *in situ* combustion test – reservoir characteristics. Jour. Petroleum Technology **17**, 337–353 (1965).

Hollenshead, C. T., Pritchard, R. L.: Geometry of producing Mesaverde sandstones, San Juan basin. In: Peterson, J. S., Osmond, J. C., Eds.: Geometry of sandstone bodies, p. 98–118. Tulsa: Am. Assoc. Petroleum Geologists 1961.

Holm, D. A.: Desert geomorphology in the Arabian Peninsula. Science **132**, 1369–1379 (1960).

Hopkins, M. E.: Harrisburg (No. 5), Coal reserves of southeastern Illinois. Illinois Geol. Survey Circ. **431**, 25 p. (1968).

Horn, Dietrich: Zur Sedimentation des Dogger-beta-Hauptsandsteines im östholsteinischen Juratrog. Meyniana **14**, 21–42 (1964).

— Zur geologischen Entwicklung der südlichen Schleimündung im Holozän. Meyniana **15**, 42–58 (1965).

Houbolt, J. J. H. C.: Recent sediments in the southern Bight of the North Sea. Geol. Mijnbouw **47** 245–273 (1968).

Hoyt, J. H.: Barrier island formation. Geol. Soc. America Bull. **78**, 1125–1136 (1967).

— Chenier versus barrier, genetic and stratigraphic distinction. Am. Assoc. Petroleum Geologists Bull. **53**, 299–306 (1969).

— Henry, V. J., Jr.: Significance of inlet sedimentation in the recognition of ancient barrier islands. Wyoming Geol. Assoc. Guidebook, 19th Field Conf., p. 190–194 (1965).

Hoyt, W. V.: Erosional channel in the middle Wilcox near Yoakum, Lavaca County, Texas. Gulf Coast Assoc. Geol. Soc. Trans. **9**, 11–50 (1959).

Hrabar, S. V., Potter, P. E.: Lower West Baden (Mississippian) sandstone body of Owen and Greene counties, Indiana. Am. Assoc. Petroleum Geologists Bull. **53**, 2150–2160 (1969).

Hutchinson, C. A., Jr., Dodge, C. F., Polasek, T. L.: Identification, classification, and prediction of reservoir inhomogeneities affecting production operations. Jour. Petroleum Technology **13**, 223–230 (1961).

Jizba, Z. V., Campbell, W. C., Todd, T. W.: Core resistivity profiles and their bearing on dipmeter survey interpretations. Am. Assoc. Petroleum Geologists Bull. **48**, 1804–1809 (1964).

Jobin, D. A.: Relation of the transmissive character of the sedimentary rocks of the Colorado Plateau to the distribution of uranium deposits. U.S. Geol. Survey Bull. **1124**, 151 p. (1962).

Johnson, W. E., Breston, J. N.: Directional permeability of sandstones from various states. Producers Monthly **15**, 10–19 (1951).

— Hughes, R. V.: Directional permeability measurements and their significance. Pennsylvania State Coll., Mineral Industries Expt. Sta., Bull. **52**, 180–205 (1948).

King, F. H.: Principles and conditions of the movement of ground water. U.S. Geol. Survey, 19th Ann. Rept., 1898–1899, p. 59–294 (1899).

Kolb, C. R., Van Lopeck, J. R.: Geology of the Mississippi River delta plain, southeastern Louisiana. U.S. Army Eng. Waterways Expt. Sta., Tech. Rept. No. 3–483, **1**, 120 p. (1958).

Krueger, W. C., Jr.: Depositional environments of sandstones as interpreted from electrical measurements – an introduction. Gulf Coast Assoc. Geol. Soc. Trans. **18**, 226–241 (1968).

Krumbein, W. C.: Fortran IV computer programs for Markov chain experiments in geology. Kansas Geol. Survey, Computer Contr. **13**, 38 p. (1967).

— Statistical models in sedimentology. Sedimentology **10**, 7–23 (1968a).

— Fortran IV computer program for simulation of transgression and regression with continuous-time Markov models. Kansas Geol. Survey, Computer Contr. No. **26**, 38 p. (1968a).

— Monk, G. D.: Permeability as a function of the size parameters of sedimentary particles. Am. Inst. Mining Metall. Engineers Tech. Pub. **1492**, 153–163 (1942).

— Sloss, L. L.: Stratigraphy and sedimentation, 2nd Ed., 660 p. San Francisco: Freeman 1963.

Krynine, P. D.: The megascopic study and classification of sedimentary rocks. Jour. Geology **56**, 130–165 (1948).

Ksiazkiewiz, Marian: Life conditions in flysch basins. Rocznik Polsk, Towarz. Geol. **31**, 1–21 (1961).

Kuenen, Ph. H.: Emplacement of flysch-type sand beds. Sedimentology **9**, 203–244 (1967).

— Turbidity currents and organisms. Eclogae géol. Helvetiae **62**, 525–544 (1968).

— Perdok, W. G.: Experimental abrasion 5. Frosting and defrosting of quartz grains. Jour. Geology **70**, 648–658 (1962).

Kulm, L. D., Byrne, J. V.: Sedimentary response to hydrography in an Oregon estuary. Marine Geology **4**, 85–118 (1966).
Land, L. S.: Eolian cross-bedding in the beach dune environment, Sapelo Island, Georgia. Jour. Sed. Petrology **34**, 389–394 (1964).
— Hoyt, J. H.: Sedimentation in a meandering estuary. Sedimentology **6**, 191–207 (1966).
Lee, Wallace, Nickell, C. O., Williams, J. S., Henbest, L. C.: Stratigraphic and paleontologic studies of the Pennsylvanian and Permain rocks in north-central Texas. Univ. Texas Pub., Bull. **3801**, 252 p. (1938).
Leopold, L. B., Miller, J. P., Wolman, M. G.: Fluvial processes in geomorphology 522 p. San Francisco: Freeman, 1964.
Levorsen, A. I., Ed.: Stratigraphic type oil fields, 902 p. Tulsa: Am. Assoc. Petroleum Geologists 1941.
Lineback, J. A.: Turbidites and other sandstone bodies in the Borden Siltstone (Mississippian) in Illinois. Illinois Geol. Survey Circ. **425**, 29 p. (1968).
Logvinenko, N. V., Remizov, I. N.: Sedimentology of beaches on the north coast of the sea of Azov. In: van Straaten, L. M. J. U., Ed.: Deltaic and shallow marine deposits. Developments in sedimentology, Vol. 1, p. 244–252. Amsterdam: Elsevier Pub. Co. 1964.
Lovell, J. P. B.: Tyee formation: Undeformed turbidites and their lateral equivalents, mineralogy and paleogeography. Geol. Soc. America Bull. **80**, 9–22 (1969).
Lucchi, F. R., D'Onofrio, S.: Transportu gravitativi sinsedimentari nel Tortoniano dell'Appennino Romagnolo (Valle del Savio). Giornale di Geologia, **34**, Ser. 2a, Fasc. 1, 1–30 (1966).
Mackenzie, F. T.: Bermuda Pleistocene eolianites and paleowinds. Sedimentology **3**, 52–64 (1964).
Madigan, C. T.: The Simpson Desert expedition, 1939 Scientific reports: No. 6, Geology – the sand formations. Roy. Soc. South Australia, Trans. and Proc. **70**, 45–63 (1946).
Maksimov, M.: Issledoraniia vzaimootnoshenii prodyktivnykh gorizontov Maikopskoi legkoi nefti sviazi s genezisom Maikopskoi rukavoobraznoi zalezhi (Investigation of the interrelation of the horizons producing Maikop high petroleum in conjunction with the genesis of the Maikop elongate oil pool). Neftianoe Khoziaistvo **17**, 813–834 (1929).
Marr, J. D., Zagst, E. F.: Exploration horizons from new seismic concepts of CDP and digital processing. Geophysics **32**, 207–224 (1967).
Marshall, K. T.: A preliminary model for determining the optimum drilling pattern in locating and evaluating an ore body. Colorado School Mines Quart. **59**, Pt. A, 223–236 (1964).
Martin, B. O.: Rosedale channel – evidence for late Miocene submarine erosion in Great Valley. Am. Assoc. Petrol. Geologists Bull. **47**, 441–456 (1963).
Martin, Rudolph: Paleogeomorphology and its application to exploration for oil and gas (with examples from western Canada). Am. Assoc. Petroleum Geologists Bull. **50**, 2277–2311 (1966).
Mast, R. F., Potter, P. E.: Sedimentary structures, sand shape fabrics and permeability. II. Jour. Geology **71**, 548–565 (1963).
Maxey, G. B.: Hydrostratigraphic units. Jour. Hydrology **2**, 124–129 (1964).
McBride, E. F., Hayes, M. O.: Dune cross-bedding on Mustang Island, Texas. Am. Assoc. Petroleum Geologists Bull. **46**, 546–551 (1962).
McCubbin, D. G., Brady, M. J.: Depositional environment of the Almond Reservoirs, Patrick Draw Field, Wyoming. The Mountain Geologist **6**, 3–26 (1969).
McKee, E. D.: The Coconino sandstone – its history and origin. Carnegie Inst. Wash. Pub. **440**, Contr. Paleont. 77–115 (1933).
— Primary structures in some Recent sediments. Am. Assoc. Petroleum Geologists Bull. **41**, 1704–1747 (1957).
— Structures of dunes at White Sands National Monument, New Mexico. Sedimentology (Spec. Issue) **7**, 69 p. (1966).
— Sterrett, T. S.: Laboratory experiments on form and structure of longshore bars and beaches. In: Peterson, J. A., Osmond, J. C., Eds.: Geometry of sandstone bodies, p. 13–28. Tulsa: Am. Assoc. Petroleum Geologists 1961.
— Tibbits, G. C., Jr.: Primary structures of a seif dune and associated deposits in Libya. Jour. Sed. Petrology **34**, 5–17 (1964).
Meckel, L. D.: Origin of Pottsville conglomerates (Pennsylvanian) in the central Appalachians. Geol. Soc. America Bull. **78**, 223–258 (1967).
— Paleozoic alluvial deposition in central Appalachians. A summary. In: Fisher, G. W., and others Eds.: Studies in Appalachian geology. p. 49–68. New York: Interscience 1970.

Menard, H. W., Smith, S. M., Pratt, R. M.: The Rhone deep-sea fan. In: Whittard, W. F., Bradshaw, R. Eds.: Submarine geology and geophysics, p. 271–284. London: Butterworths 1965.

Merriam, D. F., Ed.: Symposium on cyclic sedimentation. Kansas Geol. Survey Bull. **169**, **1**, 1–380 and **2**, 381–636 (1964).

Montadert, L., Latreille, M., Groult, J., Poujol, P., Deix, H.: Étude à l'affleurement d'un réservoir gréso-argileux: Le Dévonien inferieur (Réservoirs F4-F5) du Bassin de Fort-Polignac. Inst. Francais du Pétrole, 2nd Colloque A.R.T.F.P., p. 313–327, 1965.

Nachtigal, K. H.: Über die Unterwasser-hangmorphologie vor Rantum und Kampen auf Sylt. Meyniana **18**, 43–63 (1968).

Nanz, R. H., Jr.: Genesis of Oligocene sandstone reservoir Seeligson Field, Jim Wells and Kleberg Counties, Texas. Am. Assoc. Petroleum Geologists Bull. **38**, 96–117 (1954).

Nichols, M. M.: Characteristics of sedimentary environments in Moriches Bay. In: Miller, R. L., Ed.: Papers in marine geology, Shepard commemorative volume, p. 363–383. New York: MacMillan 1964.

Noorthoorn van der Kruijff, J. F., Lagaaij, R.: Displaced faunas from inshore estuarine sediments in the Haringvliet (Netherlands). Geol. Mijnbouw **39**, (N. S. 22), 711–723 (1960).

Off, Theodore: Rhythmic linear sand bodies caused by tidal currents. Am. Assoc. Petroleum Geologists Bull. **47**, 324–341 (1963).

Oomkens, E., Terwindt, J. H. J.: Inshore estuarine sediments in the Haringvleit (Netherlands). Geol. Mijnbouw **39** (N. S. 22), 701–710 (1960).

Paine, W. R., Meyerhoff, A. A.: Catahoula Formation of western Louisiana and thin-section criteria for fluviatile depositional environment. Jour. Sed. Petrology **38**, 92–113 (1968).

Palain, Christian: Contribution à l'étude sédimentologique du Grès à Roseaux (Trias supérieur) en Lorraine. Sci. Terre **11**, 245–291 (1966).

Parker, R. H.: Macro-invertebrate assemblages as indicators of sedimentary environments in east Mississippi delta region. Am. Assoc. Petroleum Geologists Bull. **40**, 295–376 (1956).

— Ecology and distributional patterns of marine macro-invertebrates, northern Gulf of Mexico. In: Shepard, F. P., Phleger, F. B., van Andel, Tj. H., Eds.: Recent sediments, northwestern Gulf of Mexico, p. 302–337. Tulsa: Am. Assoc. Petroleum Geologists 1960.

Peel, R. F.: The landscape in aridity. Inst. British Geographers, Trans. and Papers, Pub. **38**, 23 p. (1966).

Pelletier, B. R.: Pocono paleocurrents in Pennsylvania and Maryland. Geol. Soc. America Bull. **69**, 1033–1064 (1958).

Peterson, J. A., Osmond, J. C., Eds.: Geometry of sandstone bodies, 240 p. Tulsa: Am. Assoc. Petroleum Geologists 1961.

Phleger, F. B.: Ecology of Foraminifera in southeastern Mississippi delta area. Am. Assoc. Petroleum Geologist Bull. **39**, 712–752 (1955).

— Sedimentary patterns of microfaunas in northern Gulf of Mexico. In: Shepard, F. P., Phleger, F. B., van Andel, Tj. H., Eds.: Recent sediments, northwestern Gulf of Mexico, p. 267–301. Tulsa: Am. Assoc. Petroleum Geologists 1960.

Polasek, T. L., Hutchinson, C. A., Jr.: Characterization of non-uniformities within a sandstone reservoir from a fluid mechanics standpoint. In: 7th World Petroleum Cong. Proc., Vol. 2, p. 397–407. Barking: Elsevier Pub. Co. Ltd. 1967.

Poole, F. G.: Paleowinds in the western United States. In: Nairn, A. E. M., Ed.: Problems in paleoclimatology, p. 395–405. London: Interscience 1964.

Postma, H.: Sediment transport and sedimentation in the estuarine environment. In: Lauff, G. H. Ed.: Estuaries, Pub. 83, p. 158–179. Washington: Am. Assoc. Adv. Sci. 1967.

Potter, P. E.: Facies model conference. Science **129**, 1292–1294 (1959).

— Late Mississippian sandstones of Illinois Basin. Illinois Geol. Survey Circ. **340**, 36 p. (1962).

— Late Paleozoic sandstones of Illinois Basin. Illinois Geol. Survey Rept. Inv. **217**, 92 p. (1963).

— Sand bodies and sedimentary environments. A review. Am. Assoc. Petroleum Geologists Bull. **51**, 337–365 (1967).

— Blakely, R. F.: Generation of a synthetic vertical profile of a fluvial sandstone body. Jour. Petroleum Technology, **1967**, 243–251.

— Random processes and lithologic transitions. Jour. Geology **76**, 154–170 (1968).

— Mast, R. F.: Sedimentary structures, sand shape fabrics and permeability. I. Jour. Geology **71**, 441–471 (1963).

Potter, P. E., Pettijohn, F. J.: Paleocurrents and basin analysis, 296 p. Berlin-Göttingen-Heidelberg: Springer 1963.
Pretorius, D. A.: Conceptual geological models in the exploration for gold mineralization in the Witwatersrand Basin. Univ. Witwatersrand, Johannesburg, Econ. Geol. Res. Unit, Inf. Circ. **36**, 39 p. (1966).
Reese, D. O.: Wilcox dipmeter applications. Gulf Coast Assoc. Geol. Soc. Trans. **18**, 387–399 (1968).
Reiche, Parry: An analysis of cross-lamination the Coconino sandstone. Jour. Geology **46**, 905–932 (1938).
Reineck, Hans-Erich: Sedimentgefüge im Bereich der südlichen Nordsee. Abh. senckenb. naturf. Gesell. **505**, 138 p. (1963).
Reyment, R. A.: Interstitial ecology of the Niger delta. Bull. Geol. Institutions, Univ. Uppsala **1**, 121–159 (1969).
Reynolds, D. W., Vincent, J. K.: Western Kentucky's Bethel channel – the largest continuous reservoir in the Illinois Basin. In: Rose, W. D., Ed.: Proceedings of technical sessions, Kentucky Oil and Gas Association, 29th annual meeting, June 3–4, 1965. Kentucky Geol. Survey Ser. 10, Spec. Pub. **14**, 19–30 (1967).
Riba, Oriol, Villena, J., Quirantes, Jose: Nota preliminar sobre la sedimentacion en paleocannales Terciarios de la Zona de Caspe-Chiprana (Provincia de Zaragoza). Ana. Edafologia Agrobiologia **26**, 617–634 (1967).
Rich, J. L.: Oil and gas in the Birds Quadrangle. Illinois Geol. Survey Bull. **33**, 106–180 (1916).
— Shoestring sands of eastern Kansas. Am. Assoc. Petroleum Geologists Bull. **7**, 103–113 (1923).
— Shorelines and lenticular sands as factors in oil accumulation, Vol. 1, p. 230–239. In: Dunstan, A. E., Ed.: The science of petroleum. London: Oxford Univ. Press 1938.
— Three critical environments of deposition and criteria for recognition of rocks deposited in each of them. Geol. Soc. America Bull. **62**, 1–19 (1951).
Rittenhouse, Gordon: Problems and principles of sandstone-body classification. In: Peterson, J. A., Osmond, J. C., Eds.: Geometry of sandstone bodies, p. 3–12. Tulsa: Am. Assoc. Petroleum Geologists 1961.
Robinson, A. H. W.: The submarine port morphology of certain port approach channel systems. Inst. Navigation Jour. **9**, 20–46 (1956).
Rubey, W. M., Bass, N. W.: The geology of Russell County, Kansas. Kansas Geol. Survey Bull. **10**, 104 p. (1925).
Rühl, W., Schmid, Ch.: Über das Verhältnis der vertikalen zur horizontalen absoluten Permeabilität von Sandsteinen. Geol. Jahrb. **74**, 447–462 (1957).
Rukhin, L. B.: Osnovy obshchei paleogeografii (Principles of general paleogeography): Leningrad, State Sci.-Tech. Pub. House, Oil and Conventional Fuel-Mining Literature, 628 p. 1962.
Rusnak, G. A.: Sediments of Laguna Madre, Texas. In: Shepard, F. P., Phleger, F. B., van Andel, Tj. H. Eds.: Recent sediments, northwest Gulf of Mexico, 1951–1958, p. 153–196. Tulsa: Am. Assoc. Petroleum Geologists (1960).
— Nesteroff, W. D.: Modern turbidites: terrigenous abyssal plain verses bioclastic basin. In: Miller, R. L., Ed.: Papers in marine geology, Shepard commemorative volume, p. 488–507. New York: MacMillan 1964.
Russell, R. J.: Alluvial morphology of Anatolian Rivers. Ann. Assoc. Am. Geographers **44**, 363–391 (1954).
Rutledge, J. R.: A study of fluid migration in porous media by stereoscopic radiographic techniques. Brigham Young Univ. Studies, **13**, 89–104 (1966).
Sabins, F. F., Jr.: Anatomy of stratigraphic trap, Bisti Field, New Mexico. Am. Assoc. Petroleum Geologists Bull. **47**, 193–228 (1963a).
— Computer flow diagram in facies analysis. Am. Assoc. Petroleum Geologists Bull. **47**, 2045–2047 (1963b).
— Review of petrographic criteria for Cretaceous depositional environments. Wyoming Geol. Assoc. Guidebook, 19th Ann. Field Conf., p. 27–34 (1965).
Schlee, J. S., Moench, R. H.: Properties and genesis of "Jackpile" Sandstone, Laguna, New Mexico. In: Peterson, J. A., Osmond, J. C., Eds.: The geometry of sandstone bodies, p. 134–150. Tulsa: Am. Assoc. Petroleum Geologists 1961.
Schumm, S. A.: Speculations concerning paleohydrologic controls of terrestrial sedimentation. Geol. Soc. America Bull. **79**, 1573–1588 (1968a).

Schumm, S. A.: River adjustment to altered hydrologic regimen – Murrumbidgee River and paleochannels, Australia. U.S. Geol. Survey Prof. Paper **598**, 65 p. 1968b.

Scruton, P. C.: Delta building and the deltaic sequence. In: Shepard, F. P., Phleger, F. B., van Andel, Tj. H., Eds.: Recent sediments, northwest Gulf of Mexico, 1951–1958, p. 82–102. Tulsa: Am. Assoc. Petroleum Geologists 1960.

Sedimentation Seminar: Cross-bedding in the Salem Limestone of central Indiana. Sedimentology **6**, 95–114 (1966).

— Bethel Sandstone (Mississippian) of western Kentucky and south central Indiana, a submarine-channel fill. Kentucky Geol. Survey, Rept. Inv. **11**, 23 p. (1969).

Seibold, Eugene: Geological investigation of nearshore sandtransport – examples of methods and problems from Baltic and North Seas, in Sears, M., Ed.: Progress in oceanography, Vol. 1, p. 1–70. New York: MacMillan 1963.

Selley, R. C.: Paleocurrents and sediment transport in nearshore sediments of the Sirte Basin, Libya. Jour. Geology **75**, 215–223 (1967).

Semin, E. I.: Geologicheskia nedodarodnost neftenosnykh plastov i nekotorye sposoby ee izucheniia (Geological nonuniformity of oil bearing strata and some means of its study). Trudy VNII neft', n. 34, Gostoptekhizat, p. 3–43 (1962).

Sharp, R. P.: Wind ripples. Jour. Geology **71**, 617–636 (1963).

Shaw, A. B.: Time in stratigraphy, 365 p. New York: McGraw-Hill 1964.

Shelton, J. W.: Shale compaction in a section of Cretaceous Dakota Sandstone, northwestern Black Hills. Jour. Sed. Petrology **32**, 874–877 (1962).

— Trend and genesis of lowermost sandstone unit of Eagle Sandstone at Billings, Montana. Am. Assoc. Petroleum Geologists Bull. **49**, 1385–1397 (1965).

— Stratigraphic models and general criteria for recognition of alluvial, barrier-bar, and turbidity-current sand deposits. Am. Assoc. Petroleum Geologists Bull. **51**, 2241–2461 (1967).

Shepard, F. P.: Gulf Coast barriers. In: Shepard, F. P., Phleger, F. B., van Andel, Tj. H., Eds.: Recent marine sediments, northwest Gulf of Mexico, p. 197–220. Tulsa: Am. Assoc. Petroleum Geologists 1960.

— Submarine geology, 2nd Ed., 557 p. New York: Harper and Row 1963.

— Importance of submarine valleys in funneling sediments to the deep sea. In: Sears, M. Ed.: Progress in oceanography, Vol. 3, p. 321–332. Oxford: Pergamon 1965.

— Dill, R. F., von Rad, Ulrich: Physiography and sedimentary processes of LaJolla submarine fan and fan-valley, California. Am. Assoc. Petroleum Geologists Bull. **53**, 390–420 (1969).

— Moore, D. G.: Central Texas coast sedimentation: characteristics of sedimentary environment, recent history and diagenesis. Am. Assoc. Petroleum Geologists Bull. **39**, 1463–1593 (1955).

Shiarella, N. W.: Typical oil producing structures in the Owensboro field of western Kentucky. Kentucky Univ., Bur. Mineral Topog. Survey, Bull. **3**, 14 p. (1933).

Shotten, F. W.: The lower Bunter sandstones of North Worcestershire and East Shropshire. Geol. Mag. **74**, 534–553 (1937).

Smith, H. T. U.: Eolian geomorphology, wind direction, and climatic change in North Africa, 48 p. Bedford, Mass.: Air Force Cambridge Res. Lab. No. 63–443 (1963).

Spjeldnaes, Nils: Silurian tidal sediments from the base of the Ringerike Formation, Oslo region, Norway. Norsk Geol. Tidsskrift, **46**, 497–509 (1966).

Stokes, W. L.: Fluvial and eolian sandstone bodies in Colorado Plateau, p. 151–178. In: Peterson, J. A., Osmond, J. C., Eds.: Geometry of sandstone bodies. Tulsa: Am. Assoc. Petroleum Geologists 1961.

— Multiple parallel-truncation bedding planes – a feature of wind-deposited sandstone formations. Jour. Sed. Petrology **38**, 510–515 (1968).

van Straaten, L. M. J. U.: Megaripples in the Dutch Wadden Sea and in the basin of Arcachon (France). Geol. Mijnbouw **15**, 1–11 (1953).

— Composition and structure of Recent marine sediments in the Netherlands. Leidse. Geol. Mededel **19**, 1–110 (1954a).

— Sedimentology of Recent tidal flat deposits and the Psammites du Condroz (Devonian). Geol. Mijnbouw **16**, 25–47 (1954b).

— Coastal barrier deposits in south- and north Holland in particular in the areas around Scheveningen and Ijmuiden. Mededel. Geol. Stichtung **17**, 41–75 (1965).

Stride, A. H.: Current-swept sea floors near the southern half of Great Britain. Geol. Soc. London Quart. Jour. **119**, 175–199 (1963).

Sullwold, H. H., Jr.: Turbidites in oil exploration, p. 63–81. In: Peterson, J. A., Osmond, J. C., Eds.: Geometry of sandstone bodies. Tulsa: Am. Assoc. Petroleum Geologists 1961.

Swift, D. J. P., McMullen, R. M.: Preliminary studies of intertidal sand bodies in the Minas Basin, Bay of Fundy, Nova Scotia. Canadian Jour. Earth Sci. **5**, 175–183 (1968).

Talwani, M.: A review of marine geophysics. Marine Geology **2**, 29–80 (1964).

Tanner, W. F.: Ripple mark indices and their uses. Sedimentology, **9**, 89–104 (1967).

Terwindt, J. H. J., de Jong, J. D., van der Wilk, E.: Sediment movement and sediment properties in the tidal area of the Lower Rhine (Rotterdam Waterway). Koninkl. Nederlands Geol.-Mijnb. Genoot Verh., Geol. Ser. **21-2**, 243–258 (1963).

Testerman, J. D.: A statistical reservoir-zonation technique. Jour. Petroleum Technology **14**, 889–893 (1962).

Thomas, W. A., Mann, C. J.: Late Jurassic depositional environments, Lousiana and Arkansas. Am. Assoc. Petroleum Geologists Bull. **50**, 178–182 (1966).

Thompson, W. D.: Original structures of beaches, bars and dunes; Geol. Soc. America Bull. **48**, 723–751 (1937).

Todd, D. K.: Ground water hydrology, 336 p. New York: Wiley 1960.

Twenhofel, W. H.: Principles of sedimentation, 2nd Ed., 673 p. New York: McGraw-Hill 1950.

Uchupi, Elazar: Atlantic continental shelf and slope of the United States – physiography. U.S. Geol. Survey Prof. Paper **529-C**, 30 p. (1968).

Verrien, J. P., Courand, G., Montadert, L.: Applications of production geology methods to reservoir characteristics, analysis from outcrop observations (French). In: Proc. 7th World Petroleum Cong., Vol. 2, p. 425–446. Barking, England: Elsevier Publishing Co., Ltd. 1967.

Visher, G. S.: Use of vertical profile in environmental reconstruction. Am. Assoc. Petroleum Geologists Bull. **49**, 41–61 1965.

Voinov, V. V.: Izuchenie neodnorodnosti produktiunykh plastov nekotorykh neftianykh mestorozhdenii (The study of the nonuniformity of productive strata of several oil deposits). Trudy VNII neft', n. 50, Izd-vo. "Nedra", 145–150 (1967).

Walker, R. G.: Turbidite sedimentary structures and their relationship to proximal and distal depositional environments: Jour. Sed. Petrology **38**, 1120–1154 (1967).

— Harms, J. C.: The "Catskill Delta" – A prograding muddy shoreline in central Pennsylvania. Jour. Geology **78**, 381–399 (1970).

Walther, J.: Lithogenesis der Gegenwart, 1055 p. Jena: Gustav Fischer 1894.

Weimer, R. J.: Stratigraphy and petroleum occurrences, Almond and Lewis Formations (Upper Cretaceous), Wamsutter Arch, Wyoming. In: Sedimentation of Late Cretaceous and Tertiary outcrops, Rock Springs Uplift. Wyoming Geol. Assoc., 19th Field Conf., p. 65–81, 1965.

— Hoyt, J. H.: Burrows of *Callianassa major* Say, Geologic indicators of littoral and shallow neritic environments. Jour. Paleontology **38**, 761–767 (1964).

Weir, G. W.: Borden formation (Mississippian) in southeast-central Kentucky, Field trip No. 3. In: Guidebook for field trips, 18th Ann. Meeting, Southeastern Sect., Geol. Soc. America, p. 29–41, 1969.

Werner, Friedrick: Über den inneren Aufbau von Strandwällen an einem Küstenabschnitt der Eckernförder Bucht. Meyniana **13**, 108–121 (1963).

Wiegel, R. L.: Oceanographical engineering, 532 p. Englewood Cliffs: Prentice-Hall 1964.

Witkind, J. J., Thaden, R. E.: Geology and uranium-vanadium deposits of the Monument Valley area, Apache and Navajo-Counties, Arizona. U.S. Geol. Survey Bull. **1103**, 171 p. 1963.

Wurster, Paul: Geologie des Schilfsandsteins. Mitt. Geol. Staatsinst. Hamburg **33**, 140 p. (1964).

Wyllie, M. R. J., Spangler, M. B.: Application of electrical resistivity measurements to the problem of fluid flow in porous media. Am. Assoc. Petroleum Geologists Bull. **36**, 359–403 (1952).

Young, R. C.: Stratigraphy of coal-bearing rocks of Book Cliffs, Utah-Colorado. In: Hamblin, W. K., Rigsby, J. K., Eds.: Central Utah coals: A guidebook prepared for the Geological Society of America and associated societies. Utah Geol. and Mineral. Survey Bull. **80**, 7–21 (1966).

Zagwijn, W. H.: Pleistocene stratigraphy in the Netherlands based on changes in vegetation and climate. In: Trans. Jubilee Convention, Pt. 2. Koninkl. Nederlands Geol.-Mijnb. Genoot verh., Geol. Ser., **21-2**, 173–196 (1963).

Zeito, G. A.: Interbedding of shale breaks and reservoir inhomogeneities. Jour. Petroleum Technology **17**, 1223–1228 (1965).

Zhemchuzhnikov, Yu. A.: Tip kosoy sloistosti kak kriteriy genezisa osadkov (Types of crossbedding and genesis). Gorniy Institute, Zapiski **8**, 35–69 (1926) (French summary).

— and others: Stroenie i usloviia nakopleniia asnounykh uglenosykh suit i ugolnykh plastov arednego karvona Donetskogo basseina chast' pervaia (Structure and conditions of accumulation of the principal coal-bearing formations and coal beds of the Middle Carboniferous of the Donetz Basin). Akad. Nauk. USSR. Trans. Geol. Inst. No. 15, Pt. 1, 332 p. and Pt. 2, 347 p. (1959 and 1960).

Chapter 12. Sandstones, Sedimentary Basins, and Continental Evolution

Introduction

At some stage in the geological investigation of a sandstone, we need to place the formation in a larger perspective – to see it in relation to the basin fill as a whole and even to think about it in the longer view of geologic history and continental evolution. We need to go from a provincial to a cosmopolitan outlook. Only by placing the local study in a larger framework can we really understand the sandstone in question. Hence in this chapter we consider the larger questions related to sands and sandstones.

Sands have proved to be the most useful sediment in unravelling the geologic history of the large majority of sedimentary basins on the continents. As we have seen in Chap. 8, the kind and proportions of the detrital minerals provide clues to the character of the source rocks, the location of the source area, and something about the relief and climate of that region. The paleocurrent record is largely reconstructed from the primary sedimentary structures of sandstones. Neither the shales nor the limestones shed as much light on the interrelated problems of provenance, paleocurrents, and tectonics.

There are also such broad questions as the relations between the sands of the stable cratons and those of mobile belts. Are there basic differences? What is the relation between sands, tectonics, and the architecture of the depositional basin? And what of the sands of the deep ocean basins? Can they be related to tectonics, provenance, paleocurrents, in the same ways as sands on the continents?

Study of geosynclinal sedimentary basins suggests that their filling and structural evolution proceeds in an orderly fashion. That such is so was recognized by Bertrand as early as 1897. The concepts outlined by Bertrand have been refined and elaborated on by many workers since. A modern restatement, for example, is found in Aubouin's book on geosynclines. It remained, however, for Krynine to point out in 1941 the relations between sandstone petrography and the tectonic cycle. Regardless of whether the observed changes in character of the fill of a geosyncline are related to shifting multiple source areas or to the erosion and changing geomorphology of a single source area or both, that there is an orderly cycle of development is beyond doubt. We propose to review this problem and to illustrate it by examples from the geologic record.

We might also ask whether there is a theory capable of explaining not only the role of sands in basin filling but also the diversity and distribution of sands in time. What, for example, can we say about sandstones and the geologic past? Is there any evidence that there has been a secular change in the composition of sands, of progressive impoverishment in feldspar and enrichment in quartz as recycled sand forms an ever increasing proportion of the total sand accumulation?

Or is the crust of the earth, and the sandstones contained in it, continually recycled through the upper mantle so that there is no accumulation of sand recycled through the upper parts of the crust? This problem relates to theories of global plate tectonics and sea floor spreading. Are these new theories supported by sandstone petrography or distribution? Or, conversely, what are the implications of plate tectonics and sea floor spreading for the origin of sandstones?

What was the oldest sandstone like? Was it appreciably different from modern sands? If the earth had, within recorded geologic history, a reducing atmosphere, what would sands formed under such an atmosphere be like? Is there any evidence in the character of the early sands that such conditions existed? Did the advent of vascular land plants in the Upper Silurian and Lower Devonian have any effect on the production of sand or its character?

Has the production of sand been constant in time? Has the proportion of sand to other sediments shown any significant change? Clearly not only do sands have something to say about the evolution of a given basin but they also have proved to be most useful in the study of continental development and even the evolution of our planet as a whole. And finally, is there evidence that sand, as we would ordinarily describe it, is to be found on the moon or other planets?

We shall first turn to the geologic record to see what is known about sands of the geologic past.

Continental Structure and Sand Accumulation

Major structural elements of the crust of the earth, excluding oceanic basins, are cratons and mobile belts or geosynclines. A *craton* is a large area of the continent on which relatively thin, essentially flat-lying sediments completely or partially cover a tectonically passive and stable basement. The basement is commonly of Precambrian age, although it may also be younger. The craton (or hedreocraton) thus includes the *shields* where the basement rock is exposed, and *platforms* where a basement is covered by flat-lying rocks of variable thickness. By *mobile belt* is meant a linear welt containing one or more geanticlines and deep troughs marginal to a craton. The sediments of the troughs are commonly strongly deformed.

Geosynclines and cratons, unstable and stable areas respectively, have played an important role from the beginning of known geologic history. It is not our task to trace that history nor to discuss the origin and structural behavior of these major crustal elements. We will, however, review briefly the salient features of each and summarize what is known about the relation of sand to these features – the volume and character of the sands deposited in each and related questions.

What are the relative volumes of geosynclinal and cratonic sediments? Using Kay's estimate (1951, p. 92) for area and thickness, we find that 82 percent of the sediments by volume in North America are geosynclinal accumulations (eugeosynclines 59 percent; miogeosynclines 23 percent) and only 18 percent are cratonic. If twenty-five percent of the cratonic sediment and fifteen percent of the geosynclinal accumulation is sand, it is clear that most sandstone of the geologic past has been trapped in geosynclines, the ratio of geosynclinal to cratonic sand being at least 2.5 to 1 and probably near 4 to 1. Though some (Gilluly and others, 1970) have estimated somewhat different sediment volumes, it is clear that the total volume of sandstones of geosynclines greatly exceeds that of the cratons.

Geosynclines

As summarized by Aubouin (1965, p. 104), the geosyncline normally consists of a mio-eugeosynclinal couple. From the foreland or craton to the interior, the ideal geosyncline consists of a miogeosynclinal basin, a medial ridge which separates the miogeosyncline from the eugeosyncline, and an interior tectonic land or ridge (Fig. 12-1). This geosynclinal domain may be localized at or near the margin of the continent or, as in the case of the Urals, be within the continent.

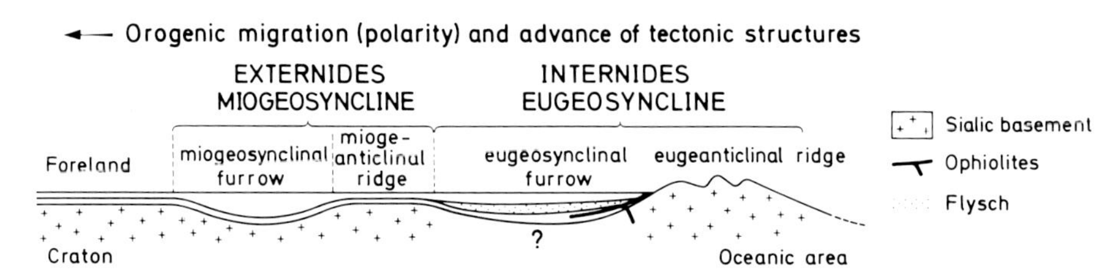

Fig. 12-1. Organizational pattern of ideal geosyncline and related tectonic elements (Modified from Aubouin, 1965, Fig. 16)

The sedimentation and volcanism within the geosyncline and its ultimate destruction by orogenesis proceeds according to a pattern. In the ideal case, the eugeosynclinal furrow and associated eugeosynclinal ridge or "tectonic land" is the first to form. It is characterized by ophiolitic emissions (pillow basalts and autointrusions now altered to greenstones) and flysch deposits. This eugeosynclinal zone is deformed early and is commonly thrust towards the craton. The miogeosynclinal domain becomes isolated from the eugeosyncline by development of the medial geanticline and generally has a non-volcanic fill derived at first from the geanticlinal ridge and later from the deformed eugeosynclinal assemblage. The miogeosynclinal assemblage undergoes later but less intense deformation and, upon uplift, sheds a great deal of coarse debris into a molasse trough, commonly, but not always exclusively, in front of the elevated mountain chain (foredeep).

The organization and evolutionary development outlined above characterizes in some form or another all the Alpine-type mountain chains of the world, including the Alps, the Carpathians, the Urals, the Appalachians, and other mobile belts of similar character. The Coronation geosyncline of Precambrian age, described elsewhere in this chapter, also belongs to this class of sedimentary basins.

What can we say about the sedimentary fill, especially the sandstones, of geosynclines? First it should be noted that geosynclines have a polarity – they are not symmetrical structures. The two sides are unlike – one being a stable cratonic area, the other being a tectonic land. Both sides generate sediment, both contribute to the fill. But the sediments, the sands in particular, furnished by the two sides are quite unlike. The sands from the cratonic side are apt to be small in volume and are apt to be mature quartz arenites, or perhaps subarkoses, but rarely true arkoses. The sands derived from the tectonic lands and medial geanticline are immature, both texturally and compositionally; they are graywackes in the ancient rocks or lithic sandstones in the younger geosynclines. Not only are the sands unlike in their petrologic character but they are unlike in their place in the evolutionary development of the belt.

In the eugeosyncline, the sands are turbidites – part of a flysch sequence. Those of the early stages in the miogeosynclinal domain are blanket type quartz arenites with the usual shallow-water structures, typically associated with shallow-water carbonates. Only in the later stages of filling does the miogeosyncline receive a flysch. In the postgeosynclinal stage, the rapid erosion of the newly uplifted range produces the molasse assemblage in the foredeep and in intermontane basins. The sands of the molasse are largely, but not entirely, alluvial lithic sandstones and arkoses.

Paleocurrent data clearly reflect the response of transport patterns to the evolutionary development of the geosynclinal domain. This is particularly true of the sands of the craton. These, which as a rule are small in volume and which appear only early in the history of a geosyncline, are limited to the cratonic margin of the miogeosyncline. Their crossbedding demonstrates a cratonic source. The flysch-sands are apt to show longitudinal as well as lateral transport – displaying, however, no transport from the craton. The only exceptions are carbonate turbidites derived from the cratonic shelf as in the case of the Dimple Limestone of the Marathon area in Texas (Thompson and Thomasson, 1969). The largely alluvial molasse shows a clear pattern of movement from the orogenic belt toward the craton. The distal edge of the molasse wedge may extend far into the craton.

The sand content of geosynclines is probably variable. In the Appalachian miogeosyncline sand forms about 23 percent of the section or about 115,000 mi^3 (1,088,980 km^3) according to Colton (1970, p. 11–12). In the Cordilleran miogeosyncline of Idaho, sandstones and other coarse clastics are estimated to constitute 30 percent of the section (Schwab, 1969, p. 156). On the other hand, the Indonesian geosyncline is believed to contain only 14 percent sand (Kuenen, 1941, p. 169).

This brief sketch of the somewhat idealized anatomy and evolutionary development of a geosynclinal tract is summarized in Fig. 12-2, but is best understood by considering some of the real examples later in this chapter.

Cratons

The consolidated and stable parts of the earth's crust were designated "cratons" by Stille according to Aubouin (1965, p. 24). In general, the term is normally restricted to tectonically passive parts of a continent and includes both Precambrian Shields and platforms consisting of essentially flat-lying late Precambrian or younger sediments. Cratonic platforms as thus defined, cover large areas and are commonly very gently warped into wide, shallow, intracratonic basins, such as the Michigan and Illinois basins in North America which are separated from one another by broad regional arches, such as the Cincinnati Arch in the Ohio

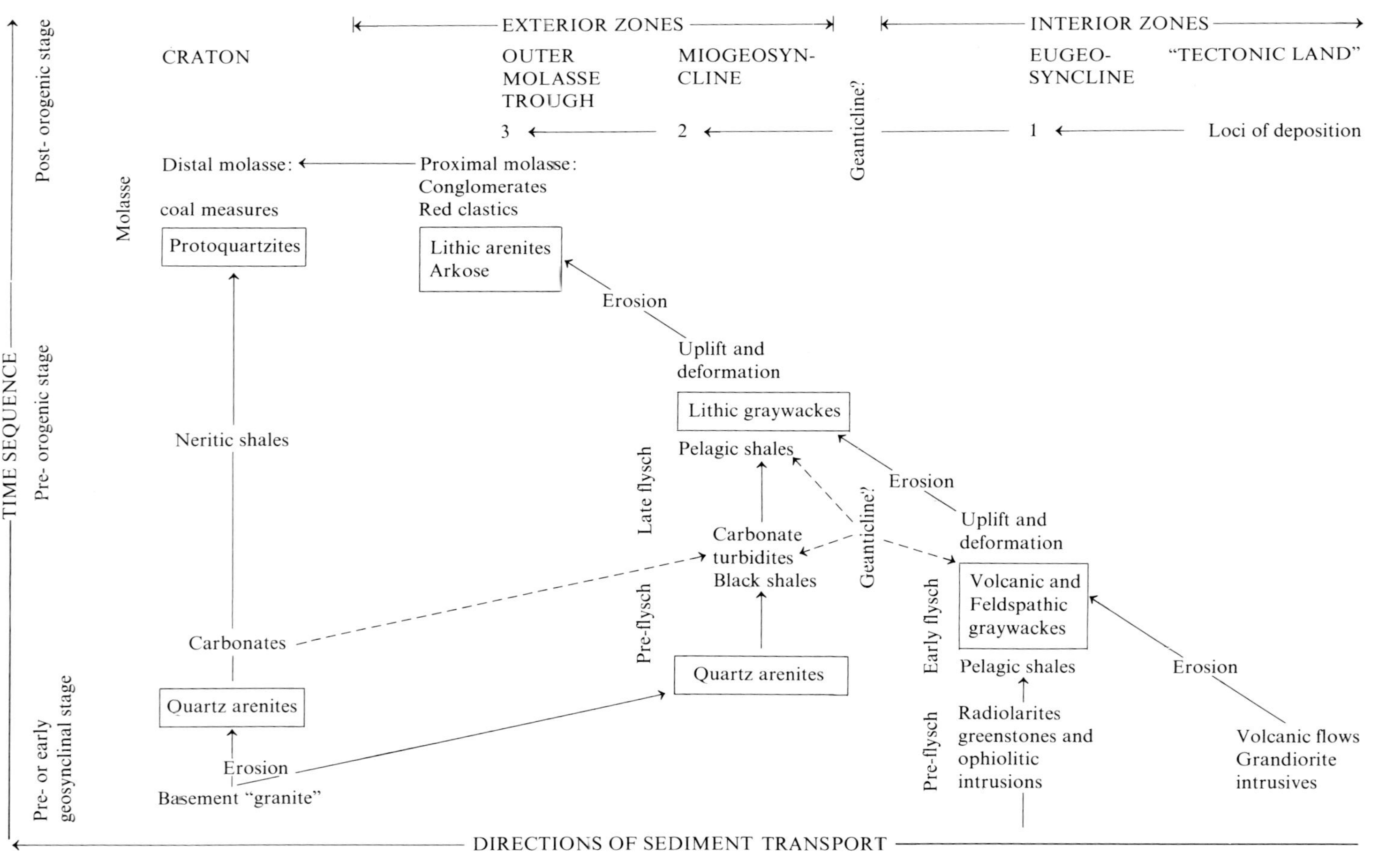

Fig. 12-2. Geosynclinal evolution and sedimentation. Solid lines indicate main paths and dashed lines less common alternative paths. See Fig. 12-1 for structural terminology

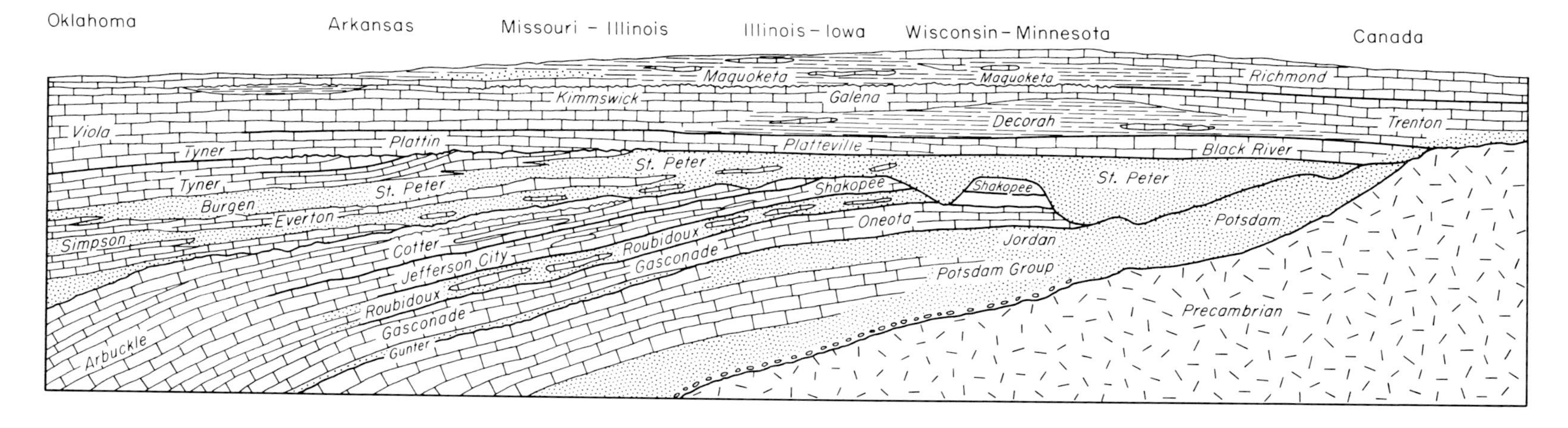

Fig. 12-3. Generalized reconstructed, north-south cross section of the early Paleozoic of the Mississippi Valley (from Potter and Pryor, 1961, Fig. 3)

Valley region. Once formed, the individual basins and arches tend to persist throughout the craton's history. The sediments overlying the concealed part of the craton vary in thickness and reach their maximum thickness in the intracratonic basins. In places they exceed 10,000 feet (3100 m).

A craton tends to be a positive area; portions of it are generally exposed even during times of maximum submergence. Much of the present shield areas had a thin sedimentary cover at some past epoch which has been stripped off by subsequent erosion. The craton is characterized by uniform and persistant paleoslopes, the latter dipping very gently from the Precambrian core toward the younger bordering marginal troughs. This persistence is demonstrated by paleocurrent data, mainly crossbedding, in sandstones deposited on the craton and also by stratigraphic relations. The cratonic platforms characteristically have a thin sedimentary cover, mainly of carbonates and sandstones, shale being subordinate. A volume of sand of about 4018×10^3 km^3 forms about 23 percent of the total volume of the sediments on the North American craton (Sloss, 1969).

Disconformities are common and widespread and become compound up slope (Fig. 12-3). Short term cyclic deposition seems to characterize some cratonic sediments, especially the cyclothems of Carboniferous strata.

Sand derived from the interior of the craton is deposited in shallow water, commonly in fluvial, littoral, and neritic environments. These sands characteristically show crossbedding and ripplemarks. Much recycling and repeated long erosional intervals, together with possible epochs of eolian action, produce mature and supermature sands. Large river systems may carry these sands long distances to bordering mobile belts.

Sand Deposition and Basin Architecture

As our understanding of the earth's history has grown, it has become increasingly clear that there is a close relation between sediments and tectonics. Beyond the self-evident relation between uplift and erosion and subsidence and sedimentation, it has become clear that the sedimentary assemblages of some areas or times differ markedly from those of other places and times and that these differences are in some way related to the architecture of the basin and/or the stage in its evolution and filling. These variations in sedimentary character are particularly well shown by sands.

O. T. Jones in 1938 noted the contrast in character of lower Paleozoic strata of Wales that rest on a Precambrian platform and those in the adjacent Caledonian geosyncline. The platform sediments are a relatively thin sequence of shallow-water carbonates and crossbedded sands, but those of the geosyncline are a very thick accumulation of dark shales with interbedded graded graywacke sands. There is a comparable difference in the Paleozoic rocks of the Ouachita geosyncline of Arkansas, Oklahoma and Texas and rocks of the same age in the Arbuckle-Ozark areas. The Ouachita facies is dominantly a clastic sequence with a great thickness of shales and interbedded turbidite sands of late Paleozoic age superimposed on relatively thin early Paleozoic strata; the Arbuckle facies, in contrast, consists of a thick, dominantly carbonate sequence of early Paleozoic age and a relatively thin late Paleozoic clastic sequence. Contrasts in sedimentary aspect related to structural site led to the concept of a "shelf" or "platform" facies in the relatively stable areas and a "geosynclinal" facies in the unstable mobile tract.

A parallel development was the recognition by Stille (1936) of different sedimentary facies within geosynclinal tracts that could be used to differentiate two tectonically distinct elements, the mio- and eugeosynclines and their absence in miogeosynclines. Glaessner and Teichert (1947) reviewed the growth of the geosynclinal concept and emphasized its importance in the understanding of sedimentary facies. Kay (1951) amplified and extended Stille's concepts and introduced many new terms, applying the name geosyncline to many kinds of sedimentary basins. King (1959, p. 57) incorporated the ideas of Stille and Kay in his analysis of the relation of sedimentation to tectonics in the evolution of North America but used a simpler terminology. All of these men, while differing in details, put as much emphasis on the differing interactions between tectonics, provenance, and sedimentary facies within different parts or kinds of geosynclines as on the differences between geosynclines and platforms.

At the same time that spatial relations of geosynclinal facies were being explored, data accumulated that seemed to indicate a time sequence in the sedimentary assemblages of the geosynclinal tracts. Bertrand (1897) was perhaps the first to note that the first sediments in the Alpine geosyncline were relatively fine grained, mainly shales (shaly flysch), that these were mildly deformed and eroded to yield a coarser sandy Flysch, and that this in turn was uplifted and redeposited as very coarse Molasse sands and conglomerates, in part red clastics. "Flysch" and "Molasse," originally Alpine stratigraphic terms, are thus sedimentary assemblages characteristic of particular stages in the filling of a geosyncline. The sands of these two assemblages were observed to be vastly different in their structures and composition – the first being graded graywackes and the second being coarser crossbedded, lithic and arkosic arenites. These concepts have been expanded and elaborated on by various workers, notably Krynine, who formulated a diastrophic or tectonic cycle (1945, p. 13; 1959, p. 746). According to Krynine the early or peneplanation stages were characterized by orthoquartzitic (quartz arenite) sands, the second stage ("geosynclinal stage") being marked by graywacke sands (including lithic arenites) and finally an "orogenic" stage of intense deformation marked by arkoses.

The concepts of tectonic control of sedimentation, especially the concept of sedimentary associations presumably related to tectonic stability of the site of deposition or of a particular stage in the filling of a geosyncline, have been the theme of a multitude of papers. The concepts outlined above have been elaborated on, developed in details, have been criticized, or even denied. Summaries of the state of our understanding, have been presented in the writings of van Andel (1958); Aubouin (1965, p. 109–144); Kay (1951, p. 85–96); Krumbein and Sloss (1963, p. 390–431); Krumbein, Sloss and Dapples (1949); and Pettijohn (1957, p. 636–644).

A critical analysis of the literature shows that some of the earlier views need considerable revision, that many earlier generalizations were premature, that the problem has been clouded by semantics, notably by lack of agreement on just what "geosynclines" are, and by failure to discriminate between the various types of sedimentary basins often lumped together as "geosynclines". The concepts are complicated further by preconceptions about continental drift as opposed to continental stability and newer conceptions of global tectonics. It is not possible in the limited space available here to trace all the ramifications of thought relative to either the tectonic or sedimentological aspects of these topics nor their relations to one another. To do so would take us far afield from our subject – sand.

Perhaps the best we can do, under the circumstances, is to review briefly what is known about the history of some particular sedimentary basins, chosen because they illustrate some contrasting types of basins, with emphasis on the sands contained in them.

Examples of Basin Development and Sand Accumulation

This section contains a synoptic description of some well-studied sedimentary basins. We have selected as examples those studies in which sandstone petrography and paleocurrents have been integrated with the usual stratigraphic and structural data. Only by this comprehensive integration can one reach a full understanding of the history of a particular basin.

Our choice of examples was further governed by a desire to illustrate several unlike basin types and show the kinds of problems that these present. We have chosen, as illustrative of the principles and methods of analyses, (1) the Coronation Geosyncline, a Canadian Precambrian example of a classic geosyncline, (2) the Paleozoic central Appalachian Basin, the "type" geosyncline, (3) the Rand Basin, a Precambrian basin of a different type, (4) the Illinois Basin, an intracratonic basin, and (5) the Mesozoic Geosyncline of California, a flysch basin.

The Coronation Geosyncline

The Coronation Geosyncline was so named from the area of most complete preservation – Coronation Gulf in the District of Mackenzie, Canada. There are four belts of Aphebian sedimentary and volcanic rocks (Great Slave, Epworth, Goulbourn, and Snare Groups) which are remnants of an arcuate north-south (convex to the west) geosyncline. This structure lies between a craton to the east and a late Aphebian orogenic belt (the Bear Province) to the west, Fig. 12-4, and was filled and deformed between 1750 and 2000 million years ago.

The strata in the geosyncline are about 32,000 feet (9760 m) thick in contrast to the 6,000–10,000 feet (1830–3050 m) recorded in the contiguous beds over the associated craton. Four stages in the depositional history are recognized:

1. a preorogenic quartz arenite-carbonate phase
2. a transitional-euxinic volcanic phase
3. an early synorogenic flysch phase
4. a late synorogenic molasse phase.

The quartz arenite sandstones of the pre-orogenic phase are derived from the craton, whereas the synorogenic feldspathic graywackes and lithic arenites are derived from the orogenic belt.

The Coronation strata are intensely deformed and metamorphosed only in the orogenic belt. Elsewhere they are indurated, but largely remain unmetamorphosed.

The pre-orogenic strata consist of a basal terrigenous sequence, a cyclical dolomite-shale assemblage, and a second terrigenous sequence restricted mainly to the craton but extending into the geosyncline in the Great Slave Lake region. Paleocurrents demonstrate that these sediments were derived from the craton to the east (Fig. 12-5A).

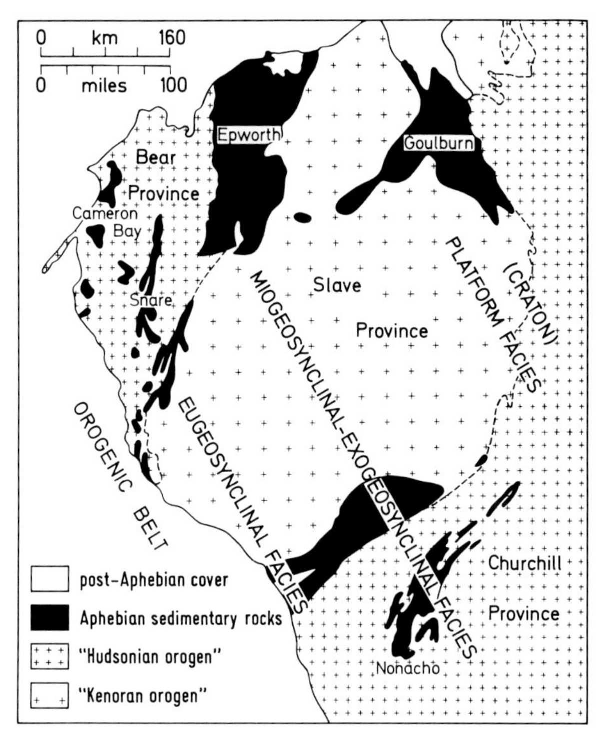

Fig. 12-4. Present outcrops and inferred regional facies belts, Coronation Geosyncline, N.W.T., Canada (Modified from Hoffman, 1969, Fig. 12). Reproduced by permission of the National Research Council of Canada

The transitional strata include a pyritic black shale deposited in a deep euxinic basin in the Coronation Gulf region. Elsewhere a thick complex of volcanic rocks was formed. Pillow lavas are absent and the volcanics are thought to have been derived from island volcanoes in an otherwise shallow water marine shelf.

The flysch phase accumulated in a deep basin in the Coronation Gulf region and consists of thousands of feet of rhythmically interbedded coarse-grained feldspathic graywacke and dark shales. The graywackes thin to the east in their down current direction and their feather edge does not extend beyond the hinge line separating basin from shelf. They were derived from tectonic lands to the west. A typical flysch phase developed slightly later in the Great Slave Lake area.

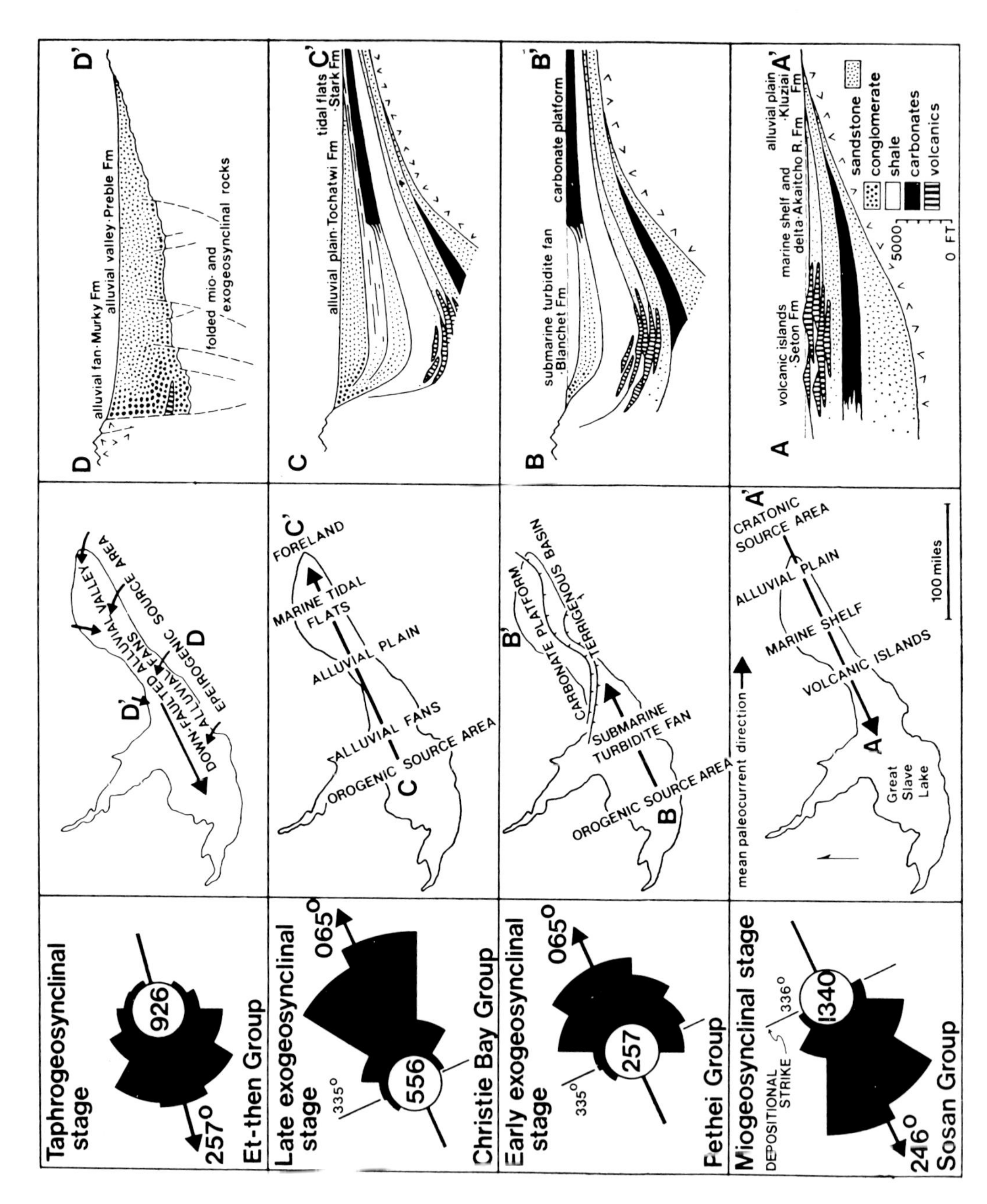

Fig. 12-5. Summary of paleocurrents in the four major sandstone units (left column) and the inferred paleogeography and cross sections during each stage of deposition. Great Slave Lake, N.W.T., Canada (from Hoffman, 1969, Fig. 11). Reproduced by permission of the National Research Council of Canada

It is noteworthy that a thick stromatolitic carbonate facies deposited on the cratonic shelf co-existed with the flysch facies in the geosyncline which lay only a few miles to the west (Fig. 12-5 B).

Above the flysch sediments and their cratonic equivalents lies a thick, broadly regressive, succession of red clastics. These consist of red siltstone grading upward into crossbedded red lithic arenites of alluvial origin. Paleocurrents demonstrate that these molasse sediments, like the flysch graywackes before them, came from the orogenic belt to the west (Fig. 12-5 C).

This molasse stage was followed, in the Great Slave Lake area, by the development of a taphrogeosyncline or half-graben filled with conglomerates and coarse sands of alluvial origin (Fig. 12-5 D).

It is clear that the depositional history of the Coronation Geosyncline may in fact be an ancestral part of the Cordilleran Geosyncline – that part of it which was initiated, filled, and deformed in Precambrian times. As commonly happens in orogenic belts, the axis of downwarp and deposition tended to shift or migrate. In Paleozoic and Mesozoic times the axis shifted westward from its Precambrian position.

The sandstones in this geosyncline belong in two major groups: the cratogenic sands which are, in general, most mature, and those which are tectonogenic, that is, derived from the orogenic belt or the tectonic source land which lay west of it.

The sands from the craton vary from the somewhat arkosic basal sands (Hornby Channel) of alluvial origin to the quartz-arenite sands of mixed coastal and eolian origin (Odjick Formation, Fig. 6-24). White quartz arenites of cratonic origin are associated with the dolomites.

The sands of the flysch facies (Blanchet and Recluse Formations, Fig. 6-11) are typical feldspathic graywackes of turbidite origin. They are poorly sorted dark rocks, typically graded, consisting of angular detritus including sedimentary and volcanic rock fragments and feldspar.

Table 12-1. *Petrography of sandstones of Coronation geosyneline of the Great Slave Lake area (Hoffman, 1969)*

Formations	Quartz	Feldspar	Rock fragments	Quartz cement	Carbonate cement	Matrix	Void space
Third Depositional Phase:							
Preble Formation	35	20	10	—	35	—	—
Upper Murky	30	2	28	—	40	—	—
Basal Murky	20	—	45	—	—	—	35
Second Depositional Phase:							
Upper Tochatwi	50	15	13	—	20	—	—
Basla Tochatwi	40	—	35	—	20	—	—
Blanchet	50	10	15	—	—	20	—
First Depositional Phase:							
Kluziai	65	12	—	20	—	—	—
Upper Hornby Channel	60	8	—	30	—	—	—
Basal Hornby Channel	65	15	5	25	—	—	—

The molasse sands (Tochatwi Formation) are red lithic arenites, texturally and compositionally immature. The sediments are coarse, in places conglomeratic, markedly crossbedded, and probably mainly fluvial as fining-upward cycles testify.

The petrographic character of the sands of the Great Slave Lake area are summarized in Table 12-1.

References:

Hoffman, P. F.: Proterozoic paleocurrents and depositional history of East Arm fold belt, Great Slave Lake, Northwest Territories. Canadian Jour. Earth Sci. **6**, 441–462 (1969).

— Fraser, J. A., McGlynn, J. C.: The Coronation Geosyncline of Aphebian age, District of Mackenzie Geol. Survey Canada, Paper 70-40, 201–212 (1970).

Central Appalachian Geosyncline

The Appalachians have been the birthplace of many of the great concepts of geology – the theory of geosynclines being the most important. We propose here to describe the sandstones of the central Appalachians and their relation to the evolution and deformation of the Appalachian geosyncline of this area.

The Appalachian Basin here described extends from the eastern Great Lakes on the north to Alabama and Georgia on the south and from the Blue Ridge on the east to the Cincinnati-Findlay Arch on the west (Fig. 12-6). It is about

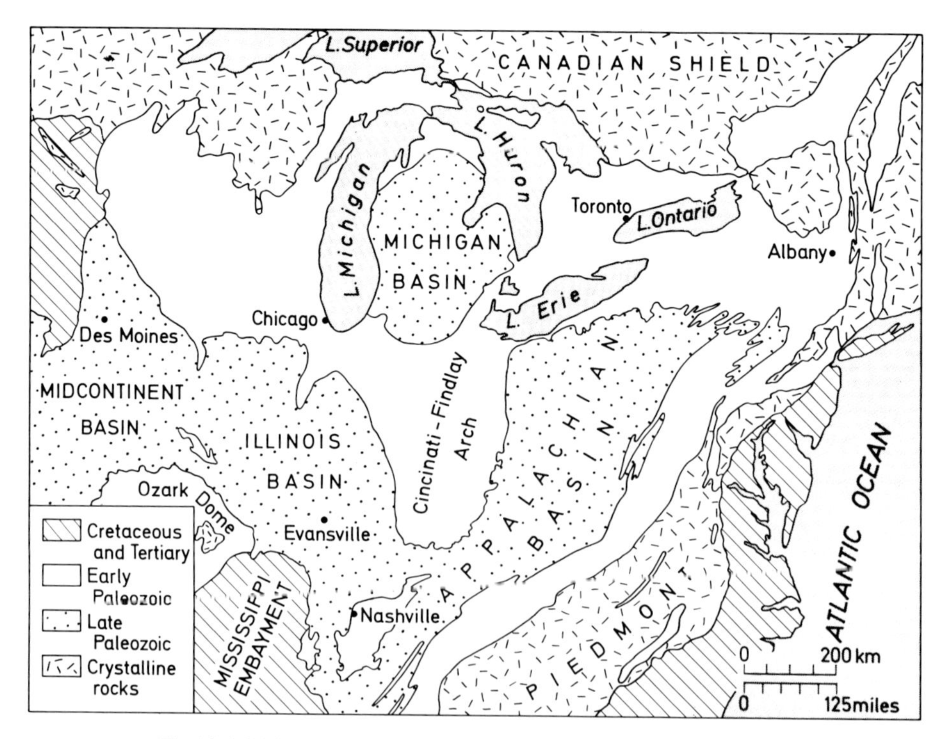

Fig. 12-6. Major structural elements in part of eastern North America

1,000 miles (1,600 km) long and 330 (530 km) wide. It is filled primarily with Paleozoic rocks, from Cambrian to Permian in age, approximately 510,000 mi^3 (2,126,000 km^3) of which 23 percent is sand (Colton, 1970, p. 11). The section thickens from less than 5,000 ft. (1525 m) in the northwest to nearly 40,000 ft. (12,190 m) along its southeastern margin.

The basin described is an asymmetrical miogeosyncline, essentially devoid of volcanic rocks. It was filled in the central and western parts mainly with carbonates of tidal flat origin, with a sparse development of quartz arenites during the early Paleozoic. A poorly known eugeosynclinal tract lay to the east of the Blue Ridge – well-developed and better studied in New England than in the central Appalachians. Muddy sediments and other clastics, with some volcanics in North and South Carolina accumulated in the eugeosyncline at this early time. It is not clear whether this eugeosynclinal tract was separated from the miogeosyncline by a geanticlinal ridge as it was in New England.

Deformation and metamorphism of the eugeosyncline in Ordovician time was followed by uplift and erosion. This uplift is recorded in the central Appalachian miogeosyncline by a late Ordovician flysch succeeded by a molasse consisting of alluvial red clastics and conglomeratic quartzites, both derived from the southeast.

This "Taconic cycle" was followed by a mid- and late-Paleozoic cycle which began with a carbonate sequence with sparse quartz arenite sands and was brought to an end by another clastic wedge derived from the southeast. These clastics consisted of a flysch-like Devonian sequence followed by a late-Devonian and Mississippian molasse. This "Acadian cycle" was followed by a third, somewhat abbreviated marine transgression which deposited a thin carbonate wedge and which was ended by final uplift and a coarse clastic wedge of southeastern derivation.

The organization of the later Paleozoic fill in the Appalachian Basin is shown schematically in Fig. 12-7.

The sandstones of this impressive fill, form, as noted above, about 32 percent of the total. The bulk of these were derived from the southeast as the paleocurrent data show (Fig. 12-8). Only a small volume came from the craton to the northwest (Schwab, 1970). The craton supplied mature quartz arenites only in the earliest Cambrian (Weverton, Antietam, and Gatesburg Formations). By mid-Ordovician time the craton was largely covered by carbonates and hence incapable of supplying clastic sediments. The late Ordovician and early Silurian clastic wedge came from the southeast. The turbidite sands (Martinsburg Shale, Fig. 6-18) of the Ordovician flysch are lithic graywackes (McBride, 1962; McIver, 1970); those of the latest Ordovician (Juniata) are lithic arenites deposited in a fluvial environment. The culmination of the molasse, the ridge-making Silurian Tuscarora, is a mixture of marine orthoquartzites, non-marine protoquartzites, and arenites (Yeakel, 1959). The paleogeography in Tuscarora time is summarized in Fig. 12-9.

The first sands of the Acadian cycle (Keefer and Oriskany Formations, Fig. 6-27), associated with the Siluro-Devonian carbonates, are small in volume and of uncertain derivation. In character they resemble the Cambro-Ordovician cratogenic sands. It is probable that they were derived by erosion and redeposition of Cambro-Ordovician quartz arenites of the continental interior – in other words are multi-cycle cratogenic sands. The clastic wedge of late Devonian and early Mississippian consists in its lower part, of turbidite sands ("Portage") of easterly derivation (McIver, 1970) and in its upper part of coarse alluvial sands (Catskill) characterized by the usual fining-upward cycles (Allen and Friend, 1968).

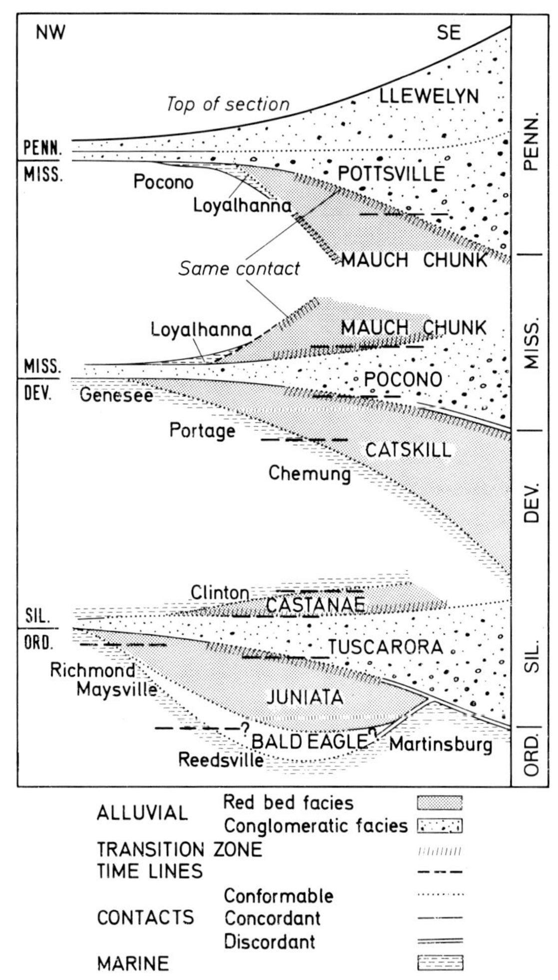

Fig. 12-7. Transverse cross sections of three alluvial wedges, central Appalachians (Modified from Meckel, 1970, Fig. 3). By permission of author and John Wiley and Sons

The culmination of this molasse is the ridge-forming Mississippian Pocono, a lithic sandstone of a nonmarine alluvial origin in the eastern areas and marine in the northwest (Pelletier, 1958). Deposition of the Mississippian clastics was followed by a brief transgression followed by another molasse-type conglomeratic deposit in the Pennsylvanian Pottsville (Meckel, 1970). See Fig. 12-10.

A rough calculation shows that the Appalachian sands are about 65 percent alluvial in origin, about 15 percent marine turbidites, and 20 percent marine cratonic "shelf" sands. The first are largely lithic arenites, the second graywackes,

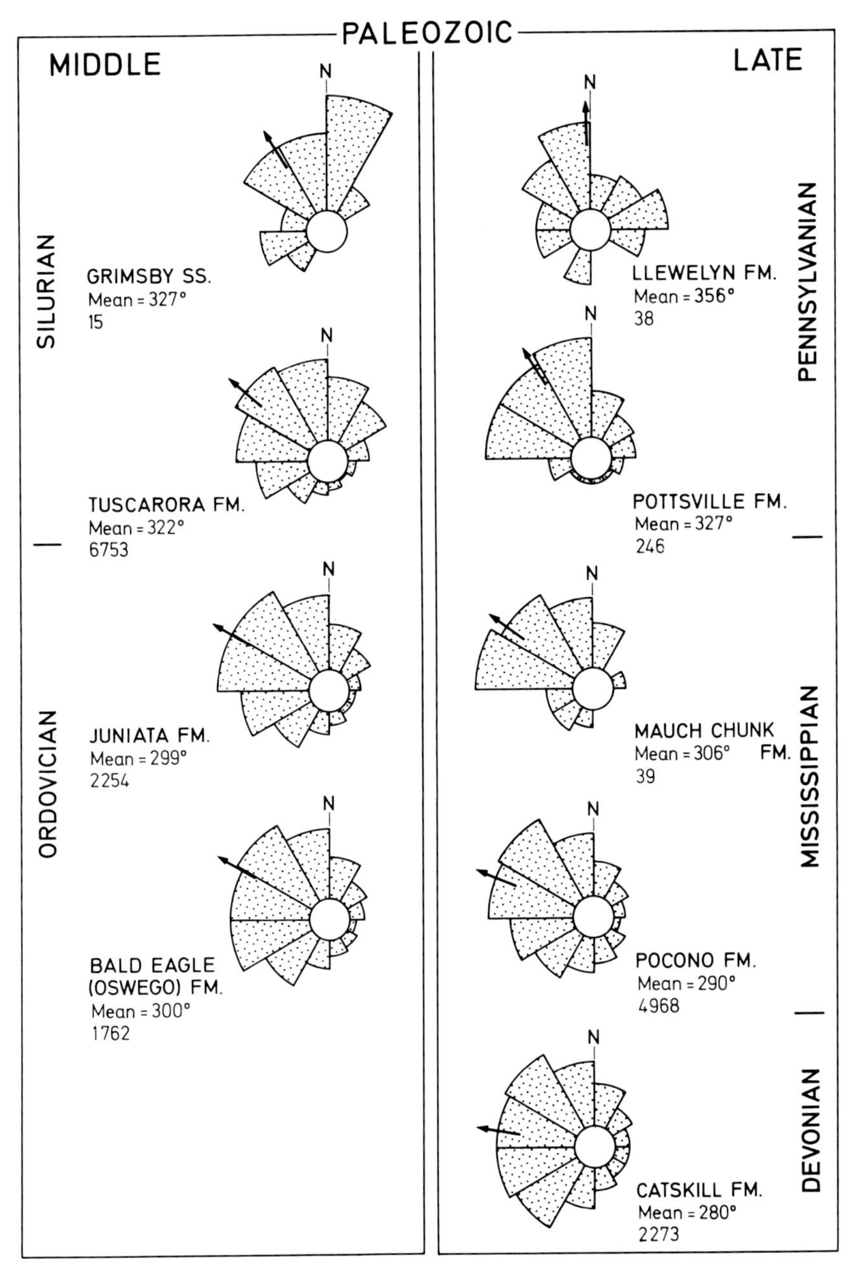

Fig. 12-8. Paleocurrents and numbers of observations in central Appalachian sandstones (Modified from Meckel, 1970, Fig. 7)

and the third, quartz arenites. The salient features of Appalachian sands are summarized in Table 12-2.

It is clear, in summary, that a good deal of the geologic history of the Appalachians can be extracted from the sandstones in this area – by careful petrographic analysis and by a study of the paleocurrent patterns deduced by mapping the primary sedimentary structures of these sands. It is clear also that the clastic fill of the Appalachian Basin is primarily of southeastern origin, that some of it was generated by uplift and erosion of previously deposited strata,

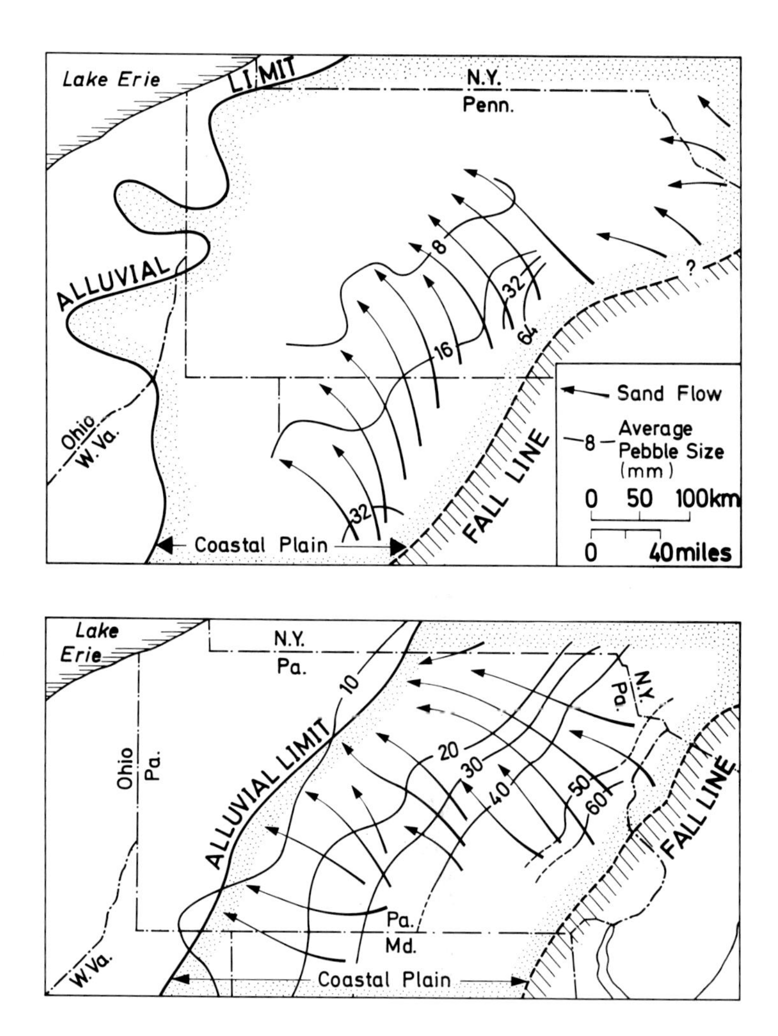

Fig. 12-9. Tuscarora (Silurian, upper figure) and Pocono (Mississippian, lower figure) paleogeographies in the central Appalachians (Modified from Yeakel, 1962, Fig. 7 and Pelletier, 1958, Fig. 16)

that very little came from the craton. The principal source land was an area of mixed low-grade metamorphic and sedimentary rocks – containing negligible volume of volcanic and plutonic igneous rocks. The eugeosynclinal fill, about which little is presently known, contained a significant proportion of volcanic materials (Sundelius, 1970).

Table 12-2. *Petrography of sandstones of Central Appalachians (based on compilation by W. R. Kaiser)*

	Quartz	Feldspar	Rock fragments	Chert	Matrix	Cement[4]
Appalachian cycle						
Molasse:						
Pottsville	67	tr	13	2	12	3
Mauch Chunk	73	7	9	1	—	—
Pocono	63	tr	28	—	6	—
Catskill	49	tr	33[1]	—	—	—
"Flysch":						
"Portage"	62[2]	1	7	—	23	6
Shelf:						
Oriskany	91	1	—	8	—	—
Keefer	74	—	1	tr	4	21
Taconic cycle						
Molasse:						
Tuscarora	54	2	20	3	12	8
Juniata	59	2	13	1	18	5
Oswego	63	2	12	2	17	4
Flysch:						
Martinsburg	35	5	24	—	37	—
Shelf:						
Gatesburg	55	16	—	—	—	29[5]
Antietam	84	5	4[3]	—	—	3

(1) Includes matrix; (2) includes chert; (3) mostly polycrystalline quartz and chert; (4) not separately determined in many sandstones; generally present but not separated from detrital quartz; (5) micrite and clastic carbonate.

References:

Allen, J. R. L., Friend, P. F.: Deposition of the Catskill facies, Appalachian region. Geol. Soc. America, Special Paper **106**, 21–74 (1968).

Colton, G. W.: The Appalachian Basin – its depositional sequences and their geologic relationships. In: Fisher, G. W., Pettijohn, F. J., Reed, J. C., Jr., and Weaver, K. N., Eds.: Studies of Appalachian geology, central and southern, p. 5–47. New York: Wiley-Interscience 1970.

McBride, E. F.: Flysch and associated beds of the Martinsburg Formation (Ordovician), Central Appalachians. Jour. Sed. Petrology, **32**, 39–91 (1962).

McIver, N. L.: Appalachian turbidites. In: Fisher, G. W., Pettijohn, F. J., Reed, J. C., Jr., and Weaver, K. N., Eds.: Studies of Appalachian geology – central and southern, p. 69–82. New York: Wiley-Interscience 1970.

Meckel, L. D.: Paleozoic alluvial deposition in the Central Appalachians: A summary. In: Fisher, G. W., Pettijohn, F. J., Reed, J. C., Jr., Weaver, K. N., Eds.: Studies of Appalachian geology – central and southern, p. 49–68. New York: Wiley-Interscience 1970.

Pelletier, B. R.: Pocono paleocurrents in Pennsylvania and Maryland. Geol. Soc. America Bull. **79**, 1033–1064 (1958).

Schwab, F. L.: Origin of the Antietam Formation (Late Precambrian? Lower Cambrian), central Virginia. Jour. Sed. Petrology **40**, 354–366 (1970).

Sundelius, H. W.: The Carolina Slate Belt. In: Fisher, G. W., Pettijohn, F. J., Reed, J. C., Jr., Weaver, K. N., Eds.: Studies of Appalachian geology – central and southern, p. 359–368. New York: Wiley-Interscience 1970.

Yeakel, L. S.: Tuscarora, Juniata, and Bald Eagle paleocurrents and paleogeography in the central Appalachians. Geol. Soc. America Bull. **73**, 1515–1540 (1959).

Mesozoic Geosyncline of California

The late Mesozoic strata of California accumulated in a geosyncline along the west margin of the North American continent. According to Bailey, Irwin, and Jones (1964, p. 21 and 123), the combined volume of the Franciscan assemblage and the Great Valley sequence, two coeval Coast Range facies derived from the Klamath-Sierra Nevada region, is about 500×10^3 mi^3 (2085×10^3 km^3). The stratigraphy of the Franciscan sequence is too uncertain for analysis but it is clearly a eugeosynclinal assemblage of graywackes (Fig. 6-14) and pillow lavas with a little radiolarian chert. The better known Great Valley sequence includes about 49,000 ft (14,945 m) of clastic sedimentary strata consisting of turbidite sandstones in an enclosing mudstone phase (Fig. 12-10).

The sandstones belong to one or another of three types: (1) sandstones derived from active andesitic volcanic sources characterized by very little or no quartz, a high ratio of plagioclase to total feldspar, and an abundance of rock fragments, mainly andesitic with some felsites; (2) sandstones derived from tectonic highlands, the detrital fraction being moderately rich in quartz, relatively low in feldspar, and rich in a diversity of rock particles; and (3) sandstones derived from deeply eroded orogens, mainly plutonic igneous and high-grade gneisses (Dickinson, 1969). These sands have high quartz and feldspar and relatively little in the way of rock particles. The composition of the major sandstone bodies in the Great Valley sequence in the Sacramento Valley area is summarized in Table 12-3.

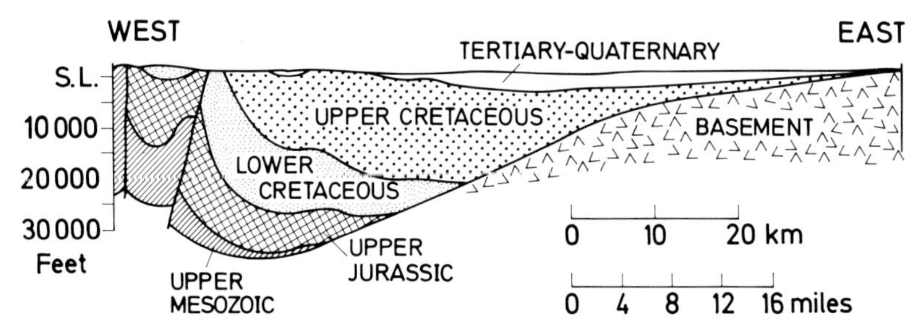

Fig. 12-10. Generalized east-west cross-section Sacramento Valley, California (Modified from Ojakangas, 1968, Fig. 3)

Table 12-3. *Approximate mean modal composition of the major sandstone bodies of the Late Mesozoic strata of the west side of the Sacramento Valley, California (Dickinson, 1969)*

Unit	System and stage	Quartz	Feldspar	Rock particles	Plagioclase/ total feldspar
Guinda	Campanian	40	40	20	0.60
Sites	Turonian	35	40	25	0.65
Venado	Turonian	30	35	35	0.70
"Brophy Canyon"	Albian	35	25	35	0.75
Unnamed	Aptian	50	25	25	0.80
"Leesville"	Valanginian?	45	30	25	0.85
"Blue Ridge"	Berriasian?	25	30	45	0.85
Unnamed	Tithonian	5	20	75	1.00

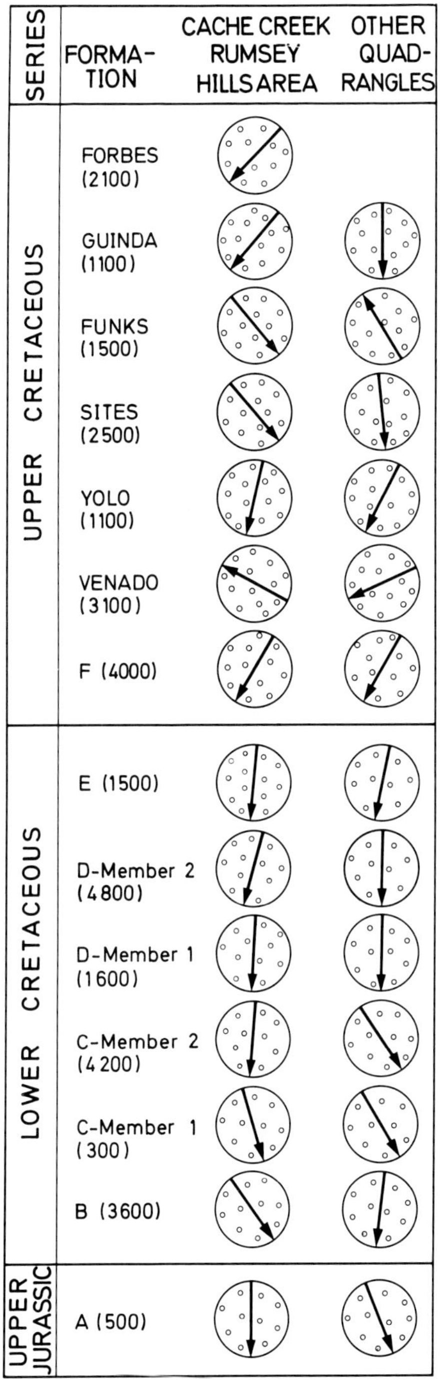

SERIES	FORMATION	CACHE CREEK RUMSEY HILLS AREA	OTHER QUAD-RANGLES
UPPER CRETACEOUS	FORBES (2100)		
	GUINDA (1100)		
	FUNKS (1500)		
	SITES (2500)		
	YOLO (1100)		
	VENADO (3100)		
	F (4000)		
LOWER CRETACEOUS	E (1500)		
	D-Member 2 (4800)		
	D-Member 1 (1600)		
	C-Member 2 (4200)		
	C-Member 1 (300)		
	B (3600)		
UPPER JURASSIC	A (500)		

Fig. 12-11. Paleocurrents of late Mesozoic sandstones, Sacramento Valley California. Numbers in parentheses are thickness in feet. Sections totals 31,900 ft. (9730 m). Modified from Ojakangas (1968, Figs. 3 and 6)

In the lower Cretaceous (Aptian and Albian) two kinds of sandstones are present, one type (quartz 55 percent, feldspar 22.5 percent, rock fragments 22.5 percent; here abbreviated as $Q_{55}F_{25.5}Rx_{22.5}$) derived largely from volcanic sources. A similar but less pronounced oscillation of plutonic and volcanic influences can be noted in the Upper Cretaceous (Turonian) where the analogous variants have compositions of $Q_{30}F_{40}Rx_{30}$ and $Q_{20}F_{30}Rx_{50}$. These variations record recurrent rejuvenation of volcanism in the source region. Some sandstones in the sequence contain recognizable volcanic debris including bipyramidal quartz and sanidine. The dominant progression, from volcanic graywackes in the Upper Jurassic to arkosic wackes in the Upper Cretaceous, records the evolution of the volcano-plutonic source from a volcanogenic superstructure to a batholithic infrastructure.

Superimposed on the reversible variations in the proportions of aphanitic and phaneritic igneous detritus, is the apparently irreversible decrease in the ratio of plagioclase to total feldspar. This trend is believed to reflect what is known of the composition of the plutons intruded at different times. Those emplaced in the late Jurassic lay west of the quartz diorite line and were mainly diorite, tonalite, and trondhjemite. The main plutons of granodiorite and adamellite were not emplaced until mid-Cretaceous time.

In summary, the sandstones are shown to faithfully record the evolutionary trend of both plutonic and volcanic igneous activity and to record the erosion and destruction of the volcanic suprastructure and unroofing and exposure of the pluton infrastructure.

Ojakangas (1968) has shown that sole-markings on the turbidite sands of the Great Valley sequence show a well-defined paleocurrent trend, from north to south, parallel to the axis of the depositional trough (Fig. 12-11). There are some divergent trends showing transport from the east to the west. The sediments are believed to have been derived from the ancestral Sierra Nevada chain on the east and the Klamath mountain area on the north.

References:

Bailey, E. H., Irwin, W. P., Jones, D. L.: Franciscan and related rocks and their significance in the geology of western California. California Div. Mines and Geology Bull. **183**, 177 p. (1964).

Dickinson, W. R.: Evolution of calc-alkaline rocks in the geosynclinal system of California and Oregon. In: Proceedings of the Andesite Conference. Oregon Dept. Geol. Min. Ind. Bull. **65**, 151–156 (1969).

Ojakangas, R. W.: Cretaceous sedimentation, Sacramento Valley, California. Geol. Soc. America Bull. **79**, 973–1008 (1968).

Witwatersrand Basin of South Africa

Both because it is one of the world's principal sources of gold and uranium and because it has so many unusual geological features, the Precambrian Witwatersrand basin (between 2.800 and 2.500 million years old) has been intensively studied, although only in recent years have sedimentological analyses been made of the basin as a whole. Much of the following is taken from the two general summaries by Brock and Pretorius (1964a and 1964b) and the publications of the Economic Geology Research Unit of the University of the Witwatersrand, Johannesburg. Another general source is Haughton (1964). Table 12-4 summarizes the characteristics of the basin.

The Witwatersrand basin covers an area of some 15,000 square miles (38,850 km^2) and lies to the south of the Bushveld basin. Elliptical arcs of tectonically reactivated granite domes border the basin which also has a series of peripheral border faults (Fig. 12-12). Along the axis of structural symmetry and

Table 12-4. *Summary of Precambrian Witwatersrand basin of South Africa*

Geometry

Sharply curved, L-shaped, probably closed, basin with asymmetrical, transverse crossection. Basin edges tend to be fault-bounded, particularly on the northwest side.

Size

Approximately 250 mi (400 km) long on a sharply arcuate depositional axis, 150 mi (240 km) wide and 25,000 ft (7,600 m) thick.

Fill

Kind: Weakly metamorphosed crossbedded quartzites, argillites and some conglomerates and lavas. Gold and uranium mostly associated with coarse clastics along basin's rim. Widespread marker beds and shallow water sedimentation. *Arrangement:* Asymmetrical filling by four alluvial fans (deltas?) on northwestern side of the basin supplied most of the coarse conglomerates which are largely limited to the basin rim. Units thin toward margins as a result of both depositional thinning and unconformities which become compound toward margins.

Current System

Locally radial along original northwestern side of basin. Some evidence of weak longshore currents.

Tectonics

Moderately dipping (30°), fault-bounded basin inset in a stable craton, surrounded by granite plutons. Deposition followed by 10,000 ft (3,050 m) of less steeply dipping lavas and clastics.

near its depositional axis the basin has been pierced by a diapiric granite body known as the Vredefort dome. The bordering granites yield ages of between 3,200 and 2,800 million years.

These neighboring deeply eroded plutons indicate that the Witwatersrand basin is inset in a stable craton known as the Kaapraal Craton, a quite remarkable crustal fragment from many points of view, constituting part of the southern African Shield. Overlying the Witwatersrand sequence are at least 12,000 ft (3,660 m) of lava and clastic sediments of the Ventersdorp sequence (2.5 to 2.2×10^9 yr) plus at least 13,000 ft (3965 m) of dolomite and clastic sediments and lavas of the Transvaal sequence (2.2 to 1.9×10^9 yr). Both these sequences have shallower dips than the Witwatersrand strata. Underlying the Witwatersrand basin, for the most part, is the Basement Complex of granites and greenstone belts (3.4×10^9 yr). Gold and uranium occur principally along the original northwest rim of the basin, the side adjacent to the main source-area of sedimentary material.

The fill of the Witwatersrand basin is almost entirely clastic: conglomerates are conspicuous locally along the original northwestern margins of the basin, where they may form as much as 8 percent of the section, but quartzites and argillites constitute most of the fill. The quartzites show a progressive change upwards from unusually pure quartz arenites at the base, through altered lithic arenites in the higher parts of the lower half, to hydrothermally altered feldspathic quartzites in the upper half. The principal sedimentary structures of the quartzites are crossbedding, ripplemark, and some scour-channels. Maximum thickness of the fill is estimated to be approximately 25,000 ft (7625 m) thick which in the

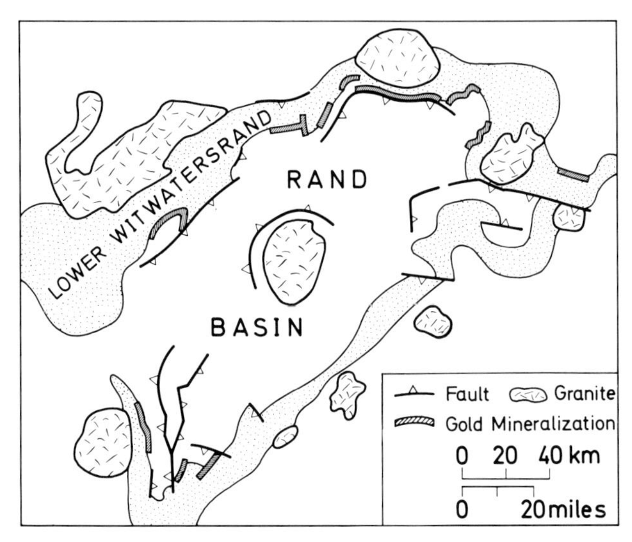

Fig. 12-12. Structural setting of Witwatersrand Basin and major areas of gold and uranium mining (Simplified from Brock and Pretorius, 1964a, Fig. 2)

central part in the basin contain some 17 marker beds. Many of these beds are comparatively thin and widespread so that the basin has an appreciable stratigraphic ordering. Because the basin progressively shrank, the preserved lower groups represent a distal facies, whereas the upper groups are essentially a proximal facies. Consequently, the conglomerate to quartzite and quartzite to argillite ratios are much higher in the upper strata than in the lower sediments. Significant gold mineralization is virtually restricted to the two upper groups, where its presence is a function of a preserved proximal facies.

Stratigraphic, sedimentologic, and structural evidence suggests that the present basin limits do not differ greatly from original ones. Along the fault-bounded north and west sides of the presently shaped basin, evidence for this conclusion includes: (1) unconformities that become compound edgeward; (2) intraformational disconformities that become angular unconformities near the basin's margin; (3) depositional thinning; (4) marginal conglomerates; and (5) four deltas or alluvial fans that served as the main sources of supply (Fig. 12-13). Although definitive evidence is lacking, these fans or deltas may have debouched into a large lake or inland sea or perhaps even into a narrow arm of the ocean. These four fans are also the principal sites of gold and uranium mineralization.

That most, if not all, of this mineralization may be of primary sedimentary origin – a placer rather than a hydrothermal deposit – is indicated by: (1) the radiating pattern of the paystreaks; (2) decrease in gold content basinward, down cross-bedding dip; (3) decrease in gold and uranium content as grain size decreases basinward (4) association of gold and uranium with basin margin unconformities; and (5) entrapment of substantial amounts of gold and uranium in algal mats, formed in quiet water environments in alluvial fan-edge and fan-base areas. With

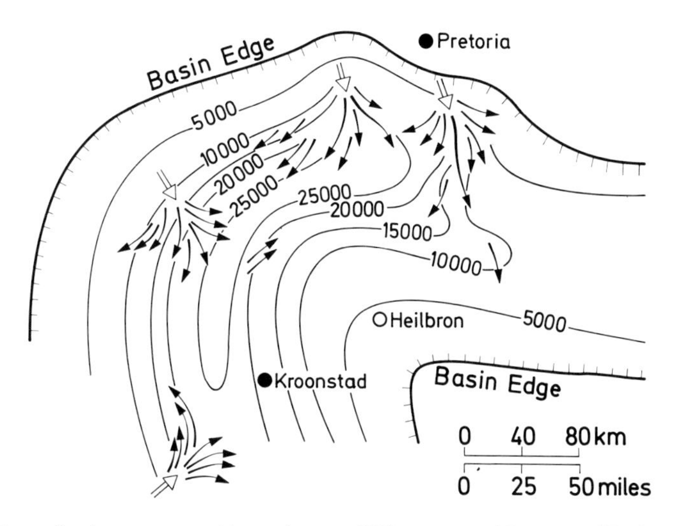

Fig. 12-13. Generalized, reconstructed isopach map of Witwatersrand System and its four major deltas or fans outlined by trend of paystreaks of gold and uranium (Modified from Brock and Pretorius, 1964b, Pls. 1 to 6)

respect to the basin proper, the marginal goldfields of the upper part of the Witwatersrand sequence are thought to be transgressive overlaps. Filling was asymmetrical with most sediment coming from the erosion of the Basement Complex to the north and west. Sediment was derived from the erosion of the domes that were uplifted beyond the major faults bounding the basin. Many of the productive sand and conglomerate beds show an unusual detrital assemblage that includes much rounded pyrite and uraninite. The presence of these two oxidizable minerals as detritus has been thought by many to signify a lack of oxygen in the earth's atmosphere at that time.

The Precambrian Witwatersrand basin is unusual in that its thick fill is largely undeformed and its margins, particularly the upper portion of the stratigraphic column, are little eroded so that basin edges with their rich placers of gold and uranium are largely preserved. The fill of the Witwatersrand and the thick overlying, gently deformed lavas, clastics and dolomites pose challenging problems, for it is not clear into what tectonic or sedimentary model the Witwatersrand fits. The Witwatersrand is unusual but is probably not unique, for the Canadian Blind River and the Brazilian Jacobina gold deposits, both Precambrian, seem to have some of the same characteristics. Is the Witwatersrand unusual among the world's basins possibly because it represents a stage of Precambrian sedimentation when oxygen was absent from the atmosphere and when the crust may have behaved somewhat differently from today?

References:

Brock, B. B., Pretorius, D. A.: An introduction to the stratigraphy and structure of the Witwatersrand goldfield. In: Haughton, S. H., Ed.: The Geology of some ore deposits of southern Africa, p. 25–61. Johannesburg: Geol. Soc. South Africa 1964a.

— — Witwatersrand Basin sedimentation and tectonics. In: Haughton, S. H., Ed.: The geology of some ore deposits of southern Africa, p. 549–599. Johannesburg: Geol. Soc. South Africa 1964b.

Haughton, S. H.: The geology of some ore deposits in southern Africa, Vols. 1 and 2, 625 and 739 p. Johannesburg: Geological Soc. South Africa 1964.

Illinois Basin and Adjacent Areas

The craton of the midwestern United States has three principal intracratonic basins filled with Paleozoic sediments (Fig. 12-6): the Illinois Basin having up to 11,000 feet (3,550 m), the Michigan Basin having a maximum thickness of 13,000 feet (3965 m), and the shallower Midcontinent Basin containing 5,000 feet (1520 m). All were sites of mild subsidence, especially in the later half of the Paleozoic; they are truly sedimentary basins, their fill thinning toward their bounding, broad, intervening arches, where thicknesses of 2,000 to 5,000 feet (610 to 1,520 m) are the rule. Southwest of the Illinois Basin lies the Ozark Dome, where some Precambrian rocks are exposed.

This part of the North American craton includes all the Paleozoic systems except the Permian. Along its southern border, in the Mississippi Embayment, there are sediments of Cretaceous and Tertiary age. Cretaceous sediments also occur in western Iowa and Minnesota.

Sandstone and limestone are the dominant lithologies covering the craton, shale being conspicuous only in the strata of Carboniferous System. There are many thin widespread interbedded carbonate units and many minor diastems as well as four major unconformities in the Paleozoic section. Generally thin but widespread cyclic coal measures of Pennsylvanian age, are present in these basins but only in the Illinois Basin have large quantities of coal been mined – over 8 billion tons. There is also a strong cyclic alternation of limestone, shale and sandstone in the underlying Chesterian (Mississippian) strata of the Illinois Basin.

Sands and sandstones have played a key role in our understanding of the history of this part of the craton. In the Midwest the Paleozoic sands of the craton's cover belong to two major suites (Table 12-5).

There is a lower pre-Mississippian suite of mature to supermature, well-rounded quartz arenites. Crossbedding and ripplemark are their dominant sedimentary structures. Most of these sands are probably of marine origin. Crossbedding and stratigraphic relations show them to have been ultimately derived from the Precambrian shield to the north, primarily as the result of unconformities that become compound northward, up paleoslope. Repeated recycling occurred. The Devonian sands are the end-product of such reworking.

Beginning in the Mississippian, sands from a new source which lay to the northeast, reached the Michigan and Illinois Basins (Fig. 12-14). These sands are less well rounded and sorted and were introduced into the area by a large river system, called the Michigan River System by Swann (1963, p. 13-15). During

Table 12-5. *Mineral composition of sandstones of Illinois Basin and adjacent areas (Potter and Pryor, 1961, Table 3)*

	Quartz	Feldspar	Rock Fragments
Tertiary	97	2	1
Cretaceous	98	1	1
Pennsylvanian			
Higher	89	6	5
Lower	97	2	1
Mississippian			
Chester	97	1	2
Osage	93	1	6
Devonian	100	0	0
Ordovician	99	1	0
Cambrian	91	8	1

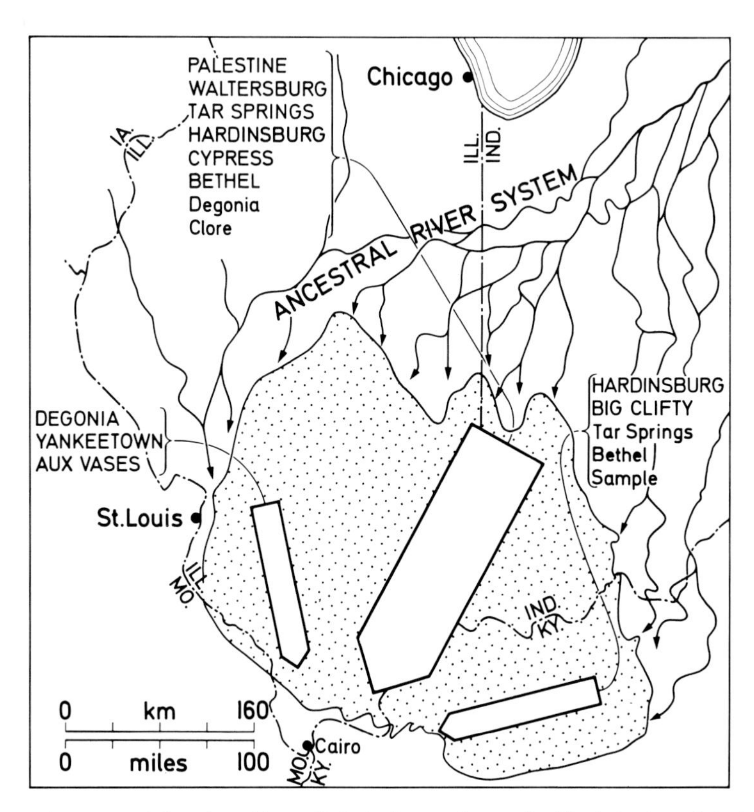

Fig. 12-14. Diagrammatic location of late Mississippian sandstones in Illinois Basin and the Michigan River System (Potter, 1962, Fig. 14)

Osage time (earliest Mississippian) immature, fine-grained, lithic arenites and even some minor turbidites were deposited in the deeper parts of the Illinois Basin, which for a short time had some aspects of a starved basin (Lineback, 1969). The overlying Chester sandstones are crossbedded and better sorted than those of the Osage (or Borden) Group and show striking deltaic patterns whereas the younger Pennsylvanian sandstones display a mixture of alluvial and deltaic patterns (Potter, 1963). Approximately 2.3×10^9 barrels of recoverable oil have

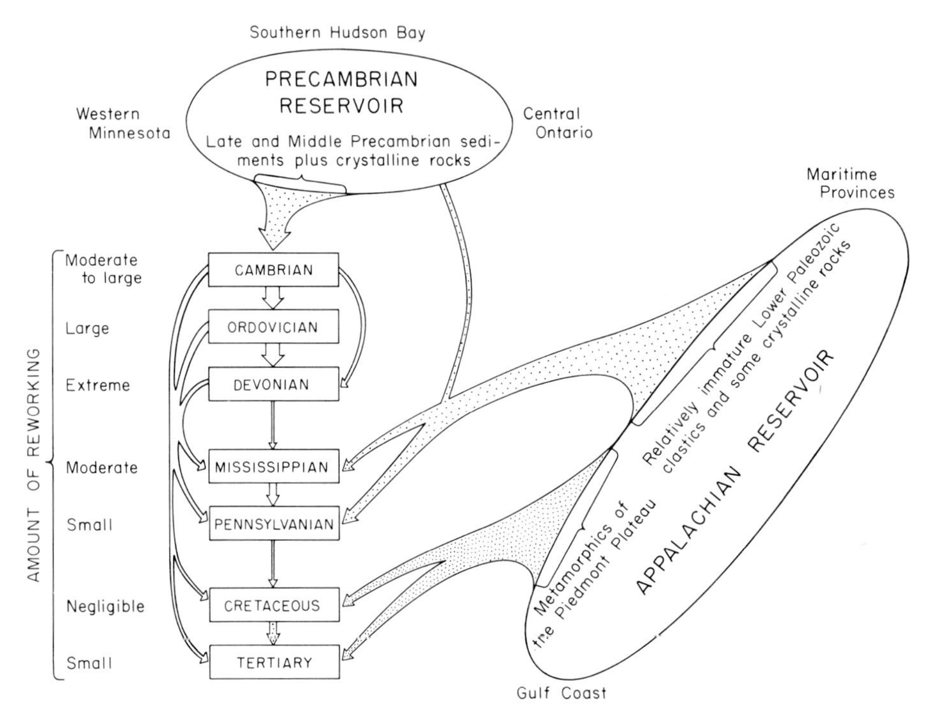

Fig. 12-15. Dispersal center and source relations of the post-Precambrian clastic sediments of upper Mississippi Valley. Stipple density is proportional to maturity of sandstones (Potter and Pryor, 1961, Fig. 14)

been found in Chesterian sandstones and approximately 0.5×10^9 barrels in Pennsylvanian sandstones. Total recoverable oil found in the basin is estimated at 3.8×10^9 barrels of which less than a third comes from carbonate rocks. A major unconformity separates Chesterian from Pennsylvanian sediments. Regional truncation of older beds to the north, northwest, and northeast together with the southwest orientation of an integrated channel system cut into the unconformity, and the crossbedding and sand dispersal pattern all indicate dominant Pennsylvanian sediment flow from the northeast and east. Cretaceous and Tertiary sands are compositionally more mature than the higher Pennsylvanian sands, but contain abundant kyanite, staurolite, and sillimanite, indicating the unroofing of the crystalline highland to the east of the Appalachian Basin (Pryor, 1960). Zircon, tourmaline and some garnet are the chief heavy minerals in all the Paleozoic sands.

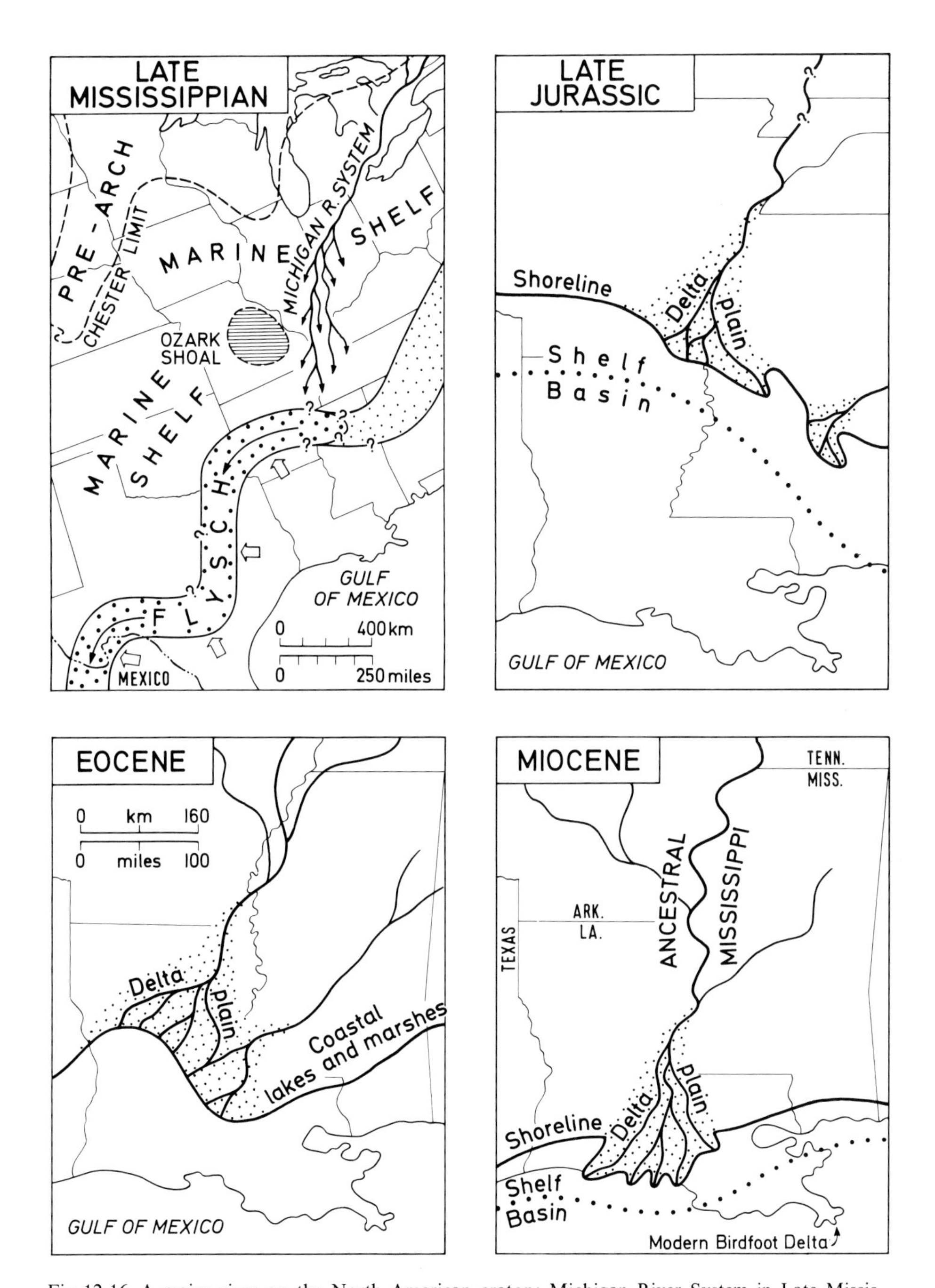

Fig. 12-16. A major river on the North American craton: Michigan River System in Late Mississippian time (Modified from Sedimentation Seminar, 1969, Fig. 15); Mesozoic and Cenozoic ancestors of the modern Mississippi (Modified from Mann and Thomas, 1968, Figs. 4, 9, and 11)

Collectively, petrology, paleocurrents, and stratigraphy suggest a Precambrian reservoir for the Devonian and earlier sands and an Appalachian Reservoir as the dominant source of Carboniferous and later sands (Fig. 12-15). The granites and gneisses of the Precambrian reservoir yield few sand-sized rock fragments, whereas the Appalachian reservoir contributed many, particularly in the Carboniferous. The Michigan River was the principal transporter of the molasse sands derived from the erosion of the Appalachians and its source regions. This river system persisted on into the Mesozoic and Cenozoic with modifications (Mann and Thomas, 1968) and was thus a forerunner of the modern Mississippi River system (Fig. 12-16). By this interpretation a major river system has, for some 250 million years, transported mud and sand across the craton, the detritus being derived from a distant, marginal mobile belt.

References:

Lineback, J. A.: Illinois Basin – sediment-starved during Mississippian. Am. Assoc. Petroleum Geologists Bull. **53**, 112–126 (1969).

Mann, C. J., Thomas, W. A.: The ancient Mississippi River. Trans. Gulf Coast Assoc. Geol. Soc. **18**, 187–204 (1968).

Potter, P. E.: Late Mississippian sandstones of Illinois Basin: Illinois Geol. Sur. Cir. **340**, 36 p. (1962).

— Late Paleozoic sandstones of the Illinois Basin. Illinois Geological Survey Rept. Inv. **217**, 92 p. (1963).

— Pryor, W. A.: Dispersal centers of Paleozoic and later clastics of the Upper Mississippi valley and adjacent areas. Geol. Soc. Am. Bull. **72**, 1195–1250 (1961).

Pryor, W. A.: Cretaceous sedimentation in Upper Mississippi Embayment. Am. Assoc. Petroleum Geologists Bull. **44**, 1473–1504 (1960).

Swann, D. H.: Classification of Genevievian and Chesterian (Late Mississippian) rocks of Illinois. Illinois Geol. Surv. Rept. Inv. **216**, 91 p. (1963).

Sandstone in the History of the Earth

In almost all of this book we have implied a uniformitarian approach to the origin of sandstones in the sense that we believe the processes that operate to produce a given sandstone type are independent of when they were deposited, Precambrian, Paleozoic, Mesozoic, or Cenozoic. We do not believe that the general interactions between provenance, tectonics, climate, environment, and diagenesis were any different in the Precambrian than they are today. Yet we also know that the earth has been going through an evolutionary process since its formation about 4.7 billion years ago. Did the change from the primitive reducing atmosphere of the early planet to the oxygen-rich atmosphere of the Phanerozoic have any effect on the nature of the sandstones produced? As the size and geography of continents and continental crust evolved through geologic time, did the average composition of sandstones change? And did the course of evolution of life and subsequent proliferation of the animal and plant kingdoms affect the character of sandstones?

Is there any evidence from mineral or chemical composition that sandstones of different ages are different? Or do the older sandstones exhibit particular textures or sedimentary structures that distinguish them from the younger sandstones? Here we briefly review the data available and then suggest ways in which sandstones may reflect some of the secular changes in the earth and thus be used to map earth history in its largest sense.

Sandstone Composition in Relation to Age

One way to check an age effect is to compare averages of mineralogical and/or chemical analyses of sandstones of different ages. Averaging the diverse compositions of sandstones of different types all formed at one time might show changes in the ratios of the major elements. Over how large or small a time interval can we or must we calculate averages? The smallest time interval will be determined by the reliability of time-stratigraphic correlations. The practical lower limit of reliability within a particular sedimentary basin is about one million years. In most cases, however, especially if we are comparing sandstones in different basins or even on different continents, the resolution of the time scale is much poorer. This is partly due to the fact that sandstones rarely contain, or are interbedded with rocks that contain, the kinds of fossils that allow precise dating. But it is also due to the limitations of the method of dating itself. Because the precision of paleontologic dating varies and is in part dependent on geologic age, the resolution varies from a few million years in the younger strata to more than a hundred million years in the older sequences (Fig. 12-17). Because of the absence of fossils, the resolution in the Precambrian is extremely poor. This is most unfortunate because we suspect that many of the most important changes in the environments of sand deposition may have taken place during the earlier stages of the earth's history.

In order to compare properly sandstones of one geologic age with those of another, be it their composition or structures, it is necessary to compare sandstones from like facies. Like must be compared with like. It is manifestly improper to compare sandstones of an Archean flysch with Paleozoic cratonic sandstones. In order to discern secular trends, if there be any, we must select an easily recognizable facies and compare the sands of this facies throughout the history of the earth. Such a facies is, for example, the molasse facies. The time interval represented by such a facies may be the order of 20-45 million years. This facies, as we have pointed out in the first part of this chapter, is a part of larger basinal or geosynclinal cycle, such as those of the Appalachians or the three cycles, Caledonian, Hercynian, and Alpine, of the Russian platform, that have been analyzed quantitatively by Ronov and others (1969). Molasse facies of ages from Precambrian to Tertiary are recognizable in the geologic record. It is likely that significant secular changes in sandstone composition, if they exist, could be brought out by appropriate sampling and averaging of these facies alone. Even narrower intervals within the geosynclinal cycle might be used – if such intervals occupy precisely the same relative position in the different cycles. Care is needed to avoid the many trends in sandstone composition within the geosynclinal cycle. It is likely, however, that significant secular changes in composition can better be demonstrated by averaging over such cycles than by taking only one of their constituent facies.

An additional complexity is the differential preservation of sandstone of different positions within the basinal cycle. For example, it is only in the youngest cycles that a significant volume of the uppermost intermontane sandstones are preserved. The matter is further complicated inasmuch as in older cycles there is a tendency for the graywacke-flysch part of the sequence to be more involved

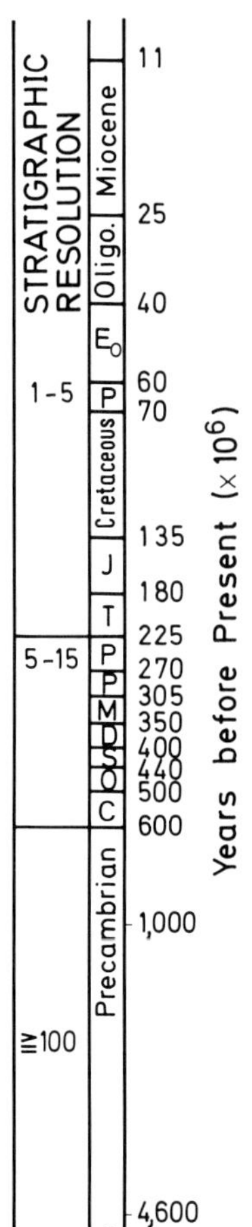

Fig. 12-17. Stratigraphic resolution of the geologic time scale for reliable geochemical or petrographic averages. The minimum number represents the best possible equivalent to stratigraphic stages; the maximum number is the poorest resolution likely with good stratigraphic work (Modified from Garrels, Mackenzie, and Siever, 1971, Fig. 1)

in high-grade metamorphism and even granitization. Sandstones belonging to this facies are thus lost for averaging purposes.

One can get some idea of the immense amount of data that must be digested to reveal such trends and how the analysis proceeds from the work on the Russian Platform of Ronov (1968) and Ronov and others (1963, 1965, 1969). So far no other comparable compilations are available, either in terms of chemical or petrographic analyses, for large areas of other continents. Until that kind of information becomes available, the best that we can do is to use the incomplete data we have to see if any obvious trends can be discerned. One way to look at chemical data is in terms of the element ratios, particularly the SiO_2/Al_2O_3 and Na_2O/K_2O ratios, that are discussed in Chapter 2.

If sandstones are distinguished only on the basis of eras (Precambrian, Paleozoic, Mesozoic, Cenozoic and Recent) without respect to petrography, there appears to be no detectable relation between age and the SiO_2/Al_2O_3 or Na_2O/K_2O ratios (Fig. 12-18). The Na_2O/K_2O ratio is affected by several

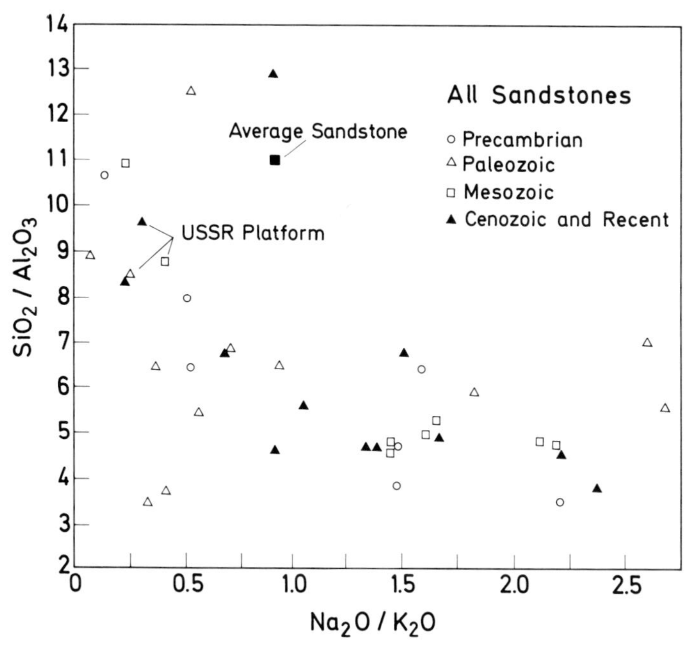

Fig. 12-18. SiO_2/Al_2O_3 versus Na_2O/K_2O ratios in sandstones. Data from various sources, mainly from Pettijohn, 1963. Average sandstone not based on these data

different mineral parameters: (1) the ratio of detrital plagioclase to detrital K-feldspar; (2) the ratio of feldspar to quartz; and (3) the ratio of illite to montmorillonite and other clay minerals. Because (1) and (2) are distinctively source factors, where (3) is a combination of environment and diagenesis – mostly the latter – sandstones have to be considered by petrographic type. If we consider only arkoses, for example, there does seem to be a dividing line at $Na_2O/K_2O = 0.6$, between Mesozoic and Tertiary feldspathic sands which are more sodic and the Paleozoic and Precambrian which are less sodic. But the differences are not great. And furthermore, the differences may not be an original attribute. Arkoses tend to be highly permeable formations and there may be some gradual dissolution of the more calcic plagioclase relative to K-feldspar that over a long time would decrease the Na_2O/K_2O ratio.

If one considers only graywackes, the sandstones that typically have the highest Na_2O/K_2O ratios, there appears to be no age effect. Some of the highest Na_2O/K_2O ratios, are those of two classic Paleozoic graywackes, the Kulm and the Tanner, and the Precambrian Knife Lake (Table 6-6). On the other hand, some of the sandy, volcanic-rich sediments of the Columbia River, shown by Whetten (1966) to be similar to graywackes in every respect but amount of matrix and that become texturally a graywacke when artificially metamorphosed (Hawkins and Whetten, 1969), have lower Na_2O/K_2O ratios than most other graywackes. The change in labile mineral composition that results from long term diagenesis – and low grade metamorphism – of graywackes is from montmorillonite, mafic glass, and plagioclase to chlorite and more albitic plagioclase. It appears that much of this change is isochemical. The differences in some graywackes may be more due to original source contributions than to diagenesis. It may be that the proportion of originally detrital volcanic debris is more important than any other factor in determining graywacke alkali metal ratios.

There is one other barrier to a proper evaluation of the analyses – the size effect. Fine grained graywackes, approach shales in composition. The coarser grained ones have the higher Na, which is undoubtedly located in the framework fraction feldspars (albites) rather than the matrix. Obviously meaningful comparisons can be made only if the rocks compared have similar grain size.

It is hard to see any significant age effect in SiO_2/Al_2O_3 ratios in sands, though no Paleozoic and Precambrian samples have ratios lower than 5.

Other ratios show some age differences. The CaO/MgO ratio appears to increase with decreasing age. The data of Vinogradov and Ronov (1956) show ratios of 0.78 in pre-Devonian sandstones, 1.39 in Devonian, 3.56 in Carboniferous, and 3.90 in Mesozoic sandstone. This change is similar to that found in carbonate rocks by Daly as long ago as 1909, and discussed in relation to other elemental ratios by Garrels and others (1971). The change in CaO/MgO is related to progressive dolomitization of carbonate cement with time. Other chemical components that show secular trends in clays and shales, such as the Fe_2O_3/FeO ratio increase with decreasing age (Nanz, 1953), do not appear to show such clear trends in sandstones.

Analyses of petrographic data on sandstones show few obvious tendencies. There is some evidence that the feldspar content of sandstones increases with decreasing age, the Paleozoic sandstones on both the North American and Russian platforms containing less feldspar than those of Mesozoic or Cenozoic age (Tables 2-2 and 2-3). Likewise, as discussed in Chapter 10, the number of heavy mineral species increases with decreasing age. The clay mineralogy of sandstones seems to show the same general tendencies as in shales, Mesozoic and Cenozoic samples being richer in smectites (montmorillonite and other expandable clays) and kaolinite and the older rocks richer in chlorite and illite (Weaver, 1967).

These apparent secular changes, if not the result of biased sampling, are probably more the consequence of long term diagenetic changes than they are of original differences in sand composition. As noted in Chapter 10, the secular variations in carbonate and silica cement, the diversity of heavy mineral suites, the changes from illite-chlorite to smectite-kaolinite clays, and the proportion of graywacke matrix can all be directly related to diagenesis.

The change in feldspar content is less easily related to gradual alteration during diagenesis. It cannot be the result of recycling of sandstones, for feldspar would decrease with increased recycling as age decreased – that is the heart of the maturity concept. Because feldspar is related to tectonism and climate (see Chapters 2 and 8) then this trend, if it were statistically substantiated, might imply a tendency for increased tectonism and more rigorous climate to the present time. Here is where we need a much sounder statistical analysis of petrographic data on which to base hypotheses. Because feldspars have differing chemical stabilities, it would be important to determine the proportion of different feldspars as a function of age to check any diagenetic effect which may be the chief cause of feldspar variation. The variations observed could, of course, be due to sample bias – the older sandstones being mainly cratogenic and the younger ones being derived from orogenic belts.

There is some evidence that the very oldest sandstones, those over 2.5 billion years old, may be different. Engel (1963) and Ronov (1964) believe that there is a higher proportion of graywackes and arkoses relative to quartz arenites or quartzose lithic arenites in such old sections. Another perplexing kind of sandstone is the gold, uraninite-, and detrital pyrite-bearing sandstones of the Witwatersrand in South Africa, the Blind River district in Canada, and the Jacobina district in Brazil. All seem to be in the time interval 1.7-2.0 billion years old and no others like them are known of any other age. Another seeming "non-uniformitarian" kind of sandstone is the extremely thick quartz arenite that seems to be widespread in the later Precambrian. Quartzites such as the Lorrain of Ontario, the Baraboo of Wisconsin, the Athabaska of Saskatchewan, and the Uinta of Utah are all very pure and well over 1000 m thick whereas Phanerozoic quartz arenites tend to be very thin, rarely over a few tens of meters thick. Are all these seemingly unique sandstones the result of special events or conjunctions of evolutionary patterns, never again to be repeated?

To summarize, most apparent secular changes that we now deduce from only a crude statistical evaluation may be traced to long term diagenetic changes but some, particularly special characters of very early sandstones may reflect evolutionary changes in the overall history of the crust.

Sandstones in Relation to Evolutionary Changes in Earth History

Have certain events in earth history been reflected in sandstone compositions? One such event already mentioned is the evolution of oxygen in the earth's atmosphere, which took place sometime in the Precambrian (Holland, 1962; Cloud, 1968). Sandstones formed in a reducing environment would be expected to contain much more detrital pyrite, magnetite, and other oxides of reduced metals. They might also be expected to have a greater abundance of detrital ferrous silicates, for part of the present rapid weathering of iron-rich olivines and pyroxenes is related to oxidation of the iron. The difficulty is that very low levels of oxygen, pressures of about 10^{-4} or 10^{-5} atmospheres, are probably sufficient to oxidize ferrous iron (Holland, 1962) and this low level might have been reached as early as 3 billion years ago. The minerals that might be found in sandstones which might be diagnostic of oxygen pressures close to or a little less than that of the present atmosphere are the opaque heavy minerals of oxidizable heavy metals. The information that we need is the precise mineralogical identification of these heavy opaques; for that one needs polished ore sections of heavy mineral concentrates – an unusual technique for the sedimentary petrologist.

Another development on the surface of the earth is the evolution of vascular land plants in the late Silurian and early Devonian. It is clear that there is no profound difference between pre-Devonian and later sandstones but may there not be some differences in erosion and denudation rates that would be reflected in sedimentation rates of sandstones? Or was there an extensive cover of primitive non-vascular plants? Or is the rate of weathering and erosion not as greatly influenced by vegetation as has been thought conventionally? Cawley and others (1969) studied chemical weathering of vegetated and non-vegetated areas in Iceland and concluded that the vegetation effect was not overwhelming. In trying to estimate rates of denudation in the distant past we may have to rely mainly on an analysis of sedimentary volumes produced by denudation of land surfaces. A good statistical comparison of this kind of data, pre- and post Devonian, difficult as it may be, will be needed to demonstrate that the evolution of the higher land plants made a significant difference. The alternative is to find some chemical tracer in sediments, as yet unknown, that would give clues to changes in the kind and/or amount of denudation.

It has been suggested that the organization of strata involving sandstones particularly the fining-upward alluvial cycles, might not be developed and that the alluvial deposits would be noncyclic sheets of coarse immature materials in the absence of a land plant cover (Schumm, 1968, p. 1583).

Has the development of a rich benthonic fauna in later times affected the character of sands and sandstones? Bioturbation, the disturbance of deposited sediment by burrowing organisms, leads to destruction of primary sedimentary features – especially laminations. Are the older sandstones generally better laminated?

How has the evolution of sialic continental crust affected sandstone compositions? Do we have any sandstones dating from the time when there was no such crust? If so, we should recognize them from their chemical and petrographic composition as having been derived solely from mafic terrains. They might have been largely volcaniclastics. And, when the sialic crust was thin, might we have still have had much more of a contribution from mafic rocks and less quartz-rich detritus? Sandstones of that stage could be recognized by an average composition intermediate between those expected from the earliest stage and all later sandstones derived mainly from sialic crust. Or the earliest sialic material might have been volcanic rather than plutonic and so be found in volcaniclastics.

Given this kind of crustal evolution, were tectonic styles different in early stages when continents were small and/or thin? Might tectonic activity been much more generally distributed over the surface and thus led to a more general distribution of sandstones reflecting that tectonism, rather than the later, more familiar localization of tectonic activity along mobile belts that produces geosynclinal sandstones there and platform sandstones elsewhere? We may phrase these questions in terms of tectonic plate theory too. At an early evolutionary stage were plates very small, so that the lines along which they converged, the linear regions of subduction zones, mountain building, and volcanism, were more numerous and formed a much larger proportion of the earth than today? The answer to most of these questions lie in the petrologic study of the oldest sandstones to determine their provenance, including careful averaging of the ratios of quartz and K-feldspar to mafic minerals and calcic plagioclase, a parameter that might be used to measure mafic vs. felsic provenance.

Opinion differs on these matters, some workers interpreting the Archean sands as acid volcaniclastic materials (Ayers, 1969) whereas others consider them to be derived from sialic crust by normal weathering and erosion (Donaldson and Jackson, 1965; Walker and Pettijohn, 1971).

How have continental drift and global tectonic movements affected the distribution of sandstones and how may we use sandstones to infer that aspect of earth history? One way is suggested by Allen and Friend (1968, Fig. 11), who noted the similarity of the Catskill sandstones of North America and the Old Red Sandstones of western Europe and assumed continental drift to get a "rational" Devonian paleogeography. And Percival Allen (1969) accounted for the provenance of the Lower Cretaceous shelf deposits of western Europe by inferring a pre-drift proximity of North America and Europe and looking for sediment sources in the Piedmont metamorphics of the eastern United States. Those who have compared the Carboniferous of eastern North America and western Europe have long been aware of the many similarities and hence the pre-drift paleogeography of that time interval might show a consistent provenance and paleocurrent pattern. Thus sandstones probably have much to tell us about paleogeographic patterns that are consistent with continental drift, sea floor spreading, and global tectonics.

Can sandstones tell us more about the tectonic and sedimentation patterns developed by converging or laterally sliding crustal plates? Dickinson (1970) has suggested that many of our preconceptions of the nature of geosynclinal tracts need revison in the light of global plate tectonics theory; a good example of this kind of reinterpretation is given by Bird and Dewey (1970), who give a model for the evolution of the Appalachians belt. Dickinson (p. 1259), singles out the notion that "a single sequential orogenic progression ... assumed to be the norm ... cannot be expected to remain part of the plate tectonic model of orogeny." It is our feeling that basinal sequences such as those described earlier in this chapter stand by themselves as descriptions of the record of "what's there" and that there is indeed a cyclic nature to many of the sequences. Plate tectonics is unreasonable or inconsistent with the facts of the rock record only *if* theorists insist that there can be only a random, non-cyclical progression of events. Much of our knowledge of ancient tectonic patterns comes from mapping of sandstones, and it is clearly necessary to rethink those paleogeographic reconstructions that are based on tectonic theory. But a good paleocurrent and provenance map is much more dependent on direct observations of sandstones and theories of weathering and transport of detritus than on tectonic theories. Thus such maps may not be changed greatly but rather can be used to work out more concretely a revised historical geology of tectonic belts in terms of the new theory.

Other secular changes in earth history are known, but have not been shown to have a great effect on the character and distribution of sandstones. One such effect is the gradual slowing down of the earth's rotation rate in response to the tidal friction induced by the moon, the quantitative measurement of the variation in length of the day and month being based on growth patterns in corals, molluscs, and stromatolites (Pannella and others, 1968). The more rapid rotation rate and higher tides from a closer moon might have produced a greater effectiveness of tidal currents. Would the marine crossbedding related to such currents be on a significantly larger scale than that of later times? An analysis of paleotidal sedimentation by Klein (1970) led him to conclude that the tidal range and frequency of tidal deposits has not changed much through geological time, though much remains to be done in the quantitative analysis of such sediments.

Mention of the moon raises the question of whether there might be sandstones on the moon to interpret its history. The results of the Apollo 11 and Apollo 12 missions to the moon do not indicate any deposits that would be sensibly called sand, though there are abundant fragmental materials of all sizes (Levinson, 1970). Our conventional definition of sand requires that materials of sand size (1/16-2 mm) constitute a major proportion of the deposit. Neither the "soils" nor breccias found on the moon meet these requirements. There is the possibility of discovering ash falls and even flows, and we could consider them in the same way as volcaniclastics, differing only in the value of gravity and the dependence of any current on the entrapment of enough gas to provide a fluid transport medium. But the results of lunar exploration thus far lead us to believe that those interested in sandstones will probably be happier studying the earth.

Conclusions and a Look at the Future

In conclusion what can we say about the usefulness of sandstones for the study of the larger aspects of the earth? To the degree that the minerals of sandstones reflect the chemistry of the atmosphere and oceans, we may get some better information on their evolution. More important, though, is the use of sandstones as indicators of provenance and paleocurrents, for these are the keys to paleogeography and ultimately to paleotectonics. But in study of larger elements of paleotectonics we must map large regions, even continents. We hope that sandstone petrologists in the future will put together paleogeographic and provenance maps of many different basins over several continents or the whole world. Just as the 50's saw sandstone petrologists study whole basins, their descendants in the 70's and 80's will need to put together continents, at the same time that they refine ways of analyzing the details of individual basin development.

At the same time much more work on statistical averaging of petrographic data, chemical analyses, and relative abundances of different sandstone types will have to be done for us to get a clearer picture of long term trends, within and between basinal sequences, and over the whole history of the earth.

Finally, the interpretation of the compositional data will have to be based on better quantitative data on chemical stabilities and reaction kinetics. The analysis of minerals that might reflect the oxidation potential of primitive earth environments will help us only if we can answer the following question. How fast does pyrite oxidize as a function of oxygen pressure? What is a reliable value for the oxygen pressure needed to convert magnetite to hematite? What chemical species influence the kinetics of the magnetite-hematite conversion?

As in drawing conclusions on other aspects of the geology of sands, we again state our conviction that the answers to most of the questions we have posed in this book will come not from unimaginable new techniques but from the intelligent application of methods and tools that we now have at our disposal. We rely on the coupling of quantitative field and laboratory analysis – primarily but not exclusively microscopy – with the ideas and data of chemistry on the one hand and fluid dynamics on the other to give us the solutions. But because we are concerned with the geology of sand – and so with the actual distribution of sands on earth, we use one of the most quantitative techniques ever developed by geologists – *the map. The decision to map implies an intelligent choice of parameters to map, and the ability to measure them quantitatively. We believe that there is an order to the earth and that through sagacious mapping we can discern much of it.*

References

Allen, J. R. L., Friend, P. F.: Deposition of the Catskill facies, Appalachian region: with notes on some other Old Red Sandstone basins. In: Klein, G. de V., Ed.: Late Paleozoic and Mesozoic continental sedimentation, northeastern North America. Geol. Soc. America Spec. Paper **106**, 21–74 (1968).

Allen, Percival: Lower Cretaceous sourcelands and the North Atlantic. Nature, **222**, 657–658 (1969).

van Andel, Tj. H.: Origin and classification of Cretaceous, Paleocene, and Eocene sandstones of western Venezuela. Am. Assoc. Petroleum Geologists Bull. **42**, 734–763 (1958).

Aubouin, Jean: Geosynclines. Developments in Geotectonics 1, 335 p. Amsterdam: Elsevier 1965.

Ayers, L. D.: Early Precambrian metasandstone from Lake Superior Park, Ontario, and its implications for the origin of the Superior Province (abs.). Geol. Soc. America, Abstracts with Programs for **1969**, part 7, 5.

Bertrand, Marcel: Structure des alpes francais et recurrence de certain facies sedimentaires. Compt. rend. Congr. Intern. geol., 6th sess., **1897**, 1–3–177.

Bird, J. M., Dewey, J. F.: Lithosphere plate-continental margin tectonics and the evolution of the Appalachian orogen. Geol. Soc. America Bull. **81**, 1031–1060 (1970).

Cawley, J. L., Burrus, R. C., Holland, H. D.: Chemical weathering in central Iceland: an analog of pre-Silurian weathering. Science, **160**, 729–736 (1969).

Cloud, P. E., Jr.: Atmospheric and hydrospheric evolution on the primitive earth. Science **160**, 729–736 (1968).

Colton, G. W.: The Appalachian Basin – Its depositional sequences and their geologic relationships. In: Fisher, G. W., Pettijohn, F. J., Reed, J. C., Jr., Weaver, K. N., Eds.: Studies of Appalachian geology – central and southern, p. 5–48. New York: Interscience 1970.

Daly, R. A.: First calcareous fossils and the evolution of the limestones. Geol. Soc. America Bull. **20**, 153–164 (1909).

Dickinson, W. R.: Global tectonics. Science **168**, 1250–1259 (1970).

Donaldson, J. A., Jackson, G. D.: Archean sedimentary rocks of North Spirit Lake area, northwestern Ontario. Canadian Jour. Earth Sciences **2**, 622–647 (1965).

Engel, A. E. J.: Geologic evolution of North America. Science, **140**, 143–152 (1963).

Garrels, R. M., Mackenzie, F. T., Siever, Raymond: Sedimentary cycling in relation to the history of the continents and oceans. In: Robertson, E. C., Ed.: The nature of the solid earth, 93–121. New York: McGraw-Hill 1971.

Gilluly, James, Reed, J. C., Jr., Cady, Wallace: Sedimentary volumes and their significance. Geol. Soc. America Bull. **81**, 353–376 (1970).

Glaessner, M. F., Teichert, Curt: Geosynclines, a fundamental concept in geology. Am. Jour. Sci. **245**, 587–589 (1947).

Hawkins, J. W., Jr., Whetten, J. T.: Graywacke matrix minerals: Hydrothermal reactions with Columbia River sediments. Science **1 66**, 868–870 (1969).

Holland, H. D.: Model for the evolution of the earth's atmosphere, In: Engel, A. E. J., James, H. L., Leonard, B. F., Eds.: Petrologic studies, a volume to honor A. F. Buddington, p. 447–477. New York: Geol. Soc. America 1962.

Jones, O. T.: On the evolution of a geosyncline. Geol. Soc. London Proc. **94**, lx-cx (1938).

Kay, Marshall: North American geosynclines. Geol. Soc. America, Mem. **48**, 143 p. (1951).

King, P. B.: The evolution of North America, 189 p. Princeton, New Jersey: Princeton Univ. Press 1959.

Klein, G. de Vries: Paleotidal sedimentation (abs). Geol. Soc. America, Abstracts with Programs **2**, no. 7, 598 (1970).

Krumbein, W. C., Sloss, L. L., Dapples, E. C.: Sedimentary tectonics and sedimentary environments. Am. Assoc. Petroleum Geologists Bull. **33**, 1859–1891 (1949).

— — Stratigraphy and Sedimentation, 2nd Ed., 660 p. San Francisco: Freeman 1963.

Krynine, P. D.: Differentation of sediments during the life history of a landmass (abst.). Geol. Soc. America Bull. **52**, 1915 (1941).

— Sediments and the search for oil. Producers Monthly, **9**, 12–22 (1945).

— A critique of geotectonic elements. Am. Geophysical Union Trans. **32**, 743–748 (1951).

Kuenen, Ph. H.: Geochemical calculations concerning the total mass of sediments in the earth. Am. Jour. Sci. **239**, 161–190 (1941).

Levinson, A. A., Ed.: Apollo 11 Lunar Science Conference Proc., Vol. 1, Mineralogy and petrology; Vol. 2, chemical and isotope analyses; Vol. 3, Physical properties, 2492 p. Geochim. et Cosmochim Acta, Suppl. 1 (1970).

Mann, John C., Thomas, W. A.: The ancient Mississippi River. Trans. Gulf Coast Assoc. Geol. Soc. **18**, 187–204 (1968).

Nanz, R. H., Jr.: Chemical composition of Precambrian slates with notes on the geochemical evolution of lutites. Jour. Geology **61**, 51–64 (1953).

Pannella, G., MacClintock, C., Thompson, M. N.: Paleontologic evidence of variations in length of synodic month since Late Cambrian. Science **162**, 792–796 (1968).

Pettijohn, F. J.: Sedimentary Rocks, 2nd Ed., 718 p. New York: Harper 1957.

Ronov, A. B.: Common tendencies in the chemical evolution of the earth's crust, ocean, and atmosphere. Geokhimiya **8**, 715–743 (1964).

— Probable changes in the composition of sea water during the course of geological time. Sedimentology **10**, 25–43 (1968).

— Girin, Y. P., Kazakov, G. A., Ilyukhin, M. N.: Comparative geochemistry of the platform and geosynclinal sedimentary rocks. Geokymiya **8**, 961–979 (1965).

— Migdisov, A. A., Barskaya, N. B.: Tectonic cycles and regularities in the development of sedimentary rocks and paleogeographic environments of sedimentation of the Russion platform (an approach to a quantitative study). Sedimentology **13**, 179–212 (1969).

— Mikhailovskaya, M. S., Solodkova, I. I.: Evolution of the chemical and mineralogical composition of sandy deposits. In: Vinogradov, A. P., Ed.: Chemistry of the earth's crust, 1, p. 201–252. Moscow: Akad. Nauk S.S.S.R. 1963.

Schumm, S. A.: Speculations concerning paleohydrologic controls of terrestrial sedimentation. Geol. Soc. America Bull. **79**, 1573–1588 (1968).

Schwab, F. L.: Geosynclines: What contribution to the crust? Jour. Sed. Petrology **39**, 150–158 (1969).

Sedimentation Seminar: Bethel sandstone (Mississippian) of western Kentucky and south-central Indiana, a submarine-channel fill: Kentucky Geol. Sur. Rept. Invs. **11**, 24 p. (1969).
Sloss, L. L.: Sedimentary volumes on the North American Craton (abst.). Geol. Soc. America Program 1968 Ann. Meetings, p. 281 (1968).
Stille, Hans: Present tectonic state of the earth. Am. Assoc. Petrol. Geol. Bull. **20**, 849–880 (1936).
Thompson, Alan, Thomasson, M. R.: Shallow to deep water facies in the Dimple Limestone (Lower Pennsylvanian) Marthon Region, Texas. In: Freidman, G. M. Ed.: Depositional environments in carbonate rocks. Soc. Econ. Paleon. and Mineralogists Spec. Publ. No. 14, p. 57–77 (1969).
Vinogradov, A. P., Ronov, A. B.: Composition of the sedimentary rocks of the Russian platform in relation to the history of its tectonic movements. Geokhimiya **6**, 3–24 (1956).
Walker, R. G., Pettijohn, F. J.: Archean sedimentation: analysis of the Minnitaki basin, northwestern Ontario. Geol. Soc. America Bull. **82**, 2099–2130 (1971).
Weaver, C. E.: Potassium, illite, and the ocean. Geochim. et Cosmochim. Acta **31**, 2182–2196 (1967).
Whetten, J. T.: Sediments from the lower Columbia River and origin of graywacke. Science **152**, 1057–1058 (1966).

Appendix: Petrographic Analysis of Sandstones

Introduction

The petrographic analysis of sediments, sandstones in particular, became an organized discipline with the development of thin-section techniques and the polarizing microscope – a development attributed to Henry Clifton Sorby. Sorby was making thin sections in 1849, published a paper on the microscopical structure of "calcareous grit" in 1851, and his paper, in 1880, on the noncalcareous stratified rocks, was a major milestone in the thin-section analysis of sandstones. Folk (1965) has ably summarized Sorby's petrographic contributions. Until shortly before World War II, most students of sedimentary rocks, unlike those of igneous and metamorphic rocks, failed to "follow through" on Sorby's auspicious beginnings. One brilliant exception was Lucien Cayeux. In more recent years, however, the thin-section analysis of sedimentary rocks has become commonplace and is now a fully exploited tool for research.

The prime object of the study of a thin section is, or should be, the reading of *rock history*. The microscope is the most useful tool for close study of the mineral composition, fabric, and general make up of a rock. Such close study is a necessary complement to field studies in interpreting the origin of sands and sandstones.

The study of sand and sandstone in the laboratory has proceeded in other directions also and there has been a proliferation of methods for study of grain size, grain shape and roundness, porosity and permeability, and the like. Many of these methods, however, are applicable only to unconsolidated deposits and, useful as they may be, the thin-section approach remains the single most effective means of investigation of sandstones in the laboratory. Its effectiveness, however, depends on the imagination and skill of the operator.

Rock Description and Analysis

Rock description and analysis are based on study of outcrops, hand specimens, and thin sections. Emphasis is here placed on hand-specimen and thin-section study.

The art of rock description and analysis is learned by doing and the study of published examples. Good descriptions of rocks of all types have been published in Bulletin 150 of the U. S. Geological Survey. A number of abbreviated descriptions are given in Grout (1932, p. 22–28).

The use of a well designed petrographic form develops a regular pattern of description and so maximizes the efficiency and effectiveness of any microscopic study. The level of description, however, may vary widely – from a concise paragraph based largely on qualitative observation and semiquantitative visual estimates to a 3 or 4 page typed report based on counts of 200 to 500 or more

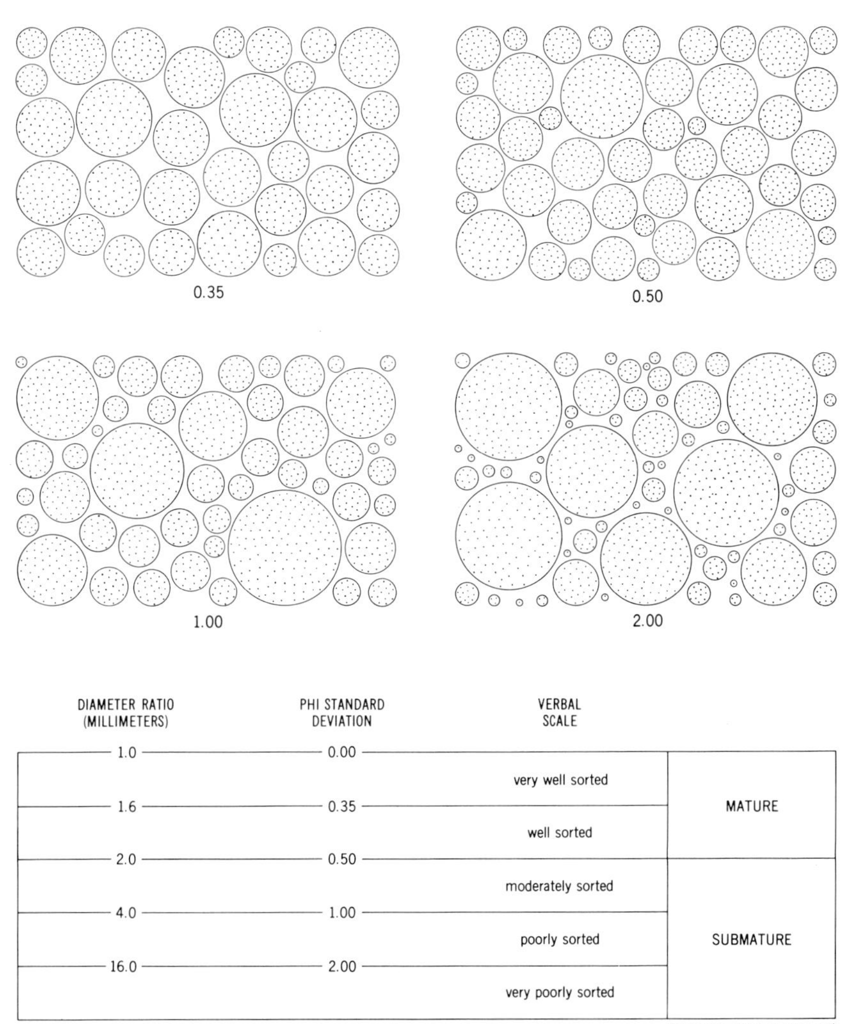

Fig. A-1. Comparison chart for sorting and sorting classes (Modified from Folk, 1968, p. 102)

grains. Or in place of a written report, the data may be directly entered on a punch card for subsequent computer processing.

Choice between semiquantitative and quantitative estimates depends on the investigator's objectives, his judgement, and the available time. If very large numbers of samples are to be studied and little time is available for the task, semiquantitative estimates must suffice. Mineral percentages can be estimated by comparison charts with a reticle (Terry and Chilingar, 1955). Sorting can also be estimated by comparison charts (Fig. A-1). Roundness of individual grains is always so estimated (Fig. A-2). Counts of 50 to 100 grains have usually been made to estimate either average grain roundness or percentage of angular grains.

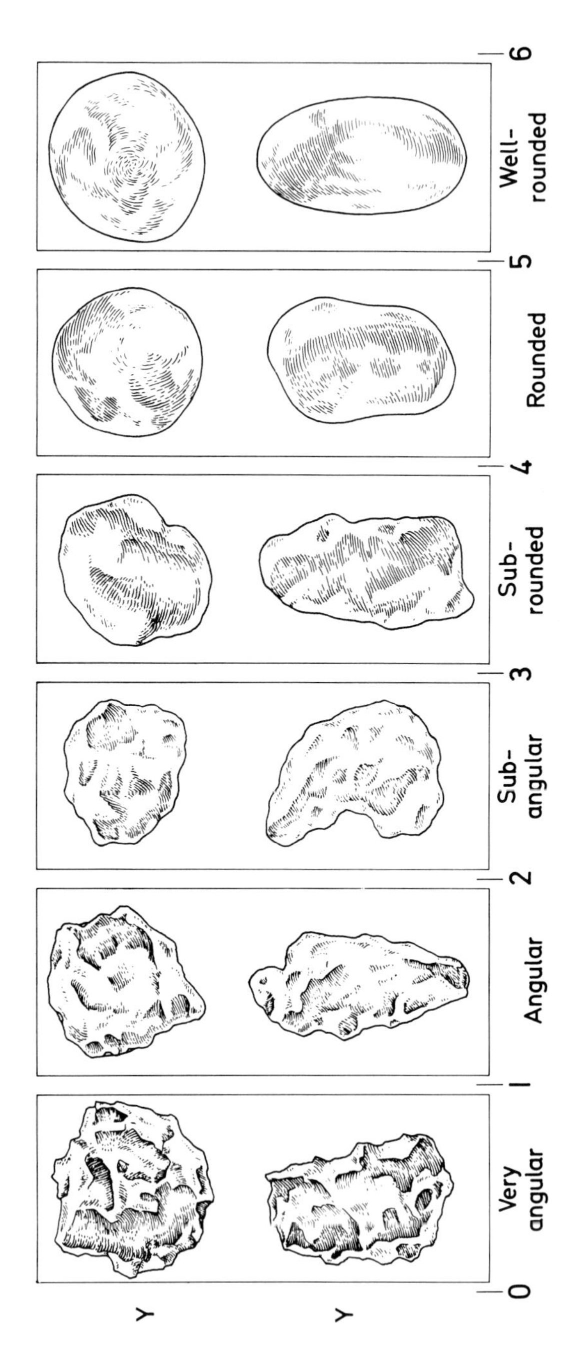

Fig. A-2. Roundness images and classes. Columns show grains of similar roundness but different sphericity (Redrawn from Powers, 1953, Fig. 1)

But the earmark of the modern petrographer is the point count of 200 to 500 grains per slide for estimation of composition. Quantitative estimates are needed for many petrographic classifications and for most subsequent statistical analysis. By using binomial and Poisson confidence charts (Pearson and Hartley, 1954, Tables 40 and 41) one can determine the reliability of an estimate based on a

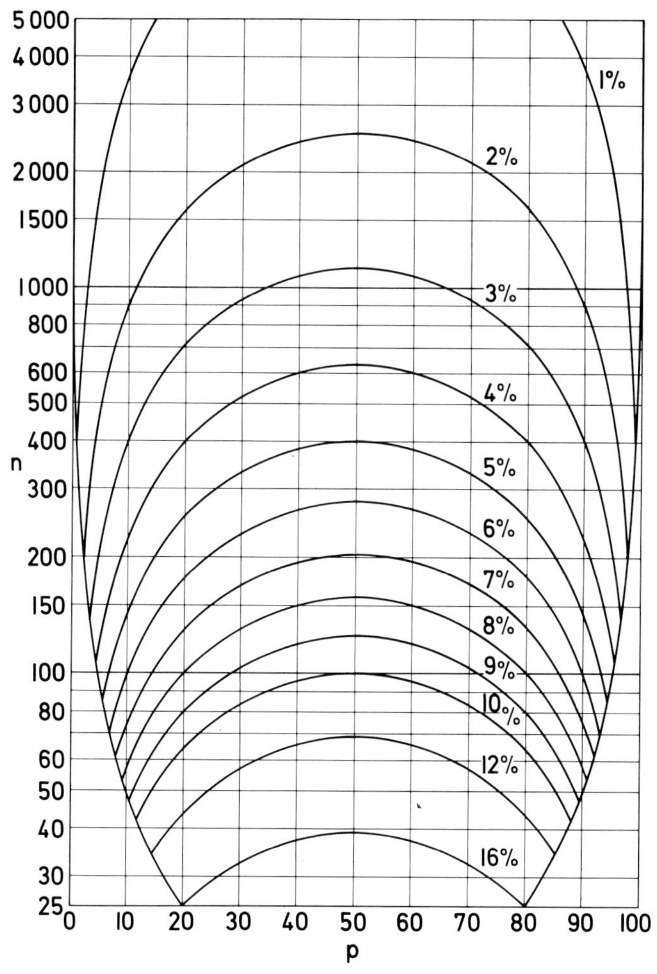

Fig. A-3. Ninety-five percent confidence limits for mineral proportions, where n is total number of grains counted and p is the *estimated* proportion of a particular mineral. Curved contours in percent give confidence limits. Worked example: n is 500, p is 28 percent and the confidence limit is 4 percent so that in repeated sampling the *true* proportion will lie within 24 and 32 percent (Modified from van der Plas and Tobi, 1965, Fig. 1)

given number of counts. Van der Plas and Tobi (1965) provide a comparable chart (Fig. A-3). Dennison (1966) also has useful charts, all of which are based on the normal and binomial distributions. In practice, it is not uncommon for a petrographer to utilize semiquantitative estimates for some of the petrographic variables and reserve systematic point counting for the one or two most significant ones. A mechanical recorder is needed for efficient point counting.

Sorting, angularity, and clay content define *textural maturity* which should be specified. Table A-1 gives a flow chart for this procedure. Sorting can be estimated by comparison with Figure A-1 or one can estimate it by determining

Table A-1. *Textural maturity flow chart (modified from Folk, 1968, p. 102)*

STEP 1	Clay content (micaceous material less than 30 μ, excluding authigenic material) a) If greater than 5 percent, sand is *immature*. b) If less than 5 percent, determine sorting.
STEP 2	Sorting (See Fig. A-1) a) If sorting is greater than 0.5 ϕ (diameter ratio over 2.0), sand is *submature*. b) If sorting is less than 0.5 ϕ, determine roundness.
STEP 3	Roundness (See Fig. A-2) a) If sand size grains are subangular to angular (3.0 or less on the Powers scale, Table 3-3), sand is *mature*. b) If roundness exceeds 3.0, sand is *supermature*.

Table A-2. *Short petrographic report for sandstones*

HAND SPECIMEN

Color, grain size, sorting, induration, bedding, etc. and field name.

THIN SECTION DESCRIPTION

Abstract:	Digest and condense all the petrography and summarize in 25 words.
Texture:	Modal size, sorting and nature of grain-to-grain contacts. Bedding. Percent sand-silt-and clay. Roundness.
Mineralogy:	Give percent of terrigenous, orthochemical (precipitated cement), and allochemical (transported grains formed within the basin of deposition) material plus description and amounts of different type of terrigenous debris. A brief paragraph for each constituent.
Interpretation:	Character of source area plus type of transportation and character of depositional basin, if possible. Diagenesis.

GENERAL COMMENT

Always keep description and interpretation separate. At times you may estimate abundances with 100 point counts or by using comparison charts.

the ratio in phi units of two representative grain diameters: the diameter in millimeters of a grain that has one-sixth of the grains by area smaller than itself (84th percentile) and the diameter of a grain one-sixth of the grains by area larger than itself (16th percentile). Conversion to phi units, subtraction, and division by two yields the sorting. Clay of authigenic origin should be ignored when clay content is determined.

To facilitate petrographic analysis, we have included two petrographic forms. Table A-2 is a skeletal form and Table A-3 a very detailed one modified from

Table A-3. *Detailed petrographic report form (modified from Folk, 1968, p. 133–138)*

I. SAMPLE IDENTIFICATION

Formation name, age, and precise geographic location.

II. FIELD RELATIONS

Outcrop thickness, position of outcrop with respect to formational boundaries, associated lithologies, bedding characteristics, sedimentary structures, fossils, deformation and mineralization.

III. HAND SPECIMEN DESCRIPTION

Concise, simple description including a field name consistent with petrographic analysis. Include color, grain size, sorting, roundness, mineral composition, fossils, induration, sedimentary structures, bedding, and tectonic deformation and weathering.

IV. THIN SECTION DESCRIPTION

A. Abstract

Brief comments, perhaps 50 words. Prepare only after all other aspects of the report are complete. Include rock name, summarized modal analysis, interpretation, possible economic significance and relevance to scientific problems.

B. Texture

1. Fabric
   Grain to grain relations, grain orientation, cementation and porosity.
2. Grain size
   Specify mean, median, and range, sorting and percent of gravel, sand, silt, and clay. Plot cumulative on log-probability paper, if 100 or more grains are counted.
3. Angularity and sphericity
   Describe and comment on how they vary with size.
4. Textural maturity.

C. Mineralogy

Separate into terrigenous, allochemical and orthochemical constituents and give percent of each. Reproduce modal analysis in a compact table.

Describe the terrigenous minerals in the following order: quartz and its varietal types, chert, feldspar, rock fragments, mica, detrital clay and heavy minerals. For each mineral record its percentage, distribution and orientation in slide, grain size (estimate median and range), morphology, overgrowths, clastic nuclei, inclusions and alteration products. Summarize each mineral species in concise, well written sentences, no more than a paragraph for each one.

For the allo- and orthochemical constituents follow an essentially similar procedure. Consult appropriate manuals for identification of biotic constituents.

V. INTERPRETATION

Here one integrates the data gained from the thin section with all other evidence: field observations, chemical analysis and the literature. Remember that the best interpretation is one that uses all the relevant facts and exploits their significance to the fullest – stopping just short of the point of unjustified conclusions.

A. Source area

Estimate lithologic composition and maturity of source sandstones, tectonic state and weathering, location, and distance from depositional site. How many sedimentary cycles are involved? Did more than one source region contribute?

B. Depositional basin

Estimate nature and strength of currents and water depth. Identify environment of deposition as fully as evidence permits. If marine, how far from shore? Utilize biotic constituents and Lebenspuren as much as possible.

Table A-3 (continued)

C. Diagenesis

State and interpret diagenetic history. Are diagenetic affects major or minor?

VI. ECONOMIC IMPORTANCE

Discuss economic importance and give industrial name. Comment on possible market value and problems of development.

VII. BEARING ON SCIENTIFIC PROBLEMS

How does the interpretation of this sample relate to the historical development of the sand body or the sedimentary basin? Does the thin section contain any features that contribute to general methodology or principles?

VIII. REFERENCES CITED

Folk (1968, p. 133–138). We recognize, of course, that in many studies all the items in these report forms will not be appropriate and that there may be special comments that we have not included. The detailed form emphasizes, however, the comprehensive nature of a full description. The analysis of the Trivoli sandstone follows this form.

We have included a few amplifying remarks for the most effective use of these forms. After looking at the hand specimen with the naked eye, a hand lens or binocular, it is best to scan the thin section with low power to appraise its general characteristics. This should be followed by a grain size analysis of its terrigenous components under medium or high power. Size analysis is commonly the best way to become acquainted with both the texture and mineralogy of the section. Counts of 100 grains are sufficient for rough estimates but counts of 300 to 500 using quarter phi units are needed for accurate ones.

Modal mineralogical analysis follows. Counts of 200 to 500 grains are generally sufficient for all but minor constituents (less than 1 percent) using high or medium power. It is best to estimate proportions of varietal types by separate counts. The modal analysis should be always supplemented by a qualitative observation of each component: its median and size range, angularity, inclusions, alteration products, etc. Such descriptions should avoid useless detail, such as, for example, enumeration of ordinary optical properties, and instead be directed to useful special features. It is, for example, pointless to note that the quartz is uniaxial but it may be significant to observe that quartz is composite and well-rounded. The relation between grain size and composition and the distribution of cementing agents within the section should always be observed. The textural relations of one mineral to another are most important and should be recorded.

Any comprehensive study of thin sections should be supplemented by X-ray analysis of interstitial matrix and microscopic heavy mineral analysis. If a chemical analysis is available, it should certainly be cited and related to petrographic features.

Good report forms clearly separate description from interpretation and one should always strive to maintain this separation. But how does one make the interpretation? Table A-4, modified from van Andel (1958, Table 1), relates

Table A-4. *Objectives of study, relevant petrographic properties and their interpretation (modified from van Andel, 1958, Table 1)*

OBJECTIVE	PROPERTY	REMARKS
Source	Roundness	Generally modified but little in a single cycle and hence useful in assessing character of source rocks. Rounded quartz generally, but not always, implies recycling.
	Directional structures	Regional mapping outlines current system in basin and thus helps locate source region. May also have some environmental significance.
	Mineralogical maturity	Mature mineralogy commonly reflects cratonic source, recycling, and appreciable weathering; immature mineralogy indicates uplift and rapid erosion of crystalline rocks. Absence of polycrystalline quartz indicates pre-existing sediments.
Transportation	Grain size	Generally not diagnostic of environment except for presence or absence of gravel; thick conglomeratic sections indicate strong gradients and proximity to source, but pebbly sands may be transported hundreds of miles by large streams. Vertical size profile may contain environmental information.
	Mineralogy	Abrasion minimal for sand in all but steepest mountain streams.
Depositional environment	Associated sediments	Knowledge of lateral equivalents plus preceding and following units essential for maximum interpretation.
	Fossils	Establish environment of deposition in many cases.
	Mineralogy	Argillaceous rock fragments may be eliminated on beaches with high wave energy; authigenic minerals.
	Sand-body shape and orientation	Map pattern of sand accumulation tells much about process and environment of sand dispersal in basin.
	Textural maturity	Clay content and framework sorting reflect final depositional environment. Rounding mostly inherited.
Post-depositional history	Authigenic minerals	Those that develop after appreciable lithification. Difficult to separate from early diagenetic effects.
	Permeability	A rough overall measure of the extent of diagenesis.
	Pressure solution	The most important key to porosity and permeability.

petrographic and other properties to the major objectives of the study of sand and sandstone. This table summarizes the essential information of much of the previous material of the book. Table A-4 emphasizes that the fullest interpretation requires some knowledge of the size, shape, and orientation of the sandstone body, its associated sediments, and position of the sandstone body in the basin. All of which underscores the fact that petrographic interpretation is greatly enhanced by other information about the sandstone body from which the sample was obtained.

Table A-4 also underscores another important point – most petrographic properties are the response to the joint effects of both inheritance and depositional environment. For example, angularity of quartz may be related to the maturity of the sandstones in the source area as well as the effectiveness of rounding in the last depositional environment. In short, a *history* is involved and the petrographer's problem is to decipher where and when the particular effect took place: in the source area, in transport, at the site of deposition, or afterward?

Although quantitative petrography is essential, there is no substitute for penetrating qualitative observation. It alone is the key to what may be worthwhile to count.

A Comprehensive Petrographic Analysis: The Trivoli Sandstone of Southern Illinois

I. Sample Identification

Trivoli sandstone member of the Modesto Formation, McLeansboro Group, Pennsylvanian System. Basal member of the Trivoli Cyclothem and named as such by Wanless (1931). Outcrops in parts of western Illinois, especially Peoria and adjacent counties, and southern Illinois, especially Williamson and adjacent counties. Known in the subsurface over wide areas of the Illinois Basin. Description is a composite from samples collected from outcrops and the subsurface in the Illinois Basin.

II. Field Relations

Maximum sandstone thickness is 157 ft (49 m) in Franklin County. Thick or channel phase generally varies from 40–100 ft (12–31 m) and mainly in erosional valleys cut in underlying beds. Thin or sheet phase, ranging from 0–40 ft (0–13 m), is distributed more uniformly. Sandstone has a sharp unconformity separating it from the gray marine shale of the West Franklin Limestone of the underlying cyclothem. Grades upward into silty shale and then to the underclay below the Chapel (No. 8) coal bed. Basal part of sandstone may be conglomeratic with locally derived pebbles.

On the outcrop the sandstone weathers into large slabs and flags and is uniform in appearance. Ripplemark and crossbedding are characteristic throughout. Crossbedding varies from planar to large troughs. Mean direction of crossbedding is 158° based on 81 observations (Andresen, 1961, p. 25). Plant fossils are characteristic, large fragments of abraded stems and twigs occur in the conglomerates, smaller particles, not easily identifiable, in the finer grained beds. Beds are gently deformed, conformable with the structure of other Pennsylvanian beds in the basin. Dips are so low that they are rarely accurately measurable in the field.

III. Hand Specimen Description

The sandstone, in outcrop, is brown to reddish-brown, but in the subsurface it is light gray. It is uniformly fine grained, well sorted, with shaly, micaceous (muscovite), and carbonaceous partings. It can readily be identified as a lithic arenite by the abundance of non-quartz grains and the lack of much matrix. Cementation is normally moderate (can easily be disaggregated without breaking grains) but some specimens show extensive cementation (disaggregation difficult and breaks grains). Grains appear angular; sparkling facets of euhedral quartz overgrowths and calcite cement crystals hide detrital outlines. Small scale cross-bedding and ripples can be seen in some specimens.

IV. Thin Section Description

A. Abstract

A typical moderately well-sorted, sublithic to lithic arenite characteristic of many continental Coal Measures (Fig. A-4). Mono- and polycrystalline quartz are subangular to subrounded and commonly form about 50 percent of the sand,

Fig. A-4. Typical appearance of Trivoli sandstone in thin section; partially crossed nicols, × 100

feldspar averages 4 percent, and rock fragments about 9 percent. Clay and badly squashed rock fragments (7 percent) plus calcite and ferroan dolomite (11 percent) bind the framework together. Outcrop and extensive subsurface mapping indicate that the Trivoli Sandstone was mostly derived from the Appalachian mobile belt and was deposited in the Illinois Basin as a coastal plain-deltaic complex.

B. Texture

1. Fabric is typical of lithic arenites. Most of the rock is supported structurally by the quartz-feldspar-chert framework, but deformation of argillaceous rock fragments has squashed the latter to "clay matrix" which intrudes into pore space and surrounds competent framework grains. Grain contacts are more numerous than expected for a well-sorted sand, indicating some compaction and rearrangement of grains accompanied the squashing of rock fragments.

Because most grains tend to be equant, a preferred shape orientation is not obvious but orientation of large detrital mica flakes is excellent and parallel to

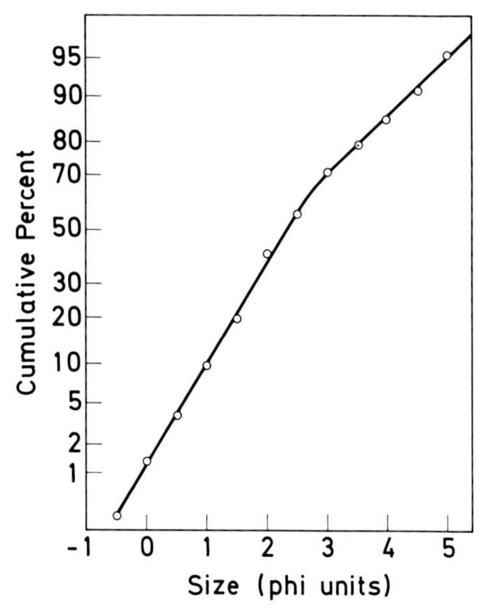

Fig. A-5. Grain size analysis of Trivoli sandstone as determined by thin section (300 counts). Note two populations on arithmetic probability paper

bedding. Quantitative studies of a similar sandstone lower in the Pennsylvanian, the Pleasantview of western Illinois (Rusnak, 1957), show that there is both a general preferred direction of elongate grains in a section parallel to bedding and an imbrication in a plane normal to the bedding.

The sandstone is cemented by a combination of the clay matrix with precipitated mineral cements of calcite, ferroan dolomite, siderite, quartz, carbonate, and, on the outcrop, hematite and limonite. Cement, most of which is carbonate, averages about 20 percent and this, coupled with clay matrix, reduces porosity to about 20 percent from an original porosity that must have been about 40 to 45 percent.

2. Grain size distribution is that of a moderately well-sorted, fine- to medium-grained sand (Fig. A-5). Size distributions are skewed to fine sizes because of the abundance of clay. The break in the size curve suggests a mixture of two size distributions. If most of the clay were assumed to come from the degradation of argillaceous rock fragments and removed, the sediment would be very well-

sorted, fine- to medium-grained sand, corresponding to the fraction coarser than about 3ϕ in Figure A-5.

3. Large numbers of framework grains are subrounded to rounded, but roundness is difficult to estimate in many grains because of lack of preserved border between detrital grain and overgrowth or replacing cement. Average visual roundness of quartz is estimated at 0.25; the feldspar average is 0.5. Sphericity is in the range .60–.85 and is highly variable because of the heterogeneous composition. Coarser samples tend to be better rounded and more spherical.

4. The Trivoli is texturally semi-mature as indicated by the sorting and roundness. If most of clay matrix is not counted as detrital, then many samples would be mature.

C. *Mineralogy (Table A-5)*

1. Terrigenous detritus

a) Quartz is separable into mono- and polycrystalline, chert and secondary overgrowths. Cathodo-luminescence shows that some of the monocrystalline quartz has low luminescence and may be of relatively low temperature origin – sedimentary, low grade metamorphic, or low temperature hydrothermal. Authigenic overgrowth quartz exceeds 5 percent in only a few samples; in most it is about 1 percent. Second-cycle grains are present in some samples.

b) Chert is present in small or trace amounts in most samples. A few grains show faint suggestions of fossil outlines.

c) Feldspars present are sodic varieties of plagioclase, orthoclase, microcline, and a few grains of anorthoclase. Some grains are kaolinized but it is not certain that they have been altered in place. More microcline grains are altered than

Table A-5. *Some typical modal analyses of the Trivoli sandstone (Andresen, 1961, Table 2)*

MINERALS	SAMPLES		
	1E	C15B	C21A
Quartz			
Monocrystalline	23	38	24
Polycrystalline	28	23	27
Chert	—	T	T
Feldspar			
Microline	T	T	T
Plagioclase	—	1	—
Untwinned	1	5	2
Mica	3	3	1
Rock fragments			
Metamorphic	4	7	8
Sedimentary	4	2	5
Clay matrix	7	5	5
Cement	23	16	21
Average quartz size (mm)	0.08	0.18	0.10

plagioclase. Most orthoclase is untwinned. The ratio of plagioclase to K-feldspar is 1.7 based on X-ray diffraction patterns of the $>2\mu$ fraction.

d) Rock fragments are dominantly argillaceous varieties, both low grade metamorphic and sedimentary, which grade into clay matrix. Many appear deformed and have corroded edges. Many siltstone and shale fragments look like Pennsylvanian rocks lower in the section and may be of local origin. A few limestone fragments were found.

e) Mica is almost all muscovite, with small amounts of biotite and chlorite. It occurs in shreds and plates, many bent and broken, many deformed around quartz grains. Mica flakes are all oriented parallel to bedding, and are very abundant in thin shaly partings.

f) Clay minerals were analyzed by X-ray diffraction after separation of the $<2\mu$ fraction. This fraction probably includes not only matrix but some of the rock fragment clays as well. Clays present are kaolinite, illite, chlorite, and a mixed layer clay. Kaolinite is well crystallized and some can be seen to be an authigenic pore filling. A scanning electron micrograph of a sample of pore kaolinite shows a flaky white appearance. Under cathodo-luminescence kaolinite appears a bright blue indicating a low temperature origin. Chlorite is poorly crystallized and appears to be an iron-rich variety on the basis of relative intensities of different orders of basal spacing (first order low, second order high). Illite shows typical grading into mixed layered varieties. The mixed layered varieties are abundant and seem to be randomly interlayered. Based on peak intensities, the ratio of kaolinite to chlorite is about 10; chlorite to illite about 0.2, and quartz to illite about 2.2. Analyses of ground samples of the $>2\mu$ fraction shows a higher ratio of kaolinite to chlorite, 12.3; a higher ratio of quartz to illite, 4.9; and the same ratio of chlorite to illite, 0.2.

g) Heavy minerals are dominantly zircon, tourmaline, and rutile with lesser amounts of apatite and garnet. The ZTR index is very high, indicating high maturity. Opaque heavy minerals are mainly leucoxene with lesser amounts of pyrite, hematite, and limonite. Roundness of the heavy minerals varies: garnet is angular, rutile is rounded, tourmaline and zircon have both rounded and angular varieties. Some tourmalines show small spikes of authigenic overgrowths.

h) Conglomerate pebbles at base of sandstone in channel deposits are almost entirely sideritic clay concretion fragments. These are the same type as found in shales overlying coal beds in the Illinois Basin. Other pebbles include hematite and limonite cemented sandstone similar to other Pennsylvanian sandstones in that part of the section, a few limestone pebbles, and rarely a coal pebble.

2. Chemical Constituents

a) Calcite is present as a clear untwinned mosaic of interlocking crystals that fills pore space and replaces detrital grains of quartz and feldspar and, much more rarely, rock fragments or clay matrix. The only inclusions are of the quartz or other framework grain replaced. Calcite is present either in small amounts, about 5 percent, as irregular patchy areas or, in a few samples, as an abundant constituent making up more than 20 percent of the rock. Calcite not only replaces detrital quartz, but is itself replaced by some sharply euhedral authigenic quartz.

b) Iron carbonates include two varieties, siderite and a ferroan dolomite, both present as small individual or clusters of rhombohedra which replace calcite in some areas and appear to be intergrown in other areas. *X*-ray diffraction of ground samples indicates both siderite and dolomite; the slight stain caused by oxidation of the iron that is ubiquitous in all of the rhombohedra in outcrop samples implies the presence of iron in the dolomite.

c) Authigenic quartz is present as secondary overgrowths deposited in optical continuity with detrital grains. Some replace calcite. This quartz variety is shown by cathodo-luminescence to be more abundant than estimates based on ordinary microscopy and in one section is about 8 percent. The difference in estimates is traceable to the fact that the detrital outlines of original grains are not always distinguishable and so anhedral overgrowths are not recognized.

d) Authigenic kaolinite is present as well as crystallized, small, vermicular aggregates in pore spaces. A few grains of glauconite, probably authigenic but possibly detrital, are present in a few samples. Colorless authigenic tourmaline overgrowths are rare. Anatase is idiomorphic and probably is authigenic.

V. Interpretation

The Trivoli is one of the best known mappable sand bodies of the upper part of the Pennsylvanian section of the Illinois Basin, primarily from the subsurface but also from its outcrop belt (Fig. A-6). It was the first Pennsylvanian sandstone of the basin to be given systematic petrographic study, and one of the few individual sand bodies for which there is a large and useful body of stratigraphic and petrologic data. The only other Pennsylvanian sand body to be studied in such detail, and a useful comparison, is the Anvil Rock Sandstone lower in the section in the Carbondale Formation of the Kewanee Group (Hopkins, 1958).

The composition and texture of the Trivoli are consistent with the general interpretation of Pennsylvanian sandstones of the Illinois Basin as part of an alluvial-deltaic complex derived primarily from a sedimentary and metamorphic highlands of the Appalachian mobile belt far to the east and northeast by a major river system (See Chap. 12, Illinois Basin). The lack of much feldspar and the abundance of metamorphic and sedimentary rock fragments imply the absence of any large area of eroding igneous rock, either intrusive or extrusive. The many sedimentary rock fragments that can be identified easily with rocks of the Illinois Basin section, such as the clay ironstones and coal, coupled with the abundance of relatively soft pelitic rock fragments, demonstrate that some material was being supplied by local source areas within the basin. Some of the material is multicycled as indicated by the roundness of quartz and heavy minerals and the presence of a few second-cycle grains; but how much of this is from nearby and how much from distant sources is not possible to say. It is possible that a careful comparison of samples from the minor tributaries mapped by Andresen (1961, p. 26) with those of major channels would show differences that would be interpretable as coming from local as opposed to distant sources. The Trivoli has much the same composition everywhere; local variance introduced by sampling sheet- or channel-phases or coarse and fine grained beds is much greater than regional variance. This homogeneity indicates a well mixed contribution from distant sources, such as would be characteristic of a large river system, with a contribution from homogeneous nearby rocks lower in the section – rocks that we know show little regional variance.

The direction and paleogeography of the source lands comes from paleocurrent and subsurface channel studies that indicate a major source to the east and northeast in the Appalachian mobile belt and a lesser source to the north and

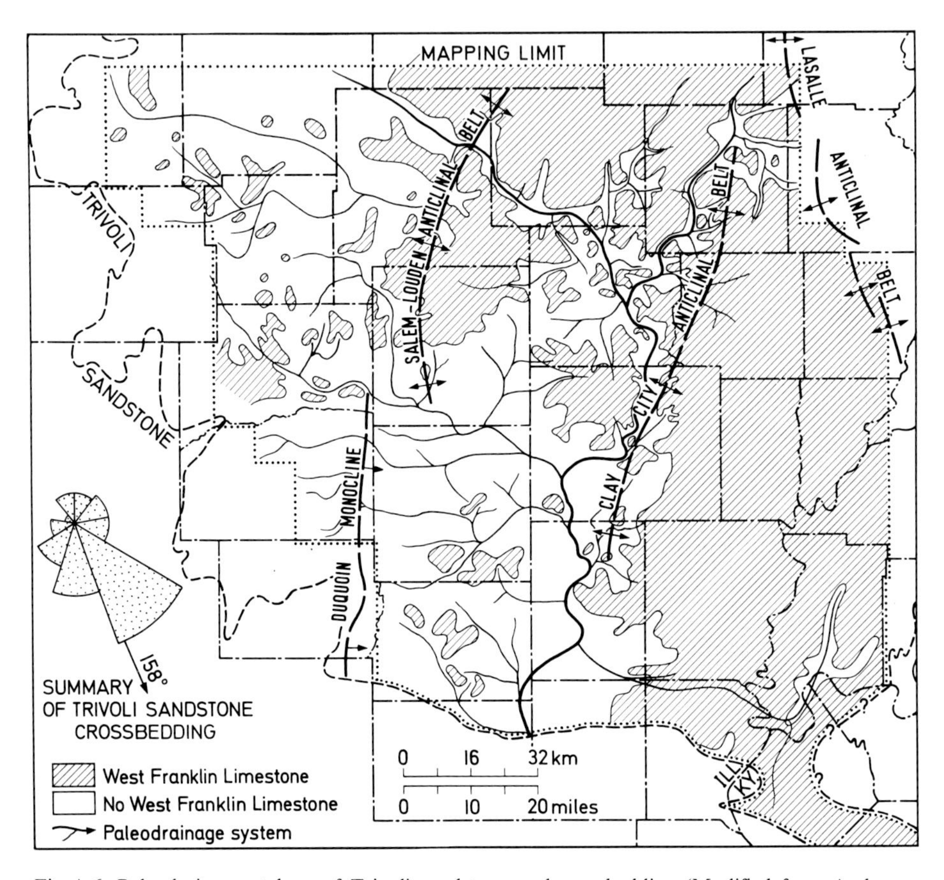

Fig. A-6. Paleodrainage at base of Trivoli sandstone and crossbedding (Modified from Andresen, 1961, Fig. 9)

northwest, from the Transcontinental Arch and Canadian Shield (see Chap. 12).

B) All of the characteristics of the Trivoli indicate a dominantly alluvial origin – the plant remains, the crossbedding, the valley system of the channel sandstones, the sorting of the detritus, and the lack of any marine fossils. The shoreline must have been to the south but the record of constant marine transgressions and the increasingly marine nature of the section to the southwest in Missouri and Kansas indicate that shoreline to have been no more than a few hundred miles away at any time and probably much closer at many times. Thus the sedimentary framework of the Trivoli is that of a deltaic-coastal plain complex.

The difference between coarser channel and finer sheet phases is explainable in terms of higher current shear stress in the high discharge channels and lower shear stress in the lower discharge streams, perhaps braided, of the sheet phase that followed the aggradation of the channels.

C) Two major changes in the sand took place as a result of diagenesis, the transformation of many soft rock fragments into matrix and the precipitation of secondary quartz and carbonate cement. The alteration of the rock fragments was mainly a mechanical process rather than chemical, for the composition of the matrix is much the same as the rock fragments. Implied is a more or less plastic flow of soft argillaceous fragments around and between the rigid, competent framework quartz and feldspar grains.

The precipitation of the cement follows the order: calcite, perhaps contemporaneous with iron carbonates but perhaps earlier, followed by secondary quartz. After being brought close to or at the surface by uplift and erosion, hematite and limonite cement were added. The calcite may have been precipitated following sand deposition when it was buried only by the overlying silt and clay and a coal swamp [Chapel (No. 8) coal]. The probable groundwater movement at this time would have been a circulation through the Trivoli sand upward to the swamp – a typical pattern for such swamps – perhaps carrying with it some of the waters still being squeezed from the underlying thick marine shale sequence. Those waters, originally meteoric, may have become acid and depleted in oxygen as a result of passage through strata containing abundant pyrite, leading to the dissolution of carbonate fossils and parts of limestone beds and to the disappearance of pyrite. The resulting water, passing through the sand and mixing with other less carbonate saturated waters might have thus become less acid and so precipitated carbonate, including iron-rich varieties. Alternatively, some of the calcite may simply have been detrital or fresh water fossil calcite redistributed by dissolution and reprecipitation in the immediate local area where the cement now is. This alternative is suggested by the irregular patchy distribution of some cement.

The authigenic quartz overgrowths may owe their origin to a more general pressure-solution process operating later in the sand's history, when it was under an overburden of later Pennsylvanian beds. Because of post-Pennsylvanian erosion, it is difficult to estimate that overburden. If some of the clay minerals were altered in such a way as to liberate silica, as, for example, from a more siliceous montmorillonite to an illite, then that silica may have been contributed to the ground water too, later to be precipitated as the groundwater slowly equilibrated with quartz.

VI. Economic Importance

The Trivoli itself is not an economic resource. Its composition makes it unsuitable for molding or foundry sands. It shows no evidence of any oil or gas, now or at any former time. Yet its study has economic significance in relation to exploration for and exploitation of coal beds in the Illinois Basin, for sands such as the Trivoli act as cutouts of coal beds and interfere seriously with mining operations.

VII. Bearing on Scientific Problems

The Trivoli is very typical of many sandstones in the Illinois Basin – and of many alluvial sandstones in other places, particularly Coal Measure sandstones. The many kinds of information available: outcrop and subsurface core samples, extensive subsurface stratigraphic data, and surface paleocurrent mapping, make its origin one of the best supported interpretations we have. Problems, such as the relation of composition to near and distant sources and any correlation between that and minor tributaries and major channels, remain to be studied. A detailed analysis of the history of the sequence in relation to diagenetic effects has yet to be made. Such detailed studies will give an inductive base for future general models.

References

van Andel, Tj. H.: Origin and classification of Cretaceous Paleocene and Eocene sandstones of western Venezuela. Am. Assoc. Petroleum Geologists Bull. **42**, 734–763 (1958).

Andresen, M. J.: Geology and petrology of the Trivoli sandstone in the Illinois Basin. Illinois Geol. Survey Circ. **316**, 31 p. (1961).

Dennison, J. M.: Graphical aids for determining reliability of sample means and an adequate sample size. Jour. Sed. Petrology **32**, 743–750 (1962).

Folk, R. L.: Henry Clifton Sorby (1826–1908), the founder of petrography. Jour Geol. Education **8**, 43–47 (1965).

— Petrology of sedimentary rocks, 170 p. Austin: Hemphill's Bookstore 1968.

Grout, F. F.: Petrography and petrology, 552 p. New York: McGraw-Hill 1932.

Hopkins, M. E.: Geology and petrology of the Anvil Rock Sandstone of southern Illinois. Illinois Geol. Survey Circ. **256**, 49 p. (1958).

Pearson, E. S., Hartley, H. D.: Biometrika tables for statisticians, Vol. 1, 238 p. New York: Cambridge Univ. Press 1954.

Powers, M. C.: A new roundness scale for sedimentary particles. Jour. Sed. Petrology **23**, 117–119 (1953).

Rusnak, G. A.: A fabric and petrologic study of the Pleasantview Sandstone. Jour. Sed. Petrology **27**, 41–55 (1957).

Siever, Raymond: Trivoli sandstone of Williamson County, Illinois. Jour. Geology **57**, 614–618 (1949).

Sorby, Henry Clifton: On the microscopical structure of the calcareous grit of the Yorkshire coast. Geol. Soc. London, Quart. Jour. **7**, 1–6 (1851).

— On the structure and origin of the non-calcareous stratified rocks. Geol. Soc. London Proc. **36**, 46–92 (1880).

Terry, R. D., Chilingar, G. V.: Summary of "Concerning some additional aids in studying sedimentary formations" by M. S. Shvetsov. Jour. Sed. Petrology **25**, 229–234 (1955).

van der Plas, L., Tobi, A. C.: A chart for judging the reliability of point counting results. Am. Jour. Sci. **263**, 87–90 (1965).

Wanless, H. R.: Pennsylvanian cycles in western Illinois. In: Papers presented at the quarter centennial celebration of the Illinois State Geological Survey. Illinois Geol. Survey Bull. **60**, 182–193 (1931).

Author Index

Numbers in *italics* denote complete citation in the References Cited and Annotated Bibliographies

Subject Index

Formation names are mentioned numerous times in the text. These are indexed *only* if mineralogical or chemical analyses are given or if a photomicrograph is included.

39*